SPRINGER SERIES IN SURFACE SCIENCES 51

SPRINGER SERIES IN SURFACE SCIENCES

Series Editors: G. Ertl, H. Lüth and D.L. Mills

This series covers the whole spectrum of surface sciences, including structure and dynamics of clean and adsorbate-covered surfaces, thin films, basic surface effects, analytical methods and also the physics and chemistry of interfaces. Written by leading researchers in the field, the books are intended primarily for researchers in academia and industry and for graduate students.

Please view available titles in *Springer Series in Surface Sciences* on series homepage http://www.springer.com/series/409

Gianangelo Bracco · Bodil Holst
Editors

Surface Science Techniques

Springer

Editors
Gianangelo Bracco
Department of Physics
University of Genoa
Genoa, Italy

Bodil Holst
Department of Physics and Technology
University of Bergen
Bergen, Norway

Series Editors:

Professor Dr. Gerhard Ertl
Fritz-Haber-Institute der Max-Planck-Gesellschaft
Berlin, Germany

Professor Dr. Hans Lüth
Institut für Schicht- und Ionentechnik Forschungszentrum Jülich Gmbh
Jülich, Germany

Professor Douglas L. Mills, Ph.D.
Department of Physics, University of California
Irvine, CA, USA

ISSN 0931-5195 Springer Series in Surface Sciences
ISBN 978-3-642-34242-4 ISBN 978-3-642-34243-1 (eBook)
DOI 10.1007/978-3-642-34243-1
Springer Heidelberg New York Dordrecht London

Library of Congress Control Number: 2012955761

Printed on acid-free paper

Springer is part of Springer Science+Business Media (www.springer.com)

To Luca and Sarah

Preface

The very first idea of a book on Surface Science techniques came at the 26th European Conference on Surface Science (2009, Parma, Italy) during the session on new experimental techniques. After this first burgeon, it took several months of 2010 for a complete plan to appear and be shaped to a definite form with a list of authors and a table of contents. The task was not easy since the first part of the editorial work was to decide the type of scope: a collection of chapters in which important results are described or a book more oriented towards the description of the experimental aspects of techniques.

Since the experience and the inclination of the Editors are on experiments and the construction of instrumentation we were naturally keen on the latter approach. Moreover we think that for many readers, who are not experts in a particular field, it is important to understand what is going on behind important results and, in a normal scientific article, the experimental details are sometimes not so explicit. The need for a more in depth knowledge of techniques is especially important for students involved in experimental research, as a support to help them understanding the literature and planning successful experiments with the right techniques. In fact, this is our greatest hope: If this book can assist the readers in finding the right technique for answering their scientific questions, then it has certainly justified its existence.

After the choice of the approach, the second step was the choice of the techniques which should appear as representative for the study of surfaces. The number of Surface Science techniques is huge and, for a single book, a selection had to be made.

The concept of surface involves both macroscopic as well as microscopic issues therefore, although microscopic techniques form the main core of the book, a few macroscopic techniques have been included. For the microscopic techniques, penetration depth is a particular crucial issue. Some techniques are strictly not penetrating while others probe several atomic/molecular layers. We decided to include also very penetrating techniques because the study of the interfaces is an important topic which cannot be neglected in Surface Science. The environment of the surface is another issue of crucial importance. Surfaces in ultra high vacuum can be kept clean for long periods of time and this is the traditional environment for

Surface Science techniques. However, industrial processes require high pressures, electrochemical investigations are performed in a liquid solution and surfaces of biological interest generally need a moist environment. For these reasons, techniques that can be applied in these realistic conditions are included in the book and we have decided to split the atomic force microscopy chapter in two chapters dealing with ultra high vacuum and liquid environment cases, respectively. Finally, due to the obvious relationship between Surface Science and Nanotechnology, topics related to nanostructuring of surfaces are also presented.

We believe that our choice of techniques is broad enough to allow the reader to understand how a study of surface (and interface) properties can be performed and how the various experimental problems might be tackled. In these choices we got very good support from several colleagues who suggested improvements in the schemes of the chapters. Surface Science is an interdisciplinary field and only a collaborative work can give rise to successful result: the same applies to this book.

The chapters are organized in parts: macroscopic techniques (contact angle and single-crystal adsorption calorimetry), microscopic techniques (optical and X-ray techniques for photons, charged (electrons and ions) and neutral particle (atoms, molecules and neutrons) techniques), and scanning probe microscopies.

Finally, we would like to thank all the participants to this project: first of all the authors who have accepted our invitation with enthusiasm and did most of the real work, then the following colleagues for their helpful suggestions: Giacinto Scoles (Udine, Italy and Princeton U.,USA), William Allison (Cambridge U., UK), Jochen Stahn (ETH Zurich and Paul Scherrer Institut, CH), Peter O'Toole (U. of York, UK), Kim Lefmann (Niels Bohr Institute, Copenhagen U., DK), Salvatore Iannotta (CNR-IMEM, Italy), Gang-yu Liu (U. of California Davis, USA), Neil Curson (U. of Nottingham, UK), Larry Scipioni (Carl Zeiss SMT), Miquel Salmeron (Lawrence Berkeley National Lab, USA), Enrico Gnecco (IMDEA Nanociencia, Madrid, Spain), Giorgio Benedek (U. of Milano-Bicocca, Italy), Charles T. Campbell (U. of Washington, USA) and David C. Joy (U. of Tennessee, USA), and last but not least, a thanks to Dr. Claus Ascheron and the staff of Springer who provided the necessary support for the accomplishment of the book.

We also gratefully acknowledge Trond Mohn, Bergen Research Foundation and the Michelsen Centre for Industrial Measurements Science and Technology for their generous support to the scientific careers of the Editors. The book would not have been possible without them.

Genoa (Italy) Gianangelo Bracco
Bergen (Norway) Bodil Holst

Contents

Part VI Scanning Probe Techniques

Contributors

David Allara Depts. of Chemistry and Materials Science and Engineering, Pennsylvania State University, University Park, PA, USA

T. Beuvier LUNAM Université, IMMM, Faculté de Sciences, Université du Maine, Le Mans Cedex 9, France

Gianangelo Bracco Department of Physics, University of Genoa, Genoa, Italy

Maurizio Canepa Department of Physics, University of Genoa, Genoa, Italy

M.S. Chebil LUNAM Université, IMMM, Faculté de Sciences, Université du Maine, Le Mans Cedex 9, France

Giovanni Costantini Department of Chemistry, The University of Warwick, Coventry, UK

Ada Della Pia Department of Chemistry, The University of Warwick, Coventry, UK

Juan de la Figuera Instituto de Química-Física "Rocasolano", CSIC, Madrid, Spain

Hans-Jörg Drescher Fachbereich Chemie, Philipps-Universität Marburg, Marburg, Germany

Vladimir A. Esaulov Institut des Sciences Moléculaires d'Orsay, CNRS and Université Paris-Sud, Orsay Cedex, France

Daniel Farías Departamento de Física de la Materia Condensada and Instituto Nicolás Cabrera, Universidad Autónoma de Madrid, Madrid, Spain

Luca Floreano CNR-IOM, Laboratorio TASC, Trieste, Italy

Jayne C. Garno Department of Chemistry, Louisiana State University, Baton Rouge, LA, USA

A. Gibaud LUNAM Université, IMMM, Faculté de Sciences, Université du Maine, Le Mans Cedex 9, France

J. Michael Gottfried Fachbereich Chemie, Philipps-Universität Marburg, Marburg, Germany

Bodil Holst Department of Physics and Technology, University of Bergen, Bergen, Norway

Andrew Jardine The Cavendish Laboratory, Cambridge, UK

Jacek J. Kolodziej Centre for Nanometer-Scale Science and Advanced Materials (NANOSAM), Institute of Physics, Jagiellonian University, Krakow, Poland

Rickmer Kose Sentys Inc, San Francisco, CA, USA

Franciszek Krok Centre for Nanometer-Scale Science and Advanced Materials (NANOSAM), Institute of Physics, Jagiellonian University, Krakow, Poland

Peter Krüger ICB, UMR 6303 CNRS-Université de Bourgogne, Dijon, France

T. Randall Lee Department of Chemistry, University of Houston, Houston, TX, USA

Venetia D. Lyles Department of Chemistry, Louisiana State University, Baton Rouge, LA, USA

Ole Lytken Lehrstuhl für Physikalische Chemie II, Universität Erlangen-Nürnberg, Erlangen, Germany

Diederik J. Maas TNO Nano-instrumentation, Delft, The Netherlands

Kevin F. McCarty Sandia National Laboratories, Livermore, CA, USA

Marina Minniti Departamento de Física de la Materia Condensada, Universidad Autónoma de Madrid, Madrid, Spain

Rodolfo Miranda Departamento de Física de la Materia Condensada and Instituto Nicolás Cabrera, Instituto Madrileño de Estudios Avanzados en Nanociencia (IMDEA-Nanociencia), Universidad Autónoma de Madrid, Madrid, Spain

Masashi Nakamura Department of Applied Chemistry and Biotechnology, Graduate School of Engineering, Chiba University, Inage-ku, Chiba, Japan

Alexei Nefedov Institute of Functional Interfaces, Karlsruhe Institute of Technology, Eggenstein-Leopoldshafen, Germany

Frédéric Ott Lab. Léon Brillouin CEA/CNRS, bat 563 Centre d'Etudes de Saclay, Gif sur Yvette Cedex, France

Lee J. Richter Material Measurement Laboratory, National Institute of Standards and Technology, Gaithersburg, MD, USA

Mario Rocca Dipartimento di Fisica dell'Università di Genova, Genoa, Italy

Osami Sakata Synchrotron X-ray Station at SPring-8, National Institute for Materials Science (NIMS), Sayo, Hyogo, Japan

Letizia Savio CNR-IMEM Unità operativa di Genova, c/o Dipartimento di Fisica Università di Genova, Genoa, Italy

Wilson K. Serem Department of Chemistry, Louisiana State University, Baton Rouge, LA, USA

Josh Stapleton Materials Characterization Laboratory, Pennsylvania State University, University Park, PA, USA

Bartosz Such Centre for Nanometer-Scale Science and Advanced Materials (NANOSAM), Institute of Physics, Jagiellonian University, Krakow, Poland

Marek Szymonski Centre for Nanometer-Scale Science and Advanced Materials (NANOSAM), Institute of Physics, Jagiellonian University, Krakow, Poland

Raoul van Gastel Physics of Interfaces and Nanomaterials, MESA+ Institute for Nanotechnology, University of Twente, Enschede, The Netherlands

Luca Vattuone Dipartimento di Fisica dell'Università di Genova, Genoa, Italy

Alberto Verdini CNR-IOM, Laboratorio TASC, Trieste, Italy

Christof Wöll Institute of Functional Interfaces, Karlsruhe Institute of Technology, Eggenstein-Leopoldshafen, Germany

Jing-Jiang Yu Nanomeasurements Division, Agilent Technologies, Inc., Chandler, AZ, USA

Yuehua Yuan Department of Chemistry, University of Houston, Houston, TX, USA

Jörg Zegenhagen European Synchrotron Radiation Facility, Grenoble, France

Part I
Macroscopic Techniques

Chapter 1
Contact Angle and Wetting Properties

Yuehua Yuan and T. Randall Lee

Abstract This chapter highlights a variety of techniques that are commonly used to measure contact angles, including the conventional telescope-goniometer method, the Wilhelmy balance method, and the more recently developed drop-shape analysis methods. The various applications and limitations of these techniques are described. Notably, studies of ultrasmall droplets on solid surfaces allow wetting theories to be tested down to the nanometer scale, bringing new insight to contact angle phenomena and wetting behavior.

1.1 Introduction

The topic of wetting has received tremendous interest from both fundamental and applied points of view. It plays an important role in many industrial processes, such as oil recovery, lubrication, liquid coating, printing, and spray quenching [1–6]. In recent years, there has been an increasing interest in the study of superhydrophobic surfaces, due to their potential applications in, for example, self-cleaning, nanofluidics, and electrowetting [7–12]. Wettability studies usually involve the measurement of contact angles as the primary data, which indicates the degree of wetting when a solid and liquid interact. Small contact angles ($\ll 90^\circ$) correspond to high wettability, while large contact angles ($\gg 90^\circ$) correspond to low wettability.

This chapter will begin with an introduction of the fundamental science behind wetting and contact angle phenomena, followed by a comprehensive description of the various techniques used to measure contact angles, as well as their applications and limitations in terms of the geometric forms of solid samples. Most of the techniques can be classified into two main groups: the direct optical method and the indirect force method. Calculations based on measured contact angle values yield an important parameter—the solid surface tension, which quantifies the

Y. Yuan · T.R. Lee (✉)
Department of Chemistry, University of Houston, 4800 Calhoun Road, Houston, TX 77204-5003, USA
e-mail: trlee@uh.edu

Y. Yuan
e-mail: yyuan2@uh.edu

G. Bracco, B. Holst (eds.), *Surface Science Techniques*,
Springer Series in Surface Sciences 51, DOI 10.1007/978-3-642-34243-1_1,

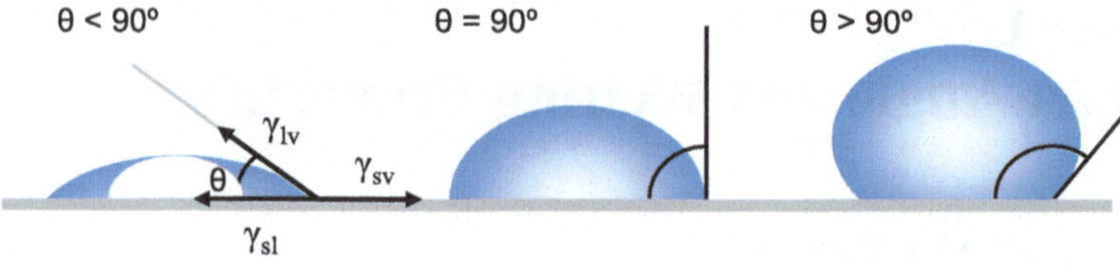

Fig. 1.1 Illustration of contact angles formed by sessile liquid drops on a smooth homogeneous solid surface

wetting characteristics of a solid material. The criteria of calculating solid surface tension based on experimental contact angle values will be discussed. Finally, the most up-to-date contact angle measurement techniques will be presented and discussed.

1.2 Theoretical Background

1.2.1 Surface Tension and Contact Angle—Young's Equation

Consider a liquid drop resting on a flat, horizontal solid surface (Fig. 1.1). The contact angle is defined as the angle formed by the intersection of the liquid-solid interface and the liquid-vapor interface (geometrically acquired by applying a tangent line from the contact point along the liquid-vapor interface in the droplet profile). The interface where solid, liquid, and vapor co-exist is referred to as the "three-phase contact line". Figure 1.1 shows that a small contact angle is observed when the liquid spreads on the surface, while a large contact angle is observed when the liquid beads on the surface. More specifically, a contact angle less than 90° indicates that wetting of the surface is favorable, and the fluid will spread over a large area on the surface; while contact angles greater than 90° generally means that wetting of the surface is unfavorable so the fluid will minimize its contact with the surface and form a compact liquid droplet. For example, complete wetting occurs when the contact angle is 0°, as the droplet turns into a flat puddle. For superhydrophobic surfaces, water contact angles are usually greater than 150°, showing almost no contact between the liquid drop and the surface, which can rationalize the "lotus effect" [13]. Furthermore, contact angles are not limited to the liquid-vapor interface on a solid; they are also applicable to the liquid-liquid interface on a solid.

Ideally, the shape of a liquid droplet is determined by the surface tension of the liquid. In a pure liquid, each molecule in the bulk is pulled equally in every direction by neighboring liquid molecules, resulting in a net force of zero. However, the molecules exposed at the surface do not have neighboring molecules in all directions to provide a balanced net force. Instead, they are pulled inward by the neighboring molecules (Fig. 1.2), creating an internal pressure. As a result, the liquid voluntarily contracts its surface area to maintain the lowest surface free

Fig. 1.2 Surface tension is caused by the unbalanced forces of liquid molecules at the surface

energy. From everyday life, we know that small droplets and bubbles are spherical, which gives the minimum surface area for a fixed volume. This intermolecular force to contract the surface is called the surface tension, and it is responsible for the shape of liquid droplets. In practice, external forces such as gravity deform the droplet; consequently, the contact angle is determined by a combination of surface tension and external forces (usually gravity). Theoretically, the contact angle is expected to be characteristic for a given solid-liquid system in a specific environment [14].

As first described by Thomas Young [15] in 1805, the contact angle of a liquid drop on an ideal solid surface is defined by the mechanical equilibrium of the drop under the action of three interfacial tensions (Fig. 1.1):

$$\gamma_{lv} \cos\theta_Y = \gamma_{sv} - \gamma_{sl} \tag{1.1}$$

where γ_{lv}, γ_{sv}, and γ_{sl} represent the liquid-vapor, solid-vapor, and solid-liquid interfacial tensions, respectively, and θ_Y is the contact angle. (1.1) is usually referred to as Young's equation, and θ_Y is Young's contact angle.

1.2.2 Contact Angle Hysteresis

From Young's equation applied to a specific liquid-solid system, three thermodynamic parameters γ_{lv}, γ_{sv}, and γ_{sl} determine a single and unique contact angle θ_Y. In practice, however, there exist many metastable states of a droplet on a solid, and the observed contact angles are usually not equal to θ_Y. The phenomenon of wetting is more than just a static state. The liquid moves to expose its fresh surface and to wet the fresh surface of the solid in turn. The measurement of a single static contact angle to characterize wetting behavior is no longer adequate. If the three-phase contact line is in actual motion, the contact angle produced is called a "dynamic" contact angle. In particular, the contact angles formed by expanding and contracting the liquid are referred to as the advancing contact angle θ_a and the receding contact angle θ_r, respectively (Fig. 1.3). These angles fall within a range, with the advancing angles approaching a maximum value, and the receding angles approaching a minimum value. Dynamic contact angles can be measured at various rates of speed. At a low speed, it should be close or equal to a properly measured static contact angle. The difference between the advancing angle and the receding angle is called the hysteresis (H):

$$H = \theta_a - \theta_r \tag{1.2}$$

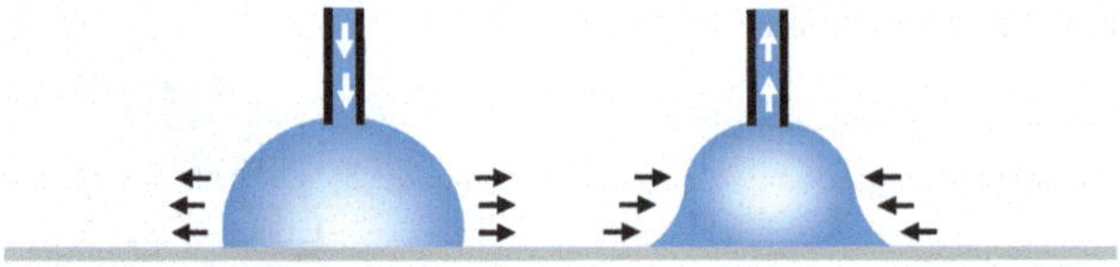

Fig. 1.3 Illustration of advancing and receding contact angles

The significance of contact angle hysteresis has been extensively investigated [16–20], and the general conclusion is that it arises from surface roughness and/or heterogeneity. For surfaces that are not homogeneous, there exist domains that present barriers to the motion of the contact line. For example, hydrophobic domains will pin the motion of the water front as it advances, causing an increase in the observed contact angle; the same domains will hold back the contracting motion of the water front when the water recedes, thus leading to a decrease in the observed contact angle. In cases that surface roughness plays the role of generating hysteresis, the actual microscopic variations of slope on the surface create barriers that pin the motion of the contact line and alter the macroscopic contact angles. Interpreting such contact angle data in terms of Young's equation can be misleading because the equation fails to consider surface topography.

Due to the complexity of contact angle phenomena, the experimentally observed contact angle might or might not be equal to Young's contact angle θ_Y [21, 22]. However, on ideal solid surfaces, there is no contact angle hysteresis, and the experimentally observed contact angle is Young's contact angle θ_Y. On smooth but chemically heterogeneous solid surfaces, the experimentally observed contact angle might not be equal to θ_Y. Nevertheless, the experimental advancing contact angle θ_a might be expected to be a good approximation of θ_Y [21, 22], while the experimental receding angle θ_r is usually found to have less reproducibility due to liquid sorption or solid swelling [23]. On rough solid surfaces, there is usually no correlation between θ_a and θ_Y. All contact angles on rough surfaces are largely meaningless in terms of Young's equation [21, 22]. The thermodynamic equilibrium contact angles on rough and heterogeneous surfaces are called Wenzel [24] and Cassie-Baxter angles [25], respectively. They are not equivalent to Young's contact angle.

There are as yet no general guidelines regarding how smooth a solid surface must be for surface roughness not to have an obvious impact on the contact angle. It is therefore recommended that the solid surface should be prepared as smooth as possible, and as inert to the liquids of interest as possible. Several techniques for the preparation of smooth homogeneous solid surfaces have been developed; these include: heat pressing [26], solvent casting [27, 28], self-assembled monolayers [29, 30], dip coating [31, 32], vapor deposition [33, 34], and surface polishing [35, 36].

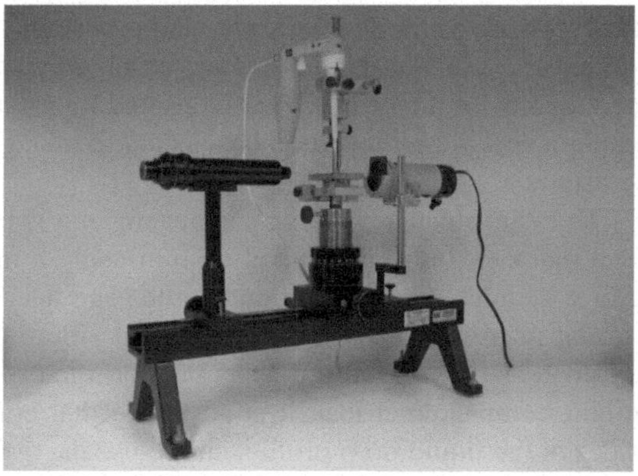

Fig. 1.4 A ramé-hart contact angle telescope-goniometer

1.3 Experimental Setup and Method

1.3.1 Direct Measurement by Telescope-Goniometer

The most widely used technique of contact angle measurement is a direct measurement of the tangent angle at the three-phase contact point on a sessile drop profile. Bigelow et al. [37] set up a simple and convenient instrument, which they referred to as a "telescope-goniometer" to measure contact angles of various liquids on polished surfaces. Later, the first commercial contact angle goniometer, designed by W.A. Zisman, was manufactured by ramé-hart instrument company in the early 1960s (Fig. 1.4).

The equipment consists of a horizontal stage to mount a solid or liquid sample, a micrometer pipette to form a liquid drop, an illumination source, and a telescope equipped with a protractor eyepiece. The measurement was achieved by simply aligning the tangent of the sessile drop profile at the contact point with the surface and reading the protractor through the eyepiece. Over the years, modifications of the equipment have been made to improve the accuracy and precision. A camera can be integrated to take photographs of the drop profile so as to measure the contact angle at leisure [38]. The use of relatively high magnifications enables a detailed examination of the intersection profile [39]. A motor-driven syringe can be used to control the rate of liquid addition and removal to study advancing, receding, or dynamic contact angles [40].

This direct optical method is advantageous because of its simplicity and the fact that only small amounts of liquid (a few microliters) and small surface substrates (a few square millimeters) are required. On the other hand, there is a relatively higher risk/impact of impurities due to the small size of the liquid and substrate. As for accuracy and reproducibility, the measurement relies on the consistency of the operator in the assignment of the tangent line, which can lead to significant error and inconsistency between multiple users. It is necessary to establish general guidelines for operators to follow. It is suggested that the telescope be tilted down slightly (1 to 2°) off the horizon so that the near edge of the sample stage (out of focus) is out of the line of sight, and a portion of the profile reflected by the substrate surface is brought into focus, which prevents forming a fuzzy liquid-substrate contact line in the profile. A background light is always used to assist observation, while a specific light source is selected to avoid undesired heating of the liquid or substrate.

To establish an advancing contact angle, it is best to slowly grow the sessile drop to a diameter of approximately 5 mm using a micrometer syringe with a narrow-gauge stainless steel or Teflon needle. The needle must remain in the liquid drop during measurement to avoid undesired vibration. The needle diameter should be as small as possible so it does not distort the drop profile shape. Because, the drop might be unsymmetrical, it is advisable that contact angles be measured on both sides of the liquid drop profile, and to use the averaged result. For a relatively large substrate, contact angles should be measured at multiple points to give an average value that is representative of the entire surface.

The direct goniometer method suffers from another serious limitation because small contact angles (below 20°) cannot be accurately measured due to the uncertainty of assigning a tangent line when the droplet profile is almost flat. Also, the imaging device only focuses on the largest meridian section of the sessile drop, which means the profile image reflects only the contact angle at the point in which the meridian plane intersects the three-phase line. In addition,the dependence of the contact angle on the drop size causes a systematic problem [41, 42]. Surface heterogeneity or roughness could well cause variations of the contact point along the three-phase contact line. Despite all of these issues, the goniometer method is considered to be the most convenient method if high accuracy is not required [43]. It is generally recognized that the direct measurement of sessile drop contact angles with a telescope-goniometer can yield an accuracy of approximately ±2° [43, 44].

Ideally, contact angle measurements should be made inside an enclosed chamber to exclude airborne contamination and establish an equilibrium vapor pressure of the liquid tested, which is especially preferable when the test liquid is volatile. It has been observed that evaporation can cause the liquid front to retract, and that a retreating or an intermediate contact angle is recorded unintentionally. However, the inherent inaccuracy of the direct measurement technique and the use of liquids with high boiling points make the enclosed chamber unnecessary in many cases.

The measurement of the contact angle hysteresis has been recommended as a means to assess the quality of the substrate surface. A "tilted plate" method (also referred to as the "inclined plate" method) was introduced by McDougall and Ockrent [45] (this is not to be confused with the "tilting plate" method in Sect. 1.3.3), who modified the sessile drop method and obtained both advancing and receding contact angles by tilting the solid surface until the drop just begins to move. The contact angles obtained at the lowest point θ_{max} and the highest point θ_{min} are considered as the advancing and receding contact angles, respectively (Fig. 1.5). This method was used by Extrand and Kumagai [46, 47] to study the contact angle hysteresis of liquids on a variety of polymer surfaces, including silicon wafers and elastomeric

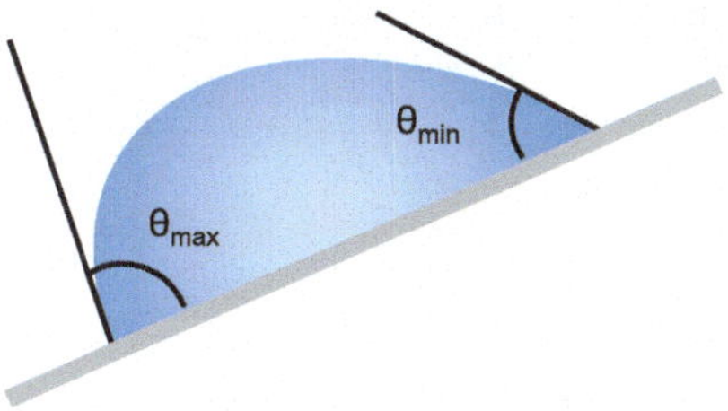

Fig. 1.5 Illustration of the "tilted plate" method, where θ_{max} and θ_{min} are assumed to be θ_a and θ_r, respectively when the drop just starts to move

surfaces. However, this particular association between the advancing/receding angles and the maximum/minimum contact angles must be used with caution because sometimes they can be quite different [48, 49].

In the early history of contact angle measurement, a platinum wire was used by Zisman and his co-workers [50] to form sessile liquid drops on solid surfaces. A fine platinum wire (about 8 cm long and 0.05–0.10 mm in diameter) was first cleaned by heating to red in a Bunsen burner, then dipped into the liquid and gently flicked to form a pendant drop hanging from the tip of the wire. The drop was then slowly brought into contact with the solid surface, and it flowed off the wire, forming a sessile drop. Although a reproducibility of $\pm 2°$ was claimed [26], there are concerns about the kinetic energy associated with the flowing droplet and the deformation that occurs when removing the platinum wire, which might lead to metastable contact angles.

Phillips and Riddiford [51], as well as McIntyre [52] have analyzed sessile drop profile photographs with a "tangentometer", which consists a mirror mounted at the baseline of the droplet. The mirror is positioned normal to the photograph at the drop tip, and is rotated until the curve of the drop shape forms a smooth, continuous curve with its reflection image in the mirror. Thus, the straight edge becomes the tangent line, and contact angle value is indicated by the protractor that is attached to the mirror. However, this technique still suffers inaccuracy due to the inherent subjectivity of tangentometers [53]. Fisher [54] has acquired a series of contact angle data less than 30° by applying the mass of the drop and the radius of the contact area into a semi-empirical formula that calculates the contact angle value. Langmuir and Schaeffer [55] used the specular reflection from a drop surface to measure the contact angle. Here, a light source above the three-phase contact line is rotated around the sessile drop, until to the height that small changes make the reflected light from the drop disappear or appear abruptly. The degree of rotation indicates the contact angle. This method was later refined by Fort and Patterson [56, 57] and has been used with both sessile drops and menisci on flat plates or inside tubes, with an accuracy of $\pm 1°$.

In the preceding sections, we have focused on techniques that involve measuring a sessile liquid drop on flat solid surfaces, which is the most common system for contact angle measurements. The following sections describe other systems and methods, including measurements of contact angles on solid samples that have different geometric forms (e.g., plates, fibers, and powders).

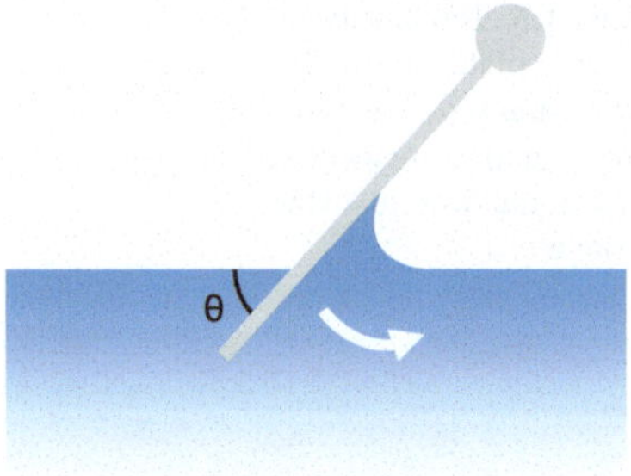

Fig. 1.6 Illustration of the tilting plate method

1.3.2 Captive Bubble Method

Instead of forming a liquid sessile drop above the solid sample, an air bubble can be formed beneath the solid sample, which is immersed in the testing liquid. The contact angle formed by the air bubble in liquid can also be directly measured. The technique was introduced by Taggart et al. [58], and is now widely referred to as the "captive bubble method". Typically, a small amount of air (about 0.05 ml) is injected into the liquid of interest to form an air bubble underneath the solid surface. Similar to the sessile drop method, the needle should remain in the bubble so as not to disturb the balance of the advancing angle and also to keep the bubble from drifting over the solid surface in case the plate is not perfectly horizontal. The captive bubble method has the advantage of ensuring that the surface is in contact with a saturated atmosphere. It also minimizes the contamination of the solid-vapor interface from sources such as airborne oil droplets. Furthermore, it is much easier to monitor the temperature of the liquid in the captive bubble method than with sessile drops, which makes it possible to study the temperature-dependence of contact angles. On clean smooth polymeric surfaces, good agreement has been observed between sessile drop and captive bubble contact angles [59]. However, due to the nature of the method, the captive bubble method requires far more liquid than the sessile drop method. It can also be problematic when the solid swells after immersion into the liquid, or a film on the solid is dissolved by the liquid.

1.3.3 Tilting Plate Method

The tilting plate method developed by Adam and Jessop [60] was once favored due to its simplicity and relatively lesser dependence on the operator's subjectivity. In this method, a solid plate with one end gripped above the liquid is rotated toward the liquid surface until the end of the plate is immersed in the liquid, forming a meniscus on both sides of the plate. The plate is then tilted slowly until the meniscus becomes horizontal on one side of the plate (Fig. 1.6). The angle between the plate and the horizontal is then the contact angle of interest. An error of $\pm 5^\circ$ was reported, which was attributed to liquid contamination. The disturbance of the liquid by the

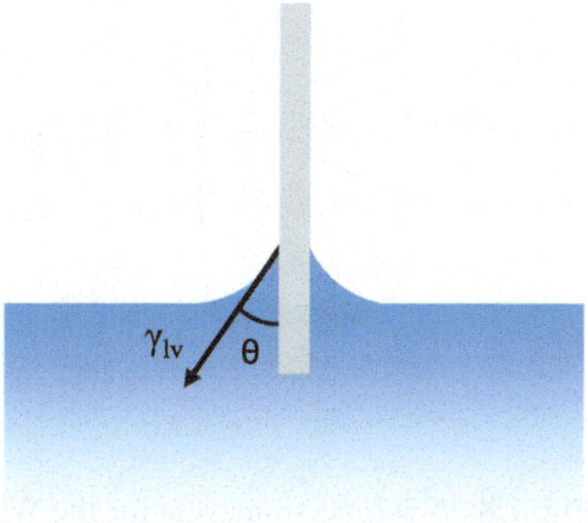

Fig. 1.7 Illustration of the Wilhelmy balance method

rotating solid plate and the requirement of considerable skills remain the major difficulties in the measurement. Fowkes and Harkins [61] improved the accuracy of the method by using glass barriers to clean the surface and a film balance to detect the presence of impurities on the liquid surface. They also used a microscope with an eyepiece to ensure the edge of the solid-liquid intersection lay on the axis of rotation. This method has also been used to measure small contact angles (less than 10°) [62]. Smedley and Coles [63] employed a scanning laser beam with the tilting plate technique to study the moving contact line, with the intention of making accurate measurements of the contact angle. The technique demonstrates high accuracy and potential to determine the velocity-dependence of dynamic contact angles automatically. Bezuglyi et al. [64] substantially enhanced the precision and reproducibility of the tilting plate technique by applying a high sensitivity thermocapillary (TC) response to the static curvature of the liquid meniscus.

1.3.4 Wilhelmy Balance Method

The Wilhelmy balance method [65] is a widely used technique that indirectly measures contact angle on a solid sample. When a thin, smooth, vertical plate is brought in contact with a liquid, the change in its weight is detected by a balance. The detected force change on the balance is a combination of buoyancy and the force of wetting (the force of gravity remains the same). The wetting force f is defined as (Fig. 1.7):

$$f = \gamma_{lv} p \cos\theta \tag{1.3}$$

where γ_{lv} is the liquid surface tension, p is the perimeter of contact line (i.e., the same as the perimeter of solid sample's cross-section) and θ is the contact angle. Consequently, the total detected force change F on the balance is:

$$F = \gamma_{lv} p \cos\theta - V \Delta\rho g \tag{1.4}$$

where V is the volume of the displaced liquid, $\Delta\rho$ is the difference in density between the liquid and air (or a second liquid), and g is the acceleration of gravity.

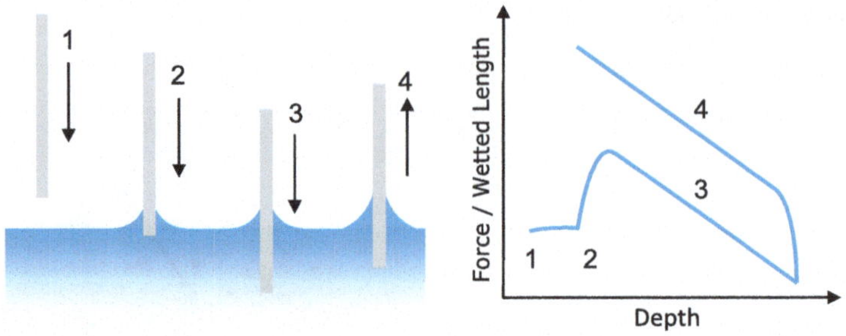

Fig. 1.8 A submersion cycle for the Wilhelmy balance measurement: (*1*) The sample approaches the liquid, and the force/length is zero. (*2*) The sample is in contact with the liquid surface, forming a contact angle $\theta < 90°$; the liquid rises up, causing a positive wetting force. (*3*) The sample is immersed further, and the increase of buoyancy causes a decrease in the force detected on the balance; the force is measured for the advancing angle. (*4*) The sample is pulled out of the liquid after having reached the desired depth; the force is measured for the receding angle

Thus, as long as the liquid surface tension and the solid perimeter are known, the contact angle value can be readily calculated. In rare cases, when the contact angle is zero and the perimeter is known, the measured force is related directly to the liquid surface tension. A technique developed by Princen [66] enables obtaining a zero contact angle in order to measure liquid surface tension by Wilhelmy balance method.

As the solid sample is pushed into or pulled out of the liquid, an advancing or receding contact angle can be established, respectively. The entire process will appear as illustrated in Fig. 1.8. The Wilhelmy balance technique is an indirect force method. It has several advantages over conventional optical methods. First, the task of measuring an angle is reduced to the measurements of weight and length, which can be performed with high accuracy and without subjectivity. Second, the measured force at any given depth of immersion is already an averaged value. Although this feature does not help determine the heterogeneity, it does automatically give a more accurate contact angle value that reflects the property of the entire sample. In addition, the graph produced by this technique (Fig. 1.8) is useful for studying dynamic contact angles and contact angle hysteresis at different wetting speeds. The smoothness of the curve indicates the heterogeneity of the solid sample. It is even possible to study absorption or surface reorientation by repeating the submersion circle. However, the method also suffers from several drawbacks. The solid sample must be produced with a uniform cross section in the submersion direction. Rods, plates, and fibers with known perimeters are ideal samples, but it is sometimes difficult to measure the perimeter and the wetted length precisely. Other than regular geometries, the sample must have the same composition and topography at all sides, which might be difficult to meet, particularly if one wants to investigate films or anisotropic systems. Also, a sufficient quantity of liquid must be used, which might cause the solid sample to swell and/or absorb vapor unintentionally.

1.3.5 Capillary Rise at a Vertical Plate

When a liquid comes in contact with a vertical and infinitely wide plate, it will rise due to the capillary effect. The height of capillary rise h can be determined by the integration of the Laplace equation [67]:

$$\sin\theta = 1 - \frac{\Delta\rho g h^2}{2\gamma_{lv}} \tag{1.5}$$

where $\Delta\rho$ is the difference in density between liquid and vapor, g is the acceleration due to gravity, and γ_{lv} is the liquid surface tension. Plates that are about 2 cm wide satisfy the theoretical requirement of being "infinitely" wide. The Wilhelmy balance method can be modified to measure the capillary rise h, in order to determine the contact angle θ [68, 69]. Dynamic contact angles are achieved by moving the plate up or down. This method has been widely used, and has proved to be particularly suitable for measuring the temperature-dependence of contact angles [70]. The method has been automated by Budziak and Neumann [71], and Kwok et al. [72]. For a specially prepared surface that forms a straight meniscus line, an accuracy of $\pm 0.1°$ can be obtained. This technique also inherited most of the advantages and disadvantages of Wilhelmy balance method. In addition, (1.4) and (1.5) can be combined through the relation $\sin^2\theta + \cos^2\theta = 1$, making it possible to determine both contact angle and liquid surface tension at the same time [73–75].

1.3.6 Individual Fiber

The direct measurement of contact angles on fibers has been attempted by Schwartz and co-workers [76, 77]. They suspended an individual fiber horizontally in the field of a microscope, and used a goniometer eyepiece to measure the contact angles of drops deposited on the fiber. Rotating the fiber along its longitudinal axis yielded approximate advancing and receding contact angles accordingly. Bascom and Romans [78] improved this method by placing a small platinum ring to hold the liquid drop, and passed a glass filament vertically through the center of the drop. Both advancing and receding contact angles were measured with a microscope as the filament was pulled through the stationary drop. Roe [79] has computed the equilibrium shapes of drops resting on fibers having different diameters. It was shown that the apparent contact angle can become misleading if the drop diameter is much larger than that of the fiber. Due to the small dimensions, large uncertainties exist because the drop curvature and the weight of the drop distort the profile where the fiber and liquid intersect. The equilibrium meniscus near a floating fiber on a liquid can also be used to calculate the contact angle [80–82]. The reflection method developed by Fort and Patterson has also been used to measure contact angles on fibers [56]. However, in practice, the relatively small depth of immersion can make it difficult to obtain good accuracy.

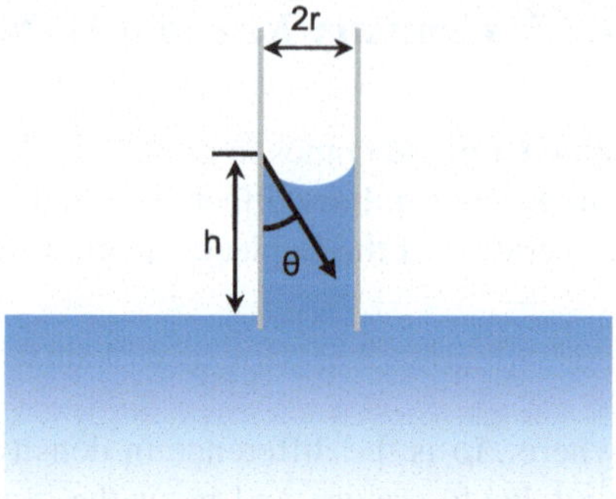

Fig. 1.9 Illustration of a capillary tube immersed in a liquid

The above mentioned Wilhelmy balance method is probably the most reliable technique for measuring contact angles on individual fibers of known diameter. The precise value of the fiber diameter can be determined by using a liquid of known surface tension to wet the fiber completely (i.e., zero contact angle). Given that $\cos\theta = 1$, the perimeter p of the fiber can be calculated from (1.4). Similar to the solid plate used in Wilhelmy balance method, a continuous immersion circle of the fiber in the liquid can also be used to test the homogeneity of the fiber surface.

1.3.7 Capillary Tube

In circumstances when both the inside and outside surfaces of the capillary tube are made of the exact same material, the Wilhelmy balance method can be used to measure the contact angle. The perimeter p of the capillary tube should be the sum of the inner and outer perimeters. In general, the Wilhelmy balance method can be applied to a wide range of plates, rods, wires, tubes, and capillaries.

For a vertical capillary with a sufficiently narrow circular cross section, the meniscus might be considered as spherical, and the capillary rise, h, is given by (Fig. 1.9):

$$h = \frac{2\gamma_{lv}\cos\theta}{\Delta\rho g r} \tag{1.6}$$

where r is the capillary radius, g is the acceleration constant of gravity, and $\Delta\rho$ is the difference in density between the liquid and vapor. The contact angle can be calculated using the values of experimentally measured h and r. If r is too small, it can be calculated from the length of the capillary occupied by a known mass of mercury. This relationship between contact angle, capillary height, and radius is known as the Jurin rule, named after James Jurin, who studied this effect in 1718. For relatively wide and transparent tubes, the above-mentioned Langmuir-Schaeffer reflection technique might also be used to directly measure the contact angle [83].

1.3.8 Capillary Penetration Method for Powders and Granules

The wetting behavior of powders and granules also involves contact angle phenomena, but it is complicated by the presence of a porous architecture. Although alternative methods have been employed to characterize porous architectures, contact angle remains the primary tool.

Direct measurement of contact angles on an individual powder or granule is almost impossible by conventional methods. Instead, the powders or granules are compressed into a flat cake, to which liquid drops are applied, and contact angles are measured [84–89]. However, due to the inherent porous architecture of compressed powder cakes, "spontaneous" liquid penetration might occur if the actual contact angle between the liquid and the solid is lower than 90°. Moreover, the actual rate of penetration might vary largely depending on the actual wettability and the porous structure. It might occur slowly enough that reproducible results can be obtained through direct measurement; or it might happen rapidly and give unstable results. Microscopic examination shows that the surfaces of compressed powders are porous, and it has been demonstrated thermodynamically that the contact angle on a porous surface is higher than on a smooth surface with the same composition [90]. The data can also be influenced by surface roughness [91, 92], particle swelling [55], and tablet porosity [61, 93]. Furthermore, during sample preparation, the topmost powder particles are likely to undergo plastic deformation by the compression, which might give different results than the uncompressed powders [61]. In addition, if the powder is slightly soluble in the measuring liquid, drops of solution saturated with the powder are used instead of the pure liquid [94].

The capillary penetration method was developed by Washburn [95], who monitored the rate at which a liquid penetrates into a compressed powder cake. The measurement was achieved by recording the depth of the liquid front intrusion as a function of time. The contact angle can then be deduced according to Washburn theory:

$$l^2 = \frac{rt\gamma_{lv}\cos\theta}{2\eta} \tag{1.7}$$

where l is the depth of liquid intrusion, γ_{lv} is the liquid surface tension, θ is the contact angle, η is the liquid viscosity, t is the time required for penetration, and r represents the pore radius. Numerous qualitative measurements have been performed, and the method has been developed theoretically [96–98].

Static measurements were originally proposed by Bartell and co-workers [99–101], whose theory has been significantly extended by White [102]. The wetting liquid penetrates upward vertically through a compressed powder cake, until it reaches a height at which the capillary pressure balances the weight of the liquid in the column. This method involves measuring the pressure necessary to balance the Laplace pressure, which drives liquid into a capillary bed:

$$\Delta P = \frac{2\gamma_{lv}\cos\theta}{r} \tag{1.8}$$

where ΔP is the pressure difference, γ_{lv} is the liquid surface tension, θ is the contact angle, and r is the pore radius.

Both Washburn's and Bartell's methods suffer from the common drawback that the effective pore radius r is not constant from point to point in the powder bed. White [102] defined the effective capillary radius r_{eff} in the following way:

$$r_{eff} = \frac{2(1-\phi)}{\phi \rho A} \tag{1.9}$$

where ϕ is the volume fraction of solid in the packed bed, ρ is the density of the solid material, and A is the specific surface area per gram of solid. Combining (1.8) and (1.9) yields the Laplace-White equation, which is a strict thermodynamic expression for ΔP in porous media:

$$\Delta P = \frac{\gamma_{lv} \cos\theta \phi A \rho}{1-\phi} \tag{1.10}$$

For clean and chemically treated smooth glass beads, a broad agreement between White's theory and experimental data has been obtained [103]. However, it can be rather problematic to determine r_{eff} directly due to the uncertainty in measuring the specific wetted area of the particles.

Diggins and co-workers [104, 105] used a second liquid that fully wet the powder, and compared its wetting behavior with the liquid of interest. Cyclohexane, with a low surface tension of 25.5 mN/m, is the most commonly used reference liquid. Prestidge and Ralston [106] reported that the success of the method relies on the correct choice of the reference liquid.

Kwok and Neumann [107] stated that, although the contact angles obtained on powder surfaces are usually not identical to Young's contact angles, the actual measured angles, along with the surface tension of the specific contacting liquid, determine the Laplace pressure ΔP, which indicates the capillary penetration. In other words, it is the actual contact angle that determines the wetting behavior of this system.

Alternatively, the so-called h–ϵ method proposed by Kossen and Heertjes [108, 109] can be used to obtain contact angles on compressed powders. This method is based on the assumption that the powder consists of identical spheres. After the powder is compressed into a cake, it is saturated with the probe liquid, and a drop is placed on the surface. The contact angle is calculated from the height of the drop, the cake porosity, and the density and surface tension of the liquid.

Subrahmanyam and co-workers [110] compared the contact angles of quartz by the capillary penetration method and the captive bubble method. These studies found that the capillary penetration method gave results that showed a better correlation with floatability than the captive bubble method.

1.3.9 Capillary Bridge Method

Restagno et al. [111, 112] developed a high-precision contact angle measuring technique, which they referred to as the "capillary bridge method". In their experiment,

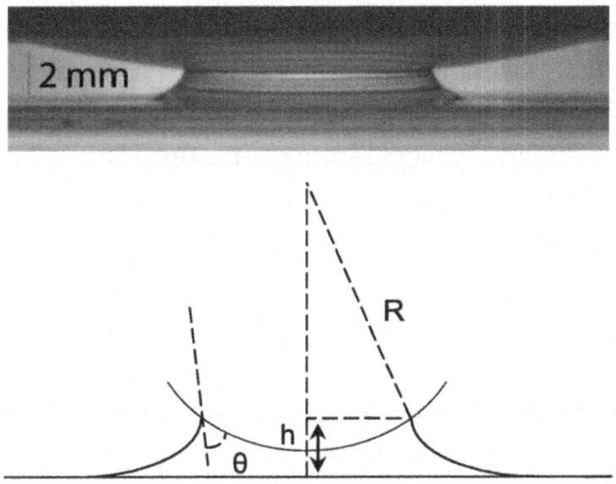

Fig. 1.10 Photograph of a "capillary bridge", illustration of contact angle calculation. Reproduced from Ref. [111]. Copyright 2009 American Chemical Society

a spherical solid surface (usually a watch glass) is put in contact with a large liquid bath. Due to the capillary effect, a meniscus or "capillary bridge" forms around the contact line, which defines the wetted area on the solid surface. The shape of this "capillary bridge" between the solid surface and the liquid changes as the solid is slowly moved up or down to give a systematically varying wetted area. By monitoring the changes of the wetted area and the distance that solid surface moves, the dynamic contact angles can be quantitatively determined through numerical resolution of the Young-Laplace equation or by a simplified approximated relation (Fig. 1.10):

$$A = 2\pi R\left(k^{-1}\sqrt{2(1+\cos\theta)} - h\right) \tag{1.11}$$

where A represents the wetted area, h is the distance of solid surface from liquid bath surface, R is the radius of the sphere surface, and k^{-1} is the capillary length, which is known for a given liquid. The contact angle θ can be deduced from the experimentally determined $A(h)$ curve.

The capillary bridge method offers several advantages. Both advancing and receding contact angles can be established by slowly pulling the surface away from the liquid or pushing back toward the liquid; thus, the dynamic contact angles can be studied. Due to its high sensitivity, the technique can be used to characterize surfaces with low contact angle hysteresis (on the order of 1°). The technique has shown great potential in discriminating low friction surface coatings [113]. However, evaporation of the test liquid can cause uncertainty of the measured distance h, which will induce a small deviation in the experimental $A(h)$ curve [111]. Moreover, to measure the wetted area, the method requires transparent films to be coated on a transparent spherical surface, which can cause serious limitation and inconvenience in sample preparing [114]. This problem might be solved by introducing a new and better means to measure the wetted area.

1.4 Applications of the Technique

One of the most important applications of contact angle measurement is the determination of solid surface tension. The assessment of solid surface tension has been of much fundamental and practical interest for decades [115–120]. However, solid surface tension cannot be directly measured since most existing techniques for surface tension measurement rely on surface deformation, which is difficult in the case of a solid. Several indirect methods have been developed for the assessment of solid surface tension [121], among which contact angle measurement is considered to be the simplest.

Young's equation (1.1) reveals the correlation between contact angle and interfacial tensions. It suggests that the observation of the equilibrium contact angle of liquid on solid might be a starting point for investigating the solid surface tension. However, there are only two measurable quantities in Young's equation: the contact angle θ and the liquid surface tension γ_{lv}. To determine γ_{sv} and γ_{sl}, an additional relation between these quantities must be established.

The pioneering work of interpreting contact angles in terms of solid surface tension was carried out by Zisman and co-workers [122], who conducted numerous studies about contact angles of high-energy liquids on low-energy solid surfaces. They found out that for a given solid, the measured contact angles do not vary randomly upon change of the testing liquid. Instead, the change of $\cos\theta$ versus liquid surface tension γ_{lv} falls into a linear trend for a homologous series of liquids such as alkanes on Teflon. In addition, extrapolation of the linear curve to where $\cos\theta = 1$ gives an important parameter, the critical surface tension γ_c. Theoretically, a liquid with a surface tension equal or less than γ_c would wet the solid surface completely ($\theta = 0°$).

Subsequent to Zisman's work, two main groups of thoughts have developed: the surface tension components theory and the equation of state theory. Numerous methods of determining solid surface tension have been developed following these considerations. For example, Neumann and co-workers [123] measured the contact angles of a large number of liquids on solid surfaces, from which they acquired a smooth curve by plotting γ_{lv} versus $\gamma_{lv}\cos\theta$. The curve moves in a regular way when γ_{sv} is changed, indicating $\gamma_{lv}\cos\theta$ is dependent on γ_{lv} and γ_{sv}, which gives the equation of state:

$$\gamma_{lv}\cos\theta = f(\gamma_{lv}, \gamma_{sv}) \tag{1.12}$$

Combining (1.12) with Young's equation (1.1), we have:

$$\gamma_{sl} = \gamma_{sv} - f(\gamma_{lv}, \gamma_{sv}) = f'(\gamma_{lv}, \gamma_{sv}) \tag{1.13}$$

As mentioned previously in Sect. 1.2.2, many experimentally accessible contact angles are not equal to Young's contact angle, and are therefore meaningless in terms of interpreting surface tension through Young's equation. Kwok and Neumann [121] developed experimental procedures and general criteria for measuring and interpreting meaningful contact angles in terms of Young's equation (1.1). Their work emphasizes that obtaining meaningful contact angles for the determination of solid surface tension depends heavily on how contact angles are measured and whether or not widely made assumptions have been violated. The assumptions are [121] (directly from the original text):

1. All approaches rely on the validity and applicability of Young's equation for surface energetic from experimental contact angles.
2. Pure liquids are always used; surfactant solutions or mixtures of liquids should not be used, since they would introduce complications due to preferential adsorption.
3. The values of γ_{lv}, γ_{sv}, and γ_{sl} are assumed to be constant during the experiment, e.g. there should be no physical/chemical reactions between the solid and the liquid.
4. The surface tensions of the test liquids should be higher than the anticipated solid surface tension.
5. The values of γ_{sv} in going from liquid to liquid are also assumed to be constant, e.g. independent of the liquids used.

In summary, the solid surfaces are required to be rigid, smooth, homogeneous, and inert to the testing liquids. Advancing angles should be used instead of receding angles, so as to minimize the possible swelling and physical/chemical effects. Kwok and Neumann [121] also pointed out that many results from the literature, such as non-rigid gels [124] and non-smooth biological surfaces [125], are open to question due to violations of the assumptions.

1.5 Recent Developments of the Technique

1.5.1 Drop Shape Analysis

Numerous methods have been developed to determine the liquid surface tension and contact angle from the shape of a sessile drop, pendant drop, or captive bubble. Ideally, the shape of a liquid drop depends on the combined effects of interfacial and gravitational forces. Surface tension tends to minimize the surface area by making the drop spherical, while gravity deforms the drop in two ways: (1) by elongating a pendant drop and/or (2) flattening a sessile drop. This balance between surface tension and external forces (such as gravity) is reflected mathematically in the Laplace equation of capillarity, which offers the possibility of determining surface tension by analyzing the drop shape.

During the early years of contact angle measurement, the $\theta/2$ method was once widely used to analyze the profile of a sessile drop. In this analysis, the liquid drop is assumed to be part of a sphere. Geometrically, the contact angle can be calculated by measuring the drop diameter and the height of the apex (Fig. 1.11):

$$\frac{\theta}{2} = \tan^{-1}\left(\frac{h}{d}\right) \tag{1.14}$$

This method yields reasonable results when the liquid drop is extremely small. However, the spherical shape assumption cannot be applied if the drop shape is large enough to be affected by gravity.

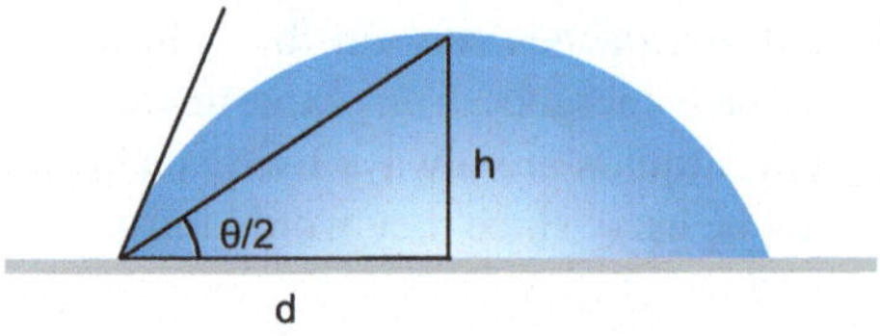

Fig. 1.11 Demonstration of the $\theta/2$ method according to (1.14)

Bashforth and Adams [126] were the first to use the Laplace equation to analyze the shape of droplet profiles. They manually generated a collection of sessile drop profiles according to different values of surface tension and the radius of curvature at the drop apex. Consequently, the task of determining surface tension became simple interpolation from their tables. Their tremendous contribution led to booming development in the area. Blaisdell [127], as well as Tawde and Parvatikar [128] extended the Bashforth and Adams tables. Fordham [129] and Mills [130] generated equivalent tables for pendant drops. Ever since digital computers became popular, drop shape analysis has been greatly improved, and many new methods have been developed [131–141].

1.5.2 Axisymmetric Drop Shape Analysis (ADSA)

During the past three decades, significant improvements were made in hardware design and computational technology, which has led to considerable development of drop shape analysis for surface science. The axisymmetric drop shape analysis (ADSA) method that was developed by Rotenberg et al. [142], improved by Spelt et al. [143], Cheng at al. [144], Río [145], and Kalantarian et al. [146] is believed to be one of the most accurate techniques for high precision contact angle measurement. With a reproducibility of $\pm 0.2°$, the ADSA method has been shown to improve the accuracy of contact angle measurement by essentially an order of magnitude, compared to the reproducibility of $\pm 2°$ by direct tangent measurements.

The basic principle of the ADSA method is to find the best theoretical profile that matches the drop profile extracted from an experimental image, from which the surface tension, contact angle, drop volume, and surface area can be computed. There are two main assumptions in the ADSA method: (1) the experimental drop is Laplacian and axisymmetric, and (2) gravity is the only external force. Surface tension is used as an adjustable parameter, and the algorithm searches for the specific value of surface tension that produces the best theoretical profile that fits the experimental drop profile. The first generation of the axisymmetric drop shape analysis-profile (ADSA-P) method was developed by Rotenberg et al. [142]. In this method, an objective function is defined as the sum of the squares of the normal distances between experimental profile points and corresponding theoretical profile points. The matching is achieved by minimizing the objective. It should be noted that extracting drop interface coordinates was performed manually in the first generation

ADSA. Cheng et al. [144] improved the first generation ADSA-P method by implementing a computer-based edge operator, Sobel [147], to extract the drop interface profiles automatically. Optical distortion correction techniques were also incorporated into the ADSA-P program to achieve better accuracy. The first generation ADSA-P method was found to give accurate results except for large and flat sessile drops, where the program failed due to the flatness of the apex.

Rio and Neumann [145] developed the second-generation ADSA-P method by integrating more efficient algorithms. They also used the curvature at the apex instead of the radius of curvature at the apex as a parameter, so as to overcome the apex limitation in the first generation ADSA-P. The vertical location of the solid surface was determined to pixel resolution from the digitized drop image.

Another branch of the ADSA method—the axisymmetric drop shape analysis-diameter (ADSA-D)—was developed to measure extremely low contact angles (less than 20°) or contact angles on non-ideal surfaces [148, 149]. Quite different from the ADSA-P method, the ADSA-D program analyzes a top view image of the drop and measures the contact diameter. Once the contact diameter, liquid surface tension, and volume of the drop are known, the contact angle can be determined from numerical integration of the Laplace equation of capillarity. It has been claimed that the contact angle results obtained by the ADSA-D and ADSA-P methods closely agree, varying no more than $\pm 0.4°$, for which ADSA-D gives a higher precision for low contact angles [148, 149]. However, Rodriguez-Valverde and co-workers [150] reported that quantification of the effects of surface roughness in contact angle measurements is unwieldy and controversial. They asserted that the equilibrium contact angle cannot be measured due to the existence of metastable states and a strong dependence on drop size.

A new drop shape analysis method called theoretical image fitting analysis (TIFA) was introduced by Cabezas et al. [151, 152]. The main difference between TIFA and ADSA is that TIFA fits the whole two-dimensional theoretical projection to the experimental image, instead of ADSA fitting only a one-dimensional theoretical profile curve to the experimental image. TIFA operates without using edge detection algorithms. Further, it employs an error function that measures pixel-by-pixel differences between the theoretical and experimental images, which are then minimized so as to fit theoretical image to the experimental image.

Both ADSA and TIFA suffer from a common limitation: the necessity of using the apex of the drop as part of the drop image analysis. For this reason, the most widely used system of a liquid drop with the needle immersed at the apex cannot be applied. Instead, drops must be formed from below the solid surface through a hole. This complication has led to the development of new versions of both methods: the TIFA-AI (theoretical image fitting analysis for axisymmetric interfaces) method [153] developed by Cabezas et al., and the ADSA-NA (axisymmetric drop shape analysis-no apex) method developed by Kalantarian et al. [146]. With TIFA-AI, the geometry of the interface at a reference level different from the apex is used to solve the Laplace equation. The radius of the profile and its inclination at the reference level are defined as two additional optimization parameters. ADSA-NA is also able to analyze the shape of axisymmetric interfaces without using apex coordinates. A recent study shows that the contact angle data obtained using these two new methods agree within $\pm 0.1°$ [146].

1.5.3 Contact Angle Measurement of Ultrasmall Droplets

Wetting phenomena have been extensively investigated and well understood at the macroscale (millimeters); while the wetting behavior at the micro- or nanoscale (micro- or nanometers) has yet to be thoroughly studied, and many issues remain unresolved. The study of ultrasmall droplets on solid surfaces allows wetting theories to be tested down to the nanoscale, where the wetting behavior is significantly influenced by line tension and liquid evaporation (both are usually negligible in macroscale studies). These studies have relevance to many industrial applications, such as friction in microelectromechanical systems (MEMS) devices [154], flotation in mineral recovery [155], and wastewater treatment [156].

At the micro- or nanoscale, it is possible to take advantage of the fact that the influence of gravity is negligible compared to the influence of surface tension; consequently, the droplet can be approximated as part of a sphere [157]. There are, however, complications due to the small size. As the drop becomes smaller, the effect of the line tension at the three-phase contact line becomes larger compared to the effect of surface tension [158, 159]. For micro- or even nanometer-sized droplets, the line tension can significantly affect the measured contact angle [160]. Thus, at micro- and nanoscales, the classical Young's equation must be modified, which is usually done by adding a second term to take the line tension effect into account [159–161]:

$$\cos\theta_\infty = \frac{\gamma_{sv} - \gamma_{sl}}{\gamma_{lv}} \tag{1.15}$$

$$\cos\theta_R = \cos\theta_\infty - \frac{\sigma\gamma_{lv}}{R} \tag{1.16}$$

In these equations, θ is the contact angle, and the subscripts, R and ∞ indicate the radius of the droplet. The quantity γ is the surface tension, σ is the line tension of the three-phase system, and R is the radius of the drop at the surface. The subscripts s, l, and v indicate the solid, liquid, and vapor phases, respectively.

Another problem associated with small droplets is the effect of evaporation, which is especially problematic when working with atomic force microscopy (AFM), because AFM requires at least 15–30 min to produce a stable image. Picknett and Bexon [162] studied the changes of droplet profile during evaporation. They found that there exist two stages: a first "constant contact area" phase dominates until the contact angle decreases to certain value, at which point a second "constant contact angle" phase dominates thereafter. This model has been supported by Soolaman and Yu [163], who reported that the evaporation of water microdroplets progresses from the pinning stage (decreasing contact angle, constant contact area)

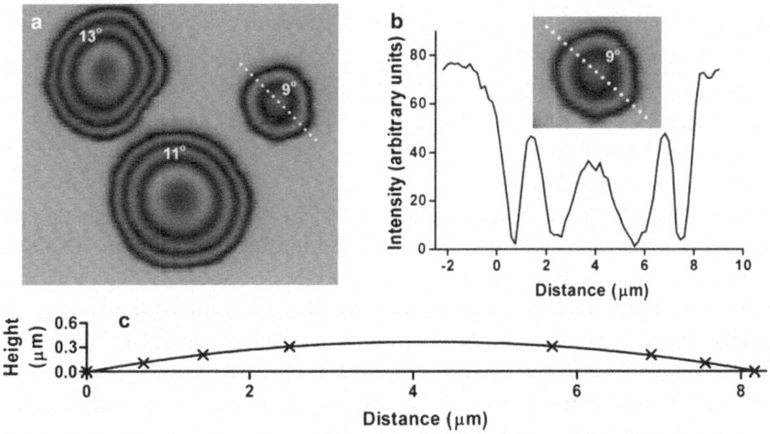

Fig. 1.12 Interference method for measuring microscale contact angles. (**a**) Interference fringes in water droplets on ethanol-cleaned glass (image size 129 × 115 μm^2); the contact angles indicated for the droplets. (**b**) Intensity profile along the *dotted line* in (**a**) and *inset* that shows a magnification of the droplet analyzed (image size 44 × 38 μm^2). (**c**) Drop profile with circular curve fitted to the data from the *dashed line* in (**a**). Reproduced from Ref. [174]. Copyright 2007 Elsevier Science

to the shrinking stage (constant contact angle, decreasing contact area). Therefore, the evaporation effect must be taken into consideration when interpreting micro- and nanodroplets profiles.

Several methods for measuring the contact angles of ultrasmall droplets have been reported thanks to the availability of advanced imaging techniques such as interference microscopy, confocal microscopy, environmental scanning electron microscopy (ESEM), and AFM to establish the droplet profiles [164–171]. Interference microscopy method utilizes the fringe patterns formed by the interfering beams reflected from the solid-liquid and the liquid-vapor interfaces to calculate the contact angle value. It is best suited for droplets with low contact angles (less than 30°), in which case individual interference fringes are readily resolved. Jameson and Del Cerro [172] used this technique for cases in which the underlying phase is transparent. Fischer and Ovryn [173] improved the technique by developing a geometrical-optics-based model to interpret the optical path length through a thick refractive specimen.

Confocal microscopy, on the other hand, is best suited for droplets with higher contact angles (30–90°) since such droplets have greater heights, thus allowing the collection of a larger number of image slices from which the drop profile can be reconstructed. Sundberg et al. [174] recently used a combination of interference microscopy and confocal fluorescence microscopy to study the contact angles of micrometer-sized nucleated water droplets (Fig. 1.12). The method was designed for measuring both low ($< 30°$) and high (30–90°) contact angles. For droplets displaying low contact angles, the profiles can be reconstructed from the interference patterns, whereas for contact angles too high to allow resolution of interference fringes, a low concentration of fluorescent dye was mixed with water to allow the recording and analysis of 3D image stacks. The results were compared with contact angles of macroscopic (millimeter scale) droplets, and a dependence of contact angle on droplet sizes was found. At small radii (below 10 μm), the contact angles obtained are larger than the macroscopic contact angles, indicating a positive line tension effect. When the radii of the droplets are more than 10 μm, the average contact angles on microscopic droplets are largely the same as their macroscopic counterparts. In general, interference and confocal methods can measure droplets with diameters of about 10 to 100 μm. New techniques with higher resolution are needed for nanoscale measurements.

A number of methods have been developed to deposit micro- and nanoscale liquid droplets on surfaces. The simplest way is to use an air sprayer [175], which generates ultrasmall droplets by mixing macroscale droplets with a jet of compressed air. Commercially available atomizers are able to spray ultrafine droplets with a wide range of sizes. Hong et al. [176] used micropipettes with submicron orifices to create submicron-sized droplets. Jayasinghe and Edirisinghe [177] used electrospraying to generate charged liquid droplets with diameters less than 1 μm. Similar methods have also been used by Paine et al. [178] and by Adhikari and Fernando [179]. Fine-emulsions and nano-emulsions formed by two immiscible phases can generate significantly smaller droplets (between 100 and 1000 nm) [180]. Zhang and Ducker [181] successfully observed ultrasmall interfacial oil nanodroplets of decane (height 2–50 nm) formed at the solid/liquid interface by direct adsorption from an emulsion. A syringe pump has also been used to create microdroplets [182]. However, with all of the aforementioned methods, control over the size of the droplet presents a challenging problem.

To address this problem, Meister et al. developed a nanoscale dispensing (NADIS) technique by modifying a commercially available silicon nitride (Si_3N_4) AFM probe tip and using it to transfer liquid from the tip to the surface by direct contact [183, 184]. Here, the size of the deposited droplet is controlled by the aperture width of the hollow AFM tip. The NADIS technique is able to create droplets with controlled sizes to measure the contact angle at micro- or nanoscales [185]. Jung and Bhushan [186] have used the NADIS technique to deposit micro- and nanodroplets of a glycerol/water mixture on different surfaces, followed by measuring the contact diameter, thickness, and volume of the droplet with an AFM to determine the contact angles. Fang et al. [187] have shown that tips with an aperture diameter of 35 nm were able to deposit nanodroplets of glycerol-based liquids with diameters down to 70 nm and to form regular arrays on silica surfaces with different hydrophilicities. Importantly, fine control of the droplet volume is also possible.

As mentioned previously, the capillary penetration technique based on Washburn's theory is the most frequently used technique for determining contact angle of particles. However, there exists an inherent limitation—it only gives an averaged contact angle value and thus cannot describe the wetting behavior of individual powders. In recent years, AFM has been used to study the contact angle of individual particles by measuring the interaction forces between a spherical colloidal particle and a bubble in aqueous solution [188–192]. The contact region of the AFM force curve is used to establish the position of zero force, which gives the depth of the particle penetration into the bubble. The data can then be used to calculate the contact angle. In this way, the intrinsic hydrophobic properties of an individual particle can be revealed. It is also worth mentioning that soft surfaces can sometimes deform during AFM force measurements [193–195]. Furthermore, Nguyen and co-workers [196] observed that the contact angle measured by AFM changes with the speed of the piezoelectric translator, showing that the measurements were generally dynamic. However, it cannot be decided whether or not the measured contact angle on a single particle indicates the hydrophobicity of the solid particle surface, since the position of the three-phase contact area relative to the particle is unknown. In the case of pinning of the three-phase contact, the determined contact angle does not entirely characterize the hydrophobic properties of the particle surface. Notably, work by Jung and Bhushan [186] using the AFM-based NADIS technique to measure the contact diameter, thickness, and droplet volume gave contact angle values that were lower than those of corresponding macrodroplets.

Checco et al. [197] measured the contact angles of micro- and nanoscale alkane droplets on a model substrate using non-contact AFM. They connected an enclosed chamber to the AFM head in order to mitigate evaporation. The large range of droplet sizes accessible allowed them to determine the contact line curvature dependence of the contact angles. The contact angle values were found to decrease when the droplet size decreased from macroscopic scales. Furthermore, they found that the modified Young's equation fails to describe correctly the experimental data at a sufficiently large range of droplet sizes, and used surface heterogeneity to explain this shortcoming, which was further supported by numerical simulations. More recently, Checco et al. produced a nanoscale wettability pattern on a surface consisting of hydrophilic COOH-terminated and hydrophobic CH_3-terminated lines (Fig. 1.13) [198]. The use of an evaporation-condensation chamber allows control of both the temperature and the vapor concentration. By measuring the contact angle of ethanol nano-channels formed on the surface, they found that ethanol selectively wetted the hydrophilic lines with an apparent contact angle at the stripe's boundary of $12^\circ \pm 1^\circ$, a value significantly lower than that measured macroscopically for an ethanol droplet on the same surface ($30^\circ \pm 1^\circ$).

Dupres et al. also used AFM to study the local wetting of human hair [199]. For nonvolatile liquids, small droplets were directly deposited on the solid by using a microsyringe; volatile liquids were deposited in a condensation chamber. Most of the liquids tested have a higher surface tension γ_{lv} than the critical surface tension γ_c; consequently, partial wetting is expected. However no spreading or drop nucleation was observed. Interestingly, more polar liquids such as water and glycerol failed to wet any region of the cuticle, although water seemed to penetrate into the bulk of the hair fiber. Hydrocarbon liquids preferentially wetted the more polar cuticle edges. In a more recent report, Dupres and co-workers studied the changes in wettability of the cuticle when different covalently and non-covalently bound fatty acids present at the cuticle surface were selectively extracted [200].

Wang and co-workers [201, 202] used AFM to image sprayed water microdroplets on polished metal surfaces. They found large differences between

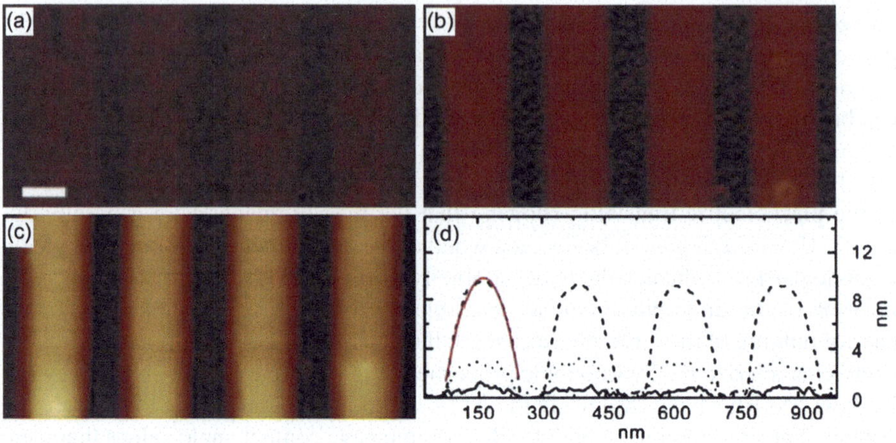

Fig. 1.13 Condensation of ethanol onto COOH nanostripes for ΔT equal to: (**a**) 10 °C, (**b**) 0.5 °C, and (**c**) −15 °C (lateral scale 100 nm). (**d**) Cross-sectional profiles corresponding to (**a**) (*solid line*), (**b**) (*dotted line*), and (**c**) (*dashed line*). Reproduced from Ref. [198]. Copyright 2006 Elsevier Science

macroscale and microscale wettability, even on the same surface. While the contact angles of macroscale water droplets were found to be larger than those of microscale water droplets, the scatter in the values of the microscale contact angles was larger than those of the macroscale contact angles. Accordingly, the shapes of the droplets could not be maintained when the droplets become too small. The influence of organic contaminants might contribute to this phenomenon. Organic contaminants in general increase surface hydrophobicity, and their distribution might be non-uniform at the micro- and nanoscale. When droplets are obtained by condensation, there should be areas with little or no organic contamination, where water preferentially condensed, leading to reduced nanoscale contact angles. This rationalization is consistent with the previously mentioned results by Checco et al. [197], who also suggested the possibility that small surface heterogeneities were responsible for the observed decrease in contact angle when the droplet size was decreased from 2 mm to 200 nm. ESEM has also been used as an imaging tool in the studies of microwetting. Stelmashenko et al. [203] performed fundamental studies on the imaging of water droplets nucleated onto different standard surfaces. Lau et al. [204] used ESEM to verify their creation of a stable, superhydrophobic surface using the nanoscale roughness inherent in a vertically aligned carbon nanotube forest together with a thin, conformal hydrophobic poly(tetrafluoroethylene) (PTFE) coating on the surface of the nanotubes. Spherical water microdroplets formed by condensing vapor onto the surface can be suspended on top of the nanotube forest. The results confirmed that the superhydrophobic characteristic is retained at the microscale. In a related superhydrophobicity study, Cheng and Rodak [205] reported that lotus leaves lost their superhydrophobic character ("lotus effect") when microscale water droplets were condensed on the leaf.

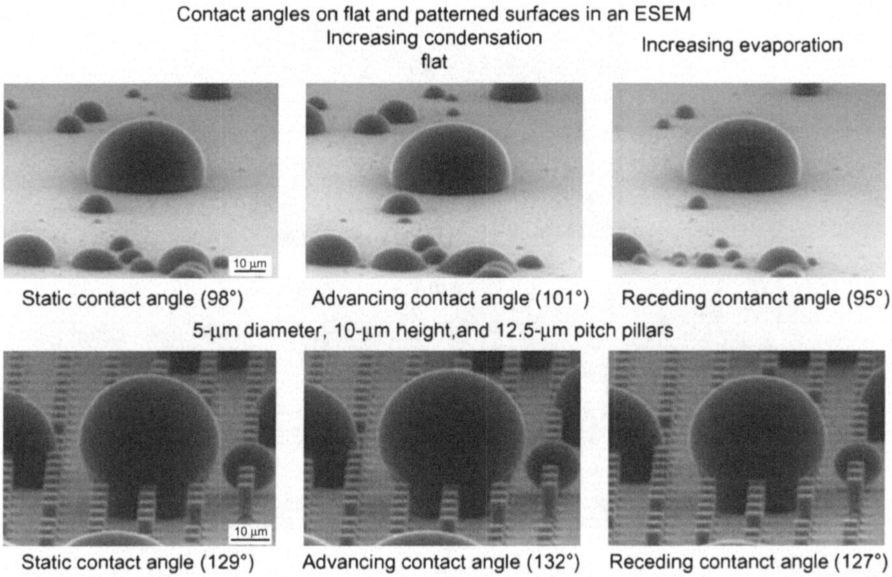

Fig. 1.14 ESEM images of static, advancing, and receding contact angles of microscopic droplets of water on flat and patterned Si surfaces. Reproduced from Ref. [206]. Copyright 2008 The Royal Microscopical Society

Recently, Jung and Bhushan [206] reported an ESEM study of the dynamic wetting of superhydrophobic surfaces. Static contact angles were measured once a dynamic equilibrium between condensation and evaporation was achieved. Advancing contact angles were obtained by cooling the substrate so as to increase the condensation and grow the drop size. Conversely, receding angles were achieved by heating the substrate. Thus, the hysteresis was measured, which showed no obvious difference compared to macroscopic droplets. The researchers found, however, that the hysteresis depended on the geometric characteristics of the patterned surface (Fig. 1.14).

ESEM has specific advantages for characterizing micro- and nanofibers, as fibers have two dimensions at the micro/nanoscale and one dimension at the macroscale. Conventionally, the Wilhelmy balance technique was used to evaluate the wetting properties of fibers. A major limitation of this technique lies in the difficulty of precise measurement of the fiber perimeters, especially in the case of micro- and nanofibers. The use of ESEM can provide additional information at the sub-single-fiber level (Fig. 1.15) [207, 209].

Wettability studies at the micro- and nanoscale have paved the way for the development of wetting patterns. For example, it is possible to deposit droplets onto surfaces to which a chemical or topographical micro- or nanoscale pattern is imposed [208–216]. Wetting patterns, especially those obtained by patterning organic molecules, are currently of great interest in the area of molecular-scale devices [217, 218]. Figure 1.16 shows work done by Seemann et al. [216], who created

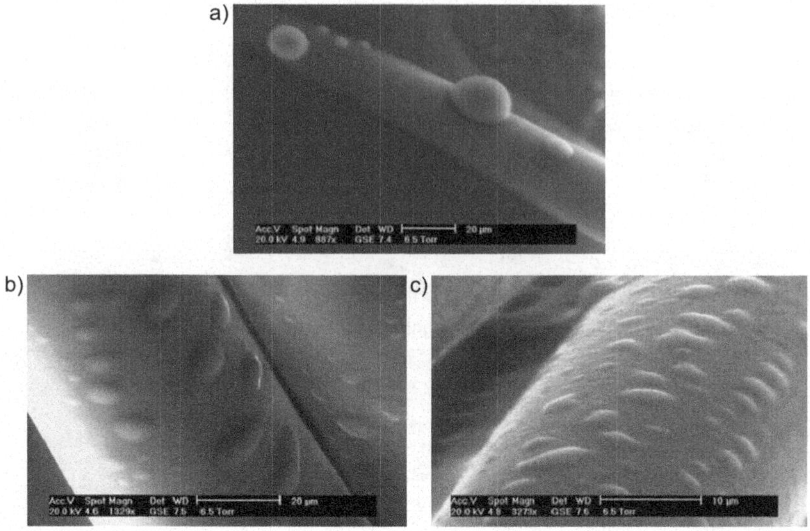

Fig. 1.15 ESEM images of water microdroplets on polypropylene (PP) fibers: (**a**) untreated; (**b**) oxygen-plasma treated for 30 s; (**c**) oxygen-plasma treated for 60 s. Reproduced from Ref. [209]. Copyright 2004 Elsevier Science

grooves with rectangular cross sections in Si by photolithographic methods with depths between 100 and 900 nm, and widths between 400 nm and 3 mm. These studies introduced a global morphology diagram of clustering droplet geometries that depends on the aspect ratio of the groove geometry and the contact angle of the substrate. To sum up, these studies on wetting patterns have revealed the occurrence of morphological wetting and dewetting transitions, which has promising applications in electrowetting, microfluidics, and nanodevice development.

1.6 Summary

In this report, the basic and advanced techniques used to characterize the wettability of materials at the macro-, micro-, and nanoscale have been described. As the importance of wettability in both fundamental and applied fields has long been recognized, innovative techniques for measuring the wettability of surfaces, powders, and fibers continue to emerge. Nevertheless, there are still many open questions regarding wetting mechanisms and for manipulating the wetting behavior at the micro- and nanoscales. AFM and ESEM are currently the most relevant techniques for imaging ultrasmall droplets on surfaces. AFM provides high resolution at the nanometer scale, but is limited by long measurement times. For example, measurements by AFM routinely require 15–30 min to acquire a stable image, which can be a great disadvantage because some nanodroplets evaporate in seconds. ESEM, on the other hand, collects images rapidly (in seconds), which circumvents evaporative

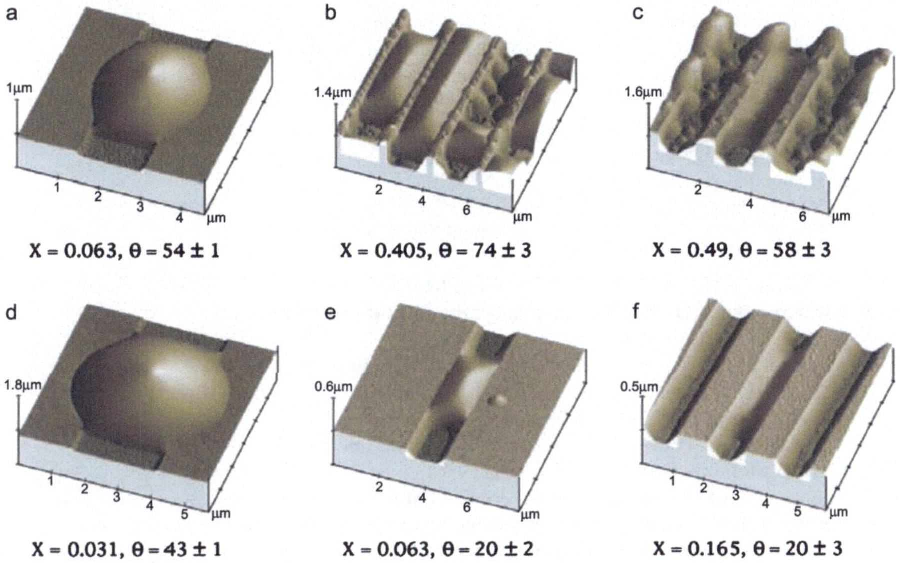

Fig. 1.16 AFM images of liquid structures in grooves with rectangular cross section. Reproduced from Ref. [216]. Copyright 2005 National Academy of Science, USA

problems and allows the collection of dynamic contact angles. However, the resolution of ESEM is only at the micrometer scale. The use of a field emission gun (FEG) in ESEM is expected to diminish beam effects and enhance the resolution to the nanometer range. As for AFM, there are current developments in both the electronic and the data acquisition systems, as well as in the design of AFM probes optimized for high-speed scanning (especially small cantilevers) [219]. Hopefully, we will witness these significant advances applied to future wettability studies and practices.

Acknowledgements We acknowledge generous financial support from the National Science Foundation (DMR-0906727), the Robert A. Welch Foundation (Grant No. E-1320), and the Texas Center for Superconductivity at the University of Houston.

References

1. K.N. Prabhu, P. Fernandes, G. Kumar, Mater. Des. **2**, 297 (2009)
2. X. Zhao, M.J. Blunta, J.J. Yao, Pet. Sci. Technol. Eng. **71**, 169 (2010)
3. Y.Q. Wang, H.F. Yang, Q.G. Hang, L. Fang, S.R. Ge, Adv. Mater. Res. **154–155**, 1019 (2010)
4. M. Sakai, T. Yanagisawa, A. Nakajima, Y. Kameshima, K. Okada, Langmuir **25**, 13 (2009)
5. Y. Son, C. Kim, D.H. Yang, D.J. Ahn, Langmuir **24**, 2900 (2008)
6. J. Perelaer, C.E. Hendriks, A.W.M. de Laat, U.S. Schubert, Nanotechnology **20**, 165303 (2009)
7. H. Yang, P. Jiang, Langmuir **26**, 12598 (2010)
8. B. Bhushan, Y.C. Jung, K. Koch, Langmuir **25**, 3240 (2009)

9. M. Rauscher, S. Dietrich, Annu. Rev. Mater. Sci. **38**, 143 (2008)
10. K. Fei, C.P. Chiu, C.W. Hong, Microfluid. Nanofluid. **4**, 321 (2008)
11. Z. Keshavarz-Motamed, L. Kadem, A. Dolatabadi, Microfluid. Nanofluid. **8**, 47 (2010)
12. Y.S. Nanayakkara, S. Perera, S. Bindiganavale, E. Wanigasekara, H. Moon, D.W. Armstrong, Anal. Chem. **82**, 3146 (2010)
13. A. Lafuma, D. Quere, Nat. Mater. **2**, 457 (2003)
14. J.H. Snoeijer, B. Andreotti, Phys. Fluids **20**, 057101 (2008)
15. T. Young, Philos. Trans. R. Soc. Lond. **95**, 65 (1805)
16. R.E. Johnson Jr., R.H.J. Dettre, Phys. Chem. **68**, 1744 (1964)
17. L.W. Schwartz, S. Garoff, Langmuir **1**, 219 (1985)
18. J.F. Joanny, P.G.J. de Gennes, Chem. Phys. **81**, 552 (1984)
19. J.W. Krumpfer, T.J. McCarthy, Faraday Discuss. **146**, 103 (2010)
20. L. Gao, T.J. McCarthy, Langmuir **22**, 6234 (2006)
21. A.W. Neumann, Adv. Colloid Interface Sci. **4**, 105 (1974)
22. D. Li, A.W. Neumann, in *Applied Surface Thermodynamics*, ed. by J.K. Spelt, A.W. Neumann (Dekker, New York, 1996), p. 109
23. R.V. Sedev, J.G. Petrov, A.W. Neumann, J. Colloid Interface Sci. **180**, 36 (1996)
24. R.N. Wenzel, Ind. Eng. Chem. **28**, 988 (1936)
25. S. Baxer, A.B.D.J. Cassie, Tex. Inst. **36**, 67 (1945)
26. J.K. Spelt, D.R. Absolom, A.W. Neumann, Langmuir **2**, 620 (1986)
27. B.B. Davidson, G.J. Lei, Polym. Sci. **9**, 569 (1971)
28. T.R. Krishman, I. Abraham, E.I. Vargha-Butler, Int. J. Pharm. **80**, 277 (1992)
29. C.D. Bain, E.B. Troughton, Y. Tao, J. Eval, G.M. Whitesides, R.O. Nuzzo, J. Am. Chem. Soc. **111**, 321 (1989)
30. C.D. Bain, G.M. Whitesides, Angew. Chem. Int. Ed. **28**, 506 (1989)
31. E.F. Hare, E.G. Shafrin, W.A.J. Zisman, Phys. Chem. **58**, 236 (1953)
32. C.J. Budziak, E.I. Vargha-Butler, A.W.J. Neumann, Appl. Polym. Sci. **42**, 1991 (1959)
33. A. Dilks, B. Kay, Macromol. **14**, 855 (1981)
34. H.J. Yasuda, Polym. Sci. **16**, 199 (1981)
35. E.I. Vargha-Butler, M. Kashi, H.A. Hamza, A.W. Neumann, Coal Prep. **3**, 53 (1986)
36. H.J. Busseher, A.W.J. van Pelt, H.P. De Jong, J. Arends, J. Colloid Interface Sci. **95**, 23 (1983)
37. W.C. Bigelow, D.L. Pickett, W.A.J. Zisman, Coll. Sci. **1**, 513 (1946)
38. J. Leja, G.W. Poling, On the interpretation of contact angle, in *Proceedings of the 5th Mineral Processing Congress* (IMM, London, 1960), p. 325
39. R.W. Smithwich, J. Colloid Interface Sci. **123**, 482 (1988)
40. D.Y. Kwok, R. Lin, M. Mui, A.W. Neumann, Colloids Surf. A **116**, 63 (1996)
41. S. Brandon, N. Haimovich, E. Yeger, A. Marmur, J. Colloid Interface Sci. **263**, 237 (2003)
42. P. Letellier, A. Mayaffre, M. Turmine, J. Colloid Interface Sci. **314**, 604 (2007)
43. A.W. Neumann, R.J. Good, in *Surface and Colloid Science: Experimental Methods*, vol. 11, ed. by R.J. Good, R.R. Stromberg (Plenum Publishing, New York, 1979), p. 31
44. R.J. Hunter, *Foundations of Colloid Science*, 2nd edn. (Clarendon Press, Oxford, 2001)
45. G. MacDougall, C. Ockrent, Proc. R. Soc. **180A**, 151 (1942)
46. C.W. Extrand, Y. Kumagai, J. Colloid Interface Sci. **184**, 191 (1996)
47. C.W. Extrand, Y. Kumagai, J. Colloid Interface Sci. **191**, 378 (1997)
48. E. Pierce, F.J. Carmona, A. Amirfazli, Colloids Surf. A **323**, 73 (2008)
49. B. Krasovitski, A. Marmur, Langmuir **21**, 3881 (2005)
50. W.A. Zisman, Adv. Chem. Ser. **87**, 1 (1968)
51. M.C. Phillips, A.C. Riddiford, J. Colloid Interface Sci. **41**, 77 (1972)
52. D.E. McIntyre, A study of dynamic wettability on a hydrophobic surface, Ph.D. Thesis, The Institute of Paper Chemistry, Appleton, WI, 1969
53. W.J. Fenrick, Rev. Sci. Instrum. **35**, 1616 (1964)
54. L.R. Fisher, J. Colloid Interface Sci. **72**, 200 (1979)
55. I. Langmuir, V.J. Schaeffer, J. Am. Chem. Soc. **59**, 2400 (1937)

56. T. Fort, H.T. Patterson, J. Colloid Interface Sci. **29**, 185 (1967)
57. K.A. Johnson, D.O. Shah, J. Colloid Interface Sci. **107**, 1 (1985)
58. A.F. Taggart, T.C. Taylor, C.R. Ince, Trans. Am. Inst. Min. Metall. Pet. Eng. **87**, 285 (1930)
59. W. Zhang, M. Wahlgren, B. Sivik, Desalination **72**, 263 (1989)
60. N.K. Adam, G.J. Jessop, Chem. Soc. **127**, 1863 (1925)
61. F.M. Fowkes, W.D. Harkins, J. Am. Chem. Soc. **62**, 337 (1940)
62. A.L. Spreece, C.P. Rutkowski, G.L. Gaines Jr., Rev. Sci. Instrum. **28**, 636 (1957)
63. G.T. Smedley, D.E. Coles, J. Colloid Interface Sci. **286**, 310 (2005)
64. B.A. Bezuglyi, O.A. Tarasov, A.A. Fedorets, Colloid J. **63**, 668 (2001)
65. L.A. Wilhelmy, Ann. Phys. **119**, 117 (1863)
66. H.M. Princen, J. Colloid Interface Sci. **30**, 69 (1969)
67. J.C. Cain, D.W. Francis, R.D. Venter, A.W. Neumann, J. Colloid Interface Sci. **94**, 123 (1983)
68. M. Shimokawa, T. Takamura, J. Electroanal. Chem. Interfacial Electrochem. **41**, 359 (1973)
69. A.W. Neumann, Z. Phys. Chem. **41**, 339 (1964)
70. C.J. Budziak, E.I. Vargha-Butler, A.W. Neumann, J. Appl. Polym. Sci. **42**, 1959 (1991)
71. C.J. Budziak, A.W. Neumann, Colloids Surf. **43**, 279 (1990)
72. D.Y. Kwok, C.J. Budziak, A.W. Neumann, J. Colloid Interface Sci. **173**, 143 (1995)
73. D.D. Jordan, J.E. Lane, Austral. J. Chem. **17**, 7 (1964)
74. A.W. Neumann, W. Tanner, Tenside **4**, 220 (1967)
75. J. Kloubek, A.W. Neumann, Tenside **6**, 4 (1969)
76. A.M. Schwartz, F.W. Minor, J. Colloid Sci. **14**, 572 (1959)
77. A.M. Schwartz, C.A. Rader, Congr. Surface Activity Brussels **2**, 383 (1964)
78. W.D. Bascom, J.B. Romans, Ind. Eng. Chem. Prod. Res. Dev. **7**, 172 (1968)
79. R.J. Roe, J. Colloid Interface Sci. **50**, 70 (1975)
80. H.M. Princen, in *Surface and Colloid Science*, vol. 2, ed. by E. Matijevic (Wiley, New York, 1969), p. 1
81. A.V. Rapacchietta, A.W. Neumann, S.N. Omenyi, J. Colloid Interface Sci. **59**, 541 (1977)
82. A.V. Rapacchietta, A.W. Neumann, J. Colloid Interface Sci. **59**, 555 (1977)
83. R.J. Good, J.K. Pascheck, in *Wetting, Spreading and Adhesion*, ed. by F. Paddy (Academic Press, New York, 1978), p. 147
84. G. Zografi, S.S. Tam, J. Pharm. Sci. **65**, 1145 (1976)
85. C.F. Lerk, M. Lagas, J.P. Boelstra, P. Broersma, J. Pharm. Sci. **66**, 1480 (1977)
86. W.C. Liao, J.L. Zatz, J. Pharm. Sci. **68**, 488 (1979)
87. V. Steiner, G. Adam, Cell Biophys. **6**, 279 (1984)
88. E.I. Vargha-Butler, S.J. Sveinsson, Z. Policova, Colloids Surf. **58**, 271 (1991)
89. A.W. Neumann, D. Renzow, H. Reumuth, I.E. Richter, Colloid Polym. Sci. **55**, 49 (1991)
90. R. Shuttleworth, G.L.J. Bailey, Discuss. Faraday Soc. **3**, 16 (1948)
91. G. Buckton, J.M. Newton, Powder Technol. **46**, 201 (1986)
92. D.T. Hansford, D.J.W. Grant, J.M. Newton, Powder Technol. **26**, 119 (1980)
93. G. Buckton, J.M. Newton, J. Pharm Pharmacol. **37**, 605 (1985)
94. A. Stamm, D. Gissinger, C. Boymond, Drug Dev. Ind. Pharm. **10**, 381 (1984)
95. E.W. Washburn, Phys. Rev. **17**, 374 (1921)
96. S. Levine, J. Lowndes, E.J. Watson, G. Neale, J. Colloid Interface Sci. **73**, 136 (1980)
97. R.J. Good, N.J. Lin, J. Colloid Interface Sci. **54**, 52 (1976)
98. L.R. Fisher, P.D. Lark, J. Colloid Interface Sci. **69**, 486 (1979)
99. F.E. Bartell, H.J. Osterhof, Ind. Eng. Chem. **19**, 1277 (1927)
100. F.E. Bartell, C.E. Whitney, J. Phys. Chem. **36**, 3115 (1932)
101. F.E. Bartell, C.W. Walton, J. Phys. Chem. **38**, 503 (1934)
102. L.R. White, J. Colloid Interface Sci. **90**, 536 (1982)
103. D. Dunstan, L.R. White, J. Colloid Interface Sci. **111**, 60 (1986)
104. D. Diggins, J. Ralston, Coal Prep. **13**, 1 (1993)
105. D. Diggins, L.G.J. Fokkink, J. Ralston, Colloids Surf. **44**, 299 (1990)
106. C.A. Prestidge, J. Ralston, J. Colloid Interface Sci. **172**, 302 (1995)

107. D.Y. Kwok, A.W. Neumann, in *Surface Characterization Methods: Principles, Techniques and Applications*, ed. by A.J. Milling (Marcel Dekker, New York, 1999), p. 37
108. N.W.F. Kossen, P.M. Heertjes, Chem. Eng. Sci. **20**, 593 (1965)
109. P.M. Heertjes, N.W.F. Kossen, Powder Technol. **1**, 33 (1967)
110. T.V. Subrahmanyam, M.B.M. Monte, A. Middea, E. Valdiviezo, F.F. Lins, Miner. Eng. **12**, 1347 (1999)
111. F. Restagno, C. Poulard, C. Cohen, L. Vagharchakian, L. Léger, Langmuir **25**, 11188 (2009)
112. L. Vagharchakian, F. Restagno, L. Léger, J. Phys. Chem. B **113**, 3769 (2009)
113. C. Cohen, F. Restagno, C. Poulard, L. Léger, Langmuir **26**, 15345 (2010)
114. R. Rioboo, M. Voué, H. Adão, J. Conti, A. Vaillant, D. Seveno, J. De Coninck, Langmuir **26**, 4873 (2010)
115. R.E. Johnson Jr., R.H. Dettre, in *Surface and Colloid Science*, vol. 2, ed. by E. Matijevic (Wiley-Interscience, Weinheim, 1969), p. 85
116. A.W. Neumann, in *Wetting, Spreading and Adhesion*, ed. by J.F. Padday (Academic Press, London, 1978), p. 3
117. A.W. Neumann, R.J. Good, *Surface and Colloid Science*, vol. 11, ed. by R.J. Good, R.R. Stromberg (Plenum Press, New York, 1979), p. 31
118. A. Marmur, Soft Matter **2**, 12 (2006)
119. A. Marmur, Annu. Rev. Mater. Res. **39**, 473 (2009)
120. A. Marmur, in *Contact Angle, Wettability, and Adhesion*, vol. 6, ed. by K.L. Mittal (Brill, Leiden, 2009), p. 3
121. D.Y. Kwok, A.W. Neumann, in *Contact Angle, Wettability and Adhesion*, vol. 3, ed. by K.L. Mittal (VSP International Science, Boston, 2003), p. 117
122. W.A. Zisman, in *Contact Angle, Wettability and Adhesion: Advances in Chemistry Series*, vol. 43, ed. by R.F. Gould (ACS, Washington, 1964), p. 1
123. A.W. Neumann, R.J. Good, C.J. Hope, M. Sejpal, J. Colloid Interface Sci. **49**, 291 (1974)
124. C.J. van Oss, L. Ju, M.K. Chaudhury, R.J. Good, J. Colloid Interface Sci. **128**, 313 (1989)
125. C.J. van Oss, R.J. Good, J. Macromol. Sci. A **26**, 1183 (1989)
126. F. Bashforth, J.C. Adams, *An Attempt to Test the Theory of Capillary Action* (Cambridge, London, 1892)
127. B.E. Blaisdell, J. Math. Phys. **19**, 186 (1940)
128. N.R. Tawde, K.G. Parvatikar, Indian J. Phys. **32**, 174 (1958)
129. S. Fordham, Proc. R. Soc. Lond. **194A**, 1 (1948)
130. O.S. Mills, Br. J. Appl. Phys. **4**, 24 (1953)
131. S. Hartland, R.W. Hartley, *Axisymmetric Fluid-Liquid Interfaces* (Elsevier, Amsterdam, 1976)
132. J.D. Malcolm, H.M. Paynter, J. Colloid Interface Sci. **82**, 269 (1981)
133. C. Maze, G. Burnet, Surf. Sci. **13**, 451 (1969)
134. C. Maze, G. Burnet, Surf. Sci. **24**, 335 (1971)
135. C. Huh, R.L. Reed, J. Colloid Interface Sci. **91**, 472 (1983)
136. J.F. Boyce, S. Schürch, Y. Rotenberg, A.W. Neumann, Colloids Surf. **9**, 307 (1984)
137. S.H. Anastasiadis, J.K. Chen, A.P. Koberstein, A.P. Siegel, J.E. Sohn, J.A. Emerson, J. Colloid Interface Sci. **119**, 55 (1986)
138. R.V. Hoggs, *An Introduction to Robust Estimation* (Academic Press, New York, 1979)
139. P. Cheng, Automation of axisymmetric drop shape analysis using digital image processing, Ph.D. Thesis, University of Toronto, Toronto, 1990
140. O.I. del Río, On the generalization of axisymmetric drop shape analysis, M.S. Thesis, University of Toronto, Toronto, 1993
141. L.R. Fisher, J.N. Israelachvili, J. Colloid Interface Sci. **80**, 528 (1981)
142. Y. Rotenberg, L. Boruvka, A.W. Neumann, J. Colloid Interface Sci. **93**, 169 (1983)
143. J.K. Spelt, Y. Rotenberg, D.R. Absolom, A.W. Neumann, Colloids Surf. **24**, 127 (1987)
144. P. Cheng, D. Li, L. Boruvka, Y. Rotenberg, A.W. Neumann, Colloids Surf. **43**, 151 (1990)
145. O.I. del Río, A.W. Neumann, J. Colloid Interface Sci. **196**, 136 (1997)
146. A. Kalantarian, R. David, A.W. Neumann, Langmuir **25**, 14146 (2009)

147. R.O. Duda, P.E. Hart, *Pattern Classification and Scene Analysis* (Wiley, New York, 1973)
148. F.K. Skinner, Y. Rotenberg, A.W. Neumann, J. Colloid Interface Sci. **130**, 25 (1989)
149. E. Moy, P. Cheng, Z. Policova, S. Treppo, D. Kwok, D.R. Mack, P.M. Sherman, A.W. Neumann, Colloids Surf. **58**, 215 (1991)
150. M.A. Rodríguez-Valverde, M.A. Cabrerizo-Vílchez, P. Rosales-López, A. Pàez-Dueñas, R. Hidalgo-Àlvarez, Colloids Surf. **206**, 485 (2002)
151. M.G. Cabezas, A. Bateni, J.M. Montanero, A.W. Neumann, Appl. Surf. Sci. **238**, 480 (2004)
152. M.G. Cabezas, A. Bateni, J.M. Montanero, A.W. Neumann, Colloids Surf. **255**, 193 (2005)
153. M.G. Cabezas, A. Bateni, J.M. Montanero, A.W. Neumann, Langmuir **22**, 10053 (2006)
154. B. Bhushan, *Handbook of Nanotechnology*, 2nd edn. (Springer, Heidelberg, 2007)
155. E. Abkhoshka, M. Korb, B. Rezaic, Expert Syst. Appl. **37**, 5201 (2010)
156. K.A. Matis, E.N. Peleka, Sep. Sci. Technol. **45**, 2465 (2010)
157. J. Chatterjee, J. Colloid Interface Sci. **259**, 139 (2003)
158. A. Marmur, J. Colloid Interface Sci. **186**, 462 (1997)
159. D.Q. Li, Colloids Surf. A **116**, 1 (1996)
160. R. David, A.W. Neumann, in *Applied Surface Thermodynamics*, vol. 151, ed. by A.W. Neumann, R. David, Y. Zuo (Taylor & Francis, Boca Raton, 2011), p. 701
161. B.A. Pethica, J. Colloid Interface Sci. **62**, 567 (1977)
162. H.Z. Yu, D.M. Soolaman, A.W. Rowe, J.T. Banks, Chem. Phys. Chem. **5**, 1035 (2004)
163. D.M. Soolaman, H.Z. Yu, J. Phys. Chem. B **109**, 17967 (2005)
164. J.P.S. Farinha, M.A. Winnik, K.G. Hahn, Langmuir **16**, 3391 (2000)
165. K.W. Stockelhuber, B. Radoev, H.J. Schulze, Colloids Surf. A **156**, 323 (1999)
166. A. Dussaud, M. Vignes-Adler, Langmuir **13**, 581 (1997)
167. F. Rieutord, M.J. Salmeron, Phys. Chem. B **102**, 3941 (1998)
168. L. Kao R, D.R. Wasan, A.D. Nikolov, D.A. Edwards, Colloids Surf. **34**, 389 (1989)
169. E. Grimoldi, S. Zanini, R.A. Siliprandi, C. Riccardi, Eur. Phys. J. D **54**, 165 (2009)
170. Z. Tao, B. Bhushan, J. Phys. D: Appl. Phys. **39**, 3858 (2006)
171. Y.-F. Gao, Y. Yang, D.-Y. Sun, Chin. Phys. Lett. **28**, 036102 (2011)
172. G.J. Jameson, M.C.G. Del Cerro, J. Chem. Soc. Faraday Trans. I **72**, 833 (1976)
173. D.G. Fisher, B. Ovryn, Opt. Lett. **25**, 478 (1979)
174. M. Sundberg, A. Månsson, S. Tågerud, J. Colloid Interface Sci. **313**, 454 (2007)
175. T. Pompe, S. Herminghaus, Phys. Rev. Lett. **85**, 1930 (2000)
176. M. Hong, K.H. Kim, J. Bae, W. Jhe, Appl. Phys. Lett. **77**, 2604 (2000)
177. S.N. Jayasinghe, M.J. Edirisinghe, J. Aerosol. Sci. **33**, 1379 (2002)
178. M.D. Paine, M.S. Alexander, K.L. Smith, M. Wang, J.P.W. Stark, J. Aerosol. Sci. **38**, 315 (2007)
179. S. Adhikari, S. Fernando, Trans. ASABE **49**, 1269 (2006)
180. T.G. Mason, J.N. Wilking, K. Meleson, C.B. Chang, S.M. Graves, J. Phys. Condens. Matter **18**, R635 (2006)
181. X.H. Zhang, W. Ducker, Langmuir **24**, 110 (2008)
182. Y. Tan, V. Cristini, A.P. Lee, Sens. Actuators B, Chem. **114**, 350 (2006)
183. A. Meister, S. Jeney, M. Liley, T. Akiyama, U. Staufer, N.F. de Rooij, H. Heinzelmann, Microelectron. Eng. **67–68**, 644 (2003)
184. A. Meister, M. Liley, J. Brugger, R. Pugin, H. Heinzelmann, Appl. Phys. Lett. **85**, 6260 (2004)
185. A. Meister, J. Polesel-Maris, P. Niedermann, J. Przybylska, P. Studer, M. Gabi, P. Behr, T. Zambelli, M. Liley, J. Vörös, H. Heinzelmann, Microelectron. Eng. **86**, 1481 (2009)
186. Y.C. Jung, B.J. Bhushan, Vac. Sci. Technol. A **26**, 777 (2008)
187. A. Fang, E. Dujardin, T. Ondarçuhu, Nano Lett. **6**, 2368 (2006)
188. M. Preuss, H.-J. Butt, Langmuir **14**, 3164 (1998)
189. S. Ecke, M. Preuss, H.-J. Butt, J. Adhes. Sci. Technol. **13**, 1181 (1999)
190. M. Preuss, H.-J. Butt, Int. J. Miner. Process. **56**, 99 (1999)
191. J. Ralston, D. Fornasiero, R. Hayes, Int. J. Miner. Process. **56**, 133 (1999)
192. D. Johnson, N. Hilal, K. Waters, K. Hadler, J. Cilliers, Langmuir **25**, 4880 (2009)

193. H.-J. Butt, A. Doppenschmidt, G. Huttl, E. Muller, O.I. Vinogradova, J. Chem. Phys. **113**, 1194 (2000)
194. H.K. Christenson, Langmuir **12**, 1404 (1996)
195. J.L. Parker, P.J. Attard, Phys. Chem. **96**, 398 (1992)
196. A.V. Nguyen, J. Nalaskowski, J.D. Miller, J. Colloid Interface Sci. **262**, 303 (2003)
197. A. Checco, P. Guenoun, J. Daillant, Phys. Rev. Lett. **91**, 186101 (2003)
198. A. Checco, Y. Cai, O. Gang, B.M. Ocko, Ultramicroscopy **106**, 703 (2006)
199. V. Dupres, T. Camesano, D. Langevin, A. Checco, P. Guenoun, J. Colloid Interface Sci. **269**, 329 (2004)
200. V. Dupres, D. Langevin, P. Guenoun, A. Checco, G. Luengo, F. Leroy, J. Colloid Interface Sci. **306**, 34 (2007)
201. R. Wang, L. Cong, M. Kido, Appl. Surf. Sci. **191**, 74 (2002)
202. R. Wang, M. Takeda, M. Kido, Mater. Lett. **54**, 140 (2002)
203. N.A. Stelmashenko, J.P. Craven, A.M. Donald, E.M. Terentjev, B.L. Thiel, J. Microsc. **204**, 172 (2001)
204. K.K.S. Lau, J. Bico, K.B.K. Teo, M. Chhowalla, G.A.J. Amaratunga, W.I. Milne, G.H. McKinley, K.K. Gleason, Nano Lett. **3**, 1701 (2003)
205. Y.T. Cheng, D.E. Rodak, Appl. Phys. Lett. **86**, 144101 (2005)
206. Y.C. Jung, B.J. Bhushan, Microsc. **229**, 127 (2008)
207. Q.F. Wei, R.R. Mather, A.F. Fotheringham, R.D. Yang, Tex. Res. J. **73**, 557 (2003)
208. D. Aronov, M. Molotskii, G. Rosenman, Appl. Phys. Lett. **90**, 104104 (2007)
209. Q.F. Wei, Mater. Charact. **52**, 231 (2004)
210. D. Aronov, M. Molotskii, G. Rosenman, Phys. Rev. B, Condens. Matter **76**, 035437 (2007)
211. M. Morita, T. Koga, H. Otsuka, A. Takahara, Langmuir **21**, 911 (2005)
212. J. Léopoldś, P. Damman, Nat. Mater. **5**, 957 (2006)
213. A. Checco, O. Gang, B.M. Ocko, Phys. Rev. Lett. **96**, 056104 (2006)
214. K. Khare, S. Herminghaus, J.-C. Baret, B.M. Law, M. Brinkmann, R. Seemann, Langmuir **23**, 12997 (2007)
215. K. Fukuzawa, T. Deguchi, J. Kawamura, Y. Mitsuya, T. Muramatsu, H. Zhang, Appl. Phys. Lett. **87**, 203108 (2005)
216. R. Seemann, M. Brinkmann, E.J. Kramer, F.F. Lange, R. Lipowsky, Proc. Natl. Acad. Sci. USA **102**, 1848 (2005)
217. L. Valentini, F. Mengoni, J.M. Kenny, A. Marrocchi, A. Taticchi, Small **3**, 1200 (2007)
218. C. Joachim, J.K. Gimzewski, A. Aviram, Nature **408**, 541 (2000)
219. A. Méndez-Vilas, A.B. Jódar-Reyes, M.L. González-Martín, Small **5**, 1366 (2009)

Chapter 2
Adsorption Calorimetry on Well-Defined Surfaces

Ole Lytken, Hans-Jörg Drescher, Rickmer Kose, and J. Michael Gottfried

Abstract The focus of this chapter is single-crystal adsorption calorimetry, which is used for the direct measurement of adsorption energies on well-defined surfaces, both single- and poly-crystalline. The method was pioneered in the 1990s mainly by D.A. King and C.T. Campbell and is based on earlier work by S. Černy. Especially in recent years, the technique has seen increasing proliferation and development. In contrast to desorption-based methods, such as temperature-programmed desorption and isosteric measurements, calorimetry is also well suited for irreversible reactions. The systems studied range from simple adsorption of small molecules, such as CO and NO, to surface reconstructions, reaction intermediates, hydroxyl group formation, metal adsorption and diffusion into polymers, particle-size dependent adsorption energies on model catalysts and even electrochemical reactions on electrode surfaces in the liquid phase.

2.1 Introduction

Adsorption processes are of fundamental importance for heterogeneous catalysis, sensor technology, and electrochemical processes, but also for the production of thin films, for example in semiconductor technology. One of the most important

O. Lytken (✉)
Lehrstuhl für Physikalische Chemie II, Universität Erlangen-Nürnberg, Egerlandstr. 3, 91058 Erlangen, Germany
e-mail: ole.lytken@chemie.uni-erlangen.de

R. Kose
Sentys Inc, 182 Howard St #750, San Francisco, CA 94105, USA
e-mail: rkose@sentys.com

H.-J. Drescher · J.M. Gottfried (✉)
Fachbereich Chemie, Philipps-Universität Marburg, Hans-Meerwein-Straße 6, 35032 Marburg, Germany
e-mail: michael.gottfried@chemie.uni-marburg.de

H.-J. Drescher
e-mail: hans-joerg.drescher@chemie.uni-marburg.de

G. Bracco, B. Holst (eds.), *Surface Science Techniques*,
Springer Series in Surface Sciences 51, DOI 10.1007/978-3-642-34243-1_2,

parameters in this respect is the adsorption energy as a direct measure for the strength of the interaction between the adsorbate and the surface [1]. For example, the catalytic activation of a molecule requires a sufficiently strong interaction of the intermediate with the surface of the catalyst; the same is true for the bonding of a molecule to a sensor and for electrochemical and electrocatalytic processes. However, if the adsorptive bond is too strong, the adsorption will lead to a poisoning of the catalyst (principle of Sabatier) [2]. The precise knowledge of experimental adsorption energies is also important for the improvement of theoretical methods, especially as focus shifts toward larger molecules and more complex systems. Single-crystal adsorption calorimetry measures the heat of reaction as an adsorbate adsorbs on a surface and is a very versatile tool, able to provide information not available with any other technique. Although the technique started out with a narrow focus on single-crystal studies, it has now become a more general technique for planar, low surface area systems.

2.2 Theory Background

2.2.1 Approaches for the Measurement of Adsorption Energies

Adsorption energies can be determined by thermodynamic and, indirectly, kinetic methods. Temperature-programmed desorption (TPD, also known as thermal desorption spectroscopy) [3, 4] belongs to the kinetic methods. Here, the desorption rate constant r is measured as a function of temperature T, and the desorption activation energy E_a is then determined using the simple rate equation (also Polanyi-Wigner equation):

$$r = \nu \exp\left(-\frac{E_a}{RT}\right)\Theta^n \tag{2.1}$$

where Θ is the coverage and n the desorption order. In the case of non-activated adsorption, the activation energy for desorption approximates the negative adsorption energy. Another classical approach is based on equilibrium thermodynamics and considers the change of the equilibrium pressure p with temperature at constant coverage. These quantities are related by the Clausius-Clapeyron equation to the heat of adsorption at constant coverage (isosteric heat) [1]:

$$q_{st} = -R\left(\frac{\partial \ln p}{\partial(1/T)}\right)_\Theta \tag{2.2}$$

The major limitation of both methods is that they can be applied only in cases of fully reversible adsorption, i.e., when the molecules desorb much faster than they decompose; a classical example is carbon monoxide desorbing from a Pt(111) surface (Fig. 2.1a). Many molecules, especially large organic or biomolecules, however, dissociate on the surface and only their fragments desorb. An example is benzene on Pt(111) (Fig. 2.1b), which adsorbs intact at room temperature and below, but decomposes at elevated temperatures forming carbon (in graphene-like structures [5])

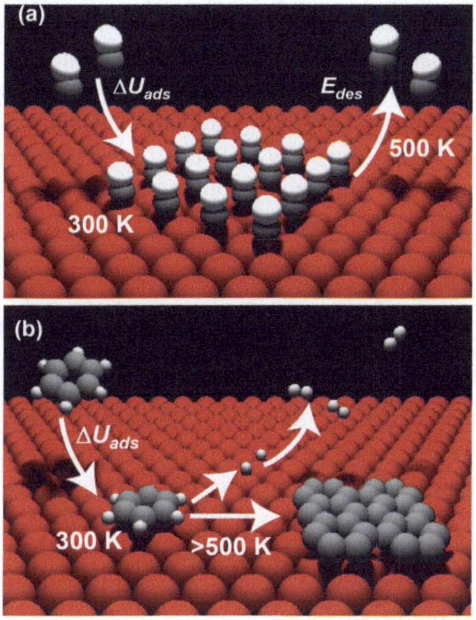

Fig. 2.1 Reversible and irreversible adsorption. (**a**) Reversible: CO on Pt(111). CO desorbs intact and the molar adsorption energy ΔU_{ads} can be estimated from the desorption activation energy E_{des}. (**b**) Irreversible: Benzene on Pt(111). Benzene adsorbs at 300 K as an intact molecule, but dissociates above 500 K forming hydrogen and carbon. In this case, calorimetry is needed to determine ΔU_{ads} (images courtesy E.K. Vestergaard)

and hydrogen, which desorbs. In this case, neither TPD nor methods based on the Clausius-Clapeyron equation (2.2) can be used. For such systems, the adsorption energy can only be determined by direct calorimetric measurement, i.e., by measuring the temperature increase due to the heat of adsorption.

2.2.2 Thermodynamics

The following considerations focus on the thermodynamic understanding of the ultrahigh vacuum-based calorimeters, especially the Campbell calorimeter (see Sect. 2.3.3). A typical experiment consists of a pulsed molecular beam striking a single-crystal sample. As each pulse strikes the sample, heat is deposited and measured as a temperature change in the sample. Simultaneously for each pulse the sticking probability (fraction of the molecules in each pulse adsorbing on the sample) is measured, and with knowledge of the amount of molecules per pulse, the number of molecules that adsorbed on the sample may be calculated. Because of the approximately isochore conditions in the ultrahigh vacuum apparatus, the heat measured (q_{cal}) is equivalent to the change of the internal energy of the sample (Δu_p). (Note that we use lowercase letters for extensive quantities and uppercase letters for intensive quantities.) Δu_p contains two contributions: one from molecules that adsorbed on the sample ($\Delta u_{\downarrow}$), and another from molecules that were reflected, but exchanged energy with the sample ($\Delta u_{\downarrow\uparrow}$):

$$q_{cal} = \Delta u_p = \Delta u_{\downarrow} + \Delta u_{\downarrow\uparrow} \tag{2.3}$$

$\Delta u_{\downarrow}$ can be expressed as the sum of three terms: (i) the actual adsorption energy Δu_{ads}, i.e., the change of the internal energy of the system gas/surface during adsorption of a gas with the sample temperature T_p, (ii) a term $\frac{1}{2}RT_{ms}$ for the energy difference between a gas *flux* and a gas *volume* at the temperature of the molecular beam source, T_{ms} [6], (iii) a contribution due to the gas temperature T_{ms} being different than the sample temperature T_p:

$$\Delta u_{\downarrow} = -\Delta u_{ads} + n_{ads}\left(\frac{1}{2}RT_{ms} - \int_{T_{ms}}^{T_p} C_v \, dT\right) \tag{2.4}$$

where n_{ads} is the adsorbed amount of gas (in mol) and C_v is the molar isochore heat capacity of the gas. Note that Δu_{ads} contributes to (2.4) with a negative sign, because the system gas/surface releases heat ($\Delta u_{ads} < 0$), but the sample considered here gains this amount of heat ($\Delta u_{\downarrow} > 0$).

The integral term in (2.4) is relevant when the temperature of the molecular beam source is different from the sample temperature and, as mentioned above, also reflected molecules can contribute to q_{cal} by exchanging energy with the sample (term $\Delta u_{\downarrow\uparrow}$ in (2.3)). Assuming that all reflected molecules leave the surface with the sample temperature T_p, i.e., that complete thermalization occurs, the following equation is obtained:

$$\Delta u_{\downarrow\uparrow} = -n_{refl} \int_{T_{ms}}^{T_p} \left(C_v + \frac{1}{2}R\right) dT \tag{2.5}$$

where n_{refl} is the reflected amount of gas. The additional term $\frac{1}{2}R$ takes into account that a flux of molecules is warmer than the corresponding volume of gas. If the volume of gas is in thermal equilibrium this corresponds to $\frac{1}{2}R$ [6].

The *adsorption enthalpy*, Δh_{ads}, at T_p equals the adsorption energy Δu_{ads} plus the volume work, which would result, under isobaric conditions, from the compressions of the gas phase (assumed to be ideal) to the negligibly small volume of the adsorbed phase:

$$\Delta h_{ads} = \Delta u_{ads} - n_{ads} R T_p \tag{2.6}$$

This quantity allows for comparison with tabulated thermodynamic standard quantities (after conversion to standard temperature, if necessary).

The *molar heat of adsorption* is normally defined as the negative molar adsorption enthalpy, $-\Delta H_{ads}$, and thus always positive. Combining (2.3)–(2.6) results in the following relation between $-\Delta H_{ads}$ and the measured heat q_{cal}:

$$-\Delta H_{ads} = \frac{-\Delta h_{ads}}{n_{ads}} = \frac{1}{n_{ads}}(q_{cal} - K_{ads} - K_{refl}) \tag{2.7}$$

with the correction terms

$$K_{ads} = n_{ads}\left[-\int_{T_{ms}}^{T_p} C_v \, dT + \frac{1}{2}RT_{ms} - RT_p\right] \tag{2.8}$$

$$K_{refl} = n_{refl}\left[-\int_{T_{ms}}^{T_p} \left(C_v + \frac{1}{2}R\right) dT\right] \tag{2.9}$$

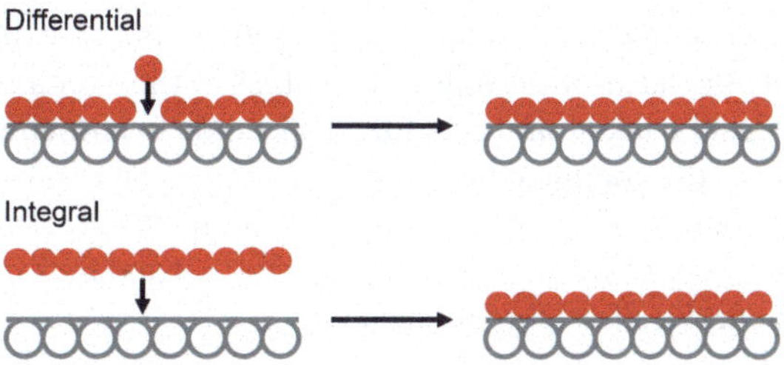

Fig. 2.2 Differential (*top*) and integral (*bottom*) heats of adsorption near saturation coverage. In the case of the differential heat of adsorption, differentially small changes of coverage are considered, while the integral heat of adsorption considers finite changes starting from $\Theta = 0$. Calorimetric experiments typically measure the differential heat of adsorption, whereas theoretical calculations typically calculate the integral heat of adsorption

$-\Delta H_{ads}$ is a differential heat of adsorption. Integration of $-\Delta H_{ads}$ over the coverage Θ yields the respective integral heat of adsorption $-\Delta H_{ads,int}$ (see Fig. 2.2):

$$-\Delta H_{ads,int} = -1/\Theta \int_0^{\Theta} \Delta H_{ads}\, d\Theta' \tag{2.10}$$

Integral and differential heats and energies of adsorption play an important role for comparisons with results from ab-initio calculations. For such comparison, it can be useful to take the changes in the internal and external degrees of freedom (DOF) of the adsorbing molecules into account. In the simplest case, a mono-atomic gas (three translational DOF), the enthalpy of the gas is given by (with $T = 0$ as the point of reference):

$$H_{gas} = U_{gas} + pV_{gas} = \frac{3}{2}RT + RT \tag{2.11}$$

For estimating the enthalpy of the adsorbed phase, H_{ads}, one needs to distinguish between mobile and localized adsorption. In the case of mobile adsorption, the molecule has two lateral translational DOF ($F_{trans} = 2$) and one vibrational DOF ($F_{vib} = 1$) perpendicular to the surface. For full excitation of all DOF, this results in (again with $T = 0$ as reference point):

$$H_{ads} = U_{ads} = -|E_0| + 2RT \tag{2.12}$$

where $|E_0|$ is the adsorbate-substrate bond energy. Thus, the following relation holds for the heat of adsorption:

$$\Delta H_{ads} = H_{gas} - H_{ads} = -|E_0| + \frac{1}{2}RT \tag{2.13}$$

In contrast, for localized adsorption ($F_{trans} = 0$, $F_{vib} = 3$), one obtains in the case of full excitation of all DOF:

$$\Delta H_{ads} = H_{gas} - H_{ads} = -|E_0| - \frac{1}{2}RT \tag{2.14}$$

$|E_0|$ can now be compared to calculated adsorbate-substrate bond energies or (in the case of non-activated adsorption) to desorption activation energies E_{des}.

The situation can become very complex, especially in the case of large molecules, because in principle the adsorption-induced changes in the excitation of internal degrees of freedom must be taken into account. This holds in particular for chemisorption, which influences the chemical bonds in the molecule, and thus, via the partition functions, the contribution of the vibrations to the total energy. However, these contributions are in the order of RT (~2.5 kJ/mol at 300 K) and thus much smaller than typical adsorption energies for large molecules.

2.3 Experimental Setup and Method

2.3.1 History

Adsorption calorimetry on thin metal wires, which serve as both substrates and resistance thermometers, were already performed in the 1930s by J.K. Roberts with the focus on a quantitative understanding of chemisorption and adsorbate-adsorbate interactions. For example, the adsorption of H_2, N_2 and O_2 on tungsten was investigated with this approach [7–9]. P. Kisliuk later improved this method by using metal ribbons instead of wires to achieve a higher surface-to-volume ratio [10]. Until the 1970s, this approach was pursued by various groups [11]. However, it became evident that the polycrystalline samples with their wide spectrum of different adsorption sites lead to problems with the reproducibility of results, especially as the wire and ribbon samples were also structurally and chemically insufficiently defined.

Around 1940, O. Beeck [12, 13] introduced the thin film calorimeter, in which vapor-deposited metal films are used. This technique was later improved and extensively used by G. Wedler [14–17] and by a number of other groups. The Wedler calorimeter (Fig. 2.3) contains a thin-walled round glass bulb (1) with a metal filament (4) inside, from which the metal is vapor-deposited onto the inner wall of the bulb as a thin closed film. Around the outside of the bulb, a thin metal wire is wound in a induction-free loop arrangement as a resistance thermometer. The gas is introduced into the evacuated bulb in small pulses, while the adsorption-induced temperature change of the bulb is measured. For thermal insulation, the glass bulb is located inside an evacuated glass cylinder (6), which is immersed in a thermostat. The heat capacity of this calorimeter is in the order of 1 J/K, and temperature changes as low as 1 μK can be detected [11, 16, 17]. With an active sample surface of 75 cm^2, absolute sensitivities in the range of 10 nJ/cm^2 are reached. For calibration of the Wedler calorimeter, the resistance thermometer is used as a resistance heater to deposit a precisely known amount of heat. This calorimeter type was used for adsorption studies of various small molecules on transition metal films [11]. A disadvantage of the technique is again the polycrystalline nature of the metal films, the dependence of the film morphology on the deposition parameters and the very limited possibilities for a structural and chemical characterization of the surface.

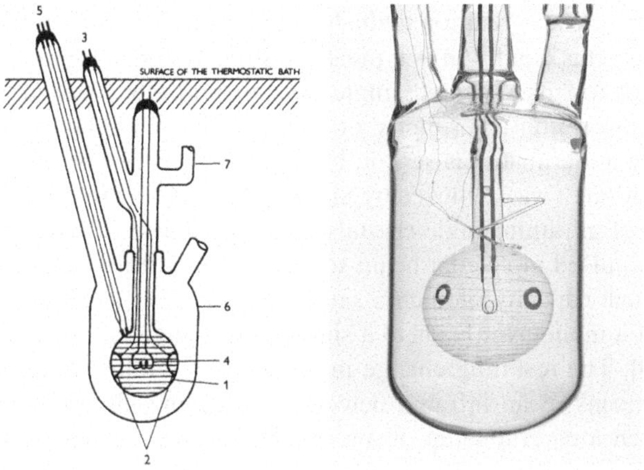

Fig. 2.3 Spherical adsorption calorimeter after Wedler [16, 17]. (*1*) Thin-walled glass bulb (∅ 5 cm, wall thickness 0.1 mm) with resistance thermometer on the outside (tungsten, length 2 m, ∅ 10 μm). (*2*) Platinum contact foils for measuring the resistance of the film and the related electrical feedthroughs (*3*). (*4*) Evaporant (metal wire). (*5*) Electrical feedthroughs for the resistance thermometer. (*6*) Evacuated glass bulb for thermal insulation (*left figure* reproduced from Ref. [11] with permission from Elsevier B.V., © 1996)

The adsorption calorimeter by S. Černý et al. [18] relies on a principle similar to that of the single-crystal calorimeters by King and Campbell, which will be described in the next section, and was developed independently around the same time. The Černý calorimeter uses a pyroelectric detector ($LiTaO_3$), onto which the sample is vapor-deposited as a thin polycrystalline metal film. The molecules are dosed with a pulsed supersonic molecular beam. For calibration, a laser beam is used, which takes the same path as the molecular beam.

Finally, we mention the micromechanical calorimeter [19–22], which uses the temperature-induced bending of a bimetallic cantilever for the detection of heats of adsorption and reaction. A typical Si/Al cantilever has a length of ~400 μm, a width of 35 μm, and a thickness of ~1.5 μm. The bending is measured similarly as in an Atomic Force Microscope. The calorimeter reaches an absolute sensitivity limit of $\sim 10^{-12}$ J, which corresponds to ~10 nJ/cm^2. Thus, the sensitivity per area is similar to that of the Wedler calorimeter, and both are limited to polycrystalline substrates.

For a more detailed description of the history of adsorption calorimetry, we refer to Ref. [11].

2.3.2 Single-Crystal Adsorption Calorimeter by D.A. King

The major challenge of adsorption calorimetry is to ensure that the heat released upon adsorption causes a measurable change in temperature, which requires the heat capacity to be low in relation to the heat of adsorption, which is proportional to the surface area. This condition is typically fulfilled in the case of powder samples [23–25], but not for macroscopic single crystals and other well-defined samples. In the 1980s, adsorption calorimetry on macroscopic Pt(111) single crystals was attempted using a thermistor as sensor, but the results were discouraging [26]. The decisive breakthrough was achieved by Sir D.A. King (Cambridge, UK) in the 1990s with the usage of ultrathin single crystals with a thickness of only 0.2 μm, in combination with a pulsed molecular beam source and an infrared detector (pyrometer) [27–29]. The heat capacity of such a sample is very low, and even the adsorption of a fraction of a monolayer leads to a significant change in temperature on the order of 1 K [30]. The resulting change in intensity of emitted infrared light can be measured by means of an infrared detector outside the ultrahigh-vacuum (UHV) apparatus. Given a careful setup, heats of adsorption of at least 1 $\mu J/cm^2$ can be measured.

For calibration of the calorimeter, a laser beam with known intensity and closely matched temporal and spatial characteristics as the molecular beam is used to deposit a known amount of heat in the sample. For this calibration, the reflectivity of the sample at the wavelength of the laser must be precisely known. Alternatively, a known heat of reaction, such as the multilayer adsorption energy of the adsorbate, may be used to calibrate the system. Although used in Schuster's electrochemical calorimeter (see below) this approach is typically not used for ultrahigh vacuum calorimeters, except to confirm the laser-based calibration.

To determine the heat of adsorption per molecule (or per mol), the number of adsorbed molecules in each pulse must be known. This number is determined by separate measurements of flux and sticking probabilities. The flux is measured with a stagnation gauge. The sticking probability is measured by a mass spectrometer, detecting the reflected molecules (King-Wells techniques [31], see Fig. 2.4). Before and after each experiment, an inert gold flag is placed in front of the sample, providing the mass spectrometer signal corresponding to 100 % reflection. Because the sticking probability is always measured by comparing with 100 % reflection, the relative uncertainty becomes larger the lower the sticking probability becomes, and unless extreme care is taken, it is generally not possible to measure reliable adsorption energies for sticking probabilities below 10 %. From the measured amounts of heat and the corresponding numbers of molecules, the molar heat of adsorption can be calculated for each pulse. In the limit of very small pulses ($\Delta\Theta \rightarrow 0$), this is a differential heat of adsorption (see Fig. 2.2).

The magnitude of the infrared signal depends strongly on the absolute sample temperature, because the radiant power, ΔP_{rad}, follows the Stefan-Boltzmann law: $\Delta P_{rad} \propto T^3 \Delta T$. This technique will, therefore, have a lower sensitivity at lower sample temperatures and a higher sensitivity at higher sample temperatures. However, cooling or heating of the sample via the support ring is impossible since lateral heat transfer is extremely ineffective. Thus, the sample temperature is dominated by radiative thermalization with the environment. To date, no attempts have been made to measure at sample temperatures above or below room temperature, e.g., by means of a cryoshroud.

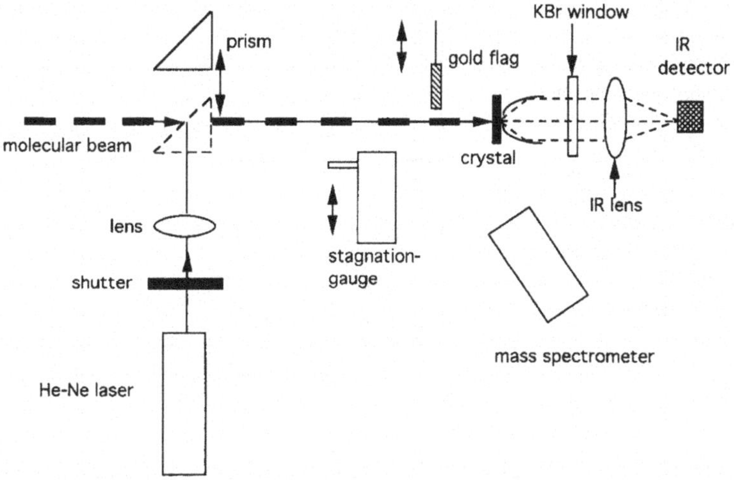

Fig. 2.4 Single-crystal calorimeter by D.A. King. The adsorption-induced temperature change of the thin single-crystal sample (0.2 μm) is measured with an IR detector outside the UHV apparatus (*right*). For dosing, a pulsed molecular beam is used (*left*). Further components: pulsed laser beam for calibration (*left*), stagnation gauge for flux measurements (*center*), and mass spectrometer and gold flag for measurements of the sticking probability after King and Wells [31] (reproduced from Refs. [27–29] with permission from Elsevier B.V., © 1996)

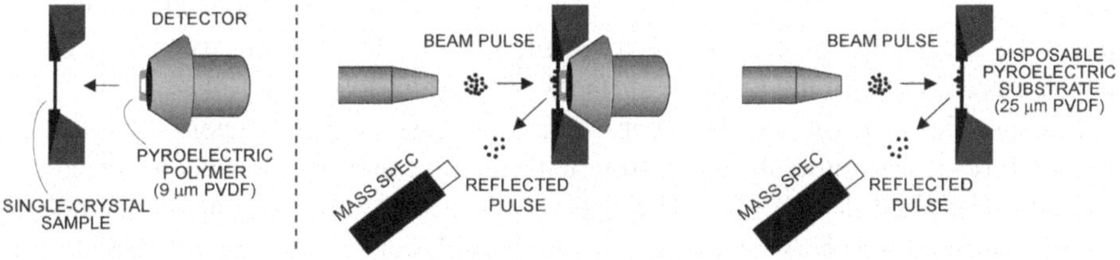

Fig. 2.5 An illustration of the Campbell calorimeter. *Left*: The setup for single-crystalline samples. A thin pyroelectric polymer ribbon is mounted in an arch and pressed gently against the backside of the sample. As a pulse of molecules adsorbs on the single-crystal sample, heat is deposited and is detected as a temperature change in the pyroelectric polymer. Simultaneously, the sticking probability of the molecular pulse is measured by a mass spectrometer, detecting the reflected fraction of the pulse. *Right*: The setup for polymer samples, organic films and other polycrystalline samples, which are deposited as thin films directly on the detector

2.3.3 Single-Crystal Adsorption Calorimeter by C.T. Campbell

The central feature of the single-crystal calorimeter by C.T. Campbell et al. is a pyroelectric detector, which gently contacts the backside of a thin single-crystal sample during the measurement (Fig. 2.5). The material of the detector is a pyroelectric polymer, typically 9 μm thick β-polyvinylidene fluoride (β-PVDF) in the

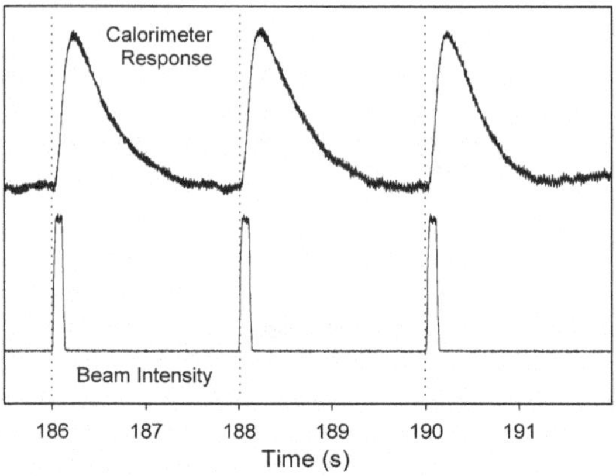

Fig. 2.6 Calorimeter signal (*top*) and molecular beam intensity (*bottom*) for the adsorption of cyclohexene on Pt(111) at 100 K. Each pulse has a length of 102 ms, contains $2.5 \cdot 10^{12}$ molecules (0.011 ML) [37] and leads to a heat input of ~250 nJ (adapted from Ref. [36] with permission from the American Chemical Society, © 2008)

form of a bent ribbon. The direct mechanical contact between sample and detector requires thicker samples than those used in the King calorimeter, typically 1 μm or thicker, but leads to an increased sensitivity, such that temperature changes in the μK range can be detected. The increased sensitivity allows samples as thick as 100 μm to be used [32–35]. This represents a considerable advantage, since 100 μm thick samples may be produced by mechanically thinning down readily available thicker single-crystal samples. The thin 1 and 0.2 μm single-crystal samples are produced as epitaxially grown films on a salt crystal and are therefore only available for certain elements and surface orientations. The smooth and mirror-like surfaces of the 100 μm samples is another advantage, because the calibration of the calorimeter requires the knowledge of the reflectivity of the sample. The reflectivity is difficult to measure accurately on the buckled surfaces of the thin 0.2–1 μm foils.

Because the heat transfer between sample and detector is efficient even at low temperatures, low-temperature measurements are possible and have been carried out with the Campbell design [36]. The upper operating temperature of the Campbell design is around 400 K, where a significant degradation of the detector will start to occur. During sample preparation, which normally includes annealing at high temperatures, the detector is retracted from the sample and does not limit the maximum annealing temperature.

The adsorbate is dosed by means of a periodically pulsed molecular beam. For gases or liquids with vapor pressures above 10 mbar at room temperature, an effusive beam is used. For metals or other solids with low vapor pressures an evaporator is used instead. The pulses have a typical length of 100 ms and a frequency of 0.5 Hz. Each pulse leads to the deposition of heat in the sample and causes a detector signal, which is proportional to the temperature increase and, thus, to the amount of heat (Fig. 2.6). Note that a pyroelectric detector is a current source, for which the current is proportional to the time derivative of the temperature, $I_{pyro} \sim dT/dt$. Calibration is achieved with a laser, as in King's setup.

In the Campbell design the flux of the molecular beam is measured before and after each experiment with a Quartz Crystal Microbalance (QCM) positioned in front of the sample. When used to measure the flux of molecules such as benzene or cyclohexene, the QCM is cooled with liquid nitrogen to allow multilayers to build up. For adsorbates with negligible vapor pressure at room temperature, such as metals, cooling of the QCM is not necessary. As in King's setup, the sticking probability is measured by a mass spectrometer, detecting the reflected molecules, see Fig. 2.5. For metals and other low vapor pressure compounds the inert flag, used to calibrate the mass spectrometer, is heated to ensure 100 % reflection.

If the molecular beam is produced by a hot source, as for example in the case of metals or compounds with low vapor pressure, then the contribution from the absorbed IR radiation to the calorimeter signal must be measured separately. For this purpose, an IR transmissive window, such as BaF_2 or KBr, is inserted in the beam path, blocking the molecules or atoms in the beam, but allowing the radiation to reach the sample.

The Campbell design has produced several offspring; one such has been constructed by R. Schäfer's group in Darmstadt [38, 39]. Schäfer's design involves a novel pulsed molecular beam, which, instead of the typical continuous beam chopped by a rotating chopper, uses an intrinsic pulsed beam. The pulses are created by a piezoelectric plunger, opening and closing to a gas containing vessel of known volume. After an experiment the gas dose per pulse can be determined by measuring the pressure drop in the gas containing volume and dividing by the number of pulses. The pressure changes are typical in the range of 10^{-4} mbar (at a starting pressure in the range of 1 mbar) and are measured by a capacitance pressure gauge.

2.3.4 Electrochemical Single-Crystal Adsorption Calorimeter by R. Schuster

All the single-crystal calorimeters discussed this far have been ultrahigh vacuum-based calorimeters, where the adsorbate is dosed from the gas phase in an ultrahigh vacuum chamber onto a single-crystal surface. However, in the group of R. Schuster the technique has been adapted to measure heat effects during electrochemical reactions, with conversions of just a few percent of a monolayer [40–43]. Similar to Campbell's design a pyroelectric polymer foil is used as the detector (Fig. 2.7). In Campbell's design the soft contact between the pyroelectric polymer detector and the sample limits the heat transfer into the detector. Schuster solved this problem by evacuating the space between the detector and the sample, allowing the ambient pressure in the electrochemical cell to press the detector and sample together, forming a tight contact. The samples are typically 12 μm thick single or polycrystalline metal foils supported by a 120 μm thick Si sheet. The Si support is used to reduce bending of the sample/detector sandwich, which would cause false signals since the PVDF-based detector is piezoelectric as well as pyroelectric.

An interesting feature of the electrochemical calorimeter is that for electrochemical reactions conducted close to equilibrium the heat effects reflect the entropy changes at the interface. Whereas the ultrahigh vacuum-based calorimeters measure reaction energies, Schuster's electrochemical calorimeter measures reaction

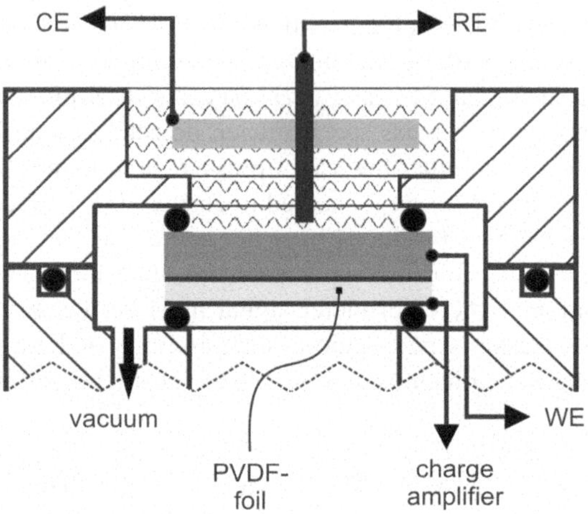

Fig. 2.7 Electrochemical cell of the calorimeter by R. Schuster et al. [40–42]. The sample (working electrode, WE) is located directly above the detector (PVDF foil). To increase the mechanical stability of thin samples, a 120 μm thick Si sheet may be inserted between the sample and the detector. A good mechanical contact between sample and detector is achieved by evacuating the volume between sample and detector. CE: counter electrode; RE: reference electrode (reproduced from Ref. [40] with permission from the American Institute of Physics, © 2010)

entropies. Schuster's calorimeter is also much smaller and simpler than the large ultrahigh vacuum calorimeters, but the interpretation of the measurements is far more complex than those for ultrahigh vacuum-based calorimeters. The sensitivity of the method is in the range of 1 $\mu J/cm^2$ and thus, despite the presence of the liquid phase, only 1–2 orders of magnitude below that of UHV calorimeters [41]. However, because of heat loss into the liquid, Schuster's calorimeter operates with much faster pulses than the ultrahigh vacuum calorimeters, typically in the range of 10 ms, see Fig. 2.8. Calibration of the calorimeter is achieved by using an electron transfer reaction, such as the $[Fe(CN)_6]^{4-}/[Fe(CN)_6]^{3-}$ redox system, depositing a known amount of heat [40]. Schuster's technique has been used to measure the deposition and dissolution of Ag and Cu as well as phase transitions in organic monolayers on electrode surfaces [40–43].

2.4 Applications of the Technique

2.4.1 Surface Phase Transitions—CO Adsorption on Pt

Surface phase transitions or reconstructions are frequently occurring phenomena and are often induced or lifted by adsorbates on the surface. An example is CO and ethylene adsorption on Pt(100). Unreconstructed Pt(100)-(1 × 1) can be prepared at 300 K, but the unreconstructed phase is metastable, and upon heating to 500 K

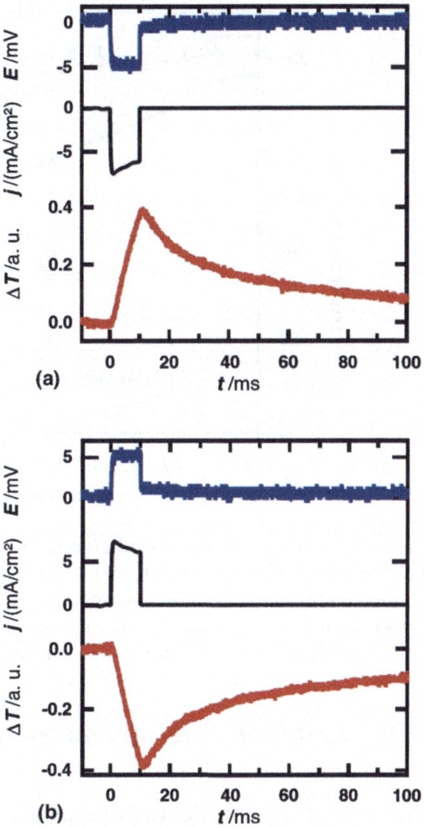

Fig. 2.8 The deposition (**a**) and dissolution (**b**) of Ag onto a 300 ML thick silver film on Au. At $t = 0$ the potential was stepped to 5 mV negative (**a**) or positive (**b**), and after 10 ms the cell current was forced to 0. During deposition the sample temperature increases with time, which can be attributed to the reduction in entropy of the electrode-solution interface by the removal of Ag ions from solution (reproduced from Ref. [40] with permission from the American Institute of Physics, © 2010)

the surface will reconstruct to Pt(100)-hex, adopting a hexagonal arrangement in the first layer. Upon adsorption of CO or ethylene at 300 K the reconstruction is lifted, and the final state of the surface is identical to the final state when CO or ethylene is adsorbed on the unreconstructed Pt(100)-(1 × 1) surface. The reconstruction energy can therefore be measured directly with single-crystal adsorption calorimetry as the difference in the integrated adsorption energies, yielding an energy difference between Pt(111)-hex and Pt(100)-(1 × 1) of 20–25 kJ/mol [44].

Single-crystal adsorption calorimetry was also used to investigate the energetics of the Pt(311)-(1 × 2) surface reconstruction [45]. This reconstruction was—albeit with some discrepancy in their findings—observed by Blakely and Somorjai [46] and Gaussmann and Kruse [47, 48] and more recently studied by means of DFT by Orita and Inada [49]. Using a combination of calorimetry and low-energy electron diffraction (LEED), it was possible to confirm the existence of the reconstruction and to show that the presence of small amounts of CO or carbon are sufficient to lift the (1 × 2) reconstruction [45]. The heats of adsorption for CO on Pt(311)-(1 × 2) and Pt(110)-(1 × 2) are shown in Fig. 2.9. For CO adsorption on Pt(311) four distinct adsorption regions can be observed, each corresponding to different adsorption structures.

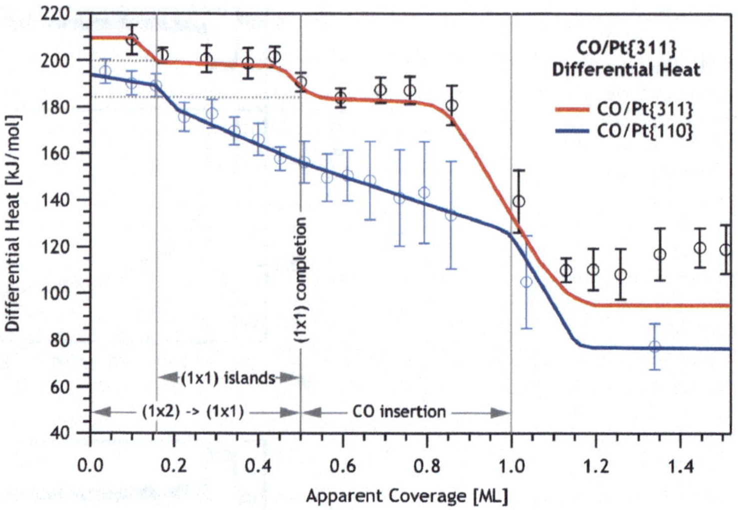

Fig. 2.9 Coverage-dependent differential heats of adsorption for CO on Pt(311) (*red lines*) and Pt(110) (*blue lines*) adsorbed at 300 K. Open circles are averaged experimental data points (error bars represent one standard deviation from averaging across several experiments), and the solid lines represent guides to the eye through these points (CO/Pt(110) data adapted from Refs. [50, 51] with permission from the American Chemical Society, © 1996 and 1998)

2.4.2 Reaction Intermediates—Cyclohexene Adsorption on Pt(111)

Platinum is an excellent hydrogenation/dehydrogenation catalyst and the adsorption of hydrocarbons on platinum has therefore received much attention in the literature. However, because many hydrocarbons readily dehydrogenate on platinum before any desorption occurs, desorption-based techniques, such as temperature programmed desorption, are very limited in their ability to study these systems. Let us consider the hydrogenation of benzene to cyclohexane. It is one of the first steps in the production of nylon, but may also be seen as a model reaction for the removal of aromatics in diesel. From a more fundamental point of view, this is a system with a wide range of interesting possible reaction intermediates that are not accessible using desorption-based techniques. An example is the adsorption of cyclohexene, a possible reaction intermediate, on Pt(111), see Fig. 2.10. At 100 K, cyclohexene adsorbs intact, but as the surface is heated, sequential dehydrogenation will occur, eventually resulting in adsorbed graphite.

Unlike temperature programmed desorption, single-crystal calorimetry does not rely on desorption and can therefore directly measure the heat of adsorption and reaction of the wealth of species formed as hydrocarbons are adsorbed on platinum. However, to control the species formed upon adsorption, it is necessary to control the temperature, at which the measurement is carried out. With the Campbell design this is possible, and Fig. 2.11 shows the experimental differential heat of adsorption of cyclohexene on Pt(111) at 100 K as an example. In the range of the first

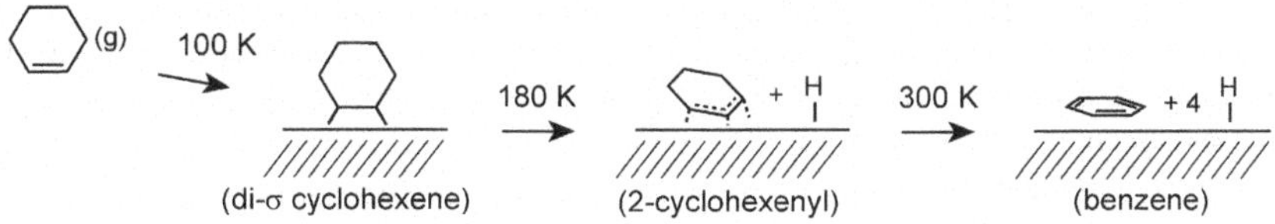

Fig. 2.10 The temperature dependent, sequential dehydrogenation of cyclohexene on Pt(111) [36]. At 100 K, cyclohexene adsorbs intact forming a di-bonded cyclohexene species. Above 180 K, cyclohexene looses an allylic hydrogen, forming a conjugated 2-cyclohexenyl intermediate. Above 300 K, three more hydrogen atoms are abstracted and adsorbed benzene and hydrogen are formed. Above room temperature, the adsorbed hydrogen will desorb, and around 400 K benzene will further dehydrogenate, eventually producing graphite at high temperatures. By controlling the adsorption temperature, it is therefore possible, using the same starting molecule, to measure the heat of formation of multiple different intermediates

adsorbed layer ($\Theta < 0.24$ ML) [37], the heat of adsorption decreases with growing coverage, which indicates repulsive interactions between the adsorbed molecules. Based on the shape of the curve, conclusions regarding the strengths of these interactions can be made [52, 53]. Using the heat of adsorption at very low coverages ($\Theta \rightarrow 0$, 130 kJ/mol), the standard enthalpy of formation of adsorbed cyclohexene (-135 kJ/mol) and the dissociation energy of the Pt–C bond (205 kJ/mol) were calculated. Both values are not accessible by any other technique than adsorption calorimetry. After saturation of the first layer ($\Theta > 0.24$ ML), the heat of adsorption has a constant value of 47 kJ/mol. This value represents the heat of condensation of cyclohexene at 100 K, since the molecules adsorb here on molecular layers and the influence of the substrate is negligible. Similar investigations were performed for benzene and naphthalene on Pt(111) [52, 53]. Based on these measurements and values for the other above-mentioned temperature ranges, a complex energy landscape for the hydrogenation of benzene to cyclohexane on Pt(111) can be derived [54]. Such data are of great importance for the quantitative understanding of heterogeneous catalytic reactions.

From the standard deviation between the data points in the multilayer range in Fig. 2.11, it can be concluded that a heat of adsorption of approximately 5 kJ/mol represents the detection limit for the calorimeter used here. This corresponds to a sensitivity limit of approximately 100 nJ/cm^2, if the active sample area is taken into account. The ability of the Campbell design to carry out experiments below room temperature has also been used to measured the heat of formation of adsorbed hydroxyl on Pt(111), by adsorbing water on an oxygen pre-covered Pt(111) surface at 150 K [55, 56].

2.4.3 Metal Adsorption on Polymers—Calcium on Poly(3-Hexylthiophene)

For some systems, desorption-based techniques are inappropriate because the adsorbate will thermally decompose before or simultaneous with desorption. For other

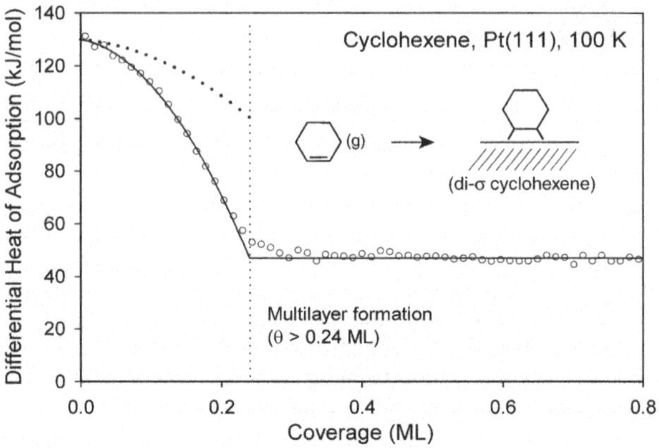

Fig. 2.11 Differential heat of adsorption of cyclohexene on Pt(111) at 100 K (*open circles*) and the integral heat of adsorption (*dotted line* in the range $0 < \Theta < 0.24$ ML), calculated using (2.10) (reproduced from Ref. [36] with permission from the American Chemical Society, © 2008)

systems, it is the substrate which will decompose. This is the case for metal adsorption on organic substrates [57–60]. Such interfaces occur in organic electronic or opto-electronic devices [61, 62]. An important parameter in this respect is the interface energy, because it determines the stability of the interface and is related to its electronic properties. For example, the charge injection rates at the interface depend on the overlap of wave functions and thus on the character of the interfacial chemical bond [63]. For polycrystalline or amorphous films, the organic film can, as shown in Fig. 2.5, be deposited directly on the detector, for example by vacuum sublimation or spin coating. The benefit of depositing the substrate film directly on the detector is that the sensitivity is increased by more than an order of magnitude.

As an example, we will discuss the adsorption of Ca on the π-conjugated, semiconducting polymer poly(3-hexylthiophene), P3HT, which leads to formation of a complex metal/organic interface. This is also a good example of how single-crystal adsorption calorimetry today is no longer restricted to single-crystal studies, but is used as a general technique for a wide range of planar, low surface area samples. The differential heat of adsorption of Ca on P3HT as a function of coverage is shown in Fig. 2.12. From the initial value is 625 kJ/mol (adsorption on defect sites), the heat of adsorption drops rapidly to 405 kJ/mol, which is still considerably larger than the sublimation enthalpy of Ca, 178 kJ/mol. This indicates a strong metal-polymer interaction or a chemical reaction at the interface. Indeed, parallel spectroscopic investigations show that Ca reacts with the thiophene units of the polymer forming calcium sulfide (CaS). This reaction occurs up to a maximum depth of 3 nm at 300 K. If the deposition is performed at lower temperatures, the reaction depth is smaller [57, 64]. This reaction dominates in the early stages of deposition, while Ca particles and finally a closed Ca film form at higher Ca coverages (> 0.5 ML). For this reason, the heat of adsorption slowly approaches the heat of sublimation of Ca and reaches this value around 5 ML. Above this coverage, all newly adsorbing Ca atoms contribute exclusively to the growth of this Ca layer.

The third competing process, the reflection of impinging Ca atoms, prevails at small coverages (initial sticking probability $S_0 = 0.35$), but becomes less important

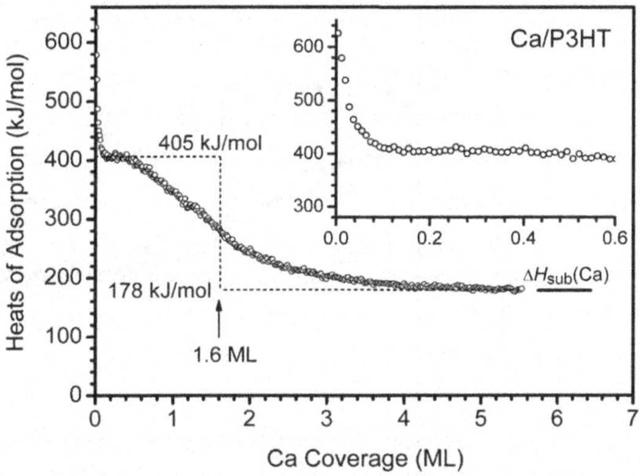

Fig. 2.12 Differential heat of adsorption of Ca on poly(3-hexylthiophene) at 300 K. A coverage of 1 ML corresponds to a closed-packed Ca(111) layer ($7.4 \cdot 10^{14}$ atoms/cm^2) (adapted from Ref. [54] with permission from the American Chemical Society, © 2010)

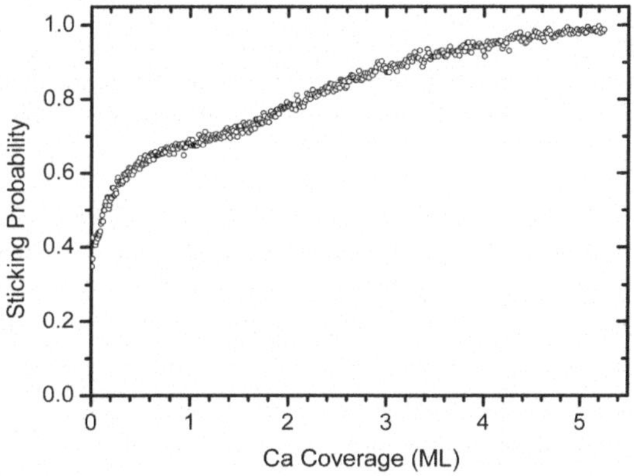

Fig. 2.13 Sticking probability of Ca atoms on poly(3-hexylthiophene) at 300 K, as measured by a modified King-Wells technique [31] (adapted from Ref. [58] with permission from the American Chemical Society, © 2010)

at higher coverages. This is evident from the dependence of the sticking probability on the coverage (Fig. 2.13).

2.4.4 Particle-Size Dependent Adsorption Energies—CO on Pd

Another offspring of the Campbell design is S. Schauermann's and H.-J. Freund's calorimeter in Berlin [65, 66]. It features several technical improvements which potentially reduce systematic errors in the calibration of the calorimeter, for example a system for *in situ* measurements of the reflectivity and an *in situ* photodiode for measuring the power of the calibration laser. Furthermore, it comprises a rotatable table for the precise alignment of sample, photodiode and stagnation gauge in exactly the same position. The construction of this table also ensures improved temperature stability and reduced susceptibility to vibrations.

The Berlin calorimeter has been used to measure the particle-size dependent heats of adsorption on planar model catalysts. For example, the adsorption of CO on Pd nanoparticles on Fe_3O_4 thin films has been studied, see Fig. 2.14. The growth

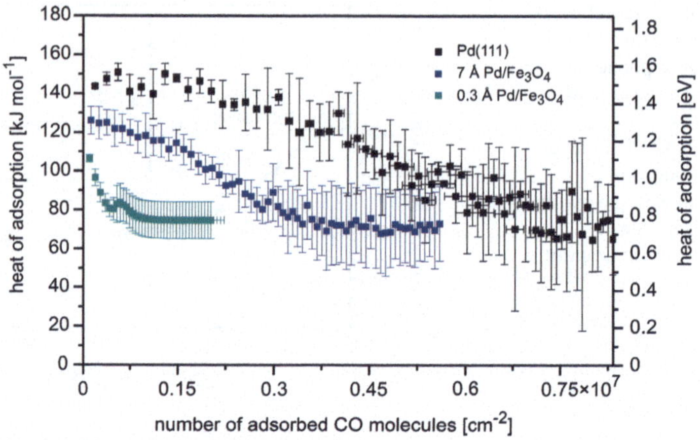

Fig. 2.14 The heat of adsorption of CO on Pd(111) and Pd/Fe_3O_4/Pt(111) model catalysts. The heat of adsorption is seen to decrease with decreasing particle size (reproduced from Ref. [66] with permission from the American Physical Society, © 2010)

of oxide films on the thin single crystals needed for calorimetry represents a major challenge in these experiments, because the sample temperature must be precisely controlled during the film growth. Since the thin samples must be mounted on a holder with a much larger heat capacity, this is difficult to achieve. During radiative heating, the edges of the thin foils are in contact with the sample holder and will remain relatively colder than the center of the foil, which is far away from the edges. Nevertheless, this challenge has been successfully overcome.

2.5 Future Developments of the Technique

Single-crystal adsorption calorimetry is a very versatile technique and has been used to study a broad range of different low-surface area systems, both single- and polycrystalline. This development is expected to continue, and in the future we may see calorimetric measurements involving adsorption of large organic or bio molecules, intercalation of metals in organic films, or even measurements of the heat of formation of metal-organic or covalent networks on surfaces. Such complex surface reactions often require a precise control of the temperature. It is therefore likely that we will see more measurements above and below room temperature and the development of more apparatus capable of operating under these conditions.

We are also likely to see more sophisticated data treatments. The standard laser calibration used today for ultrahigh-vacuum calorimeters is only valid if the heat of adsorption in the actual calorimetric measurements is deposited instantaneously. However, this is not always the case. The initial adsorption may be followed by a secondary slower reaction, such as decomposition, diffusion into the subsurface range, or desorption. This will cause the shape of the calorimeter response to change, and the new shape will contain kinetic information about the secondary reaction. As the sensitivities and signal/noise ratios of future calorimeters are expected to increase, it may become possible to deconvolute the calorimeter responses and extract this kinetic information.

Overall, we are likely to see further differentiation, because the technique will be adapted to the systems studied. There will not be one solution equally suitable for all systems. For example, different adsorbates require different molecular beam setups. Thin 1 μm single-crystal foils may be used when high sensitivity is required, e.g., for deconvolution of the calorimeter response, but thicker 100 μm crystals may be preferred for accuracy, since reflectivity measurements are easier on thicker crystals. Accuracy may also be increased by using laser wavelengths at which the samples have lower reflectivities. For example, gold and copper have reflectivities of 96 % and 92 % at 633 nm, respectively, but only 37 % and 50 % at 405 nm [67]. We may also see advanced sample mounts which will provide better temperature control for preparation of oxide thin films and model catalysts.

Radiative contributions to the calorimeter signal will be a challenge for adsorption studies with many catalytically relevant transition metals. As the evaporation temperature required increases, the heat of adsorption becomes a smaller signal on an increasing background of radiated heat. Calorimetry has already been used to measure the growth of Ag on CeO_2 and MgO [68], but attempts to move to metals with even lower vapor pressures requires refining our techniques for measuring the radiation contribution. Radiation may also represent a challenge at lower temperatures if both the evaporant and the sample have high emissivities in the far infrared region. In those cases, simple infrared transmissive windows such as BaF_2 or KBr may not transmit over a wide enough range, and alternative solutions may have to be considered, such as Au or Ag mirrors or rotating velocity filters [69], which are only transmissive for the molecules and completely eliminate the radiative contribution to the calorimeter signal.

Finally, as this technique becomes more wide-spread, we are likely to see more experimental overlaps. If the same system is investigated with more than one calorimeter, experimental accuracy and reliability will improve. Today, the best way to determine the accuracy of a calorimetric measurement is to compare it to known values, such as the heats of adsorption in the multilayer range to tabulated heats of sublimation. This approach is especially feasible for adsorption of metals. However, for many systems, such direct comparisons are not possible.

References

1. H.-J. Freund, Principles of chemisorption, in *Handbook of Heterogeneous Catalysis*, vol. 3, ed. by G. Ertl, H. Knözinger, J. Weitkamp (VCH Wiley, Weinheim, 1997), p. 911
2. M. Boudart, Principles of heterogeneous catalysis, in *Handbook of Heterogeneous Catalysis*, vol. 1, ed. by G. Ertl, H. Knözinger, J. Weitkamp (VCH Wiley, Weinheim, 1997), pp. 1–13
3. D.A. King, Surf. Sci. **47**, 384 (1975)
4. K. Christmann, *Introduction to Surface Physical Chemistry* (Steinkopff, Darmstadt, 1991)
5. T. Fujita, W. Kobayashi, C. Oshima, Surf. Interface Anal. **37**, 120 (2005)

6. G. Comsa, R. David, Surf. Sci. Rep. **5**, 145 (1985)
7. J.K. Roberts, Proc. R. Soc. Lond. A **152**, 445 (1935).
8. J.K. Roberts, Proc. R. Soc. Lond. A **152**, 464 (1935)
9. J.K. Roberts, Proc. R. Soc. Lond. A **152**, 477 (1935)
10. P. Kisliuk, J. Chem. Phys. **31**, 1605 (1959)
11. S. Černý, Surf. Sci. Rep. **26**, 1 (1996)
12. O. Beeck, Rev. Mod. Phys. **17**, 61 (1945)
13. O. Beeck, W.A. Cole, A. Wheeler, Discuss. Faraday Soc. **8**, 314 (1950)
14. G. Wedler, Z. Phys. Chem. **24**, 73 (1960)
15. G. Wedler, Z. Phys. Chem. **27**, 388 (1961)
16. G. Wedler, H. Strothenk, Ber. Bunsenges. Phys. Chem. **70**, 214 (1966).
17. G. Wedler, I. Ganzmann, D. Borgmann, Ber. Bunsenges. Phys. Chem. **97**, 293 (1993)
18. M. Kovář, L. Dvořák, S. Černý, Appl. Surf. Sci. **74**, 51 (1994)
19. J.K. Gimzewski, Ch. Gerber, E. Meyer, R.R. Schlittler, Chem. Phys. Lett. **217**, 589 (1994)
20. J.R. Barnes, R.J. Stephenson, M.E. Welland, Ch. Gerber, J.K. Gimzewski, Nature **372**, 79 (1994)
21. J.R. Barnes, R.J. Stephenson, C.N. Woodburn, S.J. O'Shea, M.E. Welland, T. Rayment, J.K. Gimzewski, Ch. Gerber, Rev. Sci. Instrum. **65**, 3793 (1994)
22. J.M. Antonietti, J. Gong, V. Habibpour, M.A. Rottgen, S. Abbet, C.J. Harding, M. Arenz, U. Heiz, C. Gerber, Rev. Sci. Instrum. **78**, 054101 (2007)
23. B.E. Spiewak, R.D. Cortright, J.A. Dumesic, Thermochemical characterization, in *Handbook of Heterogeneous Catalysis*, vol. 3, ed. by G. Ertl, H. Knözinger, J. Weitkamp (VCH Wiley, Weinheim, 1997), p. 698
24. R. Naumann d'Alnoncourt, M. Bergmann, J. Strunk, E. Löffler, O. Hinrichsen, M. Muhler, Thermochim. Acta **434**, 132 (2005)
25. L. Damjanovic, A. Auroux, Heterogeneous catalysis on solids, in *Handbook of Thermal Analysis and Calorimetry*. Recent Advances, Techniques and Applications, vol. 5, ed. by M.E. Brown, P.K. Gallagher (Elsevier, Amsterdam, 2008), p. 387
26. D.A. Kyser, R.J. Masel, Rev. Sci. Instrum. **58**, 2141 (1987)
27. C.E. Borroni-Bird, N. Al-Sarraf, S. Andersoon, D.A. King, Chem. Phys. Lett. **183**, 516 (1991)
28. A. Stuck, C.E. Wartnaby, Y.Y. Yeo, J.T. Stuckless, N. Al-Sarraf, D.A. King, Surf. Sci. **349**, 229 (1996)
29. W.A. Brown, R. Kose, D.A. King, Chem. Rev. **98**, 797 (1998)
30. R. Kose, New frontiers in single crystal adsorption calorimetry, Dissertation, Downing College, University of Cambridge, 1998
31. D.A. King, M.G. Wells, Proc. R. Soc. Lond. A **339**, 245 (1974)
32. J.T. Stuckless, N.A. Frei, C.T. Campbell, Rev. Sci. Instrum. **69**, 2427 (1998)
33. C.T. Campbell, O. Lytken, Surf. Sci. **603**, 1365 (2009)
34. S.F. Diaz, J.F. Zhu, N. Shamir, C.T. Campbell, Sens. Actuators B **107**, 454 (2005)
35. W. Lew, O. Lytken, J.A. Farmer, M.C. Crowe, C.T. Campbell, Rev. Sci. Instrum. **81**, 024102 (2010)
36. O. Lytken, W. Lew, J.J.W. Harris, E.K. Vestergaard, J.M. Gottfried, C.T. Campbell, J. Am. Chem. Soc. **130**, 10247 (2008)
37. The coverage Θ (with the unit ML) is here defined as the number of adsorbate molecules divided by the number of surface atoms (for Pt(111) 1.505 10^{15} cm^{-2}). A complete first layer of cyclohexene on Pt(111) at 100 K is reached at a coverage of $\Theta = 0.24$ ML
38. A. Schießer, R. Schäfer, Rev. Sci. Instrum. **80**, 086103 (2009)
39. A. Schießer, P. Hörtz, R. Schäfer, Surf. Sci. **604**, 2098 (2010)
40. K.D. Etzel, K.R. Bickel, R. Schuster, Rev. Sci. Instrum. **81**, 034101 (2010)
41. K.D. Etzel, K.R. Bickel, R. Schuster, Chem. Phys. Chem. **11**, 1416 (2010)
42. R. Schuster, R. Rösch, A.E. Timm, Z. Phys. Chem. **221**, 1479 (2007)
43. K.R. Bickel, K.D. Etzel, A.E. Timm, D. Nattland, R. Schuster, in *Hauptversammlung der Deutschen Bunsen-Gesellschaft für Physikalische Chemie*, Bielefeld, 13–15 May 2010
44. Y.Y. Yeo, C.E. Wartnaby, D.A. King, Science **268**, 1731 (1995)

45. R. Kose, D.A. King, Chem. Phys. Lett. **313**, 1–6 (1999)
46. D.W. Blakely, G.A. Somorjai, Surf. Sci. **65**, 419–442 (1977)
47. A. Gaussmann, N. Kruse, Surf. Sci. **266**, 46–50 (1992)
48. A. Gaussmann, N. Kruse, Surf. Sci. **279**, 319–327 (1992)
49. H. Orita, Y. Inada, J. Phys. Chem. B **109**, 22469–22475 (2005)
50. C.E. Wartnaby, A. Stuck, Y.Y. Yeo, D.A. King, J. Phys. Chem. **100**, 12483–12488 (1996)
51. W.A. Brown, R. Kose, D.A. King, Chem. Rev. **98**, 797–832 (1998)
52. H. Ihm, H.M. Ajo, J.M. Gottfried, P. Bera, C.T. Campbell, J. Phys. Chem. B **108**, 14627 (2004)
53. J.M. Gottfried, E.K. Vestergaard, P. Bera, C.T. Campbell, J. Phys. Chem. B **110**, 17539 (2006)
54. O. Lytken, W. Lew, C.T. Campbell, Chem. Soc. Rev. **37**, 2172 (2008)
55. W. Lew, M.C. Crowe, E. Karp, O. Lytken, J.A. Farmer, L. Arnadóttir, C. Schoenbaum, C.T. Campbell, J. Phys. Chem. C **115**, 11586 (2011)
56. W. Lew, M.C. Crowe, C.T. Campbell, J. Carrasco, A. Michaelides, J. Phys. Chem. C **115**, 23008 (2011)
57. F. Bebensee, J. Zhu, J.H. Baricuatro, J.A. Farmer, Y. Bai, H.-P. Steinrück, C.T. Campbell, J.M. Gottfried, Langmuir **26**, 9632–9639 (2010)
58. J. Zhu, F. Bebensee, W. Hieringer, W. Zhao, J.H. Baricuatro, J.A. Farmer, Y. Bai, H.-P. Steinrück, J.M. Gottfried, C.T. Campbell, J. Am. Chem. Soc. **131**, 13498 (2009)
59. R. Murdey, J.T. Stuckless, J. Am. Chem. Soc. **125**, 3995 (2003)
60. J. Zhu, P. Goetsch, N. Ruzycki, C.T. Campbell, J. Am. Chem. Soc. **129**, 6432 (2007)
61. R.H. Friend, Pure Appl. Chem. **73**, 425 (2001)
62. F. Faupel, V. Zaporojtchenko, T. Strunskus, J. Erichsen, K. Dolgner, A. Thran, M. Kiene, Fundamental aspects of polymer metallization, in *Metallization of Polymers 2*, ed. by E. Sacher (Kluwer Academic, New York, 2002)
63. C.H. Schwalb, S. Sachs, M. Marks, A. Schöll, F. Reinert, E. Umbach, U. Höfer, Phys. Rev. Lett. **101**, 146801 (2008)
64. F. Bebensee, M. Schmid, H.-P. Steinrück, C.T. Campbell, J.M. Gottfried, J. Am. Chem. Soc. **132**, 12163 (2010)
65. J.-H. Fischer-Wolfarth, J. Hartmann, J.A. Farmer, J.M. Flores-Camacho, C.T. Campbell, S. Schauermann, H.-J. Freund, Rev. Sci. Instrum. **82**, 024102 (2011)
66. J.-H. Fischer-Wolfarth, J.A. Farmer, J.M. Flores-Camacho, A. Genest, I.V. Yudanov, N. Rösch, C.T. Campbell, S. Schauermann, H.-J. Freund, Phys. Rev. B **81**, 241416 (2010)
67. D.R. Lide, *CRC Handbook of Chemistry and Physics*, 88th edn. (Internet Version 2008) (CRC Press/Taylor and Francis, Boca Raton)
68. J.A. Farmer, C.T. Campbell, Science **329**, 933 (2010)
69. R. Murdey, S.J.S. Liang, J.T. Stuckless, Rev. Sci. Instrum. **76**, 023911 (2005)

Part II
Optical Techniques

Chapter 3
Methods of IR Spectroscopy for Surfaces and Thin Films

David Allara and Josh Stapleton

Abstract The basic principles and experimental methods for application of infrared vibrational spectroscopy applied to characterization of surfaces and thin films are reviewed. The emphasis is on how the fundamental principles of electromagnetic radiation at interfaces dictate the experimental setups and underlie the strategies for making quantitative interpretations of the data. The discussion is directed to a broad range of applications from ultrahigh vacuum studies to the liquid/solid interface.

3.1 Introduction and Scope

Infrared (IR) spectroscopy (IRS) is a well-established, extremely powerful analytical technique for the study of thin films and surfaces. The literature is filled with examples of the analysis of the surfaces of materials in various physical forms ranging from planar single crystals for fundamental studies to powders and fibers for practical applications. Such a range is possible because of the flexibility in selecting sampling methods, which include the use of an IR beam to probe the sample via transmission, external reflection, internal reflection and diffuse reflection, but also includes even direct detection of the intrinsic thermal emission (graybody) from samples. The usefulness is bolstered by the fact that spectra can be typically acquired under conditions where the surrounding atmosphere, temperature and pressure can be widely varied. With this flexibility it is possible to study surface layers at the solid-solid, solid-liquid, solid-gas and liquid-gas interfaces. Excellent reviews which illustrate these points are available [1–3].

The dominant use of IR photons for probing surfaces and thin films is to provide vibrational spectra of the constituent substances present. Accessing vibrational

D. Allara (✉)
Depts. of Chemistry and Materials Science and Engineering, Pennsylvania State University, University Park, PA 16802, USA
e-mail: dla3@psu.edu

J. Stapleton
Materials Characterization Laboratory, Pennsylvania State University, University Park, PA 16802, USA
e-mail: jjs366@psu.edu

G. Bracco, B. Holst (eds.), *Surface Science Techniques*,
Springer Series in Surface Sciences 51, DOI 10.1007/978-3-642-34243-1_3,

information is a particular advantage for organic materials, with constituents ranging from small molecules to macromolecular and biomolecular species, all of which typically have a large number of richly varied vibrations, thus providing unique fingerprints for each type of material. The vast majority of useful vibrations is accessed in the mid-IR region of ~400–4000 cm^{-1} (equivalently, ~25–2.5 μm wavelength or ~50–500 meV energy). For example, a simple, typical polymer repeat unit with 12 atoms, including H, gives rise to ~30 different vibrations of which a large fraction, usually almost all, are accessible by IR excitation in the mid-IR. In this regard, IRS has a near unique ability to distinguish different types of organic functional groups which vary from each other only by small details in the bonding and constituent chemistry, e.g., $-(CH_2-CH-CO_2CH_3)-$ and $-(CH_2-CH-CO_2C_2H_5)-$ repeat groups in acrylate polymer films. In contrast, inorganic surfaces and thin films usually benefit far less from vibrational spectroscopy since typical inorganic materials of interest for surfaces and thin films, such as oxides, nitrides, carbides, and metal salts, have only a few characteristic vibrations (e.g., the Si–O stretch which dominates the SiO_2 spectrum) and these are often quite broad in energy due to the large numbers of ways in which the simple repeat units can synchronize their vibrations (phonon bands). At the limiting composition of pure elemental inorganic materials, such as pure metals or semiconductors, there are so few accessible vibrational features at typical IR frequencies (due to selection rules) that other methods almost exclusively are chosen for characterization. In order to view IR photon based probes in the context of the wide variety of other surface/thin film probes we show in Table 3.1 typical probes based on the general format of the particle that impacts the sample (particle in) and the particle that exits the sample (particle out) to a detector system. This leads to a general discussion of the IR technique in terms of an incoming beam of IR photons of some selected wavelength (λ) or energy ($h\nu$ or E) with an output beam from the sample. Accordingly we will need to develop the physics of the interactions in order to see how much information can be extracted by paying attention to the details of the measurements.

Notice in Table 3.1 that the various characterization methods have been categorized in terms of the types of processes that occur when the incoming particle beam strikes the sample and the effect on the outgoing particle: (1) scattering (*bouncing off*) of the incoming particle with no change in energy (**elastic scattering**), though change of direction can occur (diffraction is a typical example), (2) scattering with a loss (or sometimes gain) in kinetic energy (KE) of the recoiling particle with the energy supplied by some characteristic quantum state transition of the sample (**inelastic scattering**), (3) complete absorption of the energy of some fraction of the incoming particles (annihilating them in the case of photons) as a function of input energy (**energy absorption spectroscopy**), and (4) impact of the incoming particle beam to cause ejection or emission of a new particle from the sample which leaves with characteristic information about the sample as a function of the energy of the incoming particle (**emission spectroscopy**). In the last entry in the table also note the addition of thermally induced emission of IR radiation from the sample. Though this process does not involve injection of excitation energy into the sample by a discrete probe particle, we can think of the injection

Table 3.1 General description of particle in—particle out surface/thin film characterization techniques

Category	Particle in → Particle out	Output beam measure	Common name
elastic scattering—momentum spectroscopy	same particle; $\Delta E = 0$		
	$h\nu \rightarrow h\nu$ (x-ray);	I_{out} vs	
	$e \rightarrow e$;	scattered beam	
	atom → atom	angles	diffraction
	$h\nu \rightarrow h\nu$ (x-ray);	I_{out} vs specular reflection angles	reflectometry
	$h\nu_{in} \rightarrow h\nu_{out}$ (uv-vis-IR)	ψ, Δ vs specular reflection angles at fixed λ or vs λ at fixed angle	ellipsometry variable angle or variable λ
inelastic scattering—energy shift spectroscopy	same particle; $\Delta E \neq 0$;		
	$h\nu_{in} \rightarrow h\nu_{out}$ (uv-vis-near IR),	I_{out} vs $(\lambda \text{ or } \nu)_{out}$	Raman
	$e_{in} \rightarrow e_{out}$,	I_{out} vs ΔKE or	energy loss
	$\text{atom}_{in} \rightarrow \text{atom}_{out}$	I_{out} vs KE_{in}	
	$\text{ion}_{in} \rightarrow \text{ion}_{out}$	I_{out} vs ΔKE	ion back scattering
energy absorption spectroscopy	same particle; $\Delta E \neq 0$;		
	fraction input beam absorbed		
	$h\nu_{in} \rightarrow h\nu_{out}$ (uv, vis, IR)	I_{in}/I_{out} vs $(\lambda \text{ or } \nu)_{in}$	absorption
	$h\nu_{in} \rightarrow h\nu_{out}$ (uv, vis, IR)	ψ, Δ vs $(\lambda \text{ or } \nu)_{in}$	spectroscopic ellipsometry
emission spectroscopy	new particle ejected		
	$h\nu_{in} \rightarrow h\nu_{out}$ (uv or vis)	I_{out} vs $(\lambda \text{ or } \nu)_{out}$	fluorescence
	$h\nu_{in} \rightarrow e_{out}$	I_{out} vs KE_{out} or	photoelectron
	$(\text{uv or x-ray})_{in}$	I_{out} vs λ	emission
	$e_{in} \rightarrow e_{out}$	I_{out} vs KE_{out}	Auger
	$\text{ion}_{in} \rightarrow \text{ions}_{out}$	I_{out} vs $(\text{ion charge/mass})_{out}$	secondary ion mass spectrometry
	kT_{in} (bath phonons) $\rightarrow h\nu_{out}$ (IR)	I_{out} vs $(\lambda \text{ or } \nu)_{out}$	thermal emission

as caused by a wide spectrum of energetic impacts from contact with an adjacent thermal bath (with average energy $\sim kT$) which can induce collective motions of the atoms within the material (phonons) to a an energy sufficient to cause ejection of characteristic photons. This table covers the wide variety of typical techniques used for surface/thin film analysis, ranging across grazing incidence x-ray diffraction, electron and atom diffraction, x-ray reflectometry, and uv-vis-IR ellipsometry, all on non-absorbing samples (elastic scattering, momentum spectroscopy); Raman spectroscopy; electron, atom and ion energy loss spectroscopies (energy shift spectroscopies); standard uv-vis-IR absorption spectroscopies, including ellipsometric detection (polarization); and ejected particle spectroscopies ranging across fluorescence, x-ray and Auger electron emission, secondary ion mass spectrometry and thermal (gray-body) emission spectroscopy (emission spectroscopy). If one reflects on the type of information that can be obtained from the variety of these probes it should become clear that they cover surface/thin film features such as translational ordering; film thickness, roughness and density profiles as a function of depth; types of chemical groups and their orientation; elemental composition, also as a function of depth; and chemical interactions between components of the films and surface layer. These probes together provide measures of order parameters, thickness, morphology, orientation, intermolecular interactions, conformations and other important characteristics. So the important question is what can IR probes do relative to all this information? As a start we note that the application of IR photons (in or out) arises in each of the standard characterization categories. Thus IR photons can undergo all of the processes listed and we need to understand what can be obtained from each in detail for standard samples.

In order to investigate this aspect we need to develop the basis of the interaction of IR photons with surfaces/thin films and see what information arises from measurements. In particular it is of interest to see how much one can push the characterizations in a quantitative direction and extract the maximum information. Overall, heading in this direction, this review will focus on the general methods and approach to developing analysis strategies, with less emphasis on specific instrumentation and laboratory procedures, though a good bit of this will be presented to be practical. Of particular interest are the general experimental setups, as dictated by the overarching theory of the interactions of the samples with the electromagnetic fields of the impinging IR beams, and the types of information about the samples that can be extracted with proper interpretation, generally based on simulations grounded in the theory. So first we start with the basic theory that underlies the interaction of photons with planar surfaces and thin films at standard beam intensities. Once this basis is developed we can move briefly into the instrument components and experimental strategies needed to solve problems in surface/thin film characterization and finish with some selected examples to illustrate the range of studies possible.

3.2 Basic Principles of Plane Wave Interactions with Planar Film Stacks and Associated Constraints on Sampling Modes and Surface Characterization Attributes at IR Wavelengths

3.2.1 Basic Theory for Low to Moderate Intensity Beams (First Order Scattering Regime)

Any surface or thin film spectroscopy based on photons can be understood on the basis of two types of theory: (1) quantum theory of radiative excitation and decay between energy states [4] and (2) a classical theory of the interaction of electromagnetic (EM) waves with material media [5]. In the case of IR photons the first almost always involves transitions between vibrational energy levels with accompanying

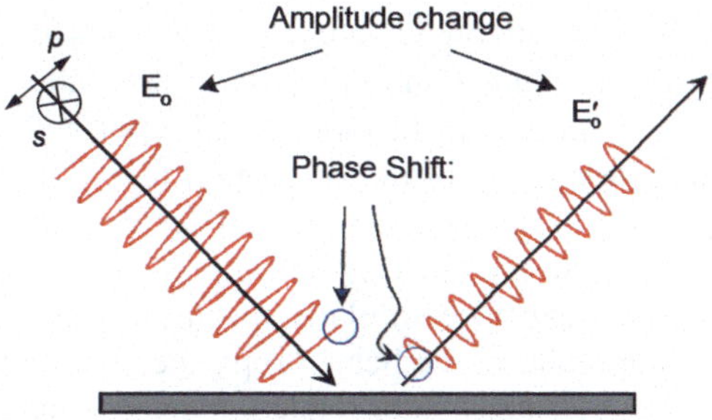

Fig. 3.1 Schematic of a reflection experiment in which a beam of polarized light is specularly reflected from a film surface. The resulting process has both an associated change in the E-field amplitude and the phase

emission or absorption of a quantum of energy manifested in the appearance or disappearance of photons, processes which can reveal features such as the types and symmetries of chemical bonds present. In the case of organic materials the bonding is further interpreted in terms of groups of atoms [6], typically defined as functional or molecular groups.

These quantum excitations can be accomplished either by intrinsic processes in the sample at a given temperature which produce excited states in the Boltzmann distribution or by an incoming light beam. Since this review is primarily focused on the use of IR beam probes we will shift to this latter aspect. For an incoming light beam to produce an oscillating electric field [E-field or E(ω), where ω is the angular frequency of the light] in the region of the sample where the film is placed one needs to properly configure the sample and the optical setup and ideally one wants to maximize E(ω) in that region. The magnitude and polarization of the E-field interacting with the medium are determined by factors incorporated in classical EM theory which treats the response of a given medium to an incoming EM wave in terms of the E-fields and propagation directions throughout the medium and across media boundaries as is appropriate. Thus the full description and quantitative interpretation of an IR spectrum in terms of the number of active IR excitation sites (typically called oscillators) per unit area or volume and the structural aspects of these sites needs both types of theory. This review will focus on the classical EM theory and related experimental aspects and treat the absorption process strictly in terms of a classical oscillator whose excitation probability in a medium of oscillators is handled in terms of a cross-section described as a simple optical absorption parameter. This approach avoids the issue of quantum mechanical effects, which then can be treated separately as needed.

The essential features of the interaction of a plane-polarized light beam with the surface of a planar film in an air or vacuum environment are shown in Fig. 3.1 for the case of specular reflection. The polarization of the E(ω)-field of the impinging beam consists of two limits, p and s, which are respectively parallel or perpendicular to the plane of incidence, which is in the plane of Fig. 3.1, where the oscillating E-field represented by the sine-wave traces is shown as p-polarized. For almost all materials of interest pure s or p polarization in the incoming beam is conserved in the outgoing beams. The vast majority of IR spectroscopy experiments are done with an unpolarized or pure s- or p-beams and only the change in beam power is measured. In the most general case, however, the incoming E-field can be elliptically polarized with both s- and p-components and under this condition the measurement of the changes in the s/p polarization ratio of the incoming and outgoing E-fields can be measured. In more detail, upon reflection the E-field undergoes a shift in the phase angle of the oscillation ($\delta\phi$) and a loss in the incoming amplitude (E_0) (see Fig. 3.1) with the magnitudes determined by the intrinsic optical properties of the film, the beam polarization and the angle of incidence. Transmission across the film interface similarly results in some corresponding shift in phase and loss in amplitude as the beam propagates into the film medium. So to be most general, in any transmission or reflection measurement there are only two parameters which can be determined. In most experiments only the change in beam power is measured and the phase and amplitude changes are lumped together. Measurement of both the phase and amplitude changes is the basis of ellipsometry [7].

For the long wavelengths of infrared light (the ~2.5–25 μm region) uniform density materials with submicron or nanometer scale density variations appear continuous with an average electron density. Under this condition no diffraction from the atomic lattice occurs due to the large relative spatial range of the each wave (covering many corrugations in the lattice electron density) and the intrinsic optical response of the bulk of a material can be described solely in terms of an average polarization of the electrons in the material. This aspect will be treated later in terms of an effective medium approximation. For materials with density fluctuations in the film interior (e.g., the presence of voids) or at the surface (rough surfaces) on a scale or larger than the wavelength of the light (e.g., large particles or very rough surfaces exceed the μm scale), diffuse scattering of the radiation can occur, often creating problems in obtaining easily interpreted IR of large particles. The treatment of such scattering effects [5] is beyond the scope of this review and only uniform materials with density fluctuations below the mid-IR wavelength scale will be considered in the theory section. Also note that at the levels of radiation power used in IR spectroscopy the polarization of the electrons in the material is linear with the electric field amplitude and is rigorously described by the frequency dependent refractive index, or equivalently a dielectric function, assigned to the medium of interest. In order to exceed the range of this behavior, very intense IR laser pulses must be used, which can lead to thermal degradation of the sample. In combination with an intense visible wavelength, synchronized pulse a non-linear effect can arise which is the basis of sum frequency generation (SFG) vibrational spectroscopy. In a general case, the refractive index and the equivalent dielectric function are defined as frequency-dependent complex quantities, respectively, in (3.1) and (3.2), with $\omega = 2\pi\nu$, where ν = frequency, and the relationship between these two quantities is given by (3.3).

$$\hat{n}(\omega) = n(\omega) + ik(\omega) \tag{3.1}$$

$$\hat{\epsilon}(\omega) = \epsilon_1(\omega) + i\epsilon_2(\omega) \tag{3.2}$$

$$\hat{n}(\omega)^2 = \hat{\epsilon}(\omega) \tag{3.3}$$

The use of the complex quantity $\hat{n}(\omega)$ allows for inclusion of both refraction effects (real part) and absorption (spectroscopic, arising from energy absorption or loss; imaginary part) effects.

The physical basis of the dielectric function follows directly. In this review we interchangeably use ω and ν to denote frequency, depending on the specific equation. The equivalent energy is given by the Einstein relationship $E = h\nu$. The frequency is related to wavelength by $v = \lambda\nu$, where $v =$ velocity of light in the medium of interest and v varies with the real refractive index n of the medium according to $v = c/n$, where $c =$ speed of light in vacuum $(n = 1) = 3.0 \times 10^8$ m/s. Finally, it is common practice in IRS to define the energy of the light in terms of the reciprocal wavelength or wavenumber as $\tilde{\nu} = 1/\lambda = \nu/v = n\nu/c$, where $\tilde{\nu}$ is typically given in cm^{-1} units for historical reasons, rather than the standard SI units of m^{-1}.

The real and imaginary parts of the complex functions $\hat{n}(\nu)$ and $\hat{\epsilon}(\nu)$ form an analytic pair of variables in the complex plane and thus are interrelated by integration over all frequency space by the well-known Kramers-Kronig (KK) integral. For the function $k_j(\nu)$ for the jth oscillator defined over the limited range ν_a to ν_b which surrounds the excitation range, a good approximation of $n_j(\nu)$ is given by the form of the KK integral in (3.4):

$$n_j(\nu) = n_j^{\infty}(\nu) + P \int_{\nu_a}^{\nu_b} \frac{k_j(\nu)\nu'}{\nu'^2 - \nu_j^2} d\nu' \tag{3.4}$$

where, $n_j^{\infty}(\nu) =$ the real refractive index outside of the oscillator absorption range (baseline refractive index), $\nu_j =$ the frequency at the maximum value of k_j and P specifies the principal part of the integral (avoiding the singularity at $\nu_j = \nu'$).

With the above background, the capabilities of IR thin film characterization can be discussed in terms of a response of the probing beam as a function of the refractive index $[n(\omega)]$ and the energy absorption or loss $[k(\omega)]$. The former carries the basis for momentum spectroscopy and the latter for energy absorption spectroscopy. This gives us a direct basis to compare with diverse techniques such as x-ray reflectometry (purely a momentum probe) and ultraviolet-visible (uv-vis) ellipsometry (both a momentum and absorption probe).

3.2.2 The Interaction of Parallel Plane EM Waves with Planar Film Stacks—Transmission, Reflection and Evanescent Wave Sampling

Beam Impinging on the Sample from a Low Refractive Index External Medium: Transmission and External Reflection Sampling

For the typical case of a propagating light beam in air or vacuum impinging onto a sample the beam will reflect and also can be transmitted through the sample if it is sufficiently non-absorbing. The process of interacting with a film stack can be understood in terms of a standing wave permeating the entire structure with the beam partitioning some of its intensity between reflection and transmission at each interface as it passes through the sample, as illustrated in Fig. 3.2 for a two layer stack.

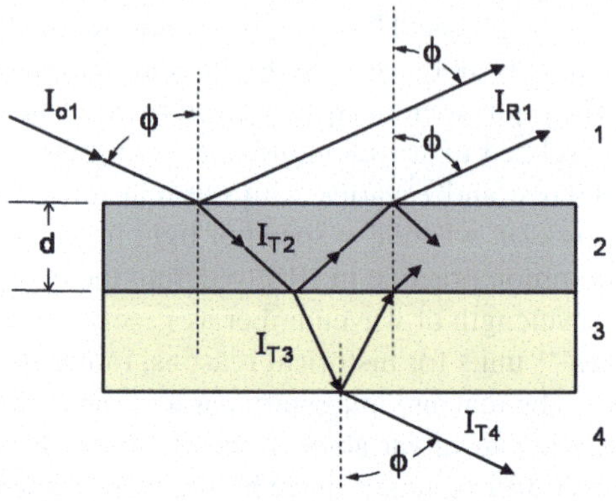

Fig. 3.2 Schematic showing a plane wave light beam at an angle of incidence ϕ interacting with a planar film of thickness d. The figure shows the ambient medium (*1*), a top layer (*2*, *gray*) and a second layer (*3*, *yellow*) which could be infinite (as for example air) or adjacent to another medium not shown

A beam of intensity I_{o1} starting in non-absorbing medium 1 (typically air or vacuum) impinges on the boundary between media 1 and 2 at an angle of incidence of ϕ to generate a reflected specular beam of intensity I_{1R} at the exit angle ϕ and also a refracted transmitted beam into medium 2 with intensity I_{T2}. If media 2 and 3 are sufficiently non-absorbing the transmitted beam will continue on and exit from the other side of the sample. The beam bouncing within the phase boundaries creates constructive and destructive interference of the E-fields, which along with absorption due to excitation of quantum states, regulates the intensity of the total output light beams (reflected or transmitted). The bending of the propagation direction at each phase boundary arises from differences in the real refractive indices between the media (Snell's law) and is a simple outcome of the change in the speed of the light in parallel rays when crossing the boundary at an angle.

With the experimental variables fixed, viz, ϕ, λ (or ν) and beam polarization, the total transmissivity and reflectivity can be related directly to the layer thickness and complex refractive indices, regardless of the numbers of layers. The basic equations, which we develop later, apply to all wavelengths, from x-rays to radio frequencies. The important point to note is that the E-field exists everywhere in the sample unless it is damped by absorption during passage, resulting only in a reflected beam.

Beam Impinging on the Sample from a High Index External Medium: Internal Reflection Sampling

If the conditions of the above experiment are arranged to have the beam source propagating into a medium with a higher real refractive index than the top film, viz, $n_1 > n_2$, and with the incoming beam direction adjusted so $\phi > \phi_c$, where ϕ_c is some critical angle determined by the ratio n_1/n_2, the beam does not propagate into the adjacent medium but rather sets up a stationary EM field which penetrates into the medium and decays exponentially with distance. This type of bounded EM field, termed an evanescent wave, is constrained to penetrate into the lower refractive index medium to a depth typically slightly less than λ before it decays away [5]. The evanescent wave decay at a media boundary is illustrated schematically in Fig. 3.3

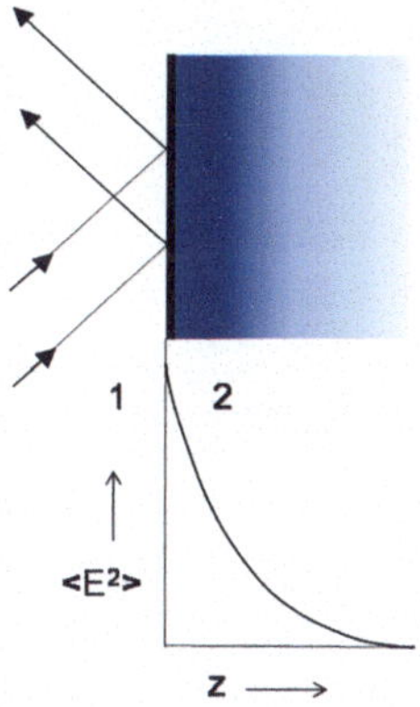

Fig. 3.3 Illustrations of a decaying evanescent E-field at a film surface

by the diminishing of the shading into medium 2 in the upper panel and by the associated plot of the mean square field in the lower panel. A general equation for this distance dependence is given in (3.5):

$$\frac{\langle E_z^2 \rangle}{\langle E_{zo}^2 \rangle} = \exp(-d_p z C) \tag{3.5}$$

where, the term $\langle E_z^2 \rangle / \langle E_{zo}^2 \rangle$ is the ratio of the mean square value of the electric field at a distance z away from the phase boundary interface to the value at the interface, C is a collection of constants and d_p is the penetration depth as defined in (3.6), where λ is the wavelength of the incoming light, n_1 and n_2 are the real refractive indices of the adjoining media and ϕ is the angle of incidence.

$$d_p = \frac{\lambda}{2\pi n_1} \left[\sin^2 \phi - \left(\frac{n_1}{n_2} \right)^2 \right]^{-1/2} \tag{3.6}$$

For such configurations one arranges to have the incoming and reflected beams propagating in a non-absorbing lower index medium (medium 2) contiguous to the sample of interest which allows some depth region into the film medium (3) to be probed selectively. This is typically done in two modes: (1) single reflection using an IR transparent hemispherical or cylindrical prism or (2) multiple reflection using a trapezoidal or rhomboidal prism, as illustrated in Fig. 3.4. In the figure, media 1, 2 and 3 represent ambient (air or vacuum), the prism material, typically of high refractive index (e.g., Si) and the film of interest. If the film is sufficiently thin to allow some of the evanescent field to penetrate through the film and exit into a final medium (medium 4, not shown) the refractive index of the final medium will have an effect on the reflectivity of the sample.

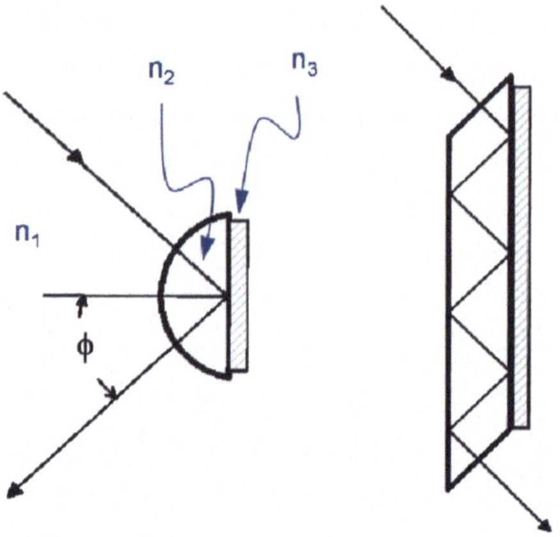

Fig. 3.4 Illustrations of common modes for internal reflectance spectroscopy. *Left* and *right*: single and multiple reflection configurations

3.2.3 General Characterization Capabilities and Features in the IR Wavelength Region

Commonly desired attributes of surface and thin film characterization techniques are:

1. *surface selectivity*: the ability to selectively sample only the surface (or interface) region of interest with signal from no other region
2. *surface sensitivity*: the ability to have sufficient signal/noise to provide good quality spectra for the region of interest
3. *non-destructive characterization*: no sample damage during analysis
4. *lateral imaging*: the ability to image laterally across the surface
5. *in-situ analysis*: directly under special sample environments

Each of these attributes is determined by the fundamental interaction of the IR beam with the sample and the limits can be understood on the basis of the simple theory in the previous section. We now discuss each of these attributes in turn.

Surface Selectivity

An important aspect of surface characterization to consider for each type of probe is the ability to selectively generate a signal from the surface region vs sampling throughout the depth of the entire sample. For example, in a totally surface selective analysis only the top row of atoms would be sampled. Generally a probe is considered as surface selective if the probe signal comes dominantly from the top ~1–2 nm of the outer sample surface. Accordingly the surface selectivity of an IRS experiment is determined by the ability to confine the overall distribution of the EM fields of the interacting light to the surface regions of the target solid.

For external reflection and transmission experiments it is clear from the basic EM theory in the previous section that the EM field permeates the entire sample and is only absent in regions where the field has been damped by the presence of strong absorption. Thus it is not possible to arrange sample configurations which can concentrate the E-field in a given region for selective analysis, e.g., at the surface or interface of a film or in the region of one particular film in film stack.

On the other hand for an internal reflection experiment using an IR transparent prism with a film of interest on the back surface, the evanescent field at the film-prism interface will be localized in film at some distance according to (3.5) and (3.6), typically of the order of the wavelength of the IR beam which is ~3–25 μm. Thus the **best** that typically can be done for IR is to achieve a selectivity limited to μm scale surface regions for internal reflection geometries so overall, in terms of surface selective analysis techniques IR offers no surface selectivity. In contrast, particle in - particle out surface techniques such as x-ray photoelectron, Auger, ion scattering and high resolution electron energy loss spectroscopy can selectively probe within a few nanometers from the substrate surface.

Surface selectivity can be arranged, however, by restricting analysis to samples with a thin absorbing overlayer film on a substrate medium in which the overlayer spectrum is distinctly different than the substrate signal. In this case subtraction of the substrate signal can reveal the overlayer spectrum with good contrast. This works well for an internal reflection experiment with a thin overlayer film since the prisms are especially selected to be non-absorbing in typical IR regions of interest. For external reflection experiments metal substrates work well since they offer the advantage they are generally spectrally featureless in the IR except for any overlayer oxides, e.g., Al_2O_3 on Al metal which absorbs near 800–1000 cm^{-1}. For non-metal substrates, however, the substrate signals can be strong and cause severe problems. For example, in the case of external reflection from a μm-thickness polymer film with a thin (~1 nm) modified surface region, the bulk signal could be orders of magnitude more intense than that of the surface region and mask the much weaker surface signal.

Surface Sensitivity

The surface sensitivity depends completely upon the signal/noise of the measurement with regard to the fraction of the total signal contributed by the surface region of interest. The signal/noise in an IRS experiment using contemporary instrumentation (e.g., interferometers with Fourier transform data processing) is sufficient in some cases to detect the signal from as little as ~0.001 monolayers in the case of highly reflective substrates (e.g., gold). As expected, the problem often is simply that with no corresponding surface selectivity the surface signal contrast is low because it is buried in a bulk signal that obscures the surface contribution.

Non-destructive Characterization

One general advantage of IRS, compared to many popular ion and electron beam surface analysis techniques, is the non-destructive nature of the analysis. Atoms and ions in beams can carry sufficient momentum to damage and even remove (sputter) material by billiard-ball type collisions while electrons often create severe damage in passing into materials, particularly organics, via electron-electron collisions which tend to disrupt bonding and cause chemical changes. In contrast, photons will tend to pass through materials with no energy dissipation except for resonant excitations of energy states, which for the IR region consist of vibrational modes. For a standard IRS experiment with current instrumentation the associated power dissipation per unit volume is well below any damage threshold. Note, however, that the heating fluctuations and changes in sample temperature as a function of the excitations at each resonant frequency are sufficiently large to be detectable and, in fact, can be used as a highly sensitive means for generating spectra, e.g., photoacoustic and photothermal reflection detectors (see later section on detectors).

Lateral Imaging

One specific disadvantage of IRS is the fact that lateral imaging capabilities are limited by the diffraction properties of the light. Thus, a beam of infrared light, using normal far-field optics (the focusing optics located many wavelengths away from the sample), cannot be focused to a point much smaller that the wavelength of the light, typically of the order of several μm, without the appearance of undesirable diffraction beams at diverging angles. In comparison, the best imaging capabilities of an ion beam (e.g., from a liquid metal) for an analysis by secondary ion mass spectrometry can reach about 0.05 μm [8].

In-situ Analysis

As a final general point, the in-situ capabilities of IRS can offer a distinct advantage for cases in which the analysis of a surface is desired while the sample remains under conditions of direct exposure to liquid or gaseous environments, e.g., electrochemical, biological or catalytic systems, and other controlled environments such as vacuum and atom, molecule, ion or electron beams. It should be noted, however, that many liquids and gases have strong absorption bands in the infrared region, e.g., water, so in-situ infrared spectroscopy is limited to solvents or gases with minimal absorption and to very limited frequency ranges between absorption features or to very short path lengths which minimize absorption.

3.3 Quantitative Aspects—Spectral Simulations

In order to obtain the maximum amount of information from spectra one needs to be able analyze the absolute intensities in a quantitative way. This information can provide a means of determining the number of IR oscillators present in the sample overlayer as well as their orientations. In order to do this the instruments need to be fully calibrated with controlled optics. Typical quantitation involves transmission spectra through films of known thicknesses with interpretation done using a simple Beer's law approximation (absorption of beam power decreases exponentially with thickness penetrated) for the power absorbed. One of the problems in the quantitative analysis of thin film samples is that for structures such as an overlayer organic film on a semiconductor wafer or a polymer sheet, reflection spectra can be highly complex with both positive and negative changes in the fraction of reflected beam power in the spectral region of an intrinsic absorption (or excitation) of the sample [9]. In such cases Beer's law approximations do not work and are confounded by what may appear to be baseline undulations and artifacts. Using more sophisticated approaches, the spectra from planar, parallel layer samples can be understood quantitatively in terms of the thickness of the film and the optical properties of the constituent layers, which includes any anisotropy such as the orientational ordering of molecular groups. In this section the fundamental basis of spectroscopic intensities will be discussed followed by the explicit electromagnetic equations which control the observed reflectivities and transmissivities of an experimental spectrum of a thin film sample. These relationships are utilized to construct practical approaches to quantitative interpretation of thin film IR spectra [10].

3.3.1 Spectroscopic Intensities in Terms of the Absorption Properties of Materials

The physical mechanism of IRS is well understood for the elementary case of a photon-electric dipole excited vibrational transition of an isolated molecule in free space [4]. In the dipole approximation the power absorbed when an electromagnetic field oscillating at a given frequency interacts with a vibrating electric dipole, is given by the proportionality:

$$I \propto |\mathbf{E}_{loc} \cdot \mathbf{p}|^2 \tag{3.7}$$

where I is the power absorbed or, equivalently, the rate of the excitation, $\mathbf{E}_{loc}$ is the electric field vector in the local vicinity of the molecule and $\mathbf{p}$, in quantum mechanical terms, is the electric dipole moment transition matrix element for the given excitation at the given frequency. The unit vector $\mathbf{p}/|\mathbf{p}|$ defines the direction of the transition dipole and $|\mathbf{p}|$ is the magnitude of a single isolated oscillator. For a condensed-phase medium of a set of parallel-aligned oscillators with some excitation or resonance at frequency ν, the value of $k(\nu) = \mathrm{Im}[\hat{n}(\nu)]$ defines the intensity of the transition. It is convenient to use k, dropping (ν) for convenience, rather than

$|\mathbf{p}|$ as the basis for intensity since k often can be derived experimentally from spectroscopic measurements on the pure material whereas $|\mathbf{p}|$ is usually thought of as a property of an isolated species. Therefore, it is convenient to form a vector $\mathbf{k}$ which carries both the direction of $\mathbf{p}$ and the intensity of k. Methods for actually making such calculations will be described later. The relationship between $\mathbf{p}$ and $\mathbf{k}$ establishes the fundamental connections between the molecular structure of the film medium and the optical response properties. This information is essential for correct interpretation of thin film spectra in terms of molecular structure, particularly the orientation of molecular or inorganic groups in organized films.

3.3.2 Electromagnetic Fields in Planar Thin Film Structures

Definition of the Fundamental Electromagnetic Problem

The general reflection/transmission experiment of interest here is schematically shown in Fig. 3.5. An incident beam of collimated plane-wave light propagating in infinite medium 1 impinges onto the surface of a stack of $N-2$ semi-infinite, parallel, planar slabs, each specified by the thickness and optical functions, described as d_j and $\hat{n}_j$ or $\hat{\epsilon}_j$ for the jth layer. To be general, the optical functions, which depend upon the frequency of the light, can be expressed as scalars or tensors, depending up whether they represent oscillators that are *isotropic*, viz, distributed with no net orientation, or *anisotropic*, viz, distributed with some uniform orientation relative to the sample coordinate frame. The angle between the incoming beam and the surface normal defines the angle of incidence ϕ which is contained in the plane of incidence. The coordinate system is chosen for convenience to have the slab surface as the xy-plane and the plane of incidence in the xz-plane. By directing the incoming beam in a different direction around the xy-plane the plane of incidence will be rotated azimuthally, shown in the diagram as the angle θ from the x-axis. The state of polarization of the electric field of the incident beam is defined in terms of the direction of the electric field vector relative to the incidence plane.

As described earlier (Fig. 3.2), part of the beam is transmitted into the sample and, if the sample layers are sufficiently transparent to the radiation, some fraction of the initial beam exits from the final slab into an adjacent, infinite medium (N in Fig. 3.5). The other portion of the incident beam is reflected from the surface. The simplest experiment consists of measuring the reflectivity or transmissivity of a pure s- or p-polarized beam of light as a function of frequency. Since s- and p-light generally have different interactions with the sample, in the case of an elliptically polarized input beam with s- and p-components the polarization ellipse will change upon exit; measurement of this change is the basis for ellipsometry. The input and output power intensity or the polarization data is interpreted in terms of formal relationships between the changes in these reflectivities and material properties via EM theory. These relationships, which are the basis for quantitation of spectra, are presented below.

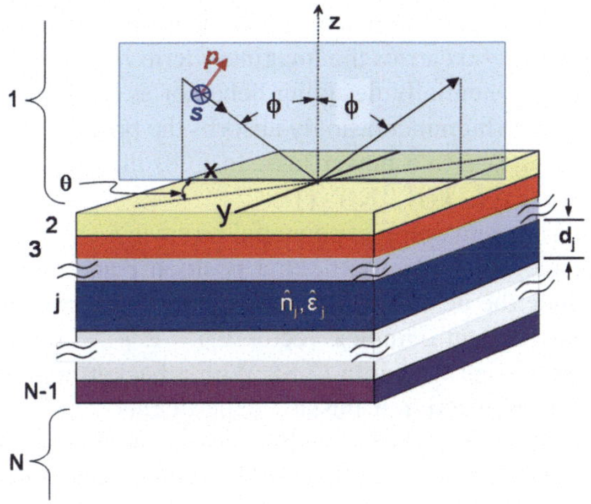

Fig. 3.5 Schematic of a N − 2 planar, parallel layer film stack with ambient media 1 and N showing polarized incoming and outgoing beams in medium 1 at incidence angle ϕ and azimuthal angle θ in the xy surface plane. Each film is characterized by a thickness d and a complex refractive index or dielectric constant

Before proceeding with the development of equations, it is necessary to briefly discuss representations. A real scalar quantity s is represented in normal typeface (s) and when complex with a "hat" $(\hat{s})$. A vector quantity v is expressed in bold typeface $(\mathbf{v})$ when real and as $\hat{\mathbf{v}}$ when complex. A real and a complex tensor of rank r are represent as $T^{(r)}$ and $\hat{T}^{(r)}$, respectively. In linear responses (first order responses to the electric field, such as are treated in this chapter) of isotropic media (e.g., liquids and amorphous solids) optical responses do not depend on the orientation of the sample, other than the trivial change due to angle of incidence, so the refractive index and dielectric function are described as simple scalars since their function is simply to scale the magnitude of the electric fields. In contrast, anisotropic media exhibit different values of the optical functions for different directions in the medium, e.g., a birefringent crystal, an oriented Langmuir-Blodgett film or a biaxial polymer film, so the refractive index and dielectric properties are expressed as second rank tensors $\underline{\hat{\epsilon}}^{(2)}$ and $\underline{\hat{n}}^{(2)}$ with their elements, which carry the directional responses, represented respectively by $\underline{\hat{\epsilon}}^{(2)}_{j,k}$ and $\underline{\hat{n}}^{(2)}_{j,k}$, essentially a 3×3 matrix description. Typically one knows when the dielectric constant or refractive index are treated as tensors (for optically anisotropic materials) so the formal rank designation is often dropped. At high incident beam powers higher order behavior can occur (2nd and even 3rd order which requires the use of higher rank tensors) but such behavior rarely occurs in typical IRS experiments so will not be considered further.

For a linearly polarized infinite, homogeneous, plane wave of vacuum wavelength λ propagating through a medium with associated refractive index $\hat{n}_1$, the time-dependent electric field vector at location $\mathbf{r}$ and time t is given by:

$$\mathbf{E}(\mathbf{r}, t) = \mathbf{E}^{\mathbf{o}} e^{-i(\omega t - \mathbf{K}\cdot\mathbf{r} + \alpha)} \tag{3.8}$$

where $\mathbf{E}^{\mathbf{o}}$ is a real vector defining the maximum amplitude and direction of the electric field, ω is the angular frequency of the light, $\mathbf{K}$ is the wavevector defined as having magnitude $|\mathbf{K}| = 2\pi\hat{n}_1/\lambda$ in medium 1 with a direction along the wave propagation axis and α is a phase angle to relate the phase of the oscillating electric field to some reference. When the refractive index is complex the absorption coefficient $k > 0$ carries the imaginary term ik into the exponent in (3.8) which generates an exponentially decaying behavior as the light wave travels through the medium. The medium increasingly absorbs the propagating wave and the energy is dissipated giving rise to a power loss. In (3.8) this dissipation arises as the real exponential term $\exp[-k(\mathbf{K}/|\mathbf{K}|) \cdot \mathbf{r}]$.

The remaining exponential terms in (3.8) are all imaginary and define the phase angle of the electric field at position $\mathbf{r}$ and time t. It is convenient, in order to include the description of ellipsometry experiments, to define the time and position independent complex vector $\hat{\mathbf{E}}^{\mathbf{o}} = \mathbf{E}^{\mathbf{o}} e^{-i\phi}$ for which $e^{-i\phi}$ carries the total phase angle contribution in (3.8). With a background established we now discuss simple reflection and transmission experiments.

In an experiment in which pure s- or pure p-polarized light reflects off the planar surface of an isotropic material with conservation of the polarization direction, the complex electric fields of the incoming and reflected outgoing linearly polarized waves at the surface are each represented by (3.8) but with different amplitude and phase factors, $|\mathbf{E}^{\mathbf{o}}|$ and ϕ.

The complex reflectivity for this special case then is defined as the time-independent ratio of the incoming and outgoing electric field vectors at the specific point of reflection:

$$\frac{|\hat{\mathbf{E}}^{\mathbf{o}}_{out}|}{|\hat{\mathbf{E}}^{\mathbf{o}}_{in}|} = \frac{|\mathbf{E}^{\mathbf{o}}_{out}|}{|\mathbf{E}^{\mathbf{o}}_{in}|} \exp\big[-i(\phi_{out} - \phi_{in})\big] = \hat{r} \tag{3.9}$$

The complex scalar $\hat{r}$, a convenient abbreviated notation for complex electric field ratios, is known as the Fresnel coefficient and is designated as $\hat{r}_s$ or $\hat{r}_p$ depending on the polarization state.

The power reflectivity, the actual quantity measured in a reflection experiment, is defined in terms of the ratio of the absolute squares of the complex electric fields, which here can be either s- or p-polarized depending upon the polarization chosen for the experiment:

$$R = \frac{|\hat{\mathbf{E}}^{\mathbf{o}}_{out}|^2}{|\hat{\mathbf{E}}^{\mathbf{o}}_{in}|^2} = \frac{|(\hat{\mathbf{E}}^{\mathbf{o}}_{out})^* \cdot \hat{\mathbf{E}}^{\mathbf{o}}_{out}|}{|(\hat{\mathbf{E}}^{\mathbf{o}}_{in})^* \cdot \hat{\mathbf{E}}^{\mathbf{o}}_{in}|} = \hat{r}^* \cdot \hat{r} = \frac{I_{out}}{I_{in}} \tag{3.10}$$

where $\hat{\mathbf{E}}^*$ denotes a complex conjugate and the I's represent the physically measurable intensities (powers) of the complex electric fields.

Similarly, one can define related quantities to describe the transmissivity:

$$T = \frac{|\hat{n}_N|}{|\hat{n}_1|} \cdot \frac{|\mathbf{E}_N|^2}{|\mathbf{E}_1|^2} = \left(\frac{\hat{n}_N}{\hat{n}_1}\right)\hat{t}^* \cdot \hat{t} = \frac{I_N}{I_{in}} \tag{3.11}$$

where the transmission Fresnel coefficients are defined as $\hat{t}_\eta = (|\mathbf{E}_N|/|\mathbf{E}_{in}|)_\eta$, η is either s or p for the polarization and the subscript N indicates the last material phase (exit phase) in the sample (Fig. 3.3).

Once the power reflectivities or transmissivities have been measured over a range of selected frequencies to generate $T(\nu)$ and/or $R(\nu)$ spectra, the next step is to relate these quantities to the material properties and the structure of the sample. One way to do this is by constructing mathematical models which represent the interaction of the light with the sample for the specific experimental conditions, fully specified by the frequency ν of the light, the beam polarization η, the angle of incidence ϕ and the azimuthal angle θ (Fig. 3.5), where the latter is chosen by setting the plane of incidence of the light relative to some important axis associated with the sample structure, e.g., a crystal axis for biaxial symmetry samples; in the case of isotropic samples all azimuthal orientations are equivalent. The standard models are based on the generalized EM theory of light interacting with an ideal, stratified-layer structure of anisotropic, optically homogeneous media. In this approach, the sample is completely described, as shown schematically in Fig. 3.5, by specifying for each layer j the associated thickness d_j and a complex refractive index tensor spectrum function $\underline{\hat{n}}_j^{(2)}(\tilde{\nu})$ which gives the values of the tensors as a function of frequency or wavenumber with the tensor elements carrying the directional properties of the sample as specified for the chosen sample geometry in the optical setup (sample coordinate frame relative to the laboratory frame). Equivalently, one can specify a complex dielectric tensor function spectrum function $\underline{\hat{\epsilon}}_j^{(2)}(\tilde{\nu})$. For the simpler and more usual case of isotropic media the tensors reduce to simple scalar function spectra $\hat{n}_j(\tilde{\nu})$ or $\hat{\epsilon}_j(\tilde{\nu})$ which are just a numerical list of the complex refractive index values $[n_j(\tilde{\nu}), k_j(\tilde{\nu})]$ or complex dielectric constant values $[\epsilon_{1j}(\tilde{\nu}), \epsilon_{2j}(\tilde{\nu})]$ for the specific frequencies.

The essential EM theory is based on the boundary value relationships of Maxwell's equations which serve to relate the complex Fresnel coefficients, defined above, to the layer thicknesses and the frequency-dependent optical response functions. The material properties of each layer are accessed by additional relationships between the optical functions and the material structure and composition. A central problem which arises in the above strategy is that although the relevant EM theory is explicitly defined in detail, in practice, for the most general experimental configurations and sample types, the theory quickly leads to such cumbersome mathematical expressions that analytical solutions are often hopelessly intractable. This problem is generally circumvented by replacing analytical solutions with numerical computation algorithms based directly on the core electromagnetic equations. This approach involves iterative comparisons of simulated data, based on trial solutions, with the experimental data and convergence based on some best-fit criterion. The computations are easily handled by commercial numerical equations solver packages or with coding in standard languages such as Fortran or C++.

The most rigorous and comprehensive approaches to spectral computation appear to be the 4×4 transfer matrix methods in which the **E** and **B** (magnetic) field vectors at any depth location in the sample are determined relative to the values at the outer boundaries of the sample by a sequential application of matrix operators across each intervening interface and through each intervening phase. A very useful approach for the present purposes has been formulated in general terms by Yeh [11]. This treatment begins by applying the continuity of the tangential components of the **E** and **B** fields across the interfaces. Further, the same functional form of the matrix transforms apply to the two fields. This is an advantage since the tangential

B-field component can be used to calculate the perpendicular **E**-field component at any point and thereby the need to consider perpendicular field components is eliminated. This simplification directly reduces the electromagnetic variables by half and leads to the 4×4 matrix formalisms of the boundary value relationships of the field vectors. An important point is that this matrix method is sufficiently general that it describes any experiment regardless of the wavelength of the incident light and once programs are written they can be used for diverse experiments ranging from x-ray reflectivity to microwave spectroscopy.

Implementation of the 4 × 4 Transfer Matrix Method for Numerical Simulations

The basic electromagnetic principles applying to these equations can be found elsewhere [11]. FORTRAN codes based on these principles have been developed for use in quantitative IRS analysis of oriented organic films [10] but are not presented here. They can be obtained from the author upon request. The solution of the electromagnetic boundary value equations leads to the following relationship between the electric fields in the final medium (N) and the first medium (1), which contains the light source:

$$\begin{pmatrix} A_s(1) \\ B_s(1) \\ A_p(1) \\ B_p(1) \end{pmatrix} = \underline{M} \begin{pmatrix} C_s(1) \\ D_s(1) \\ C_p(1) \\ D_p(1) \end{pmatrix} \tag{3.12}$$

In this equation, A, B, C, and D are the electric field intensities characterized by either s- or p-polarization directions. The A and C fields are associated with waves moving in the direction of infinite medium 1 to the exit infinite medium N and conversely the B and D fields involve waves moving in the reverse direction. The detailed nature of the 4×4 matrix operator $\underline{M}$ is explicitly defined in terms of the experimental geometry, the sample structure, the sample optical functions, and the frequency of the light. Details of the calculations of the matrix elements in terms of these quantities are given elsewhere [10].

Solution of the boundary value equations results only in setting relative values of the fields with respect to their location along the z-axis in the sample (refer to Fig. 3.3). These values are referenced to the fields in the Nth phase at the last boundary $[N/(N-1)]$ where it is required that the modal field amplitudes, $D_s(N)$ and $D_p(N)$, vanish since there can be no back reflection from this infinite phase and where the value of the outgoing fields, $C_s(N)$ and $C_p(N)$ can be set to unity for convenience. The complex reflectivities then can be computed easily by taking the ratio of the reflected amplitudes to the incident amplitudes, respectively, for a desired input polarization of the beam. For instance, the fractional complex amplitude of an s-polarized input beam which appears as a reflected s-polarized output beam can be computed as follows:

$$r_s = \left[\frac{C_s(1)}{A_s(1)}\right]_{A_p(1)=0} = \frac{m_{21}m_{33} - m_{23}m_{31}}{m_{11}m_{33} - m_{13}m_{31}} \tag{3.13}$$

where m_{jk} represents the corresponding element of the matrix $\underline{M}$ and the condition of a zero p-polarized input field is shown as a subscript condition on the second term. Similarly, the p-polarization coefficient r_p can be calculated. The power reflectivities and transmissivities then can be calculated easily from (3.9) and (3.10). The calculations are then repeated for the entire range of frequencies of interest to provide a simulated spectrum.

Generation of Optical Function Tensors

For a condensed-phase, anisotropic medium in which all the oscillators of each type are aligned in parallel, the orientations are defined by the directions of the dipole transition moments $\mathbf{p}(\nu)$. Since the collective effects of the individual oscillators are expressed in the loss term $k = \mathrm{Im}(\hat{n})$ of the medium, it is convenient to use $k(\nu)$ rather than $|\mathbf{p}(\nu)|$ as the basis for intensity since k often can be derived experimentally from spectroscopic measurements on the pure material whereas $|\mathbf{p}|$ is usually thought of as a property of an isolated species. Therefore, for the purposes at hand we form a vector $\mathbf{k}$ which carries both the direction of $\mathbf{p}$ and the intensity of k. In a real bulk sample, of course, it would need to be shown that the direction of $\mathbf{p}$ along some molecular (internal) coordinate system is not perturbed by matrix effects relative to the direction for an isolated oscillator. For a given film material the directions of the different $\mathbf{k}$ vectors (for the different oscillators of vibrational modes) are typically described in terms of some convenient material coordinate frame, generally chosen on the basis of the intrinsic material (molecular or crystal) symmetry.

The complete $\mathbf{k}(\nu)$ spectra for all the oscillators can be directly measured in a single crystal of the film material where the crystal orientation to a probing light beam can be varied but rarely is a sufficiently large single crystal available for such a measurement for most materials. Rather, the $\mathbf{k}$'s can be derived from measurements on an isotropic sample (such as a standard KBr pellet) starting with the convenient relationship:

$$|\mathbf{k}| = 3k_{iso} \tag{3.14}$$

where $|\mathbf{k}|$ represents the effective intensity response of the material when all the oscillators are aligned in the direction of the exciting electric field and k_{iso} is the value of $\mathbf{k}$ for random 3-D orientation of the oscillators in an isotropic sample. Next, using (or estimating—often simple for symmetrical molecules) the known directions of the oscillators within the material a diagonalized $\hat{\underline{k}}(\tilde{\nu})$ tensor can be created based on the x, y, z components of $\mathbf{k}$ in the chosen material coordinate frame.

A convenient way to do this is to choose an individual oscillator with some orientation in the molecular or material unit cell coordinate frame, typically chosen as the most important symmetry axis (e.g., c-axis of a unit cell), superimpose a unit vector along the z-axis of material coordinate frame), and using Eulerian transform operations, shown in Fig. 3.6(left), tilt the unit vector by angle α away from z_o and then rotate azimuthally by angle β around z_o to superimpose the unit vector on $\mathbf{k}$.

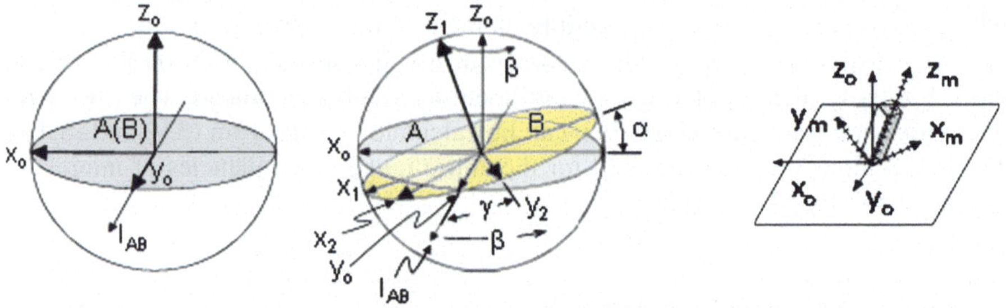

Fig. 3.6 *Left*: Transformations in a Eulerian coordinate system. Starting with a reference coordinate frame x_o, y_o, z_o (*left*) consider a second coordinate frame superimposed with xy planes A and B; initially, the axis of intersection of the two planes I_{AB} is defined to be along y_o. Three independent rotational transformations are possible that can be done in any order to achieve the same final coordinate frame reorientation provided that the proper rotation axes are always used. (1) Tilt the new coordinate frame by rotation around the I_{AB} intersection axis by angle α; this produces plane B tilted at angle α from plane A. Change the azimuthal angle of the I_{AB} axis by rotating it around the x_o, y_o plane by angle β; this rotates the B plane around the z_o axis. (3) Within the B plane rotate x and y for the plane around the z axis of the B plane by a twist angle γ. In the diagram above an initial tilt of the z axis away from z_o gives plane B tilted at α from A with axes x_1, y_o, z_1. A rotation by angle γ around z_1 gives x_2, y_2, z_1 for plane B. Finally an azimuthal rotation β will align the B plane in a new direction to give a final x_2, y_2, z_2 (not shown). *Right*: juxtaposition of the unit cell coordinate system of the film material on the sample coordinate system

The diagonal elements of the k tensor are then generated by the equations:

$$\begin{aligned} k_{xx} &= |\mathbf{k}|\left(\sin^2\alpha\right)\left(\cos^2\beta\right) \\ k_{yy} &= |\mathbf{k}|\left(\sin^2\alpha\right)\left(\sin^2\beta\right) \\ k_{zz} &= |\mathbf{k}|\left(\cos^2\alpha\right) \end{aligned} \tag{3.15}$$

The complete optical function tensor at each frequency $\underline{\hat{n}}(\tilde{\nu})$ is described by assigning each element in terms of $\hat{n}_{ee} = \mathrm{Re}(\hat{n}_{ee}) + i\,\mathrm{Im}(\hat{n}_{ee}) = n_{ee} + ik_{ee}$, where e = x, y, or z. A useful approximation is that $\mathrm{Re}(\hat{n}_{ee})$ consists of an oscillator contribution, $n_{osc}(\nu)$, which is direction and frequency dependent, and an isotropic background response, $n_\infty(\nu)$, which varies slowly or is constant with frequency. The latter quantity is essentially the constant specified in the Kramers-Kronig transform [10], and the former is the quantity given by the integral term in the transform (see (3.4)).

Once the diagonalized $\underline{\hat{n}}(\tilde{\nu})$ tensor elements are known for the oscillators with their **k** vectors specified in the material reference frame, a non-diagonal tensor for the orientation of the molecule or unit cell in the optical coordinate frame is created by a simple matrix rotation,

$$\left[\underline{\hat{n}}(\tilde{\nu})\right]_o = R^{-1}\left[\underline{\hat{n}}(\tilde{\nu})\right]_m R \tag{3.16}$$

where R is the rotation matrix defined as

$$\begin{pmatrix} \cos\alpha\cos\beta\cos\gamma - \sin\beta\sin\gamma & \cos\alpha\sin\beta\cos\gamma - \cos\beta\sin\gamma & -\sin\alpha\cos\gamma \\ \cos\alpha\cos\beta\sin\gamma - \sin\beta\cos\gamma & -\cos\alpha\sin\beta\sin\gamma - \cos\beta\cos\gamma & \sin\alpha\sin\gamma \\ \sin\alpha\cos\beta & \sin\alpha\sin\beta & \cos\alpha \end{pmatrix}$$

in the Eulerian coordinate frame (Fig. 3.6(left)).

An illustration of a material unit cell coordinate system oriented in a optical coordinate system is given in Fig. 3.6(right). The unit cell defines the intrinsic set of axes which reference the directions of all oscillators. In thin film form the unit cell assumes some orientation relative to the sample holder coordinate frame. Accordingly a vector **k** expressed as $(\mathbf{k})_m$ in the m coordinate frame transforms to $(\mathbf{k})_o$ expressed in the optical coordinate frame via $(\mathbf{k})_o = R^{-1}(\mathbf{k})_m R$. The tensor $\underline{\hat{n}}(\tilde{\nu})$ which incorporates the directions of the oscillators at each frequency similarly transforms to describe the interaction of the oscillators with the light beam in the sample holder coordinate system.

The Case of Heterogeneous Films with Uniform Nanoscale Density Variations—Application of Effective Medium Theory

For planar film media which have density variations, typically voids between crystallite regions, the refractive index n_{film} will be some average of the void medium, e.g., air or vacuum, liquid or a matrix material in a composite film, and the main film material. The standard way to specify n_{film} is through the effective medium approximation (EMA) which models the polarization response in terms of an averaging of local responses within the different density and constituent material domains. In this approximation the local domains need to be of a size scale less than the wavelength of the light, of the order of μm in the IR, in order to eliminate diffuse scattering effects. Depending on the morphology of the film two limiting equations are usually applied, the Bruggeman and the Maxwell-Garnet.

For a film with uniformly dispersed, discrete domains in contact with each other the Bruggeman equation,

$$\sum_{j=1}^{n} f_j \left(\frac{\epsilon_j - \epsilon_{film}}{\epsilon_j + 2\epsilon_{film}} \right) = 0 \tag{3.17}$$

can be used, where n is the total number of components in the multiphase system, f_j and ϵ_j are the volume fraction and dielectric constant respectively of the jth component, and film is the effective dielectric constant of the multicomponent layer [12–14].

For a film consisting of a uniform matrix of a continuous material containing dispersed domains (e.g., voids) the Maxwell Garnett equation,

$$\epsilon_{film} = \epsilon_u \left[\frac{\epsilon_1 + 2\epsilon_u + 2f_l(\epsilon_1 - \epsilon_u)}{\epsilon_1 + 2\epsilon_u - 2f_l(\epsilon_1 - \epsilon_u)} \right] \tag{3.18}$$

can be used, where ϵ_{film} represents the effective Maxwell Garnett dielectric function. ϵ_u and ϵ_l are the dielectric constants for the fractions of higher and lower composition, respectively, and f_l is the volume fraction of the lower composition.

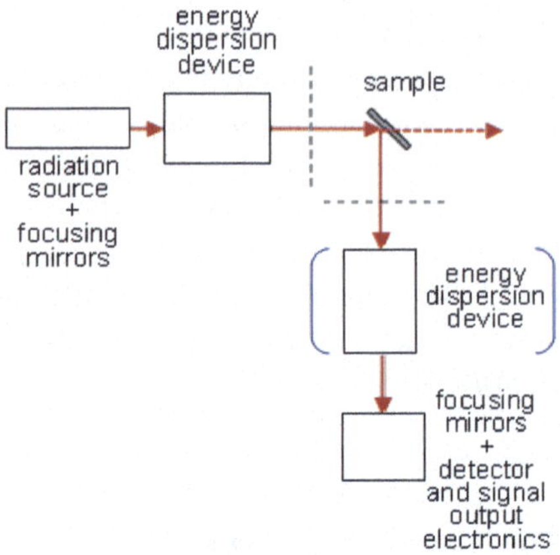

Fig. 3.7 Schematic of a general type of optical bench setup for IR spectroscopy

3.4 Instrumentation and Operating Conditions

3.4.1 Overview

An example of a highly generalized optical bench setup for IR spectroscopy of surfaces is shown in Fig. 3.7. The main components are a radiation source and associated optics which provide a broad distribution of radiation energies focused into a beam of controlled geometry, a device that provides a means to track the beam power across a small spread of energies (resolution element) at each energy in the beam energy distribution, the sample and focusing optics which direct the appropriately shape beam into a detector whose associated electronics provides a signal that is sent to a computer for recording and processing. The energy dispersion device alternatively can be placed after the beam exits the sample (shown inside the blue brackets in the figure); the choice usually depends on the dispersion mode, interferometer or grating monochromator. The diagram does not include small optical components such as beam apertures or polarizers, which are inevitably incorporated in any setup.

Since many types of surface IR experiments are done in vacuum systems dashed lines are shown in the figure to indicate the typical position of the ambient-vacuum interface. In the sections below details of the components and setups are given. The emphasis is on commercial types of instrumentation rather than home built instruments. Also discussions of sophisticated laser techniques, e.g., broad band femtosecond IR laser pulses and tunable IR sources are not treated as these are very specialized and details can be found in numerous reviews and texts. The specific types of components that are commonly incorporated in instrument setups, either home built or commercial, are shown in Table 3.2 with associated numbers as identifiers for the diagrams that follow. The environmental conditions under which each of the sampling modes is possible is given at the end of this section.

Table 3.2 Major components in typical surface IR experiments

	Component	Function	Typical type
1	IR source	generate broad λ IR radiation	ceramic glo-bar
2	variable aperture	control exit beam size and shape	circular or slit opening
3	parabolic focusing mirror	convert beams between radially divergent and collimated	metal finish
4	interferometer	generate interference pattern as function of interferometer mirror position (radiation power vs mirror stroke distance)	Michelson with beamsplitter coating for mid-IR
5	grating monochromator	disperse radiation spatially according to λ	standard IR grating
6	cylindrical mirror	convert beams between rectangular divergent and collimated	metal finish
7	ellipsoidal mirror	convert divergent beam to tight focus convergent beam	metal finish
8	linear polarizer	remove undesired linear polarization component in beam	gold wire grid on KRS-5 or similar window material
9	variable polarization phase shifter	control the exit polarization ellipse from input linear polarized beam	piezoelectric birefringent IR transparent crystal
10	single element detector	detect relative power from a single beam	HgCdTe photocurrent bolometer
11	array detector	detect relative power vs position across an incoming beam front	HgCdTe array

3.4.2 Typical Components

Energy Dispersion Devices

The heart of an IR spectrometer which uses a broadband (e.g., blackbody) source is the device that separates the response of the sample to the IR beam as a function of photon energy within some specified energy resolution, typically given in terms of wavenumber $\Delta\tilde{\nu}$ for energy or as $\Delta\lambda$ for wavelength. The most common way to accomplish this is on the basis of wavelength rather than energy by either (1) generating an interference pattern in time by passing the beam through an interferometer, which is an interference cavity with a moving mirror whose exact position is tracked by a simultaneous laser interference pattern, or by (2) using a diffraction grating to spatially fan out the beam at different angles according to wavelength. Alternatively, though not common, is the use of tunable solid state IR lasers and tunable free electron lasers in which the energy of the beam is changed stepwise with no dispersion device needed. Here we only consider the wavelength dispersion devices.

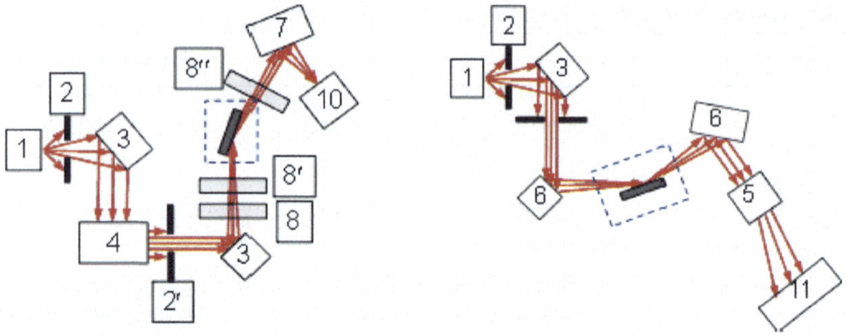

Fig. 3.8 Schematics of typical optical bench setups. The numbers of the components refer to Table 3.2. *Left*: A Fourier transform spectrometer using an interferometer for energy dispersion. *Right*: A grating based spectrometer with an array detector

Almost exclusively the basis for current commercial instrumentation involves the use of interferometry with the Michelson type of interferometer. In this case the signal from the detector is a time varying function at some base frequency in the range of ~5–100 kHz and is Fourier transformed to change from a time (or mirror distance) base to a photon wavelength base and thereby provide the desired beam power vs wavelength (or equivalently energy) trace. Thus the term Fourier transform infrared (FTIR). A simple optical setup using an interferometer is shown schematically in Fig. 3.8(left) and laid out to illustrate reflection from a sample surface. The figure is meant only to show the major components and not any optimized layout. The interferometer allows a high beam throughput (Jacquinot advantage) and the signals that arrive at the detector are strong since all wavelengths contribute to the total signal (multiplex or Fellget advantage). For surface experiments, particularly reflection at grazing angles or internal reflection experiments with multiple pass elements where specific incidence angles are required with narrow tolerances, the throughput advantage is highly compromised since the beam must be reshaped towards the limit of slit patterns with extensive loss in power since much of the initial wide diameter beam is discarded in the process. Another advantage of the interferometer is the use of the interference pattern of a visible wavelength laser line (typically the 632.8 nm HeNe line) to provide highly accurate calibrations of the IR wavelengths or frequencies.

Grating spectrometers were used widely a few decades ago but fell out of favor with the development of FTIR. The problem with the earlier grating based instruments was the throughput of the beam passing through the grating was very poor because a slit image with very small angular divergence needed to be formed, at extensive cost to the total beam power, and the resulting diffraction beams, which contained even less power, were sampled one by one with the result of very slow spectral collection and poor spectral contrast relative to the noise. For grazing incidence and internal reflection experiments, however, the slit beamshape is ideal so for these experiments this geometric factor was not the main limit in spectral quality, compared to the general noise level. For high sensitivity surface IR this could be overcome to some extent by modulating the signal, e.g., by a chopper or by applying a small sinusoidal mechanical motion to some element such as the grating and locking the electronic signal in to this frequency.

With the advent of high sensitivity, low noise array detectors of the same elements (e.g., HgCdTe) used for FTIR spectrometers, a large group of the diffracted beams from the grating can be analyzed in parallel in very short times (microsecond time scale detector response) to allow a large advantage in stability and signal averaging, thus overcoming the majority of problems. Commercial versions of the grating-based IR spectrometers are now beginning to appear on the market with abilities to collect individual scans at the rate of 100 Hz and publications of applications have appeared [15, 16]. The fast scan rates allow extensive signal averaging and noise reduction which is competitive with FTIR. A generalized example of the type of optical setup for a multichannel detector with a grating is shown in Fig. 3.8(right).

Sources and Detectors

Sources in the mid-IR typically consist of ceramic glo-bars which are heated resistively at $\sim$4 A current, held highly constant to avoid slow electromagnetic field power fluctuations which can result in serious baseline drifts. There are a variety of sources for far- and near-IR but we only consider mid-IR in this review since that by far is the region of interest for almost all studies of thin film. Specialized sources such as synchrotron radiation can also be used over various energy ranges (particularly far-IR) with the added advantage that the beam flux is highly collimated, but are rarely used because of the inconvenience of working at a special facility.

Detectors for the mid-IR region typically used for surface studies consist of semiconductor elements operating in the current mode as a bolometer and less frequently for surface work, pyrolelectric detectors. A common example of a semiconductor bolometer is the HgCdTe (MCT) detector run with a liquid nitrogen cooled element and an on-board current preamplifier. Given the constant advances in detector technology, improvements and commercial availability can be expected over time. Array versions of these detectors allow imaging measurements and the use of gratings for dispersion of the photon energy into component resolution elements.

We also note that in some cases the sample itself can serve as part of the detector if the power dissipated by relaxation of energy states excited by resonant interactions with the incoming beam is detected directly at the sample. This can be done by either using a sensitive, contiguous microphone to pick up the thermally induced acoustic fluctuations generated within the sample surface region (photoacoustic detection) or by using a visible laser to monitor the fluctuations in surface reflectivity arising from thermally induced changes in the surface refractive index (photothermal detection). These methods, though often used to advantage for special studies, are not as common as the standard far-field detectors above and fall outside the scope of this review.

Typical Optics

Optical components for forming the beam focus and steering the beam path in the mid-IR region almost entirely consist of metal-coated mirrors since metals in the IR regions can be near perfect reflectors (approaching the limit of a perfect conductor). For FTIR systems using interferometers, the mirror size is matched to the large diameter circular beams associated with the interferometer. This is not ideal, however, matching the entrance/exit requirements of internal reflection elements and for grazing incidence external reflection and some losses over what is ideal occur. On the other hand for grating based spectrometers, slit beams are needed to match the grating pattern and this is ideal for internal and external reflection but is not a good match for detector field of view requirements.

Polarizers typically consist of metal wire grids on transparent supports such as KRS5 or ZnSe windows. The wire grids block the beam propagation when the E-field vector is parallel to the grid line but pass the beam through for perpendicular E-field orientation. Brewster angle polarizers made from silicon plates are also used. At the Brewster angle of incidence, which is $\sim 74°$ for silicon in the mid-IR, s-polarization is electively blocked from transmission through the silicon plate while p-polarization is selectively transmitted.

There are tradeoffs which favor the wire grid polarizers for most applications. The Brewster angle polarizers, especially when stacked, can give extremely high separation of s- and p-polarization components from an incoming unpolarized beam but the throughput is low. On the other hand the wire grid polarizers do not completely discriminate between the two polarizations, e.g., a setting for pure p- also leaks several % or more of s-, but they give the optimum throughput of as much as $\sim$35–40 % per pass. Since low throughput tends to bring the beam power into the level of the intrinsic detector noise in most experiments on surfaces, sacrificing high polarization selectivity for beam throughput is usually a good strategy for optimum S/N.

3.4.3 Major Sampling Modes and Typical Associated Sampling Conditions and Requirements

The major sampling modes for surface studies are divided into transmission (T) and reflection with the latter subdivided into external (ER) and internal (IntR) modes. In the ER mode the incident and exit beams are both contained in a non-absorbing medium with a lower real refractive index than the reflection support (e.g., vacuum, inert gas or dry air). The converging incident beam is focused at the impact spot on the sample surface and the exit beam is at a specular (mirror) angle with diverging focal characteristics set as an inversion to the incoming beam (e.g., see the simple straight ray schematics in Figs. 3.2 and 3.5 and the ray-trace layouts in Figs. 3.7 and 3.8(left)). In the IntR mode (refer to the ray trace layouts and the schematic in

Fig. 3.4) the incident beam enters into an IR transparent prism with a higher real refractive index than the surrounding medium and undergoes total specular reflection each time an internal surface of the prism is encountered. The prisms are designed for a fixed number of internal reflections and with the ordinary (center) rays of the entrance and exit beams impinging normal to the cut faces on the element. So overall the instrument optics for the two modes are quite similar in that the beam must be delivered to and collected from the sample at different angles.

Each of the sampling modes requires a different type of material combination (e.g., IR transparent supports for T and typically metal surfaces for ER) and each has a different concentration of the electromagnetic field in the region of the sampling surface. In particular, though each of the sampling modes allows good surface sensitivity, surface selectivity is only possible with internal reflection where sampling is done in the evanescent field that decays away with the μm-scale from the internal reflection element surface. This is such a large distance compared to surface film thicknesses at the scale of nanometers or even less that it is not truly considered surface selective. On the other hand, the restricted range does allow surface sampling in the presence of low IR absorbing liquids and thus sets this mode apart from the other sampling modes which cannot tolerate the overwhelming spectral interference from surrounding liquids. A very general summary of the appropriate sampling modes for different conditions and environments is given in Table 3.3.

Some general requirements and associated aspects for the three common surface modes used for planar types of surfaces are summarized in Table 3.4. Note the relative surface sensitivity column which gives an average signal to noise (S/N) rating which is based on the relative strengths of the electromagnetic fields at the surfaces, assuming approximately equal intrinsic noise characteristics of the spectrometer electronics and detector in each mode.

3.4.4 Typical Signal/Noise Characteristics

We rate the general noise of a good quality IR spectrometer setup in terms of an FTIR spectrometer since that is by far the common instrumentation used. As a standard operating condition we use: (1) $\sim$2 cm^{-1} spectral resolution (a value several times less than the typical peak width for molecular films), (2) one polarizer, (3) a standard ceramic glo-bar source operating under extremely constant current, (4) optimized electronic parameter settings and (5) a high quality, low noise semiconductor detector (e.g., HgCdTe, liquid N_2 cooled) with response peaked at the spectral region of interest. For rating the noise we choose an $\sim$400 cm^{-1} spectral range to sample. Under these conditions the baseline noise in the mid-IR for a single scan will reach a level of $\sim 2 \times 10^{-4}$ absorbance units [peak-to-peak rms value of $\log(I/I_o) \sim 2 \times 10^{-4}$, where $I =$ beam power at the detector at a given photon energy with the signal fluctuating around a baseline power of I_o]. Typically several hundred scans are co-added which would give a typical noise level of $\sim(400)^{1/2} = 20$ times less or $\sim 1 \times 10^{-5}$ absorbance units. We can use this as

Table 3.3 Compatible environments and conditions for the major IR sampling modes

Environment or condition	Usable modes	Comments
UHV	T	use Brewster angle or wedged sample to avoid fringing
	ER	good for single crystal surfaces; for metals requires grazing incidence (>80°) optics and highly planar surfaces with p-polarization
	IntR	multiple reflection optics in UHV are difficult to adjust
Gas or vapor	T	corrections for gas or vapor spectra can be difficult over wide frequency range
	ER	corrections for gas or vapor spectra can be difficult over wide frequency range; $s \leftrightarrow p$ polarization modulation can be useful for metal surfaces although the bandwidth is restricted because of the modulator crystal range for true $s \leftrightarrow p$; note the special case of spectra from liquid surfaces which requires vibration control
	IntR	gas or vapor spectrum interference is greatly reduced by the short (∼1–5 μm) evanescent wave sampling region
Liquid (low IR absorbing)	IntR	liquid medium spectrum interference greatly reduced by the short (∼1-5 μm) evanescent wave sampling region
Variable temperature		
T > ambient	all	all modes possible if support materials are thermally stable
T < ambient	all	UHV or inert, non-condensable gas medium required

a figure of merit for stating relative values of signal/noise. With recent advances in A/D converters, stable optics and good electronics it is possible to approach the $\sim 1 \times 10^{-6}$ peak-to-peak rms noise level. Dispersive grating instruments using high quality, low noise semiconductor array detectors (e.g., HgCdTe) can give roughly the same rms $\log(I/I_o)$ noise values with ideal optics and under ideal conditions with the detector operating in the maximum response photon energy region. Regardless of the energy dispersion mechanism, the noise level values fall off as the photon energy shifts away from the maximum of the detector response and typically detectors are chosen to match their optimum response to the spectral region of interest.

In order to generate S/N values for a given sample and optical sampling method we need to state the intrinsic signal value. This consists of the peak value and the spread of the absorbed power over a spectral region, which is generally given as a peak width (or linewidth). For a narrow linewidth, the resonantly absorbed power is concentrated in a narrow spectral region which gives a large peak intensity and better S/N relative to a broad linewidth where the same power is distributed across a wide spectral region. We will use an $\sim 8\ cm^{-1}$ full linewidth at half maximum peak intensity as a figure of merit for comparing S/N values. We use transmission as a limiting example since this mode has the lowest E-field at the surface compared to reflection modes. A typical signal for a transmission through a monolayer of

Table 3.4 Some requirements and aspects of three common sampling modes for planar surface characterization

Common materials	Sample/optical requirements	Rel. surf. sensitivity	Comments
for T			
NaCl, KBr, Si, Ge, KRS5, ZnSe; also GaAs, etc.	wedged as needed to reduce fringes/straight beam path, pol. as desired	~1	simple optics, low S/N, fringing
for ER			
metals	standard/grazing angle incidence (>80°), *p*-pol	~5–10	good S/N; only *p*-pol, optics at high angles more demanding
non-metals, IR opaque (e.g., doped Si & GaAs, carbon, etc.)	standard/incidence angle typically 60–80°, pol.as desired	~1	both *s*-, *p*-pol; low S/N; peaks can be inverted; interpretation requires simulations
non-metals, IR transparent	wedged as needed/as above	~1	both *s*-, *p*-pol; low S/N; peaks can be inverted; interpretation requires simulations
for IntR			
Si, Ge, KRS5, ZnSe; also GaAs, diamond, etc.	precisely cut prisms with smooth faces/highly controlled incidence and exit angles, pol. as desired	~1–20 depending on number of reflections	increased S/N with multiple reflections; precise sample mounts and optical paths needed

a typical molecule with a typical oscillator is $\sim 0.5\text{–}1.0 \times 10^{-4}$ absorbance units, which means that for a typical noise level of $\sim 2 \times 10^{-4}$ the S/N is $\sim 1/4\text{–}1/2$ for one scan. Thus a number of scans must be co-added to signal average and raise the S/N to a level $>1/1$; e.g., typically at least ~500 scans to be in the range of S/N 10/1. These estimates are to be taken as a rough guide since noise levels will vary from one optical setup to another and vary with the spectral region and resolution.

3.5 Application Examples

This section is developed from a practical point of view in which selected application examples are presented based on the types of optical setups and sampling methods, each of which allow the surfaces of different types of samples to be characterized. The reader can refer to Tables 3.3 and 3.4 for the general background characteristics of the sampling modes.

The application examples tend to fall into two main categories: (1) study of surface structure and chemistry associated directly with the support substrate material and (2) study of a thin film which has a chemistry and structure independent of the support. In the first category a typical example would be a bare silicon substrate which is chemically modified by chemisorption of H atoms at bare Si atom sites. In the second and very common category, a native oxide-silicon support or a vapor deposited gold metal film could be used with a spun cast polymer films on the support surface with the goal of studying some chemical property of the polymer film surface unrelated to the underlying support.

3.5.1 Transmission

Transmission is often the most straightforward of all infrared measurements due mostly to the relatively simple optical setup used. If the support substrate under study is transparent in the spectral energy region of interest then a sample spectrum can be achieved which is free from support spectral artifacts. Normal incidence transmission measurements have routinely been used to study thick films (thickness $\gtrsim$10 nm). For monolayer thickness films, however, two problems must be overcome. First, the S/N is relatively low at the monolayer level (using the rms noise level given in the last section) because of the intrinsic strengths of the electromagnetic field at the surface which cannot generate sufficient absorbance to detect much less than a fraction of a monolayer for most types of monolayer films (e.g., $\sim$10 % for typical cases). The values of the E-fields differ for each type of material depending on its refractive index and on the incidence angle and beam polarization, as determined by the values given by the appropriate equations in Sect. 3.3.2. Second, intrinsic interference or fringing effects from back face reflections (for reference see the schematic in Fig. 3.2) will arise for parallel face transmission supports to create troublesome baseline artifacts that can overwhelm the spectra of interest. These interference fringes are essentially Fabry-Perot cavity oscillations for a fixed support thickness (cavity gap) with varying wavelength of the beam and can be calculated directly from the appropriate equations in Sect. 3.3.2.

The use of wedged substrates with slightly off-parallel opposite faces has long been used to eliminate or at least minimize deleterious interference fringes. This works by creating a distribution of fringe patterns with each pattern having different fringe maxima periods as a function of the location across the wedge as set by the local face-to-face thickness. For an appropriate wedge angle the superposition of all the different fringe patterns within the beam width will tend to be out-of-phase on average and cancel, thus giving a reasonably straight baseline [17].

It is also possible to reduce fringing by working at an angle of incidence equal to the Brewster angle of the transmission element (see above) and using p-polarized light [18, 19]. Under this condition the fringing associated with the selectively transmitted p-polarization is effectively quite small since the p-polarized light is transmitted with no reflection from the front and back faces, therefore eliminating the cause of interference fringes.

Two recent examples of transmission spectra of $\sim$ monolayer coverage thin films in the mid-IR are given by the studies of Chidsey and co-workers [20] and Chabal and co-workers [21]. Both groups used silicon supports and applied Brewster angle transmission to eliminate spectral fringing. In the first study, spectra were taken ex-situ after removal of the indium tin oxide-covered silicon supports from a vapor phase reaction system for deposition of monolayer alkylsiloxane coatings and the spectra used to determine to optimum conditions for preparing well-organized monolayers. In the second study, spectra obtained in situ under vacuum during thermal treatment of supported multilayer graphene oxide films, with and without water and ethanol trapped between the layers, were used to show that trapped water enhances the formation of carbonyl and hydroxyl species whereas incorporation of other species such as alcohol between graphene oxide layers can lead to different chemistries.

While silicon is a widely used substrate for transmission measurements, surface species at the monolayer level have been studied on other materials such as sapphire [22], mica [23], GaAs [24], and fused silica [25], to name a few. In each of these cases the conditions of the sample surfaces, the refractive index of the support play an important role in determining the resultant S/N for a given set of experimental optical conditions.

3.5.2 Reflection

3.5.3 External Reflection

A light beam propagating in an external medium with a lower refractive index than that of a smooth, planar sample surface will generate a specular reflected beam upon impinging on the surface (for reference, see Figs. 3.2 and 3.7). The electromagnetic field that is sustained within the reflection region at the sample will interact with materials in that region according to the complex refractive indices of the materials. The types of spectra can vary widely depending on the refractive index properties of the support and the film of interest. The major ER categories are: (1) highly reflective metals of infinite optical thickness, (2) opaque non-metals and (3) transparent non-metals.

In the first category are typical reflective metal samples such as metal single crystals and evaporated overlayer metal films of sufficient thickness to eliminate transmission. These samples then can act as supports for overlayers of interest.

The electric field vector at the surfaces of highly conductive metals, e.g., Au or Ag, is overwhelmingly perpendicular to the surface (z-axis in Fig. 3.5) since a parallel field (x, y plane) is quenched by the induced image dipole in the metal surface region. The maximum surface E-field is obtained at grazing incidence (>80°) with a *p*-polarized beam and the intensity drops off sharply at lower angles. Under this condition the spectra of overlayers is dominated by the complex part of the refractive index, viz, the loss or absorption part, and accordingly the spectra tend to look like transmission spectra with all features showing absorbed beam power, thus the common terms reflection-absorption infrared spectrometry (RAIRS) or infrared reflection-absorption spectrometry (IRRAS). For non-metals, however, the reflectivity can be dominated by changes in the real refractive index as a function of frequency which can result in inverted peaks in certain regions of the spectrum. For this reason we will drop the above acronyms in general and just use ER. Reflection IR from planar metal surfaces has been widely used to study the structure [10] and chemistry of adsorbates and films on metals [26], especially for in-situ measurements under reactive conditions [27, 28].

The second category involves materials such as semiconductor wafers which are sufficiently doped that they effectively do not transmit an IR beam and intrinsically IR opaque non-metals and weak metals. Typical examples are doped Si or GaAs and carbon materials. Films or thin wafers of these materials are opaque to IR radiation so exclusive reflection results.

The third category includes materials with IR transparent spectral regions (transmission windows). Examples include NaCl, KBr, diamond, undoped semiconductors such as Si and GaAs in the mid-IR, thin silica films in the frequency region from $\sim$1500–3000 cm^{-1} and liquids such as water, which has IR windows between the main water features near $\sim$1600 and 3000–3700 cm^{-1}. Liquids can act a supports for surface films of materials such as molecules, surfactants and polymers, provided that the mechanical vibrations of the liquid are sufficiently low to eliminate background noise in the reflected beam power.

Examples of ER studies involving in-situ vacuum conditions are given by two studies of vapor-deposited metal atoms with SAMs of H_3CO-terminated alkanethiolates on Au{111} substrates in which spectra were obtained with increasing metal atom coverage to help understand fundamental interactions with the molecules [29, 30]. The data show that whereas the $-OCH_3$ surface was expected to be moderately reactive with all the metals, as dictated by the thermodynamically favored carbide, oxide, and hydride products, in the case of Al very little chemical interaction was observed. In contrast, the deposition of Ca and Ti leads to strong chemical attack of the monolayer.

Other examples of ER characterizations include organic monolayer samples on doped semiconductors such as GaAs [31] and on materials with IR transparent spectral windows such as silica glass [10], and polymer films supported on water [16]. The latter study is noteworthy in that, in contrast to the vast majority of supported monolayer reflection studies which use FTIR instruments, this study on water was carried out using a diffraction grating based instrument.

Internal Reflection

The internal reflection element can be designed (see Fig. 3.4) to provide a single or multiple internal reflections. At each reflection from the interface the evanescent EM field extending into the exterior medium (see Fig. 3.3) can be used to probe the absorption (or more generally, the optical) characteristics of the adjoining medium within the decay length of the evanescent wave. The fraction of absorbed (or attenuated) power or, equivalently attenuated reflectance, vs energy provides a spectrum. On this basis the technique also has been termed attenuated total reflectance (ATR).

The ATR mode has been one of the most widely used reflection modes over the years and the principles have been well developed [32, 33]. This mode can be used to probe two types of surface layers: (1) a directly contiguous film or continuous medium with a sharp, uniform interface and (2) an external film which is brought into place and pressed directly in contact with the element. Examples of the first include chemical reactions arising directly on the element face, adsorbed molecular films, uniformly coated films such as spin cast polymers which fully wet the surface and immersion of the ATR element in a liquid medium. The perfect continuity of the interface allows quantitative analysis of the spectra by parallel layer simulation methods (see Sect. 3.3). The external contact mode is commonly done with materials that can be arranged to have some degree of close contact at the element face, such as powders and soft condensed matter films. Given the considerable voids or gaps of unknown spacing of the external medium at the element surface, quantitative analyses is rendered highly problematic since the exact locations of contact and gap distances from the ATR element face to an absorbing object are unknown.

Typical ATR elements for surface science studies are made from Si or Ge, robust, high refractive index materials (e.g., $n \sim 3.42$ and 4, respectively, in the mid-IR), or lower index materials such as ZnSe ($n \sim 2.42$), which are not as chemically robust. Examples of fundamental studies and applications include the formation of surface Si–H bonds on bare silicon [34], catalytic reactions at the solid/liquid interface [35], and mechanical wear-induced chemical changes (tribochemical) on the surfaces of metal coated, hard ATR elements such as Ge [36]. ATR can be used with both s- and p-polarization and the differences in the spectral intensities for an overlayer film sample can be used to determine orientations of the oscillators in anisotropic films. For the case of the contiguous film ATR mode quantitative simulations of the spectra can be carried out using the methods in Sect. 3.3.

An interesting example of a hybrid ATR-ER mode involves the use of an ATR reflection element with an internal beam propagating at a near grazing incidence within the element (termed GATR) with the element face pressed in tight contact (meaning the overall gap $\ll \lambda$) against a planar metal film sample. Under these conditions the evanescent E-field can setup a relatively intense electric field at the metal surface while also providing the opportunity for multiple reflections, in contrast to the single refection that one obtains in an external reflection mode with no ATR element involved. This combination can provide up to $\sim$2 orders of magnitude increase in signal over what the absorbance would be in transmission geometry and $\sim$1 order of magnitude over a single grazing angle specular reflection [37, 38]. Reported examples of GATR include the study of monolayers on common substrates such as gold and even silicon [39, 40].

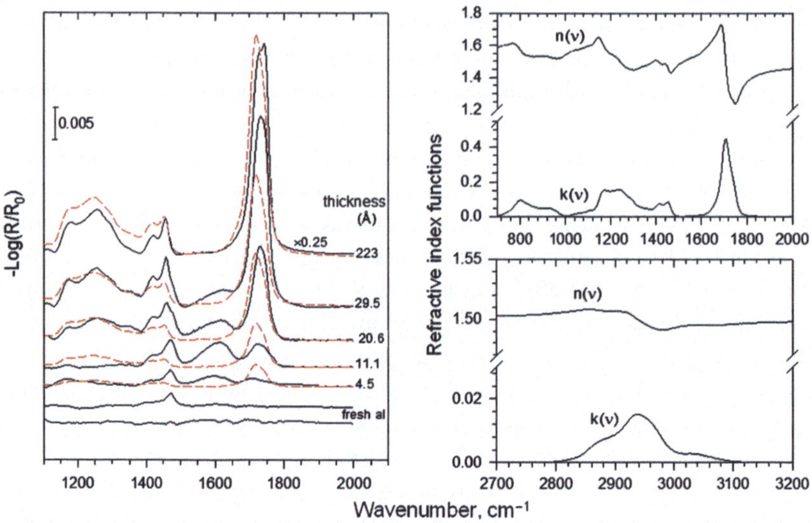

Fig. 3.9 *Left*: Grazing angle IR reflection spectra in the low frequency region for ultra thin PAA polymer films on native oxide-covered aluminum metal substrates for different polymer thicknesses (labeled on the *right*). The *black* spectra represent experiment and the *red-dashed* spectra theory simulations for isotropic films based on measured optical function spectra. *Right*: experimentally determined PAA optical function spectra in the low and high IR frequency regions

3.5.4 Examples of Spectral Simulations and Quantitative Analyses of Spectra

Solution Coated Ultrathin Polymer Films on Aluminum Substrates

In a study of the formation and structures of thin films of PAA [poly(acrylic acid); $(CH_2CHCO_2H)_n$] on vapor-deposited aluminum metal film substrates, simulations of IR reflection spectra were used to help understand the polymer film structures as a function of their thicknesses [41]. The reflection spectra taken with *p*-polarization at a grazing incidence angle of 86° are shown in Fig. 3.9, along with spectra simulated using the experimental optical setup parameters, the independently measured PAA film thicknesses and PAA optical functions. The close match of experimental and simulated reflection spectra (left) for the thickest film of 223 Å implies a fully isotropic bulk structure. The mismatch at the minimum thickness of a single polymer chain is consistent with strong $-CO_2H$ chemisorption at the oxide/polymer interface to produce carboxylate ($-CO_2^-$) ions. The vibrational features from these ions dominate the spectra, whereas the simulations based on pure PAA show only features from CH_2 and $-CO_2H$ groups. Since the intensities of these ion modes in bulk samples are highly sensitive to the counterions it was not possible to simulate these modes in the thin film samples.

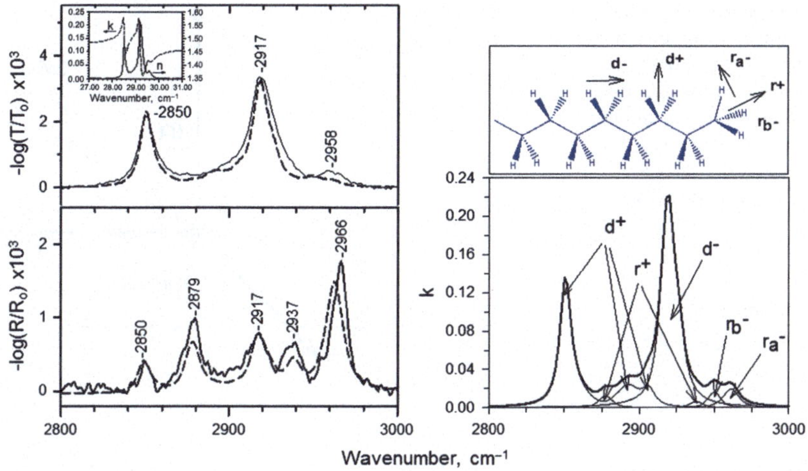

Fig. 3.10 *Left*: high frequency region IR spectra for ODS monolayer SAMs taken in the transmission mode for native oxide-covered silicon substrates (*lower*) and in the grazing angle reflection mode on vapor-deposited gold substrates (*upper*). The *solid lines* are experimental spectra and the *dashed* represent simulations based on measured optical functions for the ODS molecules in an ordered, polymerized form. *Right*: Optical function spectra of the polymerized ODS molecules. The spectra have been decomposed into the component fundamental molecular C–H stretching modes of the alkyl chains. *Upper*: Schematic of an alkyl chain showing the transition moment directions of the various C–H vibrational modes

Comparison of T and ER Spectra for Monolayer Structure on 2 Substrates

The example above illustrates a simulation based on isotropic optical functions. An example illustrating simulations based on anisotropic optical tensors is given by the study of oriented octadecylsiloxane [ODS; $CH_3(CH_2)_{17}SiO_xH_y$] monolayers on two entirely different planar substrates, native oxide-covered silicon and gold [42]. Normal incidence transmission spectra with unpolarized IR light and 86° grazing incidence with *p*-polarization were used to obtain spectra from the two types of samples, respectively, in order to determine whether the two monolayers had the same or different types of packing and alkyl chain orientations, given the entirely different types of substrates for attachment of the siloxy head groups. To answer the question simulations were made of each type of spectra using the independently measured film thicknesses (from single wavelength ellipsometry) and the known optical function properties of SiO_2, Si and Au. An isotropic complex optical function spectra $\hat{n}(\nu)$ was obtained from transmission spectra of pressed KBr pellet disks with known concentrations of polycrystalline samples of the polymer of ODS [poly(ODS) or PODS] prepared by polymerization of $CH_3(CH_2)_{17}SiCl_3$ in water [43]. The isotropic n, k spectra were converted to a tensor form for the C–H stretching region by assigning the main peaks in the isotropic spectrum to specific vibrational modes and further assigning the transition moment directions relative to the molecular backbone, as shown in the right hand side of Fig. 3.10.

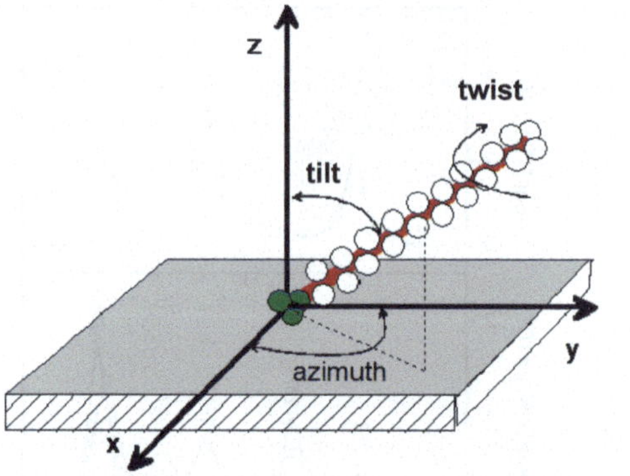

Fig. 3.11 Schematic of an alkyl chain showing the chain axis tilt from the surface normal vector, the chain twist around the chain axis and the azimuthal orientation of the chain around the surface in-plane coordinate axes

The transition moment directions were used to construct diagonalized optical function *k* tensors for each mode from the isotropic *k* spectra. Combining this with KK transforms generated n values to give a final diagonalized $\hat{n}(\tilde{\nu})$ tensor for each mode. Models of different chain orientations were constructed and simulations were done for the transmission and reflection spectra until best fits were obtained, as shown in Fig. 3.10 (left hand side). Though the experimental spectra appear strikingly different, the best fits for each sample were identical within the stated errors to give a common structure of a chain tilted $10(\pm 2)^\circ$ tilt from normal (z axis) and the C–C–C backbone plane twisted $46(\pm 2)^\circ$, essentially halfway between the xz and yz planes, as defined in Fig. 3.11, which also shows the azimuthal angle which does not apply in the present case since the films are uniaxial (no net orientation in the xy plane).

This example demonstrates the value of quantitative analysis of IR spectra using simulations based on experimentally obtained optical functions. The simulations are most valuable when combined with independent information of the sample surface or thin film structure, particularly film thickness, and in the case of oriented films if independent information is available on molecular tilt angles, for example, this allows extraction of more accurate values of twist angles. The major challenge in these simulation methods is the availability of accurate n, k data which specifically apply to the structures of the thin films. A common problem is that optical functions are obtained from a bulk polycrystalline (or even liquid) material but the film structure may pack in a slightly different way giving rise to perturbations in the mode frequencies and transition moment intensities relative to the bulk reference material. It is a common experience that occasionally new peaks appear that are not accounted for in the pure reference materials and often these peaks can be assigned to highly perturbed modes that arise when the molecules are tightly packed on the surface with a different type of ordering arrangement than in the bulk polycrystalline material. Discerning these types of effects may require sophisticated calculations of the ensembles of molecules using quantum chemical theory.

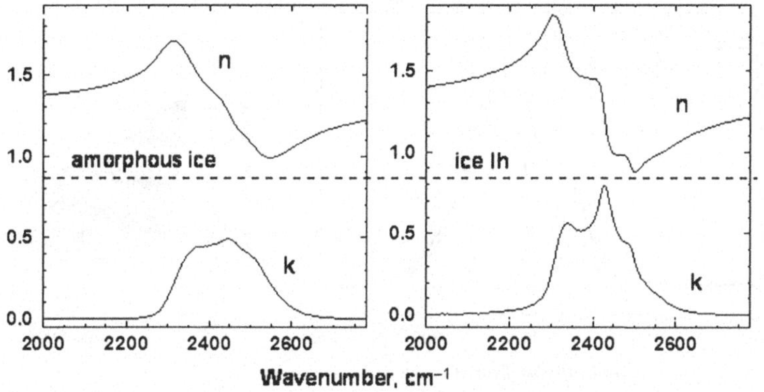

Fig. 3.12 Optical constant spectra for amorphous and crystalline Ih ice in the high frequency IR region

Application of the Effective Medium Approximation to Treat Voids in Ice Films

A final example involves the use of effective medium theory in the simulations for cases in which thin film samples contain voids. Examples of the application of the effective medium approximation (EMA) to spectral simulations can be found in previous reports of the study of ice overlayers on SAMs [44] and reconstruction of SAMs by induced chemical reactions on the SAM substrate surface [45]. Here we discuss the first example.

It is common in the case of deposited films that the full density of the film is not reached and voids are formed during the deposition. This can be seen in the following example of ice thin film structures formed by dosing D_2O vapor onto on a self-assembled monolayer of a $CH_3(CH_2)_{15}S$–/Au{111} SAM at low temperatures in vacuum [46]. The phases of the solid water exhibit different IR spectra in the O-D stretching region, as seen in the optical constant spectra in Fig. 3.12, so grazing incidence IR reflection can be applied to determining the structures. Note the use of D_2O eliminates interference of C–H stretching modes on O–H stretching modes. The isotropic optical constant spectra of amorphous (Ia) and crystalline (Ih) ice provide simulations of ice spectra, based a four medium model of vacuum, ice, SAM and Au substrate, as shown in Fig. 3.13(left), overlaid on the experimental spectrum for a particular dosing condition of $\sim$3 monolayers of water at 140 K [47].

Simulations of fully dense ice of either pure amorphous or a pure crystalline phases give poor fits to the experimental spectrum shape but a mixture of 60/40 mixture of Ia/Ih give improved fits, though with residual errors. In order to obtain closer fits the presence of nanometer scale voids was considered. Treating these using the Maxwell-Garnett model (see (3.18)) gives better fits, as shown in Fig. 3.13(left) where the best fit is obtained for a 10 Å mixed phase film with 50 % voids with the global fitting error surface for both phase mixture and void content shown in Fig. 3.13(right). Note the strong effects of the presence of voids on the band shape in Fig. 3.13(left).

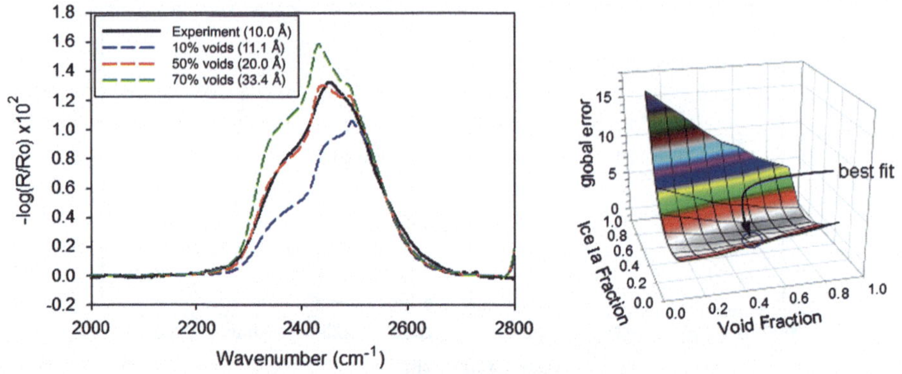

Fig. 3.13 *Left*: O–D stretching mode spectrum for a 10 Å D_2O mixed phase ice film. *Dashed lines*: simulated spectra for isotropic films of different thicknesses and void contents. The simulations were based on the optical constant data in Fig. 3.12 and the Bruggeman EMA equation. *Right*: Correlation plot of the global error for fitting simulated spectra with varying void volume and Ice Ia fractions

3.6 Future Directions

The application of infrared spectroscopy to both practical and fundamental studies of surfaces and thin films has been a popular and proven technique over many years and is expected to continue to fill this role well in the future. The instrumentation and theory of the different sampling modes are well developed, allowing generation of good signal/noise spectra well into the submonolayer level with the ability to simulate spectra for interpretation and quantitation of structural parameters. The main directions for future advancement would appear to be in improvements in energy dispersion through the use of diffraction gratings in combination with high sensitivity, low noise array detectors which will provide an alternate to the now common and mature technique of Fourier transform instruments based on interferometers. The array detector based instruments are capable of pushing the spectral acquisition times another order of magnitude or two in the near future to allow real time monitoring of surface processes in the millisecond scale with improved instrument stability to also allow extensive signal averaging for time stable samples. In addition, the use of array detectors will allow signal modulation at frequencies that would typically interfere with interferometer data frequencies (kHz), for example, rotating polarizer elements for spectroscopic IR ellipsometry experiments. It is expected there will be a number of specialized advances in areas outside of the scope of this chapter such as pulsed, tunable free electron lasers, synchrotron IR sources, supercontinuum mid-IR lasers, IR transparent optical fibers for incorporation in IR optics, and various non-linear IR techniques. Overall we expect the standard mid-IR techniques will remain extremely useful for surface and thin film applications well into the future.

References

1. V.P. Tolstoy, I.V. Chernyshova, V.A. Skryshevsky, *Handbook of Infrared Spectroscopy of Ultrathin Films* (Wiley-Interscience, Hoboken, 2003), p. 710
2. C.J. Hirschmugl, Surf. Sci. **500**(1–3), 577–604 (2002)
3. Y.J. Chabal, Surf. Sci. Rep. **8**(5–7), 211–357 (1988)
4. I.N. Levine, *Molecular Spectroscopy* (Wiley, New York, 1975), Chap. 3
5. M. Born, E. Wolf, *Principles of Optics*, 5th edn. (Pergamon, New York, 1975)
6. E. Wilson, J. Decius, P. Cross, *Molecular Vibrations* (Dover, New York, 1955)
7. H.G. Tompkins, E.A. Irene (eds.), *Handbook of Ellipsometry* (William Andrew Publishing, Norwich, 2005)
8. N. Winograd, Anal. Chem. **65**, 622A (1993)
9. M.D. Porter, T.B. Bright, D.L. Allara, T. Kuwana, Anal. Chem. **581**, 2461 (1986)
10. A.N. Parikh, D.L. Allara, J. Chem. Phys. **96**, 927 (1992)
11. P. Yeh, *Optical Waves in Layered Media* (Wiley-Interscience, New York, 1988)
12. D.A.G. Bruggeman, Ann. Phys. (Leipz.) **24**, 636 (1935)
13. D.E. Aspnes, Thin Solid Films **89**, 249–262 (1982)
14. G.E. Jellison Jr., Thin Solid Films **234**, 416–422 (1993)
15. Y. Liu, A.T. Kalambur, J.F. Rabolt, D.B. Chase, Langmuir **20**, 816–822 (2004)
16. Y.S. Kim, C.M. Snively, Y. Liu, J.F. Rabolt, D.B. Chase, Langmuir **24**, 10791–10796 (2008)
17. A.N. Parikh, D.L. Allara, I.B. Azouz, F. Rondelez, J. Phys. Chem. **98**(31), 7577–7590 (1994)
18. Y.J. Chabal, M.A. Hines, D. Feijoo, J. Vac. Sci. Technol., A, Vac. Surf. Films **13**(3), 1719–1727 (1995)
19. N.J. Harrick, Appl. Spectrosc. **31**(6), 548–549 (1977)
20. R.D. Lowe, M.A. Pellow, T.D.P. Stack, C.E.D. Chidsey, Langmuir **27**(16), 9928–9935 (2011)
21. M. Acik, C. Mattevi, C. Gong, G. Lee, K. Cho, M. Chhowalla, Y.J. Chabal, ACS Nano **4**(10), 5861–5868 (2010)
22. C.E. Taylor, D.K. Schwartz, Langmuir **19**(7), 2665–2672 (2003)
23. D.A. Guzonas, M.L. Hair, C.P. Tripp, ACS Symp. Ser. **447**, 237–250 (1991)
24. S. Ruthstein, R. Artzi, D. Goldfarb, R. Naaman, Phys. Chem. Chem. Phys. **7**(3), 524–529 (2005)
25. A.S. Lagutchev, K.J. Song, J.Y. Huang, P.K. Yang, T.J. Chuang, Chem. Phys. **226**(3), 337–349 (1998)
26. J.J. Stapleton, T.A. Daniel, S. Uppili, O.M. Cabarcos, J. Naciri, R. Shashidhar, D.L. Allara, Langmuir **21**(24), 11061–11070 (2005)
27. M. Trenary, Annu. Rev. Phys. Chem. **51**, 381–403 (2000)
28. E.L. Wilson, W.A. Brown, J. Phys. Chem. C **114**(15), 6879–6893 (2010)
29. G.L. Fisher, A.V. Walker, A.E. Hooper, T.B. Tighe, K.B. Bahnck, H.T. Skriba, M.D. Reinard, B.C. Haynie, R.L. Opila, N. Winograd, D.L. Allara, J. Am. Chem. Soc. **124**(19), 5528–5541 (2002)
30. A.V. Walker, T.B. Tighe, B.C. Haynie, S. Uppili, N. Winograd, D.L. Allara, J. Phys. Chem. B **109**(22), 11263–11272 (2005)
31. C.L. McGuiness, G.A. Diehl, D. Blasini, D.M. Smilgies, M. Zhu, N. Samarath, T.N. Weidner, M. Ballav, M. Zharnikov, D.L. Allara, ACS Nano **4**, 3447–3465 (2010)
32. N.J. Harrick, *Internal Reflection Spectroscopy* (Wiley, New York, 1967)
33. F.M. Mirabella (ed.), *Principles, Theory and Practice of Internal Reflection Spectroscopy* (Dekker, New York, 1993)
34. R. Jakob, Y.J. Chabal J. Chem. Phys. **95**, 2897–2909 (1991)
35. J.-M. Andanson, A. Baiker, Chem. Soc. Rev. **39**, 4584–4671 (2010)
36. F. Mangolini, A. Rossi, N.D. Spencer, J. Phys. Chem. C **116**, 5614–5627 (2012)
37. M.E. Mulcahy, S.L. Berets, M. Milosevic, J. Michl, J. Phys. Chem. B **108**(5), 1519–1521 (2004)
38. M. Milosevic, V. Milosevic, S.L. Berets, Appl. Spectrosc. **61**(5), 530–536 (2007)
39. Y. Qiao, D. Wang, J.M. Buriak, Nano Lett. **7**(2), 464–469 (2007)

40. S.A. Swanson, R. McClain, K.S. Lovejoy, N.B. Alamdari, J.S. Hamilton, J.C. Scott, Langmuir **21**(11), 5034–5039 (2005)
41. E. Koo, S. Yoon, S.V. Atre, D.L. Allara, Langmuir **27**, 3638–3653 (2011)
42. A.N. Parikh, D.L. Allara, F. Rondelez, Langmuir **11**, 2357–2360 (1995)
43. A.N. Parikh, M.A. Schivley, E. Koo, K. Seshadri, D. Aurentz, K. Mueller, D.L. Allara, J. Am. Chem. Soc. **119**, 3135–3143 (1997)
44. M. Östblom, R. Valiokas, P. Konradsson, S. Svensson, B. Liedberg, M. Garrett, D. Allara, J. Phys. Chem. B **110**, 1830–1836 (2006)
45. Y.-T. Tao, G.D. Hietpas, D.L. Allara, J. Am. Chem. Soc. **118**, 6724–6735 (1996)
46. I. Engquist, I. Lundström, B. Liedberg, A.N. Parikh, D.L. Allara, J. Chem. Phys. **106**, 3038–3048 (1997)
47. M. Garrett, M. Östblom, B. Liedberg, D.L. Allara, Unpublished results

Chapter 4
A Surface Scientist's View on Spectroscopic Ellipsometry

Maurizio Canepa

Abstract Nowadays, a broad choice of instruments, including dedicated synchrotron radiation beamlines, allows to exploit Spectroscopic Ellipsometry (SE) to investigate the thickness and the dielectric properties of thin films, from the terahertz down to the VUV wavelength range. Instruments combining fast parallel detection, precision, accuracy, are pushing forward *real time* and *in-situ* applications, to monitor the dynamics of processes such as e.g. film growth, oxidation, polymerization, electrochemical processes, with a diverging spectrum of scientific and industrial applications in the fields of nano-electronics, coatings, solar cell materials, polymer technology, bio-sensing, photonics, just to name a few. This chapter, beyond presenting the essentials of principles and instrumentation of SE, is intended to place this thin-film technique in the perspective of the surface scientist, through the selection of applications to ultra-thin films and nanostructures. Emphasis is placed on reflection experiments, in the 190–1700 nm wavelength range covered by high-quality commercial instruments, although some infra-red (IRSE) and far UV experiments are also discussed.

4.1 Introduction

Spectroscopic (or Multi-wavelength) Ellipsometry (SE) is sometimes superficially mistaken with single-wavelength ellipsometry (SWE). SWE, since the early modern work of Rothen [1], in turn based upon fundamental principles established by Drude [2] and the seminal work of K. Blodgett and I. Langmuir [3],[1] is universally considered as the method for *measuring the thickness* of films. Consequently, the potential of SE as a true spectroscopic tool is often underestimated. Nowadays, a broad set of instruments, including dedicated synchrotron radiation beamlines, allows to perform SE measurements from the terahertz down to the far UV wavelength range, to investigate both the thickness *and* the dielectric properties of thin

[1] Cf. [4, 5] for synthetic historical accounts.

M. Canepa (✉)
Department of Physics, University of Genoa, via Dodecaneso 33, 16146 Genoa, Italy
e-mail: canepa@fisica.unige.it

G. Bracco, B. Holst (eds.), *Surface Science Techniques*,
Springer Series in Surface Sciences 51, DOI 10.1007/978-3-642-34243-1_4,

and ultra thin films, with thickness sensitivity below the nanometer limit [6]. The development of SE instrumentation in the last three decades has been impressive. On one side, automated instruments, conjugating fast response with high accuracy and precision, have pushed forward in-situ applications, to monitor the dynamics of processes such as e.g. film growth and oxidation, polymerization, electrochemical processes, of importance in many technological contexts. On the other side, the development of instruments able to measure the Mueller matrix coefficients, together with advances in data interpretation, allow to address systems endowed with complex morphology and functionality (metamaterials), such as biaxial and magnetic films, porous films, multi-layers [6–8].

As documented in the next section there are excellent books and reviews covering the great deal of SE applications to stratified media and interfaces. This chapter has been conceived to place SE, a thin-films method, in the possible perspective of a surface scientist or, better, of a surface nano-scientist. I made a selection of a few topics where SE could most effectively integrate the other *surface science* methods presented in this book in the belief that the full potential of SE can be exploited in combination with other characterizations methods, e.g. microscopy tools (such as electron and scanning probe *nanoscopies*), structural probes (such as XRD or electron diffraction methods), and other techniques providing elemental information such as e.g. photoelectron spectroscopies or ion scattering methods. The selection necessarily implied some arbitrariness, reflecting my personal views and interests. The focus is mainly on the application of SE to ultra-thin films (thickness approximately below 10 nm), including inorganic, organic and biologic materials, and to surface nanostructures. Emphasis is placed on reflection ellipsometry experiments in the 190–1700 nm range, covered by several high-quality table-top commercial instruments; however, on specific points infrared (IRSE) and far UV experiments are also considered.

Basic information on principles and instrumentation is presented in Sects. 4.2 and 4.3, respectively. Sections 4.4 and 4.5 present a selection of applications and a few final notes.

4.2 Basics on Principles, Methods and Systems

Principles, methods and instruments of SE are presented in great detail in several books. A surely incomplete list should probably begin with the *classical* book of Azzam and Bashara [9] and with the more recent, comprehensive *Handbook of Ellipsometry* (from now on HoE) [6]. The recent book of Fujiwara [10] beyond a rigorous tutorial part, includes a compact yet exhaustive treatment of instrumentation and methods, and is useful for beginners and students as well as for specialists. Both the HoE and the Fujiwara's book dedicate ample space to the presentation of the *Stokes* and *Jones* vector formalisms, necessary to comprehend instrument design and operation, and to perspectives into several advanced topics such as the field of *Generalized Ellipsometry* [9, 11, 12]. Users, students, beginners (but not only) will appreciate the rigorous conciseness of the guide written by Tompkins and

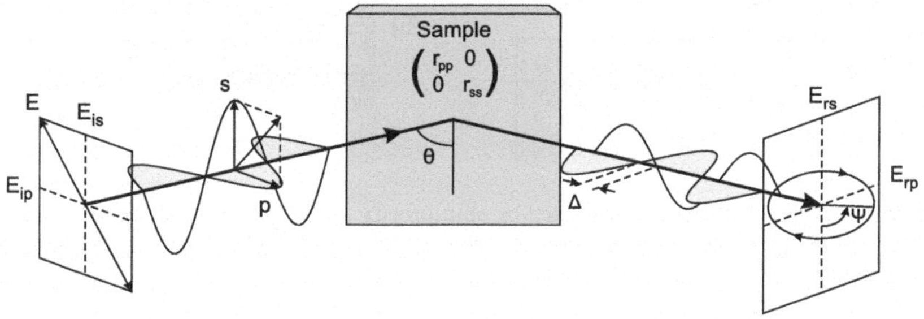

Fig. 4.1 Ellipsometry determines the change of light polarization upon reflection on a sample. The plane of incidence contains the incident beam and the normal to the sample. θ is the angle of incidence. The sample as an optical element is described by its Jones's matrix. Δ and Ψ are the ellipsometric angles (see (4.1)) [19]

McGahan [13], which describes many applications under an operative perspective. A concise presentation of SE basics was also reported in Ref. [14]. IRSE principles and applications are comprehensively treated in Refs. [15, 16] and in a dedicated chapter of the HoE [17]. Finally, updates on SE advances can be found in the proceedings of the International Conference on Spectroscopic Ellipsometry (ICSE).[2]

4.2.1 The Fundamental Relation of Ellipsometry: Ψ and Δ

In this section we summarize basic notions regarding SE. Focus, for simplicity, is placed on so-called *standard* measurements on samples consisting of homogeneous, uniform and isotropic layers.

It is always useful to premise that ellipsometry does not *measure* the thickness or the index of refraction of films; it determines the change of light polarization upon reflection on a sample (Fig. 4.1). The output of a *standard* ellipsometry measurement is represented by the *ellipsometric angles* Ψ and Δ, defined in the *fundamental* relation of ellipsometry:[3]

$$\tilde{\rho} = \frac{\tilde{r}_p}{\tilde{r}_s} = \tan\Psi \exp(i\Delta) \tag{4.1}$$

where $\tilde{\rho}$ is the complex reflection coefficient of the sample under investigation and $\tilde{r}_p$, $\tilde{r}_s$ are the Fresnel reflection coefficient for p and s polarized light, respectively, i.e. the non-vanishing diagonal elements $\tilde{r}_{pp}$, $\tilde{r}_{ss}$ of the so-called Jones matrix [19]:

$$\mathbf{J} = \begin{pmatrix} \tilde{r}_{pp} & 0 \\ 0 & \tilde{r}_{ss} \end{pmatrix} \tag{4.2}$$

which describes the sample as an optical element immersed in a isotropic medium called the *ambient*, preferably highly transparent in the wavelength range of the

[2]The proceedings of the last ICSE-V have been published in Ref. [18].

[3]Cf. [19] for an accurate discussion of the equation also in relation to the coordinate system and different conventions in use for Fresnel equations.

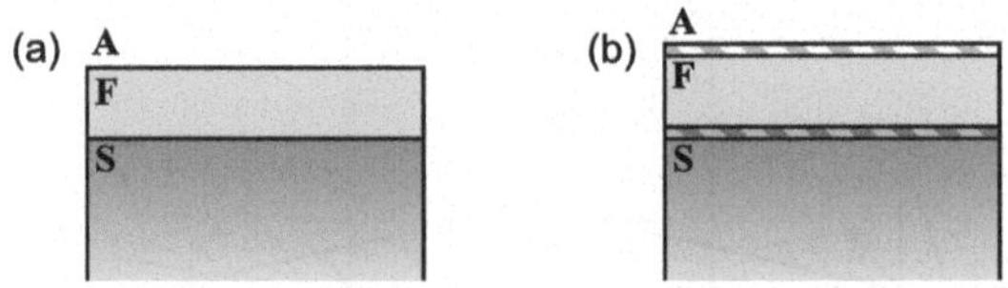

Fig. 4.2 Prototypical systems investigated by ellipsometry. (**a**) Three-phase ideal system: A (ambient)/F (film)/S (substrate). Interfaces are chemically and geometrically sharp. (**b**) A more realistic picture including the presence of interface layers

experiment. Note that as intensities ratios are measured, the calibration procedures are simplified with respect to e.g. reflectivity measurements. The Ψ and Δ representation was historically derived from single-wave ellipsometry. Modern ellipsometers derive Ψ and Δ from a Fourier analysis of the detector signal as briefly outlined in Sect. 4.3.1.

For bulk substrates in a given ambient of known index N_A, under the assumption of a chemically and geometrically sharp material/ambient interface, inversion of Ψ and Δ from data taken at angle of incidence θ would directly provide the complex dielectric function, avoiding the Kramers-Kronig (KK) analysis necessary for solely intensity (e.g reflection or transmission) measurements.

$$\tilde{\epsilon} = \sin^2(\theta)\left[1 + \tan^2(\theta)\left(\frac{1-\tilde{\rho}}{1+\tilde{\rho}}\right)^2\right] \tag{4.3}$$

In practice, ideal substrates do not exist and therefore the dielectric function derived by (4.3) must be always intended as a so-called *pseudo*-dielectric function ϵ_{pseudo}, that represents the optical response of the system as a whole. Inversion of (4.3) from data taken at different angle of incidence (or in different ambients) provides then a quick criterion to evaluate the quality of the substrate surface. If an overlayer is present, or the morphology is appreciably rough, ϵ_{pseudo} shows a significant angle-dependence.

4.2.2 The Prototypical System: Isotropic Thin Film on a Substrate

Figure 4.2(a) schematically presents the simplest system investigated by ellipsometry: an optically isotropic film of complex refractive index $\tilde{N}_F$ and thickness d_F (a few nm up to the micron scale) on a macroscopically thick (ideally semi-infinite)[4] substrate of refractive index $\tilde{N}_S$. The substrate/film and the film/ambient interfaces are in the first instance assumed chemically sharp. $\tilde{\rho}$ can be simply calculated as [10, 13]:

$$\tilde{\rho} = \frac{\tilde{r}_p^{AF} + \tilde{r}_p^{FS}\exp(-j2\beta)}{1 + \tilde{r}_p^{AF}\tilde{r}_p^{FS}\exp(-j2\beta)}\,\frac{1 + \tilde{r}_s^{AF}\tilde{r}_s^{FS}\exp(-j2\beta)}{\tilde{r}_s^{AF} + \tilde{r}_s^{FS}\exp(-j2\beta)} \tag{4.4}$$

[4]In practice, the backside of transparent substrates is prepared to be not reflective; see for example [20].

with

$$\beta = 2\pi \frac{d_F}{\lambda} \tilde{N}_F \cos(\theta_{AF}) \tag{4.5}$$

where $\tilde{r}^{ij}_{p,s}$ are the Fresnel coefficients pertinent to the involved interface and θ_{AF} is the refraction angle at the ambient/film interface. $\tilde{\rho}$ is a function of the angle of incidence θ, the wavelength λ, the thickness d_F and of $\tilde{N}_S$, $\tilde{N}_F$ and N_A. N_A and $\tilde{N}_S$ should be preferably known,[5] or accurately pre-determined with dedicated measurements. Elementary considerations on Fresnel reflection rules suggest that (i) a large substrate/film and film/ambient refractive index mismatch and (ii) an incidence angle close to the principal angle of the system (simply the Brewster's angle in the case of a dielectric film), increase the sensitivity to the film.

In order to analyze SE data a parametric model of the film refraction index $\tilde{N}_F(\lambda)$ $(n_F(\lambda), k_F(\lambda))$ is usually needed. Widely used models are reviewed in any good book on optics of thin films [21]. Popular models for dielectric films in the transparency spectral range exploit the well-known Sellmeier or Cauchy dispersion relations [9, 10]. Free-carrier absorption is accounted for by the Drude model, taking into account the effective mass of carriers [9, 10]. Absorption bands near the edge in the visible/UV region can be described by refinements and/or generalization of Kramers-Kronig consistent multi-resonance models [22], preferably backing on band structure data [23]. May belong to this category the so-called Tauc-Lorentz [24, 25], Cody-Lorentz [26, 27] and the parametric Herzinger-Johs methods [28]. Some hints on these flexible parametric models can be provided on the example of the Cody-Lorentz formula, able to describe the absorption threshold e.g. of amorphous dielectrics. The imaginary part of the dielectric constant in the Cody-Lorentz model reads as

$$\epsilon_2(E) = \begin{cases} \frac{E_t G(E_t) L(E_t)}{E} \exp\left(\frac{E-E_t}{E_u}\right), & 0 < E \le E_t \\ G(E)L(E) = \frac{(E-E_g)^2}{(E-E_g)^2+E_p^2} \frac{A E_0 \Gamma E}{(E^2-E_0^2)^2+\Gamma^2 E^2}, & E > E_t \end{cases} \tag{4.6}$$

While a thorough discussion of all parameters occurring in (4.6) can be found in Ref. [27], their role can be intuitively appreciated in Fig. 4.3: roughly, a Lorentz oscillator $L(E)$ (of amplitude A, position E_0 and intrinsic broadening Γ) is modified, above an energy threshold E_t, by a variable band edge function $G(E)$, to describe interband transitions and below E_t to approach the exponential *Urbach tail* of the absorption coefficient, accounting for defect-induced edge broadening [29].

Interpolation schemes based on polynomials, e.g. cubic splines, have been also proposed [30]. Recently, Johs and Hale discussed a flexible, Kramers-Kronig consistent method based on the recursive properties of B-Splines [31].

The dielectric constant parameters and the thickness d_F, are generally assumed as *free* parameters in a regression analysis of data looking for the best agreement

[5]The software of commercial instruments includes libraries with optical properties of many materials. These data are generally accurate but should be always considered with caution since optical properties of substrates and films may depend significantly on the preparation, post-growth treatments, ageing and contamination effects.

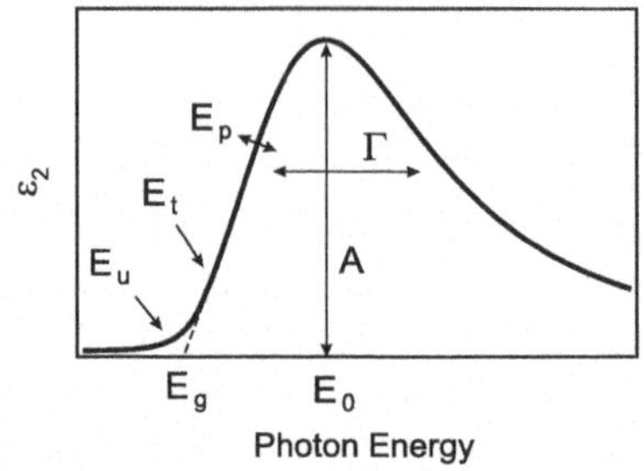

Fig. 4.3 Imaginary part of the dielectric function in the Cody-Lorentz model [27], used to describe the absorption threshold of amorphous dielectrics. E_g: band-gap; E_0, A, Γ: energy position, strength and width of Lorentz oscillator; E_p: transition energy; E_t and E_u determine the onset and extension of the Urbach tail [29]

between experimental Ψ and Δ and simulations based on the Fresnel relations. Measurements for several values of θ and d_F may improve the analysis reliability [28, 32]. In case of relatively thick dielectric films presenting a sharp absorption threshold, a diffuse practice is to determine d_F using data in the transparency range and then, given this thickness, determine n and k in the absorption region.

Best fit routines exploit the minimization, carried out e.g. by Levenberg-Marquardt or Newton methods, of suitable biased estimators [33, 34]. A popular function is the *Mean Squared Error*:

$$MSE^2 = \frac{1}{2N-M}\sum_{i=1}^{N}\left[\left(\frac{\Psi_i^{exp}-\Psi_i^{calc}}{\sigma_{\Psi,i}^{exp}}\right)^2+\left(\frac{\Delta_i^{exp}-\Delta_i^{calc}}{\sigma_{\Delta,i}^{exp}}\right)^2\right]=\frac{1}{2N-M}\chi^2 \tag{4.7}$$

where N is the number of channels, M the number of adjustable parameters and $\sigma_{\Psi,i}^{exp}$, $\sigma_{\Delta,i}^{exp}$ are the standard deviations of the experimental data point.

An important issue in data analysis is the check of the mathematical and physical correlations between fit parameters [28]. An example of such correlations is the thickness/refraction index correlation for very thin films [10, 13, 35, 36]. A combination of film and substrate with similar index of refraction in the transparency range, a situation encountered e.g. for organic and biologic films on glass [37], is another example of severe correlation. Smart methods have been proposed to circumvent the problem, combining SE and an IRSE investigation of molecular vibrations [37]. Good routines for best fit and statistical evaluation, such for example Uniqueness Test [34], are a factor of quality of the data analysis software of commercial instruments.

Transition Layers. Effective Medium Approximations Models

A careful interpretation of SE results must generally account for the presence of surface and interface layers. The optical system sketched in Fig. 4.2(b) is a multilayer. Multi-layered structures made of homogeneous uniform and isotropic layers can

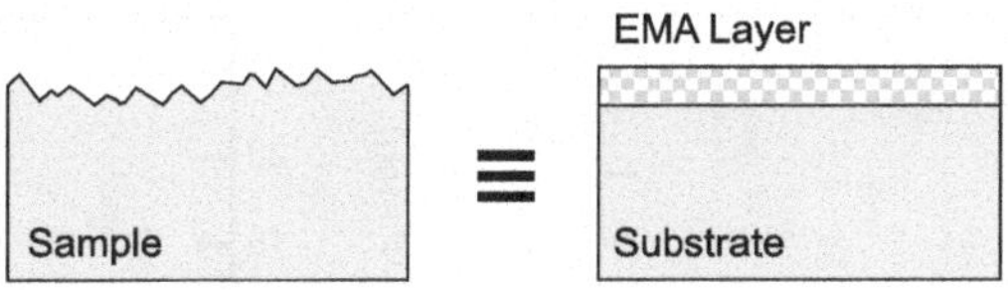

Fig. 4.4 Surface Roughness Layer. A well-established approach considers a Bruggemann-EMA [38, 39] surface layer which weights the optical properties of the ambient and of the film [41]. In the simplest approach a 50 % percentage of so-called *voids* is assumed

be treated within the framework of Fresnel formalism through recursive or matrix methods [9, 10]. Each layer is then described by an optical complex function $\tilde{N}_j(\lambda)$ $(n_j(\lambda), k_j(\lambda))$ and thickness d_j that can be eventually used as adjustable parameters in fitting routines.

Interface layers represent *per se* an example of nano film of *complex* morphology. Their dielectric properties are often described through a reasonably limited number of parameters by Effective Medium Approximation (EMA) models, in which the optical properties of an *effective* film are constructed from the optical properties of the constituent materials [38]. One of the most popular EMA model is due to Bruggemann (B-EMA) [39], in which the dielectric function of the effective film, ϵ_e, can be obtained by solving the implicit equation

$$\sum_{i=1} f_i \frac{\epsilon_i - \epsilon_e}{\epsilon_i + 2\epsilon_e} \tag{4.8}$$

where ϵ_i and f_i are the dielectric functions and volume fractions of the constituents, the latter being usually adopted as fitting parameters. A thorough introduction to the Bruggemann and other EMA models (such as Lorentz-Lorenz and Maxwell-Garnett models)[6] can be found in general references [6, 10, 21] where one can find many relevant applications.[7]

Regarding *surface roughness* for example, a well-established approach considers a B-EMA layer with thickness d_{SR} (see Fig. 4.4) which weights the ambient and film optical properties [38, 41]. Information coming from a knowledge of the film deposition process and from results of other techniques may help a-priori to optimize EMA models, and in the a-posteriori analysis stage, eventually contributing to limit correlation conflicts. An example is provided by wide-scan AFM measurements, useful to check the surface roughness [42].[8] SE determinations on the substrate/film interface layer can be usefully compared to cross-sectional TEM (destructive!) measurements [44].

[6]The work of Bruggemann is in German. The reader interested in original sources may consider the reading of the splendid paper of J.C.M. Garnett [40].

[7]EMA models play a key role in optical studies of systems of complex morphology such for example films with not-uniform density and films with intrinsic granularity or porosity. More details in Sect. 4.4.3.

[8]For a critical discussion of the surface roughness modeling, also in relation to AFM determinations see the recent Ref. [43].

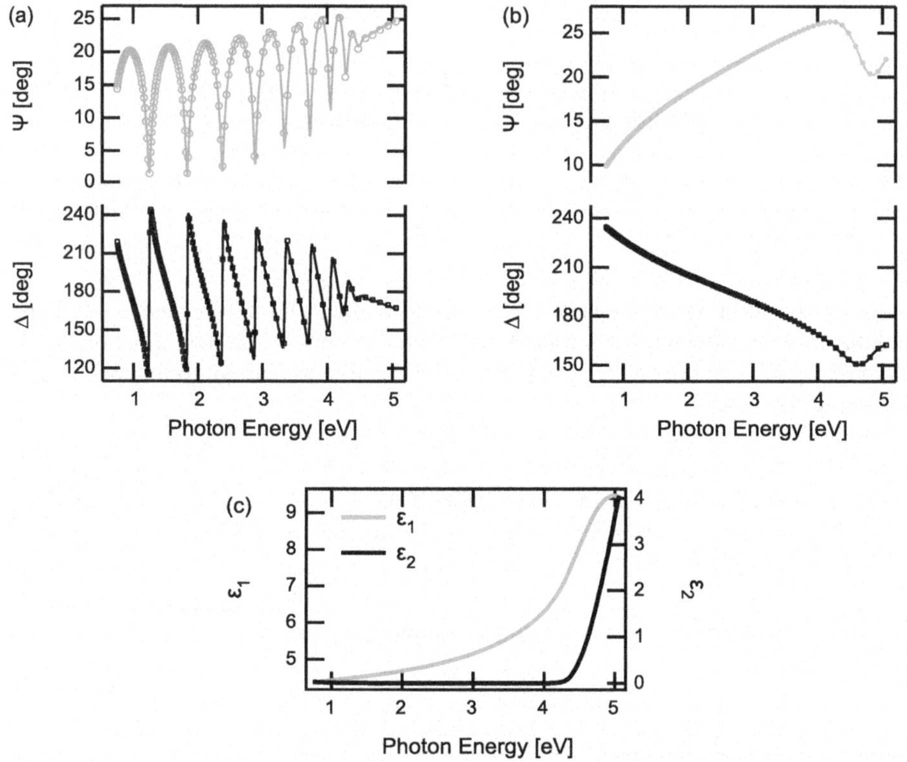

Fig. 4.5 Typical output of standard SE measurements and analysis on thin films on the example of amorphous Ta_2O_5 deposited by ion sputtering methods on fused silica substrates [46]. (**a**) Film thickness ~500 nm. Symbols: data. Note the strong damping of oscillations (analogous to interference fringes in reflectivity/transmission measurements) at the absorption threshold (about 4.2 eV). Continuous lines: best fit curves (MSE = 12) computed for a 4-layer model (ambient/surface roughness/film/substrate). The film is modeled through a Maxwell-Garnett EMA between a Cody-Lorentz oscillator, describing the Ta_2O_5 absorption threshold, and a low density of empty pores. (**b**) The same for a 40 nm-thick film (MSE = 3.2). (**c**) Complex dielectric function of the film resulting from the best fit

With a proper choice of the ambient/film index matching, one can obtain the *optical removal* of the film, thereby increasing the sensitivity of the measurement to the interface layer. This condition is exploited in the *immersion* ellipsometry technique [45].

Figure 4.5 shows the output of standard SE measurements on the example of thin amorphous tantala films deposited on fused silica substrates [46].

Anisotropic Thin Films

Systematic application of SE to anisotropic materials is relatively recent [47, 48] and requires measurement and analysis methods typical of so-called *Generalized*

Ellipsometry (GE) [49–52]. A typical GE system consists of a biaxial or uniaxial film, preferably deposited on an isotropic substrate, in an isotropic ambient. Strategies, modes of experimental implementation and analysis methods of GE are accurately presented in Ref. [12]. Within the Jones's formalism, for non depolarizing samples,[9] the four elements of the not-diagonal matrix of anisotropic samples are actually defined by six independent ellipsometric angles ($\Psi_{pp}\Delta_{pp}$, $\Psi_{ps}\Delta_{ps}$, $\Psi_{sp}\Delta_{sp}$) to be experimentally determined. The analysis is greatly simplified for suitable optical axis orientations with respect to the plane of incidence [53]. For uniaxial systems where the optical axis is parallel to the plane of incidence the off-diagonal elements of the Jones matrix vanish and meaningful results can be obtained also with appropriate analysis of standard SE measurements [54].

Application fields of GE include e.g. crystalline inorganic films [55], polymers [56–58], Langmuir-Blodgett films [59], magnetic films [60, 61]. Recent applications to photonics materials [62, 63] will be discussed in Sect. 4.4.3.

4.2.3 Ultrathin Films and Monolayers

It is easy to verify experimentally that the Ψ, Δ changes induced by ultra thin films (a few nm thick) such for example organic self-assembled monolayers (SAMs), may be very small, of the same size of the sample-to-sample Ψ, Δ variability of common substrates, and even of zone-to-zone variability on an individual sample. A convenient way to emphasize the film -induced changes can be represented by so-called difference spectra (simply $\delta(\Psi, \Delta) = (\Psi, \Delta)_{film} - (\Psi, \Delta)_{Sub}$ or $\delta\langle\epsilon\rangle = \langle\epsilon\rangle_{film} - \langle\epsilon\rangle_{substrate}$) [9, 64–66].[10] As an example we report in Fig. 4.6 δ-spectra calculated for a transparent film of variable thickness on Au and Si substrates. In the simulations $\delta\Psi$ values are very small, yet above the sensitivity limits of good instruments, whereas Δ changes are well measurable (see Sect. 4.3.1). Therefore, specially thanks to the tremendous Δ sensitivity, and in case of favorable film-substrate optical mismatch, SE can be sensitive to films with subnanometric thickness [65, 66, 68]. Figure 4.6 helps visualizing that δ-spectra depend on the film as well as on the substrate optical properties.[11] Recently, we have shown that by difference spectra it is possible to detect sharp UV-VIS absorption features of organic molecules SAMs (see Fig. 4.12) [69, 70].

The detection of such small changes is not particularly difficult in in-situ measurements, where the beam probes the same zone before and after the film deposition, as it is shown later on in Fig. 4.8 on the example of underpotential electrodeposition of a Cu atomic phase on gold [71].

[9] In the most general description the polarization state of a light beam is described by the Poincaré sphere and the Stokes formalism is adopted, leading to the concept of 4×4 Mueller matrix [12].

[10] In the IRSE community however, it is sometimes preferred to look at spectra normalized to some convenient reference spectrum [67].

[11] A detailed account, including a discussion on the dependence on the ambient, can be found in Ref. [66].

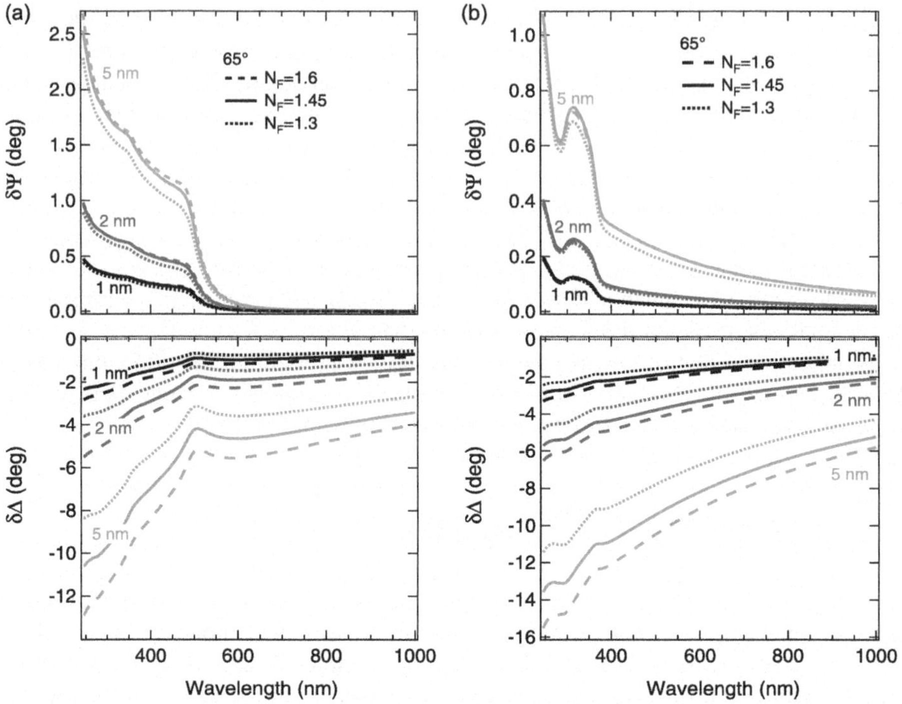

Fig. 4.6 SE difference spectra calculated at 65° incidence angle for a transparent film of variable refractive index ($N_F = 1.60, 1.45, 1.30$) and thickness (1, 2, 5 nm) on (**a**) Au and (**b**) Si substrates. The ambient is vacuum. Note that δ-spectra depend on the film as well as on the substrate optical properties. For ultrathin transparent films of given refractive index, $\delta\Delta$ is proportional to the thickness

For ex-situ measurements the uniformity of the substrate surface becomes a crucial parameter and an inaccurate choice of the reference may lead to severe systematic analysis errors as discussed in Ref. [64]. Tight experimental protocols for substrate preparation and characterization, that involve suitable averaging procedures over many zones and many samples (before and after deposition) are usually necessary to get reliable results [66, 69]. Note finally that in case of tiny $\delta(\Psi, \Delta)$ changes, good MSE values can be obtained even for models which reproduce only qualitatively the experimental difference spectra. The comparison between simulated and experimental difference spectra therefore represents the essential check to assess the quality of models [66, 70].

Still considering differential approaches, it is worthwhile to mention the so-called *d-parameter* formalism. Introduced years ago [72] in chemisorption studies, this formalism has been more recently applied to the analysis of SAMs [73] (see Sect. 4.4.2).

Towards the ultra-thin film limit, the mentioned problems of fit parameter correlation, e.g. between refraction index and thickness, become increasingly important. In favorable cases, one can assume a priori a *reasonable* value of the refractive index, e.g. by extrapolating bulk values of the same or closely related substance, and obtain fair thickness estimates. This approach has been often adopted, for example in the popular case of alkanethiols SAMs on gold (see Ref. [66] and references therein). In a different approach, an independent estimate of thickness can be provided by other techniques, as e.g. in Ref. [54] where the thickness of a Langmuir-Blodgett monolayer (about 2.5 nm) was determined by combining SE, AFM and X-ray reflectivity results. Very recently an approach that allows for simultaneous in-situ SE and quartz microbalance measurements, has been proposed to determine both thickness and porosity of ultrathin organic and biologic films [74]. In the study of ultrathin molecular films, Total Internal Reflection Ellipsometry (TIRE) (or plasmon enhanced ellipsometry) combines the advantages of ellipsometry and the surface plasmon resonance (SPR) method [9, 75–77] (see Sect. 4.4.2).

4.3 Instrumentation and Experimental Configurations

The great development of optical SE in the last forty years is intimately connected to advances in the realization of high performance optical components (e.g. compensators, optical fibers), spectrographs (e.g. compact holographic gratings) and light detectors (e.g. ccd's) as well as, for sure, to the boom of microcomputers, which allowed for remote control and expanded the possibilities of data analysis. The output of this development is represented by the commercial affirmation of automated instruments with an increasing number of channels and high measurement speed, ultimately leading to the possibility of meaningful real-time measurements. Null ellipsometry instruments, very precise and accurate but generally much slower, are seldom used for multi-wavelength measurements [78] whereas they still find wide use in so-called *imaging ellipsometry* (see Sect. 4.3.3).

4.3.1 Basic SE Instruments

Spectroscopic ellipsometers (especially commercial instruments) can be substantially divided into the two classes of instruments that are based on a photoelastic modulator (PME) [79–82] or on rotating components, such as rotating-analyzer ellipsometers (RAE) [83–85] and rotating-compensator ellipsometers (RCE) [86, 87].[12] Each category of instruments presents its own advantages with respect to specific applications, and its own estimators. Just to give a take-home message, PM ellipsometers allow for the shortest measurement time (down to 20 μs) while rotating element instruments (almost three order of magnitude slower) guarantee a much larger number of channels (several hundreds in recent models of commercial instruments).

[12]Configurations which cannot be classified in these two categories have been also presented [88].

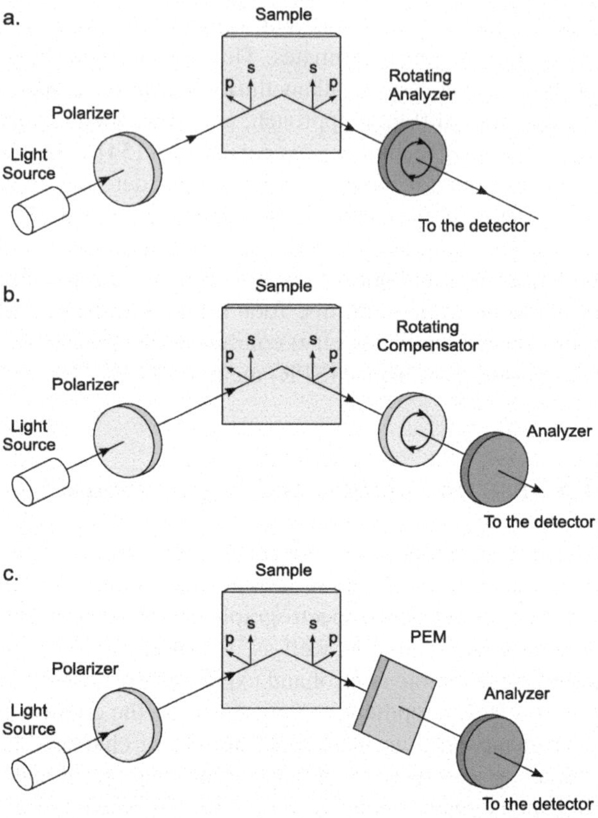

Fig. 4.7 Schemes of most common instruments. (**a**) Rotating Analyzer Ellipsometer, RAE. Note that some models combines the RA with an auto-retarder. (**b**) Rotating Compensator Ellipsometer, RCE. (**c**) Photoelastic Modulator Ellipsometer, PME. Common to all configurations are a brilliant broadband source (e.g. a xenon lamp for UV-VIS SE) and a detector stage consisting of a dispersive element (e.g. grating or prism) and a photon detector (a CCD array). The sample manipulator stage must allow for an accurate and reproducible alignment ensuring a precise identification of the plane and angle of incidence. Typical spot sizes on the sample are in the millimetric range. Suitable optics can provide so-called micro-spot options, down to a few tens of micrometers

Schemes of most common instruments are presented in Fig. 4.7. Rotating-analyzer (polarizer) ellipsometers, employing achromatic components, are probably the simplest and most employed instruments. Understanding the operation of spectroscopic ellipsometers requires the Jones's matrix analysis of the whole optical system, including the sample. This analysis [10] shows that the detected signal on each channel is a simple function of time such as, on the example of the RAE[13] (with polarizer angle P set at 45°):

$$I(t) = I_0\big(1 + \alpha \cos(2\omega t) + \beta \sin(2\omega t)\big) \tag{4.9}$$

where ω is the analyzer revolution frequency and α and β, to be determined by Fourier analysis of the signal, are connected to Ψ and Δ through the following relations [10]

$$\tan\Psi = \sqrt{\frac{1+\alpha}{1-\alpha}}\,|\tan P|, \qquad \cos\Delta = \frac{\beta}{\sqrt{1-\alpha^2}} \tag{4.10}$$

[13] A slightly more complicated formulas hold for the RCE instrument.

The Jones vector analysis of the RAE instrument response function shows that the measurement error intrinsically increases for Δ close to 0° and 180°, which can be disadvantageous in the analysis of dielectric materials [10]. Rotating-compensator instruments (and RAE instruments with additional retarder) have overcome this problem and allow accurate measurements over the full Δ range. Furthermore, the sensitivity for Ψ and Δ is uniform over all their variability range. RCE instruments allow also for partial Mueller Matrix (MM) measurements and are also capable to provide the so-called depolarization spectrum, i.e. the degree of polarization versus wavelength. A price to be paid for the great versatility and high performances of RCE is the presence of chromatic components and a more complex calibration procedure.[14] More sophisticate rotating-element instruments (e.g. dual rotating compensator ellipsometers) allow the full determination of the Mueller matrix and the complete analysis of biaxial materials [89–96].

Adopting special configurations, also PME can allow measurements of the depolarization spectrum [97]. PME can be also applied to real-time IRSE measurements, difficult with rotating-elements instruments. For PME, like for RAE, the measurement error increases in specific ranges of Ψ and Δ [10].

Measurement accuracy for Ψ and Δ of properly calibrated commercial instruments, which can be checked by straight-through measurements in air ($\Psi = 45°$, $\Delta = 0°$) reaches, depending on the quality of components and alignment, the 0.01–0.02° limit. These performances, for many substrate/film configurations, warrant a thickness sensitivity better than 0.1 nm [10, 14].[15]

The previous discussion focused on most common commercial instruments, typically covering the 190–1700 nm range. Driven by advances in lithography methods, table-top instruments have been designed and are commercially available, which operate in the VUV range, down to 140 nm [98–101]. These systems, equipped with two light sources (Xenon and deuterium lamps) and MgF_2 optics, need purging with dry nitrogen to avoid absorption of VUV light by ambient oxygen and water vapor.

A parallel development could be traced regarding instruments dedicated to mid-infrared SE [102–105] and far-infrared SE [106]. In recent years, after the first setup reported by Nagashima and Hangyo [107] a few groups have been involved in the development of ellipsometry in the terahertz range based on different designs, e.g. time- or frequency-domain instruments [108–112].

SE@Synchrotron Synchrotron radiation represents a suitable source for implementing SE experiments, coupling continuum spectral range (*white* spectrum), intrinsic polarization properties, high brilliance and beam stability. The first synchrotron SE apparatus dates back to the eighties, when a VUV (5–30 eV) beamline was built at BESSY in Berlin [113], and dedicated to the study of wide gap semiconductor, insulators and High-T_c compounds [114–116]. A VUV beamline continues this tradition at BESSY II [117] in the 2.5 eV (500 nm)—35 eV (34 nm)

[14] For commercial instruments this is accomplished by the producer.

[15] Cf. e.g. Sect. 4.4.3 of Ref. [10] for an agile discussion of the precision on film thickness and absorption coefficient determination in SE measurements.

range. Another beamline at BESSY II (IRIS) is dedicated to infra-red, far-infrared and terahertz measurements [118–121]. The two BESSY beamlines offer interesting potentiality for wide spectral range studies [122]. Early far-IRSE experiments were carried out at the National Synchrotron Light Source, (Brookhaven, USA) [123] where a new instrument with the full MM capability has been developed at the U4IR beamline [125]. Another facility for FTIR spectroscopy and ellipsometry in the mid- and far-IR spectral ranges is provided at the ANKA synchrotron (Karlsruhe, Germany) [124]. A soft x-ray polarimetry/ellipsometry end station (SXPE) is installed at the Ritsumeikan University (Shiga, Japan) [126].

4.3.2 In-situ Real-Time Measurements

In Vacuum

The input and output arms of ellipsometers can be connected to vacuum chambers through strain-free fused silica windows at convenient angles of incidence (usually in the range 60–70°). In practice, the only requisite for in-situ SE is that the substrate and the growing film are reasonably reflective. Real time in-situ SE measurements can be of great help regarding several aspects. First, SE can be used to check the substrate quality, for example after in-situ pre-deposition or post-growth surface treatments (e.g. sputtering, annealing, ion-bombardment nanostructuring [127]) as the ellipsometric response significantly depends on surface nanoscale morphology [7, 8, 128]. Second, SE provides a non perturbative monitoring of the film growth rate, of oxidation and other relevant processes [129, 130] (see also 4.4). This capability is especially appreciated when the film deposition occurs in poor vacuum reactors, and alternative control methods, such for example RHEED, are not convenient or even infeasible. Many examples can be provided in this or similar contexts e.g. reactive sputtering [131], Chemical Vapor deposition (CVD) [132, 133] or Atomic Layer Deposition (ALD) [134].

One could argue that single wavelength ellipsometry could be enough for monitoring film thickening. Note however that SE data over an extended range of wavelength bring important KK-consistent information on the growing film optical properties and, after adequate modeling, on film morphology. In this respect, spectral ellipsometry can provide dynamical insight into aspects relevant for the growth of ultrathin oxide films such as the presence of oxygen vacancies or pores [135, 136]. On the other hand, in the case of ultrathin metallic films SE can be useful to check electron-impurity scattering and size effects [8].

In-Liquid Measurements. Plasmon Enhanced Ellipsometry

Ellipsometric studies on biologic films (cf. Sect. 4.4.2) led naturally to the design of in-liquid deposition cells.[16] The design must consider several aspects [137].

[16]Two smart methods for an accurate measurement of the refractive index of the liquid ambient, i.e. *rough surface* and *prism minimum deviation* techniques, are discussed in Ref. [149].

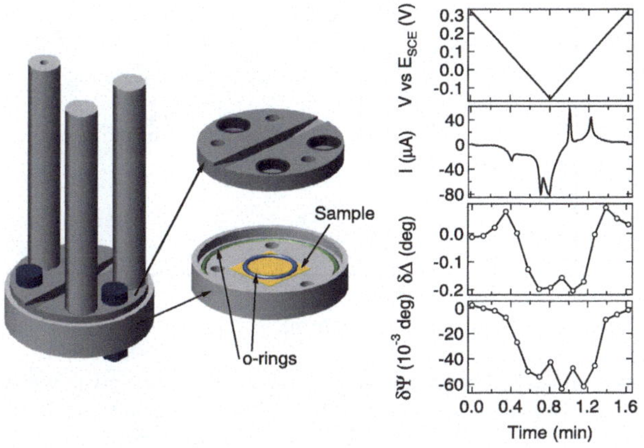

Fig. 4.8 In-situ SE monitoring of the underpotential deposition/removal of a Cu overlayer on a well-defined Au electrode; SE data have been measured with a rotating compensator instrument coupled to a home-made electrochemical cell allowing for measurements at three different angles of incidence [71]. *Left*: Scheme of sample holder (teflon). The three rods are connected to a manipulator stage providing fine z-control and tilt adjusting. The electrical contact to the sample (working electrode) is ensured by a thin wire (not shown) passing through a hollow rod down to the dry zone between the o-rings. *Right*: Evolution of potential, current and ellipsometric angles at $\lambda = 632.2$ nm during a single CV scan (rate 10 mV/s). Reproducible Δ and Ψ cyclical variations of the order of 0.3° and 0.1°, respectively, have been detected, corresponding to an effective Cu overlayer thickness of the order of 0.2 nm

Windows quality and mounting, as well as the sample manipulation stage, affect the quality of the measurement, governing e.g. the accuracy of the definition of the plane and angle of incidence. Cell volume, equipments for liquid flow, exchange and stirring are important factors for fluid-dynamics aspects. Electrochemical and temperature controls are other important issues. The variety of experiments that can be made is so wide that it is hard conceiving a really multi-purpose cell [137]. In practice, almost each group active in this field has designed a home-made cell adequate to its purposes. Examples regarding front-side SE measurements can be found in Refs. [137] and [138–141]. Some of these cells, usually presenting a single angle of incidence, may work with very reduced volume, down to a few ml, allowing measurements with expensive reagents. Other designs, functional to operation under electrochemical control and/or allowing multiple-angle of incidence and accurate definition of the incidence plane (see e.g. Fig. 4.8), may imply larger volumes [71, 142–144].

Back-side designs, in which the light probes the film passing through a transparent substrate without contacting the reagent solution, have been also successfully adopted and allowed to scale down the cell size [145–148]. Belong to this category the TIRE cells.

The TIRE method is based on the well-known process of total internal reflection. The typical set-up [76, 150], sketched in Fig. 4.9, exploits a sample mounting stage similar to the Kretschmann configuration of SPR [151]. In SPR, the energy of the p-polarized part of the light coming form a laser is efficiently transferred to

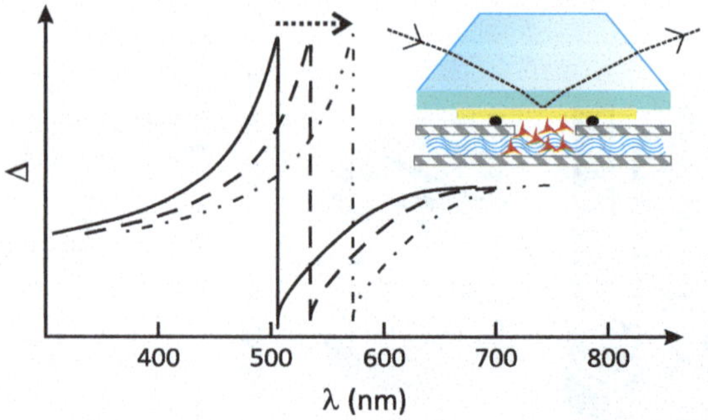

Fig. 4.9 Schematics of TIRE. The sample stage (*right*) comprises a prism which is coupled, via index matching fluid, to a glass slide coated with an ultra-thin metal film (Cr or Ti primer layer, a few nm thick, plus a 20–30 nm thick gold film, supporting the surface plasmon resonance). The sample is coupled to the flow cell with an O-ring. The angle of incidence and shape of the prism must be combined to guarantee the conditions of total internal reflection. Δ spectra (*left*) show a sharp drop near the SP resonance, which neatly shifts (*arrow*) upon molecule-sample reaction

a surface plasmon wave causing a drastic decrease of the reflectivity. Ellipsometry in addition exploits the s-polarization and phase information. The TIRE Ψ and Δ spectra present a characteristic dip and a very sharp drop, respectively, related to the surface plasmon resonance effect. These features display pronounced shifts upon interaction of gold with foreign molecules. The $\Delta(\lambda)$ behavior in particular (Fig. 4.9) ensures a sensitivity about ten times larger than conventional SPR [150]. The spectra can be fitted to the Fresnel relations adopting the same procedures outlined for conventional SE. TIRE experiments mainly addressed bio-film detection (see Sect. 4.4.2). However interesting applications were proposed also for organic films in general, as demonstrated on the example of a study on doping-induced conductivity transitions in layers of polyaniline [152], and for films for hazardous gas sensing [153].

4.3.3 Null Ellipsometry and Imaging

Imaging ellipsometry (IE) combines microscopy with the thickness sensitivity of null ellipsometry. In null ellipsometry Ψ, Δ values are estimated from the rotation angles of the polarizer and analyzer that make null the detected light intensity. It is easy to show [10] that for two orthogonal settings of the compensator ($C = 45°$ and $C = -45°$) one has a total of four settings of P and A that satisfy the null intensity condition at the detector. Accurate measurement can be performed by averaging these values (so-called four-zone averaging). A comprehensive account of early instrumental developments in this field can be found in Ref. [154] where IE was presented as a method for visualization and quantification of the lateral thickness distribution of thin (0–30 nm) transparent layers on a solid substrate. The method was based on the combined use of null and off-null ellipsometry at an incident angle close to the pseudo-Brewster angle of a high refractive index substrate (silicon). The imaging ellipsometer was basically a null ellipsometer in a PCSA configuration (Polarizer-Compensator-Sample-Analyzer). The reflected light intensity was measured by a CCD camera. The wavelength of the probe was chosen by a filter. The instrument had an accuracy of better than 0.5 nm at a lateral resolution of 5 μm in the proposed configuration; instruments claiming spatial resolution less than 1 μm have been reported [155].

Since early studies, IE was conceived as a potential technique to develop high sensitive, label-free and high throughput bio-sensors e.g. to detect the molecule mass surface concentration distribution of protein layers immobilized on a patterned surface of a solid substrate. A detailed description of the bio-sensor concept based on IE can be found in several papers of Gang Jin and collaborators [156–158]. Recent appealing developments regard the combination of IE with the TIRE configuration, employing a microfluidic reactor array for biomolecular interaction [159].

4.4 A Survey of Applications of Spectroscopic Ellipsometry to Ultrathin Films and Other Nanosystems

Attention is substantially confined to works exploiting the spectroscopic potential of the method, and to in-situ studies, in the belief they represent the most appealing field of ongoing application of SE. Ultrathin films receive more attention as two updated and extended reviews are available regarding specifically nanosystems [7, 8]. Note that sharp distinctions between the two fields are arbitrary. Ultrathin films may really consist of nanocrytallites or nanoparticles assemblies [160, 161].

4.4.1 Ultrathin Films: Inorganic Dielectrics (Mainly Oxides)

Probably, there is no category of inorganic dielectric film of technological relevance that has not been the subject of investigation through SE [7]. However, the largest part of SE papers reporting on this kind of systems dealt with relatively thick (under the perspective of this chapter) films. That is the case e.g. of the complex multilayer stack forming photovoltaic cells [162, 163]. The synthetic discussion that follows is focused on ultrathin oxide layers. The most important example is provided by the thickness metrology of SiO_2 layers, of enormous relevance in micro- and nanoelectronics (e.g. in so-called advanced gate dielectrics) [164]. This is a classical playground of SE [28, 32, 165–167] and an important point of convergence between thin film and surface sciences. SE in facts allowed the determination of the thickness and thickness-dependent optical properties of the SiO_2 layer down to the nanoscale limit [28, 32, 167], providing insight also on the morphology and extension of the interfacial SiO_2–Si layer [28, 32, 164–166].[17] The SE studies on native or ultrathin thermal oxide film on Si and other semiconductors [169, 170] represent a paradigm for any subsequent application to other insulating or wide gap semiconductor films. Indeed, the needs of the complementary metal-oxide-semiconductor (CMOS) technology continues to boost applications of SE. Scaling down of transistor size has been colliding against intrinsic limitations such as gate leakage current and oxide layer breakdown. High permittivity oxide layers may represent a viable alternative to silicon dioxide [171]. Candidates under scrutiny include binary (HfO_2, ZrO_2, TiO_2, etc.) as well as multicomponent oxides such as titanates. In facts, the thickness- and morphology-dependent band gap is a critical parameter because it is related to the band offsets between the high-κ layer and other layers in the device stack, in turn affecting the gate leakage current. SE has been systematically used to determine the thickness and optical properties of the oxide layer and of the interfacial layer with silicon [172–176]. Hafnia probably deserved the greatest attention. Recent SE works reported the thickness-dependent optical properties of ultrathin HfO_2 films up to 9 eV [177], and an accurate band gap determination [178] with an interesting comparison between several approaches to model the fundamental absorption edge. Other studies reported on sub-band-gap absorption features (so-called *trap states*) [179, 180]. While both works assigned these spectral features to oxygen deficiency, apparently conflicting conclusions were reached about their spatial localization. In facts while Ref. [179] assigned a preferential bulk character to these defects, Ref. [180] claimed a localization at the HfO_2/Si interface. Clearly, this is a challenging subject where combination of SE with structural tools could be useful and maybe necessary.

A mesoscale to nanoscale trajectory can be envisaged also in technological fields aiming at extending classical ICT, such as e.g. optoelectronics, spintronics and photonics. Perovskite and Wurtzite oxides are very popular in these fields. Regarding for example high quality (i.e. ultraflat and stoichiometric) $SrTiO_3$ (STO) films one can track along the years a definite trend towards the study of thinner and thinner films, down to the ten nm threshold [181–184]. SE has been used to provide useful information on several properties (e.g. magnetism, piezoelectricity, ferroelectricity and electron confinement) at the interface between $SrTiO_3$ itself and ultrathin perovskite films, usually deposited by laser ablation methods [185, 186]. A recent interesting IRSE application regarded the Drude response of charge carriers confined at the interface between a STO substrate and a ultrathin $LaAlO_3$ layer (LaO, thickness of 3 and 5 unit cells) [187]. The sheet concentration and effective mass of carriers have been determined in parallel with transport measurements; further, the analysis of a so-called Berreman mode resonance [188] arising in the vicinity of the highest STO longitudinal optical mode and visible as a sharp minimum in $\delta\Psi$ spectra ($\delta\Psi = \Psi_{STO+LAO} - \Psi_{STO}$) provided insight into the depth profile of the mobile carriers [187].

[17] An updated review of the vast literature reporting on the application of SE to the SiO_2–Si system can be found in a dedicated chapter of the HoE [168].

ZnO and ZnO-based alloys are another class of materials which is attracting a strong interest for their peculiar optical and electronic properties [189] with sound applications in liquid-crystal displays, plasmonics, and in photovoltaics technology as a cost-effective transparent contact layer in substitution of ITO layers [190]. SE studies on these materials are still within the thin film domain [78, 191–194] but, on the basis of recent advances in the field e.g. of transparent electronics, it is not difficult to foresee a tendency towards application of very thin films [195].

We already stressed in the previous section the importance of in-situ applications of SE. Many other examples about the real time control of in vacuum film deposition can be found in recent reviews [7, 8]. Here we mention an interesting application to the characterization of the growth of nanocrystalline diamond films prepared by CVD under different processing conditions [196]. The study obtained information both about the bulk thin film regime and on the nucleating stage, such for example the void volume fractions and the nucleation density, as well as on post-growth surface treatment by ion bombardment [127].

4.4.2 Ultrathin Films: Organic and Biologic Films

Organic Films, Polymers

Organic thin films are a strategic class of materials. The role of semiconducting and conducting polymers in organic electronics is likely the most relevant example to mention. Hybrid organic/inorganic materials find growing applications in photovoltaics, display technology, lighting, sensors. Optical spectroscopy is the *natural* tool to study the absorption behavior of any device, also thanks to the generally high absorption coefficients of organic molecules.

Though many applications of SE to organic films deal today with relatively thick films [197], also for this kind of materials one can trace a well-defined tendency to privilege thinner and thinner films [198], to contrast e.g. the limited exciton diffusion lengths and the usually inefficient charge carrier transport affecting device performances (let's think to the race to enhance hybrid solar cells efficiency). Such interface effects, even in bulk hybrid films, are appealing issues for surface scientists.

Polymers (and polymerization) attracted SE studies since the early development of fast multichannel instruments [199, 200]. The early work of Kim and co-workers dealt with the electro-polymerization of pyrrole on a gold electrode. The film growth was monitored from the monolayer regime to a thickness of $\sim$50 nm, while characterizing the evolution of film optical constants, nucleation/coalescence processes, and providing also indirect information on the film structure [199].

Polymers for Sensors and Smart Surfaces SE is particularly suited to investigate the chromic response of organic molecules to different types of stimuli (UV irradiation, heat, chemicals, and mechanical stress) [201–203]. In this field chromogenic transitions in Polydiacetylenes (PDAs) can be considered a prototypical example [202, 204]. The chromic properties of PDAs have been extensively studied in highly ordered bulk systems and ultrathin films, down to SAMs. SE, combined with fluorescence intensity measurements, has been used to study thermochromism in organized Langmuir PDA trilayer (7–9 nm thick) on SiO_2/Si [205]. The index of refraction of the film as a function of wavelength was determined through a multi-Lorentzian model, assuming the film thickness derived from AFM measurements. The authors interpreted the data in terms of a reversible thermal transition between the highly ordered initial blue form and a metastable, structurally distorted, *purple* form (between 303 and 333 K), followed by an irreversible transition to the red form after annealing above 320 K.

SE demonstrated to have the required sensitivity to allow the direct (optical) characterization of the *color* of thiolate PDAs monolayers deposited on low-roughness polycrystalline gold [69]. Difference spectra measured between data taken after and before the monolayer assembly showed specific molecular absorptions, including narrow features in the 500–700 nm wavelength range which have been interpreted as markers of the polymerization state. Polymerization mainly occurred in the so-called red phase (absorption peak around 550 nm), on very flat surface regions obtained by flame-annealing the sample. The polymeric phase was stable against moderate UV irradiance. The detection of the blue phase was questionable as specific absorptions (around 640 nm) overlap with absorptions related to the formation of the S-Au interface (see later on this section).

An example of photochromic transition in moderately thin films can be provided by diarylethene polymers spin cast on SiO_2/Si substrates. The blue-colored state is thermally stable and the backward reaction to the transparent phase is only triggered with visible light. Dynamic $\Delta(\lambda, t)$, $\Psi(\lambda, t)$ scans allow to monitor the UV irradiation-induced transition, marked by the rise of characteristic absorption features, which can be emphasized in difference spectra formed with respect to the transparent form data (Fig. 4.10).

Polyelectrolytes represent an important class of stimuli-responsive polymers useful for the development of smart surfaces with a wide range of applications including control of wettability, adhesion, friction, protein adsorption, and cell or bacteria growth. Their most attracting feature is the layer-by-layer deposition. Optical SE was used to monitor the film thickness demonstrating the regular growth under dry conditions [207] as well as in-situ [208]. More recently, SE and IRSE have been also successfully employed to investigate the pH- and temperature-sensitive swelling of ultrathin mixed brushes [67, 209, 210]. Through the detection of vibrational bands of suitable molecular groups in $\tan\Psi$ spectra referenced to data taken at the initial $pH = 2$ value, IRSE allowed to emphasize the changes in the chemical composition inside the polymer sample (COOH vs. COO-conversion) upon the controlled variation of pH, also providing information on the average concentration of ions in the brush [67].

Films for Organic Electronics SE has been extensively employed to determine the thickness and the thickness-dependent optical functions of films for organic electronics (and photovoltaics). SE investigations, often backed by structural and morphological data obtained by XRR and AFM [211, 212], embraced the in-line

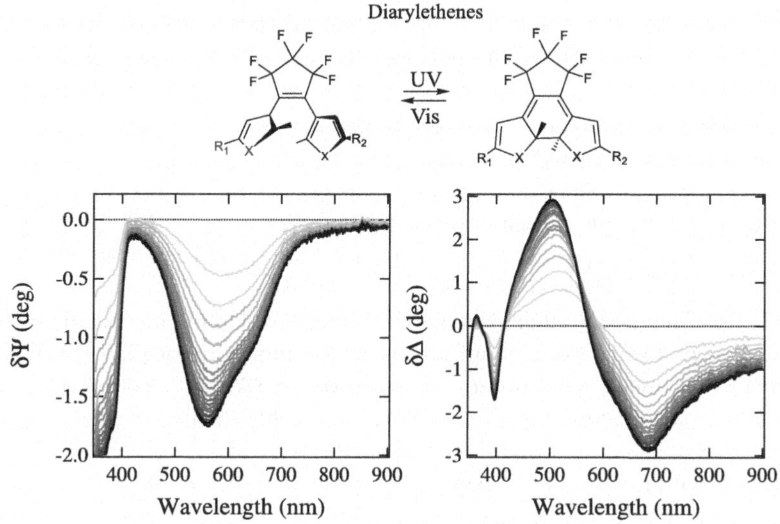

Fig. 4.10 Dynamic $\delta\Delta(\lambda, t)$, $\delta\Psi(\lambda, t)$ scans used to monitor the UV-irradiation induced photochromic transition in thin films (about 30 nm) of diarylethene polymers spin cast on SiO_2/Si substrates. Data have been measured with a rotating compensator instrument. The time interval between lines is 40 s. Absorptions (e.g. the $\delta\Psi$ feature with minimum at about 560 nm) are emphasized in difference spectra formed with respect to the transparent form data [206]. The transition is reversible after irradiation with visible light

characterization of growth processes, e.g. Organic Molecular Beam Deposition [213–215], also in a manufacturing perspective [216].

SE studies considered numerous classes of molecules (phtalocyanines, thiophenes, perylene-derivatives, oligoacenes, rubrene etc.) [213, 215, 217, 218] deposited on isotropic substrates (silicon oxide, glass sheets) and flexible polymer supports [216]. A pronounced uniaxial anisotropy was generally observed, generated by the different lateral and vertical molecular packing [217–220].

Real-time experiments investigating the evolution of dielectric functions as a function of thickness appear particularly interesting for the optimization of devices [213, 221, 222]. The optical response of films is expected to show pronounced changes from the sub-monolayer regime to bulk-like layers, due to the interplay between different sources of electronic coupling (substrate-first layer, intra-layer and interlayer molecular interactions) and their dependence on the film morphological properties. A significant example of this kind of study is a synchrotron-based UV (4–9.5 eV photon energy range), investigation of so-called Alq_3 (tris8-hydroxyquinoline-aluminumIII) layers deposited in UHV onto H-passivated silicon [213]. The observed blue-shift of molecular absorptions with respect to bulk-like layers was assigned to the effect of reduced intermolecular interaction in the sub-monolayer regime [213]. Interlayer coupling was also the subject of a recent study combining Differential Reflectance Spectroscopy (DRS)[18] and SE in a real-time

[18] Cf. [223, 224] for details on this method.

study of several organic semiconductors (pentacene, perfluoropentacene and diindenoperylene and mixed films) on SiO_2 as a function of thickness [221, 225].

Self-assembled Monolayers Self-assembled monolayers, notably those formed by organosulphur compounds on metal surfaces, are well-recognized components of nano-science [226]. Since early developments in this field, ellipsometry has been routinely used to provide estimates of the SAM thickness [227, 228]. Some general aspects related to the application of SE to this kind of systems has been already treated in Sect. 4.2.2. Here we add some more details.

While IRSE has been rather systematically employed to obtain specific information about the molecular orientation and adsorption configuration [229–234], the use of optical SE to obtain spectroscopic information on SAMs dielectric properties is comparatively scarce. Shi and coworkers were probably the first to present a SE in-situ study of the dielectric properties of alkanethiols SAMs of different thickness on gold polycrystalline films [65]. Backing on difference spectra, they also addressed the SAMs-substrate interface (about 0.3 nm thick) optical properties, important to comprehend the transport properties across the interface, and generally left aside in the routine analysis of ellipsometry data. Some years later, Richter and collaborators applied the *d-parameter* formalism to the analysis of SAMs of aromatic thiols on gold [73]. This work is particularly interesting as the model allowed to extract, under some assumptions, the polarizability tensor of the adlayer molecules, providing information on the important issue of intermolecular interactions; no evidence was found for significant perturbation of the neutral electronic transitions by interaction with the substrate. A few years ago, our group revisited the approach of Ref. [65] on alkanethiols films. High quality difference spectra allowed to disentangle spectral features related to the alkyl chain and to the interface layer. $\delta\Psi$ spectra in particular showed a well-defined transition, at about 500 nm, from positive to small yet reproducible negative values in the red-NIR region, not accounted for in models that assume a sharp film/substrate interfaces (see Fig. 4.11). This feature, with minor intensity variations, was found also for other thiolate species, including aromatic molecules [235, 236], while it was ignorable in the case of loosely bound SAMs [237]. Comparison of difference spectra with simulations based on several effective models allowed to identify the main optical features related to the interface, i.e. an absorption band whose spectral weight increases regularly from about 500 nm toward the IR, in agreement with reflectivity measurements [238]. This band was tentatively assigned to a reduction of the mean free path of *free* electrons in the near surface region likely related to nanoscale morphological modifications following the formation of the S–Au bond [238, 239].

In the field of SAMs, gold was likely the most investigated substrate. Examples of fruitful application of SE are available also for SAMs on Si, Pt [228, 240] and InAs substrates [241]. Finally we mention that SE has been used, in combination with other techniques, to investigate the interaction of a metal deposit (Au, Ag) on alkane SAMs attached to Si substrates, an issue which is of great relevance to the understanding of metal/SAMs contacts, e.g. in molecular electronics [242].

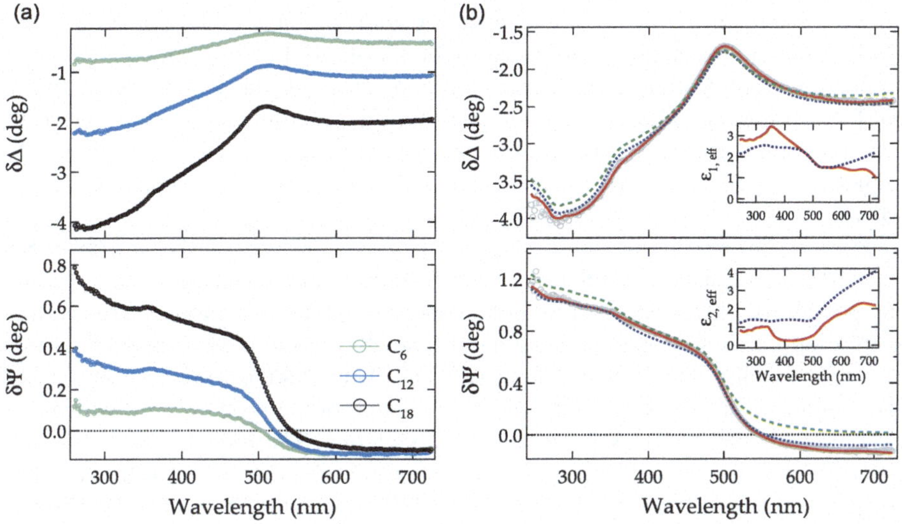

Fig. 4.11 (**a**) Ex-situ SE difference spectra measured for SAMs of alkanethiols of different alkyl chain length (C18, C12, C6) on gold polycrystalline substrates. Angle of incidence 65°. Note the $\delta\Psi$ transition to negative values, not accounted for in models that assume a sharp film/substrate interface (cf. Fig. 4.6). (**b**) Comparison between experimental (symbols; angle of incidence 70°) and simulated (*lines*) difference spectra for C18 SAMs. *Dashed lines* (*green*): sharp interface between a transparent layer and the substrate. *Dotted lines* (*blue*): interface layer modeled by a B-EMA model [249]. *Continuous lines* (*red*): interface layer modeled by a multi-Lorentzian model. *Insets*: interface layer effective dielectric constant obtained with the B-EMA (*dotted line*) and multi-Lorentzian (*continuous line*) models, respectively. For details cf. [66]

Biological Films

Beyond the repeatedly mentioned thickness sensitivity, capability of process monitoring and possibility to look directly at the solid-liquid interface, ellipsometry shows another feature specifically appealing for the investigation of biological films, i.e. the gentle and label-free character, to be exploited in bio-sensing applications, eventually in the imaging mode (see Sect. 4.3). The potential of ellipsometry to investigate bio-films was in facts well established within the vast community of scientist exploiting SWE instruments [243–247] and increased with the advances in SE and IRSE instrumentation [68, 137, 248].

Reinvigorating the tradition of SWE, most SE studies focused on the adsorption (and desorption) kinetics of protein films on a variety of substrates, including bio-functionalized and nanostructured surfaces [138, 141, 147, 249–258] and complex multi-layers systems [259, 260]. Powerful models, such as e.g. the de Feijter's [243] or Cuyper's formulas [243, 244], allow, under specific approximations, a direct translation of $\Psi(t)$ and $\Delta(t)$ scans obtained during film deposition into the dynamic quantification of the so-called surface mass density [147]. The reader mostly interested to this kind of experiments is addressed to specific review articles [137, 248] and to some recent papers [255, 256, 261] on the argument.

Note that in this field, SE is in competition with the Surface Plasmon Resonance (SPR) technique, sometimes preferred by biologists. In this respect, as mentioned in Sect. 4.3, application of TIRE can ensure a sensitivity far larger than ordinary SPR. Relevant examples are reported in Ref. [75], regarding an antibody layer, in Ref. [76] addressing a ferritin ultrathin layer and in Ref. [150] dealing with single strand DNA adsorption and hybridization and detection of low molecular weight toxins.

Here, it's worth noting another aspect where the spectroscopic character of SE (and IRSE) can make the difference with SPR and other techniques. This regards the possibility [237, 262, 263] to look at specific optical transitions bearing information on the conformational state and eventually on the preservation of molecular functionality upon adsorption at surfaces. This approach requires rigorously the building of a suitable model dielectric function of the protein layer, which can become a challenging task in the monolayer regime. The model generally needs a Cauchy (or Sellmeier) contribution, representing the effect of resonances out of the probed range and a multi-Lorentzian series to represent electronic (UV-Vis SE) and vibrational (IRSE) transitions [68]. Note that useful information may already be derived by simple inspect of spectroscopic data, eventually of spectra referenced to the bare substrate. Example of application of IRSE in this context can be found in Refs. [147, 264, 265]. In Ref. [264] the measurements allowed to determine the thickness and index of refraction of a fibrinogen ultrathin (4.5 nm) layer on a Au substrate, and to clearly detect *fingerprints* such as amide I and amide II bands.

The high sensitivity of ellipsometric UV-VIS spectroscopy was recently exploited by our group in the in-situ (buffer ambient) investigation of Yeast Cytochrome c monolayers (a prototypical system for protein/surface studies) deposited on low-roughness [70] and nano-grainy [258] Au substrates. Difference SE spectra (see Fig. 4.12) showed sharp features typical of the so-called heme group of this kind of metallo-protein. Specific simulations based on a Multi-Lorentzian model presented in Ref. [70] allowed an accurate characterization of the main absorptions of adsorbed molecules and a reliable estimate of film thickness. The position and shape of the Soret band, very sensitive to the molecule environment, matched transmission results in solution, suggesting that the adsorbed molecules under proper wet conditions likely preserved their native structure [70]. Instead, preliminary measurements on a hemoglobin film of comparable thickness on the same kind of substrates show a significant deviation from the native spectral properties possibly indicating a denatured state of at least part of the molecules.

4.4.3 Nanostructures

Nanophase materials fall within the broad class of *metamaterials*, which derive functional properties from their structure and nanoscale morphology. In many technologically-relevant cases, that is the case e.g. of materials for plasmonics and

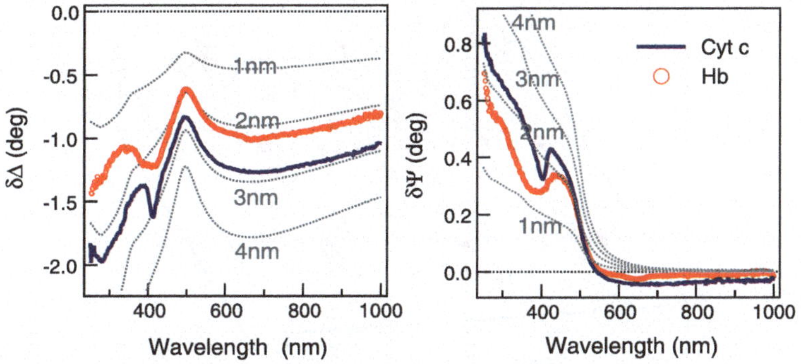

Fig. 4.12 In-situ SE difference spectra (Sect. 4.4.1) for Yeast Cytochrome (Cyt c) and hemoglobin (Hb) ultrathin films of similar thickness deposited on gold. *Thin lines* represent indicative spectra calculated for Cauchy transparent layers ($N_F = 1.5$) and for several values of thickness. Specific simulations based on a Multi-Lorentzian model are presented in Ref. [70]. The position and shape of the Soret band (the sharp dips at about 410 nm), related to the heme group of Cyt c matched transmission results in solution, suggesting that the adsorbed molecules preserved their native structure [70]. Measurements on Hb show a significant deviation from the native spectral properties

photonics, the functionality coincides with the optics [7, 8]. Here we discuss selected examples of application of SE to a broad class of materials comprising several types of assemblies of small (typical size in the 1–10 nm range) nanoparticles (NPs).

Nanoparticle Arrays for Plasmonics and Photonics

Depending on fabrication conditions, the NP mean size, aspect ratio and interparticle spacings can be controlled, allowing to achieve a variety of array morphology and plasmonic response [266–272]. We note in advance that interpreting the results of optical experiments on plasmonic systems requires the building-up of robust EMA models, that accounts for single-particle and collective properties, morphological disorder, and substrate effects (cf. Fig. 4.13) [62, 63, 272–274]. Insitu SE, like reflectivity measurements, was used to obtain real-time information on the building-up and evolution of the plasmon resonances and the reaching of the percolation limit of nanostructures [275–280]. As an example, in Ref. [278] SE was applied to the study of nucleation and percolation of plasma-deposited Ag NPs on SiO_2 at different growth temperatures. The dielectric function was determined along the growth process exploiting the popular Lorentz oscillator model and the so-called Arwin-Aspnes inversion method [36]. The two methods gave comparable, satisfactory outcomes below the percolation threshold, well characterized by ellipsometry when compared with resistance measurements [278]. The Lorentzian form described efficaciously the resonance line shape only in the low coverage phase where the particle sizes satisfy the quasi-static approximation (<20 nm).

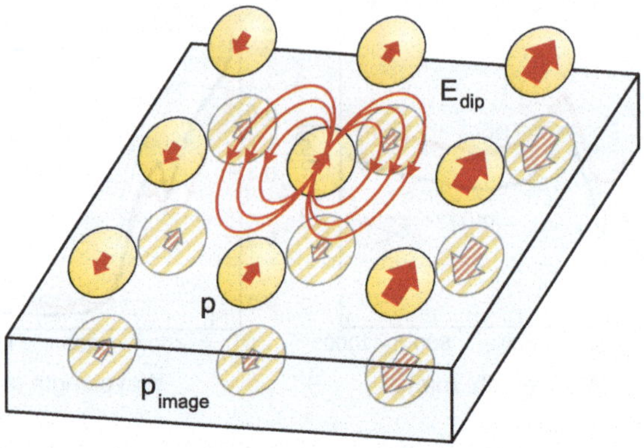

Fig. 4.13 Schematics of an array of metallic NP supported on a dielectric substrate. The particles illuminated by a light wave behave as oscillating dipoles **p** (the thickness of *arrows* represent dipole intensity). The local field acting on each particle is the sum of the incident field and of the total dipolar field. The latter includes components originating from other NPs, and components reflected at the substrate interface, equivalent to the dipolar fields of image dipoles ($\mathbf{p}_{image}$). The optical response of the array as a whole is therefore strongly affected by the dipolar interactions between the NP, i.e. by their arrangement, shape and proximity to interfaces [272]

In this context, the added value of SE with respect to bare reflectivity measurements resides in the exploitation of the phase information [160, 281–283]. It is worth discussing two recent papers which applied non-standard methods to investigate the anisotropic optical properties of arrays of nanoparticles [62, 63]. Reference [62] applied both SE and GE to Ag NP (spheres or rods) arrays formed onto flat and rippled Si–SiO_2 substrates. Ellipsometry results were correlated to the structural information coming from electron microscopy. Ripples, induced by low-energy ion bombardment, were used to template island arrays with defined periodicity. Random island films showed uniaxial anisotropy, while arrays of aligned particles were biaxial. A significant depolarization was observed near the plasmon resonances. In templated arrays, a negative refraction range was also observed between plasmon resonances, in the visible region, potentially interesting for the design of devices.

In Ref. [63] Mueller-matrix methods have been applied to a sub-wavelength square array of holes fabricated by electron beam lithography in a ultrathin, semitransparent gold layer on glass. This isotropic system exhibits a complicated mixing of polarization states upon reflection. In specific directions the rotary power is orders of magnitude larger than that observed e.g. for chiral molecules, where optical activity is a consequence of magnetic interactions. After analysis of the Mueller Matrix data, taking into account the complete three-dimensional boundary conditions similarly to the case of photonic crystals, the effect mimicking chirality was assigned to spatial dispersion leading to a strongly **k**-dependent optical response.

Other Nanophase Films

SE, exploiting EMA models, is a technique able to investigate, in an undestructive way, the structural properties of a heterogeneous material at the nanoscale through the determination of macroscopic optical properties. This is relevant for many technological issues which deal with controlled inclusion of nanoparticles (or alternatively *voids*) in a solid matrix. Relevant examples in this context are provided by porous ultrathin films [284–287] and nanocomposite systems [288–290].

Nanoporous and Nanogranular Materials Characterizing porosity of thin films by ellipsometry is a rather well-established method, that usually exploits the modeling of adsorption-desorption vapor isotherms from the system [291, 292]. Here, we concentrate on a different approach applied to nanogranular films, prepared by deposition from colloidal suspensions [160], dealloying [280] or by in-vacuum deposition methods [289, 293]. These materials exhibit typical grain and pore size of a few nanometers. Our group recently carried out experiments about optical and morphological properties of cluster-assembled nanoporous Au films deposited on a silicon wafer from a low-energy cluster beam. The nanogranular medium was investigated before [161] and after interaction with molecular solutions [294, 295]. The optical response of the system was modeled, within the framework of B-EMA [39], by explicitly introducing a variable degree of porosity in the film and finite-size effects due to the cluster size (a few nanometers) [160], obtaining quantitative information on the thickness, roughness, depth-resolved degree of porosity of the film and on the mean effective size of the nano-grains [161]. The cluster-size distribution obtained from SE matched very well AFM data, thereby suggesting the occurrence of limited cluster coalescence in the film at the deposition [161]. Upon immersion in ethanol, a clear correlation between the relative density of the films and the corresponding fraction of pores that could be filled by the liquid was obtained: Low-density structures exhibited abundant open pores, accessible by molecules from the ambient, whereas for higher-density structures the fraction of accessible pores sharply dropped [294]. Following the deposition of octadecanethiol molecules (C18) from ethanol solution we have been able to define the conditions that either favor in-depth diffusion of molecules into the pores or promote the formation of a SAM restricted to the film surface. In the presence of abundant open pores, C18 molecules strongly diffuse within the film interior and bind to the pore walls, whereas for less abundant open pores the molecules tend to remain confined to the surface region, adopting a SAM-like configuration [295].

Nanocomposites An example of nanocomposite material, still within the context of granular materials, is provided by the embedding of Ag nanoparticles in a polymer matrix, such as polystyrene [296] or poly(vinyl alcohol) [297] and subsequent extrusion after suitable annealing. The polymer helps to protect the particles and inhibits coalescence. Real time SE allowed to monitor the NPs deposition process and the evaporation of the polymer at intermediate temperatures eventually resulting in nanoparticle exposure at the surface [297]. The optical properties of the composite were modeled through the Maxwell-Garnett EMA, using a parametric Cauchy model for the transparent polymer matrix, and a Drude model with modified electron relaxation frequency to account for size effects for NPs. The particle size determined by SE, through the Maxwell-Garnett EMA, compared well to results obtained with SEM. A red-shift of the plasmon resonance was observed after the *liberation* of NPs from the polymer [297].

4.4.4 Graphene

Graphene, a 2D carbon-based nano-material [298] has known in the last few years a peak of tremendous interest (and a Nobel prize) due to unique physical, chemical, and mechanical properties and more than promising applications e.g. in nanoelectronics. The various forms of this material attracted several SE studies (mainly with micro-spot options) to extract their spectral properties [299–303]. Experiments generally backed on data on bulk and thin films of graphite [304] and theoretical predictions [305, 306].

Experiments on exfoliated graphene flakes supported on Si wafers have been reported by Kravets et al. [299] and Weber et al. [300]. In Ref. [299] a uniaxial anisotropic model was assumed, with pre-defined thickness of 0.335 nm (graphite interlayer spacing). The authors proposed a formula which derives the electronic dispersion of a thin symmetric 2D system from the optical absorption spectrum. Applying this formula to graphene flakes they claimed the existence of linear spectrum of quasiparticles. A pronounced absorption peak in the UV at 4.6 eV was reported, whose asymmetric shape and position were attributed to a van Hove singularity in the density of states, likely modified by strong excitonic effects. In the study of Ref. [300], in the 210–1000 nm wavelength range, the authors claim the simultaneous determination of the optical constants, parametrized by KK-consistent B-splines, and the thickness, which was fitted as 3.4 Å. The optical constants determined with SE were used to simulate transmittance for free-standing graphene, finding good agreement with measured data in the visible range. References [301] and [302] reported on hydrocarbon-based chemical vapor deposited (CVD) graphene on metal substrates (Cu and Ni). In the former [301] two instruments covered a wide spectral range (150–1700 nm). A KK-consistent isotropic multi-Lorentzian oscillator was chosen, the thickness set to 3.35 Å. The authors proposed an interesting comparison with the optical constants obtained in previous works [299, 300]. The three examined data sets present evident discrepancies regarding the UV range, dominated by excitonic effects, while the agreement becomes better going towards the IR region. In the visible region, values are reasonably close to those generally accepted for bulk graphite ($n \sim 2.6$ and $k \sim 1.3$) [305]. The reasons for UV discrepancies are not clear [302, 303] . They could be partly ascribed to the different models adopted; factors related to the preparation and transfer procedure of flakes are also conceivable. The influence of foreign molecules on the graphene surface, as noted in a recent IRSE study [307], and in the gap between graphene and the substrate cannot be excluded. Finally the influence of graphene rippling is not clear. In this respect very recent real time studies [302] addressing the kinetics of graphene growth and clarifying the steps involved to improve preparation appear particularly promising.

4.5 Concluding Remarks

The spectrum of scientific and industrial applications in the fields e.g of semiconductor and dielectric technology, coatings, solar cell materials, polymers, organic electronics, bio-materials, nano-materials for photonics, is so impressive and diverging that is not surprising that Spectroscopic Ellipsometers have become part of the core instrumentation of any important materials science laboratory. I hope to have shown that spectroscopic Ellipsometry, born and developed in the realm of thin solid films, can be a powerful tool in problems typical of *surface nano-science*.

As a final note I shall not omit an aspect which sometimes tends to generate *repulsion* towards SE. Popular tools such as photoemissiom or Raman spectroscopy, or ion scattering spectroscopy, may offer *at glance* information (height, position, width of *peaks*). It is generally hard to obtain *at glance* information from SE spectra. Conversely, SE can often provide fast and precious information by application of simple models, involving only the knowledge of basic notions of optics such as the Fresnel relations. Obviously, going further requires in-depth knowledge of the technique principles and optics of materials, adequate training to modeling and fitting, all that involving a not ignorable time investment. This necessity of higher level competences does not seem different from what is requested by the other techniques in modern materials science presented in this book.

Acknowledgements I thank Renzo Mattera and Silvana Terreni for their continuous support to my activity and all the people who collaborated along the years to the research on the optical properties of ultrathin films, and in particular Mirko Prato, Francesco Bisio, Ornella Cavalleri. A special thank to Luca Anghinolfi, Michael Caminale and Chiara Toccafondi for their help in experiments and in the graphics artwork of this chapter.

References

1. A. Rothen, Rev. Sci. Instrum. **16**, 26 (1945)
2. P. Drude, Ann. Phys. Chem. **39**, 481 (1890)
3. K.B. Blodgett, I. Langmuir, Phys. Rev. **51**, 964 (1937)
4. K. Vedam, Thin Solid Films Phys. **313–314**, 1 (1998)
5. R.M.A. Azzam, Thin Solid Films **519**, 2584 (2011)
6. H.G. Tompkins, E.A. Irene (eds.), *Handbook of Ellipsometry* (Andrew, Norwich, 2005)
7. M. Losurdo et al., J. Nanopart. Res. **11**, 1521 (2009)
8. T.W.H. Oates, H. Wormeester, H. Arwin, Prog. Surf. Sci. **86**, 328 (2011)
9. R.M.A. Azzam, N.M. Bashara, *Ellipsometry and Polarized Light* (North-Holland, New York, 1977)
10. H. Fujiwara, *Spectroscopic Ellipsometry: Principles and Applications* (Wiley, New York, 2007)
11. R.M.A. Azzam, N.M. Bashara, J. Opt. Soc. Am. **62**, 1521 (1972)

12. M. Schubert, in *Handbook of Ellipsometry*, ed. by H.G. Tompkins, E.A. Irene (Andrew, Norwich, 2005), p. 637 (Chap. 9)
13. H.G. Tompkins, W.A. McGahan, *Spectroscopic Ellipsometry and Reflectometry* (Wiley, New York, 1999)
14. E.A. Irene, *Surfaces, Interfaces, and Thin Films for Microelectronics* (Wiley-Interscience, New York, 2008)
15. A. Röseler, *Infrared Spectroscopic Ellipsometry* (Akademie Verlag, Berlin, 1990)
16. M. Schubert, *Infrared Ellipsometry on Semiconductor Layer Structures: Phonons, Plasmons and Polaritons* (Springer, Berlin, 2004)
17. A. Röseler, in *Handbook of Ellipsometry*, ed. by H.G. Tompkins, E.A. Irene (Andrew, Norwich, 2005), p. 763 (Chap. 11)
18. H.G. Tompkins (ed.), Thin Solid Films **519** (2011) (Proceedings of the 5th International Conference on Spectroscopic Ellipsometry)
19. J. Humlíček, in *Handbook of Ellipsometry*, ed. by H.G. Tompkins, E.A. Irene (Andrew, Norwich, 2005), p. 3 (Chap. 1)
20. R. Synowicki, Phys. Status Solidi C **5**, 1085 (2008)
21. O. Stenzel, *The Physics of Thin Film Optical Spectra: An Introduction* (Springer, Berlin, 2005)
22. M. Erman, J.B. Theeten, P. Chambon, S.M. Kelso, D.E. Aspnes, J. Appl. Phys. **56**, 2664 (1984)
23. S. Adachi, T. Taguchi, Phys. Rev. B **43**, 9569 (1991)
24. J. Tauc, R. Grigorovici, A. Vancu, Phys. Status Solidi **15**, 627 (1966)
25. G.E. Jellison Jr., F.A. Modine, Appl. Phys. Lett. **69**, 371 (1996)
26. G.D. Cody, B.D. Brooks, B. Abeles, Sol. Energy Mater. **8**, 231 (1982)
27. A.S. Ferlauto, G.M. Ferreira, J.M. Pearce, C.R. Wronski, R.W. Collins, X. Deng, G. Ganguly, J. Appl. Phys. **92**, 2424 (2002)
28. C.M. Herzinger, B. Johs, W.A. McGahan, J.A. Woollam, W. Paulson, J. Appl. Phys. **83**, 3323 (1998)
29. F. Urbach, Phys. Rev. **92**, 1324 (1953)
30. J.-T. Zettler, Th. Trepk, L. Spanos, Y.-Z. Hu, W. Richter, Thin Solid Films **234**, 402 (1993)
31. B. Johs, J.S. Hale, Phys. Status Solidi A **205**, 715 (2008)
32. G.E. Jellison Jr., J. Appl. Phys. **69**, 7627 (1991)
33. G.E. Jellison Jr., Appl. Opt. **30**, 3354 (1991)
34. J.N. Hilfiker, N. Singh, T. Tiwald, D. Convey, S.M. Smith, J.H. Baker, H.G. Tompkins, Thin Solid Films **516**, 7979 (2008)
35. E.A. Irene, Sol. State Electron. **45**, 1207 (2001)
36. H. Arwin, D.E. Aspnes, Thin Solid Films **113**, 101 (1984)
37. H.G. Tompkins, T. Tiwald, C. Bungay, A.E. Hooper, J. Vac. Sci. Technol. A **24**, 1605 (2006)
38. G.A. Niklasson, C.G. Granqvist, O. Hunderi, Appl. Opt. **20**, 26 (1981)
39. A.G. Bruggeman, Ann. Phys. (Leipz.) **24**, 636 (1935)
40. J.C.M. Garnett, Philos. Trans. R. Soc. London, Ser. B **203**, 385 (1904)
41. D.E. Aspnes, J.B. Theeten, F. Hottier, Phys. Rev. B **20**, 3292 (1979)
42. P. Petrik, L.P. Biró, M. Fried, T. Lohner, R. Berger, C. Schneider, J. Gyulai, H. Ryssel, Thin Solid Films **315**, 186 (1998)
43. A. Yanguas-Gil, B.A. Sperling, J.R. Abelson, Phys. Rev. B **84**, 085402 (2011)
44. A. Franquet, J. De Laet, T. Schram, H. Terryn, V. Subramanian, W.J. van Ooij, J. Vereecken, Thin Solid Films **384**, 37 (2001)
45. Q. Liu, J.F. Wall, E.A. Irene, J. Vac. Sci. Technol. A **12**, 2625 (1994)
46. L. Anghinolfi, M. Prato, A. Chtanov, M. Gross, A. Markosyan, A. Chincarini, G. Gemme, M. Canepa (to be published)
47. M. Schubert, B. Rheinlander, J.A. Woollam, B. Johs, C.M. Herzinger, J. Opt. Soc. Am. A **13**, 875 (1996)
48. G.E. Jellison Jr., F.A. Modine, L.A. Boatner, Opt. Lett. **22**, 1808 (1997)
49. S. Teitler, B. Henvis, J. Opt. Soc. Am. **60**, 830 (1970)

50. D.W. Berreman, J. Opt. Soc. Am. **62**, 502 (1972)
51. R.M.A. Azzam, N.M. Bashara, J. Opt. Soc. Am. **64**, 128 (1974)
52. G.E. Jellison, Thin Solid Films **313–314**, 33 (1998)
53. H. Wohler, M. Fritsch, G. Haas, D.A. Mlynski, J. Opt. Soc. Am. A **8**, 536 (1991)
54. G. Gonella, O. Cavalleri, I. Emilianov, L. Mattera, M. Canepa, R. Rolandi, Mat. Sci. Eng. C **22**, 359 (2002)
55. G.E. Jellison, L.A. Boatner, J.D. Budai, B.-S. Jeong, D.P. Norton, J. Appl. Phys. **93**, 9537 (2003)
56. C. Benecke, H. Seiberle, M. Schadt, Jpn. J. Appl. Phys. **39**, 525 (2000)
57. A. Laskarakis, S. Logothetidis, E. Pavlopoulou, A. Gioti, Thin Solid Films **455**, 43 (2004)
58. M. Campoy-Quiles, P.G. Etchegoin, D.D.C. Bradley, Phys. Rev. B **72**, 045209 (2005)
59. B. Lecourt, D. Blaudez, J.M. Turlet, J. Opt. Soc. Am. A **15**, 2769 (1998)
60. M. Schubert, Thin Solid Films **313–314**, 323 (1998)
61. M. Schubert, T.E. Tiwald, J.A. Woollam, Appl. Opt. **38**, 177 (1999)
62. T.W.H. Oates, M. Ranjan, S. Facsko, H. Arwin, Opt. Express **19**, 2014 (2011)
63. B. Gompf, J. Braun, T. Weiss, H. Giessen, M. Dressel, U. Hübner, Phys. Rev. Lett. **106**, 185501 (2011)
64. K.A. Bell, L. Mantese, U. Rossow, D.E. Aspnes, J. Vac. Sci. Technol. B **15**, 1205 (1997)
65. J. Shi, B. Hong, A.N. Parikh, R.W. Collins, D.L. Allara, Chem. Phys. Lett. **246**, 90 (1995)
66. M. Prato, R. Moroni, F. Bisio, R. Rolandi, L. Mattera, O. Cavalleri, M. Canepa, J. Phys. Chem. C **112**, 3899 (2008)
67. D. Aulich et al., Langmuir **26**, 12926 (2010)
68. H. Arwin, Thin Solid Films **519**, 2589 (2011)
69. M. Prato, M. Alloisio, S.A. Jadhav, A. Chincarini, T. Svaldo-Lanero, F. Bisio, O. Cavalleri, M. Canepa, J. Phys. Chem. C **113**, 20683 (2009)
70. C. Toccafondi, M. Prato, G. Maidecchi, A. Penco, F. Bisio, O. Cavalleri, M. Canepa, J. Colloid Interface Sci. **364**, 125 (2011)
71. M. Prato, A. Gussoni, M. Panizza, O. Cavalleri, L. Mattera, M. Canepa, Phys. Status Solidi C **5**, 1304 (2008)
72. W. Chen, W.L. Schaich, Surf. Sci. **218**, 580 (1989)
73. L.J. Richter, C.S.-C. Yang, P.T. Wilson, C.A. Hacker, R.D. van Zee, J.J. Stapleton, D.L. Allara, Y. Yao, J.M. Tour, J. Phys. Chem. B **108**, 12547 (2004)
74. K.B. Rodenhausen et al., Rev. Sci. Instrum. **82**, 103111 (2011)
75. P. Westphal, A. Bornmann, Sens. Actuators B, Chem. **84**, 278 (2002)
76. M. Poksinski, H. Arwin, Thin Solid Films **455–456**, 716 (2004)
77. A.V. Nabok, A. Tsargorodskaya, A.K. Hassan, N.F. Starodub, Appl. Surf. Sci. **246**, 381 (2005)
78. K. Postava, H. Sueki, M. Aoyama, T. Yamaguchi, Ch. Ino, Y. Igasaki, M. Horie, J. Appl. Phys. **87**, 7820 (2000)
79. S.N. Jasperson, S.E. Schnatterly, Rev. Sci. Instrum. **40**, 761 (1969)
80. V.M. Bermudez, V.H. Ritz, Appl. Opt. **17**, 542 (1978)
81. B. Drevillon, J. Perrin, R. Marbot, A. Violet, J.L. Dalby, Rev. Sci. Instrum. **53**, 969 (1982)
82. W.M. Duncan, S.A. Henck, Appl. Surf. Sci. **63**, 9 (1993)
83. D.E. Aspnes, A.A. Studna, Appl. Opt. **14**, 220 (1975)
84. R.W. Collins, Rev. Sci. Instrum. **61**, 2029 (1990)
85. I. An, Y.M. Li, H.V. Nguyen, R.W. Collins, Rev. Sci. Instrum. **63**, 3842 (1992)
86. P.S. Hauge, F.H. Dill, Opt. Commun. **14**, 431 (1975)
87. J. Lee, P.I. Rovira, I. An, R.W. Collins, Rev. Sci. Instrum. **69**, 1800 (1998)
88. T. Sato, T. Araki, Y. Sasaki, T. Tsuru, T. Tadokoro, S. Kawakami, Appl. Opt. **46**, 4963 (2007)
89. P.S. Hauge, Opt. Commun. **17**, 74 (1976)
90. R.M.A. Azzam, Opt. Lett. **2**, 148 (1978)
91. R.M.A. Azzam, K.A. Giardina, A.G. Lopez, Opt. Eng. **30**, 1583 (1991)
92. D.H. Goldstein, Appl. Opt. **31**, 6676 (1992)
93. G.E. Jellison Jr., F.A. Modine, Appl. Opt. **36**, 8184 (1997)

94. E. Compain, B. Drevillon, J. Huc, J.Y. Parey, J.E. Bouree, Thin Solid Films **313–314**, 47 (1998)
95. J. Lee, J. Koh, R.W. Collins, Opt. Lett. **25**, 1573 (2000)
96. A. De Martino, E. Garcia-Caurel, B. Laude, B. Drevillon, Thin Solid Films **455–456**, 112 (2004)
97. M. Kildemo, R. Ossikovski, M. Stchakovsky, Thin Solid Films **313–314**, 108 (1998)
98. S. Peters, T. Schmidtling, T. Trepk, U.W. Pohl, J.-T. Zettler, W. Richter, J. Appl. Phys. **88**, 4085 (2000)
99. J.N. Hilfiker, R.A. Synowicki, C.L. Bungay, B. Singh, Proc. SPIE **3998**, 390 (2000)
100. P. Boher, J.-P. Piel, P. Evrard, C. Defranoux, M. Espinosa, J.-L.P. Stehle, Proc. SPIE **3998**, 379 (2000)
101. T. Wagner, J.N. Hilfiker, T.E. Tiwald, C.L. Bungay, S. Zollner, Phys. Status Solidi A **188**, 1553 (2001)
102. A. Röseler, Infrared Phys. **21**, 349 (1981)
103. R.T. Graf, F. Eng, J.L. Koenig, H. Ishida, Appl. Spectrosc. **40**, 498 (1986)
104. F. Ferrieu, Rev. Sci. Instrum. **60**, 3212 (1989)
105. A. Röseler, Thin Solid Films **234**, 307 (1993)
106. K.-L. Barth, F. Keilmann, Rev. Sci. Instrum. **64**, 870 (1993)
107. T. Nagashima, M. Hangyo, Appl. Phys. Lett. **79**, 3917 (2001)
108. Y. Ino, R. Shimano, Y. Svirko, M. Kuwata-Gonokami, Phys. Rev. B **70**, 155101 (2004)
109. T. Hofmann, U. Schade, K.C. Agarwal, B. Daniel, C. Klingshirn, M. Hetterich, C.M. Herzinger, M. Schubert, Appl. Phys. Lett. **88**, 042105 (2006)
110. T. Hofmann, C.M. Herzinger, T.E. Tiwald, J.A. Woollam, M. Schubert, Appl. Phys. Lett. **95**, 032102 (2009)
111. N. Matsumoto, T. Fujii, K. Kageyama, H. Takagi, T. Nagashima, M. Hangyo, Jpn. J. Appl. Phys. **48**, 09KC11 (2009)
112. T. Hofmann, C.M. Herzinger, A. Boosalis, T.E. Tiwald, J.A. Woollam, M. Schubert, Rev. Sci. Instrum. **81**, 023101 (2010)
113. R.L. Johnson, J. Barth, M. Cardona, D. Fuchs, A.M. Bradshaw, Rev. Sci. Instrum. **60**, 2209 (1989)
114. M. Garriga, J. Humlíček, J. Barth, R.L. Johnson, M. Cardona, J. Opt. Soc. Am. B **6**, 470 (1989)
115. J. Barth, R.L. Johnson, M. Cardona, D. Fuchs, A.M. Bradshaw, Phys. Rev. B **41**, 3291 (1990)
116. L.X. Benedicta, T. Wethkamp, K. Wilmersb, C. Cobet, N. Esser, E.L. Shirley, W. Richter, M. Cardona, Solid State Commun. **112**, 129 (1999)
117. M. Rakel, C. Cobet, N. Esser, F. Fuchs, F. Bechstedt, R. Goldhahn, W.G. Schmidt, W. Schaff, Phys. Rev. B **77**, 115120 (2008)
118. U. Schade, A. Röseler, E.H. Korte, F. Bartl, K.P. Hofmann, T. Noll, W.B. Peatman, Rev. Sci. Instrum. **73**, 1568 (2002)
119. D.M. Rosu, J.C. Jones, J.W.P. Hsu, K.L. Kavanagh, D. Tsankov, U. Schade, N. Esser, K. Hinrichs, Langmuir **25**, 919 (2009)
120. M. Ortolani, U. Schade, Nucl. Instr. Meth. A **623**, 791 (2010)
121. T. Hofmann, U. Schade, C.M. Herzinger, P. Esquinazi, M. Schubert, Rev. Sci. Instrum. **77**, 063902 (2006)
122. K. Hinrichs, S.D. Silaghi, C. Cobet, N. Esser, D.R.T. Zahn, Phys. Status Solidi B **242**, 2681 (2005)
123. R. Henn, C. Bernhard, A. Wittlin, M. Cardona, S. Uchida, Thin Solid Films **313–314**, 642 (1998)
124. C. Bernhard, J. Humlíček, B. Keimer, Thin Solid Films **455–456**, 143 (2004)
125. T.D. Kang, E. Standard, G.L. Carr, T. Zhou, M. Kotelyanskii, A.A. Sirenko, Thin Solid Fims **519**, 2698 (2011)
126. T. Imazono, K. Sano, Y. Suzuki, T. Kawachi, M. Koike, Rev. Sci. Instrum. **80**, 085109 (2009)
127. Vue Cong, R.W. Collins, R. Messier, K. Vedam, G.F. Epps, H. Windischmann, J. Vac. Sci. Technol. A **9**, 1123 (1991)

128. D.E. Aspnes, E. Kinsbron, D.D. Bacon, Phys. Rev. B **21**, 3290 (1980)
129. Y.M. Li, I. An, H.V. Nguyen, C.R. Wronski, R.W. Collins, Phys. Rev. Lett. **68**, 2814 (1992)
130. M.S. Vinodh, L.P. Jeurgens, E.J. Mittemeijer, J. Appl. Phys. **100**, 044903 (2006)
131. S. Logothetidis, I. Alexandrou, A. Papadopoulos, J. Appl. Phys. **77**, 1043 (1995)
132. J. Koh, Y. Lee, H. Fujiwara, C.R. Wronski, R.W. Collins, Appl. Phys. Lett. **73**, 1526 (1998)
133. H. Fujiwara, M. Kondo, A. Matsuda, Phys. Rev. B **63**, 115306 (2001)
134. E. Langereis, S.B.S. Heil, H.C.M. Knoops, W. Keuning, M.C.M. van de Sanden, W.M.M. Kessels, J. Phys. D, Appl. Phys. **42**, 073001 (2009)
135. O. Stenzel et al., Thin Solid Films **517**, 6058 (2009)
136. M. Vinnichenko et al., Appl. Phys. Lett. **95**, 081904 (2009)
137. H. Arwin, in *Handbook of Ellipsometry*, ed. by H.G. Tompkins, E.A. Irene (Andrew, Norwich, 2005), p. 799 (Chap. 12)
138. V. Reipa, A.K. Gaigalas, V.L. Vilker, Langmuir **13**, 3508 (1997)
139. L.G. Castro, D.W. Thompson, T. Tiwald, E.M. Berberov, J.A. Woollam, Surf. Sci. **601**, 1795 (2007)
140. T.M. Byrne, S. Trussler, M.A. McArthur, L.B. Lohstreter, Z. Bai, M.J. Filiaggi, J.R. Dahn, Surf. Sci. **603**, 2888 (2009)
141. T. Berlind, M. Poksinski, P. Tengvall, H. Arwin, Coll. Surf. B **75**, 410 (2010)
142. N. Holmström, A. Askendal, P. Tengvall, Coll. Surf. B **11**, 265 (1998)
143. V. Reipa, H.G. Monbouquette, V.L. Vilker, Langmuir **14**, 6563 (1998)
144. V.A. Gilchrist, J.R. Lu, J.L. Keddie, E. Staples, P. Garrett, Langmuir **16**, 740 (2000)
145. I. Zudans, W.R. Heineman, C.J. Seliskar, Chem. Mater. **16**, 3339 (2004)
146. N. Pantelić, C.M. Wansapura, W.R. Heineman, C.J. Seliskar, J. Phys. Chem. B **109**, 13971 (2005)
147. S. Reichelt, K.-J. Eichhorn, D. Aulich, K. Hinrichs, N. Jain, D. Appelhans, B. Voit, Coll. Surf. B **69**, 169 (2009)
148. E. Della Gaspera, S. Schutzmann, M. Guglielmi, A. Martucci, Opt. Mater. **34**, 79 (2011)
149. R.A. Synowicki, G.K. Pribil, G. Cooney, C.M. Herzinger, S.E. Green, R.H. French, M.K. Yang, J.H. Burnett, S. Kaplan, J. Vac. Sci. Technol. B **22**, 3450 (2004)
150. A. Nabok, A. Tsargorodskaya, Thin Solid Films **516**, 8993 (2008)
151. E.Z. Kretschmann, Physik **241**, 313 (1971)
152. L. Cristofolini, M.P. Fontana, P. Camorani, T. Berzina, A. Nabok, Langmuir **26**, 5829 (2010)
153. A. Nooke, U. Beck, A. Hertwig, A. Krause, H. Kruger, V. Lohse, D. Negendank, J. Steinbach, Sens. Actuators B, Chem. **149**, 194 (2010)
154. G. Jin, R. Jansson, H. Arwin, Rev. Sci. Instrum. **67**, 2930 (1996)
155. Q. Zhan, J.R. Leger, Appl. Opt. **41**, 4443 (2002)
156. Z.H. Wang, G. Jin, Anal. Chem. **75**, 6119 (2003)
157. P. Ying, Y. Yu, G. Jin, Z. Ta, Coll. Surf. B **32**, 1 (2003)
158. G. Jin, Phys. Status Solidi A **205**, 810 (2008)
159. G. Jin, Y.H. Meng, L. Liu, Y. Niu, S. Chen, Q. Cai, T.J. Jiang, Thin Solid Films **519**, 2750 (2011)
160. E.S. Kooij, H. Wormeester, E.A.M. Brouwer, E. van Vroonhoven, A. van Silfhout, B. Poelsema, Langmuir **18**, 4401 (2002)
161. F. Bisio, M. Palombo, M. Prato, O. Cavalleri, E. Barborini, S. Vinati, M. Franchi, L. Mattera, M. Canepa, Phys. Rev. B **80**, 205428 (2009)
162. S. Kohli et al., J. Sol. Energy Eng. **131**, 021009-1 (2009)
163. M. Balestrieri, D. Pysch, J.-P. Becker, M. Hermle, W. Warta, S.W. Glunz, Sol. Energy Mater. Sol. Cells **95**, 2390 (2011)
164. H.F. Okorn-Schmidt, IBM J. Res. Dev. **43**, 351 (1999)
165. D.E. Aspnes, J.B. Theeten, J. Electrochem. Soc. **127**, 1359 (1980)
166. E.A. Irene, Thin Solid Films **233**, 96 (1993)
167. Y. Wang, E.A. Irene, J. Vac. Sci. Technol. B **18**, 279 (2000)
168. E.A. Irene, in *Handbook of Ellipsometry*, ed. by H.G. Tompkins, E.A. Irene (Andrew, Norwich, 2005), Chap. 8

169. E. Aspnes, B. Schwartz, A.A. Studna, L. Derick, L.A. Koszi, J. Appl. Phys. **48**, 3510 (1977)
170. S. Zollner, Appl. Phys. Lett. **63**, 2523 (1993)
171. M. Wallace, G.D. Wilk, Crit. Rev. Solid State Mater. Sci. **28**, 231 (2003)
172. J. Price, P.Y. Hung, T. Rhoad, B. Foran, A.C. Diebold, Appl. Phys. Lett. **85**, 1701 (2004)
173. R. Puthenkovilakam et al., J. Appl. Phys. **97**, 023704 (2005)
174. G. He, Q. Fang, J.X. Zhang, L.Q. Zhu, M. Liu, L.D. Zhang, Nanotechnology **16**, 1641 (2005)
175. C.V. Ramana, S. Utsunomiya, R.C. Ewing, U. Becker, V.V. Atuchin, V.Sh. Aliev, V.N. Kruchinin, Appl. Phys. Lett. **92**, 011917 (2008)
176. A.H. Mueller, N.A. Suvorova, E.A. Irene, O. Auciello, J.A. Schultz, J. Appl. Phys. **93**, 3866 (2003)
177. C. Jorel, C. Vallée, E. Gourvest, B. Pelissier, M. Kahn, M. Bonvalot, P. Gonon, J. Vac. Sci. Technol. B **27**, 378 (2009)
178. D. Ming, E. Bersch, A.C. Diebold, S. Consiglio, R.D. Clark, G.J. Leusink, T. Kaack, J. Vac. Sci. Technol. A **29**, 041001 (2011)
179. H. Takeuchi, D. Ha, T.-J. King, J. Vac. Sci. Technol. A **22**, 1337 (2004)
180. J. Price, P.S. Lysaght, S.C. Song, H.-J. Li, A.C. Diebold, Appl. Phys. Lett. **91**, 061925 (2007)
181. G.E. Jellison, L.A. Boatner, D.H. Lowndes, R.A. McKee, M. Godbole, Appl. Opt. **33**, 6053 (1994)
182. S. Zollner et al., J. Vac. Sci. Technol. B **18**, 2242 (2000)
183. L. Pellegrino, M. Canepa, G. Gonella, E. Bellingeri, D. Marré, A. Tumino, A.S. Siri, J. Phys. IV **11**, 337 (2001)
184. A. Dejneka, M. Tyunina, J. Narkilahti, J. Levoska, D. Chvostova, L. Jastrabik, V.A. Trepakov, Phys. Solid State **52**, 2082 (2010)
185. Š. Višnovský, *Optics in Magnetic Multilayers and Nanostructures* (CRC Press, Boca Raton, 2006)
186. M. Veis et al., J. Phys. D, Appl. Phys. **42**, 195002 (2009)
187. A. Dubroka et al., Phys. Rev. Lett. **104**, 156807 (2010)
188. D.W. Berreman, Phys. Rev. **130**, 2193 (1963)
189. C. Bundesmann, R. Schmidt-Grund, M. Schubert, Springer Ser. Mater. Sci. **104**, 79 (2008)
190. Á. Németh, Cs. Major, M. Fried, Z. Lábadi, I. Bársony, Thin Solid Films **516**, 7016 (2008)
191. P.L. Washington, H.C. Ong, J.Y. Dai, R.P.H. Chang, Appl. Phys. Lett. **72**, 3261 (1998)
192. K.J. Kim, Y.R. Park, Appl. Phys. Lett. **81**, 1420 (2002)
193. R. Schmidt et al., Appl. Phys. Lett. **82**, 2260 (2003)
194. E. Bellingeri, D. Marré, L. Pellegrino, I. Pallecchi, G. Canu, M. Vignolo, C. Bernini, A.S. Siri, Superlattices Microstruct. **38**, 446 (2005)
195. H. Frenzel, A. Lajn, H. von Wenckstern, M. Lorenz, F. Schein, Z. Zhang, M. Grundmann, Adv. Mat. **22**, 5332 (2010)
196. B. Hong, J. Lee, R.W. Collins, Y. Kuang, W. Drawl, R. Messier, T.T. Tsong, Y.E. Strausser, Diam. Relat. Mater. **6**, 55 (1997)
197. M. Losurdo, G. Bruno, E.A. Irene, J. Appl. Phys. **94**, 4923 (2003)
198. J.L. Keddie, Curr. Opin. Colloid Interface Sci. **6**, 102 (2001)
199. Y.-T. Kim, R.W. Collins, K. Vedam, D.L. Allara, J. Electrochem. Soc. **138**, 3266 (1991)
200. C. Barbero, R. Klötz, J. Electrochem. Soc. **141**, 859 (1994)
201. I. Zudans, W.R. Heineman, C.J. Seliskar, J. Phys. Chem. B **108**, 11521 (2004)
202. R.W. Carpick, D.Y. Sasaki, M.S. Marcus, M.A. Eriksson, A.R. Burns, J. Phys. Condens. Matter **16**, R679 (2004)
203. C. Akerlind, H. Arwin, F. Jakobsson, H. Kariis, K. Järrendahl, Thin Solid Films **519**, 3582 (2011)
204. B. Tieke, G. Lieser, G. Wegner, J. Polym. Sci., Part A, Polym. Chem. **17**, 1631 (1979)
205. R.W. Carpick, T.M. Mayer, D.Y. Sasaki, A.R. Burns, Langmuir **16**, 4639 (2000)
206. C. Toccafondi, L. Occhi, R. Castagna, C. Bertarelli, D. Comoretto, M. Canepa (in preparation)
207. S. Schwarz, K.-J. Eichhorn, E. Wischerhoff, A. Laschewsky, Coll. Surf. A **159**, 491 (1999)

208. C. Werner, K.-J. Eichhorn, K. Grundke, F. Simon, W. Grählert, H.-J. Jacobasch, Coll. Surf. A **156**, 3 (1999)
209. E. Bittrich, M. Kuntzsch, K.-J. Eichorn, P. Uhlmann, J. Polym. Sci. B **48**, 1606 (2010)
210. Y. Mikhaylova, L. Ionov, J. Rappich, M. Gensch, N. Esser, S. Minko, K.-J. Eichhorn, M. Stamm, K. Hinrichs, Anal. Chem. **79**, 7676 (2007)
211. A. Hinderhofer, U. Heinemeyer, A. Gerlach, S. Kowarik, R.M.J. Jacobs, Y. Sakamoto, T. Suzuki, F. Schreiber, J. Chem. Phys. **127**, 194705 (2007)
212. U. Heinemeyer, A. Hinderhofer, M.I. Alonso, J.O. Ossó, M. Garriga, M. Kytkal, A. Gerlach, F. Schreiber, Phys. Status Solidi A **205**, 927 (2008)
213. O.D. Gordan, C. Himcinschi, D.R.T. Zahn, C. Cobet, N. Esser, W. Braun, Appl. Phys. Lett. **88**, 141913 (2006)
214. H. Gommans, D. Cheyns, T. Aernouts, C. Girotto, J. Poortmans, P. Heremans, Adv. Funct. Mater. **17**, 2653 (2007)
215. D. Yokoyama, C. Adachi, J. Appl. Phys. **107**, 123512 (2010)
216. S. Logothetidis, Mat. Sci. Eng. B **152**, 96 (2008)
217. M. Friedrich et al., J. Phys. Condens. Matter **15**, S2699 (2003)
218. F. Schreiber, Phys. Status Solidi A **201**, 1037 (2004)
219. M.C. Gurau, D.M. Delongchamp, B.M. Vogel, E.K. Lin, D.A. Fischer, S. Sambasivan, L.J. Richter, Langmuir **23**, 834 (2007)
220. D. Yokoyama, A. Sakaguchi, M. Suzuki, C. Adachi, Org. Electron. **10**, 127 (2009)
221. U. Heinemeyer, K. Broch, A. Hinderhofer, M. Kytka, R. Scholz, A. Gerlach, F. Schreiber, Phys. Rev. Lett. **104**, 257401 (2010)
222. M. Kytka, A. Gerlach, F. Schreiber, J. Ková, Appl. Phys. Lett. **90**, 131911 (2007)
223. J.D.E. McIntyre, D.E. Aspnes, Surf. Sci. **24**, 417 (1971)
224. R. Forker, T. Fritz, Phys. Chem. Chem. Phys. **11**, 2142 (2009)
225. K. Broch, U. Heinemeyer, A. Hinderhofer, F. Anger, R. Scholz, A. Gerlach, F. Schreiber, Phys. Rev. B **83**, 245307 (2011)
226. J.C. Love, L.A. Estroff, J.K. Kriebel, R.G. Nuzzo, G.M. Whitesides, Chem. Rev. **105**, 1103 (2005)
227. A. Ulman, Chem. Rev. **96**, 1533 (1996)
228. D.Y. Petrovykh, H. Kimura-Suda, A. Opdahl, L.J. Richter, M.J. Tarlov, L.J. Whitman, Langmuir **22**, 2578 (2006)
229. C.W. Meuse, Langmuir **16**, 9483 (2000)
230. D. Tsankov, K. Hinrichs, E.H. Korte, R. Dietel, A. Röseler, Langmuir **18**, 6559 (2002)
231. H.G. Tompkins, T. Tiwald, C. Bungay, A.E. Hooper, J. Phys. Chem. B **108**, 3777 (2004)
232. D.C. Bradford, E. Hutter, J.H. Fendler, D. Roy, J. Phys. Chem. B **109**, 20914 (2005)
233. Z.G. Hu, P. Prunici, P. Patzner, P. Hess, J. Phys. Chem. B **110**, 14824 (2006)
234. P.N. Angelova, K. Hinrichs, I.L. Philipova, K.V. Kostova, D.T. Tsankov, J. Phys. Chem. C **114**, 1253 (2010)
235. H. Hamoudi, Z. Guo, M. Prato, C. Dablemont, W.Q. Zheng, B. Bourguignon, M. Canepa, V.A. Esaulov, Phys. Chem. Chem. Phys. **10**, 6836 (2008)
236. H. Hamoudi, M. Prato, C. Dablemont, O. Cavalleri, M. Canepa, V.A. Esaulov, Langmuir **26**, 7242 (2010)
237. F. Bordi, M. Prato, O. Cavalleri, C. Cametti, M. Canepa, A. Gliozzi, J. Phys. Chem. B **108**, 20263 (2004)
238. O. Neuman, R. Naaman, J. Phys. Chem. B **110**, 5163 (2006)
239. R. Mazzarello et al., Phys. Rev. Lett. **98**, 016102 (2007)
240. N. Gergel-Hackett, C.D. Zangmeister, C.A. Hacker, L.J. Richter, C.A. Richter, J. Am. Chem. Soc. **130**, 4259 (2008)
241. D.Y. Petrovykh, J.C. Smith, T.D. Clark, R. Stine, L.A. Baker, L.J. Whitman, Langmuir **25**, 12185 (2009)
242. C.A. Hacker, C.A. Richter, N. Gergel-Hackett, L.J. Richter, J. Phys. Chem. C **111**, 9384 (2007)
243. J.A. deFeijter, J.A. Benjamins, F.A. Veer, Biopolymers **17**, 1759 (1978)

244. P.A. Cuypers, W.T. Hermens, H.C. Hemker, Anal. Biochem. **84**, 56 (1978)
245. M. Malmsten, J. Colloid Interface Sci. **166**, 333 (1994)
246. H. Elwing, Biomaterials **19**, 397 (1998)
247. P. Tengvall, I. Lundström, B. Liedberg, Biomaterials **19**, 407 (1998)
248. H. Arwin, Thin Solid Films **377–378**, 48 (2000)
249. J. Märtensson, H. Arwin, Langmuir **11**, 963 (1995)
250. C. Werner, K.J. Eichhorn, K. Grundke, F. Simon, W. Grahlert, H.J. Jacobasch, Coll. Surf. A **156**, 3 (1999)
251. T. Byrne, L. Lohstreter, M.J. Filiaggi, Z. Bai, J.R. Dahn, Surf. Sci. **602**, 2927 (2008)
252. R.J. Marsh, R.A.L. Jones, M. Sferrazza, Coll. Surf. B **23**, 31 (2002)
253. X.Q. Wang, Y.N. Wang, H. Xu, H.H. Shan, J.R. Lub, J. Colloid Interface Sci. **323**, 18 (2008)
254. S. Lousinian, S. Logothetidis, Thin Solid Films **516**, 8002 (2008)
255. M. Reza Nejadnik, C.D. Garcia, Coll. Surf. B **82**, 253 (2011)
256. D.K. Goyal, A. Subramanian, Thin Solid Films **518**, 2186 (2010)
257. J.L. Wehmeyer, R. Synowicki, R. Bizios, C.D. García, Mat. Sci. Eng. C **30**, 277 (2010)
258. C. Toccafondi, M. Prato, E. Barborini, S. Vinati, G. Maidecchi, A. Penco, O. Cavalleri, F. Bisio, M. Canepa, BioNanoSci. **1**, 210 (2011)
259. K. Spaeth, A. Brecht, G. Gauglitz, J. Colloid Interface Sci. **196**, 128 (1997)
260. A. Nemeth et al., Sens. Lett. **8**, 730 (2010)
261. M.F. Mora, M. Reza Nejadnik, J.L. Baylon-Cardiel, C.E. Giacomelli, C.D. Garcia, J. Colloid Interface Sci. **346**, 208 (2010)
262. J. Mårtensson, H. Arwin, I. Lundström, T. Ericson, J. Colloid Interface Sci. **155**, 30 (1993)
263. J. Mårtensson, H. Arwin, H. Nygren, I. Lundström, J. Colloid Interface Sci. **174**, 79 (1995)
264. H. Arwin, A. Askendahl, P. Tengvall, D.W. Thompson, J.A. Woollam, Phys. Status Solidi C **5**, 1438 (2008)
265. G. Sun et al., Phys. Status Solidi B **247**, 1925 (2010)
266. A. Sugawara, M.R. Scheinfein, Phys. Rev. B **56**, R8499 (1997)
267. T.W.H. Oates, A. Keller, S. Facsko, A. Mücklich, Plasmonics **47**, 2 (2007)
268. S. Camelio, D. Babonneau, D. Lantiat, L. Simonot, Eur. Phys. Lett. **79**, 47002 (2007)
269. F. Cuccureddu, S. Murphy, I.V. Shvets, M. Porcu, H.W. Zandbergen, Nano Lett. **8**, 3248 (2008)
270. J.M. Flores-Camacho, G. Weidlinger, N. Saucedo-Zeni, L.D. Sun, M. Hohage, P. Zeppenfeld, Appl. Phys. A **98**, 499 (2010)
271. A. Toma, D. Chiappe, D. Massabò, C. Boragno, F. Buatier de Mongeot, Appl. Phys. Lett. **93**, 163104 (2008)
272. L. Anghinolfi, R. Moroni, L. Mattera, M. Canepa, F. Bisio, J. Phys. Chem. C **115**, 14036 (2011)
273. R.G. Barrera, M. del Castillo-Mussot, G. Monsivais, P. Villaseor, W.L. Mochán, Phys. Rev. B **43**, 13819 (1991)
274. R.G. Barrera, J. Giraldo, W.L. Mochán, Phys. Rev. B **47**, 8528 (1993)
275. I. An, H. Oh, J. Korean Phys. Soc. **29**, 370 (1996)
276. T.W.H. Oates, D.R. McKenzie, M.M.M. Bilek, Phys. Rev. B **70**, 195406 (2004)
277. T.W.H. Oates, A. Mücklich, Nanotechnology **16**, 2606 (2005)
278. T.W.H. Oates, L. Ryves, M.M.M. Bilek, Opt. Express **16**, 2302 (2008)
279. P.C. Wu Tong-Ho Kim, A.S. Brown, M. Losurdo, G. Bruno, H.O. Everitt, Appl. Phys. Lett. **90**, 103119 (2007)
280. M.C. Dixon, T.A. Daniel, M. Hieda, D.M. Smilgies, M.H.W. Chan, D.L. Allara, Langmuir **23**, 2414 (2007)
281. H. Wormeester, A.-I. Henry, E.S. Kooij, B. Poelsema, M.-P. Pileni, J. Chem. Phys. **124**, 204713 (2006)
282. A.J. de Vries, E.S. Kooij, H. Wormeester, A.A. Mewe, B. Poelsema, J. Appl. Phys. **101**, 053703 (2007)
283. M. Hövel, B. Gompf, M. Dressel, Phys. Rev. B **81**, 35402 (2010)
284. N. Layadi, P. Roca i Cabarrocas, B. Drevillon, I. Solomon, Phys. Rev. B **52**, 5136 (1995)

285. F. Ferrieu, A. Halimaoui, D. Bensahel, Solid State Commun. **84**, 293 (1992)
286. C. Wongmaneroda, S. Zangooieb, H. Arwin, Appl. Surf. Sci. **172**, 117 (2001)
287. M.T. Othman, J.A. Lubguban, A.A. Lubguban, S. Gangopadhyaya, R.D. Miller, W. Volksen, H.-C. Kim, J. Appl. Phys. **99**, 083503 (2006)
288. R. Serna, J.C.G. de Sande, J.M. Ballesteros, C.N. Afonso, J. Appl. Phys. **84**, 4509 (1998)
289. F.K. Urban, A. Hosseini-Tehrani, A. Khabaria, P. Griffiths, C. Bungay, I. Petrov, Y. Kim, Thin Solid Films **408**, 211 (2002)
290. J. Toudert, D. Babonneau, L. Simonot, S. Camelio, T. Girardeau, Nanotechnology **19**, 125709 (2008)
291. M.R. Baklanov, K.P. Mogilnikov, V.G. Polovinkin, F.N. Dultsev, J. Vac. Sci. Technol. B **18**, 1385 (2000)
292. C. Boissiere, D. Grosso, S. Lepoutre, L. Nicole, A. Brunet Bruneau, C. Sanchez, Langmuir **21**, 12362 (2005)
293. E. Barborini, P. Piseri, P. Milani, J. Phys. D **32**, L105 (1999)
294. F. Bisio, M. Prato, O. Cavalleri, E. Barborini, L. Mattera, M. Canepa, J. Phys. Chem. C **114**, 17591 (2010)
295. F. Bisio, M. Prato, E. Barborini, M. Canepa, Langmuir **27**, 8371 (2011)
296. T.W.H. Oates, Appl. Phys. Lett. **88**, 213115 (2006)
297. T.W.H. Oates, E. Christalle, J. Phys. Chem. C **111**, 182 (2007)
298. K.S. Novoselov, A.K. Geim, S.V. Morozov, D. Jiang, Y. Zhang, S.V. Dubonos, I.V. Grigorieva, A.A. Firsov, Science **306**, 666 (2004)
299. V.G. Kravets, A.N. Grigorenko, R.R. Nair, P. Blake, S. Anissimova, K.S. Novoselov, A.K. Geim, Phys. Rev. B **81**, 155413 (2010)
300. J.W. Weber, V.E. Calado, M.C.M. van de Sanden, Appl. Phys. Lett. **97**, 091904 (2010)
301. F.J. Nelson, V.K. Kamineni, T. Zhang, E.S. Comfort, J.U. Lee, A.C. Diebold, Appl. Phys. Lett. **97**, 253110 (2010)
302. M. Losurdo, M.M. Giangregorio, P. Capezzuto, G. Bruno, J. Phys. Chem. C **115**, 21804 (2011)
303. G. Isić et al., J. Nanophotonics **5**, 051809 (2011)
304. W. Wang, M. Balooch, C. Claypool, M. Zawaideh, K. Farnaam, Solid State Technol. **52**, 18 (2009)
305. M. Bruna, S. Borini, Appl. Phys. Lett. **94**, 031901 (2009)
306. L. Yang, J. Deslippe, C.H. Park, M.L. Cohen, S.G. Louie, Phys. Rev. Lett. **103**, 186802 (2009)
307. J.W. Weber, K. Hinrichs, M. Gensch, M.C.M. van de Sanden, T.W.H. Oates, Appl. Phys. Lett. **99**, 061909 (2011)

Chapter 5
Nonlinear Vibrational Spectroscopy

Lee J. Richter

Abstract Second order nonlinear spectroscopies such as sum frequency generation (SFG) are intrinsically interface sensitive and enable sub-monolayer sensitivity to be achieved at buried, optically accessible interfaces. Vibrationally-resonant SFG allows bond (functional group) selective in-situ spectroscopy of interfaces as diverse as solid-solid, liquid-liquid and liquid-gas. This chapter summarizes the theoretical underpinnings of vibrationally-resonant SFG and its practical implementation.

5.1 Introduction

With the advent of the laser, the practical study and use of the nonlinear optical response of materials became possible. As high intensity sources have become robust and affordable, nonlinear spectroscopies have moved from the laser physics lab to be routinely employed for the characterization of materials. Second order nonlinear spectroscopies such as second harmonic generation (SHG) and sum frequency generation (SFG) are well suited to the study of surfaces and interfaces and, since their introduction in the 1980's [1–5], are becoming routine tools for the surface scientist. This chapter is intended as an introduction to the technique of vibrationally-resonant SFG, a powerful molecular spectroscopy with intrinsic interface selectivity. A quick survey of the literature as of early 2011 indicates that over 2000 papers have been published based on vibrationally-resonant SFG. The focus of this chapter is on defining the fundamental underpinnings of the technique, the common experimental systems employed in its execution and the physical information that can be obtained. It is intended to be self contained, but is far from a complete discussion of the field. Excellent texts exist for a more thorough introduction to nonlinear optics [6–10]. Recent reviews exist of the application of SFG to specific systems including the study of the interfaces of liquids [11–14], polymers [15] and solids [16].

L.J. Richter (✉)
Material Measurement Laboratory, National Institute of Standards and Technology, Gaithersburg, MD 20899, USA
e-mail: lee.richter@nist.gov

G. Bracco, B. Holst (eds.), *Surface Science Techniques*,
Springer Series in Surface Sciences 51, DOI 10.1007/978-3-642-34243-1_5,

5.2 Fundamentals

5.2.1 Nonlinear Polarization

Nonlinear optical mixing can be equivalently treated in either the time or frequency domain. As vibrationally-resonant SFG (henceforth shortened to SFG) fundamentally is a functional group specific spectroscopy, a steady-state frequency-domain treatment is the most natural. Neglecting magnetic interactions, a general expression for the second order volume polarization $\boldsymbol{P}$ at the sum frequency ω_Σ in terms of applied fields $\boldsymbol{E}$ is (MKS units, repeated indices indicate summation)

$$\begin{aligned} P_i(\omega_\Sigma = \omega_2 + \omega_1) = \epsilon_0 \big[& \chi^2_{ijk}(\omega_\Sigma = \omega_2 + \omega_1) E_j(\omega_2) E_k(\omega_1) \\ & + 3\chi^3_{ijkl}(\omega_\Sigma = \omega_2 + \omega_1) E_j(\omega_2) E_k(\omega_1) E_l(0) \\ & + \chi^{\nabla_1}_{ijkl}(\omega_\Sigma = \omega_2 + \omega_1) E_j(\omega_2) \nabla_k E_l(\omega_1) \\ & + \chi^{\nabla_2}_{ijkl}(\omega_\Sigma = \omega_2 + \omega_1) E_j(\omega_1) \nabla_k E_l(\omega_2) \\ & + \nabla_j \chi^{Q}_{ijkl}(\omega_\Sigma = \omega_2 + \omega_1) E_k(\omega_2) E_l(\omega_1) + \cdots \big]. \end{aligned} \quad (5.1)$$

The power of second order nonlinear spectroscopies as interface probes comes from the symmetry constraints on the third rank χ^2 tensor. For a parametric process, all odd rank tensors vanish in centrosymmetric media and the only χ^2 contributions come from interfaces where centrosymmetry is intrinsically broken. We note that, in general, the fourth rank tensor terms, $\chi^3 : E(\omega = 0)$ and the nonlocal terms, are not negligible. Extensive work has been performed, mostly in the arena of SHG, in establishing the relative contributions of the bulk terms with respect to the interface terms. For high symmetry interfaces, such as the vacuum/solid interface of Si(100), the nonlocal terms are the origin of azimuthally anisotropic response [17]. They can also contribute significantly for molecular solids such C_{60} [18, 19]. On semiconductors, where near surface depletion fields are present, the $\chi^3 : E(\omega = 0)$ term is significant and gives rise to Electric Field Induced Second Harmonic Generation (EFISH). EFISH has been leveraged to study built in fields at the semiconductor/dielectric interface of silicon field effect transistors [20]. The near-surface EFISH contributions can even be isolated from the bulk response in intrinsically non-centrosymmetric materials such as GaAs [21]. These contributions are also important to molecular systems and have been used to study induced charge in both molecular [22] and polymer [23] based organic transistors.

For the treatment of the interface specific χ^2 terms, it is conventional to introduce an interface polarization sheet density $\wp$ and the corresponding χ

$$\begin{aligned} P_i(\omega_\Sigma) &\equiv \wp_i(\omega_\Sigma)\delta(z - z_0) \\ &= \epsilon_0 \chi^{int}_{ijk}(\omega_\Sigma = \omega_2 + \omega_1) E_j(\omega_2) E_k(\omega_1) \delta(z - z_0). \end{aligned} \quad (5.2)$$

Note that χ^{int} has units $m^2\,V^{-1}$. For a 1d model system (films thin compared to the beam cross section and plane-parallel) and a given set of incident plane waves (see Fig. 5.1), the macroscopic fields throughout the system can be expressed in terms

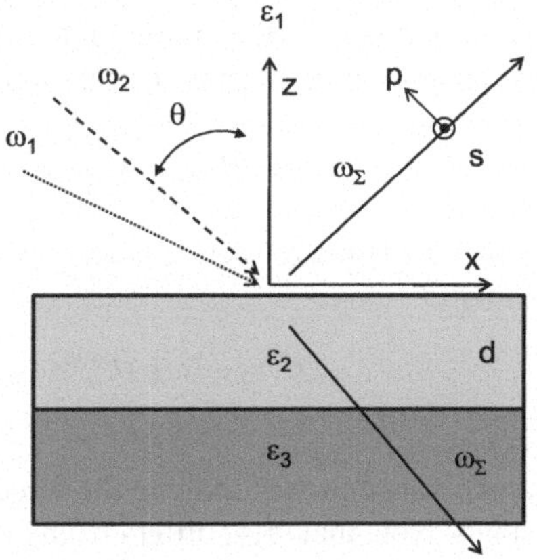

Fig. 5.1 Schematic diagram of a general SFG experiment, beams incident from a medium with dielectric constant ϵ_1 on to a film (ϵ_2) of thickness d on a substrate described by ϵ_3

of Fresnel tensors that are easily calculated via a transfer matrix formalism [24]. The radiated electric field at $\omega_\Sigma = \omega_2 + \omega_1$ can then be calculated from the induced polarizations given by (5.1). We note that this separation of the problem is only an approximation appropriate to the weak conversion limit. For macroscopic systems, where significant energy conversion into higher frequencies occurs, one must account for both depletion of the driving fields and photon occupancy in the generated fields [6–8]. This physics was initially captured in the Manley-Rowe equations for frequency conversion [25]. The calculation of the outgoing field in terms of the induced polarization can be done by boundary condition matching [7, 26] or by transfer matrix formalisms [27, 28].

For the general case of Fig. 5.1, the radiated field at ω_Σ becomes

$$\begin{aligned} E_l(\omega_\Sigma) = \epsilon_0 2\pi\omega_\Sigma \Big[& F_{li}^{out}(\omega_\Sigma; z=0)\chi_{ijk}^{top}(\omega_\Sigma = \omega_2 + \omega_1) \\ & \times F_{jm}(\omega_2; z=0)E_m(\omega_2)F_{kn}(\omega_1; z=0)E_n(\omega_1) \\ & + \int_0^d dz\, F_{li}^{out}(\omega_\Sigma; z)\chi_{ijk}^{bulk}(\omega_\Sigma = \omega_2 + \omega_1) \\ & \times F_{jm}(\omega_2; z)E_m(\omega_2)F_{kn}(\omega_1; z)E_n(\omega_1) \\ & + F_{li}^{out}(\omega_\Sigma; z=d)\chi_{ijk}^{bot}(\omega_\Sigma = \omega_2 + \omega_1) \\ & \times F_{jm}(\omega_2; z=d)E_m(\omega_2)F_{kn}(\omega_1; z=d)E_n(\omega_1) \Big] \end{aligned} \tag{5.3}$$

where we define the film depth (z) to be 0 at the 1–2 interface. In order to calculate the linear Fresnel factors F, one must define the location of the interface sheets. To maintain a common formalism for both the surface and bulk transfer tensors, we embed the nonlinear polarization in the linear dielectric constant of the medium and include the local field terms (see below) in the definition of χ. The detected signal is proportional to the emitted intensity: $I = \frac{\epsilon_0 c\sqrt{\epsilon}}{2}|E|^2$. Simplifying to the case of a single nonlinear sheet at the top of the film, this is expressed as

$$I = \frac{n_1(\omega_\Sigma)\omega_\Sigma^2}{8\epsilon_0 c^3 n_1(\omega_2)n_1(\omega_1)}\left|\chi^e\right|^2 I(\omega_2)I(\omega_1) \tag{5.4}$$

where n_1 is the refractive index ($\sqrt{\epsilon_1}$) of the incident media, the effective susceptibility is

$$\begin{aligned}\chi^e &= e_l(\omega_\Sigma)F_{li}^{out}(\omega_\Sigma; z=0)\chi_{ijk}^{top}(\omega_\Sigma = \omega_2 + \omega_1)\\ &\quad\times F_{jm}(\omega_2; z=0)e_m(\omega_2)F_{kn}(\omega_1; z=0)e_n(\omega_1)\end{aligned} \tag{5.5}$$

and e_i is a unit vector along the direction of the incident (and detected) polarization states. Note that (5.4) differs from the usual equation [29] for a polarization sheet in medium 1

$$I = \frac{\omega_\Sigma^2}{8\epsilon_0 c^3 n(\omega_\Sigma)n(\omega_2)n(\omega_1)\cos\theta_\Sigma}\left|\chi^e\right|^2 I(\omega_2)I(\omega_1) \tag{5.6}$$

by the inclusion of a $1/n\cos\theta$ term in the definition of the outgoing Fresnel factor, where both the refractive index and angle are evaluated for ω_Σ in medium 2.

5.2.2 The Susceptibility

Properties of χ

The material information content of an SFG experiment is contained in χ. In principle, χ contains a daunting set of 27 nonzero, complex elements. However, spatial symmetry can significantly simplify the form of χ. Shown in Table 5.1 are the allowed elements for the non-centrosymmetric point groups appropriate to a surface in the xy plane [30]. For the commonly encountered case of a rotationally symmetric, non-chiral surface ($C_{\infty v}$), only 4 terms are present. By definition, χ will exhibit intrinsic permutation symmetry [8]. Away from resonances one often applies Kleinman symmetry (independent exchange of frequencies). This is not a strict symmetry and can be problematic [31]. As χ arises from a causal response function, it will conform to a Kramers-Kronig relationship. However the proper relationship links the processes of sum-frequency generation and difference-frequency generation and thus is of little utility except in the degenerate case of SHG [32].

Calculation of χ

For a dilute molecular system, the interface susceptibility can be derived from the molecular hyperpolarizability β as

$$\chi_{ijk} = N_s L_{il}\langle\beta_{lmn}\rangle L_{mj}L_{nk} \tag{5.7}$$

Table 5.1 Symmetry constraints on χ^2

Point group	Unique elements	Symmetry constraints
C_1	27	None
C_s	14	zzz; xxz; yyz; xzx; yzy; zxx; zyy; xxx; xyy; yxy; yyx; zxx; xzx; xxz
C_2	13	zzz; xxz; yyz; xzx; yzy; zxx; zyy; xyz; yzx; zxy; xzy; zyx; yxz
C_{2v}	7	zzz; xxz; yyz; xzx; yzy; zxx; zyy
C_3	9	zzz; xxz = yyz; xzx = yzy; zxx = zyy; xxx = −yyx = −yxy = −xyy; yyy = −xxy = −xyx = −yxx; yxz = −xyz; yzx = −xzy; zyx = −zxy
C_{3v}	5	zzz; xxz = yyz; xzx = yzy; zxx = zyy; xxx = −yyx = −yxy = −xyy
C_4, C_6, C_∞	7	zzz; xxz = yyz; xzx = yzy; zxx = zyy; yxz = −xyz; yzx = −xzy; zyx = −zxy
$C_{4v}, C_{6v}, C_{\infty v}$	4	zzz; xxz = yyz; xzx = yzy; zxx = zyy

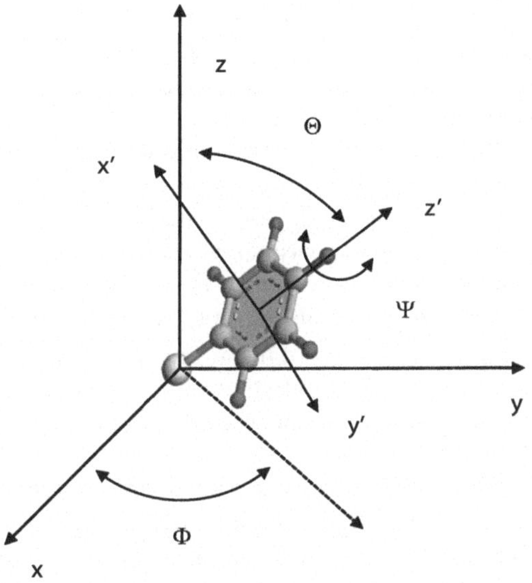

Fig. 5.2 Euler angles Θ, Φ, Ψ defining the transformation from the molecular frame (x′, y′, z′) to lab frame (x, y, z)

where N_s is the molecular surface density, L are local field terms to correct the macroscopic fields in (5.1), and $\langle\rangle$ indicates an average over the orientation distribution f of the chromophores:

$$\langle \beta_{lmn} \rangle = \iiint \sin\Theta \, d\Theta \, d\Phi \, d\Psi \, \beta_{ijk} R_{il} R_{jm} R_{kn} f(\Phi, \Psi, \Theta) \tag{5.8}$$

where Φ, Ψ, and Θ are the usual Euler angles (see Fig. 5.2) and R is a unit rotation matrix. The simple orientation average is appropriate for static disorder. In the case of dynamic disorder, deviations from the above can occur [33]. The local field factors can be estimated from continuum models, in the spirit of Lorentz-Lorenz [34] or from coupled dipole models [35, 36]. The susceptibility β can be calculated via perturbation theory from a sum over states expression [4, 37, 38]. When only vibrational resonances are relevant to the frequency range of interest

$$\beta_{lmn}(\omega_\Sigma = \omega_{VIS} + \omega_{IR}) = \beta_{lmn}^{NR} + \sum_{j=1}^{M} \frac{1}{2\epsilon_0 \omega_j} \frac{\partial \alpha_{lm}}{\partial Q_j} \frac{\partial \mu_n}{\partial Q_j} \frac{1}{\omega_{IR} - \omega_j + i\Gamma_j} \tag{5.9}$$

where β_{NR} represents the slowly varying contributions from resonances outside the region of interest, $\frac{\partial \alpha_{lm}}{\partial Q_j}$ and $\frac{\partial \mu_n}{\partial Q_j}$ are the Raman and IR transition tensors respectively and $\omega_1 \equiv \omega_{IR}$ is resonant while $\omega_2 \equiv \omega_{VIS}$ is not. In molecules with centrosymmetry, the mutual exclusion rule applies (all vibrational modes are exclusively either Raman or IR active) and β vanishes as expected.

The form of the resonant contribution in (5.9) greatly simplifies the analysis of SFG results. For example, in C_{2v} symmetry, the A_1 symmetry symmetric stretch of a CH_2 group will have an IR transition moment strictly along z and a diagonal Raman transition tensor. The result is that, of the seven distinct symmetry allowed elements in C_{2v}, only three are non-zero: xxz, yyz, and zzz. Similarly, for the B_1 symmetry asymmetric stretch, the IR transition moment is strictly along x and the Raman transition tensor is asymmetric. The result is only one distinct element for β: xzx = zxx. Compilations of such symmetry class factorings can be found in Refs. [39] and [40].

In general, the experimental determination of β is intractable, as it requires a known and narrow orientation distribution. Therefore the extraction of maximum information from an SFG experiment requires the estimation of the underlying β from (5.9). As experimental determination of the full Raman tensor is also infrequently done, the estimate is usually based on a theoretical treatment of the necessary transition tensors. In the early literature, $\frac{\partial \alpha_{lm}}{\partial Q_j}$ and $\frac{\partial \mu_n}{\partial Q_j}$ were estimated from simple calculations of the normal modes and bond-additive dipoles and polarizabilities [29, 40–43]. Modern *ab initio* quantum chemistry calculations have advanced to the point that they can be used to both assign spectra and estimate transition tensors [44, 45].

From (5.7) it is clear that the intensities in SFG spectra convolve the area density, orientation distribution, and intrinsic transition strength in a complex fashion. Strictly speaking, SFG cannot be used for determination of absolute coverage as the technique is only sensitive to the fraction of the population that has a polar orientation. For quantification of the oriented fraction, experimental conditions can be optimized to minimize the sensitivity to the specifics of the orientation distribution [44, 46]. In general, the sensitivity of SFG to the orientation is leveraged to provide insights into the underlying distribution function f. This requires great care, as accurate knowledge of both the general Fresnel factors and the local field corrections are necessary for quantitative analysis. An excellent example is the analysis of mechanically aligned polyvinylalcohol surfaces by Prof. Shen's group [29]. An advantage of SFG relative to linear dichroism techniques such as near-edge X-ray absorption fine-structure or IR absorption is the sensitivity of the second order technique to odd moments of the distribution function, allowing the determination of orientation. An exciting opportunity lies in the recent ability to compare complete orientation distributions from full molecular dynamics simulations with SFG spectra [47, 48].

5.3 Experimental Considerations

5.3.1 Experimental Configuration

There are nominally sixteen distinct configurations for an SFG experiment. If one considers the IR beam only and the case where $\epsilon_1 < \epsilon_3$ (Fig. 5.1), the IR beam can be incident from the low index side (reflection) or the high index side where total internal reflection (TIR) can occur. The visible beam can be in a co- or counter-propagating configuration on either the same or the opposite side of the sample as the IR beam. Finally, the generated sum frequency can be collected from either the ϵ_1 or ϵ_3 side. Specific calculations for many of these configurations have appeared in the literature [34, 49–51]. Most experiments are conducted with both incident beams in a common plane with the surface normal. This simplifies the determination of sample azimuthal symmetries. In a strict co-propagating configuration, alignment is simple as the sum and visible co-propagate. Independent polarization control of the IR and visible beams and spectral filtering of the sum are both difficult in this configuration but modern Raman notch filters combined with a good spectrometer provide adequate spectral contrast. Most experiments are in a near-co-propagating configuration. This facilitates independent optimization of the IR and visible beam size and polarization. As the sum will travel close to the visible, alignment is also simplified. Counter propagating configurations facilitate the spatial filtering of the sum from the intense visible beam [5, 52]. Additionally it increases the angular dispersion in the sum signal [53]. Initial alignment of the system is more complicated, but can be facilitated by use of visible tracer beams configured to follow the phase-matched sum path.

In Fig. 5.3 are shown calculations for $|\chi^e|^2$ (see (5.5)) as a function of the angle of incidence of the IR (2900 cm^{-1}) and visible (800 nm) beams (assuming a mutual plane of incidence) for two common cases: external reflection on a metal (Au) and external reflection on a dielectric (fused silica). χ_{ijk} is taken to be appropriate for $C_{\infty v}$ with all allowed terms unity. The polarizations of the beams are specified in the order Sum, VIS, and IR. General trends are clearly apparent in the results. On a high dielectric constant substrate, Fresnel factors for the z component of the field will optimize at grazing incidence while those for the s-field will optimize at normal incidence. This results in a strong preference for co-propagating configurations for ppp, as all three beams (IR, visible, and resultant sum) are grazing. In the counter-propagating configuration, the sum is closer to normal and χ^e is suppressed. For ssp, sps, and pss configurations the interesting result is obtained that the signal is optimized with one incident beam grazing and one normal. Systems with independently variable angles of incidence are difficult to construct. Thus compromise fixed angle

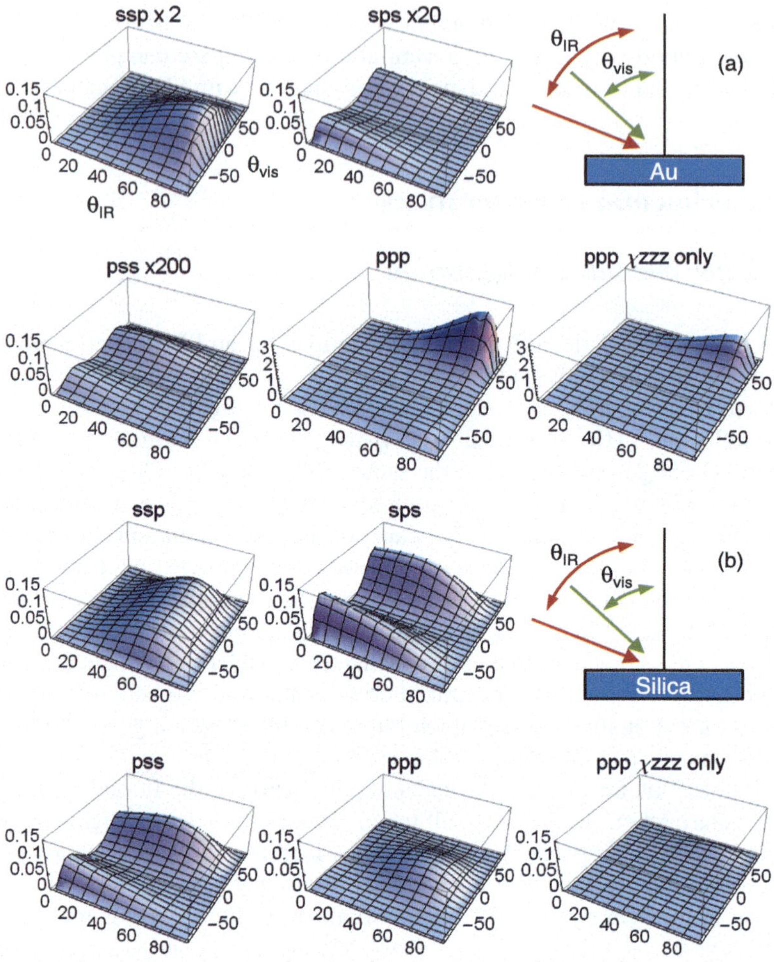

Fig. 5.3 $|\chi^e|^2$ for detection of the sum in the upper half space as a function of the angle of incidence of both the IR and visible beams; (**a**) on a gold substrate; (**b**) on a silica substrate

of incidence systems are generally employed, typically with the IR more grazing than the visible. On a weak dielectric (fused silica) substrate, the ppp configuration is intrinsically dim due to a 'Brewster-angle-like' suppression of the z field. Additionally, the xxz and zzz components in a ppp spectrum contribute with opposite sign, leading to partial cancellation.

The normal gamut of optical techniques can be applied to the engineering of the linear Fresnel factors that contribute to χ^e. As the signal is the square of the triple product, these can lead to significant enhancements. The most common practice is to employ total internal reflection (TIR) to enhance the otherwise weak signal from dielectric interfaces. This is only applicable when the substrate is transparent to at least one beam and requires tight control of the angles of incidence [11, 50, 54, 55].

Variable-angle TIR can be used to selectively probe individual interfaces in multilayer stacks of appropriate index materials [56, 57]. Simple cavities can be constructed to selectively enhance the signal from specific interfaces [58]. Waveguide effects, such as 2D surface plasmon resonance can be employed [59] as can localized surface plasmons [60].

5.3.2 Laser Sources

General Considerations

Early implementations of SFG spectroscopy employed pulsed lasers with duration (ns to 10 s of ps) long compared to the sample vibrational response and appropriate to the preceding steady state analysis. The signal level is proportional to the integral of the intensity at ω_Σ from (5.4) over space and time. In general, the duration of the visible and IR pulses are similar as they are derived from a common master pulse. Typically the respective beam sizes are slightly mismatched to facilitate spatial overlap. Assuming Gaussian spatial profiles, the signal (detected photons N) becomes:

$$N \propto \frac{1}{\hbar\omega_\Sigma} \iint I(\omega_\Sigma)\, dt\, dA\, f_{rep} \propto \left[|\chi^e|^2 \frac{\mathcal{E}_2}{A_2\tau} \frac{\mathcal{E}_1}{A_1\tau} \frac{A_1 A_2}{A_1 + A_2} \tau f_{rep} \right] \tag{5.10}$$

where $\mathcal{E}$ is the pulse energy, A, the pulse area, the integral is over the pulse duration τ and f_{rep} is the system repetition rate. Notice that the signal varies as the inverse of the sum of the two beam areas. Scaling with $1/\tau$ (τ pulse duration) favors short pulse lasers (consistent with spectral resolution constraints). Scaling with $1/A$ favors small spots. In practice, the spot size is limited by the sample damage threshold, thus $\mathcal{E}/A$ for one beam can be replaced by a safety factor η times the damage fluence F_D and (5.10) becomes

$$N \propto |\chi^e|^2 \frac{\eta F_D}{\tau} \frac{\mathcal{E}_< f_{rep}}{1+\rho} \tag{5.11}$$

where ρ is the ratio of the beam areas. Thus we see that, for an optimized system (beam size chosen consistent with F_D) the maximum signal becomes linear in the average power ($\mathcal{E}_< \times f_{rep}$) of the least damaging beam. When damage is a thermal process, F_D will scale as $\sqrt{\tau}$ so system performance improves as $1/\sqrt{\tau}$ [61]. Achieving optimal performance requires independent control of both beam sizes. Care must be taken to assure that the, typically high, fluences employed do not perturb the intended experiment. Resonant absorption of the IR cannot be neglected, as it can cause heating in the monolayer [62] in addition to molecular subphases [63].

Broad Band Systems

The practical limit to system improvement via decrease in pulse duration is reached when the frequency resolution (determined by the Fourier transform of the pulse envelope) significantly exceeds the vibrational linewidth. For Gaussian pulses, the time-bandwidth-product (tbp) is 14.7 ps cm^{-1}. With a target resolution of ≈4 cm^{-1}, pulses must be longer than ≈4 ps. A number of SFG systems have been constructed within these parameters, with repetition rates between ≈ (10 Hz and 1000 Hz). Early implementations have been based on custom made laser systems. Recently, commercial laser sources, optimized for IR generation appropriate to SFG have appeared [64]. Narrow band IR pulses have been generated by Raman shifting, difference frequency mixing, or parametric generation. Tunability in early systems was typically achieved with dye lasers; now it is almost exclusively achieved by use solid-state parametric oscillators or amplifiers.

With the advent of reliable, sub-ps, tunable laser systems, alternative routes to SFG have been developed that operate partially or totally in time domain vs frequency domain. The simplest and most widely adopted is broad-band SFG (BB-SFG), originally demonstrated using sub-ps pulses from a free electron laser [53] and then demonstrated with table top sources [65, 66]. In BB-SFG, the IR pulse is deliberately chosen to be significantly broader in frequency than the underlying vibrational resonance (significantly shorter than the polarization response time). When mixed against a narrow-band visible pulse (significantly longer than the polarization response time) a broad-band spectrum is generated that can be multiplex detected with an imaging spectrograph and scientific grade CCD detector. This approach has advantages over the conventional scanning of a narrow-band IR source in terms of ease of execution. Additionally, it is more robust to shot-to-shot amplitude fluctuations of the laser system. However, BB-SFG systems, when compared to scanned, narrow-band systems with comparable total IR pulse powers, will achieve comparable signal levels: the multiplex advantage of the BB system just compensates for the reduced power-per-unit-bandwidth. The performance and relative merits of BB vs narrow-band SFG systems closely parallel those of multiplex [67] and narrow-band Coherent Anti-Stokes Raman (CARS) systems. For spectrally dilute and imaging applications, narrow-band instruments have advantages [68]. For complex hyperspectral applications, broad-band systems have advantages. Additionally, BB-SFG systems are well suited to the study of surface dynamics, [69, 70] but ps, narrow-band sources have also been successfully used in SFG studies of dynamics [71, 72].

The details of a BB-SFG experiment are best considered in the time domain. The induced, nonlinear polarization is given by [9]

$$\wp_l(t) = \int_0^\infty d\tau_1 \int_0^\infty d\tau_2\, S^{int}_{lmn}(\tau_2, \tau_1) E_m(t - \tau_1 - \tau_2) E_n(t - \tau_1) \tag{5.12}$$

where S is the quadratic response function and is related to the Fourier transform of the frequency susceptibility χ as can be seen by comparison of (5.2) and (5.12). Causality requires that S vanish when either τ_1 or τ_2 is < 0. (5.12) contains all possible second order mixing processes. For the simple case of $E = E^{vis}(t - \tau_D) + E^{IR}(t)$ where E^{vis} is nonresonant and τ_D is the time delay between the two pulses, the visible interaction can be treated as instantaneous and (5.12) simplifies to [73]

$$\wp_l(t) = E^{vis}(t - \tau_D) \int_0^\infty d\tau_1\, S^{int}_{lmn}(0, \tau_1) e^{vis}_m E^{IR}_n(t - \tau_1). \tag{5.13}$$

If we focus on just one resonant term with ω_0 and line half width Γ_0 and assume the IR pulse is sufficiently short compared to the vibrational decay that it can be approximated by a delta function we obtain the common expression [74–76]

$$\wp_l(t) \propto \frac{\partial \alpha_{lm}}{\partial Q}\frac{\partial \mu_n}{\partial Q} E_m^{vis}(t-\tau_D) E_n^{IR} \theta(t) e^{-i\omega_0 t} e^{-t\Gamma_0} \tag{5.14}$$

where θ is the unit step function and E_n^{IR} the amplitude of the delta function. From (5.14) it is clear the essential nature of the visible pulse envelope on the measurement, as it is a window function for the free induction decay, FID, of the vibrational resonance. The signal in a homodyne detected BB-SFG experiment is obtained from the sum frequency electric field derived from Fourier transform of the polarization $I(\omega) \propto |\wp(\omega)|^2$.

The importance of the visible pulse envelope in BB-SFG has been explored by many authors [74–76] and is depicted in Fig. 5.4, which contrasts the behavior of two typical visible pulse shapes: a Gaussian pulse, representative of the output of a ps parametric source, and an exponential (in time) pulse, representative of the output of an etalon filter [75, 76]. In Fig. 5.4a and b the free-induction decay of a single vibrational resonance (no nonresonant background) with $\Gamma = 0.94\ \text{ps}^{-1}$ (10 cm^{-1} intensity FWHM) is contrasted to the visible pulse temporal envelope for pulses with an intensity frequency FWHM of 5 cm^{-1} (2.94 ps FWHM Gaussian, 0.714 ps FWHM exponential). In Fig. 5.4c and d we show the signal (at the peak of the line) as a function of time delay τ_D between the visible and IR pulses. One notes that the symmetric Gaussian pulse produces the most signal slightly delayed from the IR pulse [76], while the abrupt leading edge of the exponential pulse results in the peak signal occurring at zero delay. In Fig. 5.4e and f the resultant SFG spectrum is compared (scaled) to the vibrational susceptibility at the optimal τ_D. We see that, for visible pulses of matched spectral width, the Gaussian pulse produces a more faithful representation of the vibrational resonance. This is due to the particularly small time-band width product of exponential pulses (tbp $= 3.68\ \text{ps}\,\text{cm}^{-1}$); the exponential pulse is (relative to the Gaussian) temporally short and more significantly windows the FID.

In Fig. 5.5 we compare the evolution of the spectra for the two pulses as a function of τ_D. With the Gaussian pulse, there is a distinct narrowing of the line with increasing delay [74, 77]. The origin of this is clear from Fig. 5.4: at $\tau = 0$, the FID gates the visible pulse, allowing only $\frac{1}{2}$ the temporal duration to contribute and thus broadening the spectrum. At long delays, the FID is relatively constant over the Gaussian envelope, and the resultant spectrum is the resolution of the visible pulse (and all information on the line width is lost). The most faithful representation of the real lineshape occurs with a slight delay of the Gaussian [76]. This line narrowing was first observed in CARS with impulsive preparation pulses [78]. For the exponential pulse, the resolution is constant with time delay for delays ≥ 0.

In Fig. 5.6 we compare the two visible pulses for the more complex case of two closely spaced transitions in the presence of a non resonant background. The parameters for the susceptibility (in frequency domain) are:

$$\beta \sim 3 + \frac{1i}{\nu_{IR} - 2085 + i5} + \frac{2i}{\nu_{IR} - 2115 + i2.5} \tag{5.15}$$

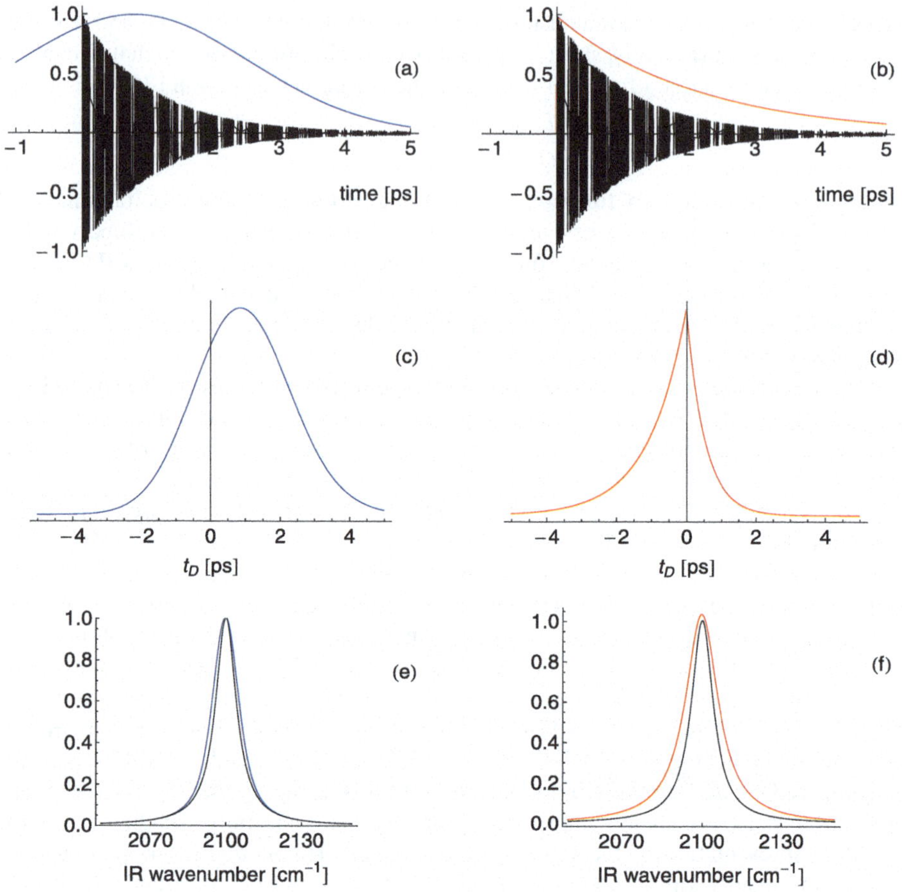

Fig. 5.4 Comparison of BBSFG performed with a Gaussian (*blue*) and exponential (*red*) visible pulse. (**a**), (**b**) Free induction decay (*black*) and visible pulse envelope (*color*). (**c**), (**d**), peak spectral intensity vs VIS-IR pulse delay. (**e**), (**f**) normalized spectra at maximum intensity (*color*) and true spectrum (*black*)

where ν_{IR} is the wavenumber in cm^{-1}. In Fig. 5.6 c and d are the τ_D dependence at the two resonance positions. The nonresonant (NR) term dominates the response at early times. In Fig. 5.6e and f are shown spectra at $\tau_D = 0$ and at $\tau_D =$ FWHM of the visible pulse (2.94 ps and 0.73 ps respectively). Also shown in Fig. 5.6e and f is the 'ideal' steady state frequency spectrum including and removing the NR term. The NR background significantly contributes to the delayed spectrum with the Gaussian probe, but is absent in the delayed spectrum with the exponential probe. The ability to fully suppress the NR background is an advantage of the exponential pulse [75]. However, it is clear that time delayed spectra must be interpreted with great care. When spectra contain features with different vibrational lifetimes, the τ_D spectra convolve both line width and intensity, suppressing spectrally broad

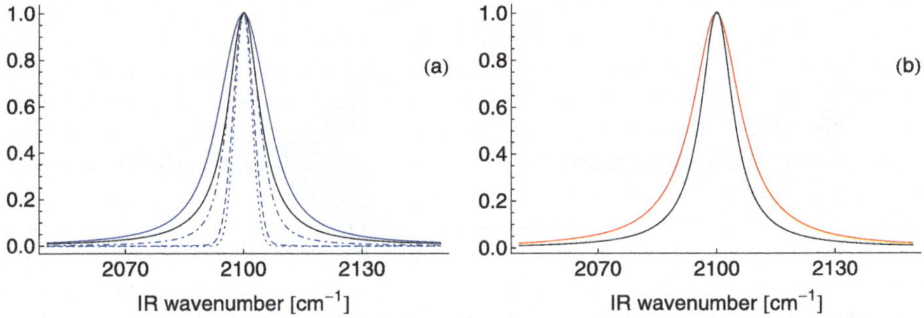

Fig. 5.5 Line profiles as a function of VIS-IR time delay τ_D: four time delays are shown: 0 and 1, 2 and 3× the visible pulse intensity FWHM (*solid*, *dot-dashed*, *dashed*, and *dotted* respectively). (**a**) Gaussian (*blue*) pulse with FWHM 2.94 ps and (**b**), exponential (*red*) pulse with FWHM 0.73 ps. Both visible pulses have 5 cm^{-1} intensity FWHM

features. Additionally, the sharp leading edge of the exponential pulse results in a phase distortion of the time delayed spectra [79]. This can be seen in Fig. 5.6f: the underlying susceptibility should exhibit destructive interference between the trailing edge of the low frequency feature and the leading edge of the high frequency feature, leading to improved resolution of the two lines as exhibited in the Gaussian and ideal spectra. However, in the time delayed exponential spectrum, the lines are blurred.

Consideration of the above implies that, for robust spectral analysis, it is important to choose relatively slow visible pulses to faithfully preserve the full information of the FID. This will be done at the expense of the efficiency of the system, as significant visible pulse energy is lost converting the weak tails of the FID. For experiments where the spectra are simple or absolute fidelity is not required, significant advantages can be obtained with short visible pulses [75]. We also note that, while scanning τ_D is a powerful tool for developing spectral diversity and refining spectral analysis by altering the relative contribution of the NR background and can thus resolve ambiguities in lineshapes [77], it provides no additional dynamic information. The full content of the experiment is contained in the FID. Scanning τ_D allows for improved reconstruction of the FID, but contains no more information than the associated steady-state frequency response [80]. Separation of the linewidth into homogeneous and inhomogeneous contributions requires higher level (pump-probe) mixing experiments [71, 80].

Recently, SFG spectra have been acquired by combining full electric field measurement (through spectral interferometry with a local oscillator) with complete time delay sampling of an ultrafast visible pulse in a manner analogous to full NMR sampling [79]. This extends earlier, homodyne detected time-domain studies [81]. While complex, the system enables the complete reconstruction of the FID and thus unbiased spectra. In analogy with conventional BB-SFG where high fidelity spectral require long (inefficient) visible pulses there is a trade-off. High resolution of the reconstructed spectra requires extensive sampling of the FID, decreasing the overall spectral acquisition time/system efficiency.

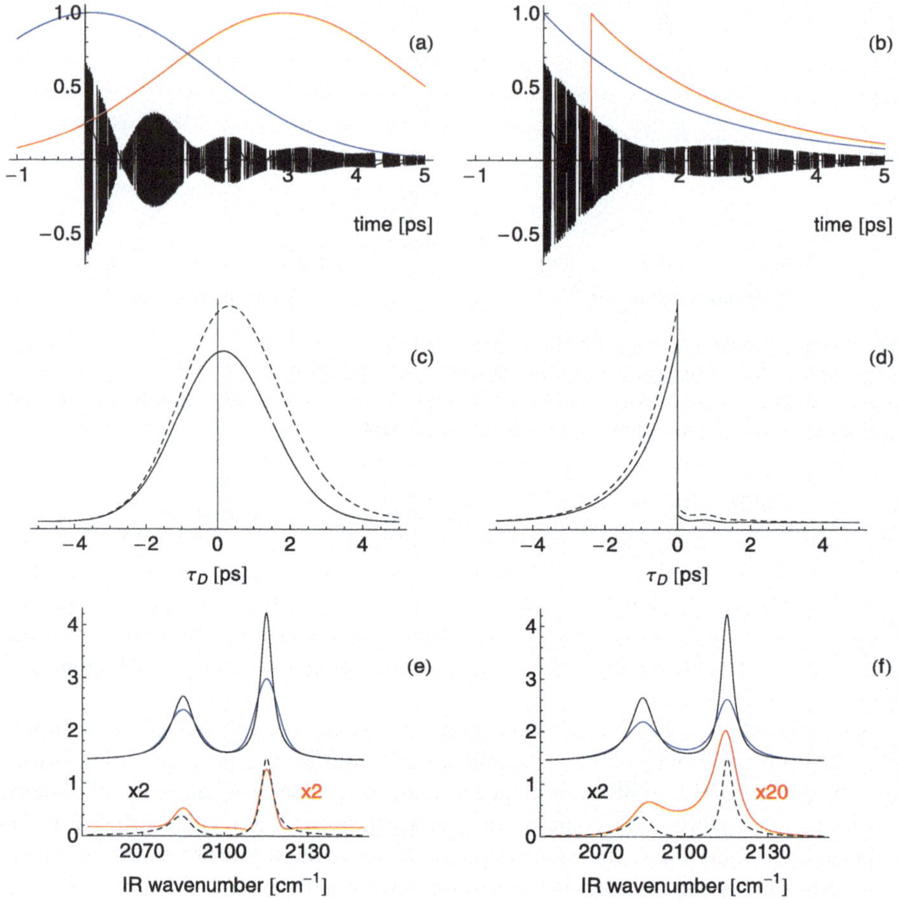

Fig. 5.6 Comparison of BB-SFG performed with a Gaussian (*blue*) and exponential (*red*) visible pulse. (**a**), (**b**) Free induction decay (*black*) and visible pulse envelope (*color*). (**c**), (**d**), peak spectral intensity (*solid* at 2085 cm^{-1}, *dashed* at 2115 cm^{-1}) vs VIS-IR pulse delay. (**e**), (**f**) normalized spectra at 0 τ_D (*blue*) and τ_D = FWHM (*red*) compared to the true spectrum (*black*) and spectrum with no NR term (*dashed*)

5.4 Data Treatment

5.4.1 Normalization

Most current implementations of SFG are done in a single beam, polarametric mode, akin to linear reflectometry. Source noise is reduced by normalizing the sample signal to either the IR and visible beam energies, or SFG from a reference sample. The normalization can be done by diverting a fraction of the incident beams onto the reference, or by replacing the sample with the reference. Diversion can be done by beam splitters (allowing shot-to-shot normalization) or by deflection mirrors (allowing shot-to-every-other-shot normalization). In charge coupled device (CCD) based detection systems, the normalization can be acquired as a second track on the 2D detector. For moderately focused, BB-SFG systems, the dominant sources of error in the normalization lie not in the counting statistics or low frequency fluctuations of the sources, but in the spatial/temporal chirp of the beams and sensitivity to beam overlap/pointing. In this case, normalization to a reference at the sample position is preferred. For studies at solid/solid interfaces (such as polymer films) this is easily achieved by implementation of multiple sample carousels. For studies at solid-liquid interfaces, we have successfully implemented dual channel (sample and reference) fluid cells [82]. In general, studies at liquid-liquid interfaces require alternative normalization (diversion) schemes.

It is essential that the dispersion of the χ^e for the reference film be weak over the entire spectral range. As adventitious hydrocarbons are ubiquitous, it is also necessary to thoroughly and frequently clean the reference. In the stretching region of the IR (1600 to 3500) cm^{-1} z-cut quartz is often used as a reference [29, 83]. Z-cut quartz has the advantage that χ^e is of comparable magnitude for all 4 typical polarization configurations. Additionally, it enables the absolute determination of the sample χ^e, based on the calibrated susceptibility of quartz [84]. Quartz is not appropriate for use in the fingerprint region (800 to 1300) cm^{-1} due to the strong dispersion from the substrate phonons [83, 85]. Vapor deposited Au films are often used as qualitative references, due to the strong signal and smooth dispersion. Simple coatings (hydrogenated or perdeuterated thiol self-assembled monolayers and or polymers) can be applied to the Au to nominally match the optical conditions of the sample and to reduce the surface free energy and thus contamination rate. The reproducibility of the linear and nonlinear properties of vapor deposited Au is such that each Au reference should be calibrated to quartz in the stretching region if absolute susceptibilities are desired.

5.4.2 *Spectral Analysis*

The typical, homodyne detected SFG spectrum arises from the complex mutual interference of both nonresonant and resonant terms as described in (5.9). In general, the absolute signal is useful only for the most qualitative of analysis. Accurate extraction of spectroscopic quantities (line positions, widths, and amplitudes) requires numerical analysis. Typically this is done by fitting a forward simulation of the signal to the general form:

$$S \propto \left| A_{NR} + \sum_j \frac{A_j e^{i\phi_j}}{\omega_{IR} - \omega_j + i\Gamma_j} \right|^2 \tag{5.16}$$

or, more commonly, to the more restricted form in which the phase ϕ_j is assumed to be the same for all vibrational features. As with all forward simulation exercises, care must be taken to include all relevant terms, but to not over fit. The primary approximation of (5.16) is the assumption of a Lorentzian (homogeneous) line shape.

This phenomenological lineshape is not necessarily well suited to condensed phase systems. For systems such as the OH stretch mode in the presence of hydrogen bonding, the Lorentzian line shape is inappropriate [86]. The coherent analog to a Voight is then commonly used [51]:

$$\int_{-\infty}^{\infty} \frac{A_j e^{i\phi_j} e^{\frac{-(\omega-\omega_j)^2}{2\sigma_j^2}}}{\omega_{IR} - \omega_j + i\Gamma_j} d\omega \qquad (5.17)$$

although the justification for a Gaussian distribution is, again, strictly phenomenological. Both the Lorentzian and pseudo-Voight functions assume no experimental broadening of the spectrum.

Intrinsically, as the signal is the square magnitude of the underlying field, information on the phase of the underlying electric fields is lost. For the simple case of (5.16), it has been shown that there are actually 2N indistinguishable, distinct sets of A_j and ϕ_j consistent with a spectrum of N resonances [87]. This can be qualitatively understood in terms of the simple case of a single line interfering with a NR background. For a spectrum with a resonance that appears as a emission line there are two possible underlying interferences: constructive interference between a weak feature and the NR term or destructive interference between a strong feature and the NR term. Thus emission-like spectra do not definitively define the relative phase of the NR and vibrational features. For absorption-like spectra, there are still two cases: destructive interference between a weak or stronger line. This 2 fold amplitude/phase ambiguity for a single feature gives rise to the 2N solutions for a complex mixture. It is even possible for ghost features (lines with finite amplitude but no visual impact to the final spectrum) to be added to the spectrum [87]. It is therefore essential that care and physical reasoning be applied to the selection of an appropriate fit. It should be noted that, while there are 2N amplitude and ϕ combinations consistent with a spectrum, the line positions, widths, and magnitude of the NR background are uniquely determined (assuming the correct line shape and number of lines are selected).

In the case of complex spectra, either empirical or *ab initio* estimates of the hyperpolarizability tensors, combined with model orientation distributions, can guide the selection of fit parameters between equivalent sets. At a minimum, the linear IR and Raman spectra are guides for the selection of features to include. Often it is assumed that all vibrational features have a single, common ϕ. This reduces the number of distinguishable solutions. However, this assumption requires that all vibrational features have a common spatial origin, that the Fresnel and local field corrections have negligible dispersion, that the NR susceptibility has negligible intrinsic dispersion, and that a single χ^e element contributes to the spectrum. For the simple case of the ppp spectrum of a monolayer on a metal, the common phase approximation is not necessarily valid as each of the χ^e elements (zzz, zxx, xzx, xxz for an isotropic system) have distinct Fresnel coefficients with distinct phase offsets. In general, quantitative calculation of the relevant F's (see (5.3)) is required to assess the more restrictive phase assumption [88]. Phase retrieval algorithms, such as maximum entropy methods can be used to guide the construction of fit functions [89]. However, they require a priori expectation on the evolution of the phase.

It is best practice to solve the above ambiguities by increasing the diversity of the spectral data set. In the frequency domain this can be done by acquiring spectra with mixed polarizations [90]. With BB-SFG systems this can be done by changing the IR-VIS time delay, varying the relative weight of the NR background and vibrational features although, as discussed earlier, care must be taken in quantitative analysis of such spectra [76, 91]. In both frequency domain and broadband systems, heterodyne techniques can be employed to measure the complex signal electric field and extract the underlying phase, removing ambiguities in fitting [92]. In a heterodyne measurement, a reference electric field is combined with the experimental field to generate the detected signal

$$S \propto |E_{ref} + E|^2 = |E_{ref}|^2 + |E|^2 + 2Re\big[E_{ref}E^*\big]. \tag{5.18}$$

Precise knowledge of the complex reference field, combined with control of the relative phase between E_{ref} and E allows the reconstruction of the complex E [93]. This can be done elegantly in BB-SFG systems through the implementation of Fourier transform spectral interferometry [94, 95]. Phase measurements place significant constraints on system stability, but are becoming increasingly common [96, 97].

5.5 Examples

We will show two examples from studies at the National Institute of Standards and Technology performed with early generation BB-SFG systems. For both systems, the visible pulse was a near-Gaussian pulse with time-bandwidth product of $\approx$2. The pulse duration (FWHM) was $\approx$7 ps. This assured that little spectral distortion occurs in the spectra presented.

One of the primary advantages of BB-SFG systems is the speed/simplicity with which multiple spectra can be acquired. Thus a variety of multivariate experiments (time, temperature, etc.) can be performed. The first example demonstrates the ability of SFG to simultaneously probe monolayer structure and coverage in a kinetic study of the adsorption of perdeuterated dioctadecyldisulfide, dODDS, on Au from an ethanol solution [98]. The self assembly of alkane thiol monolayers on Au has been extensively studied [99, 100]. It is well established that at full coverage, systems with alkane chain lengths greater than 10 to 12 methylene units form dense ($\approx 4.5 \times 10^{14}$ cm^{-2}) well-packed layers of nominally upright (average tilt angle of $\approx 30°$) nearly all trans chains [99, 101, 102]. During vapor deposition, scanned probe studies establish that adsorption proceeds through the formation of a series of ordered striped phases consisting of arrays of nearly all-trans molecules laying down on the surface [103]. Much less is known about the structural evolution during solution phase deposition.

The data was acquired in a simple flow cell created by modification of a commercial cell for microscopy applications. The cell consisted of a CaF_2 or BaF_2 flat entrance window, a 15 µm to 25 µm gasket/spacer and a glass substrate. The Au film was vapor deposited on to the glass substrate. The substrate contained machined

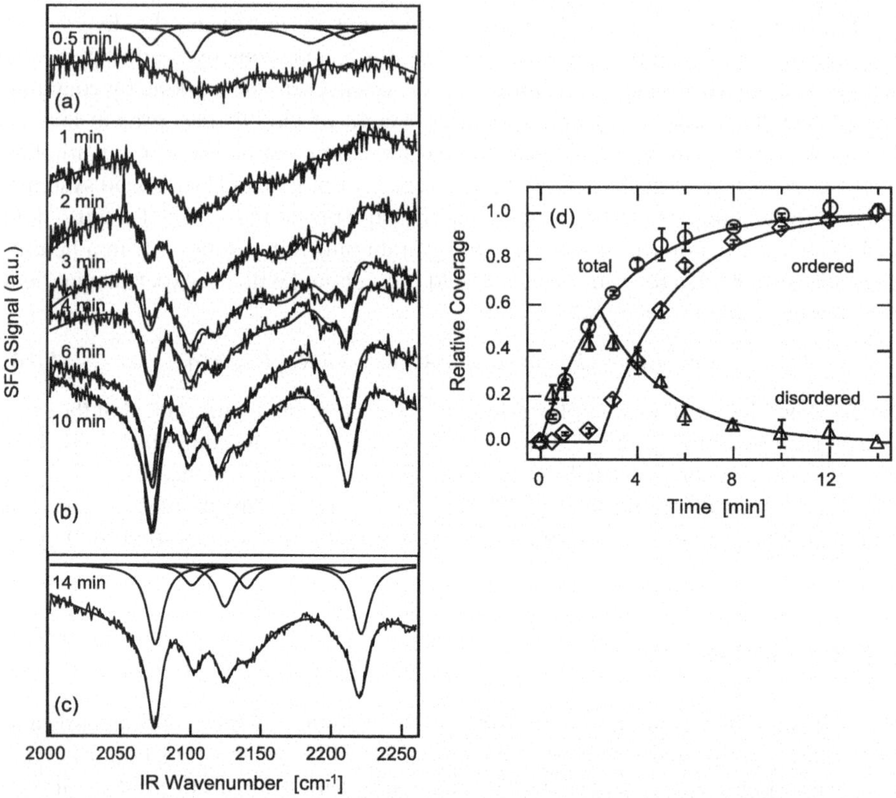

Fig. 5.7 ppp SFG spectra as a function of exposure time to flowing solution. (**a**) SFG spectrum at 0.5 min and the fit to (5.16) representing the disordered phase. Also shown are the imaginary parts to the fit. (**c**) SFG spectrum at 14 min and the fit to (5.16) representing the ordered phase. (**b**) SFG spectra at intermediate times fit to a linear superposition of the two end points. (**d**) The extracted weights of each phase (disordered and ordered), the normalized coverage from the decay of the nonresonant background, and the predictions of a simple phase coexistence model. The error bars represent the difference between two independent experiments

grooves that provided a laminar flow sheet of $\approx$12 mm width and 20 mm length over the Au film. In solution kinetics studies it is critical to control the hydrodynamic conditions of the experiment to separate transport/diffusion effects from the true surface kinetics.

Figure 5.7 reports the ppp SFG spectra of a cleaned Au film exposed to a flowing ethanol solution containing dODDS as a function of time following the introduction of dODDS in to the solution. Spectra were acquired at 30 s intervals. At early times the signal is dominated by the strong nonlinear background from the Au surface. Also present are the (destructive interference) resonant features of the dODDS. The nonlinear vibrational spectroscopy of linear alkanes has been extensively studied and assignments are known. As time evolves, there are two clear changes in the spectra: the first is a monotonic decrease in the strength of the nonlinear background.

The second is the evolution of the vibrational features from a dominance of the symmetric and antisymmetric CD_2 features at 2103 cm^{-1} and 2200 cm^{-1} at early times to a dominance by the CD_3 features at 2073 cm^{-1}, 2130 cm^{-1} and 2220 cm^{-1} at late times. Clean Au surfaces exhibit a strong hyperpolarizability [104] that is quenched by the adsorption of thiols. This has been the basis of SHG studies of thiol adsorption [105, 106]. Due to the centrosymmetry of an all-trans alkane chain, the CD_2 transitions are nominally symmetry forbidden. Thus the limiting well ordered films are dominated by the terminal methyl group. This makes SFG a powerful tool for the study of chain disorder as the introduction of gauche defects creates SFG active CD_2 (CH_2) features. This has been leveraged in many studies of order-disorder transitions in alkane containing monolayers [4, 82, 107, 108]. Qualitatively, the SFG spectra suggest initial adsorption into a highly disordered state, with ordering accompanying film densification. A simple analytical model was developed to treat the data: spectra at both early times and at completion were fit to (5.16) to develop the spectral shape of the resonant contributions to χ. All intermediate spectra were then fit to a model with three adjustable parameters: the magnitude of the nonlinear background and the magnitude of the 'disordered' and 'ordered' contributions to the spectra. The results of the fitting are shown in Fig. 5.7 and agree well with the data. The good agreement establishes that the relative phases for the initial spectral analysis were correct. A_{nr} can be approximated as linearly proportional to the total thiol coverage, $A_{nr} = A_{nr0}(1 - CN_s)$, while the spectral weights are taken to be proportional to the area density of the two conformations. Shown in Fig. 5.7d are the three fit parameters. Two trends are clear: the total thiol coverage appears to obey simple Langmuir adsorption kinetics: $N_s = N_{s0}(1 - \exp[-kt])$ and the ordered state exhibits a distinct incubation period. A kinetic model was developed, based on a proposed phase equilibrium between ordered and disordered phases above a critical coverage. Shown in Fig. 5.7d is the excellent agreement between this model and the extracted system parameters.

A second example comes from a study of the buried interface between polystyrene (PS) films and glass that established a correlation between molecular structure with adhesion [44] and highlights the ability of SFG to probe buried interfaces inaccessible to most techniques. When studying thin films, it is essential to clarify the origin of the signal: free surface, bulk, or buried interface [109, 110]. In this study a cavity technique was used to selectively study the buried interface [58]. The film of interest was deposited on a dielectric layer deposited on a good reflector, in this case Au. The thickness of the polymer and dielectric were chosen to provide enhanced sensitivity for either the free or buried interface, while simultaneously suppressing the signal from the opposite interface, thus enhancing both signal and contrast. Shown in Fig. 5.8 are ssp SFG spectra of 128 nm PS films deposited on either a 198 nm or a 330 nm dielectric film on Au. The spectra were recorded in the region of the aromatic CH stretches of the pendant phenyl ring. In Fig. 5.8a (198 nm dielectric) the system is selective for the free interface while in Fig. 5.8b the system is selective for the buried interface. The selectivity is demonstrated by damaging the free surface with a brief exposure in a UV-ozone cleaner. The degradation of the free surface signal, presumably due to oxidation of the phenyl rings, is clearly evident

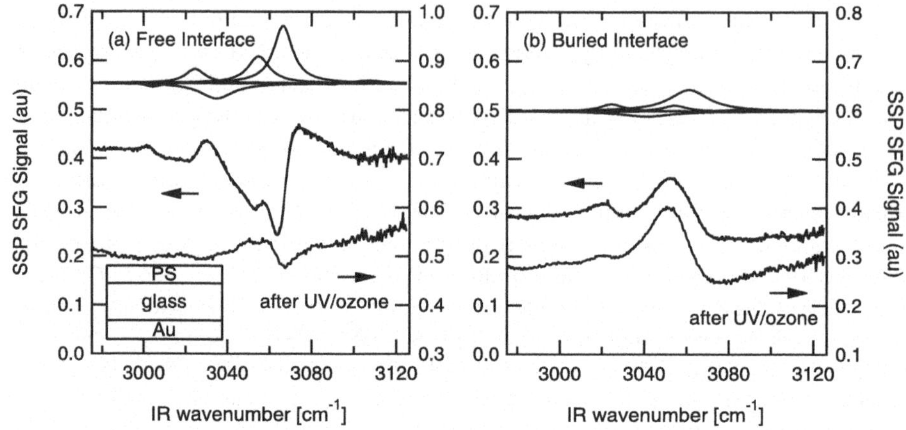

Fig. 5.8 ssp SFG spectra of matched polystyrene films spun cast onto dielectric cavities (see *inset*) to enhance: (**a**) the free surface and (**b**) the buried interface. Also shown are the same surfaces after treatment in a UV-ozone cleaner. Figure derived from Ref. [44]

as is the preservation of the buried interface. As in the study of the dODDS self assembled monolayer, the Au mirror produces a significant nonresonant background in the spectra. Control experiments established negligible background contributions from either the glass or PS itself [111]. The optical thickness between the enhanced interface and the substrate is nominally the same in the two stacks, thus the sign of the interference in the two spectra directly gives the relative orientation of the phenyl groups. Clearly the orientation is in opposite directions, relative to the Au mirror, at the two interfaces. An earlier SFG study has established that the phenyl ring at the free surface is oriented away from the film, towards the region of low excluded volume, in accord with molecular dynamics simulations [112]. Thus, the phenyl group at both interfaces is oriented away from the bulk of the film. The orientation of the phenyl group is consistent with the vibrational spectrum: all frequencies for the buried interface are in nominal accord with those found by IR and Raman spectroscopy of bulk PS. However, the frequencies at the air interface are all blue shifted by 7 cm^{-1} [111].

An advantage of the dielectric cavity technique is the ease with which multiple samples can be prepared via conventional processing (spin coating, etc.) and studied. The dielectric used in Fig. 5.8 was a spin-on-glass that created a hydrophobic, Si–H terminated surface with a water contact angle of ~95°. By selective treatment with UV-ozone, the surface could be turned hydrophilic (water contact angle <5°) by titration of the Si–H, presumably with Si–OH. The adhesion of the PS film was found to correlate with UV-ozone treatment. Films were weakly adhering on the hydrophobic surface and strongly adhering on the hydrophilic interface. Shown in Fig. 5.9 are ssp SFG from the buried interface of films as a function of dielectric treatment. Notice the dramatic differences in the spectra, indicating a significant reorientation of the phenyl ring. Of particular note in the spectrum of the hydrophilic interface is the strong intensity of the B_1 symmetry ν_{20b} mode (3028 cm^{-1}) relative

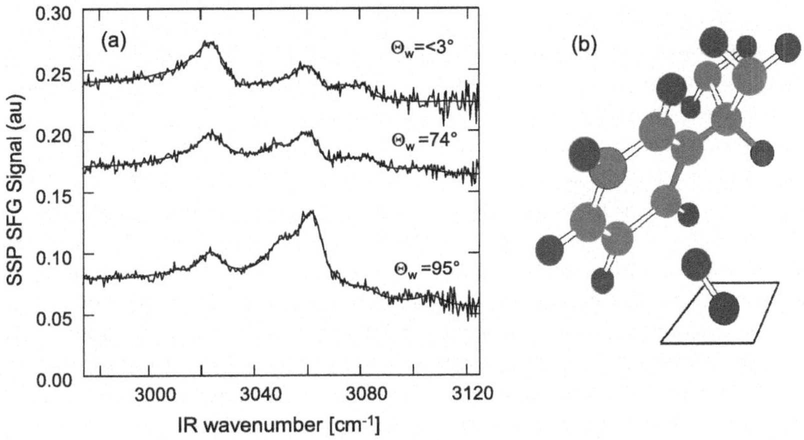

Fig. 5.9 (**a**) ssp SFG spectra of the buried interface between polystyrene and a treated dielectric as a function of the dielectric water contact angle. (**b**) molecular schematic of the phenyl ring torsion at the surface due to the development of a pi hydrogen bond with a surface hydroxyl. Figure derived from Ref. [44]

to the ν_2 mode (3065 cm^{-1}). The ν_2 mode is strong in both the Raman and IR spectra of PS and typically dominates the SFG spectra of PS [109, 111, 113] and analogous phenyl systems [42, 111, 114]. No orientation distribution of a C_{2v} phenyl group could be found to be consistent with the full spectrum of the hydrophobic interface. However, *ab initio* calculations of the hyperpolarizability of the model molecule isopropyl-benzene indicated that torsion of the phenyl ring against the polymer backbone distort the normal modes, allowing the mode at 3028 cm^{-1} to develop A character. The calculated spectra were in good accord with those observed for torsions of $\sim$30°. The torsion was attributed to the development of a pi hydrogen bond between the pendent phenyl and the substrate OH. The SFG spectra thus provided a molecular view of the improved adhesion at the interface.

5.6 Future

SFG has become an established spectroscopy for the characterization of simple, planar interfaces. Modern laser systems enable sufficient sensitivity to allow real time studies of sub-monolayer kinetics and complex, pump-probe studies of ultrafast dynamics. In the future, it is anticipated that the technique will broaden in terms of both experimental complexity and application diversity.

A logical extension of SFG is to imaging. As the diffraction properties of the radiated field are determined by the sum wavevector, one should be able to achieve visible diffraction-limited resolution with vibrational contrast and sub-monolayer sensitivity. IR-reflection microscopy is severely limited in resolution, while conventional Raman microscopy does not have monolayer sensitivity. The current competing surface sensitive imaging technique is soft X-ray microscopy [115, 116]. In principle, microscopy can be performed by either spot-scanning or imaging. The primary constraint is signal level. As discussed above, the signal in an optimized system scales linearly with system power. This assumes the spot size is adjusted to use all available power without damage. In a spot scanning system, the spot size is set by the desired resolution. Thus the beam energy must be attenuated (at constant power) by increasing the system repetition rate. Current generation BB-SFG systems based on 1 kHz amplifiers can acquire a full spectrum in the order of 10 s with spot sizes of order 100 μm. Simple scaling implies that a 300 kHz system would achieve the same spectrum from a 6 μm area. The alternative to a high repetition rate system is to image a larger illumination area [117, 118]. Such a system requires a line tuned source. A recent implementation, based on a commercial ps SFG system, has successfully demonstrated 5 cm^{-1} spectra at 5 μm resolution, acquiring a full hyperspectral image stack in order of an hour [119]. The system has been applied to the study of both catalysis and corrosion [120].

Many critical interfaces do not conform to the ideal, planar assumptions on which the preceding modeling of SFG is based. Recently SFG is being applied to non-planar systems. Following early work in SHG, SFG has been applied to rough surfaces from powders [121] and nanoparticle films [122]. Total internal reflection geometries are advantageous, in terms of minimizing multiple scattering interferences [123, 124]. Of particular note is the application of SFG to the study of colloids and inclusions. These systems are readily understood in terms of the nonlinear generalization of Mie scattering [125, 126]. Just as planar cavities can both enhance and suppress SFG signals, similar resonances occur as a function of particle size and shape. The mathematical foundations for this have been laid and recently applied to the study of coatings on colloids and buried inclusions in biological systems [127–129].

The experimental underpinnings of SFG are reaching maturity. Scientific grade detectors exist with near unity quantum efficiency that can reach the shot noise limit over a wide dynamic range. Laser sources, both tunable ps and broadband fs, are increasingly stable and powerful. As a tool uniquely suited to the non-destructive, in-situ characterization of optically accessible interfaces, it is anticipated to see even greater application as time unfolds.

References

1. Y.R. Shen, Annu. Rev. Phys. Chem. **40**, 327 (1989)
2. J.H. Hunt, P. Guyot-Sionnest, Y.R. Shen, Chem. Phys. Lett. **133**, 189 (1987)
3. X.D. Zhu, H. Suhr, Y.R. Shen, Phys. Rev. B **35**, 3047 (1987)
4. P. Guyot-Sionnest, J.H. Hunt, Y.R. Shen, Phys. Rev. Lett. **59**, 1597 (1987)
5. A.L. Harris, C.E.D. Chidsey, N.J. Levinos, D.N. Loiacono, Chem. Phys. Lett. **141**, 350 (1987)
6. R.W. Boyd, *Nonlinear Optics*, 3rd edn. (Academic Press, Amsterdam, 2008)
7. Y.R. Shen, *The Principles of Nonlinear Optics*. Wiley Classics Library (Wiley, New York, 2003)
8. P.N. Butcher, D.D. Cotter, *The Elements of Nonlinear Optics* (Cambridge University Press, Cambridge, 1990)

9. S. Mukamel, *Principles of Nonlinear Optical Spectroscopy* (Oxford University Press, London, 1999)
10. J.-C. Diels, W. Rudolph, *Ultrashort Laser Pulse Phenomena: Fundamentals, Techniques, and Applications on a Femtosecond Time Scale* (Academic Press, San Diego, 1996)
11. G.L. Richmond, Chem. Rev. **102**, 2693–2724 (2002)
12. K.B. Eisenthal, Chem. Rev. **106**, 1462 (2006)
13. P.B. Miranda, Y.R. Shen, J. Phys. Chem. B **103**, 3292 (1999)
14. Y.R. Shen, V. Ostroverkhov, Chem. Rev. **106**, 1140 (2006)
15. Z. Chen, Y.R. Shen, G.A. Somorjai, Annu. Rev. Phys. Chem. **53**, 437 (2002)
16. F. Vidal, A. Tadjeddine, Rep. Prog. Phys. **68**, 1095 (2005)
17. H.W.K. Tom, T.F. Heinz, Y.R. Shen, Phys. Rev. Lett. **51**, 1983 (1983)
18. B. Koopmans, A.-M. Janner, H.T. Jonkman, G.A. Sawatzky, F. van der Woude, Phys. Rev. Lett. **71**, 3569 (1993)
19. D. Wilk, D. Johannsmann, C. Stanners, Y.R. Shen, Phys. Rev. B **51**, 10057 (1995)
20. J.I. Dadap, X.F. Hu, M.H. Anderson, M.C. Downer, J.K. Lowell, O.A. Aktsipetrov, Phys. Rev. B **53**, R7607 (1996)
21. T.A. Germer, K.W. Kolasinski, J.C. Stephenson, L.J. Richter, Phys. Rev. B **55**, 10694 (1997)
22. T. Manaka, E. Lim, R. Tamura, M. Iwamoto, Nat. Photonics **1**, 581 (2007)
23. T.C. Anglin, D.B. O'Brien, A.M. Massari, J. Phys. Chem. C **114**, 17629 (2010)
24. P. Yeh, *Optical Waves in Layered Media* (Wiley, New York, 1988)
25. J.M. Manley, H.E. Rowe, Proc. IRE **44**, 904 (1956)
26. N. Bloembergen, P. Pershan, Phys. Rev. **128**, 606 (1962)
27. J.E. Sipe, J. Opt. Soc. Am. B **4**, 481 (1987)
28. D.S. Bethune, J. Opt. Soc. Am. B **6**, 910 (1989)
29. X. Wei, S.-C. Hong, X. Zhuang, T. Goto, Y.R. Shen, Phys. Rev. E **62**, 5160 (2000)
30. T.F. Heinz, in *Nonlinear Surface Electromagnetic Phenomena*, ed. by H. Ponath, G. Stegeman (Elsevier, Amsterdam, 1991), p. 353
31. C.A. Dailey, B.J. Burke, G.J. Simpson, Chem. Phys. Lett. **390**, 8 (2004)
32. N. Bloembergen, *Nonlinear Optics*, 4th edn. (World Scientific, River Edge, 1996)
33. X. Wei, Y.R. Shen, Phys. Rev. Lett. **86**, 4799 (2001)
34. X. Zhuang, P.B. Miranda, D. Kim, Y.R. Shen, Phys. Rev. B **59**, 12632 (1999)
35. P. Ye, Y.R. Shen, Phys. Rev. B **28**, 4288 (1983)
36. M. Cho, C. Hess, M. Bonn, Phys. Rev. B **65**, 205423 (2002)
37. S.H. Lin, A.A. Villaeys, Phys. Rev. A **50**, 5134 (1994)
38. A.J. Moad, G.J. Simpson, J. Phys. Chem. A **109**, 1316 (2005)
39. A.J. Moad, G.J. Simpson, J. Phys. Chem. B **108**, 3548 (2004)
40. H.-F. Wang, W. Gan, R. Lu, Y. Rao, B.-H. Wu, Int. Rev. Phys. Chem. **24**, 191 (2005)
41. C. Hirose, N. Akamatsu, K. Domen, Appl. Spectrosc. **46**, 1051 (1992)
42. R.N. Ward, D.C. Duffy, G.R. Bell, C.D. Bain, Mol. Phys. **88**, 269 (1996)
43. R. Superfine, J.Y. Huang, Y.R. Shen, Chem. Phys. Lett. **172**, 303 (1990)
44. P.T. Wilson, L.J. Richter, W.E. Wallace, K.A. Briggman, J.C. Stephenson, Chem. Phys. Lett. **363**, 161 (2002)
45. J. Guthmuller, F. Cecchet, D. Lis, Y. Caudano, A.A. Mani, P.A. Thiry, A. Peremans, B. Champagne, Chem. Phys. Chem. **10**, 2132 (2009)
46. G.J. Simpson, K.L. Rowlen, Anal. Chem. **72**, 3399 (2000)
47. A. Morita, J.T. Hynes, J. Phys. Chem. B **106**, 673 (2002)
48. A. Perry, H. Ahlborn, B. Space, P.B. Moore, J. Chem. Phys. **118**, 8411 (2003)
49. B. Dick, A. Gierulski, G. Marowsky, G.A. Reider, Appl. Phys. B **38**, 107 (1985)
50. J. Löbau, K. Wolfrum, J. Opt. Soc. Am. B **14**, 2505 (1997)
51. C.D. Bain, P.B. Davies, T.H. Ong, R.N. Ward, M.A. Brown, Langmuir **7**, 1563 (1991)
52. J. Miragliotta, R.S. Polizzotti, P. Rabinowitz, S.D. Cameron, R.B. Hall, Chem. Phys. **143**, 123 (1990)
53. E.W.M. van der Ham, Q.H.F. Vrehen, E.R. Eliel, Opt. Lett. **21**, 1448 (1996)
54. S.R. Hatch, R.S. Polizzotti, S. Dougal, P. Rabinowitz, Chem. Phys. Lett. **196**, 97 (1992)

55. J.C. Conboy, M.C. Messmer, G.L. Richmond, J. Phys. Chem. **100**, 7617 (1996)
56. K.S. Gautam, A.D. Schwab, A. Dhinojwala, D. Zhang, S.M. Dougal, M.S. Yeganeh, Phys. Rev. Lett. **85**, 3854 (2000)
57. G. Li, A. Dhinojwala, M.S. Yeganeh, J. Phys. Chem. C **115**, 7554 (2011)
58. P.T. Wilson, K.A. Briggman, W.E. Wallace, J.C. Stephenson, L.J. Richter, Appl. Phys. Lett. **80**, 3084 (2002)
59. E.W.M. van der Ham, Q.H.F. Vrehen, E.R. Eliel, V.A. Yakovlev, E.V. Alieva, L.A. Kuzik, J.E. Petrov, V.A. Sychugov, A.F.G. van der Meer, J. Opt. Soc. Am. B **16**, 1146 (1999)
60. Q. Li, C.W. Kuo, Z. Yang, P. Chen, K.C. Chou, Phys. Chem. Chem. Phys. **11**, 3436 (2009)
61. R.M. Wood, *Laser Damage in Optical Materials*, 1st edn. (Taylor & Francis, London, 1986)
62. Y. Goto, N. Akamatsu, K. Domen, C. Hirose, J. Phys. Chem. **99**, 4086 (1995)
63. X. Wei, P.B. Miranda, Y.R. Shen, Phys. Rev. Lett. **86**, 1554 (2001)
64. J.P. Smith, V. Hinson-Smith, Anal. Chem. **76**, 287 (2004)
65. L.J. Richter, T.P. Petralli-Mallow, J.C. Stephenson, Opt. Lett. **23**, 1594 (1998)
66. M. Bonn, C. Hess, S. Funk, J.H. Miners, B.N.J. Persson, M. Wolf, G. Ertl, Phys. Rev. Lett. **84**, 4653 (2000)
67. A. Zumbusch, G.R. Holtom, X.S. Xie, Phys. Rev. Lett. **82**, 4142 (1999)
68. J.-X. Cheng, X.S. Xie, J. Phys. Chem. B **108**, 827 (2004)
69. H. Arnolds, M. Bonn, Surf. Sci. Rep. **65**, 45 (2010)
70. J.A. Carter, Z. Wang, D.D. Dlott, Acc. Chem. Res. **42**, 1343 (2009)
71. P. Guyot-Sionnest, Phys. Rev. Lett. **66**, 1489 (1991)
72. A. Bandara, J. Kubota, K. Onda, A. Wada, S.S. Kano, K. Domen, C. Hirose, J. Phys. Chem. B **102**, 5951 (1998)
73. A.L. Harris, L. Rothberg, J. Chem. Phys. **94**, 2449 (1991)
74. T. Ishibashi, H. Onishi, Chem. Phys. Lett. **346**, 413 (2001)
75. A. Lagutchev, S.A. Hambir, D.D. Dlott, J. Phys. Chem. C **111**, 13645 (2007)
76. I.V. Stiopkin, H.D. Jayathilake, C. Weeraman, A.V. Benderskii, J. Chem. Phys. **132**, 234503 (2010)
77. M. Bonn, H. Ueba, M. Wolf, J. Phys. Condens. Matter **17**, S201 (2005)
78. W. Zinth, M.C. Nuss, W. Kaiser, Phys. Rev. A **30**, 1139 (1984)
79. J.E. Laaser, W. Xiong, M.T. Zanni, J. Phys. Chem. B **115**, 2536 (2011)
80. R.F. Loring, S. Mukamel, J. Chem. Phys. **83**, 2116 (1985)
81. D. Star, T. Kikteva, G.W. Leach, J. Chem. Phys. **111**, 14 (1999)
82. N.A. Anderson, L.J. Richter, J.C. Stephenson, K.A. Briggman, Langmuir **22**, 8333 (2006)
83. D.K. Hore, M.Y. Hamamoto, G.L. Richmond, J. Chem. Phys. **121**, 12589 (2004)
84. R.J. Pressley, *Handbook of Lasers* (Chem. Rubber Co., Cleveland, 1971)
85. W.-T. Liu, Y.R. Shen, Phys. Rev. B **78**, 024302 (2008)
86. M.G. Brown, E.A. Raymond, H.C. Allen, L.F. Scatena, G.L. Richmond, J. Phys. Chem. A **104**, 10220 (2000)
87. B. Busson, A. Tadjeddine, J. Phys. Chem. C **113**, 21895 (2009)
88. R. Braun, B.D. Casson, C.D. Bain, E.W.M. van der Ham, Q.H.F. Vrehen, E.R. Eliel, A.M. Briggs, P.B. Davies, J. Chem. Phys. **110**, 4634 (1999)
89. M. Sovago, E. Vartiainen, M. Bonn, J. Phys. Chem. C **113**, 6100 (2009)
90. J. Wang, M.L. Clarke, Z. Chen, Anal. Chem. **76**, 2159 (2004)
91. A.D. Curtis, S.B. Reynolds, A.R. Calchera, J.E. Patterson, J. Phys. Chem. Lett. **1**, 2435 (2010)
92. V. Ostroverkhov, G.A. Waychunas, Y.R. Shen, Phys. Rev. Lett. **94**, 046102 (2005)
93. L. Lepetit, G. Chériaux, M. Joffre, J. Opt. Soc. Am. B **12**, 2467 (1995)
94. I.V. Stiopkin, H.D. Jayathilake, A.N. Bordenyuk, A.V. Benderskii, J. Am. Chem. Soc. **130**, 2271 (2008)
95. S. Nihonyanagi, S. Yamaguchi, T. Tahara, J. Chem. Phys. **130**, 204704 (2009)
96. C.S. Tian, Y.R. Shen, Proc. Natl. Acad. Sci. USA **106**, 15148 (2009)
97. X. Chen, W. Hua, Z. Huang, H.C. Allen, J. Am. Chem. Soc. **132**, 11336 (2010)
98. C.S.-C. Yang, L.J. Richter, J.C. Stephenson, K.A. Briggman, Langmuir **18**, 7549 (2002)

99. F. Schreiber, Prog. Surf. Sci. **65**, 151 (2000)
100. J.C. Love, L.A. Estroff, J.K. Kriebel, R.G. Nuzzo, G.M. Whitesides, Chem. Rev. **105**, 1103–1170 (2005)
101. C.D. Bain, E.B. Troughton, Y.T. Tao, J. Evall, G.M. Whitesides, R.G. Nuzzo, J. Am. Chem. Soc. **111**, 321 (1989)
102. A.N. Parikh, D.L. Allara, J. Chem. Phys. **96**, 927 (1992)
103. D.K. Schwartz, Annu. Rev. Phys. Chem. **52**, 107 (2001)
104. D.A. Koos, G.L. Richmond, J. Phys. Chem. **96**, 3770 (1992)
105. M. Buck, F. Eisert, J. Fischer, M. Grunze, F. Träger, Appl. Phys. A **53**, 552 (1991)
106. O. Dannenberger, M. Buck, M. Grunze, J. Phys. Chem. B **103**, 2202 (1999)
107. J.C. Conboy, M.C. Messmer, G.L. Richmond, J. Phys. Chem. **100**, 7617 (1996)
108. J. Liu, J.C. Conboy, Biophys. J. **89**, 2522 (2005)
109. K.S. Gautam, A.D. Schwab, A. Dhinojwala, D. Zhang, S.M. Dougal, M.S. Yeganeh, Phys. Rev. Lett. **85**, 3854 (2000)
110. Z. Chen, Prog. Polym. Sci. **35**, 1376 (2010)
111. K.A. Briggman, J.C. Stephenson, W.E. Wallace, L.J. Richter, J. Phys. Chem. B **105**, 2785 (2001)
112. T.C. Clancy, J.H. Jang, A. Dhinojwala, W.L. Mattice, J. Phys. Chem. B **105**, 11493 (2001)
113. M. Oh-e, S.-C. Hong, Y.R. Shen, Appl. Phys. Lett. **80**, 784 (2002)
114. E.L. Hommel, H.C. Allen, Analyst **128**, 750 (2003)
115. B. Watts, C.R. McNeill, Macromol. Rapid Commun. **31**, 1706 (2010)
116. C. Hub, S. Wenzel, J. Raabe, H. Ade, R.H. Fink, Rev. Sci. Instrum. **81**, 033704 (2010)
117. D.M.P. Hoffmann, K. Kuhnke, K. Kern, Rev. Sci. Instrum. **73**, 3221 (2002)
118. M. Flörsheimer, C. Brillert, H. Fuchs, Mater. Sci. Eng. C **8–9**, 335 (1999)
119. K. Cimatu, S. Baldelli, J. Phys. Chem. B **110**, 1807 (2006)
120. K.A. Cimatu, S. Baldelli, J. Phys. Chem. C **113**, 16575 (2009)
121. G. Ma, H.C. Allen, J. Am. Chem. Soc. **124**, 9374 (2002)
122. T. Kawai, D.J. Neivandt, P.B. Davies, J. Am. Chem. Soc. **122**, 12031 (2000)
123. S.B. Waldrup, C.T. Williams, Catal. Commun. **8**, 1373 (2007)
124. M.S. Yeganeh, S.M. Dougal, B.G. Silbernagel, Langmuir **22**, 637 (2006)
125. N. Yang, W.E. Angerer, A.G. Yodh, Phys. Rev. Lett. **87**, 103902 (2001)
126. J.I. Dadap, J. Shan, K.B. Eisenthal, T.F. Heinz, Phys. Rev. Lett. **83**, 4045 (1999)
127. S. Roke, Chem. Phys. Chem. **10**, 1380 (2009)
128. S. Roke, W.G. Roeterdink, J.E.G.J. Wijnhoven, A.V. Petukhov, A.W. Kleyn, M. Bonn, Phys. Rev. Lett. **91**, 258302 (2003)
129. A.G.F. de Beer, H.B. de Aguiar, J.F.W. Nijsen, S. Roke, Phys. Rev. Lett. **102**, 095502 (2009)

Part III
X-Ray Techniques

Chapter 6
Grazing Incidence X-Ray Diffraction

Osami Sakata and Masashi Nakamura

Abstract The principles of grazing incidence X-ray diffraction (GIXD) are discussed. A sample of a crystalline material is composed of a surface region including its top layer and a bulk part. The effect of the surface region on the intensity of surface X-ray diffraction cannot be generally disregarded. With the grazing configuration this small intensity is optimized and the structural parameters of surfaces, interfaces, and thin films can be determined through the comparison between the estimated, or experimental, and calculated structure factors. For the estimation, the experimental procedures to measure GIXD profiles around reciprocal lattice points with the necessary corrections are presented. A synchrotron X-ray source and diffractometers employed to perform GIXD experiments are briefly described. We conclude with two examples of systems investigated by means of GIXD: an electrochemical interface of Ag(1 0 0) and an epitaxial thin film of $Bi_4Ti_3O_{12}$ grown on a TiO_2 (1 0 1) single crystal.

6.1 Introduction

Structural analysis of surfaces and interfaces is important for understanding their physical properties and chemical functions. In the early stages, the study of the surface structures progressed by using techniques based on electron probe such as low energy electron diffraction (LEED), then various surface analytical techniques based on other probes have been developed during the last decades.

X-ray diffractometry is one of the most standard methods for the determination of a crystallographic structure at atomic resolution. Since X-ray weakly interacts

O. Sakata (✉)
Synchrotron X-ray Station at SPring-8, National Institute for Materials Science (NIMS), Kouto 1-1-1, Sayo, Hyogo 679-5148, Japan
e-mail: SAKATA.Osami@nims.go.jp

M. Nakamura
Department of Applied Chemistry and Biotechnology, Graduate School of Engineering, Chiba University, Yayoi-cho 1-33, Inage-ku, Chiba 263-8522, Japan
e-mail: mnakamura@faculty.chiba-u.jp

G. Bracco, B. Holst (eds.), *Surface Science Techniques*,
Springer Series in Surface Sciences 51, DOI 10.1007/978-3-642-34243-1_6,

with materials, the probe can penetrate deeply in the sample and, as explained in the following, special scattering geometries are necessary to reduce the penetration and maximize the faint diffraction signal of a surface with respect to the diffraction of the bulk. On the other hand, the penetration allows the application of surface X-ray diffraction to study solid-solid and solid-liquid interfaces as well as exposed surfaces under an ultra-high vacuum condition. Moreover, an analysis based on a single scattering (kinematical) approximation is valid due to the weak interaction.

A first application of X-ray diffraction was the study of the GaAs–Al interface by Marra et al. [1]. They found that the surface sensitivity is considerably enhanced at the grazing angle of incident X-rays. The geometry at the grazing angle has some advantages for surface X-ray diffraction. The greatest merit is to reduce the background intensities from the thermal diffuse scattering and disordered area. Moreover, grazing incidence limits the penetration of the X-rays into the bulk. Since X-ray diffraction from surface is extremely weak, the use of synchrotron radiation is necessary for the acquisition of reliable data and the second-generation synchrotron facilities, which were established in the 1980s, have exponentially advanced the field of surface X-ray diffraction. The structures of surface reconstruction, such as Ge(0 0 1)2×1, Au(1 1 0)1×2, and InSb(1 1 1)2×2 were studied under ultra-high vacuum conditions [2–4].

Robinson proposed an analytical method for the determination of the out-of-plane structure such as surface relaxation and surface roughness based on the measurement of the diffraction profiles in the direction perpendicular to the surface, the so called crystal truncation rods (CTRs) [5].

In the 1990s, highly-brilliant X-rays from the third generation synchrotron radiation have been available for surface X-ray diffraction and recent investigations are focused on the analysis of nanostructures including light elements[1] at interface as well as on a detailed determination of surface structures. Currently, synchrotron-based X-ray diffraction is a quantitative and precise probe for obtaining structural information on surfaces, interfaces, and thin films.

This chapter is organized in the following way: in the first part, after a brief description of the reciprocal lattice and crystal truncation rods, we present the calculations which are necessary to obtain the structure factor. Then we explain the procedure from data acquisition to structural analysis in experiments carried out with grazing incidence X-ray diffraction. In the second part, the experimental setup and X-ray source for surface X-ray diffraction are described. Finally, as an application of the technique, we present recent results for an electrochemical interface and a thin film structure.

[1] Since X-ray diffraction intensity is proportional to electron density, low atomic number elements are weak scatterers.

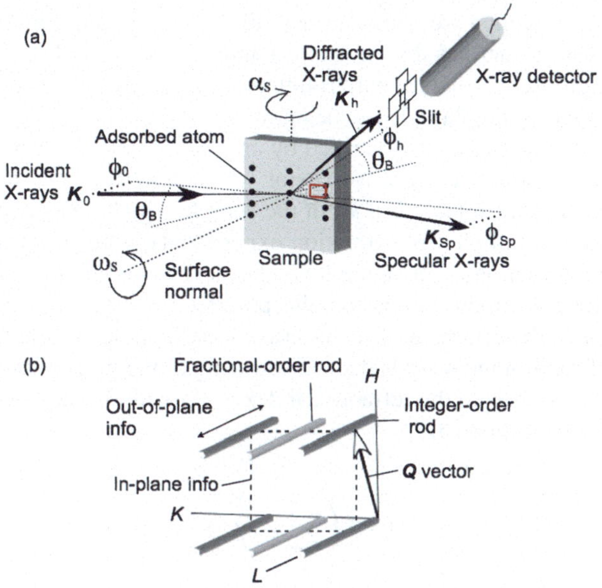

Fig. 6.1 Geometry for surface X-ray diffraction. (**a**) The flat crystal surface is shown and the *red rectangle* represents a bulk unit cell in the real space truncated on the surface. The *dots* represent atoms adsorbed on the surface and arranged with a 1×2 periodicity. Incident X-rays are monochromatic and the incident (ϕ_0) and exit (ϕ_{Sp} and ϕ_h) angles are as small as the critical angle for total external reflection. X-rays are diffracted from lattice planes perpendicular to the surface at a Bragg angle θ_B. (**b**) Crystal truncation rods in the reciprocal-lattice space. Fractional-order rods appear at half-order positions along the K direction. $\boldsymbol{Q}$ is defined as $\boldsymbol{K}_h - \boldsymbol{K}_0$

6.2 Theory and Background

6.2.1 Real-Space Geometry and Reciprocal-Space Scans Used for Surface X-Ray Diffraction

Incident X-rays with a photon energy E_p in the range between 10 and 20 keV are typically employed for structural studies using surface X-ray diffraction since the wavelength $\lambda = hc/E_p$, where h is the Planck constant and c is the speed of light, is in the range 0.6–1.2 Å therefore its value is similar to the size of the lattice parameter of materials, a condition necessary for diffraction.

Figure 6.1(a) shows a schematic geometry used for surface X-ray diffraction. The sample is mounted on a goniometer which allows four rotational degrees of freedom to change the orientation of the sample with respect to the incident beam and the detector axis: rotation of the sample for changing an angle with respect to a set of lattice planes perpendicular to the sample surface (ω_S) and around an axis in the surface plane perpendicular to the incident beam for changing an incident angle with respect to the sample surface (α_S), while the detector can be rotated around two different perpendicular axes δ_S and γ_S (not shown in Fig. 6.1). We will introduce two diffractometers which we name surface diffractometer and film diffractometer in Sect. 6.3.3. The former has the same axis names as we use in Fig. 6.1. The latter, however, has a different structure with rotation axes ω_F, χ_F, ϕ_F, and δ_F.

A monochromatic and plane X-ray beam, incident on a flat crystal surface of the sample at a grazing angle ϕ_0, is described by an electric field $\boldsymbol{E}_0 \exp(-2\pi i \boldsymbol{K}_0 \cdot \boldsymbol{r})$, where $\boldsymbol{K}_0$ is the incident X-ray wave vector whose length K ($\equiv |\boldsymbol{K}_0|$) is defined as $1/\lambda$. There is a transmitted wave $\boldsymbol{k}_0$ in the sample which gives rise to diffracted waves $\boldsymbol{k}_h$ inside the sample. In diffraction experiments, a beam linearly polarized in the horizontal direction is often employed because an undulator source of a synchrotron ring usually produces horizontally polarized X-rays. If a scattering plane is perpendicular to the direction of the polarization, the polarization factor is unity. Otherwise a diffraction intensity is decreased owing to the polarization effect.

At the X-ray wavelength the refractive index n of matter is very close to unity. In fact, neglecting absorption [6]

$$n = 1 - \delta_n,$$
$$\delta_n = 2.7 \times 10^{-6} \left(\sum_j Z_j \Big/ \sum_j A_j \right) \rho \lambda^2 \tag{6.1}$$

where $\sum_j Z_j$ is the sum on atomic charges in the unit cell and A_j are the atomic masses, ρ is the mass density in g/cm^3 and λ is in Å, hence δ_n is of the order of 10^{-5}. Accordingly the effects of refraction and reflection at the vacuum interface are generally negligible at an incident angle larger than several degrees. On the other hand, since the index is slightly less than unity, it is possible that the total reflection in the medium with the higher index of refraction, i.e. the vacuum, takes place. The critical angle ϕ_c for total external reflection is defined as an incident angle at which the transmitted wave is propagated parallel to the sample surface, i.e. $\cos\phi_c = n$, and for small angles

$$\phi_c = \sqrt{2\delta_n}. \tag{6.2}$$

Therefore a specular X-ray wave $\boldsymbol{K}_{Sp}$ occurs when ϕ_0 is close to the critical angle ϕ_c, which is inversely proportional to the incident photon energy value E_p and is proportional to the number of electrons in a unit cell volume. For reference, ϕ_c for a Si crystal is $0.1°$ at $E_p = 17.8$ keV.

Diffracted waves constructively interfere under a Bragg condition. The equation of the Bragg condition gives the relation of λ and a Bragg angle θ_B between the incident X-ray beam and interested lattice planes of a lattice spacing d. The relation for a 3D crystal can be written as $\lambda = 2d \sin\theta_B$. When we introduce the reciprocal lattice, i.e. the Fourier transforms of the crystal structure, the Bragg condition for the 3D crystal can be written as $\boldsymbol{K}_h = \boldsymbol{K}_0 + \boldsymbol{h}$. Here, a reciprocal-lattice vector $\boldsymbol{h}$ connects the origin of the reciprocal space and the reciprocal point corresponding to the lattice planes. In other words, the reciprocal point is called Bragg point of index HKL. The components (H, K, L) of $\boldsymbol{h}$ are expressed using integers H, K, and L. In addition, $|\boldsymbol{K}_h| = |\boldsymbol{K}_{Sp}| = |\boldsymbol{K}_0| = K$ which represents energy conservation, i.e. all the incident and diffracted vectors have the same length, and in a graphical representation the tips of the diffracted vectors emanating from the common origin belong to a sphere, the Ewald sphere.[2] For the surface case, it is useful to choose a reference frame with the $(H, K, 0)$ directions in the (xy) plane of the surface and the $(0, 0, L)$ direction along z, the surface normal. Therefore the reciprocal vectors $(H, K, 0)$ describe the 2D periodicity along the surface of the system. If we consider a diffraction associated to a set of lattice planes perpendicular to the sample surface having such a reciprocal vector, a diffracted wave $\boldsymbol{K}_h$ emanates from the surface at a grazing exit angle ϕ_h. Consequently $\boldsymbol{K}_0$ and $\boldsymbol{K}_h$ are in a skew configuration for grazing incidence because of the momentum transfer caused by the specular reflection. When a bulk crystal is truncated to expose the surface to be studied, there is a loss of periodicity along the z direction therefore the Bragg condition is strictly valid only in the (xy) plane for the 2D reciprocal lattice of the surface structure while the condition is relaxed for the z component of $\boldsymbol{Q}$ $(\equiv \boldsymbol{K}_h - \boldsymbol{K}_0)$ therefore any value is allowed. This corresponds to a change in the graphical representation of the bulk Bragg points which become elongated along the normal of the surface of the truncated crystal. These lines perpendicular to the surface are called crystal truncation rods or Bragg rods.

As an example, let us consider an adsorbed system grown on the surface. The system often exhibits surface superstructures, i.e. an additional periodicity with respect to the substrate. In this case, the Bragg condition in the reciprocal-lattice space has to be fulfilled both for integer-order rods, related to the substrate periodicity, and for fractional-order rods only related to the additional periodicity of the adsorbed system. The specific case of a doubling in the periodicity along the horizontal (K) direction is shown in Fig. 6.1(b): due to the presence of the adsorbed atoms, half-integer rods are present in between integer-order rods.

In the investigation of the system, a position on a Bragg rod is chosen by selecting the angles α_S and ω_S of the sample as well as putting an X-ray detector at a proper angular position. Then a diffraction intensity profile is recorded by rocking the sample around the ω_S axis (ω_S rocking scan). This procedure is repeated around as many points on the Bragg rod as possible and for different rods. Then the intensity of each rocking scan is integrated to obtain the intensity of the measured Bragg rods as function of Q_z. In fact these rod intensity curves contain the structural information of the system and they will be compared with the calculated structure factors of the sample system. The structural information along the out-of-plane and in-plane directions in the real space is parallel and perpendicular to the rods, respectively.

Generally a small L value is used for in-plane measurements and this corresponds to a small $Q_z = K(\sin\phi_h + \sin\phi_0)$. The electric field intensity on the sample surface is function of ϕ_0 and has a maximum peak at ϕ_c with a trend that generally shows a gentle decrease up to about three times of ϕ_c. Hence we can gain diffracted intensities using an incident angle $\phi_0 = \phi_c$. However, the incident angle will be doubtless

[2] This sphere is superimposed on the reciprocal space of a crystal with a radius of $1/\lambda$. An incident wave vector starts from the center of the sphere to the origin of the reciprocal space. The sphere is used for determining the directions in which an incident X-ray or other beam will be diffracted by the crystal.

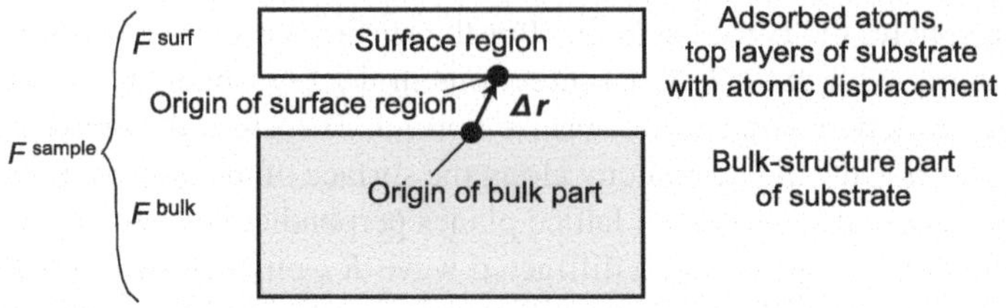

Fig. 6.2 Schematic structure of a sample. The surface region consists of adsorbed atoms and top layers of the sample substrate with atomic displacements such as lattice relaxation and in-plane superstructure. The lower part is a bulk structure of the substrate

changed due to experimental misalignment of the sample and small error in the positioning of the diffractometer when the azimuthal angle ω_S is changed to go to a different position in the reciprocal-lattice space. Therefore many researchers use ϕ_0 in the angular range from 3 to 5 ϕ_c.

For the interested reader, X-ray diffraction phenomena, such as wave fields and penetration depths under grazing incidence, were theoretically discussed on the basis of a distorted-wave approximation [7]. On the other hand, grazing incidence X-ray diffraction for a perfect crystal was theoretically considered for a large [8] and for small exit angles ϕ_h [9–11] using the dynamical theory. Furthermore, an extended dynamical theory were applied to the in-plane structural analysis of a heteroepitaxial film grown on a crystal substrate [12].

6.2.2 Structure Factor

To model the diffraction process, a sample can be treated as composed of a surface region and a bulk crystal part (Fig. 6.2). In this way the bulk crystal has the structure of the perfect 3D lattice, while, in the surface region, the change in coordination number can cause atomic displacements from the perfect lattice (for instance a lattice-relaxation among the lattice planes parallel to the surface) and/or an adsorbed layer can introduce a new periodicity as well as new chemical species. As in Sect. 6.2.1, a reference frame is defined with the positive z axis directed into the crystal.

The diffracted intensity is proportional to the square of the modulus of the structure factor F^{sample} of the sample which is the sum of the structure factor F^{bulk} of the bulk part and F^{surf} of the surface region. When we use independent coordinates for the bulk crystal part and the surface region, we can write

$$F^{sample} = F^{bulk} + \exp(-2\pi i\, \boldsymbol{Q} \cdot \Delta \boldsymbol{r}) F^{surf}. \qquad (6.3)$$

$\Delta \boldsymbol{r}$ is the difference between the origin of the bulk part and that of the surface region. The vertical component $|\Delta \boldsymbol{r}_\perp|$ of $\Delta \boldsymbol{r}$ is the interface distance of the bulk part and the surface region.

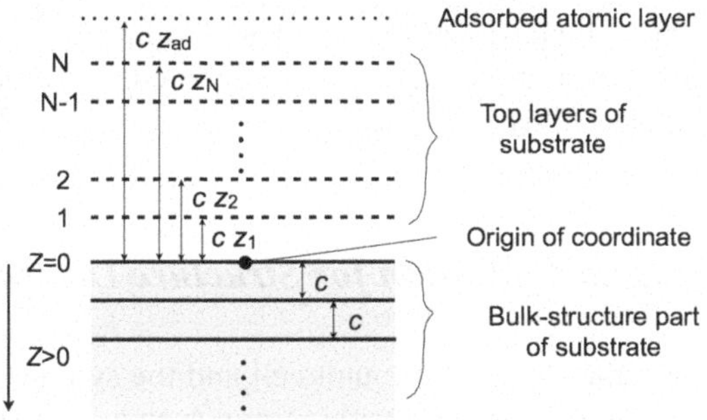

Fig. 6.3 Structural model of an adsorbed sample system. An atomic layer is adsorbed on a crystal substrate. In general, atoms on top layers of the substrate are shifted from bulk atomic positions. The origin of the coordinate for the sample system is located on the topmost layer of the bulk-structure part of the substrate

Let us consider an adsorbed sample system as an example (Fig. 6.3). We assume a single coordinate system with the origin of the coordinate located on the top atomic layer of the bulk part. The system consists of the bulk crystal part of the substrate, top layers of the substrate, and an adsorbed atomic layer. The structure factors can be expressed as follows:

$$\begin{aligned} F^{sample} &= F_{HKL}^{bulk} + F_{HKL}^{surface\ layers} + G_{HK}^{adsorbed} \exp(2\pi i L z_{ad}), \\ F_{HKL}^{bulk} &= \sum_{n=1}^{\infty} G_{HK}^{bulk} \exp(-2\pi i L n), \\ F_{HKL}^{surface\ layers} &= \sum_{n=1}^{N} G_{HK}^{surface\ layers} \exp(2\pi i L z_i), \end{aligned} \tag{6.4}$$

where G_{HK}^{bulk}, $G_{HK}^{surface\ layers}$, and $G_{HK}^{adsorbed}$, here abbreviated to G_{HK}, are structure factors of a periodic unit along the in-plane direction which can be written as

$$G_{HK} = \sum^{in\text{-}plane\ unit\ cell} f_j \exp[-2\pi i (H x_j + K y_j)] T_j. \tag{6.5}$$

Since G_{HK}^{bulk} has also a periodicity along the z direction as well, the unit cell can include one or more atomic layers. For the atom j, f_j is the scattering factor, x_j and y_j are its in-plane positions normalized by the unit cell parameters $\boldsymbol{a}$ and $\boldsymbol{b}$, while z_i and z_{ad}, for the atoms in the top layers and the adsorbed atoms, are the vertical positions normalized by the vertical lattice length c shown in Fig. 6.3. The in-plane unit cell should be identical in the bulk-structure part, in the top layers of the substrate, and in the adsorbed atomic layer. T_j is the Fourier transform of the probability distribution caused by thermal vibration and it is usually known as atomic Debye-Waller factor or atomic temperature factor. For the three-dimensional isotropic harmonic oscillator, T_j is expressed as follows [13, 14],

$$T_j = \exp[-2\pi^2 \langle u_j^2 \rangle |\boldsymbol{Q}|^2], \tag{6.6}$$

where $\langle u_j^2 \rangle$ is the three-dimensional mean square displacement. We note that, in the above expressions, absorption of X-rays is ignored. A similar approach was discussed by R. Feidenhans'l [6].

6.2.3 Outline of Data Collection for Structure Determination

It is easy to determine the shape of the unit cell and the symmetry for the surface structure because the simple crystallographic form, such as the periodicity and size of the unit cell, can be inferred from the position of each diffraction spot in the reciprocal lattice as in the observation of a LEED pattern. However, the detailed atomic structure within the unit cell must be determined from the accurate analysis of the structure factors of diffracted X-rays. There are two types of data collection for surface structure determination: in-plane and rod measurements. An in-plane measurement gives information on the atomic positions parallel to the surface, for instance it is useful in the case of surface reconstruction. Instead, out-of-plane structure can be determined by using the rod measurement along the surface normal direction. Since, in many cases, the atoms in a few layers below the surface relax along the surface normal direction, our experience suggests that rod measurements as well as the in-plane measurements must be carried out for accurate three dimensional determination of the surface structure.

These type of measurements are described in many reviews and textbooks for X-ray scattering and diffraction from a surface [6, 15, 16]. Herein we will explain the analysis procedure and data correction for the Rod measurement of the crystal truncation rod (CTR) with some recent examples using the single point detector. Figure 6.4 shows the flow chart from data correction to structure determination.

6.2.4 From Measured Intensities to Structure Factors

After the setup of the diffractometer and the determination of the crystal orientation matrix (UB matrix) are carried out, the integrated intensities of CTR are measured by ω_S rocking scan. For data acquisition, there are various measurement modes, such as the fixed incident angle mode and the incident angle equal to diffracted angle mode. The incident angle often used is several times the critical angle. Integrated intensity must be corrected by the following factors:

1. Polarization factor: Thomson scattering amplitude is dependent on the polarization direction of the incident X-ray and the scattering angle,
2. Lorentz factor: geometrical correction in integration volume,
3. Interception correction: Integration range of the rod,
4. Transmission correction: X-ray path through an absorbing medium such as Be window and liquid layer,
5. Area correction: Active area dependent on the incident beam and slits.

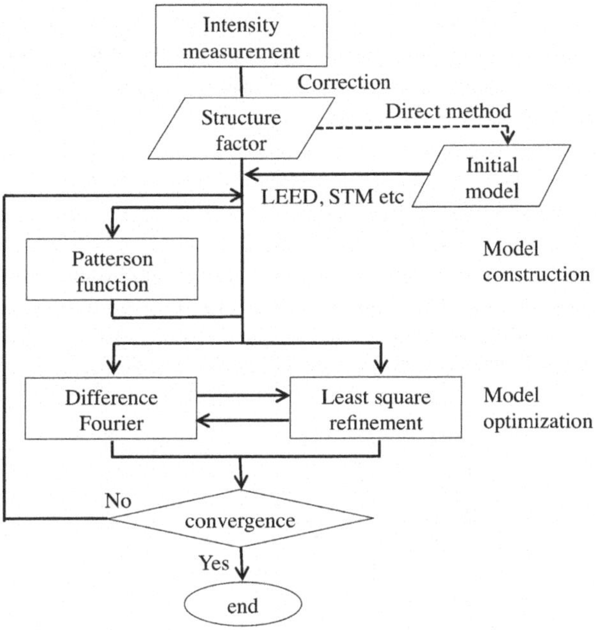

Fig. 6.4 Flow chart of structural analysis. Two input data, structure factors and an initial model are needed for standard analysis. The initial model is conventionally constructed by using the results of other experiments such as LEED and scanning tunneling microscopy (STM). A recent direct method is helpful for the construction of the initial model (*dashed line*) [17–21]

After making these corrections, we will obtain experimental structure factor $|F_{HKL}^{exp}|$ to be compared with calculated one $|F_{HKL}^{calc}|$ that is the same as $|F^{sample}|$ in Sect. 6.2.2.

These corrections depend on the measurement mode and the geometry of diffractometer [22, 23]. All corrections are not necessarily needed. For example, the vertical polarization component is often neglected for a special measurement mode using synchrotron radiation with the horizontal polarization. For the area correction, we often use the Soller slit in front of the detector to decrease axial divergence. If the beam is irradiated within the area of the sample surface, the area correction is not necessary.

6.2.5 Structural Analysis

It is important for structural analysis to select an appropriate initial model. In many cases, various surface analytical techniques such as scanning tunneling microscopy (STM) and electron diffraction will give some hints for the construction of the initial model. If the structure of the initial model is unknown, Patterson methods are available. Patterson function P is defined by the following formula:

$$P(x, y, z) = \frac{1}{V} \sum_{HKL} |F_{HKL}^{exp}|^2 \cos[2\pi i(Hx + Ky + Lz)] \tag{6.7}$$

where V is the volume of the unit cell. A contour map of the Patterson function using the structure factor amplitudes shows interatomic vectors. Atomic correlation can be obtained without the phase problem. In fact the measured intensity is proportional to the square modulus of F^{sample} and the information on its phase is lost preventing its univocal determination from experimental intensities. A study on the InSb(1 1 1)-(2 × 2) reconstructed surface is a good example about the use of the Patterson methods [4]. Recently, direct methods have been introduced for "model free analysis" [17–21].

The agreement between the experimental data and optimized model can be estimated using the χ^2 that is defined by the following formula:

$$\chi^2 = \frac{1}{N-p} \sum_{HKL} \left(\frac{|F_{HKL}^{calc}|^2 - |F_{HKL}^{exp}|^2}{\sigma_{HKL}} \right)^2 \tag{6.8}$$

where N is the number of structure factors, p is the number of free parameters in the model, i.e. the atomic positions, the occupancy factors, the Debye-Waller factors, and the roughness factors. σ_{HKL} includes photon statistics and systematic errors. The error of the diffraction intensities can be estimated from the measured reproducibility of symmetry equivalent reflections. The refinement of the parameters in the model is performed by the least-squares method to minimize χ^2. User friendly softwares for structural analysis of surface X-ray diffraction, such as ANA-ROD, are available [24].

If the reduction of χ^2 is insufficient, the model may be not appropriate or atoms may be missing in the model. In the latter case, the difference Fourier synthesis method is available. The residual electron densities originating from missing atoms can be found from a difference Fourier map. We show an example of a difference Fourier synthesis of missing water molecule on $p(2 \times 2)$–Ni(1 1 1)–O [25]. It is known that oxygen atom is preferentially adsorbed at the face-center cubic site of Ni(1 1 1), whereas the structural information for water oxygen weakly interacting with surface is unknown. The incident angle to the surface was set to be 0.7°, which is about four times as large as the critical angle for total external reflection of Ni ($\phi_c = 0.18°$ at 20 keV). Integrated intensities were obtained with the incidence angle of 0.7° using the surface diffractometer as described in Sect. 6.3.3 and were corrected for Lorentz factor $[1/(\sin\delta_S \cos\alpha_S)]$ [23] and polarization factor $[1 - (\sin^2\gamma_S \cos^2\delta_S)]$ [23, 26]. Here δ_S around the horizontal axis and γ_S around the vertical axis are the angles of rotation for the detector, while α_S around the vertical axis is for sample rotation. 13 structural parameters of the model, which includes the three Ni layers and two oxygen atoms, were determined using 137 nonequivalent reflections of the 2 × 2 fractional order rod with the p3 space group. The origins of the in-plane coordinates and the out-of-plane axis is taken to be atop of adsorbed oxygen and the third Ni layer, respectively. The atoms below 4th layer were fixed as a semi-infinite bulk crystal. Each substrate layer includes four Ni atoms locating at one independent (general position) and three equivalent atoms (special position) in a (2 × 2) unit cell. Atoms at the general position and the special position have one (z) and two (x and z) degree of freedom, respectively. In the first step, atomic positions of the initial model without water oxygen were optimized using the least-squares

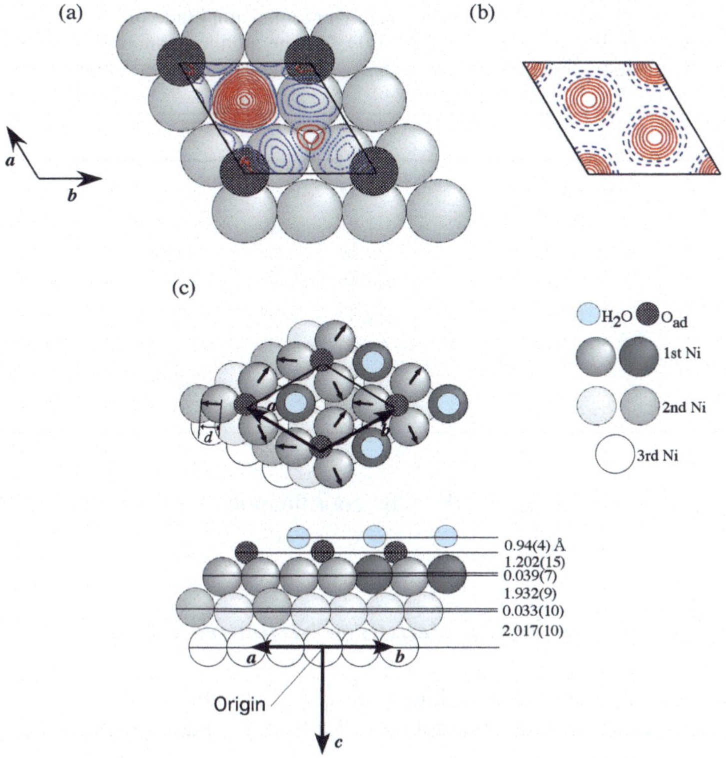

Fig. 6.5 (**a**) In-plane contour map plot of difference electron densities calculated by the difference Fourier synthesis method at the height of 2.0 Å above $p(2 \times 2)$–Ni(1 1 1)–O surface at 140 K. The *continuous lines* and *dashed lines* indicate positive and negative electron densities, drawn at every 0.05 eÅ^{-3}, respectively. (**b**) The Patterson function for the $p(2 \times 2)$ unit cell calculated from the measured in-plane structure factors at $L = 0.3$. The *solid lines* and *dashed lines* indicate positive and negative contour level, respectively. (**c**) The surface structure based on the optimum parameters for water adsorbed $p(2 \times 2)$–Ni(1 1 1)–O after annealing to 140 K. The *arrows* indicate the direction of atomic shift from the bulk position. Reproduced Fig. 1 with permission from Ref. [25] for (**a**) and (**c**). Copyright (2005) by the American Physical Society

method, then the residual electron densities $\Delta\rho$ originating from missing atoms can be calculated by the difference Fourier synthesis method as in the following equation

$$\Delta\rho(x, y, z) = \frac{1}{V}\sum_{HKL}\left(\left|F_{HKL}^{calc}\right| - \left|F_{HKL}^{exp}\right|\right)\cos\left[2\pi i(Hx + Ky + Lz)\right]. \qquad (6.9)$$

Figure 6.5(a) shows an in-plane contour map of difference electron densities at the height of 2.0 Å above $p(2 \times 2)$–Ni(1 1 1)–O. A large peak appeared at the top site of Ni atom. This peak corresponds to adsorbed water. The Patterson function also shows exactly the similar peak at threefold axis as shown in Fig. 6.5(b). Finally, all the atomic parameters with the occupancy factor and the Debye-Waller factor were

Table 6.1 Atomic positions and the mean-square vibrational amplitude $\langle u^2 \rangle$ for the optimum geometry of water adsorbed on $p(2 \times 2)$–Ni(1 1 1)–O. The atomic parameters are given in relative positions of the hexagonal coordinate system ($a = b = 4.984$ Å, $c = 6.104$ Å, $\alpha = \beta = 90°$, $\gamma = 120°$)

Atom	x	y	z	$\langle u^2 \rangle$/Å^2
O (water)	0.667	0.333	−1.028	0.08
Oad	0	0	−0.874	0.07
1st Ni(1)	0.171	0.830	−0.654	0.18
1st Ni(2)	0.667	0.333	−0.661	0.17
2nd Ni(1)	0.333	0.667	−0.338	0.19
2nd Ni(2)	0.334	0.170	−0.332	0.18
3rd Ni(1)	0	0	−0.002	0.12
3rd Ni(2)	0.500	0.500	0	0.12

determined as shown in Fig. 6.5(b). The coordinations for each atom are listed in Table 6.1.

6.2.6 Effect of Number of Measured Structure Factors

To estimate atomic parameters, a large number of structure factors is necessary: if possible, more than ten times the number of adjustable parameters. Generally, several tens of the points are measured in each rod. If the profile of diffraction rod is smooth, the interval of data points may be taken quite large. The measurement of more than a few rods is needed for full analysis of multilayer structure. Moreover, the measurement of the equivalent rods and reflections is needed to check the symmetry and systematic error. The lack of data points results in an increase of χ^2 and the standard deviation of each atomic parameter. When structural optimization is carried out for water on $p(2 \times 2)$–Ni(1 1 1)–O described in Sect. 6.2.5 using about the half number (70 points) of the experimental structure factors with twice the interval, χ^2 value increases from 1.89 to 2.12. The standard deviations also increase about $\sqrt{2}$ times.

6.3 Experimental Setup and Method

6.3.1 Beamline Components

In this section we will introduce beamline components for surface diffraction describing the beamline BL13XU as a typical example of an experimental setup for surface and interface structure studies which operates at SPring-8, a large synchrotron radiation facility in Japan. BL13XU includes an X-ray source, a front-end (FE) system, a beamline double monochromator, a pair of mirrors, and experimental facilities and instruments such as a diffractometer and ultrahigh vacuum (UHV)

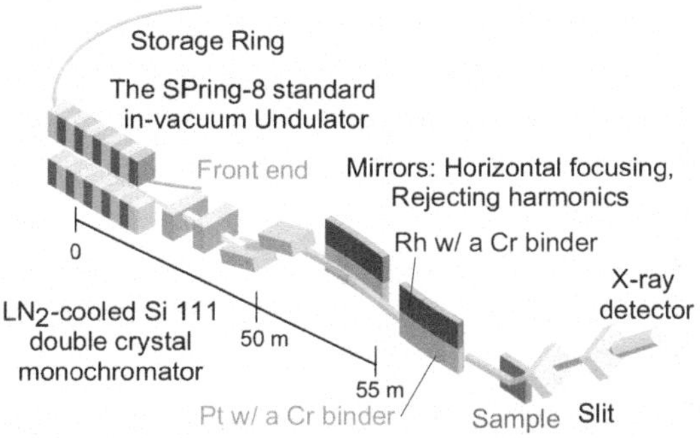

Fig. 6.6 Example of beamline layout. Main components are the SPring-8 standard undulator, which is an X-ray source, a front end, a Si double crystal monochromator, and mirrors. The monochromator and mirrors are located at an optical hutch. Researchers usually access their samples and a diffractometer in a experimental hutch. Reproduced Fig. 1 with permission from Ref. [27]. Copyright (2003) by World Scientific Publishing Co

chambers. Figure 6.6 illustrates a schematic layout of the beamline. The X-ray source is an undulator because it generates an intense beam which was strongly desired for detection of very weak diffraction peaks from surfaces and interfaces. In addition, it is suitable for diffraction measurements with an incident energy range between 10 and 20 keV. The X-ray source and many components of the front-end system are located inside the radiation shielding wall which surrounds the storage ring. An exit window at the front end, the monochromator, and mirrors are located in the optical hutch, while users perform experiments in an experimental hutch (EH).

6.3.2 X-Ray Source and Optics

BL13XU is an undulator beamline. The light source is the standard SPring-8 in-vacuum undulator (ID) [28] with a 32 mm period and 140 periods. The gap of the ID ranges from 50 mm to 9.6 mm. The fundamental energy range available is correspondingly from 18.9 to 5.5 keV. One can calculate photon flux, beam size, and beam divergence using SPECTRA [29]. The front end comprises an X-ray beam position monitor, a main beam shutter, an absorber, a graphite filter, and FE slits.

The monochromator and mirrors are installed in the optics hutch. The beamline double crystal monochromator with Si 111 reflection is cooled down with a liquid nitrogen chiller [30]. The energy resolution $\Delta E/E$ is about 10^{-4}. The stage for the second crystal is adjusted to a constant temperature using a water flow unit (de facto an heater). The two mirrors have two stripes of a rhodium (Rh) and a platinum (Pt) film with a Cr binder. The stripes are properly used depending on the incident photon energy. They reject higher harmonics of incident photons and focus the X-ray beam in a horizontal scattering geometry. The second mirror is bent for focusing the incident X-rays at the sample position.

It is interesting to estimate the performances of the system. We measured photon flux densities (Fig. 6.7) as a function of incident photon energy with a silicon pin

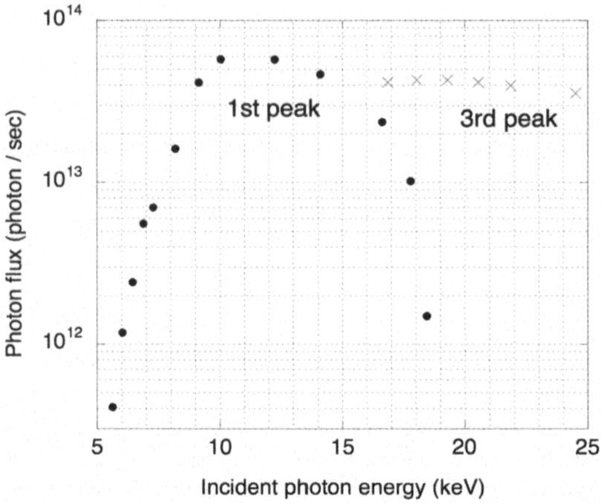

Fig. 6.7 Measurements of absolute X-ray photon flux densities. Front-end slit opening was $1 \times 0.8\ \text{mm}^2$. An undulator gap was set to get the maximum photon flux from the Si double crystal monochromator fixed for an incident photon energy. Plots were obtained by changing the combination of the gap and the angle of the monochromator. '1st' and '3rd' mean the fundamental light and third harmonics generated by the undulator. The data were taken without the mirrors. Reproduced Fig. 4 with permission from Ref. [27]. Copyright (2003) by World Scientific Publishing Co

photodiode at the entrance of EH. The data were corrected for the absorption of a 150-mm thick Be window of the photodiode and took into consideration the detection efficiency of the diode in Si 300 mm thickness. The X-ray beam passed through a 0.3 mm graphite filter and three Be windows with a total thickness of 0.75 mm in the front end. The undulator can operate with two different harmonics, 1st and 3rd harmonics, to produce the intense X-ray beam for GIXD experiments. Working in the 1st harmonic, the photon flux shows a steep increasing between 5 and 10 keV, it is constant up to 12 keV and then decreases slowly up to 17 keV. Working in the 3rd harmonic, the flux is almost constant between 17 and 25 keV and slightly less than the previous constant flux. The maximum estimated flux was 6×10^{13} photons/s with the front-end slit opening set at $1 \times 0.8\ \text{mm}^2$ and the undulator operating in the 1st harmonic around 10 to 12 keV.

6.3.3 Examples of Diffractometers

A number of UHV chambers mounted on a diffractometer have been constructed for surface structure studies at a handful of synchrotron radiation sites. They are classified into two types: a UHV chamber linked to a diffractometer [31–33] and a compact UHV chamber without surface analysis and growth tools (called a baby chamber) mounted on a diffractometer [34–36]. Our UHV system does not belong to either of the two categories. Our design is based on the idea that any of the two chambers at the user's choice can be mounted on the surface diffractometer for a

Table 6.2 Specifications of rotation axes for the surface diffractometer we used

Axis	Resolution/step	Stroke
ω_S	0.72°	±180°
α_S	0.36°	±180°
γ_S	0.36°	±180°
δ_S	0.36°	−10 to +190°

surface X-ray measurement. Chamber 1 is suitable for a structural study of a metal surface. Chambers 2 targets the determination of a semiconductor surface structure. Each UHV chamber is coupled to the surface diffractometer (3.2 × 3.2 × 2.3 m in dimension and 10 tons in weight) which is a high-precision and high-load 4-axis diffractometer with $\omega_S - \alpha_S$ goniometers for sample's orientation and $\gamma_S - \delta_S$ goniometers for detector's orientation. The sample manipulator is mounted on a base stage that has a linear translation table with three orthogonal axes for changing an in-plane position of the sample as well as a sample height and a tilting mechanism with cross swivels. The sample is rotated around a horizontal axis ω_S, which in turn rotates about a vertical axis α_S. (Note that the ω_S axis is attached to a UHV chamber. The chamber also rotates with α_S.) The detector is rotated around a horizontal axis δ_S, which in turn rotates about a vertical axis γ_S. This is why the diffractometer is classified as the S2+D2 type. The sphere of confusion of all four rotation axes is kept within 55 μm. The specifications for each axis are listed in Table 6.2.

The UHV system is composed of one of the two UHV chambers and the diffractometer (Fig. 6.8). Each UHV chamber has two 0.8-mm-thick Be windows for an incident X-ray beam and for scattered/diffracted X-rays. The Be windows for the scattered X-rays are cylindrical and cover angular ranges of 35° and 125° around a perpendicular axis and a horizontal axis for chamber 1, while they cover angular ranges of 53° and 125° for chambers 2. Sample temperature is varied in the ranges 20–2000 K and 190–930 K for chambers 1 and 2, respectively. Chamber 1 is equipped with LEED/Auger electron spectroscopy optics, a sputtering gun, a gas doser, a quadruple mass spectrometer, an electron bombardment heater, and a closed-cycle cryogenic refrigerator for sample cooling. Reflection high-energy electron diffraction (RHEED) optics, a quartz thickness monitor, Knudsen cells, a residual gas analyzer, and a pyrolytic graphite heater covered with pyrolytic boron nitride are available in chamber 2. A feature of chambers 1 and 2 is that the sample surface normal is parallel to the ω_S axis (Z sample geometry).

For chamber 1, which has no introduction chamber, the sample is attached to the top of a sample manipulator with a metal wire or spot welder in air, then the manipulator is mounted in chamber 1 and then the whole chamber is evacuated. On the other hand, chamber 2 has a sample introduction chamber. The sample is mounted on a Mo cap which has screws on its side surface and a bore on its top plate. The sample-mounted cap is put at the end of a magnetic transfer rod in the introduction chamber. After pumping the chamber, the cap is transferred to the top of the sample manipulator. Each chamber is evacuated by the combination of a 300 L/s turbomolecular pump and a 300 L/min scroll pump along with 400 L/s ion sputtering

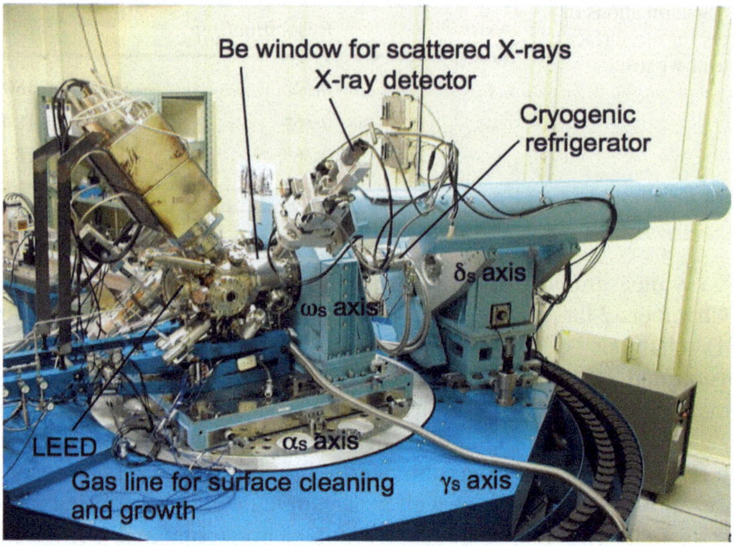

Fig. 6.8 UHV chamber 1 mounted on UHV diffractometer. A sample is put at the top of the sample manipulator parallel to the ω_S axis equipped with the UHV chamber. The sample surface is perpendicular to the horizontal plane. The chamber is rotated around a vertical α_S axis. An X-ray detector is rotated independently around a horizontal δ_S axis and a vertical γ_S axis. A surface system is made with an evaporation method after cleaning the substrate surface using a combination procedure of Ar^+ sputtering and annealing. In-situ surface X-ray measurements are performed. The photograph is viewed from a downstream perspective. The diffractometer is about 3 m in diameter. Reproduced Fig. 5 with permission from Ref. [27]. Copyright (2003) by World Scientific Publishing Co

pump plus an 800 L/s titanium sublimation pump. A sample manipulator with the ω_S axis is linked to its UHV chamber through welded bellows. The pressure gradient is maintained with a differentially pumped rotary feedthrough while the sample is rotated.

The film diffractometer is designed for structural analysis of thin films and liquid/solid and solid/solid interfaces. The instrument can be operated in either the standard four-circle or five-circle geometry. The four-circle geometry features 3 degrees of freedom (DOFs), α_F, χ_F, and ω_F, for rotating the sample and one DOF, δ_F, for rotating the X-ray detector. Axes α_F, χ_F, and ω_F, from the sample side, are mounted on the goniometer. Axes ω_F in a horizontal plane and χ_F meet at a right angle; in addition, axis χ_F is parallel to the incident X-rays for $\omega_F = 0$. Axis α_F is for changing the azimuthal angle of the sample normal. Rotation χ_F brings $\boldsymbol{h}$ into the vertical plane including the incident beam, while rotation ω_F makes an angle between the incident beam and the lattice planes investigated. All axes intersect in the center of the diffractometer within the sphere of confusion as small as 30 μm. The resolution per step of axes are 2.5×10^{-5}, 0.0002, 0.00048° for ω_F, δ_F, χ_F, and α_F, respectively. The high resolution of ω_F is required for a special use. A translation table in the horizontal and rotation axes can be attached on the detector arm; thus resulting in a five-circle geometry with 2 DOFs on the detector. The diffractometer is mounted on a stage with rotation axis α along the vertical and two translation tables X and Y. A Soller slit with an acceptance angle of 0.4° can be mounted in either geometry to improve an X-ray signal/noise ratio.

A point detector has been utilized for many surface diffraction; accordingly Fig. 6.1 is drawn for the cases. A slit put just in front of the detector enables decrease in background intensities scattered from undesirable directions and control of an angular resolution. An area detector is also available for surface X-ray diffraction measurements. For both setups (point detector or area detector), the sample has to be usually rotated however the setups with an area detector bring advantages. In particular, the PILATUS pixel detector [37] developed at Swiss Light Source is one of the promising detectors for surface X-ray diffraction. Schlepütz et al. [38] concluded that the use of the pixel detector in grazing incidence and surface diffraction has greatly improved data acquisition speed and reliability by approximately an order of magnitude.

6.4 Applications

6.4.1 Interface Atomic Structures: Electric Double Layer at the Solid Electrode

X-ray diffraction is one of the few surface science techniques that permits atomic level monitoring in electrochemical condition. It is possible to elucidate the detailed atomic structure of adsorbate and the electrode surface at an applied potential [39–41]. Typical electrochemical cells are shown in Fig. 6.9. In a thin layer cell, the electrode surface is covered by a thin film and is the conventional type for electrochemical measurement. To enhance mass transfer above the electrode surface, a drop cell is used for time resolved measurements and an electrochemical monitoring during an X-ray measurement.

At the microscopic level, ionic layers are formed above the electrode surface and form the so called electrical double layer. Models for the electrical double layer were proposed in electrochemistry [42, 43]. Specifically adsorbed ions form planar layer which is called the inner Helmholtz plane (IHP), whereas non-specifically adsorbed ions closer to the electrode surface form the outer Helmholtz plane (OHP). In an electrochemical reaction, a reactant must cross the electrical double layer to reach the electrode surface therefore the structural and physical properties of the electrical double layer are important for fundamental studies in interface science and electrochemical applications such as the development of high-performance capacitors and fuel cells. X-ray diffraction is a powerful tool to investigate the electrical double layer as well as adlayer. Recently, the structure of the electrical double layer was determined using surface X-ray diffraction [44].

Figure 6.10 shows CTR measured for 0.1 M CsBr and LiBr on Ag(1 0 0) at −0.4 V vs Ag/AgCl reference electrode. Br anion forms stable $c(2 \times 2)$ layers on

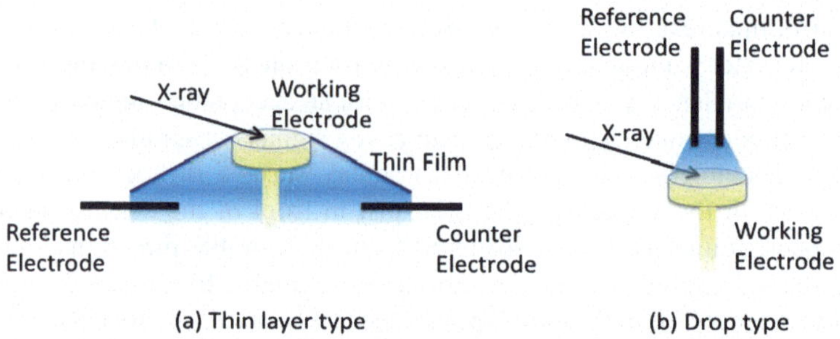

Fig. 6.9 Electrochemical cells for GIXD. (**a**) Thin layer type. Polypropylene or polyester thin film (a few μm thickness) is used. (**b**) Drop type. The reference and counter electrodes are immersed in the droplet of electrolyte on the electrode surface

Ag(1 0 0) (IHP). Integrated intensities were obtained using the film diffractometer, which were corrected for Lorentz factor $[1/(\sin 2\omega_F)]$ [45]. A body-centered tetragonal (bct) coordinate system was utilized to describe the reciprocal vector for the surface X-ray diffraction measurement, $Q = Ha^* + Kb^* + Lc^*$. Here $a^* = b^* = 2\pi/a$, $c^* = \sqrt{2}\pi/a$, $a = 2.889$ Å and L is along the direction normal to the surface. The dips at $L = 1.2$ and 3.2 of the specular CTR are due to interference between the half coverage of Br and the Ag surface. The detailed profiles around the dip in LiBr obviously differ from those in CsBr. The asymmetric dip in CsBr demonstrates the localization of electron-dense species, namely Cs^+, in the electrical double layer.

Fractional order (1/2 1/2) rod (FOR) in CsBr differs from that in LiBr, as shown in Fig. 6.10. In LiBr, the structure factor decreases monotonously along L with slight fluctuations, whereas it remarkably oscillates in CsBr due to the interference between Br and OHP Cs. This result shows that Cs in OHP is located at a specific in-plane position and the in-plane structure of OHP is affected by the Br structure. A $p(1 \times 1)$–Br model, which assumed p4mm space group, was used for the structural analysis. Lateral positional parameters of each atom are fixed due to symmetry restrictions. The origins of the in-plane coordinates and the out-of-plane axis are taken to be the top of adsorbed Br and the second Ag layer, respectively. The atoms below the third layer were fixed as a semi-infinite bulk crystal. Each substrate layer includes two Ag atoms in the $p(1 \times 1)$–Br unit cell. Three structural models were considered for the Cs position:

(i) the atop position of Br,
(ii) the bridged position,
(iii) the hollow position.

The out-of-plane and in-plane structures were refined using the CTR and FOR data, respectively. Model (iii) is in good agreement with GIXD data obtained with a χ^2 value of 1.24. Models (i) and (ii) gave χ^2 values of 2.38 and 2.43, respectively. Cs^+ in the OHP is located in the hollow site of Br with a vertical distance of 2.74(10) Å from the Br adlayer. Hydrated Cs^+ cation is accommodated in the hollow site of

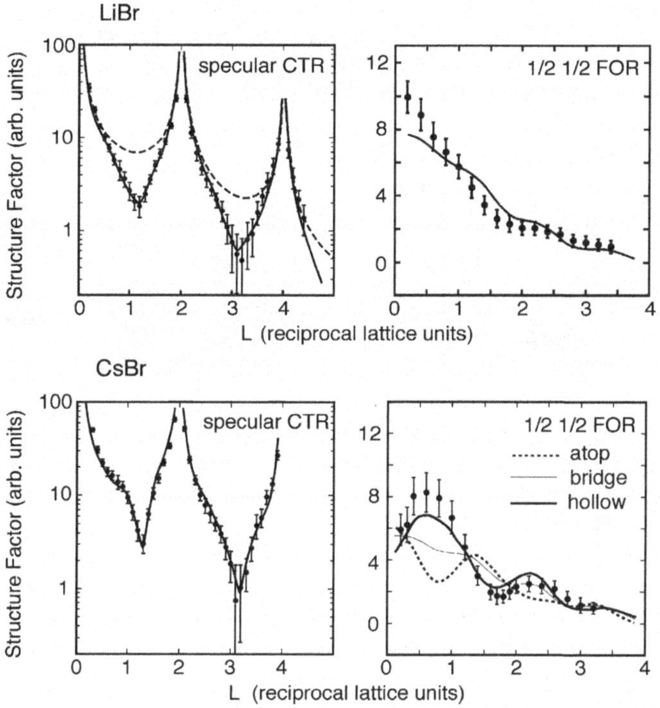

Fig. 6.10 Specular CTR and (1/2 1/2) FOR of Ag(1 0 0) in 0.1 M LiBr + 0.05 M LiOH and 0.1 M CsBr + 0.05 M CsOH at −0.4 V vs Ag/AgCl. The *solid lines* are structure factors calculated from an optimized model. The *dashed lines* in LiBr are structure factors calculated from the bulk terminated Ag(1 0 0) surface. The *dotted* and *thin lines* in CsBr are structure factors calculated from Cs atop and Cs bridge models, respectively. Reproduced Fig. 4 with permission from Ref. [44]. Copyright (2011) by Wiley

adsorbed Br layer as shown in Fig. 6.11. The coverage of Cs^+ (θ_{Cs}) per Ag topmost atom is 0.10(1), which was estimated from the occupancy factor. Cs^+ randomly occupy 20 % of the hollow sites of the Br layer. The interlayer spacing between the OHP Cs^+ and adsorbed Br increases with the increase of the potential, whereas θ_{Cs} decreases. The Cs^+ layer moves upward from the surface with the increase of the potential due to the repulsive electrostatic forces. The coordinates for respective atoms are listed in Table 6.3.

6.4.2 Quick Evaluation of Thin Film Structure: Identification of Crystal Symmetry of an Epitaxial Thin Film

The conventional surface diffraction is a very useful tool to obtain accurate structure factors of a crystallographic surface: it gives not only atomic positions in a unit

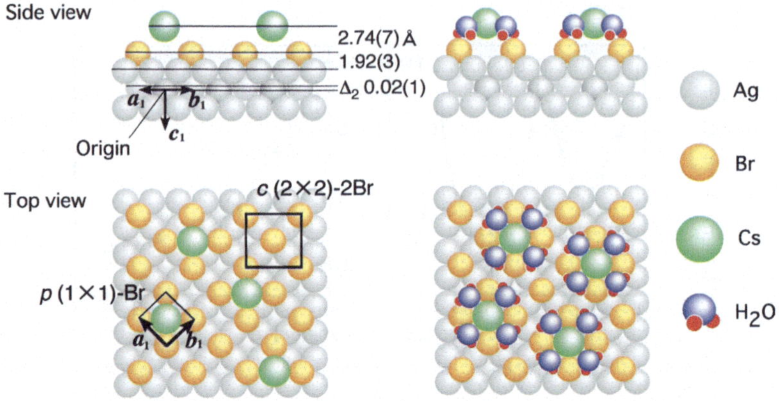

Fig. 6.11 Schematic model of Cs^+ in the OHP layer above $c(2 \times 2)$–Br on Ag(1 0 0) at −0.4 V determined using GIXD. The *square* indicates the $c(2 \times 2)$ unit cell. Hydration structure was determined using density functional theory (DFT) calculations. Reproduced Fig. 5 with permission from Ref. [44]. Copyright (2011) by Wiley

Table 6.3 Atomic positions and the mean-square vibrational amplitude $\langle u^2 \rangle$ for the optimized model of the Ag(1 0 0) electrode in CsBr at −0.4 V vs Ag/AgCl. The atomic parameters are given in relative positions of the body-centered cubic coordinate system ($a_1 = b_1 = c_1 = 4.086$ Å, $\alpha = \beta = \gamma = 90°$)

Atom	x	y	z	$\langle u^2 \rangle / \text{Å}^2$
Cs	0.5	0.5	−1.644	5.48
Br	0	0	−0.973	1.49
1st Ag(1)	0.5	0	−0.503	2.07
1st Ag(2)	0	0.5	−0.503	2.07
2nd Ag(1)	0	0	−0.006	1.37
2nd Ag(2)	0.5	0.5	0	1.37

cell but also structural information such as a crystallographic coordinate system and the space group. To achieve this, it is necessary to measure at least several 10 independent Bragg diffraction points and this means that the sample and the detector are adjusted to their appropriate angular positions, then the sample is rocked around the position of Bragg diffraction to collect the diffraction intensity for each point. Such a precise X-ray measurement is lengthy and slightly troublesome. Thus, an easier measurement method would be desired to get an overall image of the diffraction intensity profile.

To this aim, the three following methods use a sample and an area detector in angular fixed positions. The first used high-energy X-ray diffraction with a transmission-reflection scheme [46] for deeply buried interfaces and uses the transmission geometry for characterizing a composition-spread thin film [47]. An Ewald sphere for higher-energy X-rays is larger. Accordingly, the high-energy X-ray

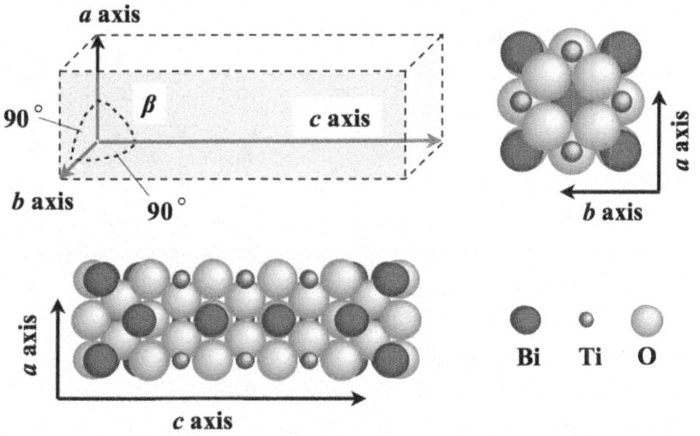

Fig. 6.12 Schematic structure of a $Bi_4Ti_3O_{12}$ crystal. $\beta \neq 90°$ for monoclinic with a $B1a1$ space group and $\beta = 90°$ orthorhombic with a $B2cb$ space group

diffraction carries the advantage that angular regions for mapping the reciprocal space are smaller. The second method for quick characterization is dispersive CTR measurements proposed by Matsushita et al. [48–50]. The third uses a grazing incidence geometry and is similar to RHEED; however, RHEED is not applicable for structure determination of a nanostructure that exists in air, interface, or solution. On the other hand, measurements in the X-ray reflection geometry were performed for characterization in UHV [51, 52], air/solid [53], solid/solid [54],[3] and liquid/solid interfaces [55]. All the methods were possible because highly brilliant synchrotron X-rays were available.

The method was also applied to identification of crystal symmetry of $Bi_4Ti_3O_{12}$ (BIT) epitaxial films [56]. BIT films with (1 0 0)/(0 1 0) orientations (i.e. *a* and *b* domains) were grown on TiO_2 (1 0 1) single crystals by metal-organic chemical vapor deposition and had a *c*-axis in-plane alignment. It was disputed that the BIT bulk structure was either monoclinic with a $B1a1$ space group or orthorhombic with a $B2cb$ space group. The schematic structure is shown in Fig. 6.12. Shrinagar et al. [57] have determined the total energy of the three structural models experimentally proposed in [58, 59], and [60] as well as two more optimized structures. They concluded that the bulk $B1a1$ structure is the most stable structure. However, there has been little investigation into the crystal symmetry of the BIT films.

Figure 6.13(a) shows the X-ray reciprocal-space maps of a 50-nm-thick BIT film using GIXD. X-rays were incident on the sample surface at an incident angle $\phi_0 = 0.2°$ and were almost parallel to the [0 1 0] direction of the TiO_2 substrate. A photon energy E_p used was 25 keV and a two-dimensional (200 × 250 mm) detector, which was a cylindrically shaped imaging plate, was used 133 mm downstream from the sample. About 120 spots from the sample at the fixed angular position were recorded within an angular range of ±50° and 50° in the horizontal and vertical directions, respectively. Figure 6.13(b) shows intersections of H-, K-, and L-constant planes with the Ewald sphere as well as the specific projected lines on the imaging plate. Here HKL is a diffraction index; in addition, we use KLH_a for

[3]GIXD patterns clearly showed the two-by periodicity of one-dimensional Bi nano lines buried in a Si (0 0 1) crystal.

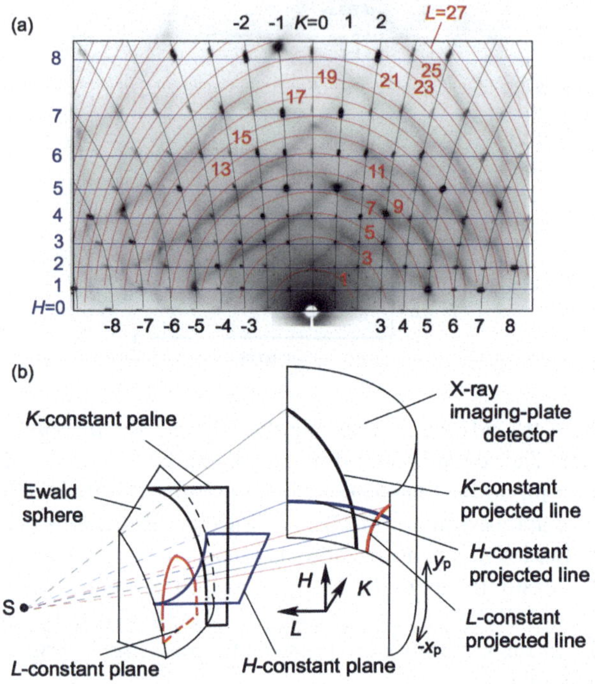

Fig. 6.13 (**a**) X-ray reciprocal-space maps of a 50 nm-thick $Bi_4Ti_3O_{12}$ film on a TiO_2 (1 0 1) single crystal substrate recorded with the sample at a fixed angular position with an exposure time of 4 min using a cylindrical X-ray imaging plate located 133 mm from the sample. The energy of the incident photon was 25 keV. The grazing incident X-rays were almost parallel to the [0 1 0] direction of the TiO_2 substrate. The incident angle ϕ_0 for the sample surface was 0.2°. Reproduced Fig. 5(a) with permission from Ref. [56]. Copyright (2011) by the International Union of Crystallography. (**b**) Calculated H-, K-, and L-constant projected lines on the imaging plate. Intersections of H-, K-, and L-constant planes with the Ewald sphere and the beginning of the incident wave vector, *point* S, define the planes. H-, K-, and L-constant projected lines are intersections between the defined planes and the imaging-plate detector. Lattice parameters used are $a = 5.436$, $b = 5.407$, and $c = 32.98$ Å. In addition, we assume $\beta = 90°$ for simplicity. The depiction here is for the a domains. Reproduced Fig. 4 with permission from Ref. [56]. The calculated projected lines are superposed on the observed spots

the a domains and HLK_b for the b domains. The projected lines are helpful to index the observed diffraction spots. The lines are superposed on Fig. 6.13(a) for the a domains. We used lattice parameters of $a = 5.436$, $b = 5.407$, and $c = 32.98$ Å when we calculated the lines. Positions (x_p, y_p) on the imaging detector is a function of diffraction index HKL.

$$x_p = -D \arctan \frac{q_x^L}{q_y^L + E_p/12.4}, \tag{6.10}$$

$$y_p = \frac{D q_z^L}{(q_x^L)^2 + (q_y^L + E_p/12.4)^2}. \tag{6.11}$$

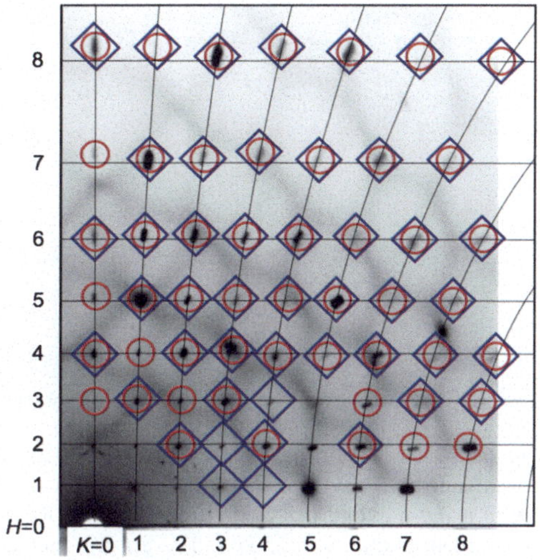

Fig. 6.14 Contributions of the a domains (*blue* ◇) and the b domains (*red* ○) on the assumption of the $B1a1$ monoclinic structure. Some diffraction spots cannot be explained using the reflection conditions of $B1a1$. Reproduced Fig. 6 with permission from Ref. [56]. Copyright (2011) by the International Union of Crystallography

Here $q_x^L = q_x^s \cos\alpha_a - q_y^s \sin\alpha_a$, $q_y^L = (q_x^s \sin\alpha_a + q_y^s \cos\alpha_a)\cos\phi_0 - q_z^s \sin\phi_0$. For the a domains, $q_x^s = -K/b$, $q_y^s = -L/c$, $q_z^s = H/(a \sin\beta)$. For the b domains, $q_x^s = H/(a \sin\beta)$, $q_y^s = -L/c$, $q_z^s = K/b$. D is the camera length. Azimuthal angle α_a is defined as 0 when incident X-rays are parallel to $\boldsymbol{c}$ of the BIT film crystal as well as the [0 1 0] direction of the TiO_2 (1 0 1) substrate and for incident angle ϕ_0 of the sample surface equal to 0.

For the a domains, the H value is decimal, whereas the K and L values are integers. This is because the reciprocal-lattice form of the thin film with the a domains elongates along the H direction from a KLH_0 Bragg point. Here H_0 is the closest integer to the H value. We tried to index all the observed spots using the $B1a1$ monoclinic structure. Figure 6.14 shows contributions of the a and b domains for the right half of Fig. 6.13(a). In general, $B1a1$ is less symmetrical than the $B2cb$ orthorhombic structure; however reflection conditions of $B1a1$ are the same as those of $B2cb$ in this region.

Most of the recorded spots can be explained using the reflection conditions of $B1a1$, while some spots are classified into forbidden reflections.[4] This means two possibilities: (1) the symmetry of the film crystal is lower than $B1a1$, or (2) the $B1a1$ structure and a structure with a lower symmetry coexist. While an intensity diffracted from an ideally sheet crystal would be homogeneous as a function of its sheet normal, an observed diffracted intensity we here discuss is the convolution of a Bragg intensity profile of the same bulk as a sample film and an external form factor, in particular the thickness, that is the Fourier transform of the external form of the film. Therefore, a diffracted intensity near a forbidden reflection from the film should be much weaker than that near an allowed reflection.

[4]While a sample crystal satisfies the Bragg condition for a forbidden reflection, diffraction does not occur because its structure factor is zero.

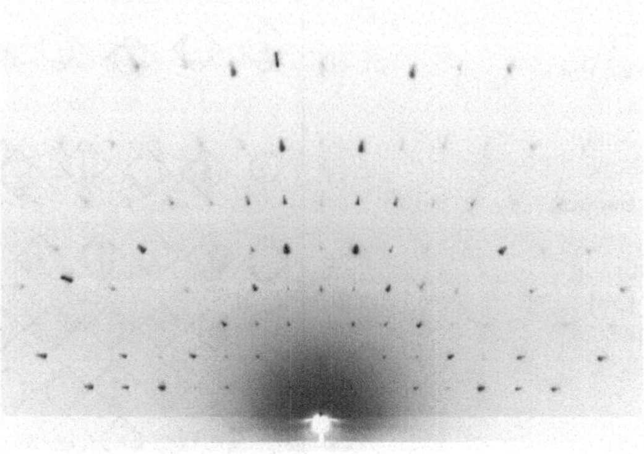

Fig. 6.15 X-ray reciprocal-space maps of a 50 nm-thick $Bi_4Ti_3O_{12}$ film on a $TiO_2(1\ 0\ 1)$ single crystal substrate recorded with the sample at a fixed angular position with an exposure time of 4 min using a cylindrical X-ray imaging plate located 133 mm from the sample. The energy of the incident photon was 25 keV. The grazing incident X-rays were almost parallel to the [0 1 0] direction of the TiO_2 substrate and ***c*** of the BIT film. The incident angle ϕ_0 for the sample surface was 0.1°. The background pattern seen in Fig. 6.13(a) is invisible. Reproduced Fig. 3(b) with permission from Ref. [56]. Copyright (2011) by the International Union of Crystallography

Grazing incidence allows us to obtain information on a depth structural profile if a sample surface is enough flat. The macroscopic curvature of the surface is typically desired to be smaller than 0.1°. The X-ray penetration depth is $\Lambda \equiv (1/\mu)\sin\phi_0$ for an incident angle larger than the critical angle ϕ_c for total external reflection of the sample. Here μ is the linear absorption coefficient.[5] The penetration depth is smaller than Λ for $\phi_0 < \phi_c$ since the total external reflection occurs [61]. The calculated penetration depth of BIT for 25 keV X-rays are 3.2 and 96 nm for incident angles ϕ_0's of 0.1 and 0.2°, respectively, when one assumes the density of 8.05 g cm^{-3}. Actually, incident X-rays for $\phi_0 = 0.2°$ reached the TiO_2 substrate and this is why the background pattern arose from the diffuse scattering from the substrate (Fig. 6.13(a)). On the other hand, Fig. 6.15 does not show such a background pattern because the X-rays at an incident angle ϕ_0 of 0.1° did not reach the substrate. This is evidence that the grazing incidence geometry is useful to investigate a structure along the depth direction.

The "obvious-at-a-glance" technique that we introduce here can be available to observe a structural change of a crystalline nanomaterial induced by an external field and to a proper sample. Moreover, such a reciprocal map will be helpful in the choice of an initial structural model in accurate structural analysis.

[5]When an X-ray beam penetrates through a material, the intensity decreases with distance x in proportion to $\exp(-\mu x)$.

6.5 Conclusions

In this chapter we have described the grazing incident X-ray diffraction which is a well established technique for the investigation of surfaces and interfaces. We have presented the principles of the technique in the kinematical approximation and introduced the concept of crystal truncation rods and structure factors. The type of measurements have been described as well as the procedure to analyze the data exemplifying also on the use of the Patterson function and difference Fourier synthesis.

We have presented measurements performed by using a single point detector as well as an area detector. The latter detector reduces the measurement time that might be very long if a large part of the reciprocal space has to be investigated. Of course this is beneficial for sensitive samples that could be damaged to prolonged irradiation with X-rays, but, on the other hand, faster detection along with an increase of the photon flux will open the possibility to time resolved diffraction experiments.

References

1. W.C. Marra, P. Eisenberger, A.Y. Cho, J. Appl. Phys. **50**, 6927 (1979)
2. P. Eisenberger, W.C. Marra, Phys. Rev. Lett. **46**, 1081 (1981)
3. I.K. Robinson, Phys. Rev. Lett. **50**, 1145 (1983)
4. J. Bohr, R. Feidenhans'l, M. Nielsen, M. Toney, R.L. Johnson, I.K. Robinson, Phys. Rev. Lett. **54**, 1275 (1985)
5. I.K. Robinson, Phys. Rev. B **33**, 3830 (1986)
6. R. Feidenhans'l, Surf. Sci. Rep. **10**, 105 (1989)
7. G.H. Vineyard, Phys. Rev. B **26**, 4146 (1982)
8. S. Kishino, K. Kohra, Jpn. J. Appl. Phys. **10**, 551 (1971)
9. A.M. Afanas'ev, M.K. Melkonyan, Acta Crystallogr. A **39**, 207 (1983)
10. P.L. Cowan, Phys. Rev. B **32**, 5437 (1985)
11. O. Sakata, H. Hashizume, Rep. Res. Lab. Eng. Mat. Tokyo Inst. Tech. **12**, 45 (1987)
12. O. Sakata, H. Hashizume, Acta Crystallogr. A **53**, 781 (1997)
13. B.E. Warren, *X-ray Diffraction* (Dover, New York, 1990), pp. 35–38
14. K.N. Trueblood et al., Acta Crystallogr. A **52**, 770 (1996)
15. I.K. Robinson, in *Handbook on Synchrotron Radiation*, vol. 3, ed. by G. Brown, D.E. Moncton (North-Holland, Amsterdam, 1991), p. 221
16. I.K. Robinson, D.J. Tweet, Rep. Prog. Phys. **55**, 599 (1992)
17. For example, J. Rius et al., Acta Crystallogr. A **52**, 634 (1996)
18. X. Torrelles et al., Phys. Rev. B **57**, R4281 (1998)
19. L.D. Marks et al., Surf. Rev. Lett. **4**, 1 (1997)
20. K. Sumitani et al., Jpn. J. Appl. Phys. **42**, L189 (2003)
21. D.K. Saldin et al., Comput. Phys. Commun. **137**, 12 (2001)
22. E. Vlieg, J. Appl. Crystallogr. **30**, 532 (1997)
23. E. Vlieg, J. Appl. Crystallogr. **31**, 198 (1998)
24. E. Vlieg, J. Appl. Crystallogr. **33**, 401 (2000)
25. M. Nakamura, M. Ito, Phys. Rev. Lett. **94**, 035501 (2005)
26. K.W. Evans-Lutterodt, M.-T. Tang, J. Appl. Crystallogr. **28**, 318 (1995)
27. O. Sakata et al., Surf. Rev. Lett. **10**, 543 (2003)
28. H. Kitamura, J. Synchrotron Radiat. **7**, 121 (2000)

29. T. Tanaka, H. Kitamura, J. Synchrotron Radiat. **8**, 1221 (2001)
30. T. Mochizuki et al., Nucl. Instrum. Methods A **467–468**, 647 (2001)
31. For example, P. Eisenberger, W.C. Marra, Phys. Rev. Lett. **46**, 1081 (1981)
32. S. Brennan, P. Eisenberger, Nucl. Instrum. Methods **222**, 164 (1984)
33. P.H. Fuoss, I.K. Robinson, Nucl. Instrum. Methods **222**, 171 (1984)
34. For example, J. Bohr et al., Phys. Rev. Lett. **54**, 1275 (1985)
35. O. Sakata, H. Hashizume, Acta Crystallogr. A **51**, 375 (1995)
36. O. Sakata et al., J. Synchrotron Radiat. **5**, 1222 (1998)
37. Ch. Brönnimann et al., Nucl. Instrum. Methods Phys. Res. A **510**, 24 (2003)
38. C.M. Schlepütz et al., Acta Crystallogr. A **61**, 418 (2005)
39. B.M. Ocko, J. Wang, A.J. Davenport, H.S. Isaacs, Phys. Rev. Lett. **65**, 1466 (1990)
40. J. Wang, A.J. Davenport, H.S. Isaacs, B.M. Ocko, Science **255**, 1416 (1992)
41. C.A. Lucas, N.M. Markovic, P.N. Ross, Phys. Rev. Lett. **77**, 4922 (1996)
42. J.O. Bockris, M.A.V. Devanathan, K. Muller, Proc. R. Soc. Lond. Ser. A, Math. Phys. Sci. **274**, 55 (1963)
43. D.C. Grahame, Chem. Rev. **41**, 441 (1947)
44. M. Nakamura, N. Sato, N. Hoshi, O. Sakata, ChemPhysChem **12**, 1430 (2011)
45. M.S. Altman, P.J. Estrup, I.K. Robinson, Phys. Rev. B **38**, 5211 (1988)
46. H. Reichert et al., Phys. B **336**, 46 (2003)
47. D.A. Kukuruznyak et al., Appl. Phys. Lett. **91**, 071916 (2007)
48. T. Matsushita et al., J. Phys. Conf. Ser. **83**, 012021 (2007)
49. T. Matsushita et al., Appl. Phys. Lett. **92**, 024103 (2008)
50. T. Matsushita et al., Eur. Phys. J. Spec. Top. **167**, 113 (2009)
51. H. Hong et al., Rev. Sci. Instrum. **71**, 3132 (2000)
52. H. Hong et al., Rev. Sci. Instrum. **73**, 1720 (2002)
53. O. Sakata et al., Appl. Phys. Lett. **84**, 4239 (2004)
54. O. Sakata et al., Phys. Rev. B **72**, 121407(R) (2005)
55. O. Sakata, M. Nakamura, Appl. Surf. Sci. **256**, 1144 (2009)
56. O. Sakata, T. Watanabe, H. Funakubo, J. Appl. Crystallogr. **44**, 385 (2011)
57. A. Shrinagar, A. Garg, R. Prasad, S. Auluck, Acta Crystallogr. A **64**, 368 (2008)
58. J.F. Dorrian, R. Newnham, D. Smith, M. Kay, Ferroelectrics **3**, 17 (1972)
59. A.D. Rae, J.G. Thompson, R.L. Withers, A.C. Wililis, Acta Crystallogr. B **46**, 474 (1990)
60. C.H. Hervoches, P. Lightfoot, Chem. Mater. **11**, 3359 (1999)
61. L.G. Paratt, Phys. Rev. **95**, 359 (1954)

Chapter 7
X-Ray Reflectivity

A. Gibaud, M.S. Chebil, and T. Beuvier

Abstract The basic principles of X-ray specular reflectivity are presented starting with the definition of the index of refraction of a material for X-rays. Applying the boundary conditions to the electric field and its derivative, one can express the reflectivity of a surface with respect to the incident angle and the index of refraction. This naturally yields the so-called Fresnel reflectivity valid for flat surfaces. The reflectivity of more complex structures such as thin layers deposited on a substrate or multilayers can be handled by the matrix technique that is then developed. The influence of interface roughness is described. When some approximations concerning refraction and multiple reflections at interfaces are made, one can extract from the so-called kinematical Born approximation a master formula which can be used for low electron density films. Experimental conditions are then discussed before finishing this chapter by a presentation of X-ray reflectivity by selected samples.

7.1 Introduction

XRR has been used for almost one century to probe the electron density profile of materials. To understand how XRR can be calculated, it is necessary to address how an electromagnetic wave interacts with matter. It is well known that this interaction is mostly governed by the knowledge of the index of refraction of the reflecting material for X-rays. Historically, the reflection coefficient for any type of electromagnetic waves was first derived by Fresnel [1]. Then, M. Von Laue in his Nobel Prize delivery speech argued in 1914 that the refractive index of materials for X-rays should be close to unity [2]. He namely quoted that "*From the fact that there was no evidence of refraction of the X-rays, it was possible to assume that, should they*

A. Gibaud (✉) · M.S. Chebil · T. Beuvier
LUNAM Université, IMMM, Faculté de Sciences, Université du Maine, UMR 6283 CNRS, Le Mans Cedex 9, 72085, France
e-mail: Alain.Gibaud@univ-lemans.fr

M.S. Chebil
e-mail: Mohamed.Chebil@univ-lemans.fr

T. Beuvier
e-mail: Thomas.Beuvier@univ-lemans.fr

G. Bracco, B. Holst (eds.), *Surface Science Techniques*,
Springer Series in Surface Sciences 51, DOI 10.1007/978-3-642-34243-1_7,

consist of a transverse wave motion- as is the case with light waves—the relevant wavelength would have to be very small, as for very small wavelengths, according to the theory of dispersion of light, the refractive index would approach unity". It seems that Compton [3] was the first to have foreseen in 1923 the possibility of totally reflecting X-rays and that Forster [4] introduced the equation giving the reflectivity of a substrate (see Eq. (7.6)). Prins [5] carried out some experiments to illustrate the predictions of this equation in 1928, using an iron mirror. He also used different anode targets to study the influence of the X-ray wavelength on the absorption. Kiessig also made similar experiments in 1931 [6] using on a nickel mirror. An account of the historical development of the subject can be found in the pioneering works of Abeles in 1950 [7] and of L.G. Parrat [8] in 1954 and the fundamental principles are discussed in the textbook by James [9]. Since then, seminal works of Névot and Croce [10], Sinha et al. [11] have complemented the bases of this technique by including the effect of surface roughness. Several interesting papers are devoted to this technique and we convey the reader to the nonexhaustive following list [12–15]. So far it is the only technique which can bring quantitative information on the electron density profile in a nondestructive manner of materials with buried interfaces. For this reason XRR has become an invaluable tool to study the structure and the organization of materials which are grown in thin films at the submicron and atomic scales. In material research on thin films, the trend is to design solid films of increasing complexity having specific properties for technical applications. The nature of the materials deposited on substrates and the techniques of deposition for such applications are extremely variable. The design of semiconductor and metallic heterostructures is well mastered by molecular beam epitaxy deposition which generally provides extremely well-crystallized materials. This technique which is very expensive is used in general for making specific materials such as quantum wells and artificial superlattices [16]. Less expensive techniques are now developed and beautiful examples of supramolecular structures can now be achieved by assembling molecules of different kinds at the surface of a substrate [17]. Well-organized structures made in such a way can provide electronic analogues, electrochromic or non-linear optical elements. Such multilayer architectures can be achieved by the LB (Langmuir-Blodgett) method and by self-assembly of the layers through covalent bonds or metal coordination. For industrial applications the sputtering technique is also widely used for coating metallic films, for making oxide thin films and for creating heterogeneous materials like cermets (ceramic metals) [18]. As can be seen there is a wealth of ways to create complex mesoscopic layered structures which in turn will be considered as interesting if their structure appears to be as perfect as possible. The perfection of mesoscopic layered super-structures is defined both by the quality of the interfaces and by the reproducibility with which one can achieve the deposition of the layers (control of the thickness, crystallinity, voids or various defects which may appear during the growth process) [19]. In particular, the roughness of the interfaces is of crucial importance for many technological applications and it is a parameter which must be determined to appreciate the quality of the interfaces. It is the main reason for which XRR is now undisputed to characterize thin films of thickness less than 200 nm (roughly speaking). Above this value, measurements are limited by the instrumental resolution. A specific paper has been devoted to a round robin test of XRR analysis that includes worldwide scientists [20]. In this chapter, we present the technique of specular X-ray reflectivity and show through various examples how it can be used to determine the electron density profile (EDP) and the roughness of the interfaces.

7.2 The Basic Principles of X-Ray Reflectivity

7.2.1 The Index of Refraction

X-ray radiation are part of the broad spectrum of electromagnetic waves. X-rays can be produced by the acceleration or deceleration of electrons either in vacuum (synchrotron) or in metallic targets (tubes). The most widely used X-ray radiation in material science is the Cu $K\alpha$ radiation. It has a typical wavelength, $\lambda = 0.154$ nm. This wavelength is associated to a very high frequency of the order of 10^{19} Hz which is at least 4 orders of magnitude greater than the eigen frequency of an electron bound to a nucleus. As a consequence the interaction of X-ray radiation with matter can be well described (in a classical way for a first approach) by an index of refraction which characterizes the change of direction of the X-ray beam when passing from air to a material. A very simple classical model (see Ref. [15]) in which an electron of the material is considered to be accelerated by the X-ray field shows that the index of refraction for X-ray radiation can be written in the following way

$$n = 1 - \delta - i\beta \tag{7.1}$$

where δ and β account for the scattering and absorption of the material, respectively. The sign of β depends on the convention of signs used to define the propagation of the electric field; here we use the convention $\exp[i(\omega t - k.r)]$. The values of δ and β (which are positive) depend on the electron density, ρ_e, and mass density, μ, of the material in the following way

$$\delta = \frac{r_e}{2\pi}\lambda^2 \rho_e = \frac{r_e}{2\pi}\lambda^2 \mu \mathcal{N} \frac{\sum_i \frac{x_i(Z_i + f_i')}{M_i}}{\sum_i x_i} \tag{7.2}$$

and

$$\beta = \frac{r_e}{2\pi}\lambda^2 \mu \mathcal{N} \frac{\sum_i x_i \frac{f_i''}{M_i}}{\sum_i x_i} \tag{7.3}$$

with $\mathcal{N}$ the Avogadro number, x_k the number of atoms k and M_k the molar mass of atom k, and, where $r_e = 2.813 \times 10^{-6}$ nm is the classical radius of the electron, V_m is the volume of the unit cell, Z_k is the number of electrons of atom k in the unit cell, f' and f'' are the real and imaginary parts of the anomalous factor for the specific energy of the incident radiation. The values of f' and f'' can be found in [21]. The sum is performed over all the atoms of the unit cell.

7.2.2 The Critical Angle of Reflection

For X-ray radiation, the refractive index of a material is slightly less than 1 [8]. Passing from air ($n = 1$) to the reflecting material ($n < 1$), it is possible to totally

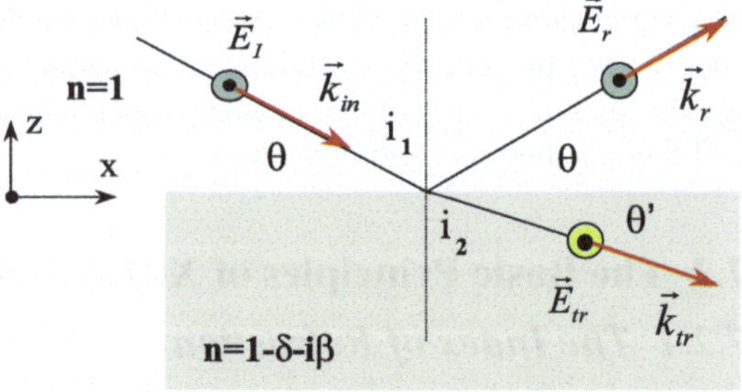

Fig. 7.1 Illustration of specular conditions for XRR measurements together with some characteristic notations used to describe the incident, reflected and transmitted waves

reflect the beam if the incident angle (which is the angle between the surface of the sample and the incident beam) is small enough. This is known as the total external reflection of X-rays. For this to occur, the incident angle must be smaller than that of the critical angle θ_c defined in absence of attenuation as

$$\cos\theta_c = n = 1 - \delta \tag{7.4}$$

Since n is very close to unity, this angle is very small and a Taylor approximation in θ_c yields

$$\theta_c^2 = 2\delta \tag{7.5}$$

7.2.3 Reflected Intensity from Ideally Flat Surface

When an X-ray beam impinges on a flat material, part of the incoming intensity is reflected and part of it is transmitted through the material. If the surface of the reflecting material is flat, the reflected intensity will be confined in a direction symmetric from the incident one and will be labeled as specular (see Fig. 7.1). The specular reflectivity is conventionally defined as the ratio, $I(\theta)/I_0$, where $I(\theta)$ is the reflected intensity at the incident angle and I_0 is the intensity of the direct through beam.

The domain of validity of X-ray reflectivity is limited to small angles of incidence where it is possible to consider the electron density as continuous, i.e. $\theta \ll \theta_B$ where θ_B is the Bragg angle of the first Bragg reflection of the material under consideration. In this approximation, the reflection can be treated as a classical problem of reflection of an electromagnetic wave at an interface. The reflected amplitude is obtained by writing the continuity of the electric field and of the magnetic field at the interface. This leads to the classical Fresnel relationships which give the reflection coefficient in amplitude for the (s) and (p) polarization [1]. The reflectivity which is the modulus square of this coefficient can be formulated in the case of X-ray radiation as [15]

$$R(\theta) = \left| \frac{\theta - \sqrt{\theta^2 - \theta_c^2 - 2i\beta}}{\theta + \sqrt{\theta^2 - \theta_c^2 - 2i\beta}} \right|^2 \tag{7.6}$$

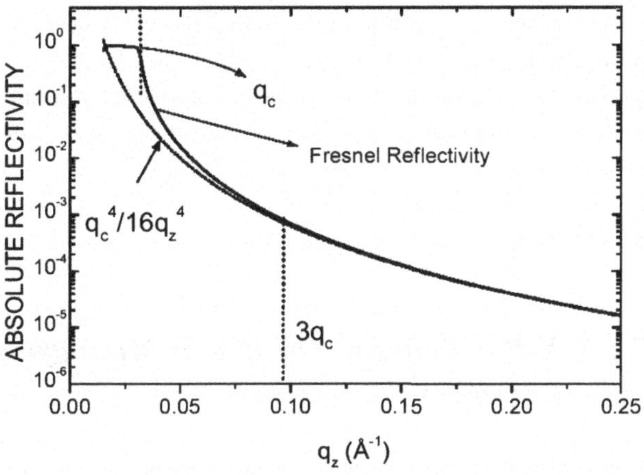

Fig. 7.2 X-ray reflectivity from a silicon wafer calculated from the Fresnel equation and in the power law regime. The critical wave vector transfer q_c is shown by an *arrow*

This expression is independent of the polarization. Since the reflectivity is only observed in specular conditions (incident angle equal to the exit angle), we obtain after introduction of the wave vector transfer $\vec{q} = (0, 0, q_z = 4\pi \sin\theta/\lambda)$,

$$R(\vec{q}) = \left| \frac{q_z - \sqrt{q_z^2 - q_c^2 - 32i\pi^2\beta/\lambda^2}}{q_z + \sqrt{q_z^2 - q_c^2 - 32i\pi^2\beta/\lambda^2}} \right|^2 \delta q_y \delta q_x \tag{7.7}$$

Figure 7.2 shows the reflectivity of a silicon wafer in the case of a complete calculation and in the power law regime which is valid when $q_z > 3q_c$ and for which $R_{flat}(q_z) = q_c^4/16q_z^4$. The deviation from unity is due to the absorption in the material which plays a major role close to $q_z = q_c = 4\pi\theta_c/\lambda$.

7.2.4 Importance of Surface Roughness

Ideally flat surfaces are fictive especially when they are analyzed with X-rays or neutron reflectometry. Such techniques are indeed extremely sensitive to any defects of flatness at the scale of tenth of nanometers. It is easy to realize that rough surfaces will be less reflecting than an ideally flat surface. It is thus important to describe the effect of roughness on the measured reflected intensity. The roughness, σ, of the surface can be apprehended statistically with help of the moments of the distribution, $P(z)$, of altitude $z(x, y)$ with respect to the mean altitude by the following relation:

$$\sigma^2 = \left\langle \left(z(x, y) - \overline{z}\right)^2 \right\rangle = \int p(z)\left(z(x, y) - \overline{z}\right)^2 dz$$

In the case of specular reflectivity, the only useful quantity is the surface roughness. Its effect is to reduce the specular reflectivity by a kind of Debye-Waller factor (this will be evidenced later on). When the correlation length of the height fluctuations is not very large then $R = R_F e^{-q_{z,0} q_{z,1} \sigma^2}$ (where $q_{z,0}$ and $q_{z,1}$ are the wave vector transfers in air and in the material) and conversely [10, 11]. A similar effect is produced by a flat graded layer in which the electron density is represented by an error function of half width σ. A more sophisticated quantity which allows to describe both the specular and the diffuse scattering is the height difference correlation function $g(X, Y)$ which correlates the two heights $z(x, y)$ and $z(x' = x + X, y' = y + Y)$ on the surface. For more details, the reader can refer to the description given in Refs. [11, 15].

7.2.5 X-Ray Reflection by Planar Multilayers with Flat and Rough Interfaces

When the wave propagates in a heterogeneous medium presenting regions of different electron densities, it is not possible to directly use the Fresnel coefficients. The calculation is performed by applying the boundary conditions of the electric and magnetic fields at each interface [7–15]. The fact that multiple reflections are taken in account in the calculation leads to the dynamical theory of reflection and the result is usually presented as the product of matrix. For this, let us consider a plane wave polarized in the direction perpendicular to the plane of incidence (s) and propagating in the medium j of a stratified material and let us choose the axes so that the wave is traveling in the xz plane (see Fig. 7.3). The electric field in medium j which propagates downwards (i.e. in the – direction) is solution of the Helmoltz's equation. It is given by

$$\mathbf{E}^- = A^- e^{+i(\omega t - k_{inx,j}x - k_{inz,j}z)}\mathbf{e}_2 \tag{7.8}$$

where $k_{inz,j}$ (resp. $k_{inx,j}$) is the z (resp. x) component of the wave vector in medium j.

The following notation will be further adopted in the derivation,

$$\begin{aligned} k_{inx,j} &= k_j \cos\theta_j \\ k_{inz,j} &= -k_j \sin\theta_j = -\sqrt{k_j^2 - k_{inx,j}^2} \end{aligned} \tag{7.9}$$

Note that the value of $k_{inx,j}$ is conserved at each interface and this condition is imposed by the Snell-Descartes law of refraction. The upward and downward traveling waves are obviously superimposed at each interface so that at a depth z from the surface the electric field in medium j is:

$$E_j(x, z) = \left(A_j^+ e^{ik_{inz,j}z} + A_j^- e^{-ik_{inz,j}z}\right)e^{+i(\omega t - k_{inx,j}x)} \tag{7.10}$$

As $k_{inz,j}$ takes a complex value, the magnitude of the upward and downward electric fields will be denoted by,

$$U_j^{\pm}(k_{inz,j}, z) = A_j^{\pm} e^{\pm ik_{inz,j}z} \tag{7.11}$$

to simplify the notation.

In addition, the quantity $k_{inz,j}$ will be replaced by $k_{z,j}$ and we shall consider this quantity as implicit in the expression of $U_j(k_{inz,j}, z)$. The conditions of continuity

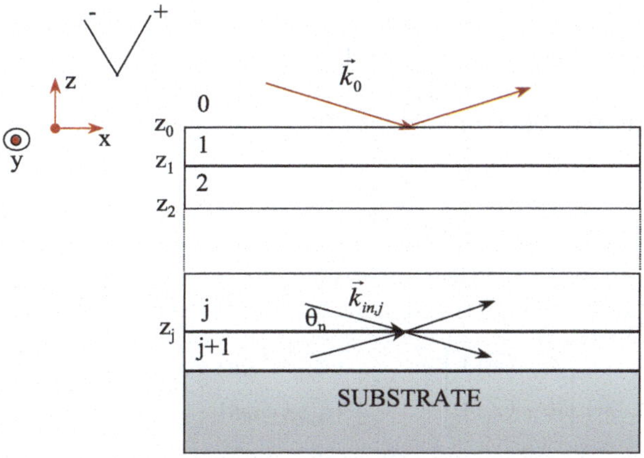

Fig. 7.3 Schematic of a layered material and of the conventions used in the text concerning the labels of the layers and of the propagation directions of the incident and reflected waves. (Note that the first air-material interface begins at altitude z_1)

of the tangential component of the electric field and the conservation of $k_{x,j}$ at the depth Z_{j+1} of the interface j, $j+1$ lead to the result that,

$$U_j^+(Z_{j+1}) + U_j^-(Z_{j+1}) = U_{j+1}^+(Z_{j+1}) + U_{j+1}^-(Z_{j+1}) \tag{7.12}$$

The tangential component of the magnetic field is also continuous provided that the first derivative of the electric field is conserved. This leads to the equality below, at the j, $j+1$ interface,

$$k_{z,j}\big(U_j^+(Z_{j+1}) - U_j^-(Z_{j+1})\big) = k_{z,j+1}\big(U_{j+1}^+(Z_{j+1}) + U_{j+1}^-(Z_{j+1})\big) \tag{7.13}$$

The combination of these two equations can be written in a matrix form, so that the magnitudes of the electric field in media j, $j+1$ at depth Z_{j+1} must satisfy,

$$\begin{bmatrix} U_j^+(Z_{j+1}) \\ U_j^-(Z_{j+1}) \end{bmatrix} = \begin{bmatrix} p_{j,j+1} & m_{j,j+1} \\ m_{j,j+1} & p_{j,j+1} \end{bmatrix} \begin{bmatrix} U_{j+1}^+(Z_{j+1}) \\ U_{j+1}^-(Z_{j+1}) \end{bmatrix} \tag{7.14}$$

with

$$\begin{aligned} p_{j,j+1} &= \frac{k_{z,j} + k_{z,j+1}}{2k_{z,j}} \\ m_{j,j+1} &= \frac{k_{z,j} - k_{z,j+1}}{2k_{z,j}} \end{aligned} \tag{7.15}$$

The matrix which transforms the magnitudes of the electric field from the medium j to the medium $j+1$ will be called the refraction matrix $\mathcal{R}_{j,j+1}$. It is worth noting that $\mathcal{R}_{j,j+1}$ is not unimodular and has a determinant equal to $k_{z,j+1}/k_{z,j}$.

In addition, the amplitude of the electric field within the medium j varies with depth z as follows,

$$\begin{bmatrix} U_j^+(z) \\ U_j^-(z) \end{bmatrix} = \begin{bmatrix} e^{-ik_{z,j}h} & 0 \\ 0 & e^{ik_{z,j}h} \end{bmatrix} \begin{bmatrix} U_j^+(z+h) \\ U_j^-(z+h) \end{bmatrix} \tag{7.16}$$

The matrix which is involved here will be denoted the translation matrix $\mathcal{T}$. The amplitude of the electric field at the surface (depth $Z_1 = 0$) of the layered material is obtained by multiplying all the refraction and translation matrices in each layer starting from the substrate (at $z = Z_{sub}$) as follows,

$$\begin{bmatrix} U_0^+(Z_1) \\ U_0^-(Z_1) \end{bmatrix} = \mathcal{R}_{0,1}\mathcal{T}_1\mathcal{R}_{1,2}, \ldots, \mathcal{R}_{N,sub} \begin{bmatrix} U_s^+(Z_{sub}) \\ U_s^-(Z_{sub}) \end{bmatrix} \tag{7.17}$$

All the matrices involved in the above product are 2×2 matrices so that their product which is called the transfer matrix $\mathcal{M}$, is also a 2×2 matrix. We thus have,

$$\begin{bmatrix} U_0^+(Z_1) \\ U_0^-(Z_1) \end{bmatrix} = \mathcal{M} \begin{bmatrix} U_{sub}^+(Z_{sub}) \\ U_{sub}^-(Z_{sub}) \end{bmatrix} = \begin{bmatrix} M_{11} & M_{12} \\ M_{21} & M_{22} \end{bmatrix} \begin{bmatrix} U_{sub}^+(Z_{sub}) \\ U_{sub}^-(Z_{sub}) \end{bmatrix} \tag{7.18}$$

The reflection coefficient is defined as the ratio of the reflected electric field to the incident electric field at the surface of the material and is given by,

$$r = \frac{U_0^+(Z_1)}{U_0^-(Z_1)} = \frac{M_{11}U_{sub}^+(Z_{sub}) + M_{12}U_{sub}^-(Z_{sub})}{M_{12}U_{sub}^+(Z_{sub}) + M_{22}U_{sub}^-(Z_{sub})} \tag{7.19}$$

It is reasonable to assume that no wave will be reflected back from the substrate if the X-rays penetrate only a few microns, so that,

$$U_{sub}^+(Z_{sub}) = 0 \tag{7.20}$$

and therefore the reflection coefficient is simply defined as

$$r = \frac{M_{12}}{M_{22}} \tag{7.21}$$

This method known as the matrix technique is general and is valid for any kind of electromagnetic wave. The above formalism can be extended to slabs presenting uncorrelated rough interfaces. For this, one can show that the coefficients $m_{j,j+1}$ and $p_{j,j+1}$ are respectively reduced by the factors $e^{-(k_{z,j+1}+k_{z,j})^2\sigma_{j+1}^2/2}$ and $e^{-(k_{z,j+1}-k_{z,j})^2\sigma_{j+1}^2/2}$ (see Ref. [15]).

7.2.6 The Refraction Matrix for X-Ray Radiation

As shown in the section above (see (7.14) and (7.15)), the refraction matrix is defined as,

$$\mathcal{R}_{j,j+1} = \begin{bmatrix} p_{j,j+1} & m_{j,j+1} \\ m_{j,j+1} & p_{j,j+1} \end{bmatrix}$$

with

$$p_{j,j+1} = \frac{k_{z,j} + k_{z,j+1}}{2k_{z,j}} \quad \text{and} \quad m_{j,j+1} = \frac{k_{z,j} - k_{z,j+1}}{2k_{z,j}} \tag{7.22}$$

(7.9) shows that $k_{z,j}$ is the component of the Wave vector normal to the surface and that it is equal to,

$$k_{z,j} = -k_j \sin\theta_j = -\sqrt{k_j^2 - k_{x,j}^2} \tag{7.23}$$

and that $k_{x,j}$ is conserved and is equal to $k\cos\theta$. As a result of this, the z component of k in medium j is,

$$k_{z,j} = -\sqrt{k^2 n_j^2 - k^2\cos^2\theta} \tag{7.24}$$

where k is the wave-vector in air. In the limit of small angles and substituting the expression of the refractive index for X-rays, this becomes,

$$k_{z,j} = -k\sqrt{\theta^2 - 2\delta_j - 2i\beta_j} \tag{7.25}$$

A similar expression can be obtained for $k_{z,j+1}$ so that the coefficients $p_{j,j+1}$ and $m_{j,j+1}$, and as a consequence, the refraction matrix $\mathcal{R}_{j,j+1}$ are entirely determined by the incident angle and by the characteristic value of δ and β in each layer.

7.2.7 Reflection from a Flat Homogeneous Material

For a homogeneous material, the transfer matrix between the air (medium 0) and the material (medium 1) is simply the refraction matrix, which means that $\mathcal{M} = \mathcal{R}_{0,1}$ so that the reflection coefficient r becomes,

$$r = r_{0,1} = \frac{U_0^+(0)}{U_0^-(0)} = \frac{M_{12}}{M_{22}} = \frac{m_{0,1}}{p_{0,1}} = \frac{k_{z,0} - k_{z,1}}{k_{z,0} + k_{z,1}} \tag{7.26}$$

or

$$r = \frac{-k\theta + k\sqrt{\theta^2 - 2\delta}}{-k\theta - k\sqrt{\theta^2 - 2\delta}} = \frac{\theta - \sqrt{\theta^2 - 2\delta}}{\theta + \sqrt{\theta^2 - 2\delta}} \tag{7.27}$$

(7.27) is of course identical to the one obtained by using the familiar expression for the Fresnel reflectivity (see (7.6)).

Similarly, the transmission coefficient is simply given by,

$$t_{0,1} = \frac{U_1^-(0)}{U_0^-(0)} = \frac{1}{M_{22}} = \frac{1}{p_{0,1}} = \frac{2k_{z,0}}{k_{z,0} + k_{z,1}} \tag{7.28}$$

7.2.8 A Single Layer on a Substrate

The transfer matrix for the case of a layer of thickness $h = Z_2 - Z_1$ (h being negative here) deposited on a substrate is given as,

$$\mathcal{R}_{0,1}\mathcal{T}_1\mathcal{R}_{1,2} = \begin{vmatrix} p_{0,1} & m_{0,1} \\ m_{0,1} & p_{0,1} \end{vmatrix} \begin{vmatrix} e^{-ik_{z,1}h} & 0 \\ 0 & e^{ik_{z,1}h} \end{vmatrix} \begin{vmatrix} p_{1,2} & m_{1,2} \\ m_{1,2} & p_{1,2} \end{vmatrix} \tag{7.29}$$

and the reflection coefficient is,

$$r = \frac{M_{12}}{M_{22}} = \frac{m_{0,1}p_{1,2}e^{ik_{z,1}h} + m_{1,2}p_{0,1}e^{-ik_{z,1}h}}{m_{0,1}m_{1,2}e^{ik_{z,1}h} + p_{1,2}p_{0,1}e^{-ik_{z,1}h}} \tag{7.30}$$

Dividing numerator and denominator by $p_{0,1}p_{1,2}$ and introducing the reflection coefficients $r_{i-1,i} = m_{i-1,i}/p_{i-1,i}$ for the two media i and $i-1$, the reflection coefficient of the electric field at the layer is then found to be,

$$r = \frac{r_{0,1} + r_{1,2}e^{-2ik_{z,1}h}}{1 + r_{0,1}r_{1,2}e^{-2ik_{z,1}h}} \tag{7.31}$$

It is worth noting that the denominator of this expression differs from unity by a term which corresponds to multiple reflections in the material, as shown by the product of the two reflection coefficients $r_{01}r_{12}$.

It is also straightforward to determine the transmission coefficient since its value is given by $1/M_{22}$; this yields,

$$t = \frac{t_{0,1}t_{1,2}e^{-ik_{z,1}h}}{1 + r_{0,1}r_{1,2}e^{-2ik_{z,1}h}} \tag{7.32}$$

In the case where the absorption can be neglected, the reflected intensity is therefore,

$$R = \frac{r_{0,1}^2 + r_{1,2}^2 + 2r_{0,1}r_{1,2}\cos 2k_{z,1}h}{1 + r_{0,1}^2 r_{1,2}^2 + 2r_{0,1}r_{1,2}\cos 2k_{z,1}h} \tag{7.33}$$

The presence of the cosine terms in (7.33) indicates clearly that the reflectivity curve will exhibit oscillations in reciprocal space whose period will be defined by the equality,

$$2k_{z,1}h \approx q_{z,1}h = 2p\pi \tag{7.34}$$

or

$$q_{z,1} = \frac{2p\pi}{h} \tag{7.35}$$

These oscillations are the result of the constructive interference between the reflected waves at interfaces *1* and *2*. The difference in path length which separates the two waves is

$$\delta = 2h\sin\theta_1 = p\lambda \tag{7.36}$$

so that

$$q_{z,1} = \frac{2\pi p}{h} \tag{7.37}$$

Figure 7.4 which corresponds to the experimental reflectivity of a copolymer deposited onto silicon substrate provides a good illustration of this type of interference phenomenon. The experimental curve is presented in open circles and the calculated one in full line. The calculation is made by the matrix technique in which we use (7.29) as the starting point of the calculation.

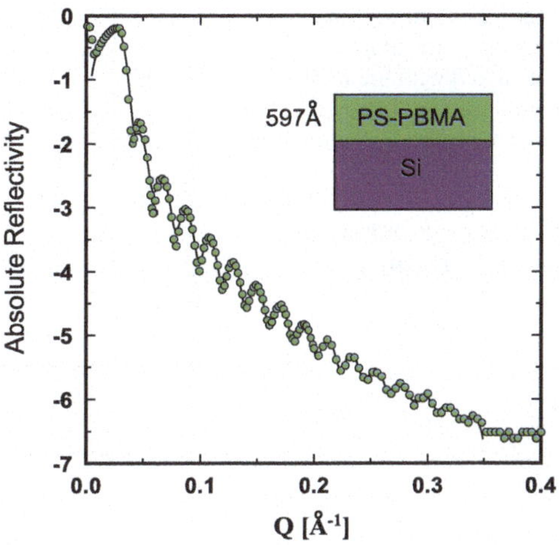

Fig. 7.4 Typical example of an X-ray reflectivity curve in which one layer of PS-PBMA was deposited on silicon. The reflectivity curve exhibits typical Kiessig fringes characteristic of the thickness of the layer

The fact that the reflectivity is less than 1 below the critical angle is related to a surface effect. At very shallow angles, it frequently happens that the footprint of the beam is larger than the sample surface so that only part of the intensity is reflected. A correction must then be applied to describe this part of the reflectivity curve. The roughness of the interfaces is also included in the calculation and this will be discussed in an other section. For two layers deposited on a substrate the analytical expression of the reflectivity becomes more tedious to express but is simple to calculate via the matrix technique with a computer.

7.3 From Dynamical to Kinematical Theory

7.3.1 The Master Equation in Kinematical Theory

The dynamical theory rigorously describes the specular reflectivity of flat surfaces but it is difficult to use in the case of rough surfaces where a significant part of the intensity is scattered off-specular. It is however possible to apply the kinematical theory which is more flexible to use providing some approximations are made. The derivation is usually made within the Born approximations for scattering processes.

The different kinds of approximations which are necessary [15, 23] will be discussed in this section. This kind of approach was first made by Rayleigh in 1912 in the context of the reflection of electromagnetic waves [22] but has since become known as the Born approximation since Born generalized it to different types of scattering processes. We will start from the dynamical expression of the reflected amplitude calculated in the previous section (see (7.33)) for a thin film of thickness

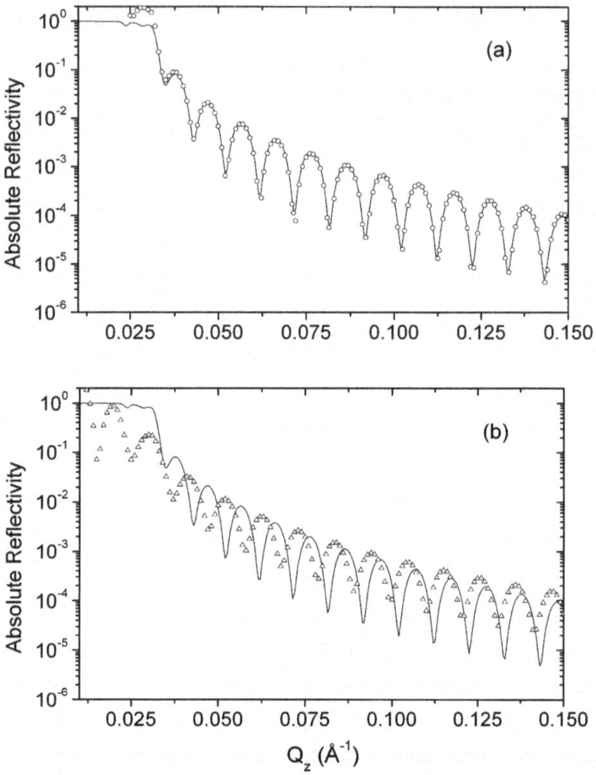

Fig. 7.5 Comparison between reflectivities calculated with the matrix technique (*full line* in (**a**) and (**b**)) with a calculation in which multiple reflections are neglected (*circles* in (**a**)) and in addition the refraction (*triangles* in (**b**))

h deposited on a substrate and follow the same approach as the one proposed by Hamley and Pedersen [23]. This yields

$$r = \frac{r_{0,1} + r_{1,2}\, e^{-2ik_{z,1}h}}{1 + r_{0,1} r_{1,2}\, e^{-2ik_{z,1}h}} \tag{7.38}$$

where the phase shift between the reflected waves on the substrate and the layer denoted by $\varphi = -2k_{z,1}h = q_{z,1}h$ can be written as a function of either k or q. The term $r_{0,1}r_{1,2}e^{i\varphi}$ in this equation represents the effect of multiple reflections in the layer and a first step in the approximations consists in neglecting this term—which is equivalent to consider that the electrons interact only with the incident field. This is justified if the wave-vector transfer is large enough, i.e. if $q_z \gg q_c$. This is illustrated in Fig. 7.5 which shows a comparison between the reflectivities calculated for a copolymer on a silicon wafer with the matrix method for the cases both with and without the multiple reflections at interfaces. It can be seen that the two curves are almost identical showing that this approximation is a valid one.

A second approximation consists in neglecting the refraction and the absorption in the material. In this case the approximation is more drastic and this can be seen in Fig. 7.5(b). It shows that the region of the curve just after the critical angle is severely affected. In addition, the location of the interference fringes is wrong.

We will therefore derive a general expression for the reflection coefficient r for a stratified material composed of n layers, according to the approximation of neglecting the multiple reflections and beginning from the equation, written as a function of q_z,

$$r = r_{0,1} + r_{1,2}e^{iq_{z,1}d_1} + r_{2,3}e^{i(q_{z,1}d_1+q_{z,2}d_2)} + \cdots + r_{j,j+1}e^{i\sum_{k=0}^{j} q_{z,k}d_k} \tag{7.39}$$

Here the ratio $r_{j,j+1}$ of the amplitudes of the reflected to the incident waves at interface j, $j+1$ is,

$$r_{j,j+1} = \frac{q_{z,j} - q_{z,j+1}}{q_{z,j} + q_{z,j+1}} \tag{7.40}$$

with the wave-vector transfer in medium j is given by,

$$q_{z,j} = (4\pi/\lambda)\sin\theta_j = \sqrt{q_z^2 - q_{c,j}^2} \tag{7.41}$$

If the effect of refraction is neglected in the phase factor we can replace $q_{z,j}$ by q_z. Further approximations are described in Ref. [23] so that if the origin of the z axis is chosen at the air interface (medium 0 at a depth of $Z_1 = 0$), we end up with

$$r = 4\pi r_e \sum_{j=1}^{n} \frac{(\rho_{j+1} - \rho_j)}{q_z^2} e^{iq_z Z_{j+1}} \tag{7.42}$$

Finally, if we consider that the material is made of an infinite number of thin layers, the sum may then be transformed into an integral over z, and the reflection coefficient r has the form,

$$r = \frac{4\pi r_e}{q_z^2} \int_{-\infty}^{+\infty} \frac{d\rho(z)}{dz} e^{iq_z z}\, dz \tag{7.43}$$

Substituting the expression for the Fresnel reflectivity into this equation,

$$R_F(q_z) = \frac{(4\pi r_e \rho_s)^2}{q_z^4} \tag{7.44}$$

shows that within the first Born approximation the reflectivity can be written as [23, 24],

$$R(q_z) = r.r^* = R_F(q_z) \left| \frac{1}{\rho_s} \int_{-\infty}^{+\infty} \frac{d\rho(z)}{dz} e^{iq_z z}\, dz \right|^2 \tag{7.45}$$

7.3.2 The Patterson Function

The above expression for $R(q_z)$ is not rigorous but it has the advantage of being easily handled in analytical calculations. In addition, if the Wiener-Kintchine theorem is applied to this result, we find,

$$\frac{R(q_z)}{R_F(q_z)} = TF\left[\rho'(z) \otimes \rho'(z)\right] \tag{7.46}$$

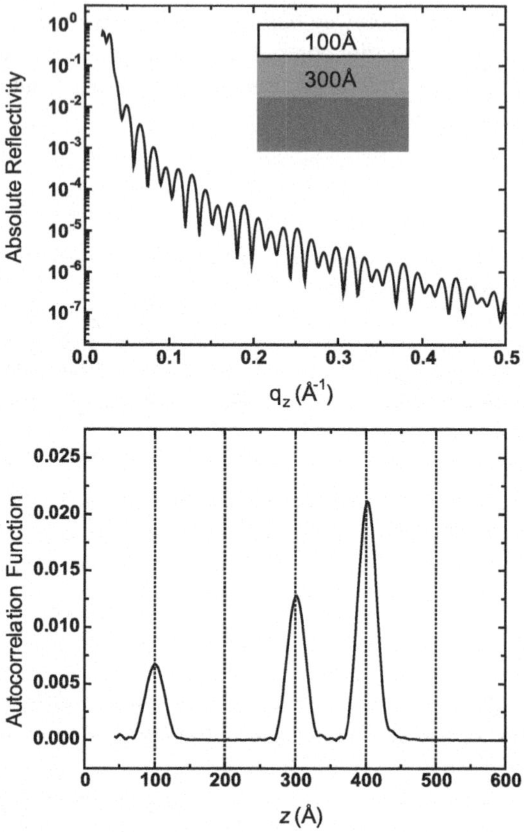

Fig. 7.6 Calculated reflectivity of a two layers system and its Fourier transform after division of the data by the Fresnel reflectivity of the substrate. In the calculation the two layers of different electron densities are 300 and 100 Å thick. The Fourier transform immediately gives the thickness of each layer without relying on any model. One can also note the expected presence of a peak at $z = 400$ Å in the autocorrelation function

so that the data inversion gives the autocorrelation function of the first derivative of the electron density [15] or the Patterson function.

Figure 7.6, illustrates the main features of this data inversion. It is based on a calculation with a model structure [15] for a sample which consists of two layers, a lower one of 300 Å and an upper one of 100 Å respectively on a substrate. The top diagram gives the calculated reflectivity curve which shows something like the "beating" effect which arises here because of the similarities between the thicknesses of the three layers of the sample. The bottom diagram gives the autocorrelation function, which has intense peaks at the interfaces where the derivative of the electron density is maximized. In an ideally flat sample these peaks would be delta functions, but for a real case their width of the depends on factors such as the roughness and degree of interdiffusion at the interfaces.

We will now consider some practical aspects encountered in XRR measurements before continuing on a set of typical examples.

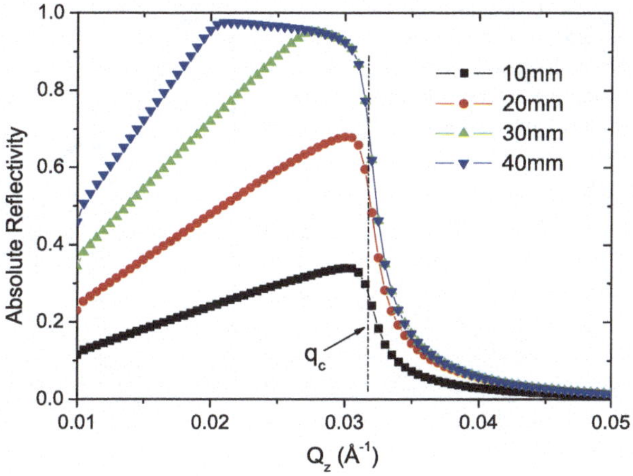

Fig. 7.7 Calculated XRR of silicon for a beam 100 μm large impinging on a sample having different lengths ranging from 10 mm to 40 mm. The footprint effect is materialized by the fact that at shallow angle the reflectivity does not reach 1 but increases linearly when q_z increases until it reaches its correct value when the footprint matches the length of the sample

7.4 Experimental Setup and Method

7.4.1 Footprint Effect

The size of the beam is an important parameter because X-ray reflectivity measurements are made at very shallow angle in particular below the critical angle of external reflection. Assuming a rectangular beam section with dimensions $t_1 t_2$ (t_2 the dimension parallel to the surface of the sample) and an incident angle α, it is straightforward to show that the footprint of the beam on the surface of the sample is

$$F = \frac{t_1}{\sin \alpha t_2} \tag{7.47}$$

At the critical angle of silicon ($\alpha \simeq 0.22°$ for $E = 8$ keV) and for $t_1 = 100$ μm, we find that the footprint along the direction of propagation of the beam is about $\frac{t_1}{\alpha} = 30$ mm. This shows that the sample size must be at least 30 mm long to totally reflect the incident beam at this angle of incidence. This condition is the minimum condition to be fulfilled to observe the plateau of total external reflection. This is illustrated in Fig. 7.7 where the Fresnel reflectivity of silicon has been calculated for a beam size of 100 μm and a sample size ranging from 10 to 40 mm. One can clearly see that below 30 mm one never reaches the plateau of total external reflection. This can even produce a severe distortion of the plateau as it is the case when the sample size is equal to 10 mm. This effect has been discussed in details in the paper of Gibaud et al. [25].

7.4.2 Instrumental Resolution

The measurement of an X-ray reflectivity curve necessitates the use of a well collimated parallel incident beam. The divergence, $\Delta\alpha$ of the incident beam needs to

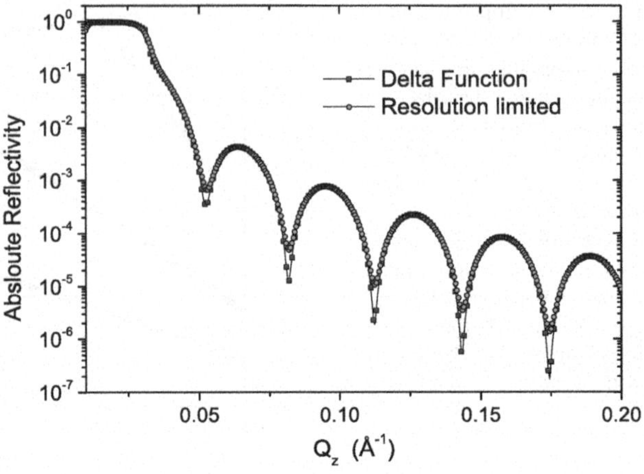

Fig. 7.8 Calculation of XRR of a 200 Å thick polystyrene film with an instrument having a perfect resolution and a resolution function presenting a width of 0.0015 Å^{-1}

be small enough to precisely define the incident angle. A similar condition stands for the divergence $\Delta\beta$ of the reflected beam which is driven by the detector aperture. If the divergences of the incident and reflected beams are quite small, one can assume that the resolution function tends towards a delta function. This is usually never the case because one clearly needs to make a compromise between a high flux and a high resolution via the beam and detector apertures. Therefore for any instrument, the resolution function is not a Dirac distribution but most of the time may be represented by a 3-dimensional Gaussian. The width (in q-space) of the Gaussian depends on the slits setting or on the monochromator and analyzer crystals used to collimate the incident and scattered beams. Therefore (7.7) must be convolved with the resolution function of the instrument to give the expression of the measured reflectivity. An illustration of the effect of the instrumental resolution is shown in Fig. 7.8.

The proper way to calculate the instrumental resolution has been reported in the paper by Gibaud et al. [25]. For this one needs to differentiate the components of the wave vector transfer with respect to the incident and exit angles. As shown in this paper one thus can express the width of the instrumental resolution along q_x and q_z scans and end up with the following relations

$$\Delta q_x = q_z \frac{w}{2} \tag{7.48}$$

$$\Delta q_z = \frac{2\pi}{\lambda} w \cos(\theta) \tag{7.49}$$

where w is the HWHM of the direct beam.

7.5 Recent Applications of the Technique

XRR is not a new technique since it was already used in the 1930's. Nevertheless recent advances in XRR are possible by achieving experiments in extreme conditions such as under high pressure or in specific environments such as humidity cells [26]. In addition, XRR experiments can be carried out at synchrotron beam lines to take advantage of the tunability of the energy so as to carry out anomalous XRR [27–29]. The trend is to be able to carry out experiments as fast as possible or as precisely as possible by increasing the flux of the sources by using mirrors at home facilities, insertion devices at synchrotron facilities or micro focus sources. In addition XRR has been recently complemented by Grazing Incidence X-ray Scattering (GISAXS) to probe the scattering in off-specular directions.

7.5.1 Contrast Matching

It is clear from (7.38) that if a thin layer has the same index of refraction as the substrate, it will be impossible to measure any Kiessig fringes from the layer since there will be no contrast between the layer and the substrate as $r_{12} = 0$. This effect is frequently used by the community of neutronists who change the contrast of a deposited layer by using deuterated samples or isotopic substitution. In this section, we show that we can achieve some contrast matching by depositing a thin film of a TiO_2 of increasing thickness on a substrate. The layer is obtained by spin coating a solution of Titanium Isopropoxide on a silicon substrate. As shown in Fig. 7.9, one can clearly see that the amplitude of the Kiessig fringes decreases when the thickness of the layer increases. This obviously shows that the film becomes denser as the thickness increases. The disappearance of the fringes is a combination of the increase of both the deposited thickness and the electron density of the layer with the one of the substrate (see inset of Fig. 7.9). As the initial sol produces a gel, this effect can be seen as the result of the increasing degree of condensation of the gel when the thickness increases.

This can be further probed by annealing the gel. Before annealing the gel has almost the same electron density as the substrate. After annealing, water molecules are lost. This produces the shrinkage of the film together with its densification (see right panel of Fig. 7.9). The densification of the film produces a shift of the critical q_c of the annealed film towards higher q_z values.

7.5.2 X-Ray Porosimetry

Porous materials are becoming increasingly popular because they are ideal to host fluids or drugs. In particular, MCM41 materials discovered by researchers at Mobil Oil Corporation have been the focus of extensive research activity [30]. These materials are made by self-assembling surfactant molecules which then serve to template an inorganic network such as silica or titania. The self assembly of surfactant is a well known topic and was used for many years to produce liquid crystalline structures of increasing complexity. More recently the concept of surfactant templating

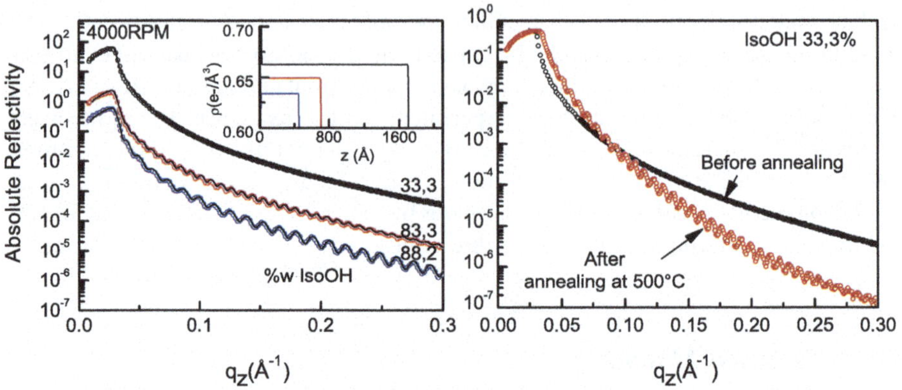

Fig. 7.9 On the *left panel*: experimental and calculated XRR of TiO_2 thin films together with a magnified view of the electron density profile (*inset* of Fig. 7.9). On the *right panel*, XRR reflectivity curves of the thicker film before and after annealing at 500 °C showing the reappearance of the Kiessig fringes together with the shift of the critical q_c of the annealed film towards higher q_z values

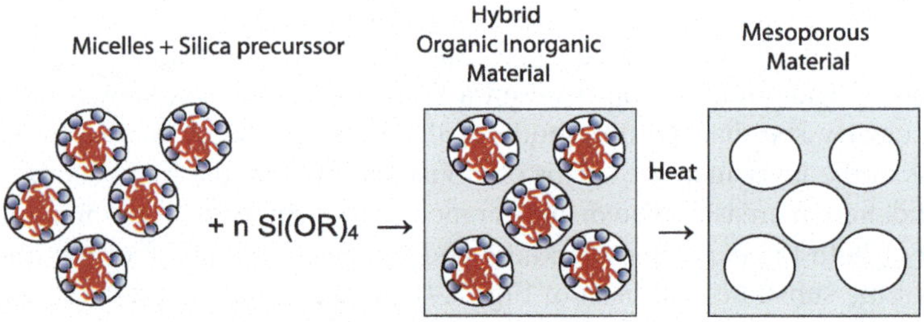

Fig. 7.10 Illustration of the design of a hybrid organic inorganic material templated by micelles of a surfactant and its transformation into a porous structure

silica was adapted to the making of thin films. This was made possible by a technique developed by J. Brinker at Sandia National Laboratory known as Evaporation Induced Self-Assembly (EISA) [31]. In this technique the surfactant is mixed to a silica precursor such as TEOS in acidic condition and diluted in ethanol. When a substrate is dipped inside such a solution and gently pulled back from it, one can get a beautiful highly organized thin film. The film is a hybrid material in which an inorganic material such as silica is templated by a surfactant. Beautiful 2D hexagonal honeycomb structures can thus be achieved by this technique [19, 32–34]. In addition the hybrid structure can be transformed into a porous mineral structure by removing the surfactant as shown in Fig. 7.10.

The main issue is then to be able to measure the porosity of such films. In many cases, porosity is determined by B.E.T., a technique which consists in exposing a porous material to a fluid and to determine the quantity of fluid which goes inside the material. This technique is very efficient for large quantity of materials but fails when it is applied to thin films with thickness of the order of 100 nm. In this case,

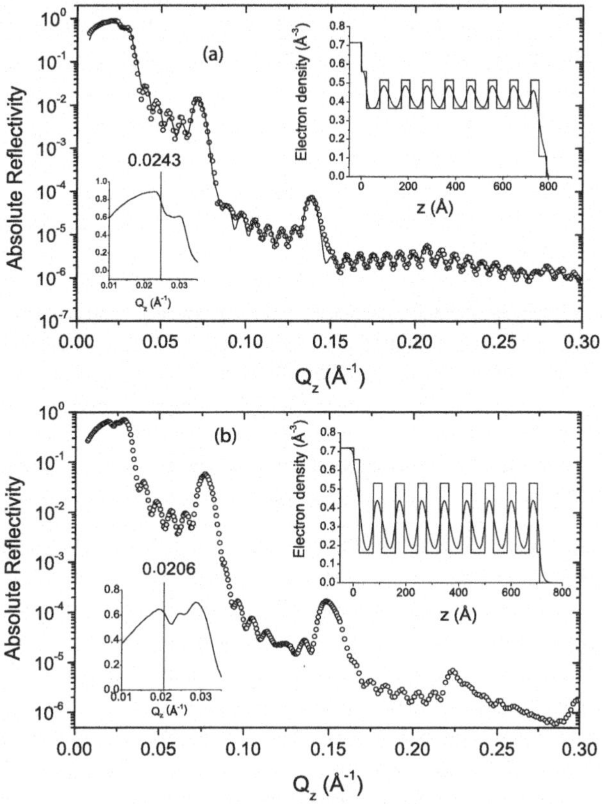

Fig. 7.11 Measured and calculated XRR curves of mesoporous silica thin films templated by P123 surfactant before (**a**) and after removing the surfactant (**b**). The *bottom inset in each figure* shows the region close to the critical angle of silicon in which one can see the critical angle of the film as well due to the fact that the film is less dense than silicon. The *top inset* shows the electron density profile corresponding to the best fit to the experimental data

XRR is a very powerful technique to extract the porosity and even to monitor how much fluid can penetrate inside the pores [35]. We show in Fig. 7.11 the X-ray reflectivity of a mesoporous silica film before and after extracting the surfactant.

The presence of two well defined critical angles in the XRR curves is typical of porous thin films deposited on silicon. Indeed, the porous film is less dense than silicon and the beam penetrates inside the film before being able to penetrate inside silicon if the incident angle is intermediate between the critical angle of the film and that of silicon. The lowest critical angle is the one of the film and it is very sensitive to whether or not surfactant molecules are in the film. When these molecules are removed, the critical angle (which let us remind you is proportional to the square root of the electron density of the film (see (7.5)) shifts towards smaller value as a signature of their removal from the silica matrix. One can show (see [35]) that the porosity of such films defined by

$$\phi = \frac{V_{pores}}{V_{total}} = \frac{V_{pores}}{V_{pores} + V_{silica}} = \frac{1}{1 + \frac{V_{silica}}{V_{pores}}} \tag{7.50}$$

can be expressed as a function of the electron densities of the film and of the silica wall defined by

$$\rho_{film} = \frac{Ne^-}{V_{total}} = \frac{Ne^-}{V_{pores} + V_{silica}} \tag{7.51}$$

and

$$\rho_{silica} = \frac{Ne^-}{V_{silica}} \tag{7.52}$$

As the number of electrons is the same in the film as in the silica walls (since there are no electrons inside the empty pores), this yields

$$\phi = \frac{\rho_{silica} - \rho_{film}}{\rho_{silica}} \tag{7.53}$$

The matrix technique allows to determine ρ_{silica} so that one can get the porosity. In this case, we end up with $\phi = 42\ \%$ since $\rho_{silica} = 0.53\ e^-/\text{Å}^3$ and $\rho_{film} = 0.300\ e^-/\text{Å}^3$. In addition, it is easy to understand that the determination of the critical angle of the film is intimately connected to what is found inside the pores. When a mesoporous film is exposed to a fluid, capillary condensation of the fluid into a liquid state usually occurs at a vapor pressure below the saturation pressure [35, 36]. The uptake of water inside the pores can be monitored by looking at the change of the critical angle versus this pressure. For this, a mesoporous film was inserted into a cell in which the relative humidity (RH) was controlled by flowing either dry or humid nitrogen. RH, defined as the ratio of the partial vapor pressure P of water to the saturation vapor pressure P_0, at ambient temperature was measured with a humidity sensor HC-610 from Ohmic instrument. XRR measurements were then measured as a function of RH. Two typical XRR curves measured at low (5 %) and high (70 %) RH are shown in Fig. 7.12.

These two curves show Bragg peaks and Kiessig fringes that are characteristic of a film having a finite measurable total thickness t. This thickness changes from $t = 66.2$ nm at RH = 5 % to $t = 67.6$ nm at RH = 70 %. This clearly shows that the silica matrix was well condensed and did not swell much upon increasing the RH. The most striking feature is the big change in the Bragg peak intensities. At RH = 5 %, the first Bragg peak is about one order magnitude stronger than at RH = 70 %. This effect is a clear signature of the water intrusion inside the mesopores of the film. Another striking feature of the RH dependence on the XRR curves is found below the critical wave vector of the silicon substrate. As shown in the bottom insets of Fig. 7.1, two different critical q_c are observed. The first one, q_{c1}, corresponds to the average electron density of the film whereas the second one, q_{c2}, is that of the silicon substrate (0.032 Å^{-1}). A comparison of the two panels clearly shows that q_{c1} strongly depends on RH while as expected q_{c2} does not vary. The shift in q_{c1} evolves from 0.02 Å^{-1} at RH = 5 % to 0.026 Å^{-1} at RH = 70 %. This corresponds in turn to a change of the average electron density from 0.28 $e^-/\text{Å}^3$ at RH = 5 % to 0.44 $e^-/\text{Å}^3$ at RH = 70 %. This increase

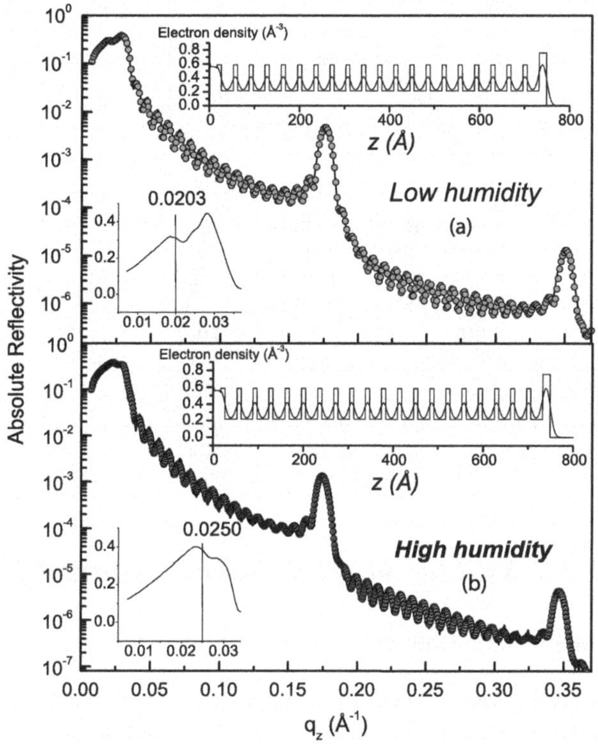

Fig. 7.12 Absolute reflectivity curves measured at low (**a**) and high (**b**) RH. The *top inset* gives the electron density profile obtained from a fit via the matrix technique to the experimental data. The *bottom inset* shows a magnified view of the region below the critical angle. The value reported along the line is the average critical wave vector of the film. The modifications induced by the water intrusion are obvious both on the electron density profiles and on the average critical wave vector

of the average electron density of the film with RH is here again the clear signature of water intrusion in the mesopores. From the determination of the electron density of the film, one can thus obtain the isotherms of condensation and desorption of the fluid inside the pores (see Fig. 7.13). The full description of this effect can be monitored during the cycling of the RH. Figure 7.13(a) shows the evolution of the critical edge versus RH while Fig. 7.13(b) inferred from the results in Fig. 7.13(a) is a plot of the average electron density of the film versus RH. This last figure is fully reminiscent of the isotherm of a mesoporous adsorbent designated as a type IV in the IUPAC classification. It presents a gradually increase at low RH as expected from a multi molecular adsorption at the material wall surface, and a hysteresis loop at RH characteristics of the capillary condensation in mesopores. Using the Kelvin equation, one can further deduce the pore size distribution, presented in the top inset of Fig. 7.13(b). The Kelvin radii derived from the desorption and adsorption branches are respectively centered at 0.9 nm and 1.4 nm.

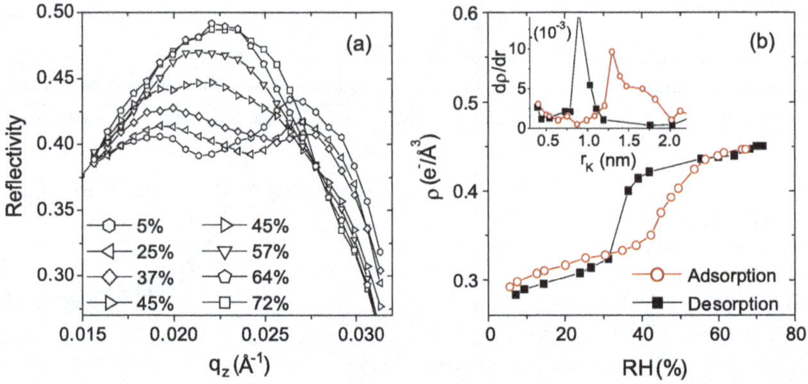

Fig. 7.13 Low q region of the XXR of a mesoporous film templated by CTAB exposed to various RH (**a**) and its electron density versus RH as deduced from the location of the critical angle of the film (**b**) during adsorption and desorption cycles. The *inset* in this figure gives the pore size distribution

7.5.3 XRR for Samples Under Pressurized Fluids

In-situ Experiments

Usually XRR is measured with a sample in contact with air. Nevertheless it can be interesting to probe the behavior of a material when it is exposed to a fluid under pressure. In this case depending on the pressure one can observe very interesting phenomena which necessitate to take into account the electron density of the fluid in contact with the material. We are going to show such an effect with a silicon substrate exposed to CO_2 under pressure. For this a silicon substrate is placed inside a cell in which CO_2 could be injected up to a pressure of 200 bars [37]. CO_2 is a gas at ambient temperature and pressure. It can be turned into a liquid by increasing at a given temperature its pressure as can be seen in Fig. 7.14.

When CO_2 pressure increases, its density increases as well as the signature of an increasing number of molecules per unit volume. This in turn implies that the electron density of CO_2 increases with pressure. As a consequence the index of refraction of CO_2 differs from unity. This has a serious consequence on the critical q_c of silicon which is usually constant and equal to 0.0318 Å^{-1}. The Snell-Descartes law dictates that when a silicon sample (medium 2) is exposed to CO_2 under pressure (medium 1)

$$k_1 \cos\theta_1 = k_2 \cos\theta_2 \tag{7.54}$$

which can be seen as the conservation of the parallel to the surface component of the wave vector. As a result, the critical angle for external reflection becomes

$$n_1 \cos\theta_{1c} = n_2 \tag{7.55}$$

which in turn yields

$$\theta_{1c} = \sqrt{2(\delta_2 - \delta_1)} \tag{7.56}$$

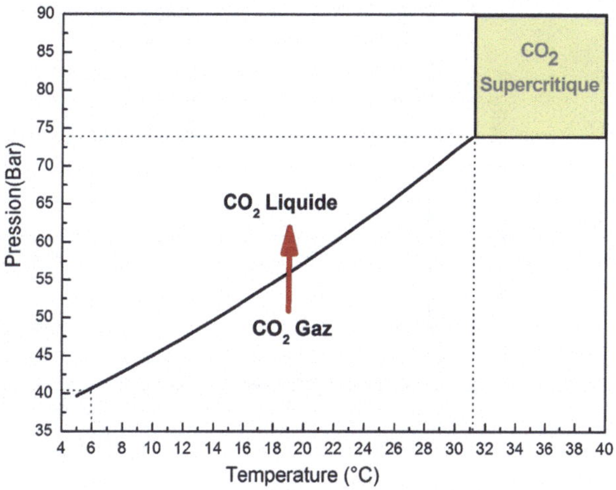

Fig. 7.14 Phase diagram of CO_2: a particularity of CO_2 is that it can be transformed into a supercritical (sc) state above 31 °C and 72 bars. In this state its density goes changes continuously from the gas phase till the sc sate. Below 31 °C the density abruptly changes (first order phase transition when the gas transforms into the liquid phase)

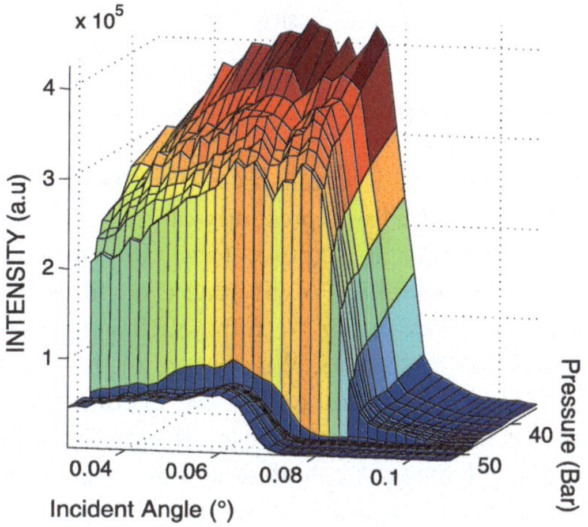

Fig. 7.15 Evolution of the reflectivity of a silicon wafer with respect to the CO_2 pressure at 15 °C. One can see that the critical q_c decreases when the CO_2 pressure increases. The effect is tremendous when CO_2 goes from the gas to the liquid phase since at this moment the density of CO_2 abruptly jumps

where δ_1 is CO_2 pressure dependent and δ_2 is related to the real part of the silicon index of refraction minus one. This last expression shows unambiguously that the critical angle of the substrate itself is affected by the presence of the pressurized gas in contact with it. It is then straightforward to show that the critical q_c will evolve as

$$q_c = \frac{4\pi\sqrt{2(\delta_2 - \delta_1)}}{\lambda} \tag{7.57}$$

A simulation of this effect can be achieved by using the density of the pressurized gas. In the case of carbon dioxide, information about the density of CO_2 at a given pressure can be obtained from the NIST data base. The change of the XRR curve of silicon in presence of pressurized CO_2 is presented in Fig. 7.15. It can be seen in this figure that the critical angle gradually decreases when the CO_2 pressure increases.

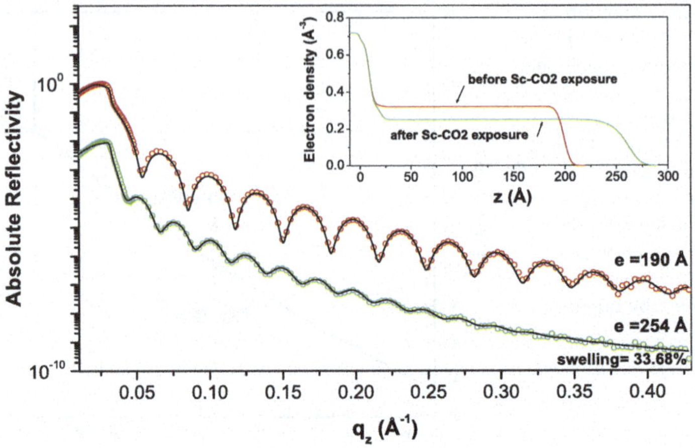

Fig. 7.16 Evolution of the reflectivity of Polystyrene (PS) film before and after exposition to sc-CO_2. The *inset* shows the electron density profile of the film showing both the swelling effect and the loss of electron density after exposure to sc-CO_2

Ex-situ Experiments

Above $T_c = 31.1$ °C and $P_c = 73.8$ bar, CO_2 is in its supercritical state (sc-CO_2). The use of sc-CO_2 is presently of great interest because it is a green solvent. When some materials are exposed to sc-CO_2, they can interact more or less with the solvent. For example, PolyStyrene (PS) films are known to swell in a nonreversible way. We show in Fig. 7.16 the XRR curves of a 20 nm thick film before and after exposition to sc-CO_2. We observe in this figure that the film swells by more than (30 %) after exposure to sc-CO_2. In conjunction with the swelling the electron density of the film decreases accordingly. This can be seen by looking at the spacing between fringes. A calculation of the total number of electrons per unit area from the electron density profile shows that this quantity is slightly increasing after exposure to sc-CO_2 compared to the one before exposure. This can be attributed to a quantity of CO_2 molecules trapped inside the film after depressurization [38]. In addition one can see that the film becomes rougher after exposure to sc-CO_2. This last example ideally shows the power of this technique to probe the electron density profile of thin films in the direction perpendicular to the substrate in a nondestructive way.

7.6 Conclusion

In this chapter we have reported the basics of specular XRR. Starting with the Fresnel equation of Reflectivity, we have shown how complex film stacking can be analyzed by the matrix technique. By simplifying the dynamical calculation thanks to some approximations, we have introduced the master formulae and its application to the determination of the autocorrelation function of the derivative of the electron density of the film. Next we have shown what kind of attention should be paid to the experimental corrections such as the footprint effect or the contribution of the instrumental resolution. Finally we have presented a selected number of examples in which we have shown how efficient is this technique to probe the electron density profile normal to the surface of a substrate. Fancy examples such as contrast matching, X-ray porosimetry and XRR determination of films exposed to pressurized fluids are finally presented.

Acknowledgements We express our deepest gratitude to our co-workers without whom this work could not have been performed: G. Vignaud, S. Dourdain, S. Fall, B. Pattier, M. Yan, J.F. Bardeau and O. Konovalov. The authors are grateful to synchrotron facilities (NSLS and ESRF) for access at beamlines X22b and ID10b respectively.

References

1. A. Fresnel, Mém. Acad. **11**, 393 (1823)
2. M. Von Laue, Nobel Lecture, November 12, 1915
3. A.H. Compton, Philos. Mag. **45**, 1121 (1923)
4. R. Forster, Helv. Phys. Acta **1**, 18 (1927)
5. J.A. Prins, Z. Phys. **47**, 479 (1928)
6. H. Kiessig, Ann. Phys. **10**, 715 (1931)
7. F. Abelès, Ann. de Physique **5**, 596 (1950)
8. L.G. Parrat, Phys. Rev. **95**, 359 (1954)
9. R.W. James, *The Optical Principles of the Diffraction of X-Rays* (Bell, London, 1967)
10. L. Névot, P. Croce, Rev. Phys. Appl. **15**, 761 (1980)
11. S.K. Sinha, E.B. Sirota, S. Garoff, H.B. Stanley, Phys. Rev. B **38**, 2297 (1988)
12. B. Vidal, P. Vincent, Appl. Opt. **23**, 1794 (1984)
13. R. Petit, *Ondes Electromagnétiques en Radioélectricité et en Optique* (Masson, Paris, 1989)
14. J. Lekner, *Theory of Reflection of Electromagnetic and Particle Waves* (Martinus Nijhoff, Dordrecht, 1987)
15. J. Daillant, A. Gibaud (eds.), *X-Ray and Neutron Reflectivity: Principles and Applications* (Springer, Berlin, 2009)
16. A. Gibaud, D. McMorrow, P.P. Swaddling, Phys. Condens. Matter **7**, 2645–2654 (1995)
17. A. Baptiste, A. Gibaud, J.F. Bardeau, W. Ke, R. Maoz, J. Sagiv, B. Ocko, Langmuir **18**, 3916–3922 (2002)
18. M. Maaza, A. Gibaud, C. Sella, B. Pardo, F. Dunsteter, J. Corno, F. Bridou, G. Vignaud, A. Désert, A. Ménelle, Eur. Phys. J. B **7**, 339–345 (1999)
19. A. Gibaud, S. Dourdain, G. Vignaud, Appl. Surf. Sci. **253**, 3–11 (2006)
20. P. Colombi, D.K. Agnihotri, V.E. Asadchikov, E. Bontempi, D.K. Bowen, C.H. Chang, L.E. Depero, M. Farnworth, T. Fujimoto, A. Gibaud, M. Jergel, M. Krumrey, T.A. Lafford, A. Lamperti, T. Ma, R.J. Matyi, M. Meduna, S. Milita, K. Sakurai, L. Shabel'nikov, A. Ulyanenkov, A. Van der Lee, C. Wiemer, Reproducibility in X-ray reflectometry: results from the first world-wide round robin experiment. J. Appl. Crystallogr. **41**, 143–152 (2008)
21. International Tables for Crystallography IV
22. J.W.S. Rayleigh, Proc. Roy. Soc. **86**, 207 (1912)
23. I.W. Hamley, J.S. Pederson, Appl. Cryst. **27**, 29–35 (1994)
24. J. Als-Nielsen, Z. Phys. B **61**, 411 (1985)
25. A. Gibaud, G. Vignaud, S.K. Sinha, Acta Cryst. A **49**, 642–648 (1993)
26. F. Venturini, S. Schöder, W.F. Kuhs, V. Honkimäki, L. Melesi, H. Reichert, H. Schober, F. Thomas, J. Synchrotron Radiat. **18**, 251–256 (2011)

27. M.K. Sanyal, S.K. Sinha, A. Gibaud, K.G. Huang, B.L. Carvalho, M. Rafailovich, J. Sokolov, X. Zhao, W. Zhao, Europhys. Lett. **21**, 691 (1993)
28. D. Vaknin, P. Kruger, M. Losche, Phys. Rev. Lett. **90**, 178102 (2003)
29. C. Park, P.A. Fenter, J. Appl. Crystallogr. **40**, 290–301 (2007)
30. C.T. Kresge, M.E. Leonowicz, W.J. Roth, J.C. Vartuli, J.S. Beck, Nature **359**, 710–712 (1992)
31. C.J. Brinker, G.W. Scherer, *Sol-Gel Science: The Physics and Chemistry of Sol-Gel Processing* (Academic Press, San Diego, 1990)
32. A. Gibaud, D. Grosso, B. Smarsly, A. Baptiste, J.F. Bardeau, F. Babonneau, D.A. Doshi, Z. Chen, C.J. Brinker, C. Sanchez, J. Phys. Chem. B **107**, 6114–6118 (2003)
33. A. Gibaud, M.J. Henderson, M. Colas, S. Dourdain, J.F. Bardeau, J.W. White, AZoNano **1** (2005) (Online Journal of Nanotechnology)
34. S. Dourdain, M. Colas, J.F. Bardeau, O. Gang, B. Ocko, A. Gibaud, Int. J. Nanosci. **4**(5), 873–878 (2005)
35. S. Dourdain, J.F. Bardeau, M. Colas, A. Mehdi, B. Smarsly, B. Ocko, A. Gibaud, Appl. Phys. Lett. **86**, 113108 (2005)
36. S. Dourdain, A. Gibaud, Appl. Phys. Lett. **87**, 223105 (2005)
37. M. Mattenet, K. Lhoste, O. Konovalov, S. Fall, B. Pattier, A. Gibaud, SRI-2009 Proc. **1234**, 111–114 (2010)
38. M. Souheib Chebil, G. Vignaud, Y. Gorhens, O. Konovalov, M.K. Sanyal, T. Beuvier, A. Gibaud, Macromolecules **45**, 6611–6617 (2012)

Chapter 8
Resonant Photoelectron Diffraction

Alberto Verdini, Peter Krüger, and Luca Floreano

Abstract A layout of a resonant photoelectron diffraction, RESPED, experiment is described from the theoretical basis to the data acquisition and analysis procedures. The theory of the resonance between the directly emitted photoelectron of a selected valence band and the electron emitted by autoionization (Auger) of the same valence band is presented within a formal frame. The critical issue of the angular symmetry and distribution of the resonating electron is discussed in connection with the current computational protocols for photoelectron diffraction, PED, analysis. A few representative applications are presented, where RESPED is shown to overcome some limitations of conventional PED thanks to the possibility of enhancing the signal of low concentration species, providing a chemical selectivity in the valence band, and discriminating the contribution from atoms of the same species, but in different structural environments or valence state.

8.1 Introduction

Photoelectron Diffraction (PED) and Resonant Photoemission (RESPES) have been widely used in the last years, thanks to the increased number of synchrotron radiation facilities. PED is a structural technique that permits to probe the local atomic arrangement around a chemically selected emitter. On the other hand RESPES is a spectroscopic technique that can provide insight into the contributions of different atomic species to the electronic states in the valence/conduction region, thus allowing to attribute a line component to one or another element in a complex system. The combination of these two techniques can provide information not only on the

A. Verdini · L. Floreano (✉)
CNR-IOM, Laboratorio TASC, Basovizza SS-14, Km 163.5, 34149 Trieste, Italy
e-mail: floreano@iom.cnr.it

A. Verdini
e-mail: verdini@iom.cnr.it

P. Krüger
ICB, UMR 6303 CNRS-Université de Bourgogne, BP 47870, 21078 Dijon, France
e-mail: peter.kruger@u-bourgogne.fr

G. Bracco, B. Holst (eds.), *Surface Science Techniques*,
Springer Series in Surface Sciences 51, DOI 10.1007/978-3-642-34243-1_8,

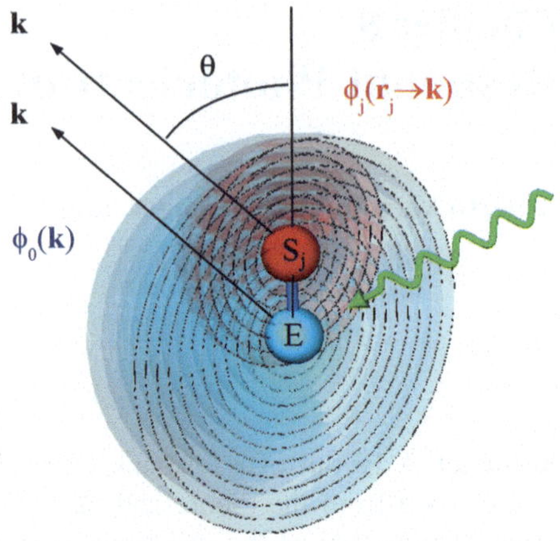

Fig. 8.1 The primary wave ϕ_0 is generated by the photoemission process of the emitter atom E. The secondary wave ϕ_j is the scattered wave by the atom S_j placed at a distance r_j from the emitter. θ is the scattering angle

electronic properties, but also on the order of the desired components. Moreover, it becomes possible to study the structure in those systems where the resonant effect can greatly enhance the signal with respect to the usual photoemission one. At variance with conventional PED, the photoemission at the resonance of selected ionization thresholds cannot be performed with a standard X-ray lamp, and the fine tuning of the photon energy provided by a Synchrotron radiation source is necessary. In this chapter we will give a brief overview of the PED, and RESPES techniques. We will address the data analysis procedure for the case of angularly diffracted photoelectrons and finally we will review a few cases where the resonant photoelectron diffraction, RESPED, technique has been applied. Additional reviews about PED and RESPES can be found in Refs. [1–3].

The basic experiment in PED involves the excitation of a core photoelectron from an atom embedded in a crystalline environment and then the measurement of the momentum modulations of the resulting peak intensities, due to the final-state scattering from the atoms (scatterers) surrounding the emitter (see Fig. 8.1). The photoelectron emission processes from two different atoms are uncorrelated in time. Therefore the measured intensity is a simple sum over the intensities of all emitting atoms.

The directly emitted photoelectron-wave exhibits interference with various scattered waves, and from this interference pattern the structural information is derived (see Fig. 8.2). The peak intensity can be monitored as a function of the emission direction or of the exciting photon energy. The diffraction pattern can be interpreted as the combination of different contributions. In particular, the photoelectrons emitted directly in the direction of the spectrometer, in the so-called forward scattering regime, produce intensity maxima when the emission occurs along a high symmetry direction because of the larger density of emitters aligned in the same directions. However, no information about the interatomic spacing is carried in forward scattering since there is no path difference between the emitted photoelectrons and the forward scattered ones (see Fig. 8.2). Moving away from the forward scattering peaks, the PED intensity stems from the interference between the primary wave (emitted

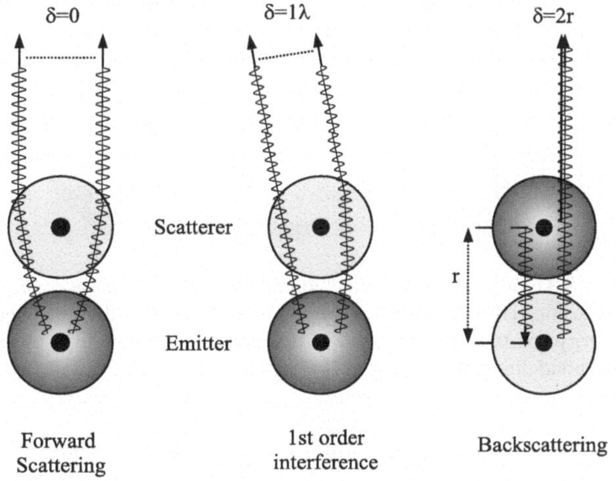

Fig. 8.2 Forward scattering cannot give information about the interatomic distances, since there is no path difference between the primary wave and the scattered waves. On the other hand interatomic distances can be obtained from the first order (or higher orders) and the backscattering interference fringes

photoelectron) and the secondary ones (scattered electrons), thus bearing the spatial information related to the path difference. In particular, the electrons emitted backward and then scattered forward bring the structural information about the interatomic spacing into the direction of forward scattering peaks. Finally, multiple scattering events also yield a complex diffraction pattern, which simulation allows the determination of the full atomic structure surrounding the emitter atom. The relative weight of these contributions depends on the kinetic energy of the photoelectrons: as shown for the case of Nickel [4, 5], forward scattering dominates the diffraction pattern at photoelectron energy higher than 500 eV (thus only allowing to detect the orientation of the crystal symmetry directions). At lower energy, also the backward scattering contribution becomes important, while higher order diffraction (thus including multiple scattering events) is relevant below 200 eV.

The mean free path for inelastic scattering of photoelectrons at the usual energies is in the range of 5–25 Å, [6] making PED a very surface sensitive diffraction technique. In conclusion, the photoelectron diffraction patterns represent the fingerprint of the local structure around a specific emitting atom lying in the first 3–5 atomic layers of a solid.

Both standard X-ray sources and synchrotron radiation can be used for excitation, with photon energies ranging from 50 eV to a few keV. Synchrotron radiation adds the capability of tuning the photon energy continuously and of studying the dependence of the diffraction on the orientation of the surface with respect to the radiation polarization. The most widespread scattering geometry in the experimental apparati (both with Synchrotron radiation and conventional X-ray laboratory sources) is characterized by a fixed angle between the photon source and the electron spectrometer (45°–55°). As a consequence, PED measurements are conventionally performed either by rotating the sample in front of the electron spectrometer or by varying the energy of the photon beam for a selected surface orientation. An azimuthal or ϕ scan, a polar or θ scan and an energy scan along a bond direction are depicted in

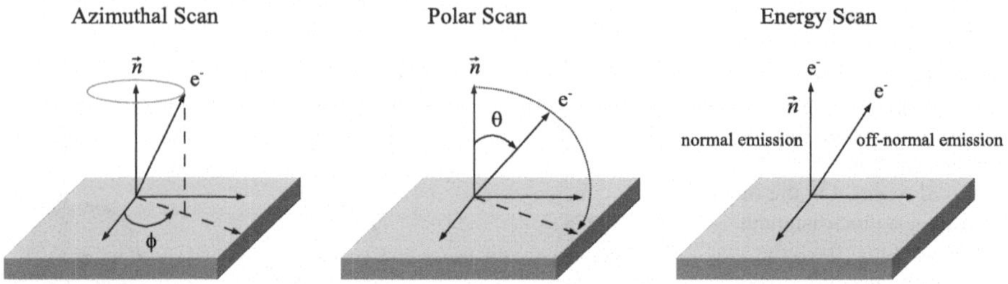

Fig. 8.3 Various experimental conditions: at fixed photon energy, the PED intensity modulations are scanned by either varying the azimuthal angle ϕ and/or the polar angle θ. By varying the photon energy and thus the electron kinetic energy, it is possible to observe intensity modulations along the bond directions

Fig. 8.3. A few experimental setups have been realized at Synchrotron radiation facilities, where the electron spectrometer is rather mounted on a goniometer, thus offering more degrees of freedom for the acquisition of PED patterns and offering the opportunity to exploit the polarization of the Synchrotron beam to enhance or suppress specific scattering contributions [7].

8.2 Theoretical Outline

Since the first theoretical paper on low energy photoelectron diffraction by Liebsch [8, 9], several discussions of the modeling of PED and Auger electron diffraction (AED) have appeared in the literature [1]. The first observations of strong diffraction effects in X-ray photoelectron emission from single crystals by Siegbahn et al. [10] and by Fadley and Bergstrom [11] were reported about 30 years ago. Ten years later, quantitative experimental studies were initiated by Kono et al. [12, 13], Woodruff et al. [14] and Kevan et al. [15]. Nowadays, photoelectron diffraction and its close relative Auger electron diffraction are widely used to study surface atomic geometries. An exhaustive treatment of the PED formalism [1] and calculation codes [16] can be found elsewhere. Hereafter we will rather emphasize details related to the resonant condition, while simply outlining the basic elements of multiple scattering calculations. The entire process of photoemission and interference between the primary wave and the secondary waves can be divided into three parts: the photoemission, the scattering due to the atoms in the solid and the surface refraction.

8.2.1 Resonant Photoemission

Resonant photoemission is the photoelectron emission process that occurs in the vicinity of a X-ray absorption threshold, that is with a photon energy around the binding energy of a core-level. Here we focus on resonant photoemission from valence states. In the photoemission process a photon is absorbed whereby one electron is excited from a valence state into the continuum and is detected at some energy ϵ_p. The process may be written as

$$|\Psi_g\rangle + \hbar\omega \xrightarrow{\text{rad}} |\Psi_f(\underline{v}), \phi_p\rangle$$

where Ψ_g is the N-electron ground state with energy E_g, $\Psi_f(\underline{v})$ is a $N-1$-electron final state with a valence hole ($\underline{v}$) and energy E_f, ϕ_p is the photoelectron wave function, and $\xrightarrow{\text{rad}}$ denotes a radiative (absorption) process. From energy conservation, we have $\epsilon_p = \hbar\omega - E_f + E_g$ and the spectral intensity is given by

$$I(\epsilon_p, \omega) = \sum_f \left|\langle\Psi_g|T(\omega)|\Psi_f(\underline{v}), \phi_p\rangle\right|^2 \delta(\hbar\omega - \epsilon_p - E_f + E_g) \tag{8.1}$$

In off-resonant photoemission, there is only the direct process, where the transition operator $T(\omega)$ corresponds to the interaction of the electrons with the photon field $V_{\text{rad}} \equiv -\frac{e}{mc}\mathbf{A}.\mathbf{p}$ where $\mathbf{A}$ is the vector potential of the light and $\mathbf{p}$ is the electron momentum. V_{rad} does not explicitly depend on the photon energy ω.

For photon energies around and above an absorption threshold, a new transition channel opens up, namely the photoabsorption from a core-level followed by the decay of the core-excited (intermediate) state through autoionization. In the latter process the excited electron and one valence electron scatter by Coulomb interaction such that one electron fills the core-hole and the other one is ejected as the photoelectron. In resonant photoemission the core-excitation and autoionization have to be considered as a coherent process which, in the case of participator decay, leads to the same final state as the direct photoemission process. A comparison between the different electron excitation channels is schematically depicted in Fig. 8.4, where the autoionization process is shown to take place along two possible decay channels.

The resonant process is

$$|\Psi_g\rangle + \hbar\omega \xrightarrow{\text{rad}} |\Psi_m(\underline{c}, k)\rangle \xrightarrow{\text{AI}} |\Psi_f(\underline{v}), \phi_p\rangle$$

where $|\Psi_m\rangle$ denotes a N-electron intermediate state with one core-hole ($\underline{c}$) and one electron in an excited state (k), and $\xrightarrow{\text{AI}}$ is the autoionization process. When the resonant process is included, the transition operator becomes, to the lowest order [19]

$$T(\omega) = V_{\text{rad}} + V_{\text{AI}} \sum_m \frac{|\Psi_m\rangle\langle\Psi_m|}{\hbar\omega + E_g - E_m + i\Gamma_m/2} V_{\text{rad}} \tag{8.2}$$

Here, $V_{\text{AI}} = e^2/r_{12}$ is the electron-electron Coulomb interaction which is responsible for the autoionization process and E_m and Γ_m are the energy and width of the intermediate state. The denominator $\hbar\omega + E_g - E_m + i\Gamma_m/2$ gives rise to a fast variation of the intensity around the threshold. Since the direct and resonant processes are added on the level of transition matrix elements (rather than intensities), they interfere, which leads to the characteristic Fano profile as a function of photon energy.

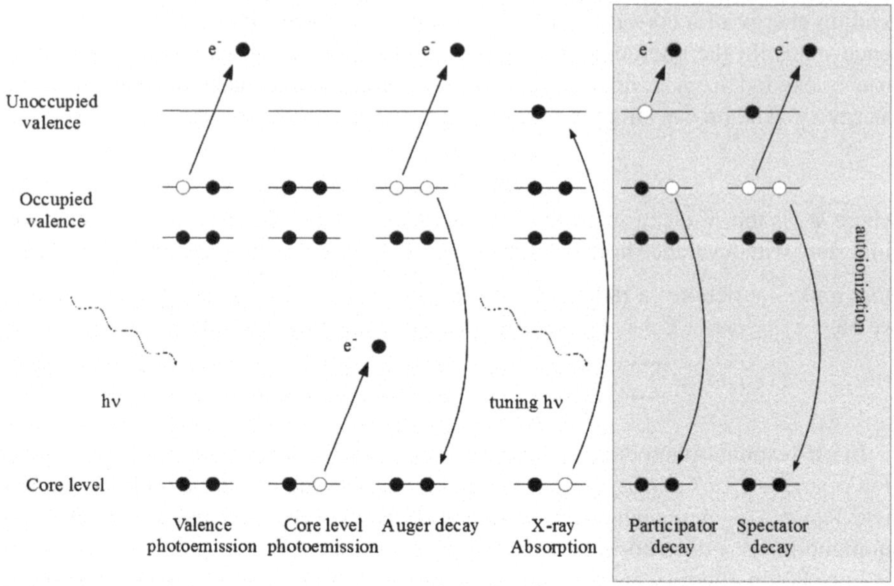

Fig. 8.4 Schematic diagram of the possible electronic excitation-deexcitation processes (also see review in Ref. [17])

Davies and Feldkamp [18] developed a theory of resonant photoemission based on the interaction between discrete states and continua. They applied the theory to the 3p-3d resonance of 3d transition metals, whose electronic structure was described by either a simple band or charge transfer model. While the calculations had strong model character, the general tendencies of the photon and binding energy dependence of resonant spectra could be explained. A much more realistic theory was presented by Tanaka and Jo [19] for resonant photoemission of transition metal oxides at the 2p threshold. The electronic structure was described with a charge-transfer cluster model which features full atomic multiplet interaction on the transition metal ion and hybridization to the oxygen ligand states. The many-electron Hamiltonian was diagonalized numerically by configuration interaction. The method was applied successfully to various transition metal oxydes [19, 20]. The theory was later extended to multiatom resonant photoemission, [21, 22] which experimental evidence is however controversial [23].

8.2.2 Angular Dependence of Resonant Photoemission

The former theories have focused on the photon and binding energy dependence of the resonant spectra but did not discuss the strong angular dependence of the valence band photoemission which manifests itself in angle-resolved photoemission spectroscopy (ARPES) and photoelectron diffraction. To account for these phenomena a precise description of the delocalized nature of the valence band states and the photoelectron continuum states is required. For non-resonant ARPES and PED, band structure and multiple scattering methods have proved very successful in this respect [24–26]. These methods are based on the independent particle approximation. When applying this approximation to resonant photoemission, all many-electron states $|\Psi_g\rangle$, $|\Psi_m\rangle$ and $|\Psi_f, \phi_p\rangle$ are taken as single Slater determinants and (8.1), (8.2) become:

$$I(\omega, \epsilon_p) = \sum_v \left| \langle \phi_p | V_{\rm rad} | \phi_v \rangle + \sum_{c,k} \frac{\langle \phi_p \phi_c | V_{\rm AI} | \phi_v \phi_k \rangle - [v \leftrightarrow k]}{\hbar\omega + \epsilon_c - \epsilon_k + i\Gamma_{ck}/2} \langle \phi_k | V_{\rm rad} | \phi_c \rangle \right|^2 \times \delta(\hbar\omega - \epsilon_p + \epsilon_v) \qquad (8.3)$$

where ($[v \leftrightarrow k]$) denotes the exchange term and all other notation is obvious from the foregoing. Cho et al. [27, 28] have calculated Ga 3d-4p resonant photoemission spectra of a GaAs(110) surface using the independent particle theory (8.3). The intermediate states ($\Psi_m(\underline{c}, k)$), were limited to the surface core exciton which was treated as an atomic excitation. Using a tight-binding scheme for the valence band states (ϕ_v), on- and off-resonant angle resolved spectra were calculated for specific points in the Brillouin zone and good agreement with experiments was achieved [28].

For direct valence band photoemission, the observed angular dependence is influenced both by the character of the initial band states and by electron diffraction of the photoelectron final state, i.e. it is a combined initial and final state effect [29]. The initial state band dispersion leads to an angular dependence within the limited solid angle range corresponding to one Brillouin zone. The angular range decreases quickly with increasing kinetic energy (e.g. 7° for 1000 eV and a lattice constant of 3 Å). At high energy, the resulting fast angular variation becomes largely smeared out by various broadening effects (finite inelastic mean free path, phonon broadening, recoil, etc.). Therefore, at high energy the angular dependence is dominated by final state photoelectron diffraction.

For strong resonances, such as 2p-3d in transition metals, the direct term can be neglected at the maximum of the resonance curve. The remaining resonant process is element selective, since it involves X-ray absorption from a core-level. Therefore, a resonant photoemission spectrum is approximately proportional to an element-projected density of states, rather than to the total density of states. This property of resonant photoemission is routinely used to "decompose" the valence band spectrum of a compound into its elemental contributions.

The physical reason for this lies in the fact that the Coulomb matrix element $\langle \phi_p \phi_c | V_{\rm AI} | \phi_v \phi_k \rangle$ in (8.3) is dominated by the on-site contribution, i.e. it is by far largest when ϕ_v and ϕ_k are located on the same site as ϕ_c. As a consequence, in resonant photoemission, the initial valence states become effectively localized on the core-level site.

The foregoing arguments indicate that the angular dependence of resonant photoemission is, especially for strong resonances and high kinetic energy, mainly due to diffraction of the outgoing photoelectron, while the band dispersion and delocalization of the initial valence states have a much weaker effect. This suggests that resonant photoelectron diffraction patterns may approximately be modeled in the same manner as standard X-ray photoelectron diffraction, i.e. as an incoherent sum over multiple scattered photoelectron waves emitted from all sites where the selected initial state is localized [26]. For core-level photoemission, this is a simple sum over all sites of the same element since the number of electrons in a core-shell is the same for all atoms of this element. For a selected valence state, however, the number of electrons depends furthermore on the crystallographic site, since the local density of states is different for inequivalent sites. So for photoelectron diffraction from valence states, the sum over emitter sites (of the same element) is to be weighted by the local density of states at the selected binding energy. This fact can be exploited to determine the charge distribution (among inequivalent sites of the same element) of specific states in the valence band.

8.3 Angular PED Analysis

In angular PED experiments, the Photoemission intensity is collected as function of the polar angle θ and/or of the azimuthal angle ϕ, as depicted in Fig. 8.3. A two dimensional pattern is obtained by spanning both the polar and the azimuthal range. Usually the azimuthal range is taken in between the main symmetry directions of the system. Particular care must be paid to the correction of any precession of the sample within the azimuthal range (with an angular precision at least one order of magnitude better than the angular acceptance of the electron spectrometer, which typically must not exceed a few degrees for PED angular measurements). The polar scan measurement can be performed by moving either the electron analyzer or the sample. The main difference is that in the first case the incidence angle, hence the illumination area, and the light polarization are fixed making simpler the data analysis. In general, the distribution of the intensity in the PED patterns is originated by two contributions: one anisotropic component χ, which is determined by the geometry of the local atomic structure, and one, slowly varying, microscopically isotropic component ISO which depends on both instrumental factors (such as sample illumination and detector angular resolution) and material dependent factors (such as atomic differential cross-section, film thickness/escape depth, surface morphology/roughness).

$$PED(\theta,\phi) = ISO(\theta,\phi)\cdot\big(1+\chi(\theta,\phi)\big) \tag{8.4}$$

8.3.1 The Isotropic Component

There are a few effective procedures to disentangle the ISO component from the χ function, which basically rely on standardized protocols to extract the signal, such as interpolation of the single polar scans (by a polynomial or a trigonometric function) or azimuthal averaging of the full angular pattern. In both cases, the strong intensity variation in correspondence with the forward scattering directions hampers a correct determination of the ISO component, thus affecting the amplitude of the oscillations of the anisotropy χ, which ultimately carries the structural information about the photoemitter environment. A more rigorous analysis of the different contributions to the ISO function has been dealt for the case of a rotatable analyzer with fixed sample orientation, which can be also applied to the reverse case of rotating the sample in front of a fixed electron spectrometer [30]. At a given azimuthal angle ϕ_0, the ISO component is represented as a function of the polar angle θ by the product of each contribution into the following functional form, where only the fitting parameters are indicated for the sake of clarity:

$$ISO = A \cdot \frac{d\sigma_{nl}}{d\Omega}(\beta) \cdot I_{IMFP}\big(D/\lambda, D'/\lambda'\big) \cdot I_R(\delta) \cdot I_{ILL}(\Gamma_{beam}, \alpha) \tag{8.5}$$

A simply represents a scale factor to be determined by the fitting of the PED polar scan. The atomic differential cross-section for the nl initial state of the photoemission process takes the analytical form of

$$\frac{d\sigma_{nl}}{d\Omega}(\beta;\gamma) \propto \left[1 + \frac{1}{2}\beta\big(3\cos^2\gamma - 1\big)\right] \tag{8.6}$$

where $\beta(\mathrm{nl}, h\nu)$ is the asymmetry parameter in the matrix elements for the photoemission process and γ is the angle between the electric field (polarization) and the photoelectron wavevector [31]; the value of γ depends on the specific geometric setup of the experimental chamber.

The inelastic mean free path I_{IMFP} takes into account the probability of escape of photoelectrons from the surface, as given by the escape depth λ. For an emitting slab of thickness D the escape photoelectron intensity is expressed by [32]

$$I_{IMFP}(D,\gamma,\theta) = \gamma\cos\theta\left[1 - e^{-\frac{D}{\gamma\cos\theta}}\right] \tag{8.7}$$

which takes the well known $\cos\theta$ form for a semi-infinite volume when $D \to \infty$. In the specific case of a non-emitting overlayer (as due to an heterogeneous thin film of thickness D'), the photoemitted intensity from a D-thick slab becomes:

$$I_{IMFP}\big(D,\lambda,D,\lambda'\big) = \gamma\cos\theta\left[1 - e^{-\frac{D}{\gamma\cos\theta}}\right]e^{-\frac{D'}{\gamma'\cos\theta}} \tag{8.8}$$

The attenuation due to the limited penetration of the photon beam can be neglected because the penetration depth is one order of magnitude larger than the electron escape depth at the typical photon energies for PED experiments (100–1000 eV). The attenuation of the photoelectron diffraction due to the surface roughness becomes relevant at large polar angle θ (grazing emission). The model of shadowing of the photoemitted electrons depends on the surface morphology. An analytical expression $I_R(\delta,\theta)$ can be found for the simplest and most common case of an isotropic normal (Gaussian) distribution of layer heights [33]. Qualitatively speaking, I_R assumes a constant unit value near normal emission, and drops to zero for $\theta \to 90°$, the steepness of the decay being determined by the amount of surface roughness. The shadowing can be completely neglected up to a maximum take-off value of the polar angle $\arctan\delta$, so that the phenomenological parameter δ can be

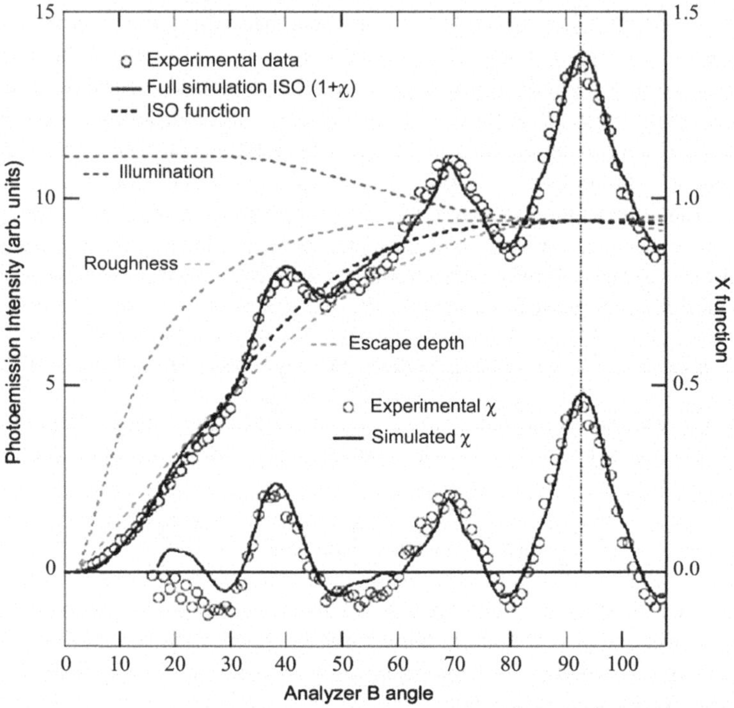

Fig. 8.5 *Upper curve*: Fe Auger LMM polar scan taken for a 36 Å Fe film grown on $Cu_3Au(001)$. The experimental AED patterns (markers) are shown together with the full simulation and the corresponding ISO function. Three additional curves are added, as obtained recalculating the ISO function by including only one factor at a time. *Bottom curve*: the anisotropy χ, after separation of the ISO function, and its simulation by multiple scattering calculations. Reprinted with permission from: F. Bruno, L. Floreano, A. Verdini, D. Cvetko, R. Gotter, A. Morgante, M. Canepa, S. Terreni, J. Elec. Spectrosc. Relat. Phenom. **127**, 85 (2002). Copyright 2002 by Elsevier

used as an effective marker of the surface roughness. Finally, the surface illumination $I_{ILL}(\Gamma_{beam}, \alpha)$ becomes a relevant component at large polar angles θ and/or small grazing angles α, where the mismatch between the projected field of view of the spectrometer and the illuminated area (projection of the beam transverse width Γ_{beam}) becomes large [30].

The individual weight of each component of the ISO function can be appreciated in the graphic of Fig. 8.5, where the anisotropy χ-function is extracted for a polar scan taken at the Fe Auger LMM line on a Fe film grown on the $Cu_3Au(001)$ surface [30]. In this specific case, the polar scan has been taken by keeping the surface at a constant grazing angle, while rotating the spectrometer in the scattering plane from the horizon to the surface normal. The opposite slope of the illumination factor with respect to both the roughness and the inelastic mean free path attenuation determines a strong deviation of the ISO function from a simple trigonometric slope close to the horizon.

8.3.2 The Anisotropy χ-Function

Since forward scattering gives rise to the major features in a PED pattern, it is possible to exploit it for a simple qualitative analysis, which yields the main symmetry directions. This simple analysis is very cost-effective for the study of heteroepitaxial growth of thin films, since it allows to follow the strain of the growing film by measuring the variation of the emission angle of the main forward scattering peaks as a function of the film thickness. In the case of a full structural determination, multiple scattering must be taken into account and a full simulation of the χ function is required. There are a few calculation codes developed by different research groups that allow the simulation of a PED pattern for a quantitative comparison. The best structural model is obtained by the minimization of a reliability factor that numerically represents the discrepancy between the experimental and simulated PED pattern, and takes the conventional form of [34]

$$R_{factor} = \frac{\sum_i (\chi_{th_i} - \chi_{exp_i})^2}{\sum_i (\chi_{th_i}^2 + \chi_{exp_i}^2)}, \tag{8.9}$$

where the χ_{th} and χ_{exp} are the simulated and experimental χ functions, and the index i runs on the single data points and corresponding simulation.

The calculated patterns typically display an amplitude of their modulations that is larger than the experimental ones, because the simulations cannot take into account the defects occurring in the real system, which smear the diffraction pattern [35]. In the numerical evaluation of the reliability factor, this inconvenient might be overcome by a modified R-factor

$$R_{factor}(A) = \min_A \frac{\sum_i (A\chi_{th_i} - \chi_{exp_i})^2}{\sum_i ((A\chi_{th_i})^2 + \chi_{exp_i}^2)}, \quad 0 \leq A \leq 1 \tag{8.10}$$

where A is a scaling factor which minimizes the R-factor for each χ_{th_i}. The A factor can be regarded as the effective portion of the XPS selected atomic species contributing to the PED pattern.

8.3.3 Simulation of PED and RESPED Angular Patterns

In the last years the PED technique has been widely used for studying structures at the surfaces and many codes for simulating the patterns, including multiple scattering, vibration effects and damping due to the mean free path of the electrons, have been proposed by several research groups: J.J. Barton and D.A. Shirley [36–38], V. Fritzsche [39–41], the MS code [42] and MSCD package [34], the EDAC code [43], the PAD code [44–46], MSPHD [47], and MS-Spec-1.0 [48]. These codes can be also used to calculate AED, and some of them are so general in the treatment of the electron scattering from the atoms that can be used to simulate even EXAFS, LEED and MEED patterns. Most of the codes include the possibility to calculate the atomic potentials for the scattering, either by Muffin-Tin approximation, or by employing non-isotropic and more realistic potentials obtained with dedicated programs [48–51]. For the RESPED technique a comment on the approximation adopted to describe the potential in the crystal is needed. In fact, the Muffin-Tin approximation, which has been developed to calculate the intensity variations in the LEED peak intensities [52] and adopted for PED, consists of assuming the potential to be spherically symmetric around each nucleus up to a certain radius and to be constant in the remaining space between atomic spheres:

$$V_{MT}(\mathbf{r}) = \begin{cases} V(|\mathbf{r}-\mathbf{R}|), & |\mathbf{r}-\mathbf{R}| < r_{MT} \\ constant, & |\mathbf{r}-\mathbf{R}| \geq r_{MT} \end{cases} \tag{8.11}$$

where $\mathbf{R}$ is a lattice vector, and r_{MT} is the chosen muffin-tin radius. This approximation may seem inadequate to describe the actual potential in a solid, which is approximately atomic-like very close to the nucleus (where the muffin-tin spherical approximation can be accurate), but follows the anisotropy of the charge distribution in the outer regions of the atom and in the crystal interstitials. Concerning PED, the muffin-tin approximation still holds as far as the kinetic energies are "high" (greater than 50 eV). This is because for high kinetic energy, electron-atom scattering is dominated by the nucleus and core-electron contribution. In the case of a typical RESPED experiment, the energies are much higher than 50 eV, for example the $L_{2,3}$ edges of transition metals for the first row of the periodic table range from 450 eV (Ti) to 950 (Cu) eV, putting the RESPED photoelectrons from the valence band far above the limits for the validity of the muffin-tin approximation.

The Muffin-Tin radius is set to half the distance between nearest neighbor atoms (d_{nn}). In fact, a slightly smaller value (90–95 %) is conveniently used in order to allow the relaxation of the crystal structure in the simulations without overlapping of neighbor atoms. If the crystal is made of different atomic species (a metal oxide for instance), the Muffin-Tin radius can be evaluated by using the atomic radius of each atomic species as a "weight":

$$r_{MT}(i) = \frac{d_{nn}}{2} \frac{r_{atomic}(i)}{\sum_i r_{atomic}(i)} \tag{8.12}$$

where $r_{MT}(i)$ is the muffin-tin radius for the i-th atom and $r_{atomic}(i)$ is the atomic radius.

The simulation of an angular distribution pattern for a quantitative comparison with the data (R_{factor} minimization), requires the calculation of the scattered intensity for a very large number of outgoing electron directions (of the order of thousand of calculated points) over clusters made of hundreds of atoms. The corresponding computation times are of the order of minutes for each angular pattern, thus requiring appropriate algorithms to optimize the minimization of the R_{factor}, such as the usual Levenberg-Marquardt method [53], or the most recent procedures based on genetic [54] and global search algorithms [55].

8.3.4 Surface Refraction

In the evaluation of the electron diffraction, the last angular contribution comes from the electron refraction at the surface in crossing the surface barrier. This barrier is introduced as a potential V_0 that changes the emission angle outside the sample following the conservation of the **k** component parallel to the surface [56]:

$$\sin\theta_{out} = \sin(\theta_{in})\sqrt{\frac{E_{out}+V_0}{E_{out}}} \tag{8.13}$$

where E_{out} is the kinetic energy outside the sample, θ_{in} and θ_{out} are the angles measured with respect to the surface normal. V_0 is treated as a parameter to be fitted and can assume values from 5 to 20 eV.

This effect may seem to be negligible in the case of RESPED measurements, where the usual Kinetic energies are much higher than the V_0. But if a PED pattern is measured with the polar angle down to 80 degrees and the V_0 is 10 eV, on the Ti L2 edge at 460 eV kinetic energy, the deviation $\theta_{out} - \theta_{in}$ starts to be 1 degree at $\theta_{in} \approx 60$ degrees, while for the Cu at 930 eV the deviation of 1 degree occurs at $\approx$73 degrees.

8.4 Application of Resonant PED

As seen before, Resonant photoemission allows to achieve chemical selectivity on the valence band of complex systems (such as alloys or molecules). In addition, the resonance condition enhances the intensity of the specific spectral line thus permitting the measurement of full angular PED patterns also on spectral features that would be normally too faint for a reliable detection with respect to the background of secondary electrons. The latter specific advantage can be exploited not only to enhance the valence band signal, whose photoemission cross-section drops quickly at the photon energy (hence electron kinetic energy) required for an useful PED study, but also to enhance the Auger transitions at the corresponding ionization edge, as may be required in the case of a low coverage of the selected atomic species, or in the case of searching specific electronic effects.

8.4.1 Resonant Valence Band PED

As seen before, when simulating the PED pattern from a resonant valence band state, one is faced with the problem of handling the angular dependence (hence symmetry) of the electron emission process. In dipole approximation, e.g. in direct photoemission, the final states depend on the l quantum number of the emitting core level. The resonating process can be thought as an electronexcitation process from an inner core level to an empty state below the vacuum level followed by the autoionization, where an electron fills the emptied core level and an electron in a filled state close to the vacuum level is emitted (Auger). In the examples shown hereafter, the resonating atom is a transition metal and the resonance process involves the electrons in the $3d$ levels (valence band). The final state (one hole in the valence band) corresponds to that of direct photoemission from the valence band, but with a much larger cross-section at the given photon energy. Currently no computational methods for PED include the autoionization process. Therefore only the direct process from a localized $3d$ level was considered. This is expected to be a fairly good approximation for $2p \rightarrow 3d$ RESPED for the following reasons. First, at resonance the initial valence states become effectively localized because the autoionization process is dominated by intra-atomic terms on the core-hole site. Second, the direct and resonant term produce similar angular photoelectron distributions, because the emitted waves have almost the same angular momenta. In fact, an electron outgoing from a $3d$-symmetry initial state ($l = 2$) has two possible angular momenta $l = 1, 3$ and the $LM_{4,5}M_{4,5}$ Auger electron (relevant to the specific transition metals considered hereafter) has three possible angular momenta $l = 1, 3, 5$; the $l = 3$ channel (f-symmetry) strongly dominates both the direct photoemission, because of the high energy behavior of the dipole matrix elements, and the Auger emission, as verified also in the condensed phase [57]. Finally, at the high kinetic energy relevant to the present cases, the PED pattern is known to be dominated by scattering effects and the angular momentum character of the emitted wave is of minor importance [58]. Hereafter we will review a few works where the RESPED from the valence band has been exploited to tackle structural issues of increasing complexity.

The Dilution of Mn into CdTe(110)

Mn-doped cadmium telluride belongs to the class of diluted magnetic semiconductors, DMS, that can be exploited for spintronic applications based on low-dimensional structures such as quantum wells and dots [59]. Because of the low dimensionality requirement, Mn doping can be achieved by simple evaporation of a small amount of Mn (a few monolayers) on the CdTe substrate and subsequent annealing to favor interdiffusion. Spectroscopic studies of the diffusion path suggested this process to take place via the incorporation of Mn atoms into the alloy crystalline matrix through the exchange with Cd [60], however a direct determination of the structural environment surrounding the Mn atoms was missing until very recently. Conventional PED is hampered by the overall low concentration of Mn. In addition, the Mn signal from a surface alloy prepared in-situ may be affected by contributions coming from clusters of metallic Mn. In principle, this aspect can be overcome by considering the signal from the Mn 2p core level that displays a characteristic splitting of the doublet associated with the crystal field of Mn in the alloy matrix (as opposed to the simple doublet of the metallic Mn). However, the

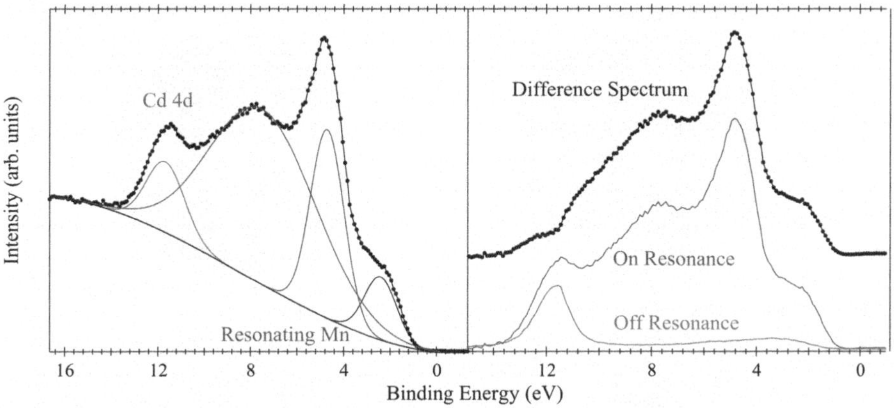

Fig. 8.6 Valence band from MnCdTe taken at the Mn L_3 edge. *Left*: a polynomial function is used to fit the background beneath the resonating spectra. *Right*: the pure resonating spectrum (top spectrum, markers) is obtained after subtraction of the background and subsequent subtraction of the off-resonance spectrum taken at a photon energy lower than the Mn 2p ionization threshold. Reprinted figure with permission from: L. Sangaletti, S. Pagliara, F. Parmigiani, A. Goldoni, L. Floreano, A. Morgante, and V. Aguekian, Phys. Rev. B **67**, 233201 (2003). Copyright 2003 by the American Physical Society

Mn 2p photoemission spectrum overlaps with the Cd 3p one, making unreliable a conventional PED study. On the contrary, resonant PED on this system can take advantage of the peculiar characteristic of DMS systems to display giant resonant photoemission.

This system displays a giant resonance at the Mn $L_{2,3}$ ionization threshold associated with the final-state interference between direct photoemission from the 3d levels $2p^63d^5 + h\nu \rightarrow 2p^63d^4 + e_f$ and the autoionization processes $2p^63d^5 + h\nu \rightarrow 2p^53d^6 \rightarrow 2p^63d^4 + e_f$, where e_f is the emitted electron [61]. Practically, the whole spectral region of the valence band between 2 and 10 eV, that is dominated by the Mn 3d states (see Fig. 8.6), is enhanced by more than one order of magnitude when measured at a photon energy of 640.5 eV, which corresponds to the maximum of the Mn 2p $\rightarrow$ 3d absorption edge [62]. Performing a resonant PED measurement yields a large signal and offers the advantage of being sensitive only to the Mn atoms corresponding to the magnetic alloy of interest.

Sangaletti and coworkers prepared the $Mn_xCd_{1-x}Te(110)$ surface by deposition of 2 ML of Mn at room temperature, without further annealing. They collected a two-dimensional angular PED pattern of the valence band measured with a photon energy tuned at the Mn L_3 edge (640.5 eV). The same angular PED was also taken for the Cd and Te 4d photoelectrons at a kinetic energy of 594.5 and 627 eV, respectively, i.e. almost the same of the resonating Mn 3d photoelectrons, thus allowing a direct comparison of the bare diffraction patterns. The angular distribution of the photoemission intensity $I(\theta, \phi)$ was measured for polar angles θ from the surface normal down to 20° above the horizon and for an azimuthal range ϕ spanning in

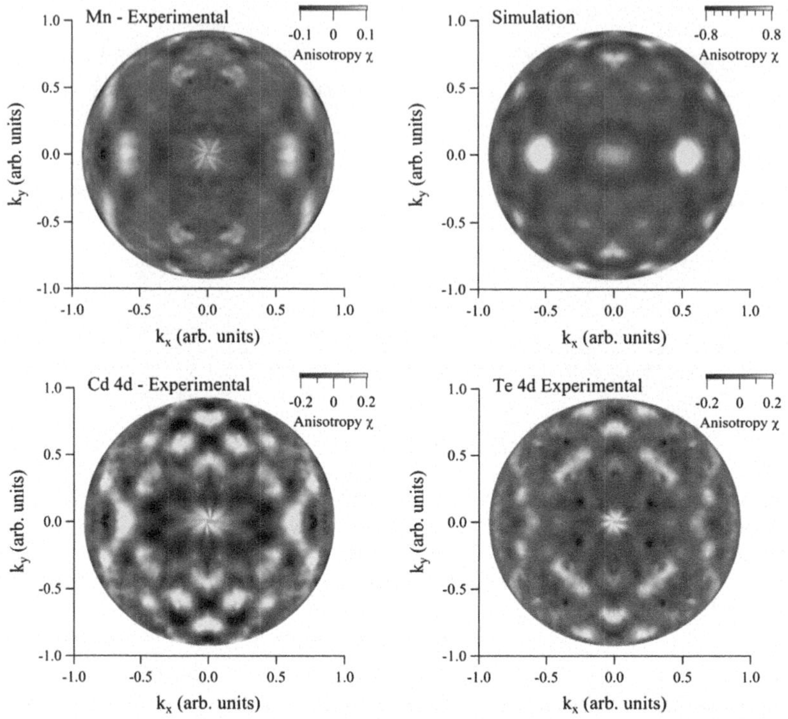

Fig. 8.7 *Top left*: measured anisotropy χ obtained by considering the signal integrated over the resonantly enhanced valence electron emission. *Top right*: calculated anisotropy for the substitutional Mn atoms distributed among 4 layers. *Bottom left*: measured anisotropy from the Cd 4d core levels (photoelectron kinetic energy = 627). *Bottom right*: measured anisotropy from the Te 4d core levels (photoelectron kinetic energy = 594.5). Reprinted figure with permission from: L. Sangaletti, A. Verdini, S. Pagliara, G. Drera, L. Floreano, A. Goldoni, A. Morgante, Phys. Rev. B **81**, 245320 (2010). Copyright 2010 by the American Physical Society

between the main symmetry directions [001] and $[1\bar{1}0]$ $\pm 5°$. The full 2π PED patterns were built by four-folding the experimental one around the main symmetry axis. The experimental patterns obtained for Mn, Cd and Te are reported in Fig. 8.7. By a simple visual inspection, one can notice the resemblance of the most prominent structures in the Mn and Cd patterns, thus indicating the effective exchange between the two species. For a quantitative comparison, simulations were performed using the MSCD package [34], where the resonant process has been approximated with the emitted wave from a Mn 3d level. For the sake of simplicity, an unrelaxed CdTe(110) crystal structure was considered. By simulating the PED patterns for slabs of increasing thickness, the authors of Ref. [62] demonstrated that already at room temperature, Mn atoms can diffuse into Cd substitutional sites down to the 4th layer beneath the surface. Calculated patterns for Mn atoms in interstitial sites (not shown) were also found to yield major discrepancies for the most prominent χ modulations, thus excluding this alternative diffusion path.

The Oxidation States of Fe into $Fe_3O_4(111)$

Magnetite Fe_3O_4 is a room-temperature ferrimagnet, whose Fe atoms display the occurrence of different oxidation states [63]. In the crystalline structure (cubic inverse spinel), the Fe cations are located either on tetrahedral sites, A, or on octahedral sites, B. At room temperature, one third of Fe atoms are located on A site as Fe^{3+} (A-Fe^{3+}), while the remaining two thirds of the Fe ions reside in B sites, equally, but randomly distributed between two different oxidation states, namely B-Fe^{2+} and B-Fe^{3+}. The magnetic moments of the A-site and B-site Fe ions are in an antiparallel alignment below 860 K, forming a ferrimagnet.

Notwithstanding the long standing research efforts on this system, the fine details of the electronic structure are still debated. An ordering of Fe^{2+} and Fe^{3+} in B sites was proposed to explain the Verwey transition [64] at about 120 K from a metallic ferrimagnet to an insulator. However, the very existence of the two B-Fe^{2+} and B-Fe^{3+} states was also questioned, in favor of an effective 2.5+ valence state common to all the B sites [65]. The discrimination among the different A/B sites and 2+/3+ valence states is hampered in photoemission by the strong overlap and large width of the corresponding components in the 2p core level [66], thus preventing a conventional structural PED study. On the contrary, magnetic circular dichroism absorption spectroscopy clearly shows sharp and distinct features in the difference spectra of the three Fe species [67]. Chen and coworkers demonstrated that the contributions to the valence band, VB, from these three species are mixed, but they can be selectively enhanced by resonant photoemission taken at the photon energy corresponding to the distinct features of the L_3 edge [20]. Magnan and coworkers exploited this ion selective enhancement of the VB to probe the crystal environment around each Fe ion by means of resonant PED [68].

In Fig. 8.8, the X-ray absorption spectrum at the Fe L_3 edge as well as the corresponding XPS spectra of the valence band are shown for the $Fe_3O_4(111)$ surface. The valence band displays a strong intensity enhancement at the Fe L_3 edge, and different VB structures are seen to resonate at different photon energies. The structure near the Fermi level (at a binding energy of $\sim$1.2 eV) is maximum at a photon energy of 707.5 eV, while the VB structures at higher binding energy near 4.4 eV and 6.6 eV resonate at higher photon energy 708.9 eV and 709.6 eV, respectively. By comparison with the calculated XAS spectrum [69], the component that resonates at 1.2 eV binding energy is predicted to stem only from the B-Fe^{2+} species [20]. On the contrary, the other two VB components (4.4, and 6.6 eV) that are mainly attributed to A-Fe^{3+} and B-Fe^{3+} respectively, are also participated by a relevant contribution of B-Fe^{2+} (see Fig. 8.9).

In order to experimentally address the nature of the VB electronic structure, Magnan and coworkers [68], performed resonant PED measurements, thus following the reversed path from the structure (assumed to be known) to the assignment of the spectroscopic features. Azimuthal scans of the VB spectra have been taken at three different photon energies of 692, 707.5, and 709 eV corresponding to off-resonance condition, to the maximum of the B-Fe^{2+} absorption, and to the maximum of the A-Fe^{3+} absorption, respectively. With this choice, the VB resonant PED patterns

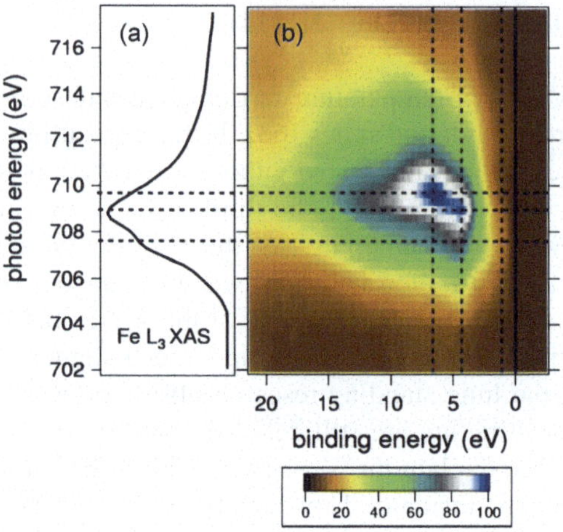

Fig. 8.8 (**a**) An experimental absorption spectrum at the Fe L3 edge recorded in partial electron yield during the acquisition of resonant VB; (**b**) 2D plot of the resonant VB spectra recorded on Fe_3O_4 in p polarization. The spectra are plotted as a function of the binding energy (horizontal scale) and for the different photon energies across the Fe L3 edge (vertical scale) with the same scale than in (**a**), the colors indicating the photoemission intensity as a function of these two variables. *Vertical dashed lines* indicate the position of 1.2, 4.4, and 6.6 eV binding energy. *Horizontal lines* indicate the position of 707.5, 708.9, and 709.6 eV photon energy. Reprinted figure with permission from: H. Magnan, P. Le Fèvre, D. Chandesris, P. Krüger, S. Bourgeois, B. Domenichini, A. Verdini, L. Floreano, and A. Morgante, Phys. Rev. B **81**, 085121 (2010). Copyright 2010 by the American Physical Society

are associated with a specific species, discriminating the A site and the B sites. For the simulations (with the EDAC package [43]), Magnan et al. used the bulk lattice parameters of magnetite. In addition, a full PED pattern from the Fe 3p taken at the same kinetic energy of the resonant VB spectra was simulated in order to tune appropriately the parameters of the simulation. A few selected scans are reported in Fig. 8.10, where the anisotropy function χ has been determined for the integrated intensity of the valence band in two energy ranges: from Fermi level to 1.5 eV (bottom spectra, FL) and from Fermi level to 8.5 eV (top spectra, VB).

The experimental data are compared with the best simulations for three structural configurations: A site only, B site only, and the sum of A and B sites. When considering the whole VB intensity, it is seen that the best simulation is obtained by considering the sum of A and B sites for the patterns taken off-resonance, as expected. Only B sites are needed, when the photon energy is tuned to 707.5 eV; while at 709 eV, both A and B sites are needed again. The simulations confirm the mixed contribution to the absorption spectrum predicted at the maximum of the A-Fe^{3+} absorption, and the prevalent B character at the maximum of the B-Fe^{2+} absorption, that dominate the whole valence band [69]. Finally, the PED simulations for the state close to the Fermi level (integrated intensity from FL to 1.5 eV) indicate that it is contributed only by Fe atoms in B sites (both off and on resonance), which is consistent with the prediction of a localization on B-Fe^{2+} atoms of the state at

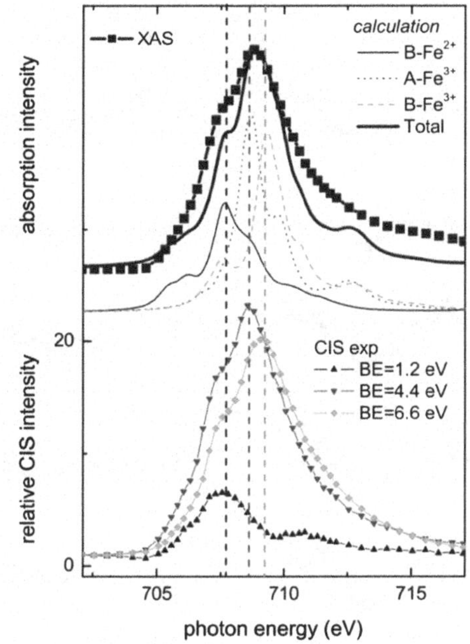

Fig. 8.9 *Top part*: experimental absorption spectrum (*filled squares*) and calculations of site-resolved contributions to the absorption spectrum of Fe_3O_4 for B-Fe^{2+} (*thick full line*), A-Fe^{3+} (*dotted line*), B-Fe^{3+} (*dashed line*), and total absorption (*full line*) [69]. *Vertical lines* indicate the position of the maximum of the three Constant Initial State, CIS, spectra. *Bottom part*: experimental VB-CIS spectra recorded on Fe_3O_4 at the Fe L_3 edge for 1.2 eV (*black up-triangle*), 4.4 eV (*gray down-triangle*), and 6.6 eV (*gray squares*) binding energies. The Constant Initial State spectra are normalized to the intensity at 705 eV. Reprinted figure with permission from: H. Magnan, P. Le Fèvre, D. Chandesris, P. Krüger, S. Bourgeois, B. Domenichini, A. Verdini, L. Floreano, and A. Morgante, Phys. Rev. B **81**, 085121 (2010). Copyright 2010 by the American Physical Society

1.2 eV [20]. The RESPED technique thus allowed to discriminate between the Fe atoms in the two A and B sites. We remark that the RESPED technique was not able to discriminate among the valence state of the Fe atoms, because of the overlap of contributions to the valence band from both A and B Fe3+ ions. In this regard, the RESPED analysis remains valid also in the case of equivalent (single valency with formal charge 2.5+) Fe ions in the B sites, as proposed in Ref. [65], and the assignments of Fig. 8.10 would simply relate to the crystallographic sites, independent of the Fe valency.

The Localization of the Defect State in $TiO_2(110)$

Within the field of electronic devices, TiO_2 is instinctively associated with the engineering of dye sensitized solar cells [70, 71], but its high dielectric constant makes it a valuable material also for the fabrication of low threshold voltage and high output

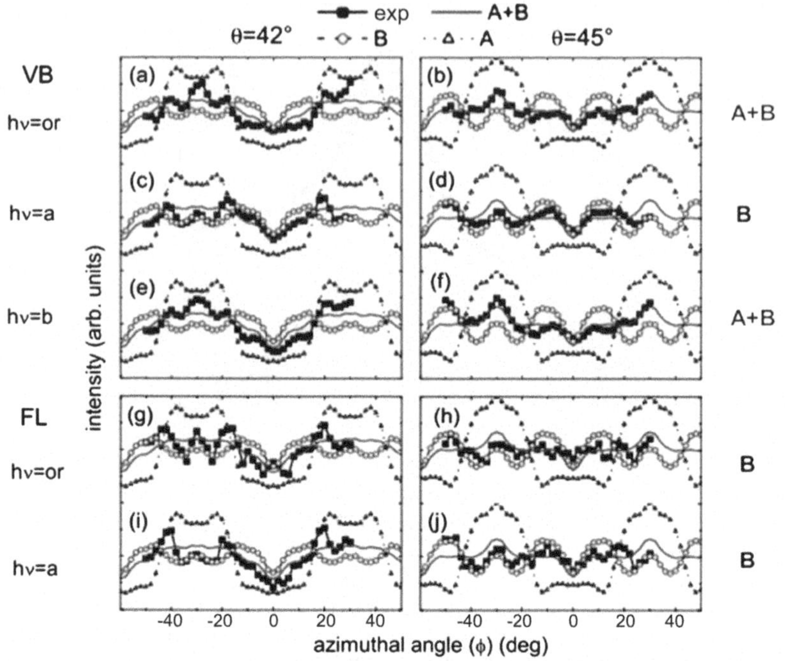

Fig. 8.10 Experimental azimuthal scans (*filled squares*) recorded at polar angle $\theta = 42°$ (*left column*) and at $\theta = 45°$ (*right column*). The photon energy is indicated on the *left*, or = 692 eV, a = 707.5 eV, and b = 709 eV. The signal is the integration of the VB intensity (**a**)–(**f**) from the Fermi level to 8.5 binding energy and (**g**)–(**j**) from Fermi level to 1.5 eV. Comparison with multiple scattering calculations for all sites of iron (*full line*), for B-sites of iron (*open circles*) and for A sites of iron (*open triangles*). The best model for each case is indicated on the *right* of the figure. Reprinted figure with permission from: H. Magnan, P. Le Fèvre, D. Chandesris, P. Krüger, S. Bourgeois, B. Domenichini, A. Verdini, L. Floreano, and A. Morgante, Phys. Rev. B **81**, 085121 (2010). Copyright 2010 by the American Physical Society

current OTFTs [72]. In particular, the rutile $TiO_2(110)$ surface attracts much attention thanks to the possibility of changing its catalytic and charge transport properties by trimming the concentration of bridge-Oxygen vacancies, O_b-vac, in the surface region (either by thermal annealing or by ion bombardment) [73]. In fact, the desorption of Oxygen atoms leads to the appearance of a new electronic state in the band gap [74], which makes the surface conductive. This electronic effect is further enhanced upon the formation of a (1×2) surface reconstruction, induced by high temperature annealing, where the surface becomes metallic [75]. Resonant photoemission experiments demonstrated that the new band gap states have mainly a Ti 3d character [76].

Within a very simple ionic model, it is possible to think that the two excess electrons made available by an O vacancy fill the 3d states of two nearby Ti atoms, giving rise to Ti^{3+} species [74]. Alternative models suggest the defect state to be associated with the occurrence of interstitial Ti atoms in the subsurface [77] or

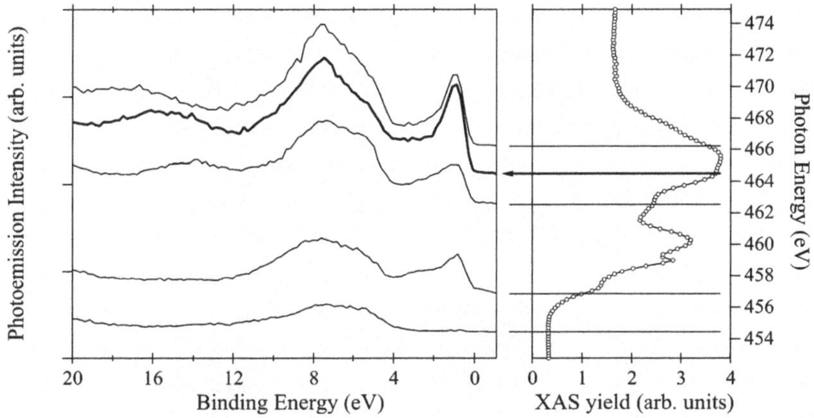

Fig. 8.11 On the *left side*, a few experimental spectra of the non-stoichiometric TiO_2 valence band are shown as taken across the Ti L_2 edge. The corresponding partial electron yield X-ray absorption spectrum is shown on the *right side*. Notice the variation in intensity of the defect state peak from off-resonance conditions (photon energy $\leq$ 454 eV) to the maximum of the L_2 edge (462–465 eV). The *arrows* associate the valence band spectra with the corresponding photon energy

even deeper layers [78], which would transfer part of their valence charge to the neighboring Ti atoms (each Ti can reduce three other Ti atoms, yielding four Ti^{3+} ions in total). Regarding the localization of Ti^{3+} ions, theory is far from consensus, since localized states in wide band gap oxides remain a challenge for first principles theory. As a matter of fact, also *ab initio* calculations that include self-interaction corrections for better evaluation of the localized band gap states yield contradicting results, where the excess electrons are expected to be localized either on surface Ti atoms next to O_b-vac [79] or on subsurface sites beneath O_b-vac [78], eventually depending on the overall concentration of O_b-vac [80, 81].

Resonant PED proved to be the best suited tool for probing the spatial distribution of the excess electrons, which yield the defect state in the band gap. As can be seen in Fig. 8.11, the photoemission from the defect state is strongly enhanced at the Ti 2p $\rightarrow$ 3d absorption edge [82], thus allowing the mapping of its emission angular distribution. For this experiment, Krüger and coworkers [83] prepared a sample by a standard protocol of Ar^+ ion bombardment followed by a mild annealing to 870 K for 45 min, which recovers the long range order over the surface, while leaving a few percent concentration of O_b-vac as point defects. The photoemission intensity $I(\theta, \phi)$ was measured for polar angles from the surface normal to 80° from it and for azimuthal angles ϕ over a range of 150° including the two symmetry directions [001] and [$1\bar{1}0$]. The data were symmetrically folded according to the symmetry of the TiO_2(110) surface. The RESPED from the defect state was recorded at the maximum of the L_2-edge resonance in correspondence of the spectral feature marked with the arrow in Fig. 8.11. In addition, a non resonant PED pattern was taken from the Ti $2p_{3/2}$ core level at the same kinetic energy of the RESPED one. The two patterns are reported in Fig. 8.12. Large differences between the two patterns can

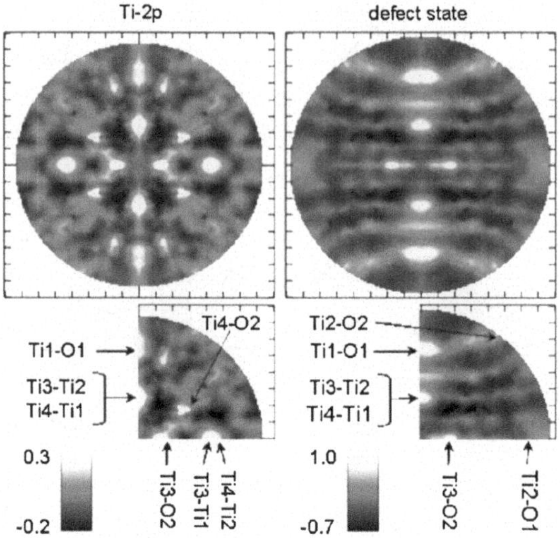

Fig. 8.12 Experimental photoelectron diffraction (PED) patterns. *Left*: Standard PED from the Ti2p$_{3/2}$ core level. *Right*: Resonant PED from the defect state. The kinetic energy is the same for both patterns (460 eV). The projection is linear in θ with the surface normal at $\theta = 0$ in the center. Ticks are drawn every 10°. The azimuth $\phi = 0$ and $\phi = 90$ are found at 3 o'clock and 12 o'clock, respectively and correspond to the $[1\bar{1}0]$ and [001] directions. Reprinted figure with permission from: P. Krüger, S. Bourgeois, B. Domenichini, H. Magnan, D. Chandesris, P. Le Fèvre, A.M. Flank, J. Jupille, L. Floreano, A. Cossaro, A. Verdini, and A. Morgante, Phys. Rev. Lett. **100**, 055501 (2008). Copyright 2008 by the American Physical Society

be detected by simple visual inspection, which already tells that the defect charge is not uniformly distributed over all Ti sites (as is the Ti 2p charge).

The resonant photoemission has been approximated by calculating the direct photoemission from localized Ti 3*d* orbitals and averaged over all the magnetic quantum numbers m_l. In order to evaluate the relative weight w_n for each Ti^{3+} contribution to the defect state charge distribution, the simulations (with the MSCD code [34]) have been performed independently for photoemitters in different Ti_n sites. Then the anisotropy function χ_{th_i} has been evaluated by a linear combination $I_{th} = \sum_n w_n I_n$ of the intensities I_n, independently calculated for each Ti_n emitting site. By minimization of the R-factor, the charge excess associated with the defect state was found to be redistributed among multiple sites. As shown in the sketch of Fig. 8.13, most of the contribution stems from the Ti_3 lattice site in the first subsurface layer, followed by the Ti_5 site in the second subsurface layer (just beneath the O_b rows) and only minor weight on surface Ti atoms. The possible occurrence of interstitial sites has been checked by simulating the RESPED pattern for any one of the four possible interstitial sites in the first and second subsurface interlayer. In any case, all of the interstitial sites yielded R-factors close to 1 when individually tested (i.e. no common features between simulation and experimental data). Moreover, when the Ti interstitial were included in the R-factor minimization, the weight was always zero for all four sites, thus the presence of interstitial Ti atoms nearby

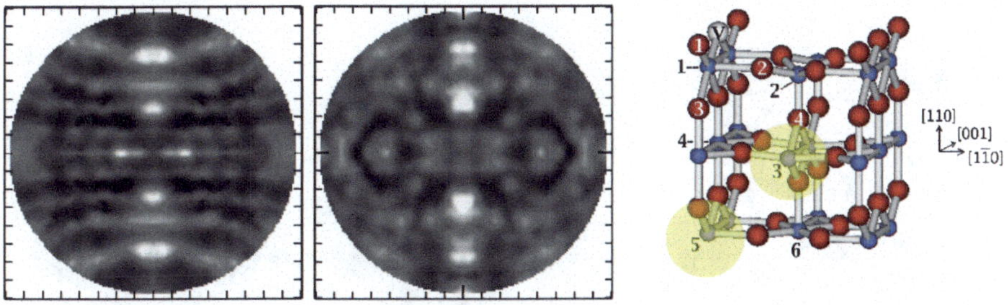

Fig. 8.13 *Left panel*: Resonant PED from the defect state recorded at a kinetic energy of 462 eV. *Central panel*: best fit MSCD simulation with multiple emitter sites. *Right panel*: structure and excess charge distribution in TiO_2. The Ti sites yielding the largest contribution to the defect state are highlighted by shadowing. Reprinted figure with permission from: P. Krüger, S. Bourgeois, B. Domenichini, H. Magnan, D. Chandesris, P. Le Fèvre, A.M. Flank, J. Jupille, L. Floreano, A. Cossaro, A. Verdini, and A. Morgante, Phys. Rev. Lett. **100**, 055501 (2008). Copyright 2008 by the American Physical Society

the surface can be excluded. Finally, the presence of O_b vacancies was found to have little influence on the simulated patterns, thus being hardly detectable by the RESPED measurements.

8.4.2 Resonant Auger ED

When exploiting resonance conditions to enhance the signal from an Auger electron, one is faced to deal with an overall symmetry of the primary wave that stems from multiple angular momentum contributions corresponding to the decay channels of the electron transition. Each angular momentum contribution brings a different scattering angular distribution, at least for low and medium ($\leq$300–500 eV) electron kinetic energy, where backward and multiple scattering events yield a large weight into the diffracted intensity [84]. While such a calculation can be performed by current diffraction packages, it is common practice to perform Auger diffraction experiment at high electron kinetic energy where forward scattering effects dominate the diffraction pattern and the emitted electron can be treated as a s wave, irrespective of the initial state symmetry [58]. This approach has been adopted also for the two following experiments of resonantly enhanced Auger electron diffraction experiments.

Intramolecular Ordering of the Endohedral Fullerene $Dy_3N@C_{80}(I_h)$ on Cu(111)

The metal-containing endohedral fullerenes represent a class of materials that allows to manipulate the metallic clusters incorporated inside the fullerene cage by simple self-assembly onto different substrates. This property can be exploited for the design

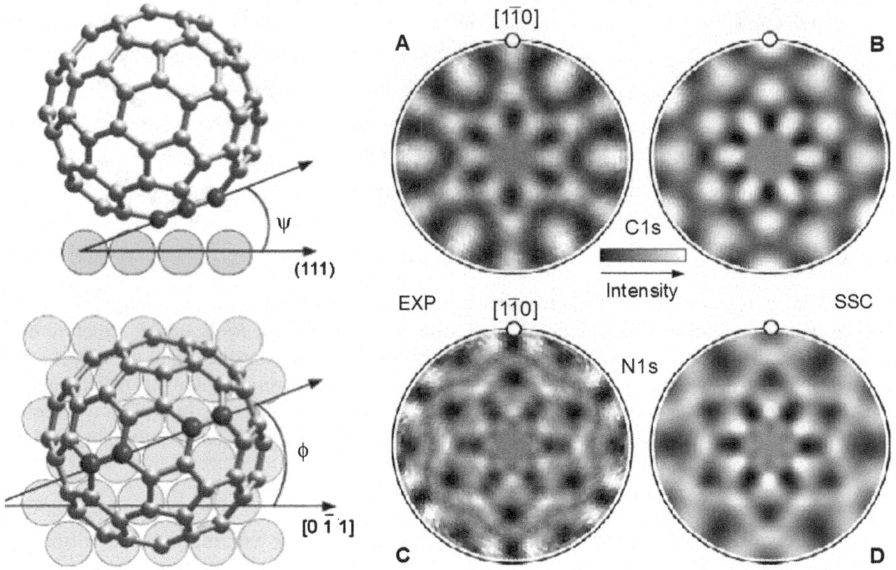

Fig. 8.14 PED patterns from the monolayer phase of $Dy_3N@C_{80}(I_h)$ on Cu(111), as measured at a photon energy of 880 eV. *Panel* **A**, Experimental C1s PED anisotropy; *panel* **B**, simulated C1s anisotropy; *panel* **C**, experimental N 1s PED anisotropy; *panel* **D**, simulated N 1s anisotropy. Reprinted figure with permission from: M. Treier, P.I. Ruffieux, R. Fasel, F. Nolting, S. Yang, L. Dunsch, T. Greber, Phys. Rev. B **80**, 081403(R) (2009). Copyright 2009 by the American Physical Society

of novel device architectures. A suitable choice of the encaged metal allows to trim the charge transfer between the metal cluster and the fullerene, thus determining the optoelectronic and electrochemical behavior [85]. In particular, the incorporation of rare-earth metals has been shown to yield unique magnetic properties, such as large magnetic moment [86]. This aspect makes them suitable candidates for the fabrication of nanoscale storage devices and the ordering of the molecules within supported layers becomes a central issue for their exploitation. Crystallographic studies have shown the emergence of a correlation between the endohedral units in the condensed phase, from which the structure, orientation and relative position of the encaged metals has been determined [87]. Understanding whether a correlation is preserved also on 2D molecular arrays is hampered by the limitation of conventional scanning probe microscopes, that can determine the orientation of the fullerene cage, but cannot directly probe the orientation of the encaged metal cluster [88, 89].

In this regard, PED is a powerful structural technique because of its chemical selectivity and it was applied to study the molecular and intramolecular orientation of the trimetallic nitride endohedral fullerene $Dy_3N@C_{80}(I_h)$ in a monolayer phase grown on Cu(111) [90]. The surface was prepared by thermal evaporation of the molecule at 770 K, thus forming a $(\sqrt{19}\times\sqrt{19})R\pm 23.4^\circ$ superlattice at completion of the first monolayer, as verified by LEED. A preliminary STM survey revealed

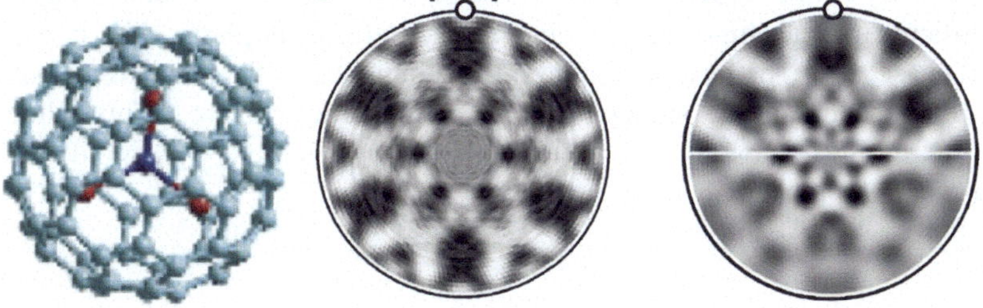

Fig. 8.15 *Left panel*: ball and stick model of the $Dy_3N@C_{80}(I_h)$ molecule, the N atom (in *blue*) at the center of the fullerene is coordinated to three metal atoms (in *red*). *Central panel*: Dy $M_5N_{6,7}N_{6,7}$ AED pattern of 1ML endohedral fullerene on Cu(111), recorded at a photon energy of 1291.5 eV (1281 eV kinetic energy). *Right panel*: best fit SSC simulation (*upper half*, see text for details) and simulation based on disordered endohedral unit (*lower half*). Reprinted figure with permission from: M. Treier, P.l. Ruffieux, R. Fasel, F. Nolting, S. Yang, L. Dunsch, T. Greber, Phys. Rev. B **80**, 081403(R) (2009). Copyright 2009 by the American Physical Society

that the molecules are facing the substrate with one of the hexagons and the cage is slightly tilted off the surface, but a quantitative evaluation was inhibited because of the difficulty of disentangling structural and electronic effects in conventional STM topography. The authors of Ref. [90] measured a full 2π PED pattern from the C 1s core level at a kinetic energy of ~600 eV (see panel A of Fig. 8.14). Although simple single scattering cluster, SSC, simulations [1] were performed and the backward scattering from the substrate was also neglected, the anisotropy of the PED pattern was large enough to determine the spatial orientation of the C_{80} cage, in particular the bottom hexagon was found to be tilted off the surface by $3° \pm 2°$. The orientation of the carbon cage was also confirmed by simulations of the PED pattern taken for the N 1s core level at a kinetic energy of 440 eV (see panels C and D in Fig. 8.14). In addition, the endohedral nitrogen atom was found to stay close to the center of the fullerene cage even in the adsorbed monolayer phase.

Contrary to the case of C and N atoms, the measurement of a PED pattern from the direct photoemission of Dy 4d (binding energy of ~154 eV) was hampered by its intrinsic very large width, accompanied by a significant overlap to the strong tail of secondary electrons from Cu 3s (binding energy of ~122 eV). In this case the authors of Ref. [90] rather measured the diffraction pattern of the Auger electrons resonantly enhanced at the Dy 3d $\rightarrow$ 4f transition, where the Auger $M_5N_{6,7}N_{6,7}$ signal at a kinetic energy of 1291.5 eV is strongly enhanced, while the signal from the Cu 3d valence band remains practically constant. When analyzing the AED pattern (bottom panel of Fig. 8.15, one must recall that, because of the ordering of the fullerene cage, an anisotropy pattern is expected also in the case of a randomly disordered distribution of the endohedral Dy atoms. In the latter case, a very low anisotropy would be expected on the basis of SSC simulations, which is not consistent with experimental data. The authors of Ref. [90] concluded that there exists a correlation between the orientation of the carbon cage and the endohedral unit, but a single orientation of the latter is not sufficient to reproduce the observed anisotropy and symmetry of the angular pattern. Because of the overall low anisotropy it was also difficult to determine whether the endohedral unit is planar, like in the condensed phase [87]. The best fit simulation was obtained as the incoherent sum from two different orientations, one with a planar endohedral structure slightly tilted with respect to the (111) substrate orientation and one with a slightly pyramidal unit, with the base parallel to the substrate surface.

Disentangling the Magnetic Dichroism from the Diffraction Pattern Rotation Induced by Photon-Electron Angular Momentum Exchange

Apart from refraction at the surface-vacuum interface, the electrons escaping from the surface can be deviated by angular momentum exchange with the incoming photon. When circularly polarized light is used to collect a PED pattern, one observes that the forward scattering peaks are (azimuthally) rotated around the surface normal according to the polarization direction. This effect can be exploited to obtain a stereographic view of the emitter structural environment by taking two PED patterns with opposite helicity of the photon beam [91]. Things become more complex when the angular patterns are collected from a magnetic system, where the symmetry of photoemission depends also on the orientation of the surface magnetization $\overline{\mathbf{M}}$ with respect to the angular momentum $\overline{\mathbf{L}}$ of the incoming photon. In an angular integrated photoemission experiment, the surface magnetization can be determined by the difference between the spectra taken with opposite helicity (magnetic dichroism), that is proportional to the scalar product $\overline{\mathbf{M}} \cdot \overline{\mathbf{L}}$, whereas angular-resolved photoemission bears the additional information coming from the structural environment of the emitter [92].

Greber and coworkers performed resonant Auger electron diffraction from Ni(111) at the Ni 2p $\rightarrow$ 3d absorption edge ($L_{2,3}$) in order to disentangle the asymmetry of the AED pattern due to the magnetic dipolar dichroism from the rotation of the forward scattering peaks due to the photon-electron angular momentum exchange [93]. By tuning the photon energy to the Ni L_2 edge (870.5 eV), the corresponding Ni Auger decay transitions L_2MM and L_3MM display a maximum of the emission dichroism at a kinetic energy of 863.8 and 846.2 eV, respectively, that is one order of magnitude larger than in off-resonance condition, thus allowing the measurements of the finest details of the angular diffraction pattern. For a given helicity $\sigma^{+/-}$ and magnetization direction, the angular diffraction pattern reveals the three-fold symmetry associated with the atomic structure of the fcc Ni(111) surface (see left panel in Fig. 8.16). When making the difference between the angular patterns taken with opposite helicity (or, equivalently, reverse magnetization direction) one obtains an asymmetric (dichroic) pattern, as shown in the right panel of Fig. 8.16, which bears the dipolar dichroism due to the surface magnetization (the dominating left-right asymmetry) and the multipolar dichroism due to the photon-electron angular momentum exchange (asymmetry of the fine structure of the pattern). The orientation of the magnetization can be easily determined by fitting the

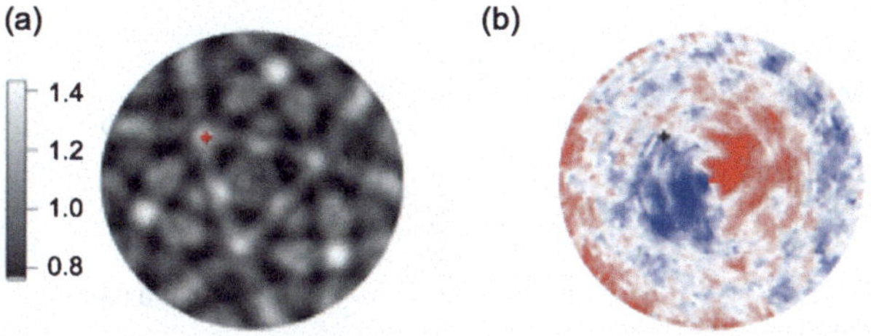

Fig. 8.16 (**a**) Resonant Auger electron diffraction data of Ni(111). The data are ϕ-averaged, i.e., normalized at each polar angle with the corresponding average intensity. (**b**) Asymmetry of two AED data sets at L_2MM measured with right and left circularly polarized light. The twofold, i.e., dipolar pattern reveals the direction of the magnetization. The *cross* marks the substrate [110] crystal direction. Reprinted figure with permission from: M. Morscher, F. Nolting, T. Brugger, T. Greber, Phys. Rev. B **84**, 140406(R) (2011). Copyright 2011 by the American Physical Society

angular pattern to a dipolar function, where the azimuthal direction is clearly indicated by the main nodal straight line between the left-right asymmetry. On the contrary, determining the amount of angular momentum exchange from the angular dichroism (opposite azimuthal rotation) of the forward scattering peaks is not trivial.

Greber and coworkers set a procedure to disentangle the contribution of the forward scattering peaks from the angular patterns by normalizing the asymmetry pattern of the L_3MM Auger to the azimuthal average of the single-helicity L_2MM pattern and viceversa (see Ref. [93] for details about the peculiar normalization of the asymmetry patterns). For each helicity $\sigma^{+/-}$, the normalized asymmetry pattern mainly bears the magnetic dipolar dichroism (with a clear nodal line) and an overall azimuthal rotation due to the photon-electron angular momentum exchange. As can be seen in Fig. 8.17, the two normalized asymmetry patterns of opposite helicity display a different orientation of the magnetic nodal line, which is solely due to the angular momentum exchange and to the distance between the emitter and the corresponding scatterer [94]. By assuming that the rotation of the angular pattern is mainly associated with the forward scattering from the 12 nearest neighbor atoms in the Ni fcc crystal, Greber and coworkers found that the Auger L_2MM electron patterns at the Ni L_2 edge have the same sense of rotation of the photon helicity, but they correspond to the transfer of $2\hbar$ angular momentum to the emitted Auger electrons, whereas a single $\hbar$ is transferred to the direct photoelectron emission [95]. Surprisingly, the Auger L_3MM patterns, as taken at the same Ni L_2 edge, display an opposite sense of rotation and a larger angular momentum transfer of $6\hbar$.

8.5 Conclusions

In summary, we have shown that RESPED can be fruitfully exploited to overcome several limitations of the conventional PED. The bare enhancement of the electron emission (both for the valence band emission and the next Auger emission) is much

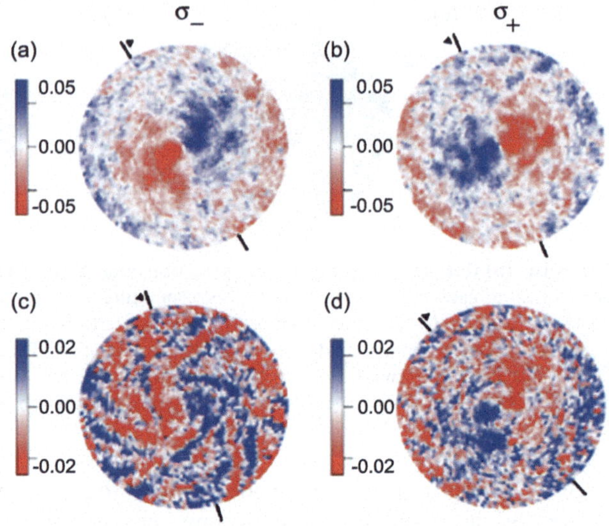

Fig. 8.17 (**a**)–(**d**) Normalized asymmetry patterns for each helicity and Auger decay channel. (**a**) Normalized asymmetry for the L_2MM Auger and helicity σ^-, (**b**) L_2MM Auger and helicity σ^+, (**c**) L_3MM Auger and helicity σ^-, and (**d**) L_3MM Auger and helicity σ^+. The ticks lie on the azimuth of the node of the corresponding dichroic dipole. Reprinted figure with permission from: M. Morscher, F. Nolting, T. Brugger, T. Greber, Phys. Rev. B **84**, 140406(R) (2011). Copyright 2011 by the American Physical Society

helpful when investigating the structural environment of atoms at very low concentration or when the corresponding core levels overlaps with other spectral features of the system under study. Beyond the element selectivity of standard PED, RESPED offers the possibility to discriminate between atoms of the same species but differing in their valence state and/or structural environment. Therefore RESPED can be used to study certain problems hitherto unaccessible to conventional structural techniques, as nicely demonstrated by the determination of the charge distribution of the TiO_2 defect state among multiple Ti sites. The production of O_b vacancies in TiO_2 can be regarded as an effective doping of the surface and lets us envisage the possibility to apply the RESPED technique for the study of semiconductor doping. This application might be particularly relevant to the case of organic semiconductors, where the carbon atoms within a molecule can carry different amount of charge, depending on the local chemistry of its ligands. A routinary use of the RESPED technique must await the implementation of the autoionization process in PED simulation packages. Meanwhile, approximate calculations may be performed by considering only the direct emission from a localized initial state. This is expected to yield reasonable results for high kinetic energy, because the resonant process is essentially localized on the core-hole site and the PED pattern is dominated by final state scattering rather than by the details of the primary wave.

References

1. C.S. Fadley, in *Synchrotron Radiation Research: Advances in Surface and Interface Science, vol. 1: Techniques*, ed. by R.Z. Bachrach (Plenum, New York, 1992) p. 421
2. C. Westphal, Surf. Sci. Rep. **50**, 1 (2003)
3. D.P. Woodruff, Surf. Sci. Rep. **62**, 1 (2007)

4. M. Sagurton, E.L. Bullock, C.S. Fadley, Phys. Rev. B **30**, 7332 (1984)
5. M. Sagurton, E.L. Bullock, C.S. Fadley, Surf. Sci. **182**, 287 (1987)
6. R. Shimizu, Jpn. J. Appl. Phys. **22**, 1631 (1983)
7. J. Wider, F. Baumberger, M. Sambi, R. Gotter, A. Verdini, F. Bruno, D. Cvetko, A. Morgante, T. Greber, J. Osterwalder, Phys. Rev. Lett. **86**, 2337 (2001)
8. A. Liebsch, Phys. Rev. Lett. **32**, 1203 (1974)
9. A. Liebsch, Phys. Rev. B **13**, 544 (1976)
10. K. Siegbahn, U. Gelius, H. Siegbahn, E. Olsen, Phys. Lett. **32A**, 221 (1970)
11. C.S. Fadley, S.A.L. Bergstrom, Phys. Lett. **35A**, 375 (1971)
12. S. Kono, C.S. Fadley, N.F.T. Hall, Z. Hussain, Phys. Rev. Lett. **41**, 117 (1978)
13. S. Kono, S.M. Goldberg, N.F.T. Hall, C.S. Fadley, Phys. Rev. Lett. **41**, 1831 (1978)
14. D.P. Woodruff, D. Norman, B.W. Holland, N.V. Smith, H.H. Farrell, M.M. Traum, Phys. Rev. Lett. **41**, 1130 (1978)
15. S.D. Kevan, D.H. Rosenblatt, D. Denley, B.C. Lu, D.A. Shirley, Phys. Rev. Lett. **41**, 1565 (1978)
16. Y. Chen, M.A. Van Hove, MSCD package user guide—simulation of photoelectron diffraction using Rehr-Albers separable representation, handled out at Trieste School on Use of Synchrotron Radiation, November 1997, and references therein. For updated information see http://electron.lbl.gov/mscdpack/mscdpack.html
17. P.A. Brühwiller, O. Karis, N. Mårtensson, Rev. Mod. Phys. **74**, 703 (2002)
18. L.C. Davies, L.A. Feldkamp, Phys. Rev. B **23**, 6239 (1981)
19. A. Tanaka, T. Jo, J. Phys. Soc. Jpn. **63**, 2788 (1994)
20. J. Chen, D.J. Huang, A. Tanaka, C.F. Chang, S.C. Chung, W.B. Wu, C.T. Chen, Phys. Rev. B **69**, 085107 (2004)
21. H. Arai, T. Fujikawa, Phys. Rev. B **72**, 075102 (2005)
22. F.J. Garcia de Abajo, C.S. Fadley, M.A. van Hove, Phys. Rev. Lett. **82**, 4126 (1999)
23. P. Feulner, M. Ecker, P. Jakob, R. Romberg, R. Weimar, D. Menzel, A. Föhlisch, W. Wurth, S.-H. Yang, C.S. Fadley, R. Larciprete, S. Lizzit, K. Kostov, G. Tyuliev, Phys. Rev. B **71**, 125409 (2005)
24. J. Pendry, Surf. Sci. **57**, 679 (1976)
25. M. Grass, J. Braun, G. Borstel, Phys. Rev. B **47**, 15487 (1993)
26. D. Sébilleau, R. Gunella, Z.-Y. Wu, S. Di Matteo, C.R. Natoli, J. Phys. Condens. Matter **18**, R175 (2006)
27. K. Cho, Y. Miyamoto, Surf. Sci. Lett. **192**, L835 (1987)
28. C. Janowitz, R. Manzke, M. Skibowski, Y. Takeda, Y. Miyamoto, K. Cho, Surf. Sci. Lett. **275**, L673 (1992)
29. P. Krüger, F. Da Pieve, J. Osterwalder, Phys. Rev. B **83**, 115437 (2011)
30. F. Bruno, L. Floreano, A. Verdini, D. Cvetko, R. Gotter, A. Morgante, M. Canepa, S. Terreni, J. Electron Spectrosc. Relat. Phenom. **127**, 85 (2002)
31. J.J. Yeh, *Atomic Calculation of Photoionization Cross-Sections and Asymmetry Parameters* (Gordon & Breach, Langhorne, 1993)
32. S. Tanuma, C.J. Powell, D.R. Penn, Surf. Interface Anal. **20**, 77 (1993)
33. A.V. Yakovenko, J. Electron Spectrosc. Relat. Phenom. **74**, 237 (1995)
34. Y. Chen, F.J.G. de Abajo, A. Chasse, R.X. Ynzunza, A.P. Kaduwela, M.A. Van Hove, C.S. Fadley, Phys. Rev. B **58**, 13121 (1998). http://www.ap.cityu.edu.hk/personal-website/Van-Hove_files/mscd/mscdpack.html
35. R.S. Saiki, A.P. Kaduwela, M. Sagurton, J. Osterwalder, D.J. Friedman, C.S. Fadley, C.R., Brundle. Surf. Sci. **282**, 33 (1993)
36. J.J. Barton, D.A. Shirley, Phys. Rev. B **32**, 1892 (1985)
37. J.J. Barton, D.A. Shirley, Phys. Rev. B **32**, 1906 (1985)
38. J.J. Barton, S.W. Robey, D.A. Shirley, Phys. Rev. B **34**, 778 (1986)
39. V. Fritzsche, Surf. Sci. **213**, 648 (1989)
40. V. Fritzsche, J. Phys. Condens. Matter **2**, 1413 (1990)
41. V. Fritzsche, J.B. Pendry, Phys. Rev. B **48**, 9054 (1993)

42. A.P. Kaduwela, G.S. Hermann, D.J. Friedman, C.S. Fadley, J.J. Rehr, Phys. Scr. **41**, 948 (1990)
43. F. García de Abajo, M.A. Van Hove, C.S. Fadley, Phys. Rev. B **63**, 075404 (2001). http://nanophotonics.csic.es/static/widgets/edac/index.html
44. X. Chen, D.K. Saldin, Comput. Phys. Commun. **112**, 67 (1998)
45. G.R. Harp, Y. Ueda, X. Chen, D.K. Saldin, Comput. Phys. Commun. **112**, 80 (1998)
46. X. Chen, G.R. Harp, Y. Ueda, D.K. Saldin, Comput. Phys. Commun. **112**, 91 (1998)
47. R. Gunnella, F. Solal, D. Sébilleau, C.R. Natoli, Comput. Phys. Commun. **132**, 251 (2000)
48. D. Sébilleau, C. Natoli, G.M. Gavaza, H. Zhao, F. Da Pieve, K. Hatada, Comput. Phys. Commun. **182**, 2567 (2011)
49. R. Díez Muinõ, D. Rolles, F.J. Garcia de Abajo, F. Starrost, W. Schattke, C.S. Fadley, M.A. Van Hove, J. Electron Spectrosc. Relat. Phenom. **114–116**, 99 (2001)
50. J.J. Rehr, R.C. Albers, Rev. Mod. Phys. **72**, 621 (2000)
51. J.J. Rehr, W. Schattke, F.J. Garcia de Abajo, R. Díez Muinõ, M.A. Van Hove, J. Electron Spectrosc. Relat. Phenom. **126**, 67 (2002)
52. J.B. Pendry, *Low Energy Electron Diffraction* (Academic Press, London, 1974)
53. W.H. Press, S.A. Teukolsky, W.T. Vetterling, B.P. Flannery, *Numerical Recipes in Fortran*, 2nd edn. (Cambridge University Press, New York, 1992)
54. M.L. Viana, R.D. Muio, E.A. Soares, M.A. Van Hove, V.E. De Carvalho, J. Phys. Condens. Matter **19**, 446002 (2007)
55. D.A. Duncan, J.I.J. Choi, D.P. Woodruff, Surf. Sci. **606**, 278 (2012)
56. C.S. Fadley, Prog. Surf. Sci. **16**, 275 (1984)
57. D.K. Saldin, G.R. Harp, B.P. Tonner, Phys. Rev. B **45**, 9629 (1992)
58. J. Osterwalder, T. Greber, S. Hüfner, L. Schlapbach, Phys. Rev. Lett. **64**, 2683 (1990)
59. L. Zutic, J. Fabian, S. Das Sarma, Rev. Mod. Phys. **76**, 323 (2004)
60. A. Wall, A. Raisanen, G. Haugstad, L. Vanzetti, A. Franciosi, Phys. Rev. B **44**, 8185 (1991)
61. L. Sangaletti, S. Pagliara, F. Parmigiani, A. Goldoni, L. Floreano, A. Morgante, V. Aguekian, Phys. Rev. B **67**, 233201 (2003)
62. L. Sangaletti, A. Verdini, S. Pagliara, G. Drera, L. Floreano, A. Goldoni, A. Morgante, Phys. Rev. B **81**, 245320 (2010)
63. N. Tsuda, K. Nasu, A. Yanase, K. Siratori, *Electronic Conduction in Oxides* (Springer, Berlin, 1991)
64. E.J.W. Verwey, K. Siratori, Nature **144**, 327 (1939)
65. J. Garcia, G. Subias, M.G. Proietti, H. Renevier, Y. Joly, J.L. Hodeau, J. Blasco, M.C. Sanchez, J. Berar, Phys. Rev. Lett. **85**, 578 (2000)
66. S. Gota, J.-B. Moussy, M. Henriot, M.-J. Guittet, M. Gautier-Soyer, Surf. Sci. **482-485**, 809 (2001)
67. P. Kuiper, B.G. Searle, L.-C. Duda, R.M. Wolf, P.J. van der Zaag, J. Electron Spectrosc. Relat. Phenom. **86**, 107 (1997)
68. H. Magnan, P. Le Fèvre, D. Chandesris, P. Krüger, S. Bourgeois, B. Domenichini, A. Verdini, L. Floreano, A. Morgante, Phys. Rev. B **81**, 085121 (2010)
69. A. Mirone, M. Sacchi, S. Gota, Phys. Rev. B **61**, 13540 (2000)
70. A. Hagfeldt, M. Grätzel, Chem. Rev. **95**, 49 (1995)
71. A. Hagfeldt, M. Grätzel, Acc. Chem. Res. **33**, 269 (2000)
72. L.A. Majewski, R. Schroeder, M. Grell, Adv. Funct. Mater. **15**, 1017 (2005)
73. U. Diebold, Surf. Sci. Rep. **48**, 53 (2003)
74. C.M. Yim, C.L. Pang, G. Thornton, Phys. Rev. Lett. **104**, 036806 (2010)
75. M. Blanco, J. Abad, C. Rogero, J. Mendez, M.F. Lopez, J.A. Martin-Gago, P.L. de Andres, Phys. Rev. Lett. **96**, 55502 (2006)
76. Z. Zhang, S.-P. Jeng, V.E. Henrich, Phys. Rev. B **43**, 12004 (1991)
77. S. Wendt et al., Science **320**, 1755 (2008)
78. A.C. Papageorgiou et al., Proc. Natl. Acad. Sci. **107**, 2391 (2010)
79. C. Di Valentin, G. Pacchioni, A. Selloni, Phys. Rev. Lett. **97**, 166803 (2006)
80. C.J. Calzado, N.C. Hernandez, J.F. Sanz, Phys. Rev. B **77**, 045118 (2008)

81. B.J. Morgan, G.W. Watson, J. Phys. Chem. C **113**, 7322 (2009)
82. P. Le Fevre, J. Danger, H. Magnan, D. Chandesris, J. Jupille, S. Bourgeois, M.-A. Arrio, R. Gotter, A. Verdini, A. Morgante, Phys. Rev. B **69**, 155421 (2004)
83. P. Krüger, S. Bourgeois, B. Domenichini, H. Magnan, D. Chandesris, P. Le Fèvre, A.M. Flank, J. Jupille, L. Floreano, A. Cossaro, A. Verdini, A. Morgante, Phys. Rev. Lett. **100**, 055501 (2008)
84. T. Greber, I. Osterwalder, D. Naumovic, A. Stuck, S. Hüfner, L. Schlapbach, Phys. Rev. Lett. **69**, 1947 (1992)
85. H. Shinohara, Rep. Prog. Phys. **63**, 843 (2000)
86. M. Wolf, K.-H. Müller, Y. Skourski, D. Eckert, P. Georgi, M. Krause, L. Dunsch, Angew. Chem., Int. Ed. **44**, 3306 (2005)
87. S. Yang, S.I. Traoyanov, A.A. Popov, M. Krause, L. Dunsch, J. Am. Chem. Soc. **128**, 16733 (2006)
88. D.F. Leigh, C. Nöremberg, D. Cattaneo, J.H.G. Owen, K. Porfyrakis, A. Li Bassi, A. Ardavan, G.A.D. Briggs, Surf. Sci. **601**, 2750 (2007)
89. D.S. Deak, F. Silly, K. Porfyrakis, M.R. Castell, J. Am. Chem. Soc. **128**, 13976 (2007)
90. M. Treier, P.l. Ruffieux, R. Fasel, F. Nolting, S. Yang, L. Dunsch, T. Greber, Phys. Rev. B **80**, 081403(R) (2009)
91. F. Matsui, T. Matsushita, H. Daimon, J. Electron Spectrosc. Relat. Phenom. **178**, 221 (2010)
92. A. Chassé, W. Kuch, M. Kotsugi, X. Gao, F. Offi, S. Imada, S. Suga, H. Daimon, J. Kirschner, Phys. Rev. B **71**, 014444 (2005)
93. M. Morscher, F. Nolting, T. Brugger, T. Greber, Phys. Rev. B **84**, 140406(R) (2011)
94. H. Daimon, Phys. Rev. Lett. **86**, 2034 (2001)
95. H. Daimon, T. Nakatani, S. Imada, S. Suga, Y. Kagoshima, T. Miyahara, Jpn. J. Appl. Phys., Part 2 **32**, L1480 (1993)

Chapter 9
Surface Structure Analysis with X-Ray Standing Waves

Jörg Zegenhagen

Abstract The most common and important application of the X-ray standing-wave technique is the analysis of surface adsorbates. Structural analysis can be performed element and even chemical species specific with high spatial resolution. The technique allows the investigation of low surface coverage and does not require long range order. In the present review I will present some examples. The fundamental principles and some variations of the technique and the experimental requirements are presented briefly. The quality and quantity of the information obtained from standing-wave measurements are discussed.

9.1 Introduction

Surface science is an important branch of condensed-matter physics and surface properties are highly relevant in industry and technology. In fact, the interest in surfaces and related low-dimensional phenomena dictated by surface properties has generally received a further boost recently last not least with the upcoming of nanoscience.

However, surfaces are still a challenge for theoretical description. The best available modern *ab initio* tools depend on the guidance by experiments. Yet, experimental surface analysis is likewise difficult, because of the dilute nature of the system under study. Fortunately, corresponding investigations have become much easier over the course of the last 20 to 30 years with the establishment of new, powerful techniques.

To understand surface properties, the knowledge of the stoichiometry, chemistry and structure is of prime importance. If the composition and structure of a surface are unknown, all other properties cannot be described correctly. This clearly calls for experimental tools which are valuable for determining accurately the stoichiometry, chemical composition, and the microscopic structure of surfaces. The X-ray standing-wave (XSW) technique [1] is a powerful tool in this respect [2].

J. Zegenhagen (✉)
European Synchrotron Radiation Facility, Grenoble, France
e-mail: zegenhagen@esrf.fr

G. Bracco, B. Holst (eds.), *Surface Science Techniques*,
Springer Series in Surface Sciences 51, DOI 10.1007/978-3-642-34243-1_9,

The XSW method employs the X-ray scattering of an X-ray interference field which serves as a ruler for the measurement of distances on the picometer scale. The characteristic scattering response of atoms within the range of the interference field reveals their spatial distribution. The X-ray standing wave is created by diffraction (or reflection). By the combination of the XSW technique with spectroscopic tools, atomic distributions are analyzed independently for each individual element present on the surface. Adsorbate or impurity concentrations may be less than 10^{13} atoms cm^{-2} and positional information can nevertheless be obtained with an accuracy of 1 pm under favorable conditions.

The XSW method is not a mainstream technique. The most common experimental tools for the determination of atomic structures of surfaces are nowadays scanning probe and diffraction methods. However, without discussing in great detail their individual strengths and weaknesses, it is fair to say that both techniques are not (or very limited) element specific. Furthermore, diffraction requires ordered superstructures and scanning microscopy images are the more difficult to interpret, the rougher and more disordered a surface is. An ordered (auto-correlated) superstructure is in principle not necessary to apply the XSW technique. In contrast to a typical diffraction experiment, one probes with an XSW experiment the correlation of the atom(s) with the X-ray interference field. An extension of the typical diffraction experiment is the so-called truncation rod X-ray scattering, which also probes the correlation of superstructures with the bulk lattice. However, combining the XSW technique with photoelectron or fluorescence spectroscopy, structural analysis of atomic distributions is carried out element specific. In contrast, X-ray diffraction yields information about electron densities, which renders element discrimination much more difficult and ambiguous. On the other hand, it is also fair to say that the analysis of clean surface structures is virtually impossible for the XSW technique.

The XSW method is occasionally categorized erroneously as diffraction technique. The difference between XSW and diffraction measurements can be highlighted when considering that in principle the adsorption site of a single atom can be determined by the XSW technique, which is not conceivable with a diffraction experiment. The XSW localization of a single atom is just a *Gedankenexperiment*, but it highlights an important difference between both techniques, which is particularly important for the application of the XSW technique in surface science. There are many cases where atomic arrangements on surfaces exhibit poor autocorrelation but pronounced correlation with the substrate bulk lattice because of a simple reason: Surface to adsorbate distances, i.e. bond lengths can only vary within a reasonable range.

Any adsorbate will be influenced by the (periodic) substrate potential in all three dimensions. Depending on the strength of interaction, it may be described as commensurate (coherent with the substrate lattice), incommensurate (incoherent with the substrate lattice) or even as a confined liquid or gas, but it will respond to the surface potential and thus be correlated with the substrate lattice. An XSW experiment will allow to accurately measure the correlation of the adsorbate with the substrate which is imposed by the corrugated substrate surface potential [3]. The structural correlation of the adsorbate with the substrate is immediately revealed by coherent fractions larger than zero. Even if we assume that an overlayer would be in-plane completely disordered, the (mean) surface-adsorbate distance can easily be determined by an XSW experiment using diffraction planes parallel to the surface. Such a determination is frequently straight forward, since the XSW technique is element specific and phase sensitive.

9.2 X-Ray Standing Wave Technique for Surface Analysis: Milestones

In 1964 Boris W. Batterman reported about an experiment [1] which provided the evidence that an X-ray interference field (= X-ray standing wave) is formed inside a crystal during Bragg reflection by the superposition of the incident and Bragg reflected X-ray beam. Furthermore, the experiment provided evidence that the wavefield shifts its position, when passing the range of the Ge(220) Bragg reflection excited by $E_\gamma = 17.4$ keV radiation from an X-ray tube. The anti-nodal planes, i.e. the intensity maxima located between the (220) lattice planes on the low angle side are moving on to the planes on the high angle side of the Ge(220) Bragg reflection. This was evident from the slightly asymmetric profile of the recorded Ge K_α fluorescence. The Ge K-fluorescence was noticeably a bit weaker on the low angle and a bit stronger on the high angle side.

The most important step toward the application of the XSW method in surface science followed in 1980. Cowan, Golovchenko and Robbins demonstrated that the X-ray interference field extends into the region above the surface [4]. They determined the position of bromine on the Si(110) surface adsorbed from a methanol solution. Using X-rays from a laboratory source, such a measurement lasted many hours (or even days). The application of synchrotron radiation (SR) a few years later [5] resulted in convincingly short measuring times ($\approx$ minutes) and low detection limits. Another technical achievement, performing XSW measurements in ultra high vacuum, (UHV) was another important step forward [6–9].

Up to the early 1980s, exclusively low order reflections of highly perfect semiconductor crystals and diffraction planes parallel to the surface had been used. Consequently, non-experts were left under the impression that XSW can only determine adsorbate positions normal to the surface and can only be applied to (perfect) semiconductor substrates. Furthermore, the formalism of the XSW analysis had never been published, and the quality of information, gained from XSW experiments, remained unclear.

Thus, it was important progress that it was shown in 1982 how to determine atomic coordinates in three dimensions [10] and the basic formalism of XSW structural analysis was eventually published in 1985 by Hertel et al. [11]. An XSW measurement, employing a diffraction vector **H**, determines the amplitude f^H and phase P^H of the Fourier component **H** of the atomic distribution under study. Traditionally, f^H and P^H are called coherent fraction and coherent position, respectively.

Shortly afterwards followed another crucial step forward for the further application of the method by using substrates other than semiconductor crystals [12]. By using Bragg reflection at near normal incidence [13, 14] even imperfect metal crystal can be used as substrates [15]. At about the same time it was shown that long period XSW produced by grazing incidence reflection from non crystalline mirror surfaces can be used to study Langmuir-Blodgett films [16].

By the end of the 1980s the most important aspects of the XSW technique were recognized or demonstrated and I reviewed its application in surface science in 1993. In the conclusion and outlook of this Surface Science Review I anticipated some further interesting developments, which I just like to quote [2]: "Next-generation SXR sources may initiate another quantum leap forward. The brilliance of these sources, and brilliance is the real figure of merit for XSW experiments, is supposed to increase by several orders of magnitude compared to present-days sources. A perfectly plane X-ray wave is ideally suited for XSW experiments. Plane wave means such a wave should be absolutely monochromatic with no divergence—an idealization of course. However, monochromatized radiation from third-generation, highly collimated SXR sources will come closer to this ideal and still deliver enough intensity. With a well monochromatized and collimated X-ray beam (a pseudo-plane-wave) a real Fourier analysis (synthesis) of the distribution function of atoms under study would be possible. While in principle possible already now, it is in practice prohibited. If we employ different sample H vectors, we usually have to change the MC crystals in order to avoid dispersive crystal arrangements. However, if we had a pseudo-PWG, an XSW measurement could be carried out rather similar to a common X-ray diffraction measurement. We would record a large number of reflections but, different from diffraction experiments, monitor inelastic scattering from the sample simultaneously. Finally, we would like to point out a few more, attractive future prospects. With sufficient energy resolution we can identify in XPS spectra lines exhibiting a shift in energy due to a different chemical environment of the emitting atomic species (chemical shift). ... Last but not least, the higher the incident intensity, the smaller could be the substrate crystal. Crystallites of—$10 \times 10 \times 10\ \mu m^3$ should be well suited for use with the XSW technique." [17].

With the help of third generation synchrotron radiation sources, all these expectations have been more than fulfilled meanwhile. With the proper instrumentation [18] XSW measurements can be carried out equivalent to diffraction experiments recording a multitude of structure factors [19], however, element specific with amplitude *and* phase. This allows to reconstruct an image [19, 20] of the position of the element under study on the length-scale of the XSW, which is typically the unit cell of the substrate crystal. XSW in combination with (hard) X-ray photoelectron spectroscopy has become a very prominent application for the analysis of soft condensed matter on surfaces. Following the first demonstration by Sugiyama et al. [21–23] there are meanwhile numerous papers published where this combination is used to structurally resolve different chemical species of the same, in particular light elements. Finally, at the end of this brief overview of the important milestones of the XSW technique I would like to mention that the brilliant beam of third generation SR sources has also allowed to perform XSW analysis on the micrometer scale [24].

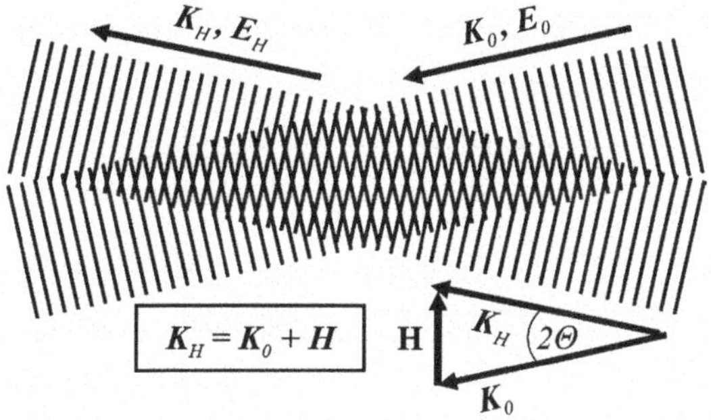

Fig. 9.1 Snapshot at fixed time of two coherent plane waves, characterized by momentum ($\mathbf{k_0}, \mathbf{k_H}$) and electric field ($\mathbf{E_0}, \mathbf{E_H}$) and traveling in different directions leading to the formation of a standing wave (*interference field*) in the overlap region

9.3 Basic Principle of the XSW Technique

The XSW technique is in essence an X-ray interference technique. The intensity of the radiative emission (photoelectrons, Auger electrons, X-ray fluorescence, ...) following the photo absorption process of specific atoms within the wavefield range signals the position of these atoms with respect to the XSW pattern. This principle is based on the fact that the probability of the photoelectric process reflects the wavefield intensity at the center of the atom. Furthermore, in order to use the XSW as a ruler requires to being able to manipulate the X-ray interference field, to shift its position in a controlled way.

We will restrict the following considerations to a planar X-ray wavefield, which is created by the superposition of two plane X-ray waves (cf. Fig. 9.1) with the electric field vectors $\mathbf{E}_0 = \mathbf{e}_0 E_0$ and $\mathbf{E}_H = \mathbf{e}_H E_H$:

$$\mathbf{E}_0 = \mathbf{e}_0 E_0 \exp\left[2\pi i(\nu_0 t - \mathbf{K}_0 \mathbf{r})\right] \;\&\; \mathbf{E}_H = \mathbf{e}_H E_H \exp\left[2\pi i(\nu_H t - \mathbf{K}_H \mathbf{r})\right]$$

as shown schematically in Fig. 9.1. Here $\mathbf{e}_0$ and $\mathbf{e}_H$ are polarization vectors and $\mathbf{K}_0$ and $\mathbf{K}_H$ are the X-ray propagation vectors with

$$|\mathbf{K}_0| = |\mathbf{K}_H| = |\mathbf{K}| = K = \lambda^{-1}$$

where λ is the X-ray wavelength related to the X-ray energy E_γ via $E_\gamma \lambda =$ 1.240 keVnm. The two waves are coherent and thus

$$\nu_0 = \nu_H = \nu$$

The two E-field vectors are related by an amplitude and a phase factor via

$$E_H = \sqrt{R} E_0 \exp(i\upsilon)$$

with

$$R = \frac{I_H}{I_0} = \frac{|E_H|^2}{|E_0|^2}$$

The momentum vectors of the two plane waves are related by

$$\mathbf{K}_H = \mathbf{K}_0 + \mathbf{H} \tag{9.1}$$

as shown in Fig. 9.1.

The XSW, which is generated in the overlap region of the two plane waves, is given by

$$|\mathbf{E}| = |\mathbf{E}_0 + \mathbf{E}_H| = (\mathbf{e}_0 \cdot \mathbf{e}_H)\big[E_0 e^{2\pi i(\nu t - \mathbf{K}_0\mathbf{r})} + \sqrt{R} E_0 e^{i\upsilon} e^{2\pi i(\nu t - \mathbf{K}_H\mathbf{r})}\big]$$

Using

$$\mathbf{E_0} \cdot \mathbf{E_0^*} = I_0$$

the wavefield intensity I can be cast into the simple form:

$$I = I_0\big|1 + \sqrt{R} e^{i(\upsilon - 2\pi \mathbf{Hr})}\big|$$

Using the cosine theorem of vector algebra, it can be converted into the more convenient expression:

$$I = I_0\big(1 + R + 2\sqrt{R}\cos(\upsilon - 2\pi \mathbf{Hr})\big)$$

The dimensionless scalar product **Hr** describes the spatial modulation of the wavefield intensity. In the direction normal to **H** the wavefield intensity is constant, but it is periodically modulated with the spacing $d_H = H^{-1}$ in the direction along **H** where $H = |\mathbf{H}|$ (cf. Fig. 9.1). We denote the particular direction, anti-parallel to **H** by the coordinate z_H.

The common way to create an X-ray interference field in the application of the XSW technique to surface structure analysis is Bragg diffraction from a single crystal in which case the XSW adopts the periodicity of the lattice spacing. Calculating reflectivity, wavefield intensities, the phase υ and other parameters requires the dynamical theory of X-ray diffraction [25–27]. We denote the spacing of a set of Miller planes (hkl) as d_M. The reflection order m of (an allowed) reflection is defined by Bragg's law

$$2d_M \sin\theta = m\lambda \tag{9.2}$$

With the help of Fig. 9.1 in can be easily seen that the Bragg equation (9.2) can be deduced from the Laue equation (9.1) via $\sin(\theta) = (H/2)/K$ considering that $H = |\mathbf{H}| = 1/d_H$. The wavefield spacing can be expressed as $d_H = d_M/m$, with $\mathbf{H} = m(hkl)$. It is also referred to as the diffraction plane spacing. The wavefield exists above the X-ray exposed surface up to a distance which is limited by the temporal coherence length of the X-ray beam, which is given by $\lambda^2/\Delta\lambda$, where $\Delta\lambda$ is the wavelength spread of the incident radiation.

Instead of using a single crystal as a generator for the XSW, one can also use an artificial "crystal", i.e. a layered synthetic microstructure (LSM) or multilayer [28]. For multilayer, the "lattice" spacing d_{ML} is at least two to three nanometer and thus much larger than for a single crystal. Thus, accordingly larger distances can be probed by the standing wave.

An X-ray interference field can also be created by the superposition of an incident X-ray wave and the total reflection from a mirror surface [16]. Below the so called critical glancing angle θ_c, an X-ray beam cannot enter a material since the refractive index for X-rays is smaller than unity. Since θ_c is small (typically a few mrad), H is also small. Thus, the wavefield spacing $d_H = H^{-1} \approx \lambda/2\theta$ becomes much larger than the X-ray wavelength. As a function of glancing angle θ it can vary from values larger than 100 nm, usually determined by the mirror quality of the surface, down to $D_c = \lambda/2\theta_c$.

9.4 X-Ray Absorption, Photoelectron Emission and Subsequent Decay Channels

Using the XSW technique for (surface) structural analysis, the photoelectric process [29] is exploited and the strength of the photo-absorption of the interference field is monitored as a function of the XSW position. When the X-ray photon has sufficient energy, a "photoelectron", is emitted from an atom with the kinetic energy E_{kin}. The probability for the photo excitation is in the dipole approximation [30] directly proportional to the amplitude of the X-ray electric field at the center of the atom. (No angular momentum is transferred to the electron by the photon.) This is the key to the excellent structural resolution of the XSW technique. To perform an XSW measurement, as a measure of the strength of the photo absorption the intensity of the photoelectron emission or subsequent decay channels of a particular element/species is detected.

The kinetic energy of the photoelectron can be expressed as $E_{kin} = E_\gamma - E_B$ where E_B is the binding energy of the photo-excited electron. Obviously, for a well defined excitation energy E_γ every photoelectron carries the information about the binding energy of the electronic state from which the electron originates and which is characteristic for the particular element. Since the binding energy slightly varies depending on the specific chemical environment of the atom, the kinetic energy of the photoelectron carries also chemical information. This was first realized by Kai Siegbahn and exploited for the well known technique coined electron spectroscopy for chemical analysis (ESCA). Figure 9.2 shows some examples for the so called chemical shift taken from a paper published by Siegbahn in 2003 [31]. The chemical shift of core levels is caused by rearrangement of the valence charge and thus it is most pronounced for electron shells with low binding energies, i.e. orbitals which are little screened by other shells.

After the core hole production, the ionized atom is in a highly excited state. It relaxes rapidly losing the excess energy by a cascade of subsequent transitions and emission processes. When the core hole is filled (very rapidly, of the order of fs) by an electron from an upper shell, the excess energy is carried away by a photon $\hbar\omega_F$ or used to kick out another electron from an upper shell, leaving the atom doubly ionized. This so-called Auger electron is emitted (to a good approximation) with the kinetic energy $E_{kin} = E_{B1} - E_{B2} - E_{B3}$, where E_{B1} is the binding energy of the photo emitted electron, E_{B2} is the binding energy of the electron, which fills this hole, and E_{B3} is the binding energy of the electron which is emitted. The process is called a B1B2B3 Auger process (e.g. KLL, KLM, LMM, etc.) according to the involved electronic shells. The non-radiative and radiative de-excitation process, i.e. the emission of an Auger electron or a fluorescence photon, are competing. For light

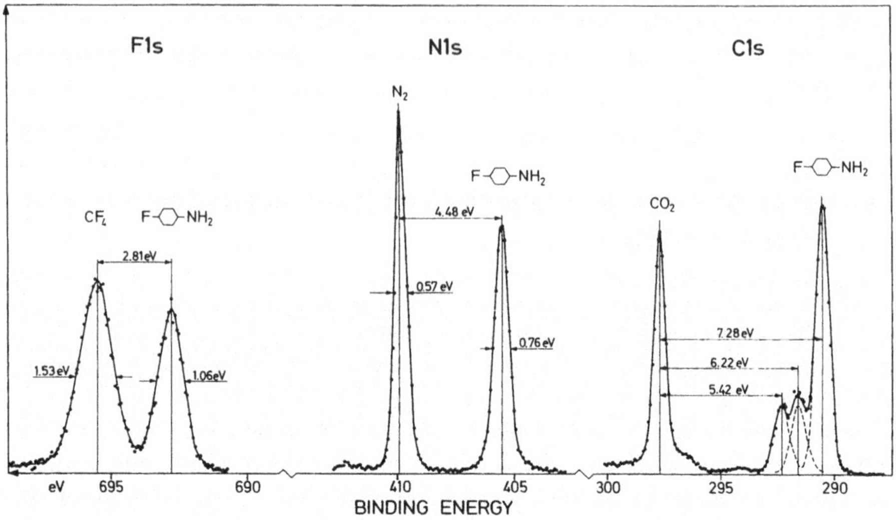

Fig. 9.2 Photoelectron peaks of fluorine 1s, nitrogen 1s and carbon 1s in different compounds. The different chemical environment leads to significantly different binding energies. The C 1s peak at the *right hand side* shows three different components for C bound to F, C bound to C, and C bound to NH_2 (taken from [31])

Fig. 9.3 The probabilities for radiative P_F (fluorescence) and non-radiative P_A (Auger electron emission) decay of K- and L_3-shell core holes for elements with atomic number Z from 1 to 100 (taken from [2])

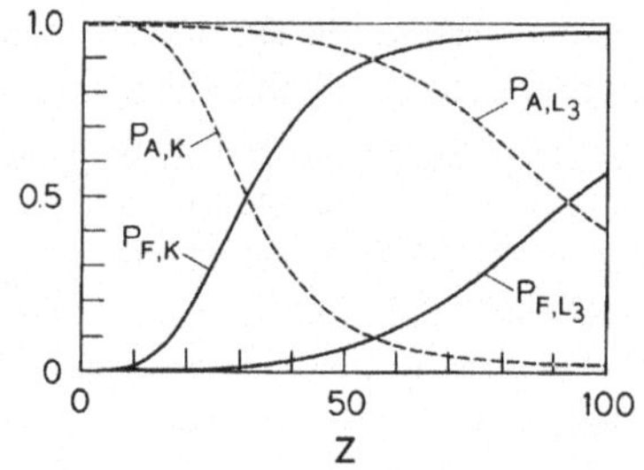

elements, the Auger process is dominant whereas for heavier atoms fluorescence is more probable (cf. Fig. 9.3). With further de-excitation of the atom, other relaxation channels open up (vibration, rotation, etc.). As for the immediately emitted photoelectron the energies of the Auger electron and the fluorescence photon are element specific. They are even chemical specific, but less pronounced than in case of the photoelectron.

For the dipole absorption process, the photoelectron emission is symmetric with respect to the E-field polarization vector. Thus, if furthermore the E-field polarization vectors of the incident and reflected wave are parallel, the photoelectron, fluorescence or Auger yield I_S excited by the interference field, i.e. $\mathbf{E_0} + \mathbf{E_H}$, is simply expressed by

$$I_S = I_{S,0}\left|1 + \sqrt{R}e^{i(\upsilon - 2\pi \mathbf{Hr})}\right|$$

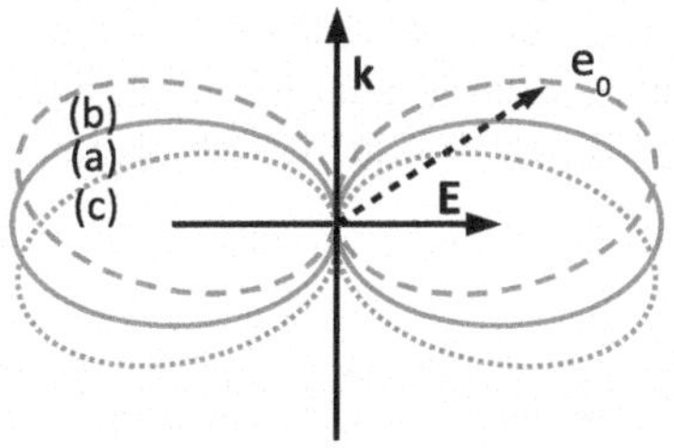

Fig. 9.4 The photoelectron emission profile is symmetric with respect to the E-field vector in case of pure dipolar emission (**a**) and is shifted along the K-vector forward (**b**) or backward (**c**) when quadrupole contributions become sizeable. Thus, in case of normal incidence Bragg reflection, the emission is shifted in one direction for the incident beam and the opposite direction for the reflected beam

The dipole approximation is largely valid even for hard X-rays in terms of the total cross section. However, higher order multipole terms can show up strongly in the differential cross section. When quadrupole terms in the photo-absorption become noticeable, the emission may be bent forward or backward with respect to the photon momentum (cf. Fig. 9.4) and thus either E_0 or E_H may contribute stronger to the photoelectron signal, depending on the position of the electron detector. As a consequence, and without going into detail, if the dipole approximation is not valid any longer, the equation describing the photoelectron emission would need to be modified (for the corresponding formalism see [32–36]), introducing three new parameter, S_R, S_I, and ψ,

$$I_S = I_{S,0}\big(1 + S_R R + S_I 2\sqrt{R}\cos(\upsilon - 2\pi \mathbf{Hr} - \psi)\big)$$

It is obvious from Fig. 9.4 that for certain directions of electron emission the one or the other X-ray wave may contribute more if the E-field vectors of the two X-ray beams are not collinear. Since the fluorescence and the Auger emission are isotropic (at least sufficiently far from resonant condition) and proportional to the total cross section, they are little affected neither by polarization effects nor by multipole contributions and thus, to a good approximation, $S_R = S_I = 1$ and $\psi = 0$.

Detecting X-ray fluorescence or electrons has advantages and disadvantages. For detecting electrons, the studied sample has typically to be kept in vacuum. X-ray fluorescence has a much larger penetration power and in most cases the sample can be in any environment. The bulk of solids, and all sorts of interfaces (solid/solid, solid/liquid, solid/gas) can be studied. However, the resolution in the spectral analysis is rather limited. The best solid state detectors, the work-horses of X-ray spectroscopy, allow an energy resolution of a few percent under favorable conditions. Crystal spectrometers allow much better energy resolution but are much more difficult to operate and, at highest resolution, cover only a small spectral range.

All three emission processes (photoelectrons, Auger electrons, X-ray fluorescence) are absolute element specific. Even more so, since the binding energy of all electronic states is influenced by the chemical environment (the more, the weaker the shielding by upper shells), the kinetic energy of the emitted electron and photon is correspondingly affected. The immediate photoelectron exhibits the strongest chemical shift. For light elements it can reach values of about 10 eV. However even if the X-ray fluorescence exhibits such a large shift in energy, which is typically not the case, it would be very difficult to resolve with common X-ray detectors.

9.5 Structural Analysis: XSW as Fourier Technique

In the following I will assume that the X-ray photon is virtually absorbed at the center of the atom, i.e. the dipole approximation is valid and that the E-field vectors are collinear.[1] In this case the X-ray excited normalized emission $Y_H = I_S/I_{S,0}$ from a single atom A at position $\mathbf{r}$ is given by

$$Y_H = 1 + R + 2\sqrt{R}\cos(\upsilon - 2\pi\mathbf{Hr})$$

For an XSW created by a Bragg reflection, the atomic structure of the substrate determines the location of the standing wave planes for each value of υ, i.e. for a specific energy and Bragg angle. The phase υ changes by π when crossing the Bragg reflection. The position of the XSW planes with respect to the unit cell is calculated using the dynamical theory of Bragg reflection. The origin of $\mathbf{r}$ corresponds to the chosen origin of the structure factor. We denoted the spatial coordinate of the atom along $\mathbf{H}$ as z_H. Obviously, the value of the cosine function and thus the yield of an atom at position z_H and $z_H + nd_H$ is identical since $|\mathbf{H}| = H = 1/d_H$ and thus $\mathbf{Hr} = z_H + n$ with integer n. This means that positions can only be determined on the length-scale of the wavefield spacing, which is referred to as modulo-d ambiguity of the XSW method. The quantity $|\mathbf{H}| = H = 1/d_H$ has the dimension of an inverse length. The dot product is dimensionless and can be expressed as $\mathbf{Hr} = P^H$, called coherent position. It corresponds to a position normal to the wavefield planes, normalized by the wavefield spacing d_H and, because of the above said, $0 \leq P^H < 1$. The graphical meaning of the coherent position is shown in Figs. 9.5 and 9.6 and the characteristic shape of the yield curve is shown for different P^H values and Fig. 9.7.

In reality a large number N_A of specific atoms A will be excited by the XSW. The total yield is the summation of the yields of all individual atoms. Normalized, it is expressed as

$$Y_H = N_A^{-1} \sum_j^{N_A} \left\{1 + R + 2\sqrt{R}\cos(\upsilon - 2\pi\mathbf{Hr}_{Aj})\right\}$$

The summation over cosine functions yields again a cosine function. Depending on the individual phases, i.e. $\mathbf{Hr}_{Aj} = P_{Aj}$, the resulting cosine function exhibits a certain amplitude $f_A^H \leq 1$ and a certain phase P_A^H i.e.

$$Y_H = 1 + R + 2\sqrt{R} f_A^H \cos\left(\upsilon - 2\pi P_A^H\right) \tag{9.3}$$

[1]The generalization is straight forward, but does not aid the understanding.

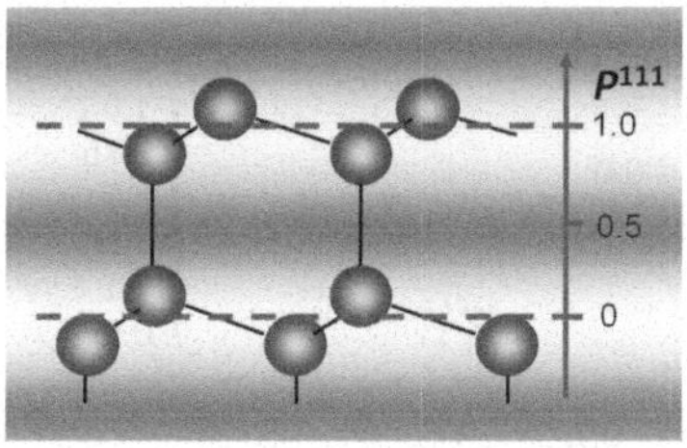

Fig. 9.5 Schematic representation of the XSW in a diamond structure crystal in cross section for a (111) reflection. The (111) diffraction planes, which are also Miller planes but no atomic planes, are marked by *dashed lines*. When passing the Bragg reflection from low to high energy or glancing angle, the wavefield maxima move in direction $-\mathbf{H}$ from being halfway between the diffraction planes to onto the planes, as shown. The coherent position P^{111} with $0 < P^{111} \leq 1.0$ is indicated

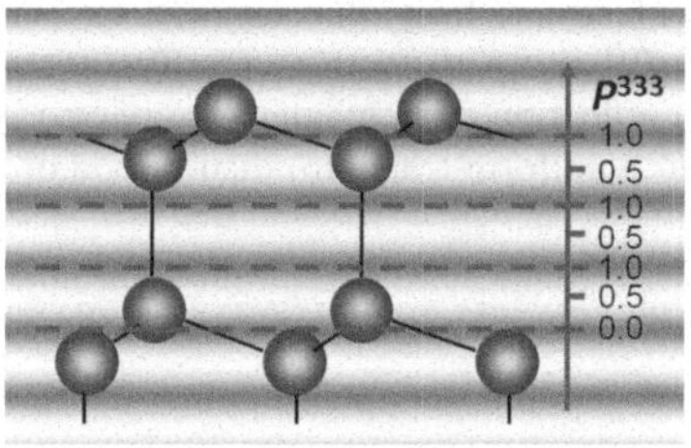

Fig. 9.6 Schematic representation of the XSW in a diamond structure crystal in cross section for a (333) reflection. The (333) diffraction planes are marked by *dashed lines*. The coherent position P^{333} with $0 < P^{333} \leq 1.0$ is indicated

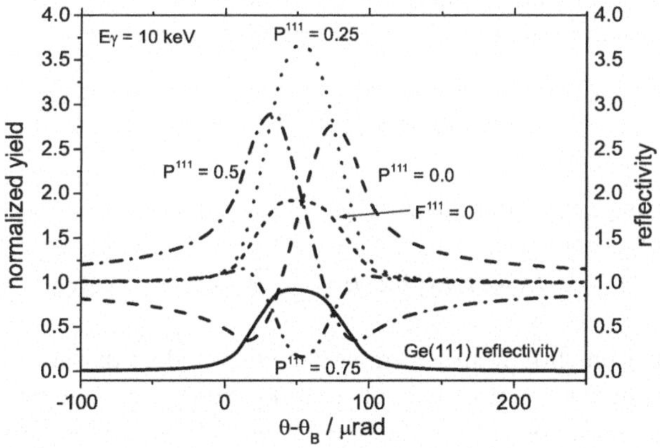

Fig. 9.7 Ge(111) reflection at 10 keV using the beam from a Si(111) monochromator. Emission yield (photoelectrons, X-ray fluorescence, ...), calculated according to (9.3). Displayed are yield curves for atoms at different position $P^H = z_H/d_H$ with respect to the Ge(111) diffraction planes. When passing the reflection curve from low to high energy or glancing angle, the wavefield moves in direction $-\mathbf{H}$ by half a diffraction plane spacing resulting in the characteristic yield profiles. The calculation of R and v requires the machinery of the dynamical theory of X-ray diffraction [25–27]

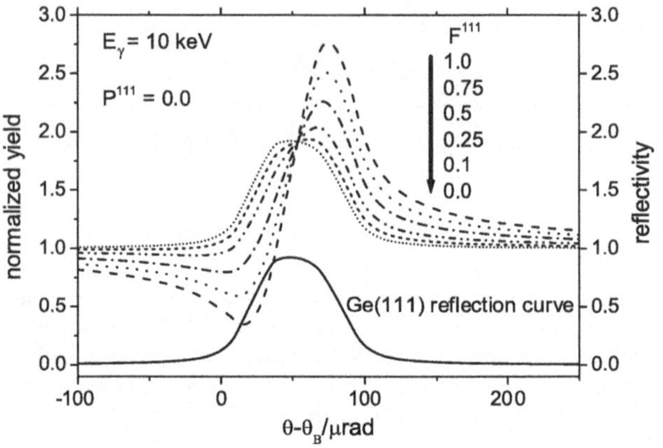

Fig. 9.8 Emission yield (photoelectrons, X-ray fluorescence, etc.) calculated using (9.3) for atoms distributed around the position $P^H = z_H/d_H = 0.0$ with different width leading to decreased coherent fractions F^{111}. Ge(111) reflection at an incident X-ray energy of 10 keV. A coherent fraction of $F^{111} = 0$ corresponds to a uniform (random) distribution of atoms along P^{111}. Otherwise as Fig. 9.7

with

$$P_A^H = Hz_{AH} = \mathbf{Hr}_A$$

The parameter f^H is called coherent fraction. Figures 9.8 and 9.9 show examples of calculated yield curves for two different coherent positions and several different coherent fractions. Since the number N_A of atoms is in reality very large, we can express the distribution of atoms by a function $G(\mathbf{r})$ given by

$$G(\mathbf{r}) = N_A^{-1} \sum_j^{N_A} \delta(\mathbf{r} - \mathbf{r}_j)$$

Substituting the summation by an integral, we can write

$$Y_H = \int_{\mathbf{r}} d\mathbf{r}\, G(\mathbf{r})\{1 + R + 2\sqrt{R}\cos(\upsilon - 2\pi\mathbf{Hr})\} \tag{9.4}$$

Casting the yield function in this form, it becomes evident that the two parameters f_A^H and P_A^H (coherent fraction and coherent position) in (9.3) represent in fact the amplitude and phase of one of the coefficients of the Fourier transform of $G(\mathbf{r})$, i.e.,

$$G(\mathbf{H}) = G^H = f_A^H e^{-i2\pi\mathbf{Hr}} = f_A^H e^{-i2\pi P_A^H}$$

A more common expression is obtained by converting the cosine function into exponentials

$$\cos(\upsilon - 2\pi\mathbf{Hr}) = \cos\left(\upsilon - 2\pi P_A^H\right) = \left[e^{-i(\upsilon - 2\pi P_A^H)} + e^{i(\upsilon - 2\pi P_A^H)}\right]/2.$$

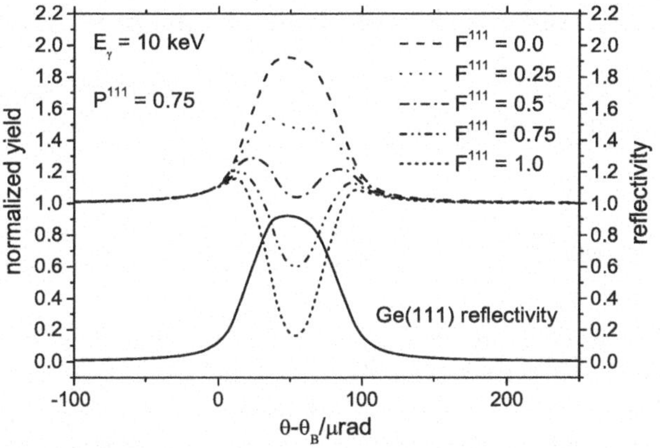

Fig. 9.9 Emission yield (photoelectrons, X-ray fluorescence, etc.) calculated using (9.3) for atoms distributed around the position $P^H = z_H/d_H = 0.75$ with different width leading to decreased coherent fractions F^{111}. Ge(111) reflection at an incident X-ray energy of 10 keV. A coherent fraction of $F^{111} = 0$ corresponds to a uniform (random) distribution of atoms along P^{111}. Otherwise as Fig. 9.7

Using $\int_{\mathbf{r}} d\mathbf{r} G(\mathbf{r}) = 1$ we convert (9.4) into

$$
\begin{aligned}
Y_H &= \int_{\mathbf{r}} d\mathbf{r}\, G(\mathbf{r})\big\{1 + R + \sqrt{R}\big[\exp\big(i(\upsilon - 2\pi\mathbf{Hr})\big) + \exp\big(-i(\upsilon - 2\pi\mathbf{Hr})\big)\big]\big\} \\
&= 1 + R + \sqrt{R} \\
&\quad \cdot \left[e^{i\upsilon}\int d\mathbf{r}\, G(\mathbf{r}) \exp i(-2\pi\mathbf{Hr}) + e^{-i\upsilon}\int d\mathbf{r}\, G(\mathbf{r}) \exp i(2\pi\mathbf{Hr})\right] \\
&= 1 + R + 2\sqrt{R}\cdot\Re\big[e^{i\upsilon}G(\mathbf{H})\big]
\end{aligned}
$$

and consequently we obtain (9.3), i.e.

$$
Y_H = 1 + R + 2\sqrt{R} f_A^H \cos\big(\upsilon - 2\pi P_A^H\big).
$$

The amplitude f_A^H and phase P_A^H are amplitude and phase of the (complex) Fourier coefficient [37] of a **particle density function** $G(\mathbf{r})$, where the specific element A is selected by spectroscopy. With sufficient Fourier coefficient an image of the analyzed distribution of atoms A can be created by Fourier back transformation.

$$
G_{XSW}(\mathbf{r}) = 1 + 2\sum_H f_A^H \cos\big[2\pi\left(P_A^H - \mathbf{Hr}\right)\big]
$$

However, the available H-values are restricted by the structure factor of the substrate and structural information beyond the unit cell dimension is not accessible by an XSW measurement when using Bragg reflection. Thus, it is principally not possible to determine the true particle density function $G(\mathbf{r})$. In contrast, by XRD we can obtain structural information on a large length scale, but we obtain only the amplitude of the Fourier transform of the **electron density** and the phase information is lost.

9.6 Structural Analysis: Simple Interpretation of XSW Results

Employing sufficient reflections and performing a Fourier analysis one can obtain structural information without any *a priori* assumptions or models. However, in particular because of the availability of phase information, XSW data of just one reflection (or very few reflections) can be sufficient to learn about the structure of surface adsorbates. In particular, since bond lengths are restricted, surface distances of adsorbates can frequently be measured accurately using just one reflection with the diffraction planes parallel to the surface.

The emission intensity I_H from N_A adsorbate atoms of a particular element A recorded during an XSW scan using the diffraction vector $\mathbf{H}$ can be written as

$$I_H = I_0 \left\{ 1 + R + 2\sqrt{R} N_A^{-1} \left[\sum_{j=1}^{N_A} \cos(\upsilon - 2\pi \mathbf{H}\mathbf{r}_{Aj}) \right] \right\}$$

The constant I_0 is proportional to the adsorbate coverage, the cross section, and the emission probabilities for the chosen element and emission line. Furthermore, it is a linear function of the incident X-ray intensity, as well as detector solid angle and efficiency. As introduced before, $\mathbf{H}\mathbf{r}_{Aj} = P_j^H$ because of the periodic nature of the wavefield and the above equation can thus be rewritten as

$$I_H = I_0 \left\{ 1 + R + 2\sqrt{R} \left[\sum_{i=1}^{n} D_i^H \cdot c_i \cos\left(\upsilon - 2\pi P_i^H\right) \right] \right\}$$

if we assume that the element A under consideration occupies a small number n of discrete coherent positions P_i^H with respect to the wavefield. Furthermore, the Debye-Waller factor $D_i^H \leq 1$ takes into account thermal vibrations (or some static disorder since N_A is usually very large). In the harmonic approximation, $D_i^H = \exp(-2\pi^2\sigma^2)$ with $\sigma^2 = \langle u^2 \rangle / d_H^2$, where $\langle u^2 \rangle$ is the mean square vibration amplitude of the considered atom and d_H is the above defined diffraction plane spacing. The factor c_i describes the fraction of the element on the position P_i^H with $\sum_{i=1}^{n} c_i = 1$. Assuming that the Debye-Waller factor is the same for all atoms of the element A regardless of their specific position P_i^H we can simplify further and write for the normalized yield Y_H

$$Y_H = I_H / I_0 = 1 + R + 2\sqrt{R} \cdot D_A^H \sum_{i=1}^{n} c_i \cdot \cos\left(\upsilon - 2\pi P_i^H\right)$$

and thus

$$Y_H = 1 + R + 2\sqrt{R} \cdot f_A^H \cos\left(\upsilon - 2\pi P_A^H\right)$$

Here, the coherent position P_A^H is now the "average" coherent position of all atoms A excited by the XSW and the coherent fraction $f_A^H = D_A^H \cdot F_A^H$ describes their distribution around this mean position. The parameters P_A^H and F_A^H can be expressed via

$$F_A^H = \left[(G_c^H)^2 + (G_s^H)^2 \right]^{1/2}$$

and

$$P_A^H = (2\pi)^{-1} \tan^{-1}\left(G_s^H / G_c^H\right) + x$$

with $x = 0.5$ if $G_c^H < 0$ and $x = 0$ otherwise.

We denote here the "geometric coherent fraction" as F_A^H. The two variables G_c^H and G_s^H are simple functions of the n occupation factors c_i and the n discrete positions P_i^H via

$$G_c^H = \sum_{i=1}^{n} c_i \cdot \cos\left(2\pi P_i^H\right)$$

and

$$G_s^H = \sum_{i=1}^{n} c_i \cdot \sin\left(2\pi P_i^H\right)$$

These relationships allow to refine models and to reproduce the XSW results in case one has some starting information about the sites occupied by the particular element under study. The coherent position, i.e. the "average" position very often provides here a very good starting point.

Frequently, f^H and P^H, coherent fraction and coherent position, are interpreted in terms of the degree of order of adatoms and their mean position. However, a particular coherent fraction f^H (i.e. a particular Fourier component) of the particle distribution could be zero by chance and the system nevertheless be well ordered. For example, two specific sites $\mathbf{r}_{A1}$ and $\mathbf{r}_{A2}$ may be occupied such that for the chosen diffraction vector $\mathbf{H}\mathbf{r}_{A2} = P_{A2} = \mathbf{H}\mathbf{r}_{A1} + 0.5 = P_{A1} + 0.5$. The XSW yield is then given by $Y_H = 1 + R + 2\sqrt{R} f^H \cos(\upsilon - 2\pi P^H) = 1/2 \sum_{j=1}^{2} \{1 + R + 2\sqrt{R} \cdot \cos(\upsilon - 2\pi P_{Aj})\}$ and adding two cosine functions which are out of phase by π leads consequently to $f^H = 0$.

9.7 Experimental

A principle experimental set-up for an XSW measurement using the X-rays from a third generation synchrotron radiation source is shown in Fig. 9.10. The beamline optics has to provide an X-ray beam which should largely resemble a plane X-ray wave. This means that the beam needs to exhibit a small energy spread and divergence. To achieve these beam properties, brilliant undulator radiation from large high energy synchrotrons such as the ESRF in Grenoble, the APS in the USA, Spring8, etc. is nowadays very well suited. While the angular collimation of undulator radiation is for a wide range of applications mostly sufficient, the energy bandwidth needs to be decreased with the help of monochromators. The primary monochromator with liquid nitrogen cooled Si(111) crystals allows to cope with the high power of the undulator radiation and transmits a bandwidth of $\Delta E/E = 10^{-4}$.

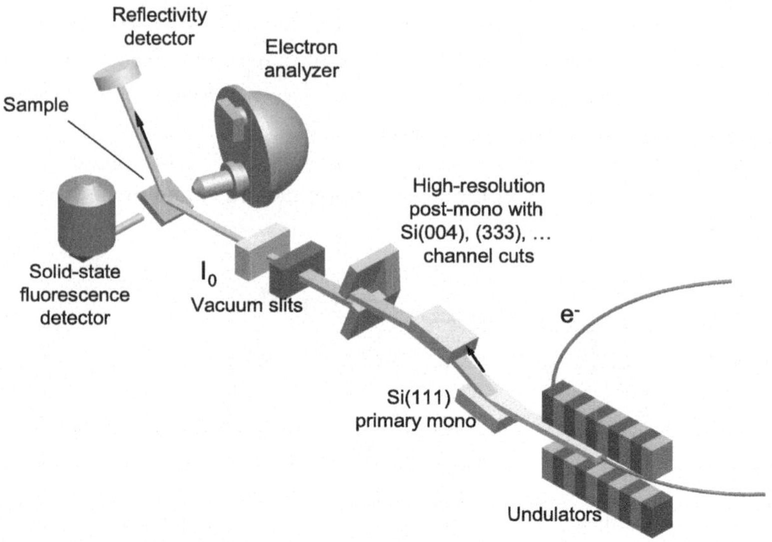

Fig. 9.10 Schematic of an XSW set-up at a synchrotron radiation source

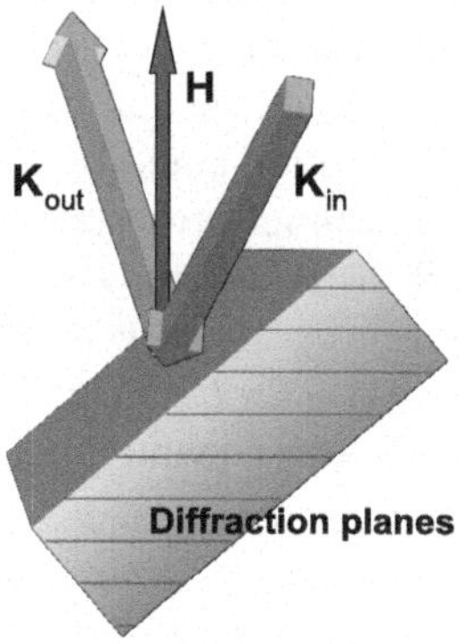

Fig. 9.11 Preferred geometry when using diffraction planes which are inclined with the surface plane. For the beam with the surface the incidence angle equals the exit angle. Furthermore, the electric field vectors (not shown) of the incident and reflected beam should lie in the diffraction plane

Better energy resolution and if necessary a better collimation than the intrinsic vertical divergence of about 10 μrad from the undulator can be achieved with high-resolution, secondary monochromators (post-mono). Other optical elements such as compound refractive lenses can be used for collimating or focusing the monochromatic beam.

The sample needs to be mounted on a kinematical stage, which allows to orient it in such a way that any desired Bragg reflection can be excited. For Bragg reflection XSW, two conditions simplify the data analysis greatly: (i) The polarization vectors of the incident and the reflected beam should be collinear, lying in the diffraction plane and (ii) the incident and exit angle with the sample surface should be equal, as shown schematically in Fig. 9.11.

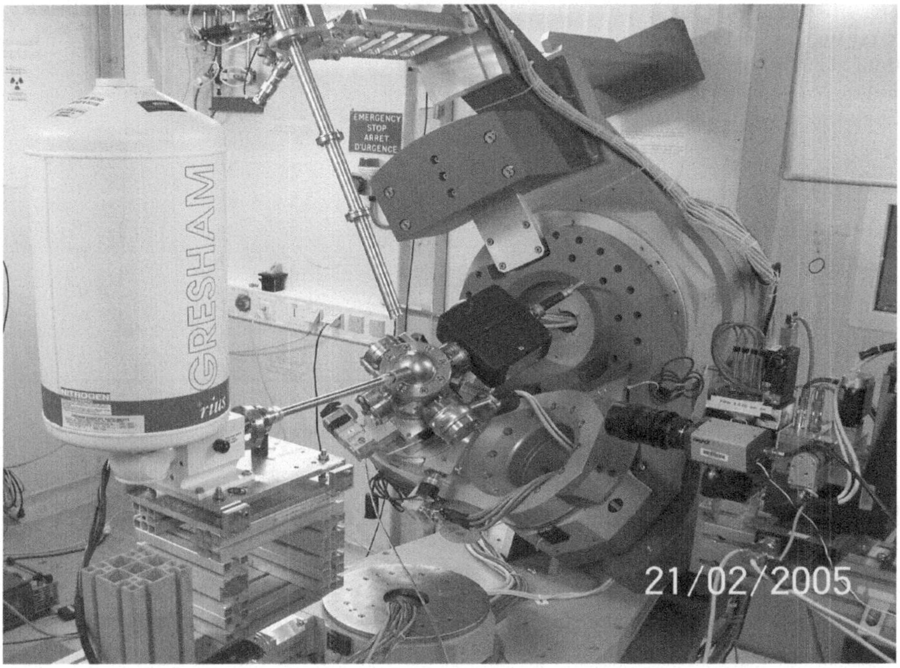

Fig. 9.12 XSW set-up on a Kappa-diffractometer for XSW "imaging" i.e. for scanning a large number of Bragg reflections. A liquid nitrogen cooled solid state detector is pointing toward the sample. A scintillation counter is mounted at the end of the flight tube also pointing at the sample. The sample is contained in a small UHV chamber, covered with a hemispherical Be window. The chamber is pumped by an ion-getter pump at the rear side (taken from [39])

XSW measurements are conveniently performed using a diffractometer such as a six-circle instrument commonly employed for standard X-ray diffraction experiments. Best suited for XSW experiments is a diffractometer where the axes of the three Eulerian circles, carrying the sample, are not orthogonal but arranged in the so-called kappa geometry as in the experimental set-up shown in Fig. 9.12. This design leaves generous space for the placement of the detectors. At least two detectors are needed for an XSW experiment. An X-ray counter is used to record the beam reflected from the sample (the reflectivity). Energy discriminating detector(s) are used to record the inelastic scattering (X-ray fluorescence or Auger-/photoelectrons) from the sample. For the experiment shown in Fig. 9.12, the sample is contained in a small ultra high vacuum (UHV) chamber with a spherical, X-ray transparent beryllium window and the X-ray fluorescence signal originating from the sample kept in UHV is detected by a liquid nitrogen cooled solid state detector. For an XSW measurement, the sample rocking curve, i.e. the Bragg reflection is scanned either by varying the beam energy or the angle of incidence on sample. Simultaneously, the reflected intensity and the inelastic scattering from the sample (electron or X-ray fluorescence yield) is recorded while traversing the Bragg reflection.

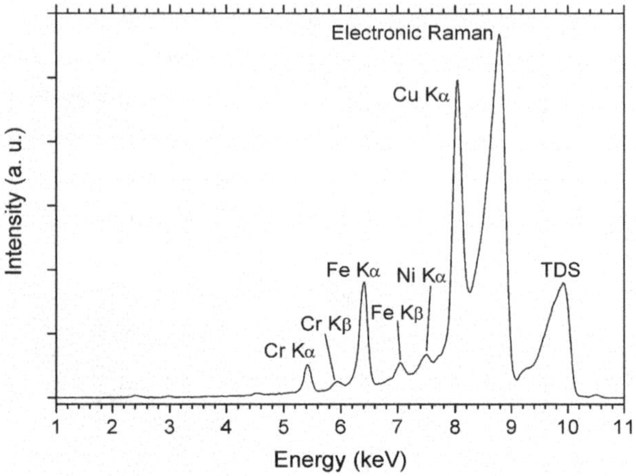

Fig. 9.13 Example of an X-ray fluorescence spectrum from a GaAs(001) crystal surface covered with 0.5 ML Cu recorded with a solid state X-ray fluorescence detector (taken from [39])

9.8 XSW Fluorescence Analysis: Electrodeposited Versus UHV Deposited Cu on GaAs(001)

In the following I describe briefly the XSW analysis of copper adsorbed on GaAs(001). The adsorption of copper deposited from an effusion cell in ultra high vacuum and electrodeposited from aqueous solution on As terminated GaAs(001) was investigated employing X-ray fluorescence spectroscopy. The analysis of the adsorbate was performed *in situ* with the samples either in ultra high vacuum or in aqueous electrolyte under potential control. More details about the experimental set-ups used and the analysis can be found for the electrodeposition and UHV deposition in Refs. [38] and [39], respectively.

The measurements on the UHV deposited and electrodeposited Cu on GaAs were performed at the ID32 beamline [18] at the European Synchrotron Radiation Facility ESRF in Grenoble and the X15A beamline [2] at the National Synchrotron Light Source NSLS at Brookhaven National Lab. in the USA, respectively. For the high resolution XSW imaging study at the ESRF, we used a Si(555) channel-cut post-monochromator delivering an energy resolution of about 20 meV at the chosen beam energy of 10 keV. In both cases, the X-ray fluorescence from the sample was detected with the help of an energy dispersive solid state detector. A spectrum is shown in Fig. 9.13. The experimental set-up at the ID32 beamline is shown in Fig. 9.12. Figure 9.14 shows the experimental set-up used in the study at the NSLS.

For the UHV study, a clean GaAs(001)-2 $\times$ 4 As-terminated surface was prepared by desorbing a protective amorphous As layer and annealing the sample at slightly below 450 °C. Subsequently about 0.5 ML of Cu was deposited by a thermal effusion cell with the sample held at room temperature. It was then transferred under UHV into a small, portable UHV chamber ("baby chamber") which was mounted on the six circle diffractometer at the ID32 beamline (cf. Fig. 9.12). Fifteen Bragg (hkl) reflections were scanned, each in small steps, recording reflectivity and a whole fluorescence spectrum for each step.

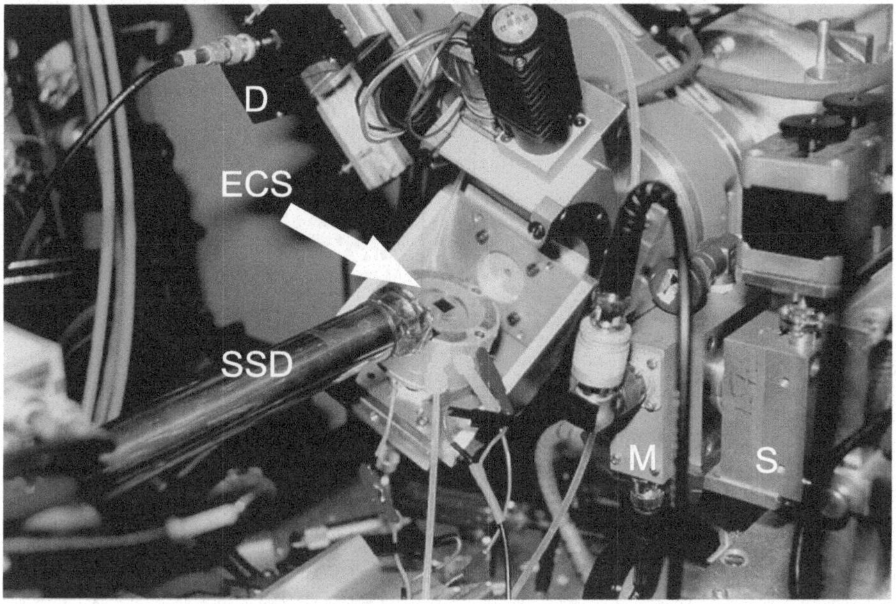

Fig. 9.14 Experimental set-up for XSW measurements at the X15A beamline for *in-situ* investigation of solid/electrolyte interfaces with slit (*S*) intensity monitor (*M*), solid state detector (*SSD*), reflectivity detector (*D*), and the electrochemical cell (*ECS*) with the 1 cm^2 GaAs sample (*small square*) in the center. The cell is filled with electrolyte and covered with a thin mylar window

For the *in situ* study of the electrodeposited Cu, we used a three-electrode electrochemical cell with a saturated calomel (SCE) reference electrode and a Pt wire as counter electrode. A sub-monolayer of Cu was electrodeposited from 0.5M $H_2SO_4 + 10^{-5}$ M $CuSO_4$ on the etched GaAs(001) surface. The concentration of Cu^{2+} ions in solution was chosen low enough to prevent any contribution from the Cu ions, which are present in the 1 μm thick sheet of solution above the sample surface, to the detected Cu X-ray fluorescence signal. The coverage was estimated from the charge during the electrochemical deposition. On GaAs(001), 1 ML is defined as 6.25×10^{14} atoms/cm^2. In total three reflections were used.

The result of two XSW measurements for GaAs(001) surfaces covered with ≈ 0.5 ML Cu, employing the (004) and (−111) reflection, are shown in Fig. 9.15 for each, the UHV and electrochemical prepared surfaces. It is obvious that the yield curves of the Cu-K fluorescence are significantly different for the UHV and electrodeposited Cu. An analytical expression given by (9.3) is fitted to the reflectivity and the Cu-K fluorescence yield. The result of each XSW measurement thus provides one complex Fourier coefficient, i.e. phase (coherent position) and amplitude (coherent fraction) for the Fourier representation of the Cu real space distribution. The coherent fractions F for the electrodeposited Cu are close to unity, proving that Cu occupies practically exclusively one lattice site. The coherent positions from three measurements, i.e. (111), (004) and (−111), correspond exactly to the Ga substitutional site.

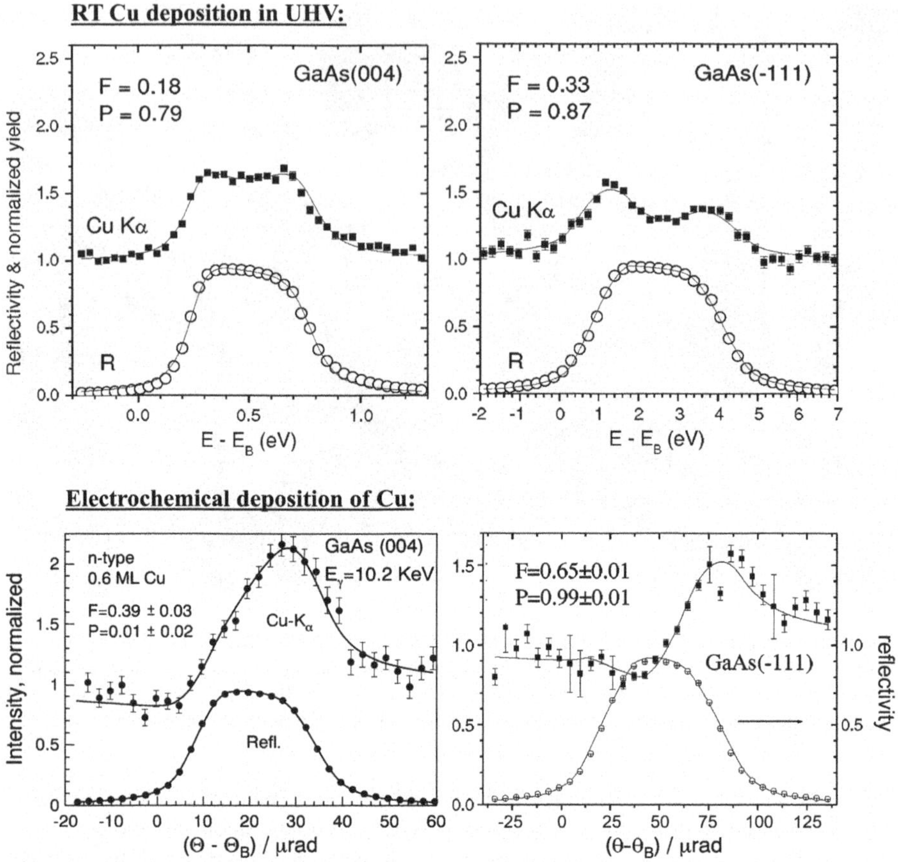

Fig. 9.15 Examples for yield curves resulting from XSW measurements for Cu on GaAs(001). The points are data and the lines are fits to the data points using the dynamical theory of X-ray diffraction and (9.3) (taken from [39])

For the UHV deposited sub-monolayer coverage of Cu the results are shown in Fig. 9.16 for all 15 (hkl) reflections. Employing the symmetry of the cubic, zinc blende GaAs lattice, this results in a total of 29 Fourier coefficients. All values of the coherent fractions for the 15 XSW scans are much smaller than unity, which proves that the Cu atoms deposited at room temperature in UHV are not occupying a unique lattice site.

By simple back transform of the amplitude f^H and phase P^H of the obtained Fourier coefficients of the Cu distribution function, we can create an image of the Cu lattice location. For the electrodeposited Cu, the resulting image in Fig. 9.17, created by the XSW result for three reflections (= five Fourier coefficients by symmetry), shows that the Cu is substituting for the Ga, as already expected. The reconstructed image of the UHV deposited Cu confirms the immediate assignment that the Cu is

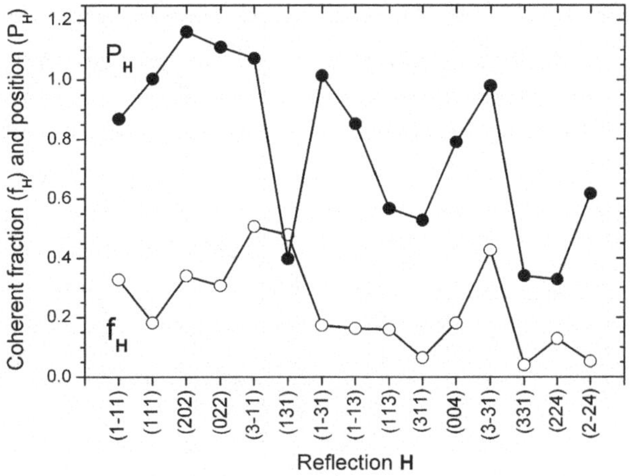

Fig. 9.16 The amplitude i.e. coherent fraction (f_H) and phase i.e. coherent position (P_H) resulting from 15 XSW scans for Cu on GaAs(001) deposited in UHV on a 2×4 reconstructed surface at room temperature (taken from [39])

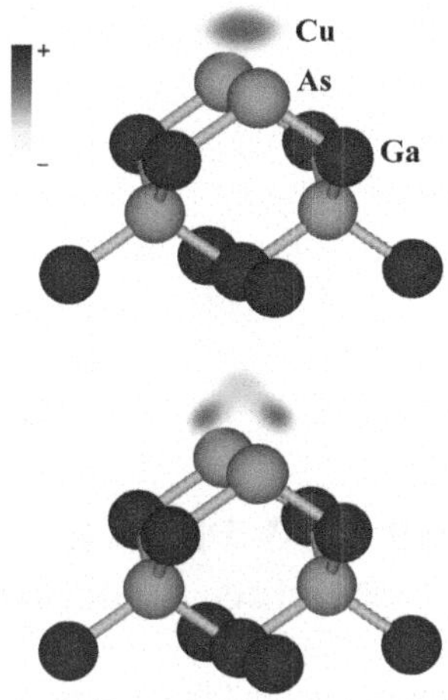

Fig. 9.17 Images created by Fourier back-transform of the Fourier coefficients determined by the XSW measurements for electrodeposited Cu (*top*) and UHV deposited Cu (*bottom*) (taken from [39])

occupying different lattice sites of lower symmetry. The Ga substitutional site may be populated by a minority of the Cu adsorbate.

In summary, XSW Fourier analysis (XSW imaging) was used to identify the adsorption sites of Cu on GaAs(001). If Cu is deposited from sulfuric acid solution on etched GaAs(001) surfaces, it assumes predominantly Ga substitutional positions.[2] Sub-monolayer amounts of copper deposited on GaAs(001)-2×4 at room temperature in UHV behave differently. Several lattice site with lower symmetry are occupied by Cu and only a minor fraction of the Cu may be located on the Ga substitutional site.

[2]However, the Cu is not located exclusively on the surface, but sub-monolayer amounts of Cu diffuse into the GaAs interface region up to a depth of several nm [40].

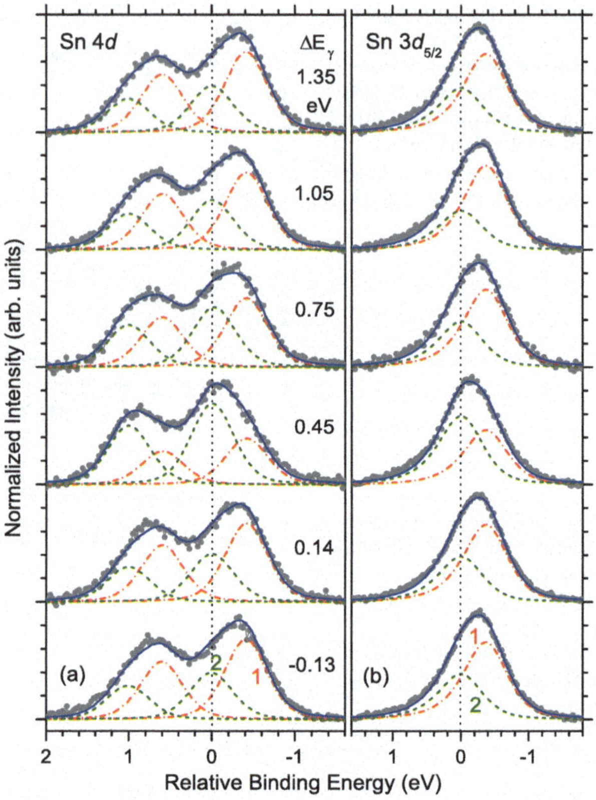

Fig. 9.18 The shape of the Sn 4$d_{3/2}$ and Sn 4$d_{5/2}$ (**a**) and 3$d_{5/2}$ (**b**) XPS peaks recorded from Ge(111):Sn-$\sqrt{3} \times \sqrt{3}$ when passing the Ge(111) Bragg reflection by scanning the incident energy. The peaks need to be fitted by two components, the ration of which is strongly dependent on the position on the rocking curve, i.e. on the wavefield position (taken from [41])

9.9 XSW Photoelectron Spectroscopy: Chemically Resolved Surface Structure of Sn on Ge(111)

Utilizing the high chemical sensitivity of photoelectron spectroscopy, i.e. the chemical shift, for XSW investigation was first demonstrated by Sugiyama et al. and shortly afterwards by Jackson et al. [21–23]. Since then mostly organic molecules on surfaces were investigated in this fashion.

We used this approach recently to investigate the room temperature Ge(111):Sn-$\sqrt{3} \times \sqrt{3}$ phase [41], which is still under discussion. Scanning tunneling microscopy images had suggested that the Sn adatoms form a flat layer, which can be described as a simple $\sqrt{3}$ structure [42–44]. However, this was in conflict with PES results which showed that Sn 4d photoelectron spectra clearly showed two components indicative of two chemically non-equivalent Sn species on the Ge(111) surface [45]. We decided to address this controversy by exploring the Sn adatom structure behind the 4d core-level shift for the Ge(111):Sn-$\sqrt{3} \times \sqrt{3}$ phase by an XSW photoelectron spectroscopy measurement.

The Sn 3d and Sn 4d photoelectron peaks from a Ge(111):Sn-$\sqrt{3} \times \sqrt{3}$ phase were recorded while scanning through the Ge(111) rocking curve by changing the

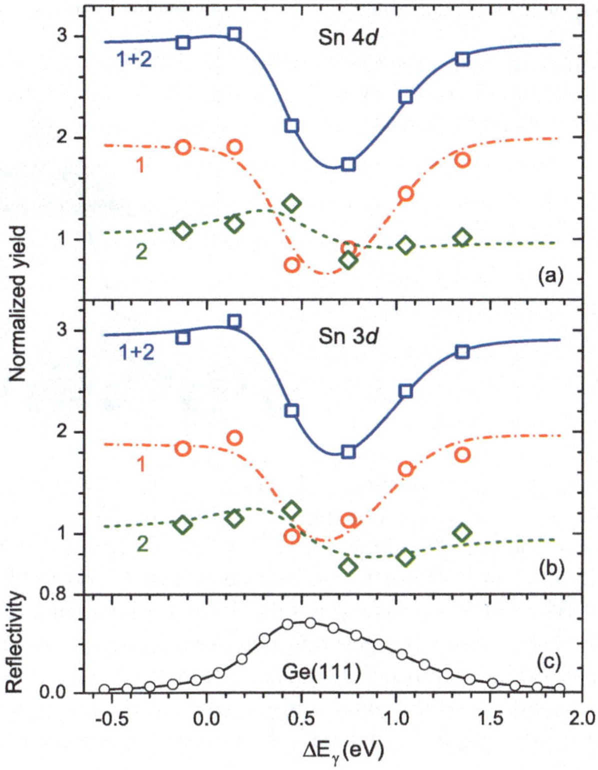

Fig. 9.19 Data (symbols) for Ge(111) rocking curve and the yield curves, i.e. the integrated intensity of the two Sn components (1, 2, and the sum of both in Fig. 9.18) as a function of excitation energy crossing the Bragg reflection; lines are fits to the data; (**a**) Sn 4*d*, (**b**) Sn 3*d*, (**c**) Ge(111) rocking curve (taken from [41])

energy of the monochromatic beam from $\Delta E = -0.13$ eV to $\Delta E = 1.35$ eV. It is obvious that the 3*d* and 4*d* peaks contain two components (intensity ratio of two to one). The intensity of the two components changes differently when scanning over the Ge(111) rocking curve (cf. Fig. 9.18). This is a clear indication that the Sn atoms, from which the two chemically shifted components originate, are located at different heights above the Ge(111) surface.

This can be seen more clearly in Fig. 9.19 where the integrated intensities of both components of the Sn 3*d* and 4*d* peaks are shown as a function of the position on the Ge(111) rocking curve. Fitting Eq. (9.3) to the yield curves, P^{111} and f^{111} are obtained. The two components with a 400 meV binding-energy difference are associated with two Sn species which reside at two different heights. A high coherent fraction is proof that either Sn species occupies a single position on the Ge(111) surface. The vertical separation amounts to exactly 23 pm. The two times stronger component, which appears on the lower binding-energy side is associated with the higher Sn position. The weaker component at the higher binding energy accounts for 1/3 of the Sn atoms which reside closer to the surface. This result can be intuitively understood in terms of simple charge transfer: A higher binding energy caused by an emptier valence state results in shorter surface distance. Conversely, the lower binding-energy caused by a more filled valence state is associated with

Fig. 9.20 Snapshot of the Ge(111):Sn-$\sqrt{3} \times \sqrt{3}$ surface. The ratio of up and down atoms is 2/1. However, the Sn atoms are flipping rapidly up and down

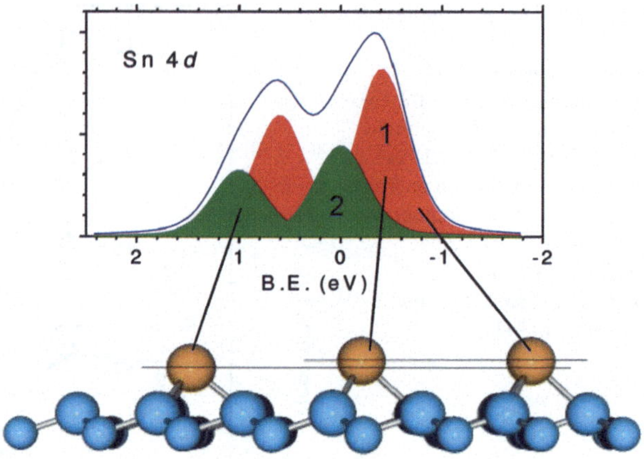

a larger surface distance. The result of the XSW analysis ("two up, one down") is shown schematically in Fig. 9.20. That the Sn associated with lower binding energy component is higher up and the Sn associated with the higher binding energy component is closer to the surface is understood by the above said. However, the XSW result is in contrast to other/earlier assignments, claiming the reverse, i.e. that the larger, lower (higher) binding energy component corresponds to Sn atoms at the lower (higher) position ("one up, two down"). This conclusion was based mostly on scanning tunneling microscopy images of the low temperature $c(2 \times 4)$ phase, but it is not intuitively understood and requires arguments, which are less straightforward.

In summary, the XSW analysis shows clearly that the two components of the Sn peaks in the XPS spectrum are associated with Sn atoms at two different heights above the Ge(111) surface (cf. Fig. 9.20). The Sn surface atoms flip randomly [46] between these two positions at a speed which is much too fast for the time scale of STM. Since the XSW analysis clearly reveals the Sn atoms with a high coherent fraction at either of the two extreme positions, the motion of the Sn atoms is not a harmonic oscillation, instead the flip transition appears to be instantaneous, supposedly facilitated by tunneling. Still, the Ge(111):Sn-$\sqrt{3} \times \sqrt{3}$ phase is not completely understood yet. A recent publication concludes that there should be three components in the Sn 4*d* line and thus there should be three different Sn species on the Ge(111) surface [47]. An XSW/XPS study with (even) better energy resolution in the XPS spectra could address this question.

9.10 Conclusion and Outlook

Spectroscopic techniques such as X-ray fluorescence and photoelectron spectroscopy are very valuable in its own right. Combined with the XSW technique, high resolution (pm) structural information is added to spectroscopic features. The XSW technique provides amplitude and phase of the Fourier decomposition of the distribution of a particular element or chemical species. In particular the available phase information can frequently be interpreted without any further assumption or knowledge as distance value. However, for Bragg reflection XSW the accessible length scale is limited to the size of the unit cell of the used substrate crystal. Thus, e.g. the information whether a structure is 1×1 or $n \times m$ cannot be distinguished and must be gathered by some other technique. On the other hand, since bond lengths are limited, surface distances of adsorbates can be determined practically immediately and unambiguously. In case of more complicated distribution functions, XSW imaging using a larger number of Bragg reflections is the method of choice.

There has been great progress over the last ten to twenty years in the quality of X-rays and the instrumentation used for XSW studies and the technique has reached a state of maturity. The brilliance of synchrotron radiation and improvements in beamline optics have, despite of increasing complexity of the equipment, rendered XSW measurements much easier such that the method has become highly attractive and feasible for scientists who are not keen on studying the dynamical theory of X-ray diffraction in detail before engaging on an XSW measurement. With the properties of nowadays X-ray sources, the incident flux is hardly a limiting factor any longer. It is in fact for many applications, in particular involving organic molecules, threatening to destroy the sample. The measuring times have thus also decreased sharply, however, in many applications unfortunately not reflecting the gain in intensity at all because of serious bottlenecks in detector technology. This is most intensely felt in fluorescence XSW, where count rates, energy resolution, and exploited solid angles are all about two orders of magnitude worse than desirable. Nevertheless, further exciting developments are on the horizon.

While the brilliance of X-ray sources cannot immediately be translated in shorter measuring times, it can be utilized already now for gaining results of higher quality, with more information. E.g. the higher available flux allows to decrease the X-ray energy bandwidth and with the help of higher resolution XPS detectors, which are meanwhile available it is practically routinely possible to differentiate chemical species in XSW measurements.

Furthermore I expect that in the future the XSW technique will be increasingly applied in combination with other spectroscopic techniques such as angular resolved photoelectron spectroscopy, synchrotron Mößbauer spectroscopy, photoelectron diffraction, inelastic X-ray scattering, and more. Last not least, with a significant leap in detector technology the time domain may eventually be exploited.

References

1. B.W. Batterman, Phys. Rev. **133**, A759 (1964)
2. J. Zegenhagen, Surf. Sci. Rep. **18**, 199 (1993)
3. G.C. La Rocca, J. Zegenhagen, Phys. Rev. B **44**, 13666 (1991)
4. P.L. Cowan, J.A. Golovchenko, M.F. Robbins, Phys. Rev. Lett. **44**, 1680 (1980)
5. G. Materlik, J. Zegenhagen, Phys. Lett. A **104**, 47 (1984)
6. S.M. Durbin, L.E. Berman, B.W. Batterman, J.M. Blakeley, J. Vac. Sci. Technol. A **3**, 973 (1985)

7. P. Funke, G. Materlik, Solid State Commun. **54**, 921 (1985)
8. J.R. Patel, P.E. Freeland, J.A. Golovchenko, A.R. Kortan, D.J. Chadi, G.-X. Quian, Phys. Rev. Lett. **57**, 3077 (1986)
9. T. Ohta, Y. Kitajima, H. Kuroda, T. Takahashi, S. Kikuta, Nucl. Instrum. Methods A **246**, 760 (1986)
10. J.A. Golovchenko, J.R. Patel, D.R. Kaplan, P.L. Cowan, M.J. Bedzyk, Phys. Rev. Lett. **49**, 560 (1982)
11. N. Hertel, G. Materlik, J. Zegenhagen, Z. Phys. B **58**, 199 (1985)
12. G. Materlik, J. Zegenhagen, W. Uelhoff, Phys. Rev. B **32**, 5502 (1985)
13. T. Ohta, H. Sekiyama, Y. Kitajima, H. Kuroda, T. Takahashi, S. Kikuta, Jpn. J. Appl. Phys. **24**, L475 (1985)
14. T. Ohta, Y. Kitajima, H. Kuroda, T. Takahashi, S. Kikuta, Nucl. Instrum. Methods A **246**, 760 (1986)
15. D.P. Woodruff, D.L. Seymour, C.F. McConville, C.E. Riley, M.D. Crapper, N.P. Prince, R.G. Jones, Phys. Rev. Lett. **58**, 1460 (1987)
16. M.J. Bedzyk, D.H. Bilderback, G.M. Bommarito, M. Caffrey, J.S. Schildkraut, Science **241**, 1788 (1988)
17. SXR = synchrotron x-radiation, PWG = plane wave generator, MC = monocromator crystal, XPS = X-ray photoelectron spectroscopy
18. J. Zegenhagen, B. Detlefs, T.-L. Lee, S. Thiess, H. Isern, L. Petit, L. André, J. Roy, Y. Mi, I. Joumard, J. Electron Spectrosc. Relat. Phenom. **178–179**, 258 (2010)
19. T.L. Lee, C. Bihler, W. Schoch, J. Daeubler, S. Thiess, M.S. Brandt, J. Zegenhagen, Phys. Rev. B **81**, 235207 (2010)
20. L. Cheng, P. Fenter, M.J. Bedzyk, N.C. Sturchio, Phys. Rev. Lett. **90**, 255503 (2003)
21. M. Sugiyama, S. Maeyama, M. Oshima, Phys. Rev. Lett. **71**, 2611 (1993)
22. M. Sugiyama, S. Maeyama, S. Heun, M. Oshima, Phys. Rev. B **51**, 14778 (1995)
23. G.J. Jackson, J. Ludecke, D.P. Woodruff, A.S.Y. Chan, N.K. Singh, J. McCombie, R.G. Jones, B.C.C. Cowie, V. Formoso, Surf. Sci. **441**, 515 (1999)
24. M. Drakopoulos, J. Zegenhagen, A. Snigirev, I. Snigireva, M. Hauser, K. Eberl, V.V. Aristov, L. Shabelnikov, V. Yunkin, Appl. Phys. Lett. **81**, 2279 (2002)
25. M.v. Laue, *Röntgenstrahl-Interferenzen* (Akademische Verlagsgesellschaft, Becker & ErlerKolm.-Ges, Leipzig, 1941) (Akademische Verlagsgesellschaft, Frankfurt, 1960)
26. B. Batterman, H. Cole, Rev. Mod. Phys. **36**, 681 (1964)
27. A. Authier, *Dynamical Theory of X-Ray Diffraction*. IUCr Monographs on Crystallography, Oxford Science Publications (2001)
28. T.W. Barbee, W.K. Warburton, Mater. Lett. **3**, 17 (1984)
29. A. Einstein, Ann. Phys. **17**, 132 (1905)
30. In the dipole approximation, the exponential $e^{2\pi i\mathbf{kr}}$ of the E-field operator in the Matrix element, describing the photoelectric process, is simply replaced by 1, the first term of its corresponding representation in a Taylor series $(1 + 2\pi i\mathbf{k}r - \pi^2(\mathbf{kr})^2 - \cdots)$
31. K. Siegbahn, Nucl. Instrum. Methods Phys. Res. A **547**, 1 (2005)
32. I.A. Vartanyants, J. Zegenhagen, Solid State Commun. **113**, 299 (1999)
33. I.A. Vartanyants, T.L. Lee, S. Thiess, J. Zegenhagen, Nucl. Instrum. Methods Phys. Res. A **547**, 196 (2005)
34. I.A. Vartanyants, J. Zegenhagen, Nuovo Cimento **19 D**, 617 (1997)
35. M.B. Trzhaskovskaya, V.K. Nikulin, V.I. Nefedov, V.G. Yarzhemsky, At. Data Nucl. Data Tables **92**, 245 (2006)
36. D.P. Woodruff, Nucl. Instrum. Methods Phys. Res. A **547**, 187 (2005)
37. N. Hertel, G. Materlik, J. Zegenhagen, Z. Phys. B, Condens. Matter **58**, 199 (1985)
38. J. Zegenhagen, F.U. Renner, A. Reitzle, T.L. Lee, S. Warren, A. Stierle, H. Dosch, G. Scherb, B.O. Fimland, D.M. Kolb, Surf. Sci. **573**, 67 (2004)
39. J. Zegenhagen, T.L. Lee, Y. Gründer, F. Renner, B.O. Fimland, Z. Phys. Chem. **221**, 1273 (2007)

40. G. Scherb, A. Kazimirov, J. Zegenhagen, T.L. Lee, M.J. Bedzyk, H. Noguchi, K. Uosaki, Phys. Rev. B **58**, 10800 (1998)
41. T.L. Lee, S. Warren, B.C.C. Cowie, J. Zegenhagen, Phys. Rev. Lett. **96**, 046103 (2006)
42. J.S. Pedersen, R. Feidenhans'l, M. Nielsen, K. Kjaer, F. Grey, R.L. Johnson, Surf. Sci. **189**, 1047 (1987)
43. M. Gothelid, M. Hammar, C. Tornevik, U.O. Karlsson, N.G. Nilsson, S.A. Flodstrom, Surf. Sci. **271**, L357 (1992)
44. O. Bunk, J.H. Zeysing, G. Falkenberg, R.L. Johnson, M. Nielsen, M.M. Nielsen, R. Feidenhans'l, Phys. Rev. Lett. **83**, 2226 (1999)
45. J. Avila, A. Mascaraque, E.G. Michel, M.C. Asensio, G. LeLay, J. Ortega, R. Perez, F. Flores, Phys. Rev. Lett. **82**, 442 (1999)
46. D. Farias, W. Kaminski, J. Lobo, J. Ortega, E. Hulpke, R. Perez, F. Flores, E.G. Michel, Phys. Rev. Lett. **91**, 016103 (2003)
47. A. Tejeda, R. Cortés, J. Lobo-Checa, C. Didiot, B. Kierren, D. Malterre, E.G. Michel, A. Mascaraque, Phys. Rev. Lett. **100**, 026103 (2008)

Chapter 10
Advanced Applications of NEXAFS Spectroscopy for Functionalized Surfaces

Alexei Nefedov and Christof Wöll

Abstract Spectroscopic techniques based on the use of synchrotron radiation have significantly contributed to a better understanding of macroscopic as well as of microscopic properties of materials. One of them is near edge X-ray absorption fine structure (NEXAFS) spectroscopy, a source of valuable information on the electronic structure and orientation of molecular adsorbates on metal surfaces. NEXAFS spectroscopy has had its largest impact in connection with an electron-based detection of the photon absorption, which turns this technique in a highly surface sensitive method. The technique has been developed over the past 30 years and presently is routinely used to study the adsorption of organic molecules on a large variety of different substrates, including metals, oxides and polymers. In the last decades, in addition to above-mentioned systems, the spectroscopic characterization of nanostructures has become an important topic. The present contribution describes the basics of NEXAFS spectroscopy and demonstrates the potential of the method by discussing several case studies.

10.1 Introduction

Reliable and precise knowledge about the geometric and electronic properties are of utmost importance for a fundamental understanding of the macroscopic and microscopic behavior of materials. For functionalized interfaces this rather general statement is even more true since the enormous importance of such properties for the general behavior of nano-, macro- and micro-scaled systems is paralleled by a huge experimental challenge to gather this information. These problems are caused by the fairly small amount of material which governs the properties of an interface. A reliable characterization of interface properties requires the utilization of highly

A. Nefedov · C. Wöll (✉)
Institute of Functional Interfaces, Karlsruhe Institute of Technology,
Hermann-von-Helmholtz-Platz 1, 76344 Eggenstein-Leopoldshafen, Germany
e-mail: christof.woell@kit.edu

A. Nefedov
e-mail: alexei.nefedov@kit.edu

G. Bracco, B. Holst (eds.), *Surface Science Techniques*,
Springer Series in Surface Sciences 51, DOI 10.1007/978-3-642-34243-1_10,

surface sensitive methods. X-ray diffraction methods, the standard tool to obtain information on geometric structure and in particular on the orientation of molecules in the bulk, cannot be applied to ultrathin layers of molecules of surfaces in a direct fashion, since sensitivity problems, a fairly small threshold for radiation damage represent major experimental challenges in this area.

Among the different tools and techniques developed for the exploration of electronic and structural properties of interfaces synchrotron based methods have played a prominent role. Synchrotrons have first been built in the 1940s, mainly for applications in nuclear physics, and synchrotron radiation (SR) was first used in the "parasitic" mode of operation. Only when—some 50 years ago—second generation machines devoted to the use of SR for the characterization of materials were built, the method started to generate a larger impact on the understanding of physical and chemical properties of surfaces. Over the last three decades, second and more recently also third generation synchrotrons have contributed significantly to a better understanding of materials on the molecular and macroscopic scale. The large impact of SR in the area of surface science mainly results from information obtained via photoelectron spectroscopy. Due to the fairly short exit depth of photoelectrons excited in a solid substrate by absorption of photons, all variants of photoelectron spectroscopy based on electron detection are highly surface sensitive.

With regard to the chemical characterization of materials X-ray photoelectron spectroscopy (XPS) carried out with laboratory sources is the method which has had the largest impact in the field; in fact, laboratory X-ray sources are used in virtually every laboratory focusing on surface science, including the area of industrial research. Being able to set the incident energy of the photons to arbitrary values—and not being limited to the rather selected values of photon energies available from standard X-ray tubes—certainly already generates a huge potential; e.g, just setting the incident energy 30–50 eV higher than the core level binding energy increases the cross section by more than an order of magnitude.

In this chapter, however, we focus on a different variant of synchrotron-based photoelectron spectroscopy, one where the incident energy of the photons is continuously varied over a certain energy range. This method has the obvious disadvantage that it cannot be used in a connection with a laboratory X-ray source, this technique clearly requires the availability of a high performance synchrotron source. The particular variant of X-ray absorption spectroscopy to be discussed here, near-edge X-ray absorption fine structure spectroscopy (NEXAFS) was devised in the 1980s. The technique, which occasionally is also referred to as X-ray absorption near edge structure (XANES), was developed in order to provide information on the electronic and geometric structure of molecules bonded to solid surfaces, mostly that of metals. The method has had its largest success when applied to low Z molecules (Z is the atomic number), for which intense absorption edges are located in the soft X-ray region (100–700 eV). This is not a severe restriction, most organic molecules consisting of light atoms such as carbon, nitrogen, oxygen and fluorine, can be investigated in a straightforward fashion by using photons in this energy region [1]. In its standard variant the techniques probes transitions from a K-edge (the deepest core shell) of an atomic species into unoccupied molecular orbitals. As a result the technique is sensitive not only to ones of intramolecular character present in the molecule before adsorption on the surface, but also to interactions between the molecule and extramolecular atoms, e.g. surface atoms. A particular advantage of the method over other photoelectron spectroscopy variants which are sensitive to molecular orbitals, in particular, over ultraviolet photoelectron spectroscopy (UPS), is that the molecular orbitals detected with NEXAFS are selected as regards contribution of a specific type of atoms, e.g. carbon or nitrogen atoms. The huge background from the metal substrate typically governing UPS data is directly removed, making the analysis of experimental data considerably more straightforward.

Compared to other techniques which are used to gather information on the electronic and geometric properties of ultrathin molecular layers supported by solid substrates, e.g. methods employing vibrational spectroscopy, infrared spectroscopy (mostly used under grazing incidence, infrared reflection adsorption spectroscopy (IRRAS)) or electron energy loss spectroscopy (EELS) as well as other photoelectron spectroscopy based methods, NEXAFS has several advantages. First, it can be applied in a rather straightforward fashion, data analysis is fast and does not require sophisticated considerations. Secondly, charging problems as often encountered in XPS and UPS applied to insulation substrates, are a problem only in rare cases. Finally, the transition dipole moment (TDM) governing the electronic excitations can be oriented both normal and parallel to a metal surface to allow for a detection of the corresponding resonance, in contrast to infrared spectroscopy (and EELS carried out in the dipole scattering regime) where only transition dipole moments orientated perpendicular to the surface can be probed. NEXAFS is also rather sensitive, investigations for small adsorbate coverages substantially below one monolayer can be carried out in a routine fashion.

Before describing the properties of NEXAFS spectroscopy in more detail we will briefly address the two most important variants of photoelectron spectroscopy which can be carried out with laboratory sources, X-ray photoelectron spectroscopy (XPS) and ultraviolet photoelectron spectroscopy (UPS). These two techniques allow to determine the binding energies of occupied states in the core-level region (XPS) and the valence-band region (UPS), respectively.

The photoelectron spectra consist of peaks corresponding to the direct, primary excitationand a background caused by secondary processes, including Auger decays within the core as well as inelastically scattered electrons. Even slight charging of the samples generates severe problems in data analysis and special precautions need to be applied to avoid spurious shifts of peaks or the appearance of artifacts, e.g. a charge compensation using the second electron source or in some cases even a second photon source. A striking advantage of NEXAFS when applied to insulating substrates such as diamond and polymers is that charging effects are, in principle, absent since the absorption of incident photons is detected. Of course, when the X-ray absorption is detected via the monitoring of the secondary electron yield (partial electron yield or Auger yield, see below) a substantial charging of the sample has to be avoided. Experience has shown, however, that there are only few substrates (e.g. mica) where charging causes severe problems. On most other insulating substrates, including thin polymer films and diamond substrates, no substantial problems are encountered, small charging of up to 10 eV do not cause major problems in the electron-yield based detection of photon absorption.

With regard to investigating organic molecules a clear advantage of NEXAFS is the very high sensitivity towards the chemical nature of intramolecular bonds. For example, the C1s level binding energies as determined by XPS for carbon atoms in either a fully saturated hydrocarbon or in an aromatic compound such as benzene differ by ~0.1 eV only. This value is often below the resolution of the electron energy analyzer available at the experimental setups. Even if this energy resolution is available, a careful calibration of the energy scale is required [2]. As a result, an unambiguous distinction between single and double bonds is a major challenge in XPS. In contrast to XPS, where the binding energy of the initial state is measured, NEXAFS is sensitive to the position of the final state as well. As a result the pronounced differences in energy of unoccupied σ^*- and π^*-orbitals cause features separated by several electron-volts in the corresponding NEXAFS spectra which makes the detection of double bonds rather straightforward.

These facts make the NEXAFS technique particularly well suited for following chemical modifications within ultrathin organic layers, e.g. as a result of a reaction with a substrate or a reaction between adsorbed molecules. Note, however, that, in contrast to XPS, NEXAFS is not quantitative, i.e. a direct determination of the sample stoichiometry or the relative ratio of σ- and π-bonded C atoms cannot be easily carried out. For that reason a combination of XPS and NEXAFS is recommended.

When it comes to a determination of the chemical nature and the orientation of the molecular species on a solid substrate the most important other technique is infrared (IR) spectroscopy and related techniques, like surface enhanced Raman spectroscopy (SERS) or sum frequency generation (SFG). The most striking advantage of these methods is that they can be employed as in-house techniques. With IR spectroscopy the vibrational spectra of adsorbed molecular species can be determined with the frequency of the vibrational bands being highly sensitive to the nature of intra- and intermolecular chemical bonds. While for metals the application of IR spectroscopy is straightforward for insulating substrates intensity problems render the detection of molecules adsorbed on single crystal surfaces rather difficult [3]. Only for normal incidence IR-spectroscopy can be applied to such semiconducting and insulating substrates in a straightforward fashion, however in such a case only vibrations with a transition dipole moment orientated parallel to the substrate can be seen [4]. Only recently a routine investigation of infrared vibrations on a solid oxide substrate has become possible [5]. When IR-spectroscopy is applied to metal surfaces, only vibrations with a transition dipole moment orientated perpendicular to the surface can be seen in IRRAS, vibrations with a transition dipole moment parallel to the substrate are completely screened, leading to the so called surface selection rule in infrared spectroscopy [6]. In several cases the lack of sensitivity to such molecular vibrations with transition dipole moments orientated parallel to the surface have made a unique interpretation of experimental IR and EELS data sometimes difficult. If, e.g., a vibration of a diatomic species (e.g. O_2) cannot be seen it might imply either that the molecules is orientated with its axis aligned parallel to surface—or not present on the surface at all. This restriction is not present in NEXAFS. As a result of the high energy of the soft X-ray photons being located well above the plasmon energies of a metal, the screening by the metal electrons is ineffective and the orientation of the incident $\boldsymbol{E}$ vector can be chosen at will.

The present contribution describes the basics of the NEXAFS technique and the corresponding experimental setup. Selected examples of applications are given to demonstrate the breadth and the strength of this particular type of spectroscopy.

10.2 Theory Background

An understanding of the various features observed in NEXAFS spectra on a solid theoretical basis is a prerequisite for the use of NEXAFS as a reliable tool to investigate electronic structures and molecular orientations. In the following, a brief and simplified description of the underlying electronic effects governing NEXAFS spectroscopy is given.

A NEXAFS spectrum reflects the dependence of the photoabsorption cross section on the photon energy for values from just below the core level ionization threshold up to around 50 eV above. For the (over)-simplified case of an electron located in a single bound state (core level) and in the absence of any other empty electronic states the variation of the photoabsorption cross section when scanning the photon energy across the ionization energy would resemble a step function (the thick dashed line in Fig. 10.1c). For photon energies below the core electron binding energy no absorption would take place, whereas for photon energies above the ionization threshold an excitation of the core electron into the continuum of final states (plane waves describing an electron traveling away from the core hole) makes an absorption of photons possible.

Even for a free single atom (Fig. 10.1a), however, the measured cross section looks considerably more complicated. This is a result of the fact that there are considerably more final states than in the oversimplified model referred to above, the presence of the electrostatic potential of the atom's positively charged nucleus gives rise to a number of empty electronic states, located below (bound states) as well as above (unbound states) of the ionization threshold. The bound states can be roughly divided into valence state localized in the proximity of the core and Rydberg states with binding energies converging to the ionization threshold. Note, however, that in some cases this distinction is problematic [7]. The unbound states mostly correspond to excitations into (molecular) antibonding states and are located well above the ionization threshold, a typical feature of NEXAFS resonances arising from excitations into such states is the large energetic half-width.

A cartoon showing schematically the origin of NEXAFS features for the case of a diatomic unit is provided in Fig. 10.1b. The effective electrostatic potential and the corresponding K-shell spectrum are also shown. Around the ionization threshold, resonant transitions are superimposed on the step-like absorption edge as shown in Fig. 10.1c. Empty molecular orbitals are labeled as σ^*- and π^*-orbitals according to their symmetry. The lowest unoccupied molecular orbital (LUMO) of a π-bonded

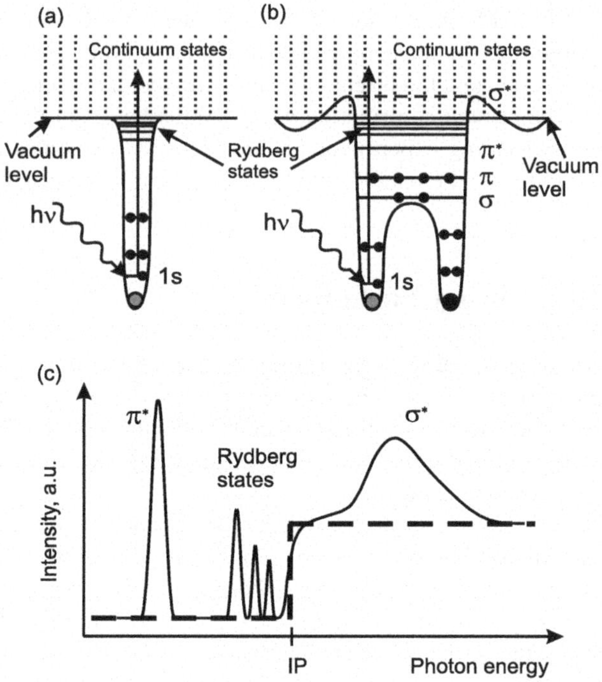

Fig. 10.1 (**a**) Schematic potential for isolated atom; (**b**) Schematic potential of a diatomic molecular (sub)group. In addition to Rydberg states and a continuum of empty states similar to those expected for atoms, unfilled molecular orbitals are present, which is reflected in the absorption spectrum (**c**). The *thick dashed line* corresponds to the photoabsorption cross section of an electron located in a single bound state (core level)

diatomic subunit of a molecule is usually a π^*-orbital, while σ^*-orbitals are found at higher energies (see Fig. 10.1c). The latter are most often found above the vacuum level for the neutral molecule. Of course, an $1s \rightarrow \pi^*$ transition or π^*-resonance can only be observed for molecules with π-bonding, i.e. double and triple bonds or aromatic systems, but not for single bonds.

The measured width of a resonance is determined by the resolution of the instrument (resulting in a Gaussian lineshape), the lifetime of the excited state (resulting in a Lorentzian lineshape), and the vibrational motion of the molecule leading to an unsymmetrical broadening [8]. Broadening due to the lifetime of a π^*-resonance is generally very small, on the order of a about 100 meV for e.g. the C1s $\rightarrow \pi^*$ excitation in benzene. Essentially, the final state lifetime is determined by the re-filling, or the decay of the core-hole pair, either by the excited electron falling back or by an Auger-transition involving other electrons. Transitions into unoccupied orbitals of σ-symmetry which are found at higher energies above the ionization threshold are energetically considerably broader, indicating strongly reduced lifetimes for such cases. This experimental finding can be rationalized by considering that the large overlap of these states with continuum states will, of course, strongly decrease the lifetime of these states. As a rule of thumb, the higher the resonance lies in the continuum the larger is its linewidth.

Rydberg orbitals give rise to sharp but weak resonances occurring below the ionization threshold. They are generally located between the π^*-resonance and the ionization potential. In the condensed phase or for strongly chemisorbed molecules, pure Rydberg resonances are quenched because of the large spatial extent of these electronic states. A clear separation is, however, not always possible. E.g. in the case of single CH bonds, a mixing of Rydberg orbitals with hydrogen-derived antibonding orbitals of the same symmetry leads to hybridization, thus increasing the intensity of the corresponding NEXAFS resonance located just before the C-K edge [9, 10]. Finally, NEXAFS spectra can also exhibit multi-electron features, such as so-called shake-up structures, which can be associated with the indirect excitation of "passive" electrons. The sudden creation of the core hole potential induced by absorption of an X-ray photon by the "active" electron may knock one or more of the "passive" electrons into excited states. It is necessary to mention that in photoemission spectra the satellites are observed at lower kinetic energy than the main active-electron peak, but in X-ray absorption shake-up features are observed at higher photon energy than the primary resonance [1].

In addition to the information on the electronic structure of molecules or molecular fragments, NEXAFS can be used to determine the orientation of a molecule relative to the substrate surface. Bonds and the corresponding molecular orbitals are highly directional and the spatial orientation of an orbital, i.e. the direction of maximum orbital amplitude on the excited atom, determines the angular dependence of the K-shell spectra. Therefore, the transition intensities depend on the orientation of the electric field vector relative to the orientation of the molecule. Exploiting the polarization characteristics of synchrotron radiation (for a simplicity, we will assume that the radiation is linearly polarized), the orientation of a molecular orbital can be found by determining the intensity of the corresponding resonance as a function of the incidence angle of the photons, θ. The quantum mechanical description of the excitation process for a single electron in the dipole approximation leads to an equation relating the initial state Ψ_i and the final state Ψ_f to the photoabsorption cross section σ_x:

$$\sigma_x \propto \left|\langle \Psi_f | \boldsymbol{e} \cdot \boldsymbol{p} | \Psi_i \rangle\right|^2 \rho_f(E) \tag{10.1}$$

with $\boldsymbol{e}$ being the unit electric field vector, $\boldsymbol{p}$ the dipole transition operator, and $\rho_f(E)$ the density of final states [1]. In the case of linearly polarized light, the angular dependence of the matrix element of interest $|\langle \Psi_f | \boldsymbol{e} \cdot \boldsymbol{p} | \Psi_i \rangle|^2 = |\boldsymbol{e}\langle \Psi_f | \boldsymbol{p} | \Psi_i \rangle|^2$ assumes a simple form and the expression $\langle \Psi_f | \boldsymbol{p} | \Psi_i \rangle$ is known as transition dipole moment (TDM). For a 1s initial state and a directional final state orbital the matrix element $|\boldsymbol{e}\langle \Psi_f | \boldsymbol{p} | \Psi_{1s} \rangle|^2$ points in the direction of the final state orbital $\boldsymbol{O}$ and the transition intensity becomes

$$I \propto \left|\boldsymbol{e}\langle \Psi_f | \boldsymbol{p} | \Psi_{1s} \rangle\right|^2 \propto |\boldsymbol{e} \cdot \boldsymbol{O}|^2 \propto \cos^2 \delta \tag{10.2}$$

with δ being the angle between the electric field vector $\boldsymbol{E}$ and the TDM direction. Therefore, the intensity of a resonance is largest when the electric field vector $\boldsymbol{E}$ lies along the direction of the final state orbital $\boldsymbol{O}$ (or the TDM direction), as schematically shown in Fig. 10.2, and vanishes when $\boldsymbol{E}$ is perpendicular to it. Note that σ^*-orbitals have a maximum orbital amplitude along the bond axis while π^*-orbitals have maximum amplitude normal to the bond direction (Fig. 10.2). In the case of molecules with π-bonds, the large π^*-resonance is most suitable for a quantitative

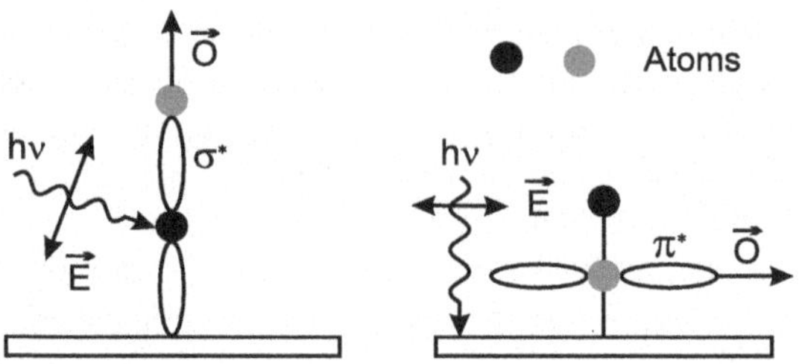

Fig. 10.2 Schematic representation of the origin of the angular dependence of NEXAFS resonances for a π-bonded diatomic molecule adsorbed with its molecular axis normal to the surface. As a result of the different overlap between the electric field vector ***E***, and the direction of the final state orbitals ***O***, the π*-resonance is maximized at normal incidence (*right*), while the σ*-resonance is maximized at grazing incidence (*left*)

evaluation of the angular dependence because of its clear separation from the ionization step and the absence of any interfering background.

In common case, the TDM direction can be described through a polar angle α and an azimuthal angle ϕ and in equations for angle dependencies of resonance intensities there is a dependence both on α and on ϕ [1]. It means for the determination of the molecule orientation it is necessary to measure at least two series of NEXAFS spectra at two different azimuthal sample orientations. Fortunately, the azimuthal dependence is eliminated in many cases by the symmetry of the surface. The surface symmetry established several equivalent in-plane chemisorption geometries which lead to the formation of adsorbate domains. Thus for molecules adsorbed on a surface with an at least three-fold symmetry, there is an averaging over the azimuthal angle ϕ and the NEXAFS resonance intensities can be expressed as:

$$I \propto P \cos^2\theta \left(1 - \frac{3}{2}\sin^2\alpha\right) + \frac{1}{2}\sin^2\alpha, \tag{10.3}$$

where P denotes the degree of polarization of the incident X-ray light. This equation is quite simple and it is routinely used for a determination of the molecular orientation.

(10.3) could be also written in the form

$$I \propto P \cos^2\theta \left(\cos^2\alpha + \frac{1}{2P}\tan^2\theta \sin^2\alpha\right) \tag{10.4}$$

and it is clearly seen that at the incidence angle $\theta = \arctan(\sqrt{2P})$ the absorption intensity does not depend on the molecule orientation. Taking in to account that $P \approx 1$, this angle θ is about 55° and it is known as "magic angle". The absorption intensity measured at this angle is very important for data analysis, because it allows to determine a scaling factor, which contains all physical and experimental constants. Usually for samples without any preferable orientation of molecules it is enough to record NEXAFS spectra only at this "magic" angle.

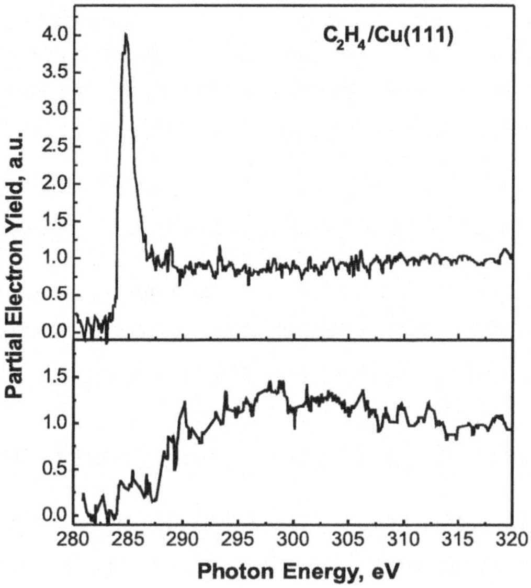

Fig. 10.3 NEXAFS spectra recorded at normal incidence (*lower panel*), and at grazing incidence (*upper panel*) for monolayers of ethylene adsorbed on Cu(111) at temperature of 100 K

Another useful consequence of (10.3) is a presentation of the difference of intensities measured at two different angles, namely at normal (90°) and gracing (20°) incidence. Taking in to account that $\cos 90° = 0$ and $\cos 20° \approx 1$ we obtained:

$$I(90°) - I(20°) \propto \left(\frac{3}{2}\sin^2\alpha - 1\right). \tag{10.5}$$

Here α is the angle between the electric field vector $\boldsymbol{E}$ and the TDM direction averaged over azimuthal angles ϕ. This difference spectra is known also as a linear dichroism and it is negative for small α and positive at $\alpha \geq 55°$. According to Fig. 10.2, if molecule have double/triple bonds in its molecular plane the negative dichroic signal of π^*-resonances corresponded to these bonds will result in a flat orientation of the molecule and positive one to upright-oriented molecules. Since σ^*- and π^*-orbitals are normal to each other, the dichroic signal of σ^*-resonances has the opposite sign than for π^*-resonances.

As an obvious case, we presented the results obtained by Fuhrmann et al. [11] on the determination of the molecular orientation of ethylene adsorbed on Cu(111). C_2H_4 was adsorbed at temperature of 100 K and the NEXAFS spectra were recorded at grazing and normal incidences. They are presented in Fig. 10.3 and the absence of π^*-resonance at E = 284.7 eV for normal incidence (Fig. 10.3, lower panel) clearly demonstrate a flat orientation of ethylene molecules adsorbed on Cu(111).

A rather reliable basis for the assignment of the features in the experimental NEXAFS spectra can be provided by a series of calculations with a quantum chemistry program package, e.g. the StoBe (Stockholm–Berlin) package [12]. Note that StoBe allows to compute and analyze the electronic structure as well as spectroscopic and other properties of molecules. A particular advantage of this program is that due to the use of density functional theory (DFT) methods it can deal with rather large molecules, e.g. hexa-peri-hexa-benzocoronene with 42 carbon atoms and 18 hydrogen atoms represents no major problem. Also atomic clusters can be handled in a straightforward fashion. The approach is based on self-consistent solutions of the Kohn–Sham equations using a linear combinations of Gaussian type orbitals and it can produce the projection intensity of the core-π^* transition according to the molecular symmetry and the sum of these projections will give rise to the final NEXAFS spectra. The theory and numerical details of the realization can be found in [12–15]. The striking advantages of using the results of such theoretical calculations with regard to the interpretation of features in experimental NEXAFS data will be demonstrated for selected cases in Sect. 10.4.

10.3 Experimental Method and Setup

The key feature of NEXAFS spectroscopy with regard to other variants of photoelectron spectroscopy is the fact that the resolution of the method is not determined by the energy analyzer used to determine the kinetic energy of the photoexited electrons. Since the measured quantity, measured either directly or indirectly, is the absorption of X-rays, the energy resolution is only determined by the employed monochromator.

The most straightforward way to measure the X-ray absorption cross section would be to measure the attenuation of a direct beam of X-ray photons transmitted through a sample using an X-ray sensitive detector. This method can be easily applied only for rather thin samples since the penetration depth of X-rays is in the μm- to mm-region (depending on incident energy). A striking disadvantage of such straightforward measurements would be the complete absence of any surface sensitivity.

In the most important variant of NEXAFS spectroscopy used today for the investigation of surface properties, the absorption of X-ray photons is measured by detecting the secondary electrons generated by the decay of the core hole emitted from the sample into the vacuum. While the most direct way to measure the total current would be to use a sensitive amperemeter to measure the total photocurrent generated by the impinging photons as a function of photon energy, in practice it has proven more effective to measure the electrons emitted into the vacuum using electron multipliers or channel plates.

A typical NEXAFS experimental setup is presented in Fig. 10.4a. First, a monochromator is used to transmit only light with predefined energy which then hits the sample substrate. The shape of the beam is defined by a set of horizontal and vertical slits. A gold grid (typical transmittance 85 %) is placed in the optical path making it possible to monitor the intensity of the X-ray photon beam incident on the sample. Although the current recorded for this gold grid cannot be used to directly determine the transmission function of the monochromator (see below) the recorded signal can be conveniently used for energy calibration purposes. The X-ray beam

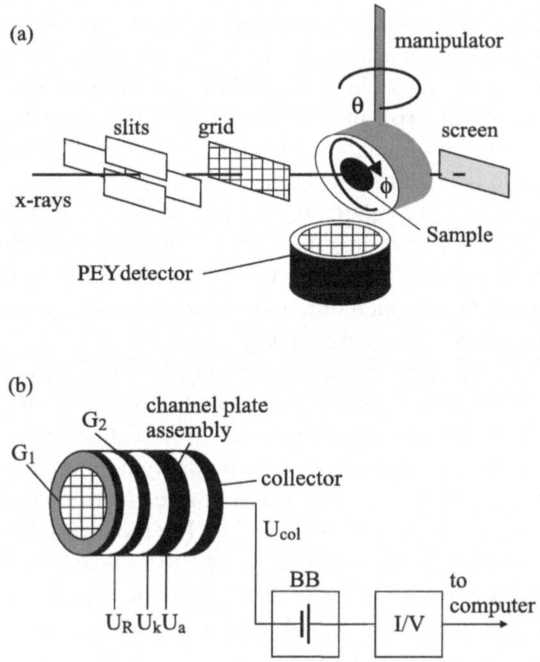

Fig. 10.4 (**a**) The scheme of the NEXAFS experimental setup; (**b**) the scheme of the PEY detector with corresponding electronics. G1 and G2—two grids, BB—battery box, I/V—current amplifier; U_R, retarding voltage, U_a and U_k—voltages on channel plates, U_{col}—voltage on collector

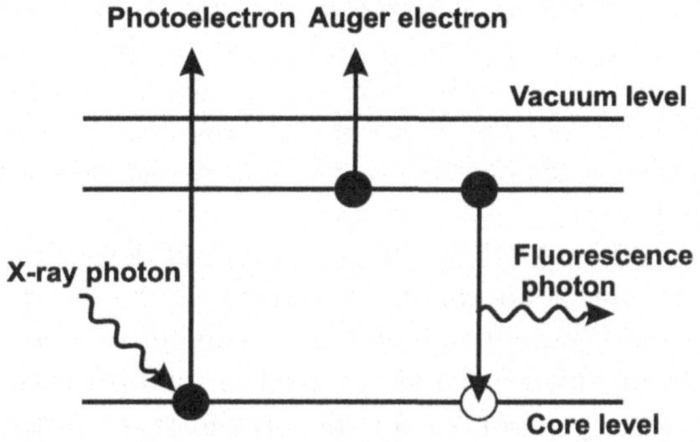

Fig. 10.5 The photoabsorption process including filling of the core hole by emission of a photon or an Auger electron

then enters the analysis chamber and hits the sample mounted on a special manipulator allowing to change sample orientation with respect to the incident beam (polar as well as in azimuthal angles).

As mentioned above the photoabsorption process taking place when the incident photons hit the sample substrate results in the creation of a photoelectron and a core hole. The hole is subsequently filled by another electron, the corresponding excess in energy is dissipated either radiatively by the emission of a fluorescent photon, or non-radiatively by the emission of an Auger electron (see Fig. 10.5). Both channels are a direct result of the core hole created in the X-ray photoabsorption process and thus provide a basis to determine the absorption cross section. In principle, either dissipation process can be used for the detection. It is necessary to note, however, that for low-Z elements (C, N, O) the Auger electron yield is much higher than the fluorescence yield (FY), making the electron yield channel better suited for low-Z molecules [16]. In addition, electron detection provides the higher surface sensitivity and in the majority of studies published in the literature this so-called electron yield detection scheme has been employed. The reason for the higher surface sensitivity is the relatively low kinetic energy of the electrons and the corresponding mean free path in solid matter, which is typically less than 1 nm for energies between 250 eV and 600 eV [17]. The inelastic scattering process leads to an electron cascade, of which only those electrons with sufficient energy to overcome the work function of the material will escape the surface. The resulting effective escape depth and, therefore, the information depth of electron yield, has been estimated to be in the range of 5 nm for metals and semiconductors, and slightly larger for insulators due to the reduced electron-electron scattering mechanism [1]. The surface sensitivity can be further enhanced by applying a retarding voltage. By suppressing lower kinetic energy electrons, only those electrons that emerge from the outermost surface region (3 nm) are detected. For the investigation of adsorbates on surfaces, this so-called partial electron yield (PEY) detection has a better signal-to-background ratio than total electron yield (TEY) detection, where all electrons that emerge from the surface are detected. A further option is Auger electron yield (AEY) detection where only elastically scattered Auger electrons are recorded. The AEY mode provides the best surface sensitivity of the three detection techniques, but requires an electron energy analyzer. In the case of fluorescence detection, an appropriate detector has to be employed. In this contribution we do not discuss the FY detection method in detail, but note that this detection method is the one of choice for liquid and bulk samples, since in such environments the limited mean free path of photo- and secondary electrons makes the application of the PEY and TEY method impossible.

In Fig. 10.4b we show the typical design for a partial electron yield detector used for monitoring the absorption of the X-ray photons. The PEY detector is assembled using two high transmission metal grids for retardation, a double or triple channel-plate assembly for electron multiplication and a collector [18]. The first grid is usually grounded and the second grid is operated at retardation voltage $-U_R$, which rejects all electrons with energies less than U_R. Typically U_R is set to a value 100–120 eV less than the energy of the corresponding Auger peaks located at around 260 eV (C), 370 eV (N) and 510 eV (O). Electrons with the energies above U_R pass through these two grids and are then accelerated towards the channel-plate assembly by a small positive voltage ($U_k \approx 50$ eV) on the contact just before the upmost channel plate—the cathode. The electron signal is amplified by channel-plate arrangement by the application of a voltage ($U_a \approx 2000$ eV) on the contact just behind the bottommost channel plate—the anode. The double channel-plate arrangement operates at a total gain of 10^7–10^8 and the electron output is collected by a collector which is connected for current measurement to a floating battery box. This battery box supplies a low-noise positive potential (U_{col}) in the 2–3 kV range. The high potential is generated by a series of batteries which are suitable insulated from each other and the battery box housing and soldered together to eliminate noise. The achievable noise level is limited by leakage currents and is about 10^{-11} A. Normally, the collector voltage U_{col} is kept somewhat ($\approx$200 eV) larger than U_a. The negative side of the battery string is connected to the input of voltage/current amplifier which generates an output voltage proportional to the input current. With a gain of 10^7–10^8 the output voltage will be in the range 0.5–5 V which fits quite well to corresponding interface card (A/D converter) of the computer. The computer must be also connected with the monochromator to control the energy of the incident X-ray photons. Note, that for a good signal-to-noise ratio of the electron yield signal the acceptance angle of PEY detector should be as large as possible. Correspondingly, it should be positioned close to the sample. An alternative is the use of channel plates with large diameter.

Before recording NEXAFS spectra it is very important to precisely adjust the spot on the sample surface illuminated by the incident X-ray photons, since contributions e.g. from the sample holder to the measured signal will make a proper analysis of the data impossible. In particular for gracing incidence the position of the sample has to be defined within a few tenth of a mm. In order to achieve such a proper alignment and to avoid contributions from unintentionally illuminated parts of the experimental setup it is quite useful to employ a phosphor-covered test sample and a phosphor-covered screen behind the sample-holder. Fluorescence in the visible regime will then allow for a good alignment of the manipulator and the whole chamber.

To extract quantitative information about adlayers on a substrate raw NEXAFS spectra recorded at different angles of incidence have to be properly normalized. We typically apply the following procedure: first, the constant background signal, present without illumination (dark currents) is subtracted from the spectra. Then the spectrum recorded for the clean substrate is subtracted from the adlayer spectra. The resulting data is then divided by a spectrum recorded for a freshly sputtered gold film to compensate the energy dependence of the transmission function of the beamline. Finally, the intensities are normalized to an edge jump of 1, i.e., intensity difference between 275 and 330 eV in a case of C K-edge. This procedure provides information exclusively on the adsorbed layer. There is also a series of another normalization procedures (see [1]) and in dependence on a sample a proper processing has to be used.

Experimental NEXAFS System The apparatus used by us to measure NEXAFS spectra is additionally equipped with complimentary surface analytical techniques, namely XPS/UPS, low energy electron diffraction (LEED) and thermal desorption spectroscopy (TDS). Especially for more complex systems the possibility to characterize samples by other methods has been shown to be crucial for providing a unique and reliable interpretation of the experimental NEXAFS data. The scheme of such rather complex, multichamber UHV-system used by us is presented in Fig. 10.6. It consists of the analysis chamber (5) for NEXAFS/XPS measurements, the preparation chamber (2) equipped by evaporators (3), ion sputter guns and LEED system (4), a sample transfer system, including the distribution chamber (7) with the

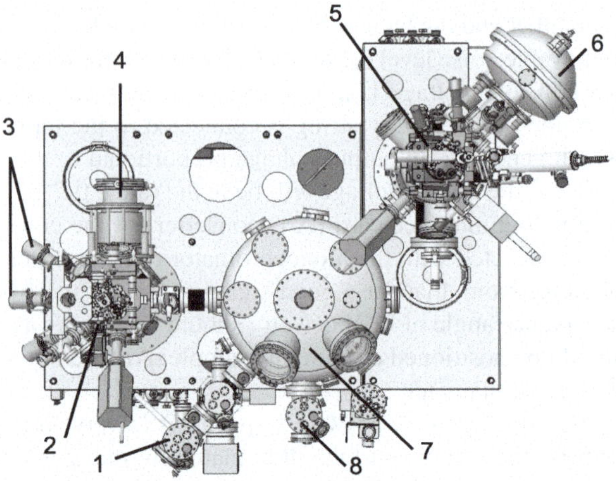

Fig. 10.6 The drawing of XPS/NEXAFS system (*top view*): *1*—load lock chambers, *2*—preparation chamber equipped by evaporators (*3*) and LEED system (*4*), *5*—analysis chamber with XPS/UPS electron analyzer Scienta R3000 (*6*), *7*—distribution chamber with park-station (*8*), which serves for keeping samples under UHV-conditions for long time

Fig. 10.7 The view of XPS/NEXAFS end-station installed on HESGM beamline of synchrotron BESSY II in Berlin

park-station (8) and two load lock chambers (1). One of the load lock chambers is designed for a use of a special transport box, which allows to transfer samples to another systems under UHV conditions. A crucial point is the sample transfer system which at the same time allows to carry out measurements in an extended temperature region. With the present setup samples can be investigated in a temperature region of at least 50–1000 K. This NEXAFS/XPS apparatus was designed and built by PREVAC (Poland) [19] and is operated at the HESGM beamline of synchrotron facility BESSY II (Berlin, Germany). A photo of the setup is presented in Fig. 10.7. Similar end-stations are installed at another synchrotron facilities, e.g. at MaxLab (Lund, Sweden), NSLS (Brookhaven, USA), ALS (Berkeley, USA), etc.

We would like to conclude this chapter by noting that the spatial resolution of NEXAFS is determined by the spot size of the X-rays on the surface and the acceptance area of the analyzer, and is usually macroscopic, i.e. the information obtained on the electronic structure and the molecular orientation will normally be averaged over a macroscopic (mm-sized) region.

10.4 Applications of NEXAFS Spectroscopy

A detailed investigation of the fine structure at the adsorption edges of light elements such as C, N and O was first carried out in the 1980s for the case of "small" molecules such as CO and NO adsorbed on single crystal metal surfaces [20]. Since then NEXAFS has been used successfully to determine the electronic structure and molecular orientation of a fairly large number of molecules on many different substrates. The seminal book by Stöhr [1] provides an excellent overview of the basic principles of this technique and the results obtained until the early 1990s. In more recent reviews one can find reports on the application of the NEXAFS spectroscopy for the study of liquids and polymers [21] as well as for a study of inorganic samples, in particular, transition metal carbides and nitrides and other interstitial compounds [22]. In the present contribution we concentrate on the application of the NEXAFS technique for studies of functionalized surfaces and organic nanostructures. Rather than providing a broad overview of experimental studies carried out in the last years we will present a number of selected examples which demonstrate the advantages of NEXAFS spectroscopy also to scientists new to this field.

10.4.1 Structural Rearrangements at Organic/Metal Interfaces

Distortion of C–H Bonds in Benzene Adsorbed on Metal Surfaces Over the past years a number of surface science studies of benzene adsorbed on a variety of close-packed metal surfaces have been reported [23–26]. The data consistently show that for coverages up to a monolayer, benzene adsorbs with its symmetry plane parallel to the substrate. In the case of transition metal surfaces with a significant fraction of the unoccupied d-bands the interaction of the benzene system with the substrate atoms results in the formation of a stable chemical adsorbate-substrate bond. The NEXAFS spectra of physisorbed and chemisorbed benzene are strongly modified with respect to gas phase spectra. These changes arise from physical (final state effects, screening of the core hole by the metal substrate) as well as from chemical (changes of electronic structure as a result of the benzene-metal bonding) interactions. The precise origin of these differences is not completely understood and has remained to some extent controversial, although the benzene-metal adsorption system is among the most extensively adsorbate systems studied. As demonstrated in a study by Weiss et al. [27] the strongest modifications in the NEXAFS spectra can be observed for the benzene π^*-resonance which is found to be significantly broadened with respect to the gas phase even in the case of physisorption on the most inert metal, Au. In this study NEXAFS spectra of benzene adsorbed on different metal

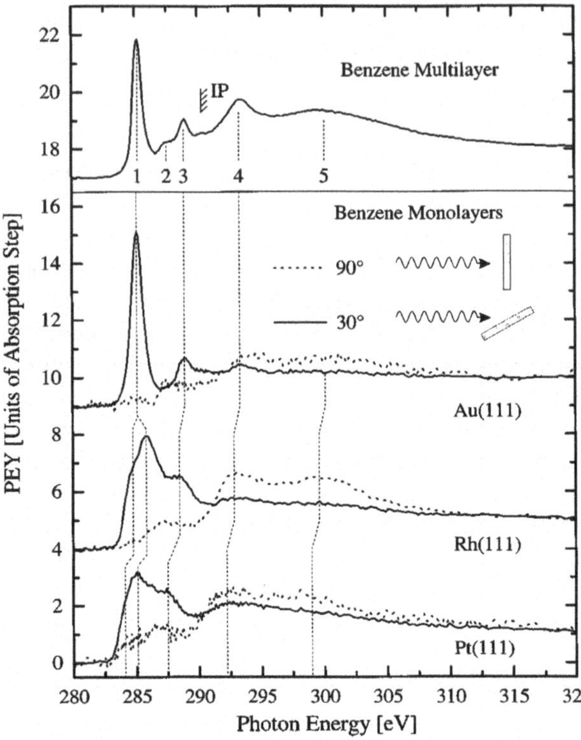

Fig. 10.8 NEXAFS spectrum of a benzene multilayer at 120 K (*top*). Polarization dependence of NEXAFS spectra recorded for benzene monolayers adsorbed on Au(111), Rh(111) and Pt(111) (*bottom*). The spectra were recorded for normal (90°) and grazing (30°) photon incidence. The hatched line at 290.3 eV indicates the ionization threshold. (Reprinted with permission from K. Weiss, S. Gebert, M. Wühn, H. Wadepohl, Ch. Wöll, Journal of Vacuum Science and Technology A **16**, 1017 (1998) Copyright (1998) American Vacuum Society)

surfaces, Au(111), Rh(111), Pt(111) and Ru(0001) were measured and it has been demonstrated that a detailed analysis of the dichroism seen for NEXAFS resonances provides direct information on the adsorption induced out-of-plane bending of C–H bonds [28].

The NEXAFS spectrum recorded at the C1s edge for a thick (d > 10 nm) benzene multilayer as shown at the top of Fig. 10.8 is very similar to the gas phase spectrum of benzene observed by Horsley et al. [23]. Peaks 1 and 3 correspond to π^*-resonances, peaks 4 and 5 correspond to σ^*-resonances. The weak feature at about 287.5 eV labeled 2 is assigned to a Rydberg resonance. For a more detailed assignment of the various resonances the reader is referred to [27].

In the lower part of Fig. 10.8 we display NEXAFS spectra recorded for benzene monolayers adsorbed on Au(111), Rh(111), and Pt(111) at normal and grazing photon incidence under the same experimental conditions. For the Au(111) surface the benzene NEXAFS spectra show a strong angular dependence. This strong linear dichroism is expected, since benzene adsorbs with its molecular plane parallel to the Au-substrate surface (Fig. 10.9, upper panel). In particular from the fact that the intensity of the π^*-resonance recorded at normal incidence its virtually zero it can be concluded that the benzene basically retains its gas phase structure, in particular the tilt of the C–H bonds away from the molecular plane must be much smaller than in the case of transition metals with only partially filled d-bands, see below.

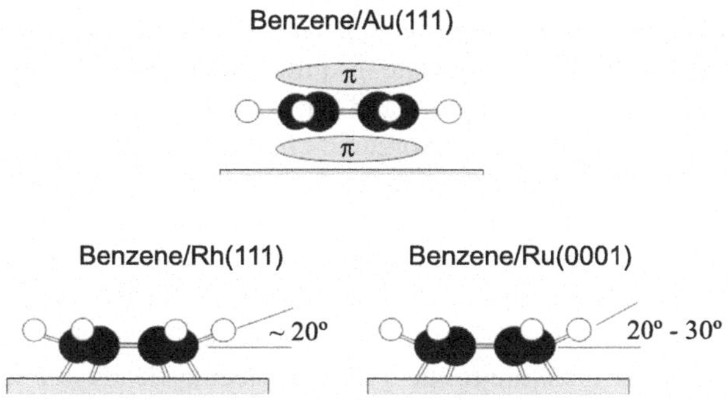

Fig. 10.9 Schematic models of benzene molecule absorbed on Au(111), Rh(111) and Ru(0001)

The spectra of benzene monolayers strongly chemisorbed on Rh(111) and Pt(111) are similar to spectra reported earlier for benzene chemisorbed on Pt(111) [23]. In contrast to the spectra recorded for benzene monolayers on Au(111) several pronounced differences can be observed between the monolayer and multilayer spectra. The most conspicuous feature is the strong broadening of the π^*-resonance. This splitting of the π^*-resonance is attributed to a hybridization of the benzene π-orbitals with metal electronic states leading to the formation of several different hybrid electronic states with contributions from both, C atomic orbitals and empty states in the metal conduction band, as discussed by Stöhr [1]. This hybridization (or mixing) of molecular (or atomic) orbitals with metal states leads to a strong decrease of the final state lifetime and thus explains the significant increase in linewidth when going from the multilayer to the monolayer.

The spectra recorded at normal incidence for the chemisorbed benzene species on Rh(111) and Pt(111) show considerable intensities for the π^*-resonance. At first this experimental finding seems to indicate a tilted adsorption geometry since for a perfectly flat orientation of the benzene molecular plane with regard to the substrate the intensity of the π^*-resonance should be strictly 0 at normal incidence, as observed for the case of Au(111) (see above). Chemical intuition as well as the results obtained by other techniques strongly suggest, however, that the CCC plane of benzene is orientated largely parallel to the surface on both, Rh(111) [29] and Pt(111) [30]. A more detailed investigation of the binding situation has revealed that the rather pronounced residual intensity of the π^*-resonance at normal incidence can be related to conformational changes resulting from the formation of chemical bonds between the carbon atoms and the transition metal substrate. The rehybridization of the carbon π^*-orbitals from sp^2 (in benzene) to sp^3 results in a pronounced change in the orientation of the transition dipole moment governing the excitation of the C1s core electron into the empty molecular orbitals. These electronic changes are accompanied by a pronounced bending of the C–H bonds away from the C–C–C plane (see Fig. 10.9, lower panel). In a very simplified picture these adsorption induced distortions can be explained in terms of interfragment interaction, since the C–H bond bending away from the metal surface maximizes the overlap between the orbitals of benzene and the electronic states of the metal. For a detailed discussion of the implications of such a distortion of the NEXAFS spectra see [27, 28].

Charge-Transfer-Induced Rearrangements of Tetracyano-p-Quinodimethane on Cu(100) Tetracyano-*p*-quinodimethane (TCNQ) is one of the strongest organic electron acceptors and has long been regarded as a prime candidate for the formation of organic/inorganic charge-transfer compounds [31]. The interface between TCNQ adsorbate and a Cu substrate can thus be considered a model system for the interfaces between the strong organic acceptors usually involved in organic light-emitted diodes or organic solar cells and the metallic contacts.

In a recent paper [32] a combination of different experimental techniques supported by theoretical calculations has demonstrated that the donation of an electron from the substrate to the TCNQ molecules leads to a molecular conformation very similar to the well-documented structure of the corresponding anion or the dianion in solution. The conformation of the negatively charged (or anionic) molecule results in a rather strong bonding between the nitrogen lone pairs and the copper atoms at the surface. A more detailed theoretical investigation revealed that this interaction is rather localized, so that the surface atoms bonded to the cyano groups are significantly lifted (or bent away) from their usual positions at the surface.

The main features that determine TCNQ acceptor functionality are the four peripheral cyano groups and the central hexagonal ring. In its neutral form the hexagonal ring is not aromatic and the molecular conformation is very rigid owing to the alternation of multiple and single bonds (*p*-quinoid character). Upon transfer of one electron, however, the central hexagon ring is aromatized which can only take place by changing the character of the bond between the hexagon ring and the dicyanomethylene group from double to single. The extra electron is thus accommodated in one of the peripheral nitrogen atoms, so that one of the molecular ends remains radical in character, whereas the other end loses the radicaloid character by accumulating one extra electron. Bond conjugation at the dicyanomethylene ends, however, remains, so that both the extra electron and the radical character are delocalized at each dicyanomethylene group. The dianion form of TCNQ is also well known: in this case the second electron transferred to the molecule is accommodated at the radicaloid dicyanomethylene end. In the anionic and dianionic forms, bond conjugation is restricted to the central ring and to the dicyanomethylene groups: the terminal carbon of bond between the hexagon ring and the dicyanomethylene group now has a higher sp^3 character and, consequently, the cyano groups bend away from the molecular plane.

As demonstrated by the angle resolved results presented in [32] NEXAFS spectroscopy is ideally suited to monitor the conformational and bonding changes accompanying the adsorption TCNQ on a Cu(100) surface. An investigation of the N1s spectra reveals that the peak assigned to the CN group is split, in contrast to NEXAFS spectra measured for TCNQ powder where only a single resonance is observed. The CN group has two orthogonal π^*-orbitals and for the neutral molecules the energies are very close thus resulting in a single peak only. However, upon adsorption on the Cu-substrate, the strong interaction of the CN groups with the Cu(100) substrate leads to a pronounced increase of this splitting, resulting in two separate peaks. Their angular dependencies of these two different contributions were analyzed by a fit procedure according to (10.3) yielding tilt-angles of 19.7°

and 10.0° for the low-energy and high-energy peaks, respectively. Together with the C1s analysis, it was concluded that TCNQ adopts a planar, flat-lying adsorption geometry on Cu(100), with the CN groups tilted away from the surface by an angle of $15 \pm 5°$.

These results exemplify the fundamental role that charge-transfer processes across metal-organic interfaces have on molecular self-assembly and subsequent crystal growth. Such a charge transfer modifies the energy level alignment at the interface, but they also affects molecular conformation and, thereby, adsorbate self-assembly. These results also demonstrate that in case of strong bonding between a metal surface at a molecular substrate the static surface approximation may no longer be applicable: substrate reconstruction and surface-mediated interactions must be explicitly taken into account. Predictions based on gas phase geometries of the molecules and corresponding intermolecular interactions may yield a poor description of the experimental results only.

10.4.2 Internal Twist Structure in Aromatic Self-assembled Monolayers

Since the first systematic investigation of self assembled monolayers (SAMs) started about 25 years ago [33], this field has developed into an important section of research in the field of nanotechnology (see [34–36] and references provided there). The huge diversity and the enormous potential of these organic thin films continue to drive a number of new developments. In earlier work mainly alkanethiols have been used for the formation of SAMs on gold surfaces to unravel fundamental aspects of film formation, structure and properties. In later years aromatic organothiols have attracted an increasing amount of attention because of the higher rigidity of their molecular backbone and also because of the more interesting electronic properties of the conjugated π-systems. Today, an important application of SAMs is to create organic surfaces exposing predefined functionalities. Attaching an appropriate moiety to the monomers from which the SAM is fabricated allows tailoring the wettability and the reactivity of the organic surfaces exposed by the SAMs. Such organic surfaces have numerous potential applications in molecular electronics, electrochemistry and biochemistry [37]. A recently established, exciting new field based on SAM-modified substrates is interface-based supramolecular chemistry, where organic monolayers are used as templating substrates to anchor and grow highly porous materials like metal-organic frameworks (MOFs) [38–40].

Pyridine-terminated SAMs represent a particularly interesting type of organic substrates. An interesting result is described in a recent paper [41], where such pyridine-terminated organic surfaces have been used to enhance the rate of heterogeneous electron transfer between electrodes and the solution-phase of biological species. Recent studies have revealed that the chemical activity of pyridine-terminated SAMs is quite complicated [42, 43] and their properties cannot be predicted in a straightforward fashion from the properties of pyridine in solution.

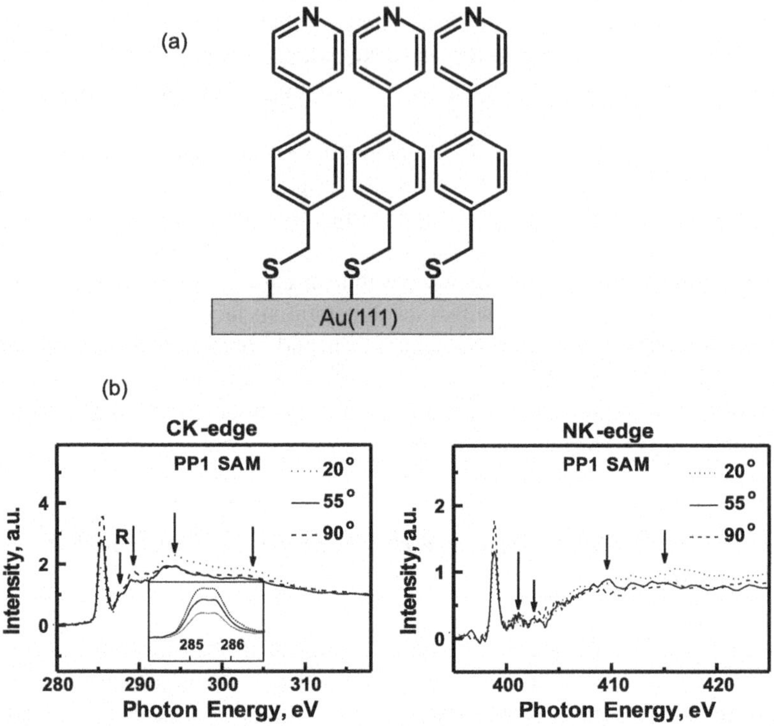

Fig. 10.10 (**a**) Schematic drawings of the PP1-SAM. (**b**) C K-edge and N K-edge NEXAFS spectra of PP1-SAM on Au(111) recorded at grazing incidence ($\theta = 20°$), the magic angle ($\theta = 55°$) and normal incidence ($\theta = 90°$) are displayed. The *inset* shows the region from 284–287 eV with the $1s \rightarrow \pi^*$ transition that consists of at least two components

NEXAFS spectroscopy is particularly well suited for the investigation of SAMs. An interesting example has been provided in [44] where it has been demonstrated the potential of the method for the case of a series of four related pyridine-terminated thiols with backbones comprising both aliphatic and aromatic parts. In this study the terminating pyridine-unit is followed by one (PPn) or two (PPPn) phenyl rings, where n = 1–3 is the number of CH_2 groups separating the aromatic group and the thiol anchor. The schematic structure of one representative of this series of SAMs, namely PP1-SAM, is sketched in Fig. 10.10a.

In Fig. 10.10b C1s and N1s NEXAFS spectra recorded for PP1-SAM at different angles of incidence are presented. All spectra exhibit a number of characteristic absorption resonances which arise from excitations of C1s core electrons into unoccupied π^*- and σ^*-orbitals localized at the aromatic rings as well as into molecular orbitals of Rydberg character. Both carbon and nitrogen K-edge NEXAFS spectra reveal a pronounced dichroism, the strongest variations of intensity with angle of incidence are observed for π_1^*-resonances. The inset in Fig. 10.10b clearly demonstrate that the $C1s \rightarrow \pi_1^*$ transition consists of at least two components. This splitting of the $C1s \rightarrow \pi^*$ excitation is typical for pyridine and has already been observed in previous NEXAFS studies on pyridine [23, 45, 46] and other organic compounds containing pyridine moieties [42, 43]. Basically, this splitting is due to differences in electronic configuration of carbon atoms forming both phenyl and pyridine rings.

In principle, a determination of the tilt angle of the pyridine moieties can be carried out by analyzing the linear dichroism observed in the experimental data with the help of (10.3). When carrying out such an analysis, however, it has to be considered that the value for the angle α obtained from such analysis corresponds to an average of the tilt angles of the TDMs of all aromatic rings of the respective molecules (will below be referred to as "average tilt angle"). Therefore, the angle α_C determined from the carbon edge is an average value, and obtaining the value for a specific phenyl ring is not possible in a straightforward fashion. In contrast, the values α_N obtained from N K-edge dichroism corresponds to the tilt angle of the TDM of the pyridine moiety only (referred to below as "pyridine unit tilt angle"), since the other phenyl rings do not contain a N atom.

The TDM tilt angles obtained for the SAMs investigated in this study were found to be different for the different resonances, the average tilt angle amounts to 64–68° whereas to the pyridine unit tilt angle a value of 58–61° was found. These differences are considered significant and strongly indicate the presence of a non-planar conformation with substantial twist-angles between adjacent aromatic rings. For a given orientation of the molecular axis (described by the tilt angle β with respect to the surface normal) the TDMs of the $1s \rightarrow \pi_1^*$ transition and, thus the average angle α, will depend on the rotation (or twist) of the aromatic ring with respect to the molecular axis, denoted by γ. Following this definition of the angles, the relationship between them is [1]:

$$\cos\alpha = \sin\beta\cos\gamma. \tag{10.6}$$

The values for α obtained from the analysis of the C K-edge and the N K-edge spectra will be different if the twist angle γ of the pyridine unit is different from the average of the aromatic moieties in the molecule, i.e. if the aromatic rings are twisted with respect to each other. The difference between the twist angles γ is equal to the internal twist angle ω of the aromatic rings. From detailed structural analysis for bulk crystals made from biphenyl and several other oligophenyls it is known that typical values of the twist angle γ between adjacent phenyl unit in oligophenyl units amounts to about 15° at room temperature [47, 48]. For the free molecules, however, these values are much larger, e.g. for gas phase biphenyl a value of $40 \pm 5°$ has been reported [49]. The large twist angle for the free molecule results from the rather pronounced steric repulsion between adjacent H atoms. The huge differences between the twist angles in the bulk and the free molecule results from the fact that a reduction of the twist angle strongly increases the packing density, therefore upon crystallization the twist angle is strongly reduced. Clearly, the gain in crystallization energy overcompensates the steric repulsion between the H atoms. Unfortunately, in the case of pyridine-phenyl units no experimental information from the bulk structure analysis is available concerning the twist angle between the pyridine unit and the adjacent phenyl rings. For this reason we have carried out a geometry

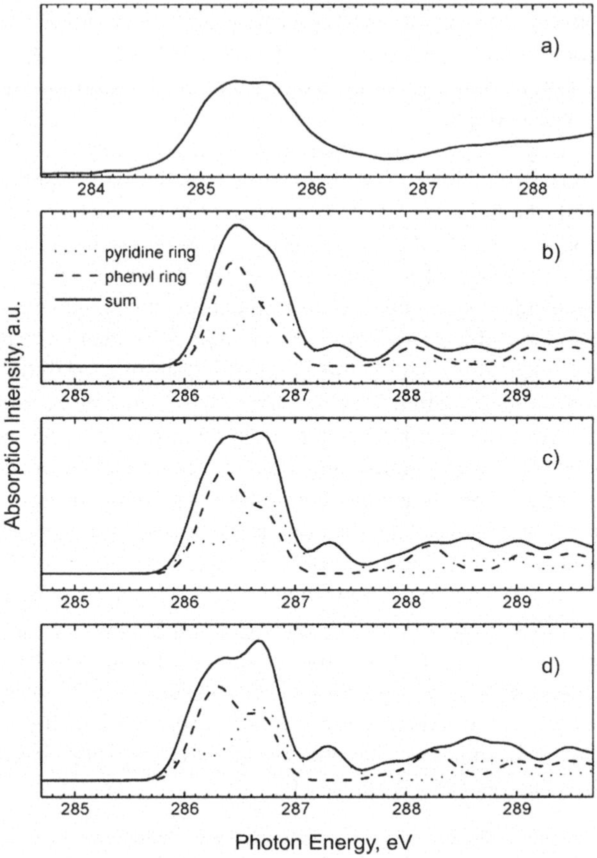

Fig. 10.11 Comparison of the experimental NEXAFS spectrum of the PP1-SAM on Au(111) recorded at the magic angle (**a**) and NEXAFS spectra of PP1 molecules calculated using the StoBe program package. Spectra have been calculated for three different internal twist angles ω of the aromatic rings: (**b**) 36°, (**c**) 18°, and (**d**) coplanar conformation ($\omega = 0°$). Both maxima of the π^*-resonance have contributions from C atoms both in phenyl and pyridine rings

optimization using the commercial quantum chemistry software package Gaussian [50]. The optimized geometry obtained from using this program package for the free PP1 molecule, yields a value of 36°, quite similar to that reported for free biphenyl molecules [49].

In principle, the presence of a non-zero twist angle could, in addition to the changes of the linear dichroism governing the intensity of the individual resonances, also have a direct effect on the NEXAFS spectra. This expectation is based on the following consideration. Starting from a hypothetical planar conformation, a torsion of the phenyl unit with regard to the pyridine unit will reduce the coupling (mixing) of the π^* molecular orbitals of adjacent aromatic rings. As a result, for a given C atom, the ratio of intensities of transitions into final states localized at the ring containing the C atom under consideration to those into a final state localized at an adjacent ring will depend on the twist angle. To demonstrate this effect, NEXAFS spectra of PP1 were simulated using the software package StoBe [12] for different values of the torsion angle ω. The corresponding results are presented in Fig. 10.11 where they are compared with an experimental NEXAFS spectrum recorded at the magic angle $\theta = 55°$. The theoretical results reveal that there is indeed a substantial variation of relative NEXAFS intensities with the internal twist angle. Comparison of experimental PP1-SAM data and theoretical spectra shows the best agreement for an intermediate torsion angle of 18°. This analysis thus makes possible an independent determination of the twist angle ω between adjacent phenyl units.

The tilt angle β of the molecular axis of PP1 in respect to the surface normal can be estimated from the experimental values α_C and α_N using (10.6) and the relationship between the twist angles of the pyridine unit and the phenyl unit:

$$\gamma_{phenyl} - \gamma_{piridyne} = \omega. \tag{10.7}$$

Assuming that the PP1 molecules chemisorbed on the Au surface actually have an internal twist angle of $\omega = 18°$ and considering that α_C is equivalent to the arithmetic average of the TDM tilt angles of the pyridine and the phenyl units, a value of 15° is obtained for the tilt angle β. This value is fully consistent with the other results obtained by other techniques (IR spectroscopy, XPS) as discussed in more detail in [44].

10.4.3 Coverage Dependent Orientation of Terephthalic Acid Molecules on $TiO_2(110)$

Chemically functionalized surfaces play an important role in many nanotechnological applications. At present, most works concentrate on the adsorption of organic molecules on metal surfaces, only few attempts have been made so far to create a chemically functionalized metal oxide surface by adsorption of organic molecules, even though metal oxide surfaces provide anisotropy and specific adsorption sites that could lead to well-ordered monolayers. Terephthalic acid (TPA) is a versatile molecular linker and has, consequently, received considerable attention with regard to the assembly of three-dimensional supermolecular frameworks stabilized by metal-carboxylate bonds [38–40], as well as in two-dimensional supramolecular architectures at surfaces [51–54]. In particular, TPA/TiO_2(110) is a promising system since one might expect the formation of an adlayer consisting of upright oriented molecules exposing an organic surface terminated by carboxyl groups. Such an organic substrate would have wide-ranging applications, similar to those of carboxyl-terminated SAMs [34]. Tekiel et al. studied the structure of a complete monolayer of TPA on TiO_2 by means of scanning tunneling microscopy and non-contact atomic force microscopy [55]. Their data was consistent with TPA adlayers with such an orientation perpendicular to the surface at a coverage of one monolayer (ML). This finding, however, is not consistent with the results of a recent ab-initio DFT calculations where it was concluded that the energetically most favorable orientation is one where the phenyl unit is oriented flat with regard to the TiO_2(110) surface [56]. Since micrographs obtained by scanning probe techniques, STM as well as AFM, do not allow for a direct determination of molecular orientation it is of pronounced interest to consider the results of a recent detailed NEXAFS investigation [57]. In this work, a coverage-dependent NEXAFS study of TPA adsorption and structure

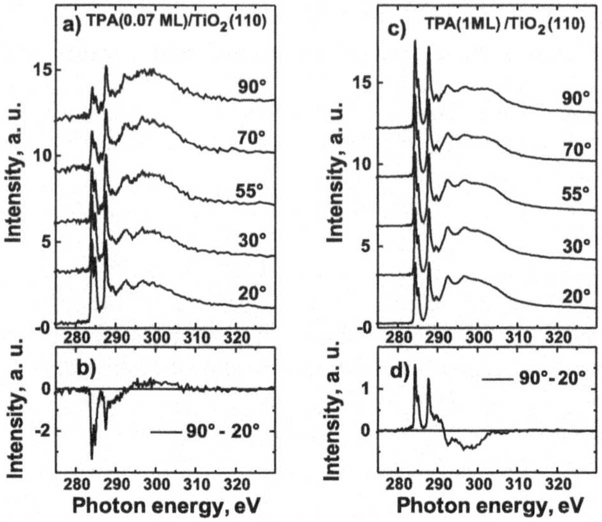

Fig. 10.12 Carbon K NEXAFS spectra for 0.07 ML (**a**) and 1 ML (**c**) of TPA coverage measured at different photon incidence angles. Difference of the spectra measured at $\theta = 20°$ and 90° for 0.07 ML (**b**) and 1 ML (**d**), correspondingly

formation on a rutile $TiO_2(110)$ surface has been presented. The orientation of the TPA molecules was determined from the dependence of the NEXAFS resonance intensities on the photon angle of incidence in case of the lowest (0.03–0.07 ML) and the highest (1 ML) coverages. In order to allow for a reliable analysis of the data, the measurements were carried out at $\theta = 20°$, 30°, 55°, 70° and 90°, and at different azimuthal orientations of the TiO_2 crystal with azimuthal angles $\phi = 0°$ and 90° with respect to the $[\bar{1}10]$ direction of the (110) surface. Figure 10.12a shows a series of carbon K-edge NEXAFS spectra taken for a $TiO_2(110)$ substrate covered with 0.07 ML of TPA molecules. The set of spectra displayed in Fig. 10.12a shows a clear dependence of the peak intensities on the incident angle θ. In particular, the pronounced negative dichroic signal of the peaks at E = 284.9 eV and E = 285.5 eV, both assigned to the phenyl ring, and of the peak at E = 288.3 eV, assigned to the carbonyl group in Fig. 10.12b, indicates almost flat lying molecules, in full agreements with the results of the theoretical calculations as discussed above.

A numerical fit of the experimental NEXAFS resonance intensities yields an average tilt angle of 18°. Ideally, for TPA molecules in a completely planar adsorption geometry, these peaks should vanish completely for $\theta = 90°$. However, TPA molecules adsorbed at step edges, defect impurities or dislocation sites might have a more tilted geometry and contribute to the NEXAFS signal as well [28]. In addition, a dynamic tilt angle resulting from the thermal occupation of the low energy vibrations of the molecule might also contribute, as was discussed for the case of benzene adsorbed on the same surface [58]. We thus conclude that the presence of a perfectly planar TPA adsorbate on the $TiO_2(110)$ surface would still be consistent with the average 18° tilt angle as determined from an intensity analysis of the corresponding NEXAFS resonances.

The NEXAFS spectra of the 1 ML sample in Fig. 10.12c show the same resonances as observed for the low-coverage 0.07 ML sample in Fig. 10.12a, however the dependence of the peak intensities on the photon angle of incidence is opposite already indicating a rather different molecular orientation of the TPA units. The difference spectrum reproduced in Fig. 10.12d demonstrates a positive signal for the π^*-resonances, a clear evidence of an upright orientation of the TPA molecules on TiO_2. Since the phenyl ring can rotate with respect to the carboxyl group, a random azimuthal orientation of the TPA molecules was assumed. On the basis of this model we obtain an average tilt angle the molecular axes with respect to the surface plane of $70° \pm 10°$. To summarize, this experimental study [57] revealed the presence of disordered, flat-lying TPA molecules at low coverages on the $TiO_2(110)$ rutile substrate. Upon increasing the coverage, a transition from flat-lying to upright-oriented molecules was observed.

10.5 Conclusions

Near edge X-ray absorption fine structure (NEXAFS) spectroscopy has contributed significantly to a better understanding of the properties of interfaces on the macroscopic as well as on microscopic scale, revealing detailed information on the electronic structure on the orientation of adsorbed molecules. As demonstrated by the examples discussed in this contribution today NEXAFS is a standard technique for obtaining direct information on the orientation and electronic structure of molecules within ultrathin layers supported by solids surfaces made from many different materials. The availability of computer codes allowing for a routine simulation of NEXAFS spectra for small to medium-sized molecules (e.g. StoBe [12]), today make possible a straightforward interpretation of NEXAFS resonances. The major drawback of the method is the fact that experiments have to be carried out at a synchrotron, thus requiring traveling to such a large scale instrument. Today, due to the success of the technique, virtually all second and third generation synchrotrons provide NEXAFS end-stations equipped with transfer-systems. These systems make it possible also for users not familiar with this particular technique to carry out experimental investigations either on samples prepared at their home laboratories or on thin films prepared in situ under UHV-conditions. In connection with further developments we expect that this particular spectroscopy method will become even more popular. These new developments include small spot NEXAFS spectromicroscopy (better than 50 nm spatial resolution) and photoelectron microscopy where a NEXAFS contrast is used. [59]. In addition, because of the high sensitivity of the method we will soon also see more time dependent studies which will allow to monitor changes within ultrathin organic layers in real time.

References

1. J. Stöhr, *NEXAFS Spectroscopy* (Springer, Berlin, 1992), pp. 1–403
2. J.L. Solomon, R.J. Madix, J. Stöhr, Surf. Sci. **255**, 12 (1991)

3. J. Kattner, H. Hoffmann, in *Handbook of Vibrational Spectroscopy*, ed. by J.M. Chalmers, P.R. Griffiths (Wiley, Chichester, 2002), p. 1009
4. J. Heidberg, E. Kampshoff, R. Kühnemuth, O. Schönekäs, M. Suhren, J. Electron Spectrosc. Relat. Phenom. **54/55**, 945 (1990)
5. Y. Wang, A. Glenz, M. Muhler, Ch. Wöll, Rev. Sci. Instrum. **80**, 113108 (2009)
6. B.E. Hayden, in *Vibrational Spectroscopy of Molecules on Surfaces*, ed. by J.T. Yates, T.E. Madley (Wiley, New York, 1987), p. 267
7. P.S. Bagus, K. Weiss, A. Schertel, Ch. Wöll, W. Braun, C. Hellwig, C. Jung, Chem. Phys. Lett. **248**, 129 (1996)
8. D. Hübner, F. Holch, M.L.M. Rocco, K.C. Prince, S. Stranges, A. Schöll, E. Umbach, R. Fink, Chem. Phys. Lett. **415**, 188 (2005)
9. K. Weiss, P.S. Bagus, Ch. Wöll, J. Chem. Phys. **111**, 6834 (1999)
10. S.G. Urquhart, R. Gillies, J. Phys. Chem. A **109**, 2151 (2005)
11. D. Fuhrmann, D. Wacker, K. Weiss, K. Hermann, M. Witko, Ch. Wöll, J. Chem. Phys. **108**, 2651 (1998)
12. K. Hermann, L.G.M. Pettersson, M.E. Casida, C. Daul, A. Goursot, A. Koester, E. Proynov, A. St-Amant, D.R. Salahub, contributed authors: V. Carravetta, H. Duarte, C. Friedrich, N. Godbout, J. Guan, C. Jamorski, M. Leböuf, M. Leetmaa, M. Nyberg, S. Patchkovskii, L. Pedocchi, F. Sim, L. Triguero, A. Vela, StoBe-deMon version 3.1, http://www.fhi-berlin.mpg.de/KHsoftware/StoBe/index.html. Cited 12 Jul 2011
13. D.R. Salahub, M.E. Castro, R. Fournier, P. Calaminici, N. Godbout, A. Goursot, C. Jamorski, H. Kobayashi, A. Martinez, I. Papai, E. Proynov, N. Russo, S. Sirois, J. Ushio, A. Vela, in *Theoretical and Computational Approaches to Interface Phenomena*, ed. by H. Sellers, J.T. Gotlab (Plenum, New York, 1995), p. 187
14. D.R. Salahub, M.E. Castro, E.I. Proynov, in *Relativistic and Electron Correlation Effects in Molecules and Solids N*, ed. by G.L. Malli. NATO ASI Series (Physics), vol. B318 (Plenum, New York, 1994), p. 411
15. W. Kohn, L. Sham, J. Phys. Rev. **140**, A1133 (1965)
16. H. Winick, in *Synchrotron Radiation Research*, ed. by H. Winick, S. Doniach (Plenum, New York, 1980)
17. G. Ertl, J. Küppers, *Low Energy Electrons and Surface Chemistry* (VCH, Weinheim, 1985)
18. J.L. Wiza, Nucl. Instrum. Methods **162**, 587 (1979)
19. PREVAC company, http://www.prevac.eu. Cited 13 Jul 2011
20. J. Stöhr, K. Baberschke, R. Jaeger, R. Treichler, S. Brennan, Phys. Rev. Lett. **47**, 381 (1981)
21. G. Hähner, Chem. Soc. Rev. **35**, 1244 (2006)
22. A.K. Ghosh, A.K. Naskar, S. Sengupta, L. Bizbiz, J. Labat-Robert, A. Alperovitch, L. Robert, The EVA-group, J.G. Chen, Surf. Sci. Rep. **30**, 1 (1997)
23. J.A. Horsley, J. Stöhr, A.P. Hitchcock, D.C. Newbury, A.L. Johnson, F. Sette, J. Chem. Phys. **83**, 6099 (1985)
24. B.E. Koel, J.E. Crowell, C.M. Mate, G.A. Somorjai, J. Phys. Chem. **90**, 2949 (1986)
25. H.-P. Steinrück, P. Heimann, W. Huber, P. Jakob, T. Pache, D. Menzel, J. Electron Spectrosc. Relat. Phenom. **52**, 91 (1990)
26. S. Aminpirooz, L. Becker, B. Hillert, J. Haase, Surf. Sci. **244**, L152 (1991)
27. K. Weiss, S. Gebert, M. Wühn, H. Wadepohl, Ch. Wöll, J. Vac. Sci. Technol. A **16**, 1017 (1998)
28. C. Mainka, P.S. Bagus, A. Schertel, T. Strunskus, M. Grunze, Ch. Wöll, Surf. Sci. **341**, L1055 (1995)
29. M.A. Van Hove, R.F. Lin, G.A. Somorjai, Phys. Rev. Lett. **51**, 778 (1983)
30. A. Wander, G. Held, R.Q. Hwang, G.S. Blackman, M.L. Xu, P. de Andres, M.A. Van Hove, G.A. Somorjai, Surf. Sci. **249**, 21 (1991)
31. J.B. Torrance, Acc. Chem. Res. **12**, 79 (1979)
32. T.-Ch. Tseng, Ch. Urban, Y. Wang, R. Otero, S.L. Tait, M. Alcamí, D. Écija, M. Trelka, J.M. Gallego, N. Lin, M. Konuma, U. Starke, A. Nefedov, A. Langner, Ch. Wöll, M.Á. Herranz, F. Martín, N. Martín, K. Kern, R. Miranda, Nat. Chem. **2**, 374 (2010)

33. R.G. Nuzzo, D.L. Allara, J. Am. Chem. Soc. **105**, 4481 (1983)
34. J.C. Love, L.A. Estroff, J.K. Kriebel, R.G. Nuzzo, G.M. Whitesides, Chem. Rev. **105**, 1103 (2005)
35. A. Ulman, *Self-Assembled Monolayers of Thiols (Thin Films)* (Elsevier, Amsterdam, 1998)
36. F. Schreiber, Prog. Surf. Sci. **65**, 151 (2000)
37. M. Kind, Ch. Wöll, Prog. Surf. Sci. **84**, 230 (2009)
38. O. Shekhah, H. Wang, M. Paradinas, C. Ocal, B. Schüpbach, A. Terfort, D. Zacher, R.A. Fischer, Ch. Wöll, Nat. Mater. **8**, 481 (2009)
39. O. Shekhah, H. Wang, T. Strunskus, P. Cyganik, D. Zacher, R. Fischer, Ch. Wöll, Langmuir **23**, 7440 (2007)
40. O. Shekhah, H. Wang, D. Zacher, R.A. Fischer, Ch. Wöll, Angew. Chem. Int. Ed. **48**, 5038 (2009)
41. S. Yoshimoto, Bull. Chem. Soc. Jpn. **79**, 1167 (2006)
42. C. Silien, M. Buck, G. Goretzki, D. Lahaye, N.R. Champness, T. Weidner, M. Zharnikov, Langmuir **25**, 959 (2009)
43. Y. Zubavichus, M. Zharnikov, Y.J. Yang, O. Fuchs, E. Umbach, C. Heske, A. Ulman, M. Grunze, Langmuir **20**, 11022 (2004)
44. J. Liu, B. Schüpbach, A. Bashir, O. Shekhah, A. Nefedov, M. Kind, A. Terfort, Ch. Wöll, Phys. Chem. Chem. Phys. **12**, 4459 (2010)
45. S. Hövel, C. Kolczewski, M. Wühn, J. Albers, K. Weiss, V. Staemmler, Ch. Wöll, J. Chem. Phys. **112**, 3909 (2000)
46. C. Kolczewski, R. Puttner, O. Plashkevych, H. Agren, V. Staemmler, M. Martins, G. Snell, A.S. Schlachter, M. Sant'Anna, G. Kaindl, L.G.M. Pettersson, J. Chem. Phys. **115**, 6426 (2001)
47. J.L. Baudour, Acta Crystallogr. Sect. B: Struct. Sci. **47**, 935 (1991)
48. J.L. Baudour, H. Cailleau, W.B. Yelon, Acta Crystallogr. Sect. B: Struct. Crystallogr. Cryst. Chem. **33**, 1773 (1977)
49. A. Almenningen, O. Bastiansen, L. Fernholt, B.N. Cyvin, S.J. Cyvin, S. Samdal, J. Mol. Struct. **128**, 59 (1985)
50. M.J. Frisch, *GAUSSIAN 03 (Revision E.01)* (Gaussian, Wallingford, 2004)
51. S. Stepanow, T. Strunskus, M. Lingenfelder, A. Dmitriev, H. Spillmann, N. Lin, J.V. Barth, Ch. Wöll, K. Kern, J. Phys. Chem. B **108**, 19392 (2004)
52. S. Stepanow, M. Lingenfelder, A. Dmitriev, H. Spillmann, E. Delvigne, N. Lin, X. Deng, C. Cai, J.V. Barth, K. Kern, Nat. Mater. **3**, 229 (2004)
53. J.V. Barth, G. Costantini, K. Kern, Nature **437**, 671 (2005)
54. M.E. Cañas-Ventura, F. Klappenberger, S. Clair, S. Pons, K. Kern, H. Brune, T. Strunskus, Ch. Wöll, R. Fasel, J.V. Barth, J. Chem. Phys. **125**, 184710 (2006)
55. A. Tekiel, J.S. Prauzner-Bechcicki, S. Godlewski, J. Budzioch, M. Szymonski, J. Phys. Chem. C **112**, 12606 (2008)
56. M. Watkins, T. Trevethan, M.L. Sushko, A.L. Shluger, J. Phys. Chem. C **112**, 4226 (2008)
57. P. Rahe, M. Nimmrich, A. Nefedov, M. Naboka, Ch. Wöll, A. Kühnle, J. Phys. Chem. C **113**, 17471 (2009)
58. S. Reiss, H. Krumm, A. Niklewski, V. Staemmler, Ch. Wöll, J. Chem. Phys. **116**, 7704 (2002)
59. A.P. Hitchcock, C. Morin, X. Zhang, T. Araki, J. Dynes, H. Stöver, J. Brash, J.R. Lawrence, G.G. Leppard, J. Electron Spectrosc. Relat. Phenom. **144–147**, 259 (2005)

Part IV
Neutral Particle Techniques

Chapter 11
Neutron Reflectivity

Frédéric Ott

Abstract Neutrons are neutral particles which have specific properties. The neutron scattering potential depends on the atomic number A in a non-monotonous way. The scattering lengths of ^1H and ^2H are both large and of opposite sign. This makes isotopic labeling a very useful tool for the study of polymers or biological systems since specific parts of a system can be highlighted by creating large contrasts between the hydrogenated and deuterated elements. The second key property of neutrons is a large magnetic interaction with the magnetic induction field which is of the same order as the nuclear interaction potential. This makes neutrons especially suited for the study of magnetic thin films. The scattering lengths of the classical 3d metal (V, Cr, Mn, Fe, Co, Ni, Cu) are quite different which makes neutrons useful for the study of metallic layers. The neutrons are also very weakly absorbed by most materials which allows performing studies on buried interfaces and in particular makes solid-liquid and liquid-liquid interface characterizations rather easy to perform.

In this chapter we will present the basics of neutron reflectivity and illustrate its use in various fields of solid state physics: polymer science, biology, metallurgy and magnetism.

11.1 Introduction

Neutron reflectometry is a relatively new technique [1, 2]. The key property of neutrons for polymer studies is their large contrast between ^1H and ^2H which allows *selective labeling* by deuteration. It has been extensively used for solving soft matter problems like polymer mixing at interfaces [3–5] or the structure of liquids at a surface [6, 7]. In the late 80's, a new field of application of neutron reflectometry emerged. Following the discovery of giant magneto-resistance in

F. Ott (✉)
Lab. Léon Brillouin CEA/CNRS, bat 563 Centre d'Etudes de Saclay, 91191 Gif sur Yvette Cedex, France
e-mail: frederic.ott@cea.fr

G. Bracco, B. Holst (eds.), *Surface Science Techniques*,
Springer Series in Surface Sciences 51, DOI 10.1007/978-3-642-34243-1_11,

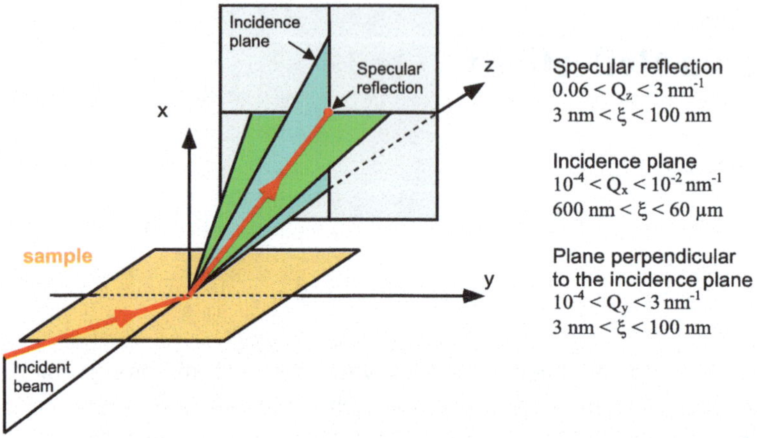

Fig. 11.1 The different surface scattering geometries. (*Red line*) specular reflectivity geometry; (*blue plane*) off-specular scattering plane, corresponding to the incidence plane; (*green plane*) GISANS scattering plane, perpendicular to the incidence plane. These different scattering geometries probe a very wide range of length-scales and directions in the sample surface

anti-ferromagnetically coupled multilayer films [8] and new magnetic phenomena in ultra-thin films, there has been an interest in the precise measurement of the magnetic moment direction in each layer of a multilayer and at the interface between layers. Owing to the large magnetic coupling between the neutron and the magnetic moment, neutron reflectometry has proved to be a powerful tool for obtaining information about these magnetic configurations and for measuring magnetic depth profiles.

At grazing incidence, it is possible to distinguish three scattering geometries (Fig. 11.1) probing different length scales ξ and directions in the sample surface. Specular reflectivity probes the structure *along the depth in the film* (3 nm $< \xi <$ 100 nm). This is the scattering geometry which is mostly used. Off-specular scattering (in the incidence plane) probes *in-plane surface features* at a micrometric scale (600 nm $< \xi <$ 60 μm). Grazing Incidence Small Angle Neutron Scattering (SANS) probes *in-plane* surface features in the range 3 nm $< \xi <$ 100 nm. These different scattering geometries allow the study of a very wide range of length-scales ξ, ranging from a few nm up to several μm.

In this chapter, we give an overview of the experimental and theoretical methods used in neutron reflectometry, focusing mainly on specular reflectivity. The corresponding theory is partly derived from previous work on X-rays, and we emphasize on the aspects specific to neutrons. The use of neutron reflectivity in the field of polymers films and of magnetic layers is then illustrated by several examples. Finally, we present the use of off-specular scattering and grazing incidence SANS applied to the study of in-plane nano-structures.

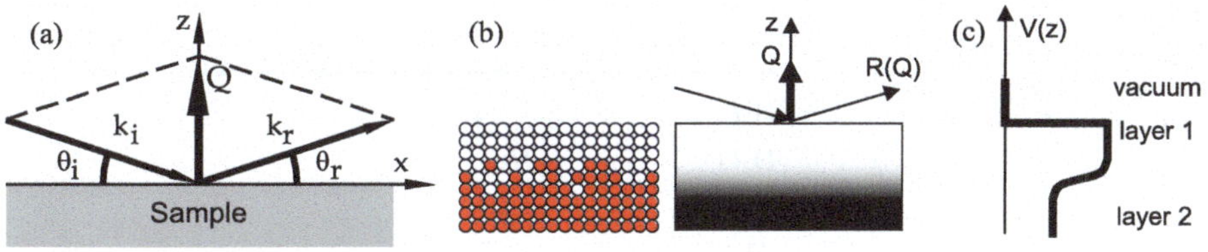

Fig. 11.2 (**a**) Specular reflectivity geometry. The reflection angle is equal to the incidence angle; the scattering wave-vector Q is perpendicular to the sample surface. (**b**) Interface between 2 surfaces. In the optical approximation, the interface is approximated as a continuous medium. (**c**) Reflection on a thin film deposited on a surface. The reflectivity measures the Fourier transform of the interaction potential $V(z)$

11.2 Principle of Specular Reflectivity

Neutrons can be reflected on surfaces in the same way as X-rays or electrons [9]. All the formalisms developed for X-ray reflectivity can be transposed to neutron reflectivity [10]. In a reflectivity geometry (Fig. 11.2a), the incidence angle θ_i on the surface is small (typically ranging from 0.5 to 5°). The reflection angle θ_r is the same as the incidence angle θ_i. As a consequence, the scattering wave-vector Q is perpendicular to the surface. The typical range of accessible scattering wave-vector $Q = k_r - k_i$ is 0.05–3 nm^{-1}. This corresponds in the real space to typical length-scales ranging between 2 and 100 nm so that neutron reflectivity does not probe structures at the atomic level. In a reflectivity geometry it is thus possible to do the optical approximation [10] and model the neutron interaction with the material as a continuous potential. The details of the atomic structure are smoothed out (Fig. 11.2b). The interaction potential V with a material is given by:

$$V = \frac{h^2}{2\pi m}\rho \quad \text{with } \rho = \frac{1}{V}\sum_i b_i$$

where h is the Planck constant and m is the neutron mass. ρ is called the *scattering length density* (SLD) and is the average of the nuclear scattering lengths b_i of the different nuclei in the material in a small volume V.

In the case of a magnetic system, the interaction between the neutron spin and the material magnetization is of the form $V = -\vec{\mu} \cdot \vec{B}$ where $\vec{\mu}$ is the magnetic moment of the neutron and $\vec{B}$ is the magnetic induction inside the thin film.

In the reflectivity geometry, the equivalent of a *neutron optical index* can be derived from the Schrödinger equation [10]. Neglecting absorption, the value of this optical index is given by the following expression:

$$n^{\pm} = 1 - \delta \mp \delta_M = 1 - \frac{\lambda^2}{2\pi}\rho \mp \frac{m\lambda^2}{h^2}\mu \cdot B$$

where δ is the nuclear contribution to the optical index, and δ_M is the magnetic contribution to the optical index. The sign of the magnetic contribution depends on the relative orientation of the neutron spin with respect to the magnetization (parallel or anti-parallel). Table 11.1 gives values of optical indexes for some typical

Table 11.1 Nuclear and magnetic optical index $n = 1 - \delta \pm \delta_M$ for some materials at $\lambda = 0.4$ nm

Element	$\delta(\times 10^{-6})$	$\delta_M(\times 10^{-6})$	σ_a (barns)
H_2O	-1.43		
D_2O	16.2		
Si	5.28		
Al_2O_3	14.5		
polystyrene-H	3.6		
polystyren-D	16.5		
Au	11.5		
Cu	16.6		
Fe	20.45	11.7	2.56
Co	5.7	10.3	37.2
Ni	24	3.7	4.49
Gd	5.0	14.5	49700

materials. One should notice that the magnetic optical index is of the same order of magnitude as the nuclear optical index. The use of polarized neutrons permits to measure both optical indexes n^+ and n^- and thus to obtain detailed information about the magnetic structure of the sample.

In a specular reflectivity measurement, the most important assumption is that the system is invariant in translation in the thin film plane, that is, there are no inhomogeneities along the film surface. Thus the interaction potential V is assumed to be only a function of the depth z in the multilayer system (Fig. 11.2c). In a first approximation, the specular reflectivity measures the Fourier transform of the optical index profile $n(z)$. However, at low incidence angles, there is total reflection up to a critical wave-vector Q_c and thus the Born approximation is not valid at small scattering wave-vectors. The Born approximation can be applied only above a scattering wave-vector of about $3Q_c$. Below this limit, one must solve the Schrödinger equation and perform a full dynamical calculation. The detailed theoretical treatment of the polarized reflectivity can be found in [1, 10–13].

Figure 11.3a presents the situation of the reflection of a neutron beam on a multilayer Si || Cu/Cr: above the critical wave-vector of total reflection, the reflected intensity decreases as $1/Q^4$. Modulations of the reflected intensities are observed. They correspond to constructive and destructive interferences of the neutron waves scattered by the different interfaces of the multilayer system. These oscillations are called Kiessig fringes. Their pattern is characteristic of the multilayer system. Figure 11.3b presents the situation of a magnetic thin film on a substrate. In this case, the optical index depends on the relative orientation of the neutron spin with respect to the thin film magnetization. The measured reflectivity is very different for neutron incident with a spin parallel to the magnetization (optical index $n^+ = 1 - \delta - \delta^+$) and for neutrons incident with a spin anti-parallel to the magnetization (optical index $n^- = 1 - \delta + \delta^-$).

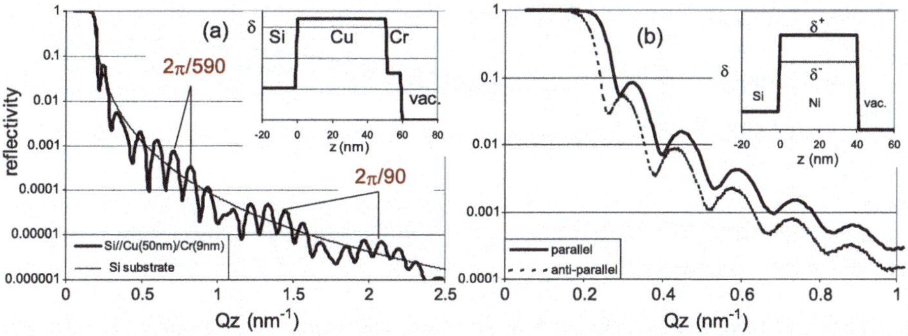

Fig. 11.3 (**a**) Reflectivity on a multilayer system Si ‖ Cu(50 nm)/Cr(9 nm). The short period oscillations are characteristic of the total thickness of the layer (59 nm). The long range modulation is characteristic of the thin Cr layer (9 nm). (*Insert*) optical index profile as a function of the depth in the film. (**b**) Reflectivity of a magnetic film Si ‖ Ni(40 nm). The reflectivity depends on the relative orientation of the neutron spin with respect to the magnetization. (*Insert*) optical index profile for both neutron polarizations (parallel and anti-parallel)

The measure of the reflectivity probes the profile of optical index $n(z)$ along the normal to the thin film system. Numerical models are then used to reconstruct the thickness of the different layers of the system as well as their individual scattering length densities ρ which is characteristic of their chemical composition. Interdiffusion and roughness at interfaces can be quantified with more detailed models. In the case of magnetic systems, information on the amplitude and the direction of the magnetization of the different layers can be obtained using polarized neutron reflectivity. One should note that polarized reflectivity is sensitive to the induction in the thin films: no difference is made between the spin and orbital magnetic moments. In practice, it is possible to measure 4 cross-sections in a polarized reflectivity experiment: 2 non-spin-flip cross sections, R^{++} (resp. R^{--}), corresponding to the number of incoming *up* (resp. *down*) neutrons reflected with an *up* (resp. *down*) polarization; 2 spin-flip cross sections, $R^{+-} = R^{-+}$, corresponding to the number of neutrons experiencing a spin-flip during the reflection on the sample. In a first approximation, the non-spin-flip cross sections probe the components of the magnetization which are parallel to the applied field; the spin-flip cross sections are sensitive to the components of the magnetization perpendicular to the applied field. Combining this information it is possible to reconstruct the magnetization direction and amplitude along the depth of the film. The depth resolution is of the order of 2–3 nm in simple systems. Polarized reflectivity is a surface technique and thus is not sensitive to paramagnetic or diamagnetic contribution from the substrate. There is no absorption. There are no phenomenological parameters. The data are naturally normalized. All these characteristics make neutron reflectivity data easy to model and interpret.

In the following, we will describe typical neutron reflectometers and illustrate some of the possibilities offered by neutron reflectivity on polymer and magnetic films. For other examples, the interested reader should refer to the following recent reviews [14–20].

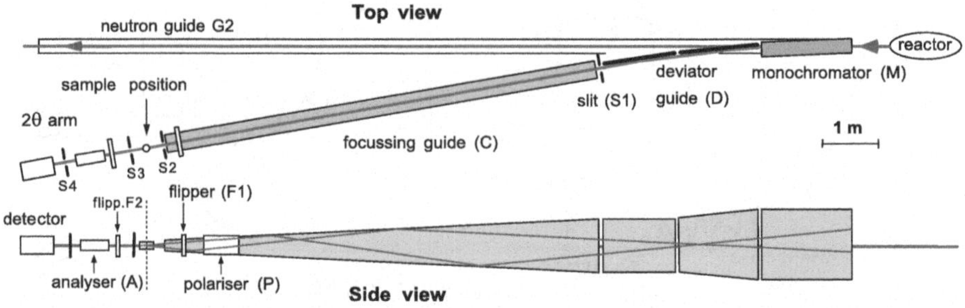

Fig. 11.4 The 2-axis reflectometer PRISM at the LLB. The beam is focused from an initial height of 100 mm down to 20 mm at the sample position

11.3 Experimental Setup and Method

The spectrometers can be divided in two different groups: time of flight reflectometers such as EROS at the Laboratoire Léon Brillouin, INTER and POLREF at ISIS, D17 and FIGARO at the ILL, and monochromatic reflectometers such as PRISM at LLB and ADAM at the Institut Laue Langevin (ILL). A list of existing reflectometers can be found in [21].

11.3.1 Monochromatic Reflectometers

Monochromatic reflectometers are basically two axes spectrometers. The wavelength is fixed (e.g. 0.4 nm for PRISM) and the reflectivity curve is obtained by changing the incidence angle θ. In this case, the sample is usually vertical. On this type of reflectometer it is easy to put a polarizer and an analyzer in order to select the spin states of the incident and reflected neutrons. The flippers can be of Mezei type (2 orthogonal coils) [22]. They allow to flip the neutron spin state from *up* to *down*. An example of two-axis spectrometer is presented on Fig. 11.4.

11.3.2 Time of Flight Reflectometers

The time-of-flight technique consists in sending a pulsed white beam on the sample. Since the speed of the neutron varies as the inverse of the wavelength, the latter is directly related to the time taken by the neutron to travel from the pulsed source to the detector (over the distance L) by $\lambda = \frac{h}{mL}t$. This relation is also written as $\lambda(\mathrm{nm}) = \frac{t\,(\mu\mathrm{s})}{2527L(\mathrm{m})}$. On a spallation source, the neutron beam is naturally pulsed and the time of flight technique must be used. On a reactor, neutron pulses are produced by a chopper. One advantage of a time-of-flight set-up is that it is very easy to change the resolution by changing the chopper and slits parameters. Another advantage is

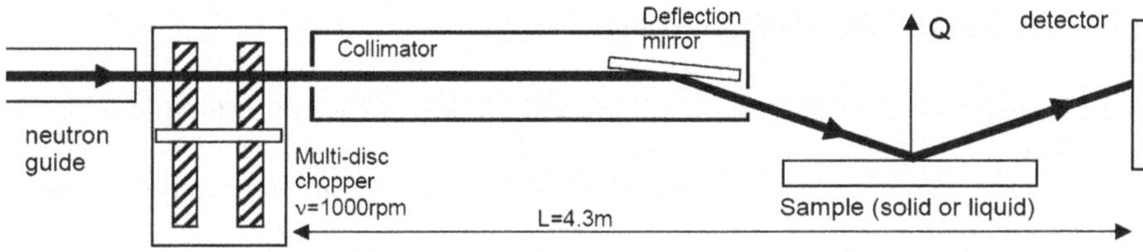

Fig. 11.5 Description of the time of flight reflectometer EROS at the LLB [23]

that the sample does not need to be moved during an experiment and thus it is easy to measure free liquid surfaces.

For a reflectivity measurement, the angle is fixed and the reflectivity curve is obtained by measuring the reflectivity signal for each wavelength of the available spectrum, each wavelength corresponding to a different scattering wave-vector. In practice, the wavelength spectrum (typ. 0.2–2 nm) is not wide enough to cover a very large Q-range in the reciprocal space. Thus, usually two or more different incidence angles are used to cover a wider Q-range. An example of time of flight spectrometer is presented on Fig. 11.5.

11.4 Applications of Specular Neutron Reflectivity

11.4.1 Polymer Systems

Isotopic Labeling

The substitution of hydrogen by deuterium in organic materials allows to strongly change the neutron optical index of the material without changing its physical or chemical properties. A very interesting possibility offered by neutron scattering is to do selective labeling by deuteration which leads to a very large contrast between deuterated ($b_D = 6.67$ fm) and protonated ($b_H = -3.7$ fm) systems [24, 25]. Such labeling is used in two ways: (i) the measurement of the conformation of polymeric chains at the interface in good solvent by using hydrogenated polymers in deuterated solvents (for example, adsorption profiles of polymers at interfaces have been measured by neutron reflectivity [26–29]). (ii) The determination of the structure of complex systems involving two polymers by mixing hydrogenated and deuterated polymers. This can be achieved with deuterated and hydrogenated chains of the same polymer (to study the inter-diffusion of chains at the interface of two molten polymers for example [30]) or of different polymers (multilayers of polyelectrolyte of opposite charges for example [31]). Combining these two advantages to determine the structure of a mixture of two different polymers in good solvent is possible by using the variation contrast method: measurements are performed in successive mixtures of hydrogenated and deuterated solvent that either match the neutron optical index of the first polymer, or match the neutron optical index of the second one. It allows to resolve the whole structure of the system [32].

Inter-diffusion Between Polymer Layers

To illustrate these two aspects we shall present a study on the conformation of dense grafted brushes of polystyrene (PS) on silicon [33–35]. Such macromolecular architectures are designed to answer the technological demand of controlled and reproducible thin polymers films. It is based on recent *grafting from* techniques that allow grafting polymers onto a surface in an efficient way. Classically, the most common method for polymer grafting is the *grafting onto* where end functionalized polymers react with appropriate surface sites. In this more promising *grafting from* method, the chains grow *in situ* from preformed surface-grafted initiators [36]. This latter approach is thus a suitable way for building high-density polymer brushes because it is not limited by polymer diffusion. It also allows a fine control of the polymer layer. This strategy has been applied by Devaux et al. [33] to realize grafted brushes of PS on silicon which have been studied by neutron reflectivity.

In order to test the homogeneity of the chains growth *during* the polymerization process, a specific chain designed for neutron reflectivity measurement has been fabricated with a two-steps process: the first part of the chains has been grown using deuterated monomers and the second part using hydrogenated monomers (Fig. 11.6a). Chemically, the polymer chains behave as a single physical unit. However, NR allows to easily distinguish between the 2 parts of the chains as shown on Fig. 11.6c that presents the reflectivity of the polymer layer at the polymer/air interface: Kiessig fringes arise from the deuterated layer, the hydrogenated layer and the whole layer. It allows a very accurate determination of the width of the interface thickness between the deuterated and the hydrogenated parts of the polymers: if the width is null we would get large oscillations from the deuterated layer and if it is too large we would only get small oscillations from the whole layer as other oscillations vanish (Fig. 11.6c presents the simulated profiles of Fig. 11.6b). Best fit shows that the interface width is limited to 25 Å for a brush of thickness 430 Å. This proves that this *grafting from* technique allows to built very well ordered polymer brushes and that the growth of the brushes is very homogeneous.

11.4.2 Solid-Liquid Interfaces

Besides isotopic labeling, one advantage of neutrons is that it is easy to set-up complex sample environments. This is especially true in the case of solid/liquid interfaces where the neutron beam can be sent through the substrate (typically Si or quartz) and probe the solid-liquid interface with negligible absorption. This is illustrated by the swelling capacities of an hydrogenated PS brush of a dry thickness of 220 Å placed in a good solvent (deuterated toluene) (Fig. 11.7a). Figure 11.7b presents the monomeric concentration profile as a function of the depth deduced from the fit of experimental reflected curves (Fig. 11.7c). It shows that the swelling of the layer is limited. The volume fraction ϕ of the polymer remains as high as 0.8 of the density of the dry polymer showing that the solvent hardly penetrates the layer. At the polymer-solvent interface, one can observe a parabolic variation of the

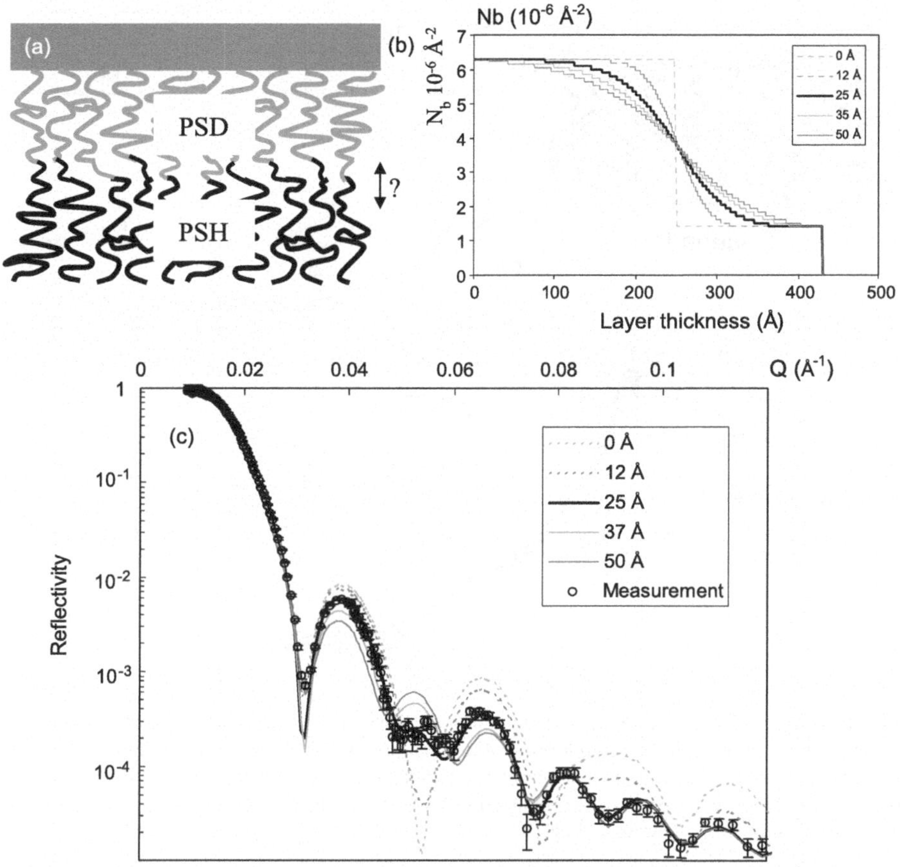

Fig. 11.6 Reflectivity on a polymer grown using a *grafting from* method. (**a**) A polymer layer grafted on a silicon substrate, half of the layer is hydrogenated, the second half of the layer is deuterated. (**b**) Optical index along the thickness of the layer assuming different interface thickness (the profile is assumed to vary as an erf function). (**c**) Reflectivity of the system and theoretical curves for the different thickness of the interfacial layer (zone separating the H and D-polymer layers). The best agreement is obtained for an interface layer of thickness 25 Å

polymer density. Three fitting methods have been tested and provide very similar results. The detailed information about the very top of the polymer layer is limited because the Q-range of the measurement was limited. The maximal stretch of the layer can nevertheless be evaluated as it roughly corresponds to the maximal extension of the profile (300 Å). It shows that the chains were already strongly stretched in their dry state ($\phi \sim 0.7$) as the layer width was 220 Å. Such initial strong stretching is due to a very high density of the grafting of the polymeric chains that explains the unusual low swelling capacities of the brushes. This example illustrates some of the unique possibilities offered by neutron reflectivity for the study of solid-liquid interfaces.

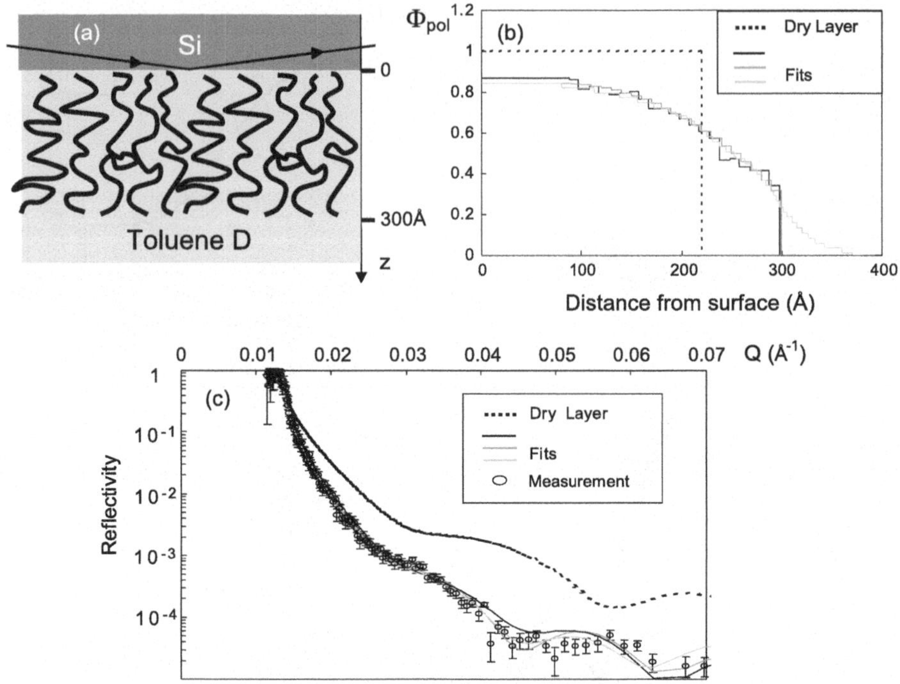

Fig. 11.7 PS layer in a good solvent (toluene). (**a**) The measurement set-up, with the neutron beam incident on the system through the silicon substrate. (**b**) Fit of the polymer density profile ϕ in deuterated toluene (3 fitting methods of the NR data have been tested and give very similar results). The polymer density α is normalized to 1 for the dry layer. (**c**) Reflectivity measurement and numerical modeling curves [33, 34]

11.4.3 Biological Model Systems

During the last decade, neutron reflectivity has started to be used for biological systems, mainly for the study of biophysical problems at solid-liquid interfaces. We illustrate the field by the example of bio-mimetic membranes [37]. Tethered membranes are systems designed for the incorporation of membrane-associated proteins. In order to be useful as a bio-membrane model, such membranes need to retain their fluid in-plane organization and at the same time remain separated from the supporting solid interface by a molecularly thin hydration layer. Neutron reflection is uniquely capable of characterizing in molecular detail the resulting membrane structures, particularly with respect to this hydration layer. A typical example is shown in Fig. 11.8. The sample preparation involves the co-adsorption of the tether lipid WC14 with a smaller thiolated *back-filler*, -mercaptoethanol, in order to dilute the lipopolymer at the interface, thus creating space for hydration. Figure 11.8b shows the SLD profile of the system discussed above in two distinct solvent contrasts: pure D_2O and an H_2O/D_2O mixture. The derived SLD profiles is obtained using slab models that is co-refined assuming that the samples are identical except for the

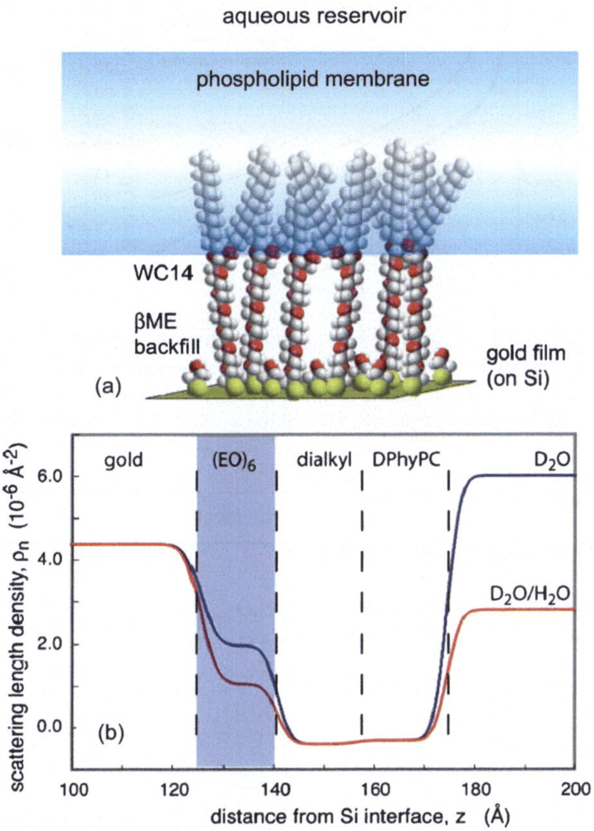

Fig. 11.8 Neutron reflection from a silicon wafer surface functionalized with a 100 Å gold film on which a lipid bilayer membrane was chemically tethered. (**a**) The chemical structure shown schematically. Because the lipopolymer WC14 was co-adsorbed with the back-filler, -mercaptoethanol, in order to reduce the density of WC14 grafted to the gold surface, we refer to this system as a sparsely-tethered membrane. (**b**) Scattering-length density profiles derived from simultaneous fits to the neutron reflectivity data. Adapted from [37]

distinct solvent contrasts. As is evident from the differences in the SLD profiles, the hexaethylene oxide spacer region $(EO)_6$ is highly hydrated while the bilayer membrane covers the substrate homogeneously. Preparations such as the one shown thus constitute biomimetic membrane systems that are very well suited for a functionalization by reconstitution with membrane-associated proteins.

For the readers interested in this field, they shall refer to the reviews of G. Fragneto-Cusani [20] and S. Krueger [38]. The following references also give good examples of what can be achieved using neutron reflectometry [39–46].

11.4.4 Oxide Layers

Neutron reflectivity can be used to probe oxide layers since the neutron optical index of oxides are usually very different from non-oxidized materials (Table 11.2) [47]. This makes neutron reflectivity much more sensitive to details in an oxide structure than X-ray reflectivity (Fig. 11.9).

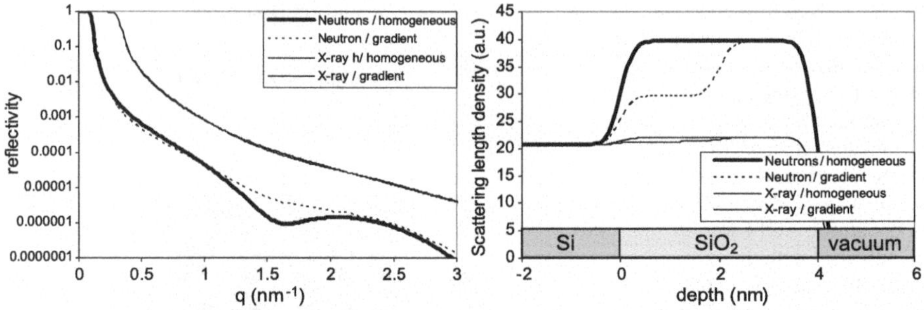

Fig. 11.9 Comparison of the neutron reflectivity and the X-ray reflectivity on a $SiO_2 \parallel Si$ structure. The contrast between the layers is much larger in the case of neutrons, leading to higher amplitude oscillations, which on the contrary are very small in the case of X-rays. Sub-structures in the oxide layer, due for example to an oxidation gradient, give measurable effects in neutron reflectivity

Table 11.2 Optical indexes of different oxides (at 0.4 nm for neutrons and 0.154 nm for X-rays). Note the large difference between δ_{Si} and δ_{SiO_2} in the case of neutrons

Element	Neutrons (10^{-6})	X-rays (10^{-6})
Si	5.3	7.6
SiO_2	10.1	8.11
SiC	13.3	10.43
Al	5.29	8.45
Al_2O_3	14.23	12.9

For example, the preparation of SiO_2 oxide films on silicon substrates by three different methods (thermal, chemical and electrochemical oxidation) have been compared by Bertagna et al. [48]. Depending on the preparation method, the obtained films give very different reflectivity results. Anodic and chemical oxides are found to be not very dense (60–75 % of the theoretical density). Thermal oxides are the densest (95 %).

Neutron reflectivity may also be useful in the case of some specific materials such as boron which is strongly absorbing neutrons (ex: the study of borophosphosilicate glass thin films used in microelectronic circuit devices [49] or titanium which has a negative scattering length (such as TiO_x coatings for glazing [50–52]). Neutron reflectometry has sometimes been used to characterize the oxidation of metallic thin films (Fig. 11.10) [53–55].

Another key advantage of neutrons is their high sensitivity to D_2O (compared to X-rays). Neutron reflectivity has for example been used to characterize the moisture transport through Al_2O_3/polymer multilayer barrier films for flexible displays [56]. It has also been used to characterize the adsorption of water on hydrophobic/hydrophilic TiO_2 surfaces under U.V. illumination. Another advantage which must be mentioned is that in neutron reflectometry it is easy to set-up complex sample environments. This is especially true in the case of solid/liquid interfaces where the neutron beam can be sent through the substrate and probe the solid-liquid

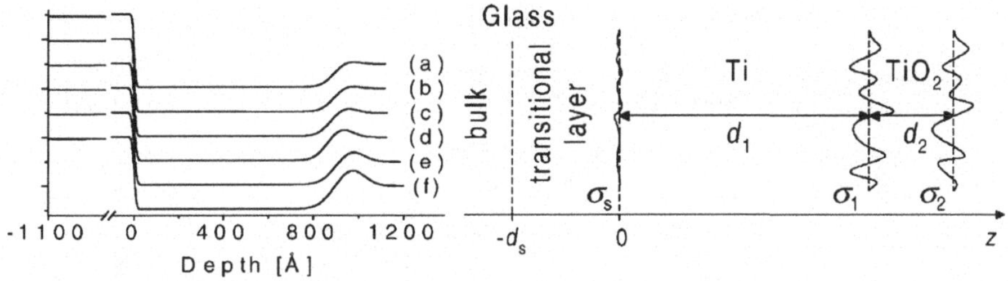

Fig. 11.10 Oxidation profile as a function of the annealing time at the surface of a Ti film. (**a**) As prepared and annealed at (**b**) 150 °C for 60 min (**c**) 200 °C for 30 min (**d**) 250 °C for 20 min (**e**) 300 °C for 15 min (**f**) 350 °C for 15 min. Adapted from [54, 55]

interface with negligible absorption to perform for example in-situ electro-chemical studies [57–59].

11.4.5 Interface Magnetism

Metallic Magnetic Layers

Metal Trilayer We present here the example of the study of a coupled trilayer system produced by S. Nerger at the Forschung Zentrum Jülich [60]. The structure of the sample is shown on Fig. 11.11a. The active region is formed by the layers FeCo/Mn/FeCo. The Ag layer is used to promote an epitaxial growth of the system. The Au layer is a simple protective capping. The presented system is $Fe_{0.5}Co_{0.5}$/Mn(8 Å)/$Fe_{0.5}Co_{0.5}$. The specificity of this system is that the magnetic couplings between Fe and Mn, and Co and Mn are of opposite sign. Ab initio calculation predicted that in such a system, contrary to a pure Fe/Mn interface, a complex magnetic behavior of the Mn layer arises. A first measurement (not shown) was performed in a saturating field of $1T$. A numerical modeling of the data shows that the magnetic moment in the $Fe_{0.5}Co_{0.5}$ layers is 2.4 μ_B/atom (as in bulk materials). A net magnetic moment of 0.8 μ_B/atom in Mn is also observed. This induced magnetization in the Mn layer was theoretically predicted for FeCo alloys by the ab-initio calculations [61, 62]. Note that similar measurements in Fe/Mn/Fe trilayers [63] did not show any magnetization in Mn, also in agreement with ab-initio calculations.

The applied field was then decreased down to 1.2 mT. The reflectivity was remeasured. In these conditions a large spin-flip signal is observed (Fig. 11.11b, yellow curve). The reflectivity data was fitted by letting the magnetization directions vary. The best adjustment was obtained when the magnetization of the FeCo layers make an angle of 45° with respect to the applied field. The two magnetic layers make an angle of 90° with respect to each other; we have a quadratic coupling (Fig. 11.11).

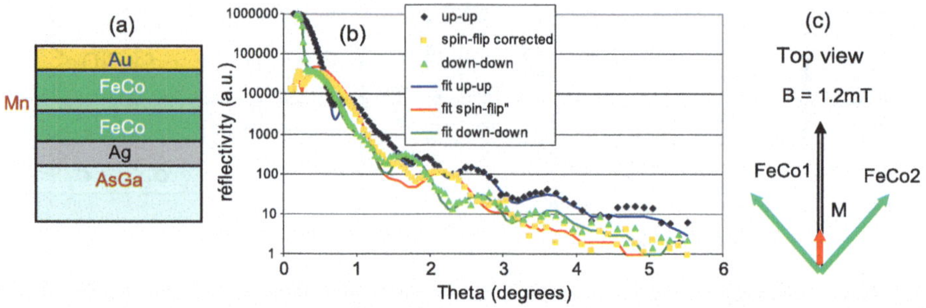

Fig. 11.11 (**a**) Trilayer system FeCo(8 nm)/Mn(0.8 nm)/FeCo(5 nm). (**b**) Reflectivity in the remanent state (1.2 mT). (**c**) Magnetic configuration as deduced from the fit performed by S. Nerger [60]

Exchange Bias—Spin-Valves The magnetic thin film system which has enjoyed the most popularity until now is the spin-valve. It consists of a stack of two magnetic layers separated by a non magnetic spacer. The electrical resistance of the system depends on the relative orientation of the magnetizations of the two layers. In industrial systems, one of the magnetic layers is pinned by a coupling with an anti-ferromagnetic material through the so-called exchange-bias mechanism. The materials which are used in such structures are numerous: Co, Fe, Ni, NiFe, Fe_3O_4, $CoFe_2O_4$, $LaSrMnO_3$ for ferromagnetic layers; Cu, Cr, V, Al_2O_3, HfO_2, $SrTiO_3$ for the spacer layers; FeMn, IrMn, CoO, NiO, $BiFeO_3$, Co/Ru/Co for the anti-ferromagnetic exchange bias layer.

Such spin valve systems have been extensively characterized [64–67] and are now well understood. However, the microscopic understanding of exchange bias has been a long standing problem for decades now. A wealth of literature is being produced on various systems [68–74]. It appears that the exchange bias mechanism combines very subtle effects. The reversal process of the coupled magnetic layer has been studied in detail. Since the origin of the phenomenon is often linked to micro-magnetic problems, reflectivity studies are often complemented with off-specular scattering which probe the underlying micro-magnetic structures. This technique is described in the following.

Magnetic Oxide

Polarized neutron reflectivity has also been used to probe the magnetism of oxide thin films (manganites [75, 76] or Fe_3O_4 [77]). For example, the hysteresis cycle of $La_{0.7}Sr_{0.3}MnO_3$ thin films shows a region with a low coercivity on which is superimposed a contribution which requires $0.3T$ to be saturated. This suggests that the films are not homogeneous and that they are composed of several phases having different coercivities. Neutron reflectivity measurements were performed on single $La_{0.7}Sr_{0.3}MnO_3$ thin films in order to probe the magnetization profiles through the depth of the films as a function of the temperature. Figure 11.12a shows the reflectivity on a 16 nm $La_{0.7}Sr_{0.3}MnO_3$. Modeling using a homogeneous magnetic

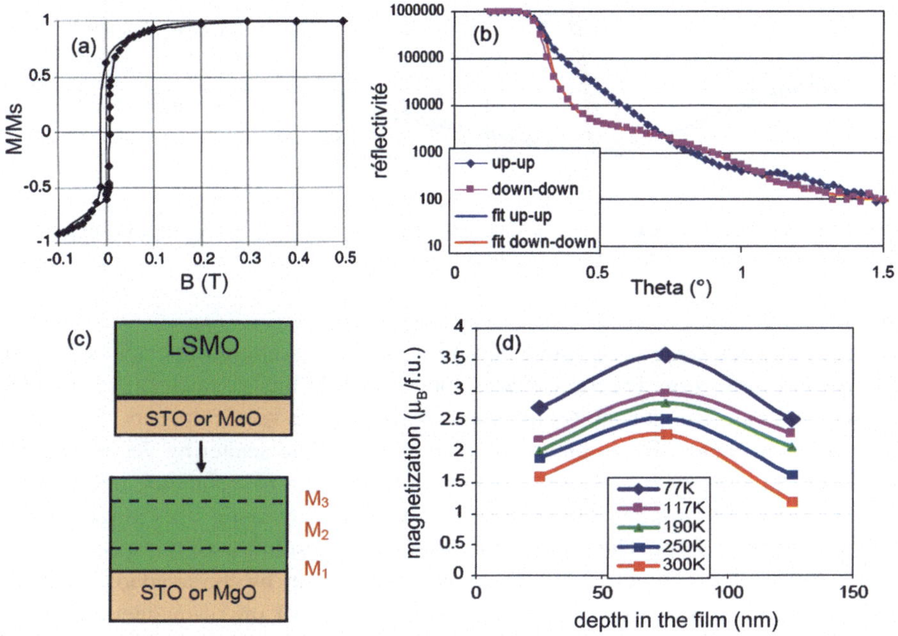

Fig. 11.12 (**a**) Hysteresis cycle of a manganite thin film. (**b**) Reflectivity of a $La_{0.7}Sr_{0.3}MnO_3$ (16 nm) film deposited on $SrTiO_3$. (**c**) Modeling of the system: (*top*) perfect system, (*bottom*) more realistic 3 layers model. (**d**) Magnetization profiles as a function of the temperature for the system $La_{0.7}Sr_{0.3}MnO_3$ (16 nm)//$SrTiO_3$

layer does not provide satisfactory fits. In order to quantitatively model the data, it is necessary to introduce a model taking into account different magnetizations at the interfaces. A 3 layers model with magnetizations M_1, M_2 and M_3 in the depth of the film is needed. Figure 11.12c shows the variations of the magnetizations M_1, M_2 and M_3 as a function of the temperature. One can note that the interface magnetization is reduced by 25 to 30 %.

11.4.6 Probing μm Scale In-Plane Structures: Off-Specular Scattering

Principle

In the case of specular reflectivity, the scattering vector **Q** is perpendicular to the sample surface and thus one only probes the structure of the sample along its depth. All the structures in the thin film plane are averaged out. However, in a number of situations, it can be of interest to probe in-plane structures. This is for example the case when there is formation of magnetic domains in the films or if the surface

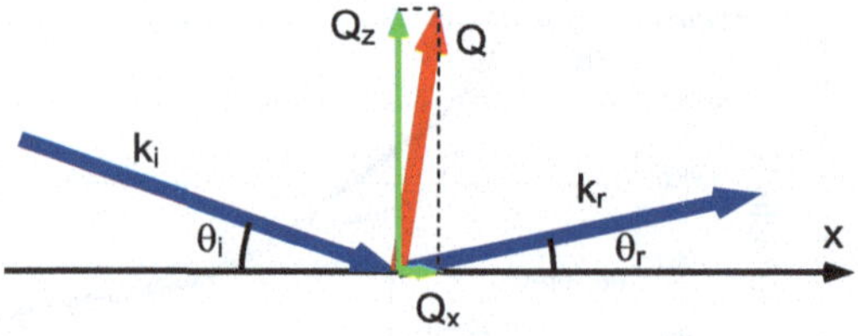

$$Q_x = \frac{2\pi}{\lambda}\left(\cos\theta_f - \cos\theta_i\right)$$

$$Q_z = \frac{2\pi}{\lambda}\left(\sin\theta_f + \sin\theta_i\right)$$

Fig. 11.13 Off-specular scattering geometry. The scattering vector **Q** is not perpendicular to the thin film plane. There is a small Q_x component in the thin film plane

is rough. In such cases, it is interesting to look at the scattered intensities around the specular reflection direction, that is for $\theta_r \neq \theta_i$. In this geometry (Fig. 11.13), a small in-plane component Q_x of the scattering wave-vector appears. Note that at grazing incidence this in-plane component Q_x of the scattering vector is very small, of the order of 0.1–10 μm^{-1}. Thus, in this scattering geometry, one will be mostly sensitive to in-plane lateral structures with characteristic sizes ranging from 50 μm down to 0.5 μm. The upper limit is set by the resolution of the spectrometer and the size of the direct beam. The lower limit is set by the available neutron flux. These sizes actually correspond to typical sizes of micro-magnetic domain structures. Thus magnetic off-specular is mostly used to probe such problems. Off-specular measurements are usually performed by using a position sensitive detector set after the sample and by measuring the scattering on the detector as a function of the incidence angle. A 2D mapping in the reciprocal space (Q_x, Q_z) can thus be obtained. The intensity scattered along the Q_z direction contains information about the depth structure of the system, the intensity scattered along the Q_x direction contains information about in-plane micrometric structures. The pioneering work in the field of off-specular scattering was presented in the early 1990s [78].

On Polymer Systems

Polymer-supported single lipid bilayers are models to study configurations of cell membranes. We illustrate the use of off-specular scattering for the study of in-plane height-height correlations of interfacial fluctuations in a lipid bilayer deposited on a polymer cushion [79] (Fig. 11.14). At 37 °C when the polymer cushion is collapsed, only weak off-specular scattering is detected (not shown), suggesting that the membrane is nearly planar (in-plane height-height correlation lengths of $\xi = 30$ μm, rms roughness of 35 Å). Cooling the system to 25 °C, the polymer layer swells and causes a dramatic enhancement of the off-specular signal, indicating the development of a short in-plane undulation of the membrane ($\xi = 11$ μm, rms roughness of about 105 Å) (Fig. 11.14a). ξ corresponds to thermally excited capillary waves and suggests that the membrane is flat at 37 °C and significantly more distorted at 25 °C (Fig. 11.14c).

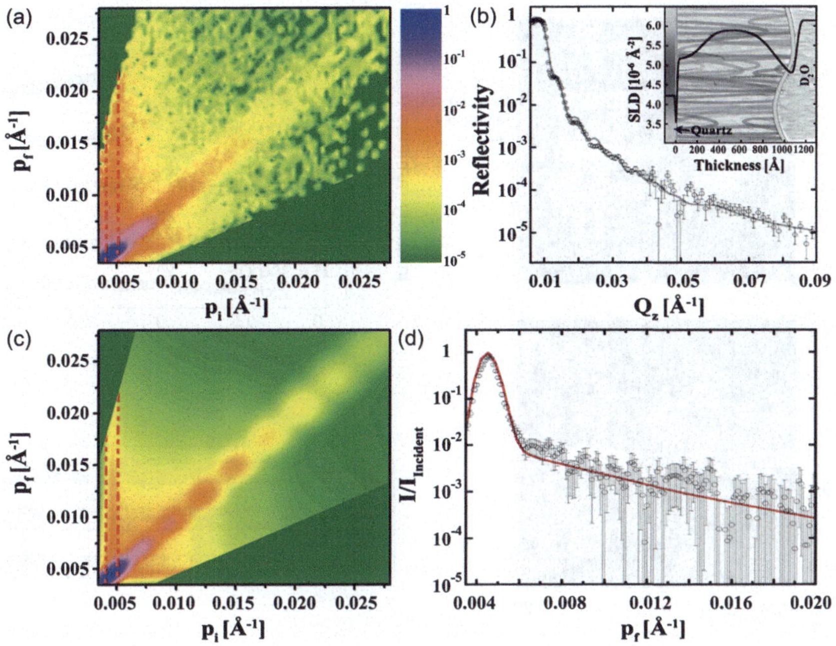

Fig. 11.14 Analysis of the neutron scattering from the polymer-membrane system at 25 °C. *Panels* (**a**) and (**c**) depict the measured and theoretical scattering maps, respectively, with a common intensity scale. *Panel* (**b**) illustrates the specular reflectivity data (*open circles*), fit (*gray line*), and corresponding real-space interpretation (SLD distribution shown in *inset*). *Panel* (**d**) compares the measured (*open circles*) and theoretical (*red line*) off-specular scattering along a Yoneda wing. Adapted from [79]

On Magnetic Systems

Figure 11.15 presents an example of the off-specular scattering from a $[Co/Cu]_{50}$ multilayer. The diffuse signal has been measured as a function of Q_x and Q_z. On Figs. 11.15a and 11.15b, one observes the structural correlation peak [001] corresponding to the chemical periodicity. At remanence, a strong diffuse scattering peak is observed at the position $[0\ 0\ \frac{1}{2}]$. Since the magnetic diffuse scattering is localized around the position $[0\ 0\ \frac{1}{2}]$, it is possible to say that the Co layers are globally anti-ferromagnetically coupled along the thickness of the layer. However, since there is a strong diffuse scattering, it is also possible to say that there exists a significant magnetic disorder in the plane of the Co layers. The width of the diffuse scattering peak around the position $[0\ 0\ \frac{1}{2}]$ (Fig. 11.15c) is inversely proportional to the magnetic domain size and gives an estimate of the mean magnetic domain size which ranges from 1 μm at remanence (30G) and grows to 6 μm at 250G.

Magnetic off-specular scattering has been mostly used to probe the magnetic domains sizes in multilayers. Detailed quantitative analysis of the magnetic off-specular scattering can be performed [81]. The effect of the micro-magnetic

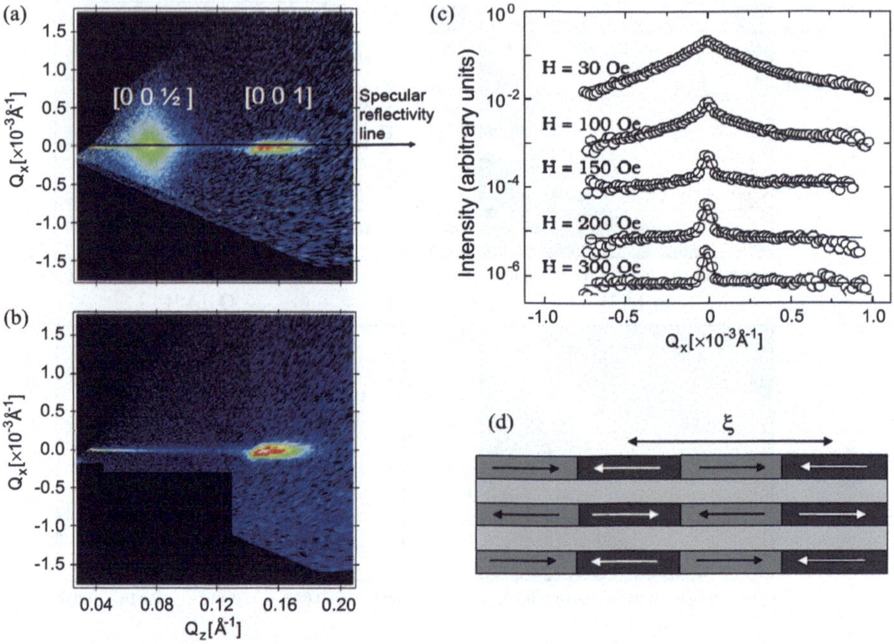

Fig. 11.15 [Co(2 nm)/Cu(2 nm)]$_{50}$ multilayers (adapted from Langridge et al. [80]). (**a**) Diffuse scattering at H = 0. One observes a strong diffuse signal at the AF position. (**b**) Diffuse scattering in a saturating field. The AF peak has disappeared. (**c**) Evolution of the AF peak as a function of the applied field (cut along $Q_z = 0.75\ \text{nm}^{-1}$). (**d**) Magnetic coupling between the layers. ξ is the lateral correlation length between magnetic domains. The Co layers are locally coupled AF but there is a strong disorder within each Co layer

structure can then be correlated with other properties such as the magneto-crystalline anisotropy (in Fe/Cr superlattice [82]) or the magneto-resistive effect (in Fe/Cr [83, 84] or Co/Cu [85] superlattices). The formation of micro-magnetic structures is very important with respect to the transport properties in magnetic sensors. The signal-to-noise ratio of Giant Magneto Resistive systems is very sensitive to the micro-magnetic structure [64]. Off-specular studies are also used to complement studies on exchange bias systems: Co/CoO [86], $Ir_{20}Mn_{80}/Co_{80}Fe_{20}$ [87]. Off-specular scattering has also been used to study the problem of the reversal process in neutron polarizing super-mirrors [88]. In some special cases, it has been shown that it is also possible to probe single interfaces (Fe/Cr/Fe trilayer [89] or waveguide structures [90]. Off-specular has also been used to study patterned or self-organized micro-structures [91–96]. In a number of studies, the influence of patterning on the exchange bias has been probed [97–100]. These studies are of interest when the magnetic heterostructures are to be integrated in large scale micro-circuits (typically for Magnetic RAMs).

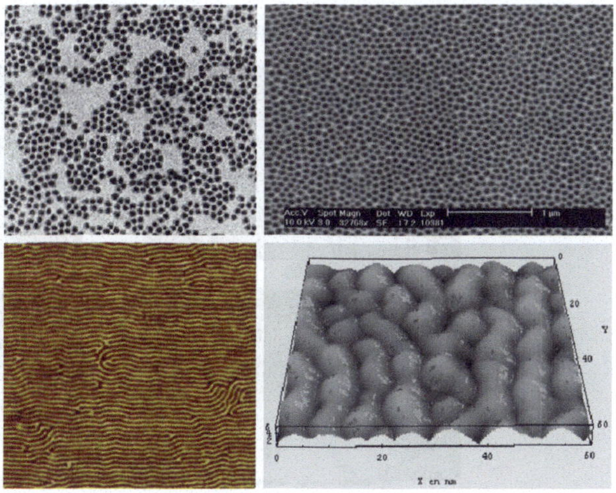

Fig. 11.16 (**a**) Small particle sitting on a surface; (**b**) arrays of nanowires; (**c**) magnetic domains self-organized in a regular structure; (**d**) magnetic-structural surface correlations

11.4.7 Probing nm Scale In-Plane Structures: Grazing Incidence Small Angle Scattering

Principle

Since nanosciences are aiming at very small scales (well below 1 μm), off-specular scattering will reach its limits since it can probe only rather large correlation lengths ($\xi > 500$ nm). This is why surface scattering has been extended to the SANS geometry. In this geometry, one looks at the scattering in the plane perpendicular to the incidence plane (Fig. 11.1, green plane). The scattering wave vector Q_y is in a range comparable to the scattering wave vectors in SANS experiments: $10^{-4}\ \text{nm}^{-1} < Q_y < 3\ \text{nm}^{-1}$. This corresponds to correlation lengths ξ ranging from 3 nm to 100 nm.

GISANS may typically be used to study small particles sitting on a surface ($\xi \sim 20$–100 nm) (Fig. 11.16a), arrays of nanowires ($\xi \sim 20$–100 nm) (Fig. 11.16b), magnetic domains self-organized in a regular structure ($\xi \sim 100$ nm) (Fig. 11.16c), magnetic-structural surface correlations ($\xi \sim 10$–20 nm) (Fig. 11.16d).

On Polymer Systems

The understanding of self assembly at the molecular scale in the vicinity of a solid substrate is a key issue for the fabrication of self-assembled nanodevices. Pluronics are well known as templates for the synthesis of ordered mesoporous materials since they self-assemble into rich patterns of ordered structures depending on the temperature, the concentration or the pressure. Figure 11.17 illustrates the use of Grazing incidence SANS to probe the structure of micelles close to solid interfaces [101, 102]. In this case, the formation of micelles of the triblock copolymer F127 at

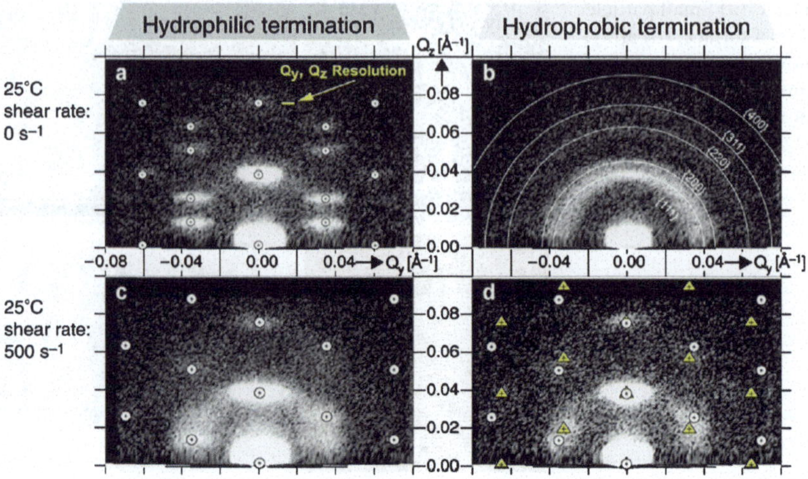

Fig. 11.17 Detector images for a 18 wt % solution of F127 dissolved in heavy water taken at different sample conditions. The *circular* and *triangular symbols* refer to calculated diffraction patterns explained in the text. Adapted from [102]

the vicinity of an interface has been studied as a function of the surface termination (hydrophillic/hydrophobic) and under planar shear.

In the case of hydrophobic coating (Fig. 11.17b) the diffraction pattern consists of sections of Debye–Scherrer cones implying a three-dimensional (3D) powder structure corresponding to a fcc structure. In case of the hydrophillic coating (Fig. 11.17a) well defined Bragg reflections appear, corresponding to the formation of a 2D powder with a fcc structure, where the cubic [111] axis is aligned perpendicular to the surface. When shear is applied, in the case of the hydrophilic termination, (Fig. 11.17c), the diffraction is globally unchanged except for a reduced Q-resolution. This is attributed to the development of a texture within the plane of the surface, while the fcc structure remains unaltered. If the texture were to become perfect, i.e. if a 3D single crystal were to form, a diffraction pattern as shown by the superimposed spots is expected. In case of a hydrophobic termination, shear modifies the 3D isotropic powder structure to a 2D ordering (Fig. 11.17d) which is best described by assuming a textured 2D fcc structure as found in the case of hydrophilic termination and a coexisting hcp structure.

Other GISANS studies have been performed during the last few years. We can mention for example the structure of a surfactant layer adsorbed at a hydrophilic solid/liquid interface [103], the lateral structures of buried interfaces in triblock copolymer films [104], the hydration of casein micelles [105].

On Magnetic Systems

We present here the first example of a Grazing Incidence SANS experiment on a magnetic thin film [106]. FePt thin film layers self organize themselves in magnetic

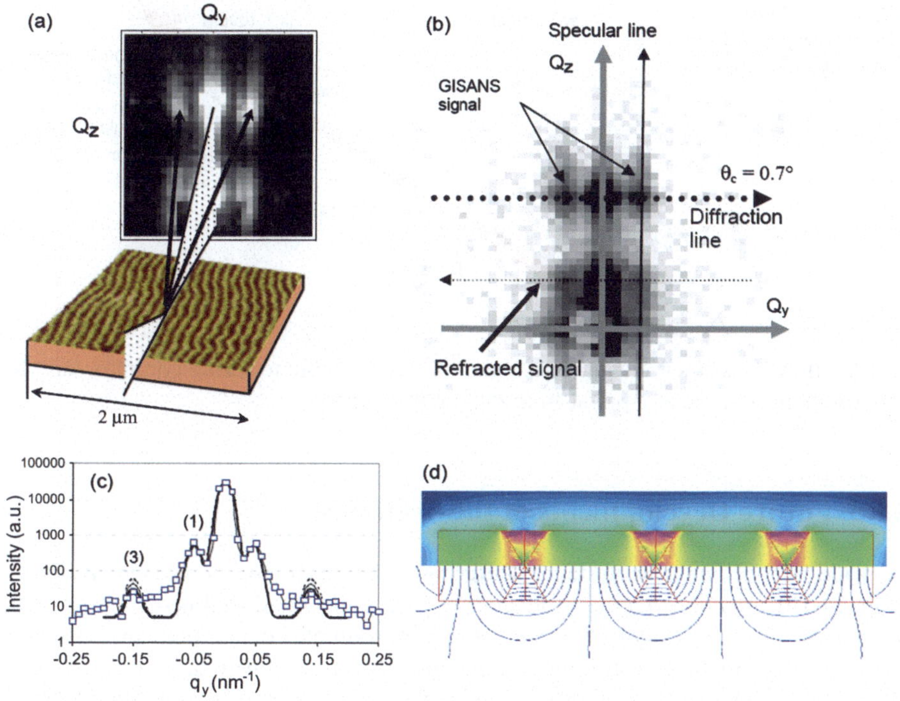

Fig. 11.18 GISANS signal from a magnetic domains nano-structure. (**a**) Magnetic Force Microscopy image of the magnetic domain and scattering geometry. (**b**) GISANS signal on the detector for $\theta_{in} = 0.7°$. (**c**) GISANS signal at constant q_z. (**d**) Distribution of the magnetic induction in the thin films

stripe domains (Fig. 11.18a). The stripes are almost perfectly ordered as a periodic pattern with a period of about 100 nm. In order to study in-depth this magnetic pattern, a Grazing incidence SANS experiment was performed on the spectrometer PAPYRUS at the LLB. The neutron beam was sent at grazing incidence ($\theta_{in} = 0.7°$) on the layer, the magnetic domains being parallel to the incidence plane. Diffraction from the magnetic domains can be observed. Figure 11.18b details the different contributions of the Grazing Incidence SANS signal. An integration at fixed Q_z has been performed and is presented on Fig. 11.18c. Three diffraction orders can be observed (the second order being extinct). In order to model the system, it is necessary to take into account the Néel caps between the magnetic stripes as well as the magnetic stray fields (Fig. 11.18d) [107].

Other systems with stripe domains have been studied (ex. Fe/FeN [108]). Systems of magnetic Fe nanodots have also been observed using Grazing Incidence SANS [109].

Compared to Magnetic Force Microscopy, which is a direct space probe, the technique permits to probe buried layers and to obtain quantitative information about the magnetization. Force microscopy only gives surface information and no quantitative

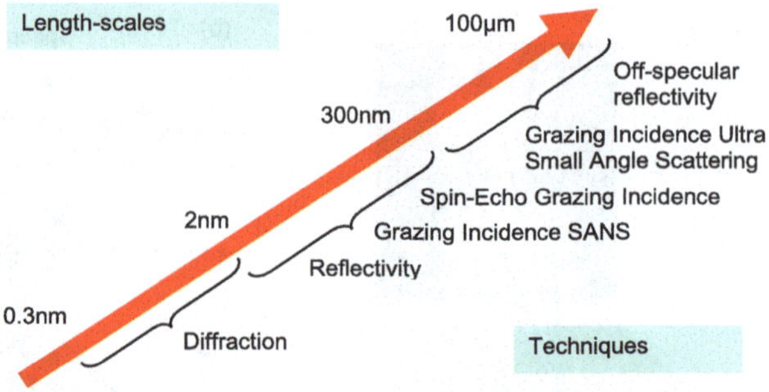

Fig. 11.19 Correlation lengths and suitable scattering techniques

information. The other advantage is that it is also possible to set-up complex sample environments (furnace-cryostat-high magnetic fields).

11.5 Future Developments of the Technique

A wide set of neutron surface scattering techniques are nowadays available (Fig. 11.19): *specular neutron reflectivity* which is operated routinely, *off-specular scattering* which is easily performed but requires complex data processing, *Grazing Incidence SANS* which is still in development, and *diffraction* on thin films which in the case of good quality systems is feasible. A very large range of correlation lengths in thin film systems can thus be probed using these different scattering techniques.

A large number of neutron reflectometers are available across the world [21]. As a whole, neutron scattering techniques are making steady progress. The instruments efficiency has been increasing by about one order of magnitude per decade during the last 30 years. This may however look as a slow pace compared to other techniques such as X-rays scattering. Nevertheless technical progress are such that this rhythm is likely to be sustained at least for another two decades. We are presently designing the instruments for the next decade and dreaming about the instruments for 2020. Flux gains ranging from 10 to 100 can reasonably be expected in the next decade through the implementation of new types of neutron reflectometers. Quantitative gains in the measuring time and in the minimum sample size will be achieved. However it is not yet clear if qualitative gains, i.e. new types of measurements besides the ones presented here, will be achieved.

Gains in flux will be achieved so as to be able to measure reflectivities down to 10^{-8} within hours. However in practice, most specular reflectivity experiments will require only to measure reflectivities down to 10^{-6}, both because of the incoherent scattering and the intrinsic roughness of most samples. This means that it will be possible to perform reflectivity measurements in a matter of minutes. Most of the effort will thus be put on providing advanced and complex sample environments, especially to perform in-situ measurements for which neutrons are perfectly suited since they are barely absorbed by any surrounding sample environment.

The case of off-specular scattering is less clear-cut. Presently the measurements of off-specular scattering are limited to very rough systems or to multilayer systems because of the low flux issue. The key problem is that even a gain of a factor 100 in flux might be insufficient to alleviate this limitation. On top of that, the technique suffers at the moment from complex data analysis and processing (especially for magnetic and time of flight measurements). Because the flux is so limited, usually only a very narrow Q-range in the reciprocal space can be explored meaning that only rather long correlation lengths (above 1 μm) can be probed. The question arises if neutron scattering is an appropriate technique for such studies (except for very specific problems).

At the moment, it is rather clear that the popular field of *nanosciences* is interested in sizes rather well below 100 nm. In this range of sizes, one should consider GISANS techniques. Dedicated instruments have recently been built (e.g. REFSANS at FRM2 in Münich) and they show impressive performances.

A large potential probably lies in the use of completely new techniques based on Larmor precession encoding. However, it is yet not clear how well these techniques will perform in everyday use. In a more general way, the next neutron sources based on the spallation technique are coming online (SNS in the USA, J-PARC in Japan and ESS in Europe). This will benefit to the construction of new high flux neutron reflectometers.

References

1. G.P. Felcher, R.O. Hilleke, R.K. Crawford, J. Haumann, R. Kleb, G. Ostrowski, Rev. Sci. Instrum. **58**, 609–619 (1987)
2. C.F. Majkrzak, J.W. Cable, J. Kwo, M. Hong, D.B. McWhan, Y. Yafet, J. Waszcak, Phys. Rev. Lett. **56**, 2700 (1986)
3. J. Penfold, R.K. Thomas, J. Phys. Condens. Matter **2**, 1369–1412 (1990)
4. T.P. Russel, Mater. Sci. Rep. **5**, 171–271 (1990)
5. T.P. Russel, Physica B **221**, 267–283 (1996)
6. L.T. Lee, D. Langevin, B. Farnoux, Phys. Rev. Lett. **67**, 2678–2681 (1991)
7. J. Penfold, E.M. Lee, R.K. Thomas, Mol. Phys. **68**, 33–47 (1989)
8. M.N. Baibich, J.M. Broto, A. Fert, F. Nguyen van Dau, F. Petroff, P. Etienne, G. Creuzet, A. Friedrich, J. Chazelas, Phys. Rev. Lett. **61**, 2472 (1988)
9. J. Lekner, *Theory of Reflection of Electromagnetic and Particle Waves* (Nijhoff, Dordrecht, 1987)
10. C. Fermon, F. Ott, A. Menelle, Neutron reflectivity, in *X-Ray and Neutron Reflectivity: Principles and Applications*, ed. by J. Daillant, A. Gibaud. Springer Lecture Notes in Physics, vol. 770 (Springer, Berlin, 2009), p. 163
11. S.J. Blundell, J.A.C. Bland, Phys. Rev. B **46**, 3391–3400 (1992)
12. C. Fermon, C. Miramond, F. Ott, G. Saux, J. Neutron Res. **4**, 251 (1996)
13. N.K. Pleshanov, Z. Phys. B **94**, 233–243 (1994)
14. C.F. Majkrzak, Physica B **221**, 342–356 (1996)
15. H. Zabel, R. Siebrecht, A. Schreyer, Physica B **276–278**, 17–21 (2000)
16. H. Zabel, K. Theis-Brohl, J. Phys. Condens. Matter **15**, S505–S517 (2003)
17. G.P. Felcher, J. Appl. Phys. **87**, 5431–5436 (2000)
18. G.P. Felcher, Physica B **267–268**, 154–161 (1999)
19. M.R. Fitzsimmons et al., J. Magn. Magn. Mater. **271**, 103–146 (2004)
20. G. Fragneto-Cusani, J. Phys. Condens. Matter **13**, 4973–4989 (2001)
21. http://material.fysik.uu.se/Group_members/adrian/reflect.htm

22. F. Mezei, Z. Phys. **255**, 146 (1972)
23. F. Cousin et al., Euro. Phys. J.+ **126**, 109 (2011)
24. T.P. Russel, A. Menelle, W.A. Hamilton, G.S. Smith, S.K. Satija, C.F. Majkrzak, Macromolecules **24**, 5721–5726 (1991)
25. X. Zhao, W. Zhao, X. Zheng, M.H. Rafailovich, J. Sokolov, S.A. Schwarz, M.A.A. Pudensi, T.P. Russel, S.K. Kumar, L.J. Fetters, Phys. Rev. Lett. **69**, 776 (1992)
26. L.T. Lee, O. Guiselin, A. Lapp, B. Farnoux, J. Penfold, Phys. Rev. Lett. **67**, 2838–2841 (1991)
27. O. Guiselin, L.T. Lee, B. Farnoux, A. Lapp, J. Chem. Phys. **95**, 4632–4640 (1991)
28. L.T. Lee, B. Jean, A. Menelle, Langmuir **15**, 3267–3272 (1999)
29. C. Marzolin, P. Auroy, M. Deruelle, J.-P. Folkers, L. Leger, A. Menelle, Macromolecules **34**, 8694–8700 (2001)
30. M. Geoghegan, F. Boué, G. Bacri, A. Menelle, D.G. Bucknall, Eur. Phys. J. B **3**, 83–96 (1998)
31. M. Lösche, J. Schmitt, G. Decher, W.G. Bouwman, K. Kjaer, Macromolecules **31**, 8893 (1998)
32. A. Glidle, A.R. Hillman, K.S. Ryder et al., Langmuir **25**, 4093–4103 (2009)
33. C. Devaux, E. Beyou, Ph. Chaumont, J.P. Chapel, Eur. Phys. J. E, Soft Matter **7**, 345–352 (2002)
34. C. Devaux, J.P. Chapel, Eur. Phys. J. E, Soft Matter **10**, 77–81 (2003)
35. C. Devaux, F. Cousin, E. Beyou, J.-P. Chapel, Macromolecules **38**, 4296–4300 (2005)
36. O. Prucker, J. Rühe, Macromolecules **31**, 592 (1998)
37. J.A. Dura, D.J. Pierce, C.F. Majkrzak, N.C. Maliszewsky, D.J. McGillivray, M. Lösche, K.V. O'Donovan, M. Mihailescu, U. Perez-Salas, D.L. Worcester, S.H. White, Rev. Sci. Instrum. **77**, 074301 (2006)
38. S. Krueger, Curr. Opin. Colloid Interface Sci. **6**, 111 (2001)
39. J. Majewski, T.L. Kuhl, M.C. Gerstenberg, J.N. Israelachvili, G.S. Smith, J. Phys. Chem. B **101**, 3122 (1997)
40. T. Charitat, E. Bellet-Amalric, G. Fragneto, F. Graner, Eur. Phys. J. B **8**, 583 (1999)
41. G. Fragneto, F. Graner, T. Charitat, P. Dubos, E. Bellet-Amalric, Langmuir **16**, 4581 (2000)
42. A.P. Maierhofer, D.G. Bucknall, T.M. Bayer, Biophys. J. **79**, 1428 (2000)
43. G. Fragneto, T. Charitat, F. Graner, K. Mecke, L. Perino-Gallice, E. Bellet-Amalric, Europhys. Lett. **53**, 100 (2001)
44. S. Krueger, C.W. Meuse, C.F. Majkrzak, J.A. Dura, N.F. Berk, M. Tarek, A.L. Plant, Langmuir **17**, 511 (2001)
45. D. Schwendel, T. Hayashi, R. Dahint, A. Pertsin, M. Grunze, R. Steitz, F. Schreiber, Langmuir **19**, 2284 (2003)
46. I. Burgess, M. Li, S.L. Horswell, G. Szymanski, J. Lipkowski, J. Majewski, S. Satija, Biophys. J. **86**, 1763 (2004)
47. J.A. Dura, C.A. Richter, C.F. Majkrzak, N.V. Nguyen, Appl. Phys. Lett. **73**, 2131 (1998)
48. V. Bertagna, R. Erre, M.L. Saboungi et al., Appl. Phys. Lett. **84**, 3816–3818 (2004)
49. H.H. Chen-Mayer, G.P. Lamaze, K.J. Coakley, S.K. Satija, Nucl. Instrum. Methods Phys. Res., Sect. A, Accel. Spectrom. Detect. Assoc. Equip. **505**, 531 (2003)
50. F. Caccavale, R. Coppola, A. Menelle, M. Montecchi, P. Polato, G. Principi, J. Non-Cryst. Solids **218**, 291–295 (1997)
51. C. Battaglin, F. Caccavale, A. Menelle, M. Montecchi, E. Nichelatti, F. Nicoletti, P. Polato, Thin Solid Films **351**, 176 (1999)
52. G. Battaglin, A. Menelle, M. Montecchi, E. Nichelatti, P. Polato, Glass Technol. **43**, 203 (2002)
53. D.G. Wiesler, C.F. Majkrzak, Physica B **198**, 181 (1994)
54. V.M. Pusenkov, K. Moskalev, N. Pleshanov, A. Schebetov, V. Syromyatnikov, V. Ul'yanov, A. Kobzev, O. Nikonov, Physica B **276**, 654 (2000)
55. S.V. Metelev, N.K. Pleshanov, A. Menelle, V.M. Pusenkov, A.F. Schebetov, Z.N. Soroko, V.A. Ul'yanov, Physica B **297**, 122 (2001)

56. B.D. Vogt, H.J. Lee, V.M. Prabhu, D.M. DeLongchamp, E.K. Lin, W.L. Wu, S.K. Satija, J. Appl. Phys. **97**, 7 (2005)
57. Z. Tun, J.J. Noel, D.W. Shoesmith, Physica B **241**, 1107 (1997)
58. Z. Tun, J.J. Noel, D.W. Shoesmith, J. Electrochem. Soc. **146**, 988 (1999)
59. J.J. Noel, H.L. Jensen, Z. Tun, D.W. Shoesmith, Electrochem. Solid-State Lett. **3**, 473 (2000)
60. S. Nerger et al., Physica B **297**, 185–188 (2001)
61. C. Cornea, D. Stoeffler, J. Magn. Magn. Mater. **198–199**, 282 (1999)
62. C. Cornea, Ph.D. Thesis, Louis Pasteur University, Strasbourg, 1999
63. E. Kentzinger, U. Rücker, W. Caliebe, G. Goerigk, F. Werges, S. Nerger, J. Voigt, W. Schmidt, B. Alefeld, C. Fermon, Th. Brückel, Physica B **276–278**, 586 (2000)
64. M. Pannetier, T.D. Doan, F. Ott, S. Berger, N. Persat, C. Fermon, Europhys. Lett. **64**, 524–528 (2003)
65. S. Moyerman et al., J. Appl. Phys. **99** (2006)
66. C. Schanzer et al., Physica B, Condens. Matter **356**, 46–50 (2004)
67. Z.Y. Zhao et al., Phys. Rev. B **71**, 1098-0121 (2005)
68. M.R. Fitzsimmons et al., Phys. Rev. Lett. **84**, 3986–3989 (2000)
69. C. Leighton et al., Phys. Rev. B **65**, 064403/1-7 (2002)
70. M.R. Fitzsimmons et al., Phys. Rev. B **65**, 134436 (2002)
71. M. Gierlings et al., Phys. Rev. B **65**, 092407/1-4 (2002)
72. P. Blomqvist et al., J. Appl. Phys. **96**, 6523–6526 (2004)
73. A. Paul et al., Phys. Rev. B **74**, 054424 (2006)
74. S. Roy et al., Phys. Rev. Lett. **95**, 047201 (2005)
75. F. Ott et al., J. Magn. Magn. Mater. **211**, 200–205 (2000)
76. R.P. Borges et al., J. Appl. Phys. **89**, 3868–3873 (2001)
77. J.-B. Moussy et al., Phys. Rev. B **70**, 174448 (2004)
78. G.P. Felcher et al., Neutron News **5**, 18–22 (1994)
79. M.S. Jablin, M. Zhernenkov, B.P. Toperverg, M.D. Dubey, H.L. Smith, A. Vidyasagar, R. Toomey, A.J. Hurd, J. Majewski, Phys. Rev. Lett. **106**, 138101 (2011)
80. S. Langridge et al., Phys. Rev. Lett. **85**, 4964–4967 (2000)
81. V. Lauter-Pasyuk et al., Phys. Rev. Lett. **89**, 167203 (2002)
82. D.L. Nagy et al., Phys. Rev. Lett. **88**, 4 (2002)
83. H. Lauter et al., J. Magn. Magn. Mater. **258**, 338 (2003)
84. V. Lauter-Pasyuk et al., J. Magn. Magn. Mater. **226**, 1694 (2001)
85. A. Paul et al., Physica B, Condens. Matter **356**, 31 (2005)
86. M. Gierlings et al., Physica B **356**, 36–40 (2004)
87. A. Paul et al., Physica B **356**, 26–30 (2005)
88. U. Rücker, E. Kentzinger, B. Toperverg, F. Ott, Th. Brückel, Appl. Phys. A **74**, S607–609 (2002)
89. E. Kentzinger, U. Rucker, B. Toperverg et al., Physica B **335**, 89–94 (2004)
90. S.V. Kozhevnikov, F. Ott, E. Kentzinger, A. Paul, Physica B **397**, 68–70 (2007)
91. F. Ott, A. Menelle, C. Fermon, P. Humbert, Physica B **283**, 418–421 (2000)
92. F. Ott, P. Humbert, C. Fermon, A. Menelle, Physica B **297**, 189–193 (2001)
93. K. Theis-Brohl, Physica B, Condens. Matter **345**, 161–168 (2004)
94. K. Theis-Brohl et al., Phys. Rev. B **71**, 4 (2005)
95. K. Theis-Brohl et al., Physica B **356**, 14 (2005)
96. S. Langridge et al., Phys. Rev. B **74**, 6 (2006)
97. K. Temst et al., Eur. Phys. J. B **45**, 261–266 (2005)
98. K. Temst et al., J. Magn. Magn. Mater. **304**, 14–18 (2006)
99. E. Popova et al., Eur. Phys. J. B **44**, 491–500 (2005)
100. K. Theis-Brohl et al., Phys. Rev. B **73**, 14 (2006)
101. M. Wolff, U. Scholz, R. Hock et al., Phys. Rev. Lett. **92**, 255501 (2004)
102. M. Walz, A. Magerl, M. Wolff, H. Zabel, Superlattices Microstruct. **41**, 185–189 (2007)
103. R. Steitz, P. Müller-Buschbaum, S. Schemmel, R. Cubitt, G.H. Findenegg, Europhys. Lett. **67**, 962–968 (2004)

104. P. Müller-Buschbaum, L. Schulz, E. Metwalli, J.-F. Moulin, R. Cubitt, Langmuir **24**, 7639–7644 (2008)
105. E. Metwalli, J.-F. Moulin, R. Gebhardt, R. Cubitt, A. Tolkach, U. Kulozik, P. Müller-Buschbaum, Langmuir **25**, 4124–4131 (2009)
106. C. Fermon et al., Physica B **267–268**, 162–167 (1999)
107. M. Pannetier, F. Ott, C. Fermon, Y. Samson, Physica B **335**, 54–58 (2003)
108. W. Szuszkiewicz et al., J. Alloys Compd. **423**, 172–175 (2006)
109. C.P. Li, I.V. Roshchin, X. Batlle et al., J. Appl. Phys. **100**, 074318 (2006)

Chapter 12
Probing Surfaces with Thermal He Atoms: Scattering and Microscopy with a Soft Touch

Bodil Holst and Gianangelo Bracco

Abstract Helium atom scattering (HAS) is a well established technique, particularly suited for the investigation of insulating and/or fragile materials and light adsorbates including hydrogen. In contrast to other beam techniques based on X-rays or electrons, low energy (typically less than 100 meV) He atoms are scattered by the tail of the electron density distribution which spill out from a surface, therefore HAS is strictly a non-penetrating technique without any sample damage. HAS has been used to investigate structural properties of crystalline surfaces, including precise determination of atomic step heights, for monitoring thin film growth, to study surface transitions such as surface melting and roughening and for determining the presence and properties of adsorbates. Energy resolved HAS can provide information about surface vibrations (phonons) in the meV range and surface diffusion. This chapter provides a brief introduction to HAS with an outlook on a new, promising surface science technique: Neutral Helium Microscopy.

12.1 Introduction

After a systematic study of the properties of molecular beams [1], in 1930 Otto Stern and collaborators observed for the first time the diffraction of He and H_2 beams from LiF and NaCl [2, 3]. These results not only confirmed the validity of the de Broglie wave hypothesis also for atoms and molecules but provided the basis for a sensitive and versatile technique for studying surface properties. Unfortunately such experiments where too advanced for the vacuum technology of the time. Moreover the study on beams were performed by using effusive sources which have a broad velocity distribution and whose intensity is not optimized for scattering experiments.

B. Holst (✉)
Department of Physics and Technology, University of Bergen, Allegaten 55, 5007 Bergen, Norway
e-mail: bodil.holst@uib.no

G. Bracco (✉)
Department of Physics, University of Genoa, via Dodecaneso 33, 16146 Genoa, Italy
e-mail: bracco@fisica.unige.it

G. Bracco, B. Holst (eds.), *Surface Science Techniques*,
Springer Series in Surface Sciences 51, DOI 10.1007/978-3-642-34243-1_12,

A breakthrough came in 1951 when Kantrowitz and Grey [4] suggested the use of a gas at high pressure in the source so that hydrodynamic effects during expansion in vacuum become important and in 1954 these sources were developed in practice [5]. For these novel sources, the frequent collisions in the exit aperture of the source, the nozzle, determine a cooling of the expanding gas in the vacuum and the resulting supersonic beam (quasi-monochromatic beam) has an increased intensity with a velocity distribution narrower than the equilibrium Maxwell distribution of the gas in the source: these properties are the necessary conditions for molecular scattering experiments.

The energy resolution of the supersonic beams was improved further and at beginning of the 80' was much better than 1 meV [6], sufficient to apply the technique of inelastic HAS (IHAS), in which the scattered beam are analyzed in energy, for the study of the surface phonon dispersion curves and in the 90' appeared also setups for quasi-elastic scattering to study surface diffusion before the development of the ^{3}He spin echo method which presents a better resolution. Since this last technique is described in Chap. 13 of this book, this chapter will be limited to the discussion of techniques based on ^{4}He.

The helium atom at thermal energies is a neutral gentle probe and HAS can be employed for the investigation of any surface of conductive or insulating samples without charging effects and chemisorbed or physisorbed systems without beam damage effects.

In the rest of this introduction an overview is given of some of the systems that has been investigated by HAS. The overview is necessarily incomplete but helps to illustrate the versatility of the techniques and to show the variety of systems that were investigated by HAS.

Insulating, ionic systems such as LiF, NaCl, and NaF were the first systems studied, because a clean (100) surface could be obtained by cleaving the crystal to expose a fresh surface. The crystal could then be quickly mounted in the apparatus, which was pumped down to reach UHV conditions. Then an annealing to high temperature before measurements was sufficient to avoid water contamination because those ionic compounds are hygroscopic. Elastic He diffraction experiments were performed by different groups [7–10] with the aim to extract both the surface periodicity from the diffracted peak position and the corrugation amplitude from the diffracted intensities. Moreover the analysis of bound state resonance effects (see Sect. 12.2.1) allowed to estimate the levels of the He–LiF interaction potential [11].

After some observation of inelastic features on LiF(100) [12, 13], the first inelastic scattering experiments with a high resolution apparatus were performed by the Toennies group [6] at the beginning of the 80'. These measurements were essentially limited to the acoustic region of the spectrum. Further measurements were performed on NaF, KCl [14, 15], KBr [16], RbCl [17]. The experiment were carried out with the in-plane geometry (the detector collects the scattered beams only in the scattering plane, i.e. the plane defined by the incident and the normal to the surface, see also 12.3) therefore only longitudinal and vertical (along the normal) polarized phonons could be detected. In the case of NaCl [18] also shear horizontal phonons were measured exploring selected regions outside the first Brillouin zone and exploiting Umklapp processes.

The IHAS was considered to be relatively insensitive to surface optical modes, because there was no evidence of surface optical phonons in the first investigations. In fact, in case of alkali halides, the counterphase motion of oppositely charged ions do not strongly modulate the electronic density at the relatively large distance of the turning points where He is bounced off from the surface. The size of He atoms further smeared out the tiny modulations. In spite of these limitations, optical phonons were measured for the first time [19] increasing the incident energy to 93 meV (He source temperature at ~450 K) and choosing NaF where the F^- ion, which has the larger coupling with He, vibrates with a larger amplitude in the optical range because it is lighter than Na^+. The reverse is true for the NaF acoustic branches and for LiF where in the optical region the less coupled and lighter Li^+ has a larger amplitude. A few years later, with a beam energy of 64 meV also the optical dispersion curves of LiF were measured [20, 21] up to 50 meV. In this case the inelastic transition were enhanced by bound state resonance effects.

Graphite is a layered crystal and the basal plane can be easily exposed by applying and peeling off adhesive tape before to transfer the sample in vacuum. The surface is relatively inert and generally an annealing is sufficient to get a clean surface. For this reasons it was studied intensively along with alkali halides. The analysis of elastic scattering measurements [22] showed that the corrugation is 0.21 Å. Moreover the measurements of bound state resonance effects performed by the Boato group allowed an estimation of the levels of the He-graphite interaction potential with band effects [23]. A further analysis can be found in Ref. [24]. Some of the features related to bound state resonances were assigned by Cantini and Tatarek to resonant processes assisted by phonons [25, 26]. The phonon dispersion curves were studied by Benedek et al. [27].

Of course the present hot topic of a single layer of graphite, graphene, has been recently investigated [28, 29]. Also the (111) surface of another semimetal, Bi, which shows a linear dispersion for electrons similar to graphene, has been recently studied [30, 31].

Much attention was devoted to the surface of semiconductors and ultrathin metal/semiconductor alloys because of the large variety of surface reconstructions and phase transitions occurring upon truncation of a bulk lattice due to the otherwise unsaturated surface bonds. The first HAS measurements of Si [32] and GaAs [33] surfaces were performed by Cardillo and Becker at Bell Labs. The (111) surface of silicon is one of the most studied systems in surface science due to its technological applications. This surface undergoes different spontaneous reconstructions, i.e. the structure of the exposed surface is different from the ideal (1×1) truncation of the crystal bulk along that plane: after cleavage it presents a 2×1 reconstruction that is transformed in the (7×7) configuration after an annealing at elevated temperatures. The saturation with H of the Si dangling bonds allows the stabilization of the (1×1) surface (see Fig. 12.5). These 3 different (111) surfaces were investigated by means of elastic and IHAS [34–36]. Recently also the structure and the dynamics of methyl-terminated Si(111) has been studied [37]. The interaction potential between He and deuterated diamond(111) was investigated [38, 39].

Thanks to its large transfer width (that can reach 100 nm), HAS proved to be a valuable tool for probing also the surface phase transitions of Ge(111) [40],

Ge(001) [41], and the renown 1/3 ML phases of Sn/Ge(111) [42]. The study of surface phonons on semiconductors by HAS was challenging because of an overall larger stiffness (higher energy) with respect to metal surfaces, placing the optical branches at the sensitivity limit of the HAS technique. Nonetheless, the phonon dynamics of Ge(111) [43], of deuterated diamond(111) [44], as well as of the InSb(110) compound [45] were investigated, too.

Due to their more reactive nature, the study of metal surfaces was more challenging. For the close-packed (111) surfaces of f.c.c. metals, the electronic density shows a very small corrugation therefore the intensity on the diffracted peaks is very low and all the intensity is essentially concentrated in the specular peak, neglecting diffuse and inelastic scattering due to surface defects and surface vibrations, respectively. At the beginning of the 70', the first system investigated was the (112) surface of tungsten [46] which is strongly corrugated in one direction. With the improvement of the HAS technique, a few years later, Ag(111) was the first close-packed surface on which diffracted peaks were observed [47]. Quantitatively, on a well prepared Ag(111) a reflectivity of 90 % was measured [48].

By exposing a metal surface, the electronic density undergoes a large perturbation which may lead to a reconstruction of the metal surfaces. Ag(111) and Cu(111) are not reconstructed but, for instance, Au(111) shows a very complex reconstruction successfully investigated by HAS [49]. HAS investigations were performed on other reconstructed surfaces such as Au(110) [50] and Rh(110) [51]. Not only the structure but also the surface vibrations were investigated [51–58]. On Pt(111) [59] a Kohn anomaly, i.e. the phenomenon that for a specific wavevector the frequency of a phonon is considerably lowered due to the electron-phonon interaction, was observed in the surface phonon dispersion curves. The same anomaly was observed more recently in films of Pb on Cu(111) [60]. Moreover, also binary alloys were investigated, such as Cu_3Au [61, 62].

Other important classes of materials are oxides. Since oxide are usually insulators the HAS technique is well suited to study them and has been used to obtain structural and dynamic information for instance on the surfaces of zinc oxide [63], magnesium oxide [64–66], α-quartz [67], nickel oxide [68]. Other compounds such as mica [69] and $KTaO_3$ [70] were also investigated.

Physisorbed as well as chemisorbed layers were investigated intensively by means of HAS. Typical systems studied with HAS were noble gases Xe and Kr physisorbed on the basal plane of graphite [71, 72] to test interatomic potentials. The study were extended to metal substrate [73–76]. Oxygen [77–79] as well as alkali [80–82] chemisorption were investigated on different systems. It is remarkable that HAS is sensitive to hydrogen whose detection is very difficult with other techniques due to the low electron density. Therefore HAS was employed to study H chemisorption [83–87]. The other technique with sensitivity to H is described in Chap. 15 on low energy ion scattering but the technique causes some damage of the sample.

The interaction of water with surface is an important topic and HAS was employed to investigate water dissociation [88], bare ice surfaces [89, 90], water-ammonia ice [91], and ice formation on alcohol films [92]. Remarkably, studies on ice require an experimental setup which can work with the sample in a pressure range up to $\sim 10^{-2}$ mbar.

Pioneering HAS experiment on self-assembled monolayers (SAM) of alkanethiols on gold were performed by the Scoles group [93–96]. The surface vibrations of SAMs were also investigated [97–99]. Other HAS experiments have focused on adsorbate layers of organic molecules such as lipid monolayers [100], acetonitrile [101], pentacene [48, 102–104], Cu-phthalocyanine [105], α-sexithienyl [106, 107] to mention a few. HAS has also been used to investigate the surface structure of very fragile organic hydrogen-bonded crystal obtained by crystal engineering [108]. After cleavage, this organic crystal of guanidinium methanesulfonate with a well-defined layered architecture can expose two distinct surfaces which are different for the presence or absence of methyl groups and that were discriminated by HAS measurements [109, 110].

All the above examples involve mainly ordered periodic system but HAS has also been used to investigate systems with a reduced crystalline order or which are disordered [111], including systems that undergo surface phase transitions. Measurements were performed on quasi-crystals [112, 113] and in 2007 HAS experiments from an amorphous solid, the silica glass, were published with the first surface measurement of the so called boson peak, a well known low-frequency characteristic of amorphous materials [114]. Study of amorphous and semicrystalline polymer films have been recently performed [115–117].

One of the most studied surface transition is the roughening transition [118, 119]. At low temperatures any surface exists in a flat configuration since the creation of steps costs some energy. Increasing the temperature the entropic gain may determine the vanishing of the step free energy hence steps can proliferate on a surface. The height-height correlation function, which has a finite limit at low temperature, diverges logarithmically with the increase of the distance across the surface and the elastically diffracted intensities evolve from a δ-function Bragg peak to power-law line shape. A similar behavior might happen for the step edges on a stepped surface: at low temperature steps are straight because the creation of a kink costs energy but step edge can meander at high temperatures for the vanishing of the corresponding free energy. Lapujoulade and coworkers started the study on high-index crystal faces of copper [120–122], and other groups investigated the roughening transition on different metals and surfaces [123–127]. Some surfaces, such as Au(110) or Pt(110), present at low temperature a surface reconstruction (1×2), but increasing the temperature, this reconstruction can be lifted and the surface may roughen [128–130]. The gas-solid transition for Xe deposited on Pt(111) was studied by the Comsa group in the low coverage regime down to 0.1 % [131].

The effect of single defects on elastic bound state resonances on NaCl [132, 133] and the cross section of isolated CO molecules adsorbed on Pt(111) [134] were investigated.

At the end of 80', with the use of low temperature sources, it became possible to measure the quasi-elastic broadening (tens of μeV) of the elastic component of the scattering caused by the diffusion of species onto surfaces. This quasi-elastic helium scattering (QHAS) was applied for studying surface melting and surface diffusion [135–137].

Elastic HAS measurements were employed to study the decay of nanostructures generated by ion sputtering. As shown by Bradley and Harper [138], a ripple phase can be generated by using grazing ion incidence. By means of a low energy He beam of ~3 meV it was possible to measure the thermal decay of ripple phases generated on Ag(110) [139] and Au(110) [140] with an average wavelength of 105 and 225 Å, respectively.

HAS is a particular excellent monitor of the large scale quality and reproducibility of thin films and monolayers also for surface processing in real time, such as sputtering or annealing, since the probe is not affected by the presence of electric or magnetic fields.

This has been exploited also for obtaining information about surface step height and growth modes. Information about the growth mode can be obtained by measuring changes in the specular reflectivity during deposition. Layer by layer growth will lead to a characteristic oscillations. This method is similar to reflection high energy electron diffraction (RHEED), but because the helium beam is inherently strictly surface sensitive, it is not necessary to go to a grazing incidence angle, and it is possible to monitor for example the growth of fragile organic films on insulating surfaces, something which can be difficult with RHEED. This RHEED-like oscillatory behavior is further enhanced when the He beam energy and scattering geometry are tuned in order to achieve out-of-phase scattering conditions for terraces separated by a monoatomic step. In the ideal case of perfect layer-by-layer growth, the specularly reflected beam would vanish at half integer coverage, at the same time, satellite peaks emerge aside the specular peak, the corresponding reciprocal momentum transfer is inversely proportional to the mean separation between equivalent steps (i.e. to the mean island separation). The intensity is then recovered when a new layer is completed. Very often, an initially 2D layer-by-layer growth regime evolves into a 3D growth because of anticipated island nucleation on top of a growing layer, before its full completion. The process gradually increases the number of layers exposed on the surface and the scattering from terraces no longer met a pure constructive/destructive interference from a perfect bilayer. As a consequence the oscillation amplitude is gradually dampened.

The information about surface step heights can in principle be readily extracted from the experimental oscillatory data, monitoring the specular signal as a function of beam energy and/or incident angle (i.e. parameterized by the momentum exchange perpendicular to the surface) and applying a Fourier-transform-based algorithm [141–143], the peaks in the Fourier transform provide an estimation of the layer spacings. On the other hand, Bedrossian et al. [144] observed that the intensity oscillations for Ag islands grown on Ag(100) depend on the size of the growing islands: the curve are shifted for small islands (average island separation ~25 Å) and the estimated step heights are different from the findings obtained on the same system by electron diffraction while the two values are equal for larger islands (~65 Å). This was interpreted as due to the smoothening of the electronic density profile for small and/or closely spaced islands probed by the He atoms. For Fe grown on Ag(100) Terreni et al. [145] observed the same dependence on the island size and performing an accurate estimation of the He interaction potential, also for the mentioned Ag on Ag(100) case, they were able to explain the shift as due to a refractive effects since the van der Waals interaction produces a local well on islands which depends on the size of the islands and which is different from the well estimated for the uncovered regions of the substrate. Although the interpretation of the shift is not unique, this shows that the analysis of the intensity oscillations requires some care to extract reliable values.

HAS proved to be well suited to study also the case of surface erosion by ion bombardment on both metals [111] and semiconductors, where Ar^+ ion sputtering at few hundreds eV was demonstrated to strictly follow an inverse-growth mechanism driven by the nucleation and growth of thermally diffusing vacancy islands [146–148].

HAS was also employed to study the spatial confinement of Fermi electrons in the case of thin metal films. In this case, the film thickness confines the electrons between the substrate and vacuum boundaries possibly leading to the manifestation of quantum size effects [149], that are associated to exotic phenomena such as "electronic growth" at very low temperature, critical thickness for layer-by-layer growth, bilayer growth [150] etc. In particular, HAS has been employed to study the layer-by-layer oscillation of the step height in the case of low temperature layer-by-layer growth of Pb on Ge(001) [149, 151].

A neutral probe is particularly useful for studying the growth of organic molecules, like aminoacids [152, 153], that can be easily damaged by electron beams. In some cases, the non-penetrating nature of the He probe may help to discriminate between the ordering of the organic overlayer and that of the inorganic substrate, which contribution to the scattering can dominate the diffraction pattern for both X-ray and electrons, specially for electron-rich transition metals [105].

At present, the complementarity with X-ray techniques has been exploited in a single synchrotron beamline [105, 154] where experiment can be performed in-situ on the same system.

Finally, there are other applications of helium atom beams that are related to surfaces but that will not discussed further in the rest of the chapter. Seeded helium beams are used for example in supersonic deposition of organic molecules on surfaces [102, 155, 156]. Other examples are the pioneering work on quantum reflection of helium from surfaces [157, 158], the very interesting work on scattering of keV helium beams from surfaces [159, 160] (the corresponding results will be briefly discussed in Chap. 15). The same applies to the work on metastable helium surface scattering, which can be used, among others, to investigate magnetic effects [161]. For a review of metastable helium surface scattering see [162].

For the interested readers, further information can be found in the following reviews on structural and dynamic investigations performed with HAS: [163–165]. A more recent review focuses on dynamic investigations [166], see also [167].

This chapter aims to introduce HAS to a new research community. Section 12.2 provides a description of the basic principles of HAS. Section 12.3 presents the typical experimental setup. Section 12.4 an application to a quartz sample.

The paper finishes with an outlook on a new, exciting application of neutral helium beams in Neutral Helium Microscopy (NEMI), Sect. 12.5.

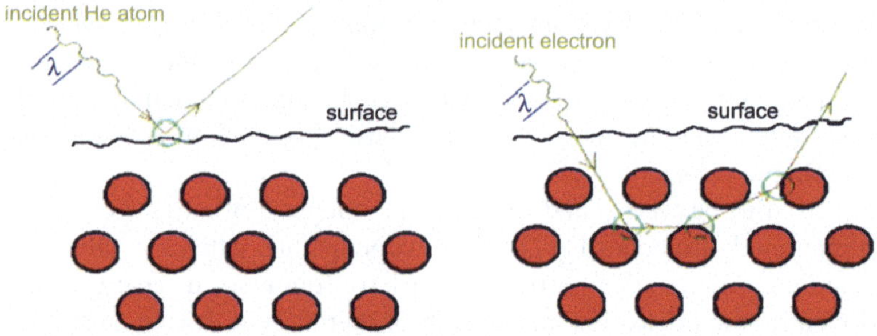

Fig. 12.1 "Helium Atom Scattering is the ideal tool to probe the topmost surface layer: While electrons can penetrate 3–5 layers below the surface, helium atoms are scattered at a surface of constant electron density, whereby the classical turning points are about 3–4 Å away from the surface atom cores". Figure and citation from [168]

12.2 The Basic Principles of HAS

Helium is a neutral atomic probe, hence without rotational or vibrational degrees of freedom, only the translational kinetic energy of the incident and scattered beam need to be analyzed in order to extract information about the surface. Moreover it has the advantage to be a noble gas, therefore it does not chemically interact with the surface and at thermal energies (1–100 meV) absolutely does not perturb the sample. For this range of energies, its de Broglie wavelength is of the order of 1 Å which matches the size of surface meshes therefore diffraction can be observed. The Pauli repulsion due the overlapping of the He closed shell and the surface electronic density is strong enough that the classical turning points are a few Å above the outermost layer of the surface, in the tail that spills off from the surface. While other beam techniques, such as electrons and X-rays, always have a certain penetration depth into the material, helium does not. This means that HAS is a unique beam probe in surface science. This is illustrated in Fig. 12.1, where the interaction of a thermal helium beam atom with a surface is compared schematically with that of an electron.

When a helium atom is far away from a surface, there will be a long range attraction between the atom and the surface due to the van der Waals interaction. For free atoms this has the r^{-6} dependence with the distance r, but in the case of the atom-surface interaction, the summation on the atoms of the crystal modify the asymptotic behavior as z^{-3}, with z the distance from the surface. At short range there is the mentioned Pauli repulsion with a quasi-exponential dependence on the distance. Figure 12.2 shows a schematic diagram of a helium surface interaction potential with the characteristic well shape caused by the long range attraction in combination with the short range repulsion. The well depth, D, is typically of the order of up to a few meV. This is just a section along z at a fixed point $\mathbf{R} = (x, y)$ on the surface, therefore this potential is parameterized by the position $\mathbf{R}$. Hence for a perfect surface, the full potential $V(\mathbf{R}, z)$ may change periodically the distance from the outermost layer, the steepness in the repulsive part, D, and the size of the attractive well. Generally the repulsive part has a stronger dependence on $\mathbf{R}$ while the

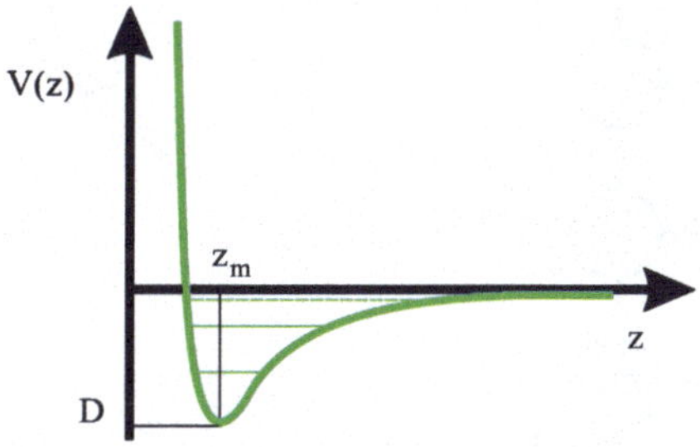

Fig. 12.2 Schematic diagram of the interaction potential between the neutral helium atom and the surface. The long range attractive van der Waal interaction and the short range, strong repulsive force give rise to a well shaped potential, with well depth D at the distance Z_m. Bound states of the well are indicated as lines. Figure after [165, 170]

attractive part has a weaker one, i.e. the repulsive part is more corrugated. Making a section at the He incident energy ($E_i = V(\mathbf{R}, z)$), this defines a profile $\xi(\mathbf{R})$ which is the corrugation function at that energy, the amplitude of its oscillation usually increases with the incident energy [169] (see Fig. 12.3).

The different scattering mechanisms of a helium atom off a surface are illustrated in Fig. 12.3. Upon interaction with a surface the helium atom will either be scattered elastically (no energy transfer) or inelastically through the excitation or deexitation of one or more surface vibrations (phonon creation or annihilation). These are essentially the only two possibilities since in HAS the atoms will never chemisorb onto the surface and D is too small so requires a low crystal temperature for the occurrence of physisorption. There are two exceptions. The electron-hole excitation is generally negligible [171, 172], but for H-covered Mo(110) and W(110) surfaces this excitation due to He scattering was observed [173]. The second possibility is the related to the selective adsorption resonance [163]. In this case helium atoms may be scattered in a metastable state in which the atom can move across the surface but the energy associated to the z motion corresponds to one of the (negative) energy levels in the potential well (see Fig. 12.2) with the total energy positive. The helium atom becomes momentarily trapped in the potential. This phenomenon, that was already observed in the pioneering experiments by Stern and coworkers and correctly interpreted by Lennard-Jones and Devonshire [174], can be exploited to study the helium-surface interaction potential.

12.2.1 Kinematics of the Scattering

During the scattering, an He atom is impinging on a crystal surface which lies in the $z < 0$ region (z is the axis normal to the mean plane of the surface) with a kinetic energy E_i and in the direction defined by the polar and azimuthal angles θ_i and ϕ_i. After the scattering, the atom leaves the surface with an energy E_f and in the direction defined by the polar and azimuthal angles θ_f and ϕ_f. The

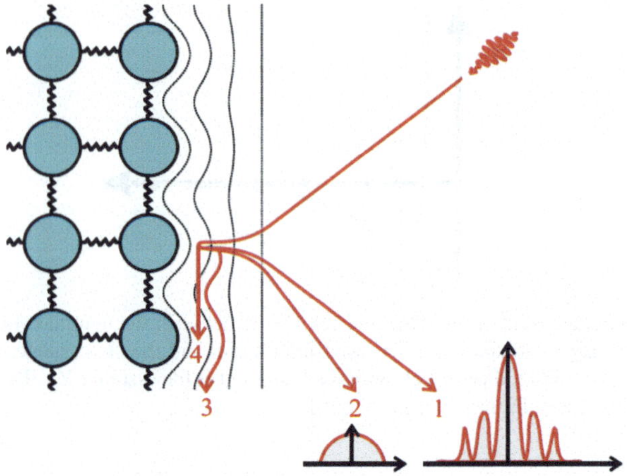

Fig. 12.3 This diagram illustrates the different scattering mechanism of the helium atom off the surface. (*1*) Elastic scattering, which for a surface with periodicity on the atomic scale will give rise to a diffraction pattern. This is normally the dominating part. (*2*) Inelastic scattering. There is an energy exchange between the helium atom and the surface giving rise to the creation or annihilation of phonons. (*3*), (*4*) The helium atom is trapped in bound states in the well. The *black lines* illustrate the decrease in potential corrugation with distance from the surface. Figure after [165, 170]

scattering geometry is drawn in Fig. 12.4. The incident He atom wavevector is denoted by $\mathbf{k}_i = (\mathbf{K}_i, k_{iz})$, where $\mathbf{K}_i$ is the wavevector component parallel to the surface and k_{iz} is the component perpendicular to the surface,[1] while for the scattered He atom the wavevector is $\mathbf{k}_f = (\mathbf{K}_f, k_{fz})$ and the wavevector exchange is $\Delta\mathbf{k} = (\Delta\mathbf{K}, \Delta k_z) = \mathbf{k}_f - \mathbf{k}_i = (\mathbf{K}_f - \mathbf{K}_i, k_{fz} - k_{iz})$. The incident and scattered He atoms are free particles hence the incident and scattered energies are $E_i = \hbar^2 k_i^2/2m$ and $E_f = \hbar^2 k_f^2/2m$ where m is the He atom mass. For multiphonon exchange, the momentum conservation law

$$\mathbf{K}_f = \mathbf{K}_i + \sum_s \mathbf{Q}_s + \mathbf{G}_{n,m} \tag{12.1}$$

and the energy conservation

$$E_f = E_i + \sum_{k,s} \big(\pm\hbar\omega_k(\mathbf{Q}_s)\big) \tag{12.2}$$

must be fulfilled, where $\mathbf{G}_{m,n} = m\mathbf{G}_{1,0} + n\mathbf{G}_{0,1}$ is a reciprocal vector of the surface mesh with $\mathbf{G}_{1,0}$ and $\mathbf{G}_{1,0}$ the primitive vectors of the 2D unit cell in reciprocal space, and $\omega_k(\mathbf{Q})$ is the frequency of a phonon of the k-th branch with momentum $\mathbf{Q}$ and the sum is over all the s exchanged phonons. It is worth noting that He may couple with both true surface phonons and phonons of the surface projected

[1] A symbol in boldface indicates a vector, whose surface component is written with the capital symbol.

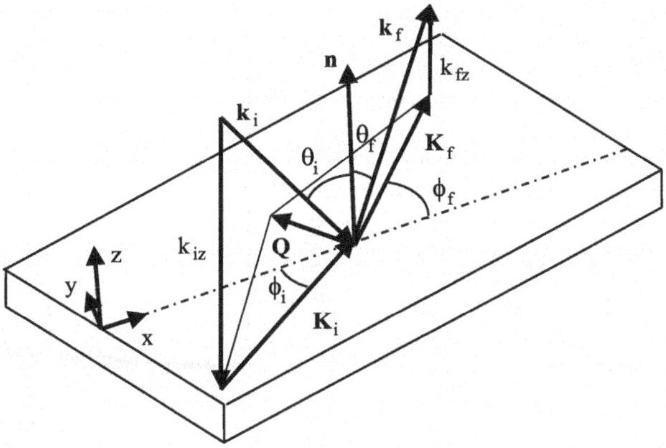

Fig. 12.4 He surface scattering geometry. **Q** is the wavevector of a surface phonon and **n** is the normal to the surface. Adapted from Ref. [175]

bulk bands. The sign + refers to the annihilation events, while the sign – refers to creation events. Typically, due to the low He mass, the low energy involved and for sample temperatures not too high, elastic scattering and, with less probability, single phonon-exchange are the predominant processes in HAS.

Elastic Scattering and Bound State Resonances

For elastic scattering $E_i = E_f$ (i.e. $k_i = k_f$), and for the in-plane geometry which is the most used geometry for the experimental setups, the momentum equation reads

$$k_f \sin(\theta_f) = k_i \sin(\theta_f) = k_i \sin(\theta_i) + G_{m,n} \tag{12.3}$$

where $G_{m,n}$ is the reciprocal vector in the scattering plane. An example of diffraction pattern is shown in Fig. 12.5. Through (12.3), the G vectors can be determined and the reciprocal surface mesh determined.

The final z component can be calculated as

$$k_{fz}^2 = k_{\mathbf{G}z}^2 = k_i^2 - (\mathbf{K}_i + \mathbf{G}_{n,m})^2 \tag{12.4}$$

The magnitude of G vectors can increase without limit therefore they are partitioned in two sets: (i) the set of open channels around the $\mathbf{G}_{0,0}$ where $k_{fz}^2 \geq 0$ is verified and these Gs are within the Ewald sphere and correspond to scattering channels associated to outgoing waves $A_f \exp[i(\mathbf{K}_f \cdot \mathbf{R} + k_{fz} z)]$, (ii) the set of closed channels for the remaining vectors which verify $k_{fz}^2 < 0$, there are no traveling waves but evanescent ones. On the other hand, if the energy associated to the z motion fulfills

$$\frac{\hbar^2 k_{\mathbf{G}z}^2}{2m} = \epsilon_j \tag{12.5}$$

where $\epsilon_j < 0$ is one of the levels of the He-surface interaction potential, the atom moves across the surface but its motion in the perpendicular direction is bound to the surface. Its total energy is positive therefore with another diffraction event the atom may be diffracted in one of the open channels leaving the surface. When an

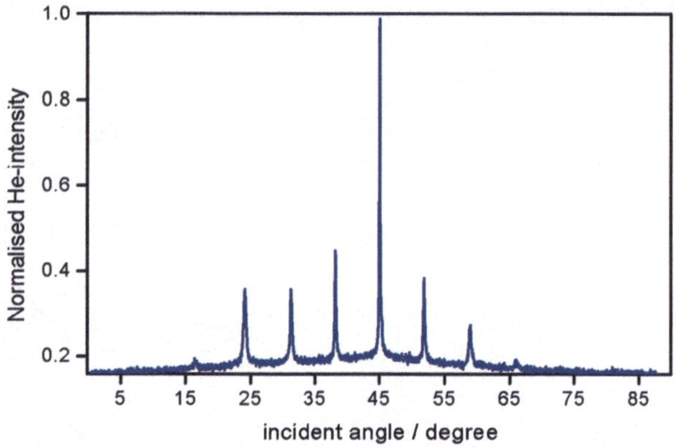

Fig. 12.5 Helium diffraction pattern from a Si(111)–H(1 × 1) surface, obtained with the instrument MAGIE, see Fig. 12.7. The pattern is obtained by keeping the source-detector geometry fixed in a 90° geometry and rotating the crystal. The incident angle signifies the angle between the incident helium beam and the surface normal

atom enters in a resonant state, the scattering probabilities change in a narrow range of the kinematical parameters. For instance measuring the specular intensity as a function of θ_i or ϕ_i, maxima or minima (or even Fano-type structures [176]) on an average monotonic trend of the intensity can be related to resonant phenomena. The position of these resonant features allow the estimation of the levels ϵ_j. It is worth noting, that the levels are actually bands because the system has a 2D periodicity. Neglecting band effects and considering the potential averaged on the unit cell (see below), phenomenological potentials whose energy spectrum is known are useful for the analysis. For instance, a very flexible potential is the potential analyzed by Mattera et al. [177]

$$V(z) = D\big[(1+\lambda z/p)^{-2p} - 2(1+\lambda z/p)^{-p}\big] \tag{12.6}$$

which depends on three parameters: the well depth D, the range λ, and the exponent p which gives the symmetry and asymptotic behavior of the potential. The spectrum of this potential can be analytically calculated with great accuracy therefore the measured levels can be used in a fitting procedure to estimate the three parameters. The value of p has to be checked to give the right asymptotic behavior ($p \sim 3$ [178]).

Any open channel is allowed kinematically but the associated intensity, proportional to the scattering probability $P_{\mathbf{G}} = (k_{\mathbf{G}z}/k_{iz})|A_{\mathbf{G}}|^2$, is determined by the interaction potential. If the potential $V(\mathbf{R}, z)$ is known, the solution of the scattering problem corresponds to the solution of the Schrödinger equation with the suitable boundary conditions: an ingoing incident wave $\exp(i(\mathbf{K}_i, -k_{iz}) \cdot (\mathbf{R}, z))$ and open as well as closed channels. Exploiting the 2D periodicity of the system and applying the Bloch theorem [179], the potential can be written in terms of its Fourier coefficient V_G which depend on z, $V(\mathbf{R}, z) = \sum_G V_G(z) \exp[i\mathbf{G}_{m,n} \cdot \mathbf{R}]$, where the term associated to $G_{0,0} = (0, 0)$ is the interaction potential averaged on the unit cell. The equation is therefore split into N($\rightarrow \infty$) equations, one for each channel. Limiting N to a finite number, the open channels and some of the closed channels, it is possible to find an approximate solution. This close coupling method can provide an accurate solution to the scattering problem but require a lot of computing power and is time consuming. In the case the potential is not exactly known and/or depends on different parameters, a fitting procedure is not feasible. Alternatively, approach based on time-dependent wavepacket calculations are reported [180].

If the repulsive part have a sufficient steepness (the potential is not soft), a useful approximation is to consider the surface as an hard corrugated wall (HCW) whose profile is given by the $\xi(\mathbf{R})$ function. With this HCW approximation, the function can be estimated from the knowledge of the size of the atoms (for instance the van der Waals radius) that are forming the surface and He. This can produce a first approximation to $\xi(\mathbf{R})$. The 2D Fourier transform of $\xi(\mathbf{R})$, truncated to a suitable number of terms, provides useful coefficients that can be used as free parameters in a fitting procedure. For the HCW methods were developed for solving the scattering problem [181–183]. The real potential presents a well which is neglected in the HCW approximation. A useful approximation is to consider the incident energy increased by the well depth D as for a square well: in this case the parallel **K** components are conserved crossing the well and the z components are increased (Beeby correction [184]).

The HCW potential is a good approximation for alkali halides or graphite but not for metal surfaces. Esbjerg and Nørskov [185, 186] showed that, in a first approximation, the He repulsive part is proportional to the electron density of the metal providing a way to calculate the potential in the case the electronic density is available. Tommasini and coworkers [187, 188] showed that in the region of interest the interaction potential can be calculated as the sum of pseudo-pairwise terms. Starting from a potential divided in repulsive and attractive parts [189], the pairwise potentials based on the atomic densities are phenomenologically modified to take into account the anisotropic rearrangement of the charge density of the surface atoms and the polarizability of the both the target atoms and the He probe.

The He-surface potential can be calculated by first principle models, such as the Density Functional Theory (DFT), but the standard exchange-correlation functionals are not able to describe the van der Waals part of the interaction. On the other hand, in the repulsive part they are accurate enough in describing the gas-metal potential. The unperturbed electron density is always expected to present maxima at top positions (above the ionic cores) and minima in between atoms (bridge) positions therefore the potential in the Esbjerg and Nørskov approximation will follow the electronic corrugation. Instead, depending on the metal surface, the classical turning points of $V(\mathbf{R}, z)$ calculated by DFT may be closer at the top than at bridge positions of the first surface atomic layer [190, 191] and this is sometimes termed as anticorrugation.

Inelastic Scattering

In a single phonon exchange scattering the momentum conservation equation reads

$$\mathbf{K}_f = \mathbf{K}_i + \mathbf{Q}_j + \mathbf{G}_{m,n} = \mathbf{K}_i + \Delta\mathbf{K} \tag{12.7}$$

and the energy conservation can be written

$$E_f = E_i \pm \hbar\omega_k(\mathbf{Q}_j) = E_i + \Delta E \tag{12.8}$$

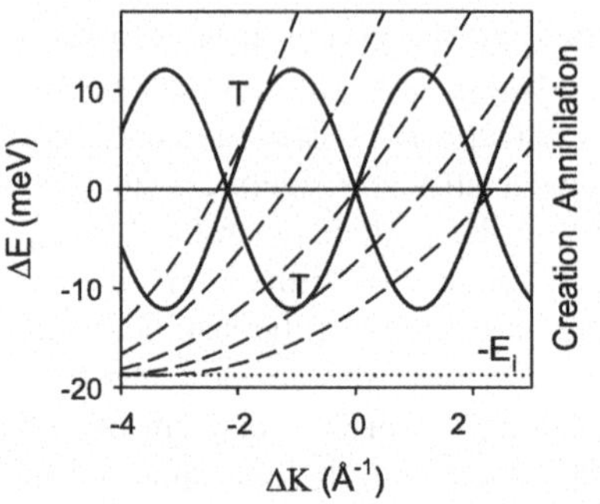

Fig. 12.6 Representative set of scan curves (*dashed parabolic curves*) with $\theta_d = 110°$ and for different incident angles (from left: $\theta_i = 75°, 65°, 55°, 45°, 35°$) and $k_i = 6\,\text{Å}^{-1}$, a typical value for a beam generated by a source cooled with liquid nitrogen. The *solid curves* represent a phonon dispersion curve $\pm\hbar\omega$ in a extended zone representation. Phonon creation ($\Delta E < 0$) can occur only for $\Delta E > -E_i$ (*dotted line*), the maximum energy available from the incident beam. Two scan curves are tangent to the phonon dispersion curve (T points)

by using both equations and solving for the phonon energy

$$\pm\hbar\omega_k(\mathbf{Q}_j) = \Delta E = \frac{\hbar^2}{2m}\left(\frac{(K_i + \Delta K)^2}{\sin^2(\theta_f)} - k_i^2\right) \tag{12.9}$$

where the function $\Delta E(\Delta K)$ gives the kinematically allowed points (scan curve) in the $\Delta K, \Delta E$ plane. The intersections of the scan curve with the phonon dispersion curves provide the allowed phonon excitations as shown in Fig. 12.6. If the scattered intensity is higher than the background, for any intersection a peak may be measured. Due to the complexity of the time-of-flight detector which requires many differential pumping stages to increase the signal to noise ratio (S/N), IHAS is performed generally with experimental setups having a constant deflection angle θ_d between the incident beam and the detector axis, i.e. $\theta_d = \theta_i + \theta_f =$ constant, therefore changing the incident angle with the sample polar rotation also the scattering angle is varied accordingly and the scan curve explores the $\Delta K, \Delta E$ plane. The phonon dispersion curves in the extended zone can be reconstructed by the measured $\Delta K, \Delta E$ positions of the inelastic structures. The scan curve can be also tangent to a phonon dispersion curve, in this case a very intense structure is measured due to the kinematical focusing [192].

If the phonon excitation is mediated by a selective adsorption in a level of the interaction potential [193], (12.4) must be modified including the phonon momentum and energy as

$$\frac{\hbar^2}{2m}k_{\mathbf{G}z}^2 = \frac{\hbar^2}{2m}k_i^2 \pm \hbar\omega_k(\mathbf{Q}) - \frac{\hbar^2}{2m}(\mathbf{K}_i + \mathbf{G}_{n,m} + \mathbf{Q})^2 = \epsilon_j \tag{12.10}$$

Resonant effects can enhance the detection of inelastic structures [194, 195]. Moreover inelastic effects related to atom-phonon bound states were investigated by Benedek et al. [196].

The calculation of inelastic intensities requires the knowledge of the surface phonon spectrum and the interaction potential between He and the surface potential which acquires a temporal dependence due to the vibrations of surface atoms.

The problem of the inelastic scattering of non-penetrating particles in solids was treated by Cabrera, Celli, and Manson [197] in a pioneering article in 1969. The total potential U was divided in a large part V and a remainder $U - V$ which contains the time dependent part of U and is considered as a perturbation. This approximation is similar to the Born approximation with the difference that the unperturbed wavefunctions are the solution of the scattering problem with the potential V, which describe the elastic interaction, therefore the asymptotic plane waves are "distorted" by V at close distance from the surface. This distorted wave Born approximation (DWBA) was used to calculated inelastic reflection coefficients to be compared with the experimental results. As mentioned for the elastic case, the interaction potential is approximately proportional to the electronic density and for the (111) surface of noble metals, Bortolani and coworkers considered the superposition of atomic densities [198, 199] which decay exponentially with the distance from the surface with a softness parameter β and expressing the differential reflection coefficient, which is the number of scattered particles in the final solid angle $d\Omega_f$ around the scattering direction defined by $\mathbf{k}_f$ and within and interval dE_f around the final energy, as

$$\begin{aligned}\frac{d^2R}{dE_f\,d\Omega_f} &= \frac{2}{MN}\sum_{\mathbf{Q},j}\frac{n(\omega_j(\mathbf{Q}))\sqrt{2mE_f}}{\omega_j(\mathbf{Q})|k_{iz}||k_{fz}|^2}\left|\mathbf{Q}\cdot\mathbf{e}(\mathbf{Q},j) - i\beta e_z(\mathbf{Q},j)\right|^2\\ &\times \exp\left(-Q^2/Q_c^2\right)\left|\langle\chi_f|\frac{\partial V}{\partial z}|\chi_i\rangle\right|^2 \exp(-2W)\\ &\times \delta(\mathbf{K}_f - \mathbf{K}_i - \mathbf{Q})\delta\left(E_f - E_i - \hbar\omega_j(\mathbf{Q})\right)\end{aligned} \tag{12.11}$$

where N is the number of atoms per unit surface area, M the surface atomic mass, $n(\omega_j(\mathbf{Q}))$ the Bose occupation number, $\mathbf{e}(\mathbf{Q}, j)$ is the polarization vector of the phonon, and $\exp(-2W)$ is the Debye-Waller factor (see below). Since for in-plane scattering geometry, $\mathbf{Q}$ is along the intersection of the scattering plane with the surface plane, the term $|\mathbf{Q}\cdot\mathbf{e}(\mathbf{Q}, j) - i\beta e_z(\mathbf{Q}, j)|$ provides a selection rule because is zero for modes perpendicular to the scattering plane, i.e. shear horizontal modes. Moreover, the coefficient shows an exponential decrease with $\mathbf{Q}$ expressed by $\exp(-Q^2/Q_c^2)$, where Q_c is the cutoff parameter, related to the ratio between the softness of the potential and the distance of turning points, which limit the detection of inelastic features for $Q > Q_c$. For another derivation of the DWBA see also [200].

The surface spectrum $\omega_j(\mathbf{Q})$ can be calculated by using a parametrization with force constants connecting crystal atoms, i.e. the Born-von Karman model, in order to construct the dynamical matrix for a slab composed of several layers parallel to the surface that is subsequently diagonalized [201–203] or considering the surface as a perturbation of a bulk crystal and solving the problem with a Green function method [204, 205].

Three body interactions have to be included to describe the inhomogeneities of the electron gas and to account for the large violation of the Cauchy relations between the macroscopic elastic constants C_{12} and C_{44} for instance present in most metals [206].

The interaction between atoms at the surface may be different from the same atoms in the bulk, hence the force constants connecting atoms in the surface region are generally varied with respect to bulk values to improve the fit with the experiments, i.e. to fit both the frequency and the corresponding intensity in the measured spectra.

In the above scheme, the scattering calculations and lattice dynamics calculations appear as two separated steps. Instead electrons have an active role in both the steps. In fact He is scattered 3–4 Å above the outermost nuclear plane with the important consequences that a vibration of ionic cores must induce a vibration of the electronic density in the region of the He turning points otherwise the vibration is not coupled with He and IHAS cannot detect it. Moreover, the electron density at the turning points is perturbed by the incoming He atom.

Phenomenological models were introduced to consider the dynamical coupling between electron and lattice vibrations. For instance, the shell model [207] that was used to study ionic insulators such as alkali halides, describes the ions as being constituted of an outer spherical shell of m electrons and a core consisting of the nucleus and the remaining electrons. Applying an electric field, the shell moves with respect to the core with an harmonic restoring force which provides a finite polarizability. The shell can be rigid or deformable and this model was applied also to study surface vibrations [208]. In a different model, the bond charge model or multipole expansion model, the electron density is expanded in multipolar terms starting the expansion in special positions between the ionic cores, the multipolar expansion coefficients act in the lattice dynamics as electronic degrees of freedom [209]. This model has been used to interpret for instance the IHAS results obtained for Cu(111). In fact a common feature of many metal surfaces, observed also for Cu(111), is the presence of two low lying acoustic modes: the Rayleigh mode which is a surface mode mainly polarized in the vertical (z) direction and a longitudinal resonant mode largely polarized in the plane parallel to the surface. Vertical surface modes are expected to be associated to intense features in IHAS spectra but generally the longitudinal resonance is detected with an intensity higher than the Rayleigh mode. In the Born-von Karman model, the HAS reflection coefficient calculated with the DWBA could fit the experiment only assuming a large and unphysical reduction of the surface force constants (50–70 %) while the electron energy loss spectroscopy (EELS, for a description of the technique see Chap. 17) results shows that the reduction is only 15 % [210] (and for other metals and surfaces it is limited to no more than 30 %). With the multipolar model, the softening of the force constants for HAS is similar to EELS analysis [211].

A different analysis of the Cu(111) case was performed by Santoro et al. [212] including phenomenologically the anticorrugation in the He-surface interaction potential and with the lattice dynamics described by a Born-von Karman model. A good agreement was obtained with a 20 % variation of the force constants.

All these results pointed out the need for first principles calculations to get a unifying picture of inelastic He scattering process taking into account the dynamic role of the electrons in the solid. In particular the electron-phonon coupling is a key parameter. In fact very recent results on bare Cu(111) [213] have shown that the longitudinal resonance is related to phonon modes localized in the second atomic layer while the study of Pb films deposited on Cu(111) [214] have shown that He can couple to phonon modes localized several layers beneath the surface. These findings provides new perspective to IHAS that it was considered a strictly surface sensitive technique due to its non-penetrating behavior.

Further information on the phonon calculations from first principles can be found on the following references on the density-functional perturbation theory [215, 216]. Recent quantum calculations of inelastic scattering of atoms from corrugated surfaces can be found in [217].

Similarly to other diffraction techniques, elastic and single phonon intensities must be multiplied by a Debye-Waller (DW) factor $\exp[-2W]$ [163], where $2W = \langle(\Delta\mathbf{k}\cdot\mathbf{u})^2\rangle$ is the thermal average of the product between the mean square amplitude $\langle\mathbf{u}^2\rangle$ of the atomic vibration and the momentum exchange $\Delta\mathbf{k}$ which includes also the Beeby correction. The interaction of He with the surface atoms is not weak or localized (He interacts strongly with more than one atoms in the unit cell), is not fast (the interaction time is not negligible in comparison with the vibrational period of phonons) therefore the simple formula used for X-ray or neutrons is not completely valid. Different studies were carried out to assess the validity of the DW factor and find suitable corrections [218–220] and a theoretical analysis of the influence of the focusing of inelastically scattered beam on the Debye-Waller factor is reported in Ref. [221].

To conclude, it is worth noting that not all the peaks measured with IHAS correspond to phonon excitations. In fact, peaks related to the elastic diffraction of the tails of the incident beam can contribute with features to inelastic spectra with an intensity similar to true inelastic peaks. Hence all the supposed inelastic peaks must be checked for the presence of these peaks, the so called deceptons [14, 222, 223].

12.3 The Experimental Setup in HAS

The typical experimental setups for HAS is identical to the experimental setups used for H_2 diffraction described in the chapter "Diffraction of H_2 from metal surfaces" by D. Farìas et al. in this book. Often the same apparatus is used for both types of experiments. Just as for H_2 experiments we distinguish between two different types of experimental setup: Fixed angle setup, where the angle between incident and outgoing reflected beam is fixed and the sample can be rotated to vary the incident angle relative to the sample and fixed incident beam setup, where the angle of the incident beam is kept fixed relative to the sample and the detector is moved. In most systems the detector can only be moved in one plane, but a few systems, including one described by Farìas et al. in this book can move out of plane. This gives for

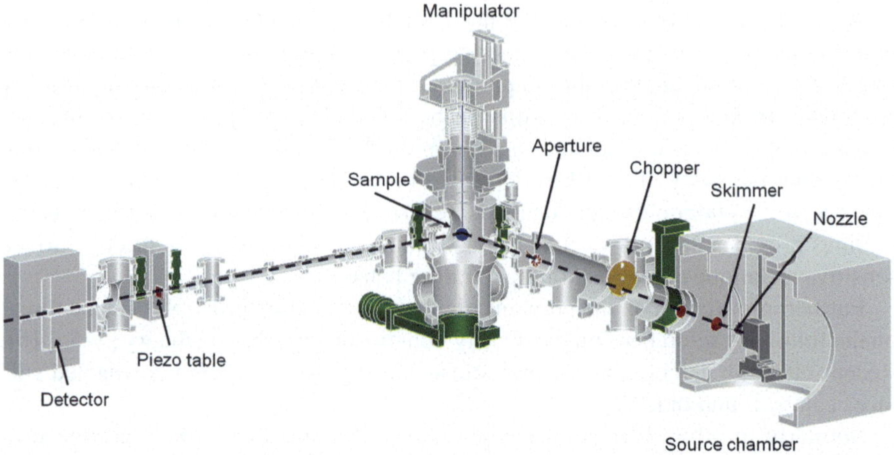

Fig. 12.7 The helium scattering instrument MAGIE at the University of Bergen. A unique feature of the instrument is that the detector arm can be rotated between 30° and 180° relative to the incident beam. The piezo table enables high resolution scans, which allow the diffraction patterns of large scale periodic structures with periodicity of up to around 1 μm to be resolved. The piezo table has also been used to perform the first neutral helium microscopy images, see Sect. 12.5. The length of each arm is about 1.5 m. A long flight path is necessary for a good energy resolution in TOF experiments. A unique feature of MAGIE is that the length of both arms can be adjusted to obtain different geometries for focusing experiments

example the possibility of obtaining a 2D diffraction pattern similar to a LEED pattern, with the difference that the whole pattern cannot be obtained at once and that the beam is not incident normal to the surface.

Figure 12.7 shows a diagram of the helium scattering instrument MAGIE at the University of Bergen. The helium beam is produced in a supersonic expansion [224]. In a supersonic expansion atoms from a high pressure helium reservoir (usually up to 200 bar) expands though a 5–10 μm nozzle (indicated in Fig. 12.7). The nozzle can be cooled or heated and the temperature T_0 of the nozzle determines the energy of the beam and thereby the velocity and hence the wavelength of the atoms. The dimensions are chosen so that the free mean path of the atoms is much smaller than the nozzle diameter. This leads to a supersonic beam rather than an effusive beam (which is what one would get if the free mean path was greater than the nozzle diameter). The atoms in the nozzle undergo the effect of the collisions and the random thermal energy in the source is converted into the beam translational kinetic energy with a decreasing temperature of the gas. The gas expand in vacuum and the diverging streamlines of the flow become straight at a distance of a few nozzle diameters. These streamlines seem to diverge from a single point, the virtual source point which is located in front of the exit of the nozzle. With the progress of the expansion, the collision frequency becomes so small that a local equilibrium cannot be kept and the gas finally is in a free molecular regime. The velocity distribution cannot evolve anymore, it is "frozen", and the gas temperature has reached a value $T \ll T_0$ which in turn means that the width of the velocity distribution $\Delta v = \sqrt{2k_B T/m_{He}}$ in the beam is much smaller than in the source, the narrowing of the distribution is the great advantage of a supersonic helium beam. Neglecting T which is in the mK range, the average velocity v_{av} of a helium atom in a supersonic beam is given as [224]

$$v_{av} = \sqrt{\frac{5k_B T_0}{m_{He}}} \tag{12.12}$$

where k_B is the Boltzmann constant and m_{He} the mass of the helium atom. The number typical used to characterize the expansion is the so called speed ratio S defined as [225]

$$S = 2\sqrt{\ln 2}\left(\frac{v_{av}}{\Delta v_{fwhm}}\right) \tag{12.13}$$

Typical values for S are 50–200; the higher the pressure, the more narrow the velocity distribution.[2] See also Fig. 2.11 in [224] and Refs. [226, 227].

When the molecular flow regime is reached, the atom trajectories can be traced back to a plane perpendicular to the beam direction containing the virtual-source point thus generating a spatial distribution function, which was labeled the virtual source by Beijerinck and Verster [228]. The size of the virtual source is an important parameter for high brightness He sources. For recent studies on it see the following Refs. [229, 230]

The central part of the beam is selected by a conical collimator, the skimmer, typically 200–400 μm in diameter for scattering experiments (those are commercially available [231]) and usually with much smaller diameter for microscopy.

Further down the beam line a collimating aperture narrows the angular spread of the beam. This can be important for high resolution diffraction studies.

For dynamic measurements it is necessary to determine the energy of the helium atoms. This is done using Time of Flight Analysis (TOF). The beam is chopped into pulses of a few μs using a chopper and the flight time of these pulses (in the ms range) measured. In Fig. 12.7 a simple chopper with two slits, leading to two pulses per rotation is shown. For some experiments where a particular high intensity is needed, a pseudo-random chopper can be used [232]. For high precision measurements a movable detector is useful to reduce the indetermination on the flight length [233]. After scattering off the sample the beam is measured in the detector. At present there are no "2-Dimensional" helium detectors and the scattered beam is detected "point wise". Typically the detector chamber is in ultrahigh vacuum where the main contribution to background is hydrogen. A ionizer convert the neutral helium to ions. This process is inefficient because for 10^5 atoms only about one ion is produced. To increase the S/N a mass spectrometer is employed to separate helium from hydrogen ions. Magnetic sector as well as standard quadrupole mass spectrometers have been employed in HAS. The mass selected ions are collected by

[2] A value $S = 1000$ was measured for a pulsed He source working up to 136 bars at room temperature with a nozzle of 0.125 mm diameter [234].

an electron multiplier or a channeltron and the output pulse current is fed to the counting electronics. For TOF measurements a multiscaler accumulates the counts in time bins to produce the TOF spectrum. Several repetitions are summed up and stored until the inelastic features are extracted from the noise. Finally, the spectrum is saved in a computer for the analysis.

For interested readers on molecular beams, further information can be found in [224, 225]. A description on the construction of various parts of a HAS apparatus can be found in [235–238].

12.4 An Application Example: Structural Properties of the α-Quartz(0001) Surface

In the introduction to this chapter we presented an overview of systems that have been studied with HAS. As a more detailed example of an application of HAS we use this section to describe a HAS study of the structure of the α-quartz(0001) surface. The work was originally published in [67]. Though the work to fully understand this surface structure is still ongoing, HAS could already now provide new insight, which had not been obtained with other techniques or predicted by theory.

Quartz (SiO_2) is one of the most common minerals on earth and it is produced industrially on a large scale, but surprisingly little experimental works on the surface structure has been carried out up till now. This may be partly due to the fact that quartz is an insulating material and hence difficult to investigate with many surface science techniques. This illustrates the power of the HAS technique in investigating insulating surfaces, including mineral surfaces. It seems likely that this work will expand in the years to come. In particular if the Neutral Helium Microscopy technique described in the last section will become really successful, since this would open for diffraction-studies on microcrystallites with focused beams.

12.4.1 Surface Preparation

The quartz sample was a synthetically grown, twin free crystal, exposing the (0001) surface with a miscut less than 0.5°. The crystal was polished to optical quality using diamond powder, and then cleaned in a soap solution (Alcanox) and rinsed three times in batches of deionized water. The sample was finally annealed to 1025 °C for 72 h under a pressure of 2 bar of oxygen. In other studies the crystal was also etched in HF but in the present study no etching was carried out. The sample morphology was checked by atomic force microscopy (AFM). The AFM images showed large terraces with an average width of about 200 nm. A characteristic feature of this surface is that the terraces are oriented with angles of 60° relative to each other. Moreover the surface presents randomly distributed pits with depths between 2 and 5 nm. The origin of those pits is presumably related to relaxation processes during the annealing.

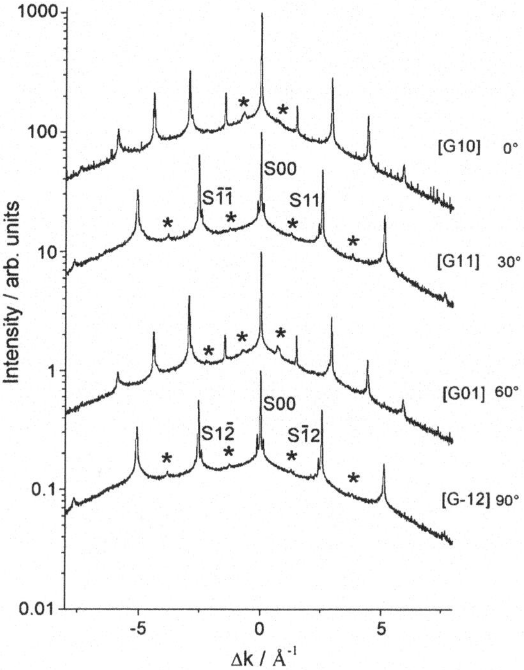

Fig. 12.8 In-plane HAS diffraction patterns from α-quartz(1000) for different azimuthal directions separated by 30°. The repeat distance can be readily determined as 0.490 ± 0.002 nm, similar to the bulk unit cell. The satellite peaks indicate an unpredicted long-scale surface reconstruction. From [67]

12.4.2 Surface Structure

As discussed in the previous sections, information about surface structure can be obtained with HAS by measuring surface diffraction patterns, similar to LEED or grazing incidence X-rays experiments, but with the advantages of strict surface sensitivity, low energy and no surface charging or damage. A drawback of HAS in comparison to the two other methods is that at present no 2-Dimensional detector for helium exists and so the measurements have to be performed point by point, typically by rotating the crystal (see Fig. 12.7). This is relatively time consuming and means that one must be very careful to avoid surface contamination.

Figure 12.8 shows a diffraction measurements from an α-quartz (1000) surface. The measurements were performed by rotating the polar angle of the crystal in a range between 10° and 80° with a step of 0.05° and with a fixed $\theta_d = 90°$. The source temperature was kept stable slightly above room temperature therefore the beam energy was 67.3 meV with an energy spread $\Delta E/E \sim 2\ \%$. The angular scale has been converted in momentum exchange by using (12.3). Each pattern has been measured along a fixed azimuthal direction ϕ and the four pattern are separated by steps $\Delta\phi = 30°$. The peak at $\Delta K = 0$ is the specular peak. Its width is 0.006 Å^{-1} which is limited by the present angular resolution of the apparatus whose transfer width is at least 200 Å. This means that the terraces do not contribute to the specular width hence their linear size is greater than the transfer width. This is consistent with the AFM measurements. It is evident that patterns measured every 60° are

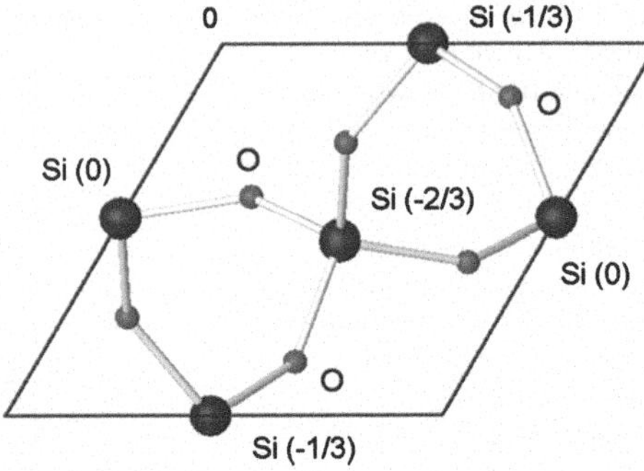

Fig. 12.9 α-quartz(1000) bulk unit cell projected unto the (0001) plane. The *large balls* represent Si-atoms and *small balls* represent oxygen atoms. The *numbers in brackets* indicate the relative positions of the Si-atoms in the bulk unit cell. From [67]

equal within the experimental incertitude. Each pattern shows a series of intense diffraction peaks which are separated by a constant momentum exchange

$$\Delta \mathbf{K} = \mathbf{K}_f - \mathbf{K}_i = \mathbf{G} \tag{12.14}$$

and the reciprocal lattice vector corresponds to a hexagonal unit cell in real space with a lattice parameter $a = 4.90 \pm 0.02$ Å, which in turn corresponds to the bulk unit cell projected on the (0001) plane as shown in Fig. 12.9.

Very weak peaks, labeled by a star, can be seen exactly half way between the main peaks in both main directions. Their presence indicates the doubling of the periodicity, i.e. a surface reconstruction. This reconstruction can either stem from a (2×2) reconstruction of a single surface, or from three domains of (2×1) reconstructions rotated 60°. The AFM image shows terraces at 60° angles, which supports the reconstruction hypothesis. Theoretical work [239, 240], proposed a (1×1) reconstruction for the surface (the so called dense structure) and suggested also a (2×1) reconstruction (the so called semi-dense) structure, which is predicted to have a slightly higher energy than the dense (1×1) phase. It appears that our diffraction pattern stems from a co-existence of the two structures. A more detailed analysis will hopeful enable us to determine the area ratio between the two structures on the surface.

The small satellite peaks, very close to the main diffraction peaks ($\Delta K = 0.13$ Å^{-1}), are very probably evidence of a very large scale reconstruction unpredicted by theory. Their origin is as yet unresolved.

12.5 Outlook: Neutral Helium Microscopy (NEMI)

The idea of exploring the advantages of helium atom scattering (neutral, low energy and strictly surface sensitive) in a neutral helium microscope (NEMI) has been around as long as the HAS technique itself. One can easily envisage an instrument, similar to a scanning electron microscope, where a focused beam of neutral helium atoms is scanned across a surface and an image is obtained by collecting the backscattered helium atoms at a particular angle. Note the difference here to a Scanning Electron Microscope (SEM) or a Scanning Helium Ion microscope (SHIM);

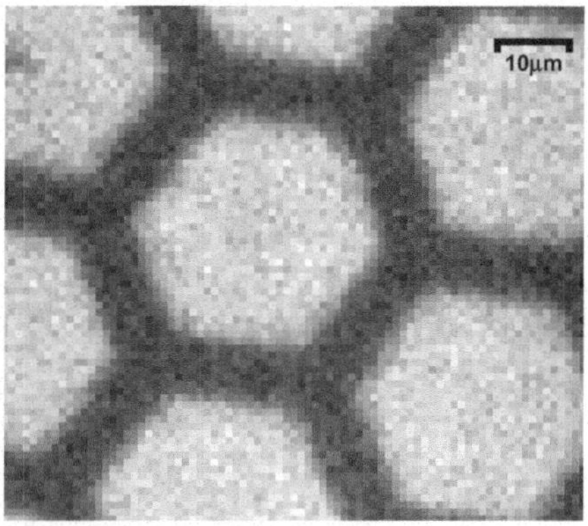

Fig. 12.10 The first 2D image obtained using neutral atoms (helium) to image a sample. The sample imaged is a hexagonal copper grating with a rod thickness of 8 μm. The image is obtained in transmission mode by scanning the focused beam across the sample (the sample is being moved). From [242]

Because of the low energy of the neutral helium atoms, there will be no secondary emitted electrons in NEMI. This is an advantage. For both SEM and SHIM the most surface sensitive measurements are generally obtained with secondary emitted electrons and this creates a particular problem with measurements at edges. The increased emission of secondary electrons at edges means that sharp features cannot be resolve with such a high precision [241].

Provided the numerical aperture is small enough, a depth of field comparable to a scanning electron microscope can be obtained for a NEMI instrument. The depth of field is an important point because it means that large aspect ratio structures with sharp features can be imaged on the nanoscale with high precision. Presently there is no technique available that can do that. Scanning Probe Microscopy (SPM) such as Atomic force microscopy and Scanning Tunneling Microscopy are also strictly surface sensitive, but they can only be used readily on flat surfaces. Another issue with SPM is that the measurements are very sensitive to the properties of the tip, which can be difficult to control, whereas a focused beam of neutral helium atoms makes a very well defined probe.

The NEMI instrument would be the perfect tool for testing polymeric nanostructures, such as they are created for example by nanoimprint or nanostructured coatings on high-aspect ratio structures and much more. Ultimately the resolution limit for such an instrument would be determined by the wavelength of the helium atoms, less than 0.1 nm, through the Abbe criterion.

The first image obtained using neutral (helium) atoms for imaging was published in 2008 [242]. The image was a shadow image obtained in transmission and can be seen in Fig. 12.10. The resolution was around 1.5 μm. The image was obtained using the apparatus MAGIE in Fig. 12.7 with the detector arm rotated to allow the beam to go straight through. The sample (a hexagonal copper grating) was mounted on the piezo table. Instead of a sample, a nanostructured zone plate with free standing zones [243, 244] was inserted as focusing element in the central chamber. The zone plate and sample distance was optimized to minimize the beam spot on the sample. The principle is simple: A beam of helium atoms with a narrow velocity distribution $\Delta v/v < 1\,\%$ is created by supersonic expansion though a 5–10 μm nozzle. The temperature of the nozzle determines the temperature of the beam and thereby the

Fig. 12.11 A reflection image obtained using neutral atoms as an imaging probe. The image shows two uncoated pollen grains. The image is created by placing a very small helium source very close to the sample and detecting the atoms scattered at a particular solid angle. The shadow effect is due to this particular setup, with the helium beam being incident perpendicular to the overall sample plane and the scattered atoms being collected at an angle relative to the overall sample plane. The detector is placed roughly in the *upper right corner* of the image, which is why we see a 'shadow' effect in the features. This is comparable to the shadow effect well known from scanning electron microscopy images obtained with secondary emission. The image resolution is similar to the transmission image in Fig. 12.10. Printed with the permission of Philip Witham, see also [245]

wavelength of the atoms as discussed previously. The central part of the beam is selected by a micro skimmer in front of the expansion hole to reduce the effect of the virtual source size, which increases with the increase of the pressure in the source. The beam is focused onto the sample using an optical element, here a zone plate. Following classical optical considerations, the resolution is determined by the geometry of the optical system (distance from object to lens (zone plate) and distance from zone plate to image plane (sample) and by the size of the object, which is being imaged onto the sample). Here the object is the atomic source, and the size is determined by the skimmer opening [230]. 1 μm skimmers are used routinely in molecular beam experiments [242, 246]. Holes down to 10 nm can easily be created with present day techniques (for example using a focused ion beam). Of course a reduction of the diameter determines a decrease of the intensity, therefore a trade off must be found.

In 2011 the first surface image obtained with neutral helium atoms was published [245]. An example of a neutral helium atom reflection image can be seen in Fig. 12.11. Also here the resolution is around 1.5 μm. The good contrast caused by the unique surface sensitivity of the helium atoms is evident. The experimental setup for the reflective images is elegant in its simplicity. The helium beam is entering perpendicular to the sample through a very small hole and the scattered atoms are collected at an angle through another small hole. The sample is very close to source and detector thus avoiding focusing elements all together.

The setup presented in [245] may in the years to come prove an interesting alternative to AFM. For a more versatile instrument, with the capability of imaging of high aspect ratios structure and with a large depth of field and zooming properties a focusing element is necessary.

So far the two big challenges in neutral helium microscopy have been the optical elements (how to focus neutral helium) and the low efficiency in the helium beam detection. This will be discussed in the following.

12.5.1 Focusing Elements

The very low polarizability of the helium atom means that they cannot be manipulated in electrostatic fields. The helium atoms can only be manipulated via their de Broglie matter-wave properties. As discussed earlier the wavelength of the helium atoms is around 0.1 nm, which puts a very high demand on the optical elements. Since helium atoms do not penetrate into materials at these energies, material lenses such as those used for visible photons are not an option and what is left are mirrors and zone plates with free standing zones. The case is similar to the focusing of X-ray. The first focusing of helium atoms with zone plates was done in the group of Mlynek and co-workers [247]. Later work includes [246]. This year a neutral beam of helium atoms was successfully focused to a submicron spot using a zone plate [248].

Zone plates are circular diffraction grating operating in transmission. The set of radially symmetric rings have a width that decreases with the radius, with a smallest value Δr at the diameter $\mathcal{D}$. The first order diffraction is focused with a focal length given by $f = \mathcal{D}\Delta r/\lambda$ and the resolution is of the order of Δr, therefore nanolithographic techniques are required for their construction. Zone plates are convenient because they can be inserted directly into the beam line but they suffer from chromatic aberrations [249], since the focal length depends on the wavelength λ, and hence the velocity spread Δv in the supersonic beam puts a limit on the effective resolution one can obtain without additional velocity selection [230]. Furthermore the intensity is divided among different diffraction orders and only ~10 % of the incident beam goes into the focus.

Mirrors on the other hand do not suffer from chromatic aberrations and there is no inherent limit as to which fraction of the incident beam is being focused. However, for the mirror to work, it must not only have the correct shape on the macroscopic level. It must also have a surface which is smooth relative to the wavelength of the incoming beam, which in the case of the helium atoms is around 0.1 nm. In practice this means that the surface needs to be crystalline. The first mirror focusing experiment used an 50 μm thick Si(111)–H(1 × 1) wafer, electrostatically bent to form a parabolic mirror, to create a focused helium spot, 200 μm in diameter [250]. In a later paper a similar wafer was bent to an ellipsoidal shape to yield a focus size of about 30 μm diameter [251]. Unfortunately, the specular helium reflectivity from a Si(111)–H(1 × 1) surface is rather low and there are issues with surface stability.

It seems however, that these problems have now been solved by using a new type of mirror. A very recent paper [252] demonstrates a graphene helium atom mirror with 20 % reflectivity. Very importantly this graphene coated surface retains the high reflectivity even after having been left in air for a month.

12.5.2 Detection

The big limiting factor for the resolution in NEMI microscopy at present is the detection efficiency of the neutral helium atoms. The low detection efficiency forced the researchers in the first microscopy experiments, see Figs. 12.10 and 12.11 to use relatively large skimmer openings, thus limiting the resolution. The detectors used in both measurements were mass spectrometers, where the helium ions are first ionized using electron bombardment and then mass selected before detection. Unfortunately the ionization efficiency of such detectors is less than 1×10^5 [245] as mentioned previously in this chapter. The fundamental problem is that space charge puts a limitation to how densely the electrons can be packed. Fortunately there are improvements in sight. Recent work indicates that an increase in the ionization volume combined with very good vacuum control can give a significant improvement in detection efficiency [253]. Field ionization has been pursued as a possible detection alternative [254–256]. Here the helium atoms are ionized by the strong field created in the vicinity of a sharp tip with an applied positive voltage. The high field cause the electrons to tunnel into the tip. The principle is known from field ionization microscopy. Field ionization would in principle be a very attractive method since it would open up for the possibility of 2D detection and hence instant imaging rather than scanning. The first promising results using an array of carbon nanotubes as a 2D field ionization detector for helium was published earlier this year [257].

With the recent improvements in neutral helium optics and neutral helium detection, there can be no doubt that the development of neutral helium microscopy (NEMI) will be very interesting to follow in the years to come.

Acknowledgements We are grateful to Giorgio Benedek, Maurizio Canepa, and Luca Floreano for fruitful discussions during the preparation of the chapter and to Anna Franchini for useful suggestions. One of the author (G.B.) acknowledges Michelsen Centre for Industrial Measurement Science and Technology (Bergen) for a generous support.

The Helium Scattering Apparatus MAGIE, now at the University of Bergen was originally constructed at the Max Planck Institute for Fluid Dynamics, Göttingen in the group of J. P. Toennies. Bergen Research Foundation Recruitment Program with founder Trond Mohn, is gratefully acknowledge for making it possible to start up a helium scattering laboratory in Bergen. Work on Neutral Helium Microscopy presented here was supported by the European Union Project INA—Imaging with Neutral Atoms.

References

1. O. Stern, Z. Phys. **2**, 49 (1920)

2. I. Estermann, O. Stern Z, Physik **61**, 95 (1930)
3. I. Estermann, R. Frisch, O. Stern Z, Physik **73**, 348 (1932)
4. A. Kantrowitz, J. Grey, Rev. Sci. Instrum. **22**, 328 (1951)
5. E.W. Becker, K. Bier, Z. Naturforsch. **9a**, 975 (1954)
6. G. Brusdeylins, R. Bruce Doak, J.P. Toennies, Phys. Rev. Lett. **44**, 1417 (1980)
7. G. Boato, P. Cantini, U. Garibaldi, A.C. Levi, L. Mattera, R. Spadacini, G.E. Tommei, J. Phys. C, Solid State Phys. **6**, L394 (1973)
8. J.R. Bledsoe, S.S. Fisher, Surf. Sci. **46**, 129 (1974)
9. G. Boato, P. Cantini, L. Mattera, Surf. Sci. **55**, 141 (1976)
10. B. Wood, B.F. Mason, B.R. Williams, J. Chem. Phys. **61**, 1435 (1974)
11. J.A. Meyers, D.R. Frankl, Surf. Sci. **51**, 61–74 (1975)
12. S.S. Fisher, M.N. Bishara, A.R. Kuhlthau, J.E. Scott Jr., in *Proc. Intern. Symp. Rarefied Gas Dynamics, 6th*, Cambridge, 1968 (1969)
13. B.R. Williams, J. Chem. Phys. **55**, 3220 (1971)
14. G. Brusdeylins, R. Bruce Doak, J.P. Toennies, Phys. Rev. B **27**, 3662 (1983)
15. R. Fatema, D.H. Van Winkle, J.G. Skofronick, S.A. Safron, F.A. Flaherty, Phys. Rev. B **77**, 024305 (2008)
16. G. Chern, J.G. Skofronick, W.P. Brug, S.A. Safron, Phys. Rev. B **39**, 12828 (1989)
17. G.G. Bishop, J. Duan, E.S. Gillman, S.A. Safron, J.G. Skofronick, J. Vac. Sci. Technol. A **11**, 2008 (1993)
18. A. Glebov, W. Silvestri, J.P. Toennies, G. Benedek, J.G. Skofronick, Phys. Rev. B **54**, 17866 (1996)
19. G. Brusdeylins, R. Rechsteiner, J.G. Skofronick, J.P. Toennies, G. Benedek, L. Miglio, Phys. Rev. Lett. **54**, 466 (1985)
20. G. Bracco, R. Tatarek, S. Terreni, F. Tommasini, Phys. Rev. B **34**, 9045 (1986)
21. G. Bracco, M. D'avanzo, C. Salvo, R. Tatarek, S. Terreni, F. Tommasini, Surf. Sci. **189**, 684 (1987)
22. G. Boato, P. Cantini, R. Tatarek, Phys. Rev. Lett. **40**, 887 (1978)
23. G. Boato, P. Cantini, C. Guidi, R. Tatarek, Phys. Rev. B **20**, 3957 (1979)
24. M.W. Cole, D.R. Frankl, D.L. Goodstein, Rev. Mod. Phys. **53**, 199 (1981)
25. P. Cantini, G.P. Felcher, R. Tatarek, Phys. Rev. Lett. **37**, 606 (1976)
26. P. Cantini, R. Tatarek, Phys. Rev. B **23**, 3030 (1981)
27. G. Benedek, G. Brusdeylins, C. Heimlich, J.P. Toennies, U. Valbusa, Surf. Sci. **178**, 545 (1986)
28. A. Politano, B. Borca, M. Minniti, J.J. Hinarejos, A.L. Vazques de Parga, D. Farias, R. Miranda, Phys. Rev. B **84**, 035450 (2011)
29. Y. Yamada, C. Sugawara, Y. Satake, Y. Yokoyama, R. Okada, T. Nakayama, M. Sasaki, T. Kondo, J. Oh, J. Nakamuara, W.W. Hayes, J. Phys. Condens. Matter **22**, 304010 (2010)
30. M. Mayrhofer-Reinhartshuber, A. Tamtögl, P. Kraus, K.H. Rieder, W.E. Ernst, J. Phys. Condens. Matter **24**, 104008 (2012)
31. A. Tamtögl, M. Mayrhofer-Reinhartshuber, N. Balak, W.E. Ernst, K.H. Rieder, J. Phys. Condens. Matter **22**, 304019 (2010)
32. M.J. Cardillo, G.E. Becker, Phys. Rev. Lett. **40**, 1148 (1978)
33. M.J. Cardillo, G.E. Becker, S.J. Sibener, D.R. Miller, Surf. Sci. **107**, 469–493 (1981)
34. U. Harten, J.P. Toennies, Ch. Wöll, L. Miglio, P. Ruggerone, L. Colombo, G. Benedek, Phys. Rev. B **38**, 3305 (1988)
35. U. Harten, J.P. Toennies, Ch. Wöll, Phys. Rev. Lett. **57**, 2947 (1986)
36. G. Lange, J.P. Toennies, P. Ruggerone, G. Benedek, Europhys. Lett. **41**, 647 (1998)
37. J.S. Becker, R.D. Brown, E. Johansson, N.S. Lewis, S.J. Sibener, J. Chem. Phys. **133**, 104705 (2010)
38. G. Vidali, D.R. Frankl, Phys. Rev. B **27**, 2480 (1983)
39. G. Vidali, M.W. Cole, W.H. Weinberg, W.A. Steele, Phys. Rev. Lett. **51**, 118 (1983)
40. C.A. Meli, E.F. Greene, G. Lange, J.P. Toennies, Phys. Rev. Lett. **74**, 2054 (1995)

41. D. Cvetko, L. Floreano, A. Crottini, A. Morgante, F. Tommasini, Surf. Sci. Lett. **447**, L147 (2000)
42. L. Floreano, D. Cvetko, G. Bavdek, M. Benes, A. Morgante, Phys. Rev. B **64**, 075405 (2001)
43. J. Lobo, D. Farìas, E. Hulpke, J.P. Toennies, E.G. Michel, Phys. Rev. B **74**, 035303 (2006)
44. A. Glebov, J.P. Toennies, S. Vollmer, S.A. Safron, J.G. Skofronick, V. Gräschus, A. Mazur, J. Pollmann, Phys. Rev. B **57**, 10082 (1998)
45. M. Buongiorno Nardelli, D. Cvetko, V. De Renzi, L. Floreano, A. Morgante, M. Peloi, F. Tommasini, Phys. Rev. B **52**, 16720 (1995)
46. D.V. Tendulkar, R.E. Stickney, Surf. Sci. **27**, 516 (1971)
47. G. Boato, P. Cantini, R. Tatarek, J. Phys. F, Met. Phys. **6**, L237 (1976)
48. M.F. Danişman, L. Casalis, G. Scoles, Phys. Rev. B **72**, 085404 (2005)
49. U. Harten, A.M. Lahee, J.P. Toennies, Ch. Wöll, Phys. Rev. Lett. **54**, 2619 (1985)
50. P. Cortona, M.G. Dondi, D. Cvetko, A. Lausi, A. Morgante, K.C. Prince, F. Tommasini, Phys. Rev. B **47**, 6705 (1993)
51. A. Bellman, A. Morgante, M. Polli, F. Tommasini, D. Cvetko, V.R. Dhanak, A. Lausi, K.C. Prince, R. Rosei, Surf. Sci. **298**, 1 (1993)
52. R.B. Doak, U. Harten, J.P. Toennies, Phys. Rev. Lett. **51**, 578 (1983)
53. G. Benedek, J. Ellis, N.S. Luo, A. Reichmuth, P. Ruggerone, J.P. Toennies, Phys. Rev. B **48**, 4917 (1993)
54. U. Harten, J.P. Toennies, C. Wöll, Faraday Discuss. Chem. Soc. **80**, 137 (1985)
55. G. Bracco, R. Tatarek, F. Tommasini, U. Linke, M. Persson, Phys. Rev. B **36**, 2928 (1987)
56. P. Zeppenfeld, K. Kern, R. David, K. Kuhnke, G. Comsa, Phys. Rev. B **38**, 12329 (1988)
57. A.M. Lahee, J.P. Toennies, Ch. Wöll, Surf. Sci. **191**, 529 (1987)
58. J. Braun, K.L. Kostov, G. Witte, L. Surnev, J.G. Skofronick, S.A. Safron, Ch. Wöll, Surf. Sci. **372**, 132 (1997)
59. U. Harten, J.P. Toennies, C. Wöll, G. Zhang, Phys. Rev. Lett. **55**, 2308 (1985)
60. J. Braun, P. Ruggerone, G. Zhang, J.P. Toennies, G. Benedek, Phys. Rev. B **79**, 205423 (2009)
61. B. Gans, P.A. Knipp, D.D. Koleske, S.J. Sibener, Surf. Sci. **264**, 81 (1992)
62. C. Mannori, G. Boato, M. Canepa, P. Cantini, L. Mattera, S. Terreni, Europhys. Lett. **45**, 686 (1999)
63. Ch. Wöll, Prog. Surf. Sci. **82**, 55 (2007)
64. D.R. Jung, M. Mahgerefteh, D.R. Frankl, Phys. Rev. B **39**, 11164 (1989)
65. F. Traeger, Chem. Phys. Chem. **7**, 1006 (2006)
66. G. Brusdeylins, R.B. Doak, J.G. Skofronick, J.P. Toennies, Surf. Sci. **128**, 191 (1983)
67. W. Steurer, A. Apfolter, M. Koch, E. Sarlat, E. Søndergård, W.E. Ernst, B. Holst, Surf. Sci. **601**, 4407 (2007)
68. W.P. Brug, G. Chern, J. Duan, G.G. Bishop, S.A. Safron, J.G. Skofronick, J. Vac. Sci. Technol. A **10**, 2222 (1992)
69. G. Brusdeylins, D. Schmicker, Surf. Sci. **331–333**, 237 (1995)
70. J.A. Li, E.A. Akhadov, J. Baker, L.A. Boatner, D. Bonart, F.A. Flaherty, J. Fritsch, S.A. Safron, U. Schroeder, J.G. Skofronick, T.W. Trelenberg, D.H. Van Winkle, Phys. Rev. Lett. **86**, 4867 (2001)
71. G. Bracco, P. Cantini, E. Cavanna, R. Tatarek, A. Glachant, Surf. Sci. **136**, 169 (1984)
72. S. Chung, A. Kara, J.Z. Larese, W.Y. Leung, D.R. Frankl, Phys. Rev. B **35**, 4870 (1987)
73. K. Kern, R. David, P. Zeppenfeld, R. Palmer, G. Comsa, Solid State Commun. **62**, 391 (1987)
74. J. Braun, D. Fuhrmann, A. Šiber, B. Gumhalter, Ch. Wöll, Phys. Rev. Lett. **80**, 125 (1998)
75. M.F.M. De Kieviet, D. Bahatt, G. Scoles, G. Vidali, M. Karimi, Surf. Sci. **365**, 789 (1996)
76. M.F. Danişman, B. Özkan, Rev. Sci. Instrum. **82**, 115104 (2011)
77. D. Farías, H. Tröger, K.H. Rieder, Surf. Sci. **331**, 150 (1995)
78. L. Yang, T.S. Rahman, G. Bracco, R. Tatarek, Phys. Rev. B **40**, 12271 (1989)
79. G. Bracco, R. Tatarek, G. Vandoni, Phys. Rev. B **42**, 1852 (1990)
80. A.P. Graham, J.P. Toennies, Phys. Rev. B **56**, 15378 (1997)

81. C. Huang, D.A. MacLaren, R.T. Bacon, W. Allison, J. Phys. Condens. Matter **23**, 355001 (2011)
82. D. Fuhrmann, E. Hulpke, W. Steinhögl, Phys. Rev. B **57**, 4798 (1998)
83. M. Patting, D. Farías, K.-H. Rieder, J. Chem. Phys. **113**, 8283 (2000)
84. R. Apel, D. Farías, H. Tröger, E. Kirsten, K.H. Rieder, Surf. Sci. **364**, 303 (1996)
85. B. Poelsema, K. Lenz, G. Comsa, J. Chem. Phys. **134**, 074703 (2011)
86. P. Jiang, M. Zappone, S.L. Bernasek, J. Chem. Phys. **99**, 8126 (1993)
87. M.F. Luo, D.A. MacLaren, I.G. Shuttleworth, W. Allison, Chem. Phys. Lett. **381**, 654 (2003)
88. M. Canepa, P. Cantini, L. Mattera, E. Narducci, M. Salvietti, S. Terreni, Surf. Sci. **322**, 271–284 (1995)
89. M.T. Suter, P.U. Andersson, J.B.C. Petterson, J. Chem. Phys. **125**, 174704 (2006)
90. P.U. Andersson, M.T. Suter, N. Markovic, J.B.C. Pettersson, J. Phys. Chem. C **111**, 15258 (2007)
91. M.T. Suter, Chem. Phys. Lett. **445**, 208 (2007)
92. X. Kong, P.U. Andersson, E.S. Thomson, J.B.C. Pettersson, J. Phys. Chem. C **116**, 8964 (2012)
93. C.E.D. Chidsey, G.-Y. Liu, P. Rowntree, G. Scoles, J. Chem. Phys. **91**, 4421 (1989)
94. N. Camillone, C.E.D. Chidsey, G.Y. Liu, T.M. Putvinski, G. Scoles, J. Chem. Phys. **94**, 8493 (1991)
95. N. Camillone, C.E.D. Chidsey, G. Liu, G. Scoles, J. Chem. Phys. **98**, 3503 (1993)
96. M.F. Danişman, L. Casalis, G. Bracco, G. Scoles, J. Phys. Chem. B **106**, 11771 (2002)
97. S. Vollmer, P. Fouquet, G. Witte, C. Boas, M. Kunat, U. Burghaus, Ch. Wöll, Surf. Sci. **462**, 135 (2000)
98. S.B. Darling, A.W. Rosenbaum, S.J. Sibener, Surf. Sci. Lett. **478**, L313 (2001)
99. A.W. Rosenbaum, M.A. Freedman, S.B. Darling, I. Popova, S.J. Sibener, J. Chem. Phys. **120**, 3880 (2004)
100. V. Vogel, C. Woell, Thin Solid Films **159**, 429 (1988)
101. I.G. Shuttleworth, Surf. Rev. Lett. **14**, 387 (2007)
102. L. Casalis, M.F. Danisman, B. Nickel, G. Bracco, T. Toccoli, S. Iannotta, G. Scoles, Phys. Rev. Lett. **90**, 206101 (2003)
103. V. Lanzilotto, C. Sanchez-Sanchez, G. Bavdek, D. Cvetko, M.F. Lopez, J.A. Martin-Gago, L. Floreano, J. Phys. Chem. C **115**, 4664 (2011)
104. E. Mete, I. Demiroğlu, E. Albayrak, G. Bracco, Ş. Ellialtioğlu, M.F. Danişman, J. Phys. Chem. C **116**, 19429 (2012)
105. L. Floreano, A. Cossaro, R. Gotter, A. Verdini, G. Bavdek, F. Evangelista, A. Ruocco, A. Morgante, D. Cvetko, J. Phys. Chem. C **112**, 10794 (2008)
106. M. Buongiorno Nardelli, D. Cvetko, V. De Renzi, L. Floreano, R. Gotter, A. Morgante, M. Peloi, F. Tommasini, R. Danieli, S. Rossini, C. Taliani, R. Zamboni, Phys. Rev. B **53**, 1095 (1996)
107. S. Prato, L. Floreano, D. Cvetko, V. De Renzi, A. Morgante, S. Modesti, C. Taliani, R. Zamboni, F. Biscarini, J. Phys. Chem. **103**, 7788 (1999)
108. M.D. Ward, Chem. Rev. **101**, 1697 (2001)
109. G. Bracco, J. Acker, M.D. Ward, G. Scoles, Langmuir **18**, 5551 (2002)
110. G. Bracco, M.D. Ward, G. Scoles, J. Chem. Phys. **118**, 8405 (2003)
111. B. Poelsema, G. Comsa (eds.), *Scattering of Thermal Energy Atoms*. Springer Tracts in Modern Physics, vol. 115 (Springer, Berlin, 1989)
112. H.R. Sharma, M. Shimoda, A.P. Tsai, Adv. Phys. **56**, 403 (2007)
113. K.J. Franke, H.R. Sharma, W. Theis, P. Gille, P. Ebert, K.H. Rieder, Phys. Rev. Lett. **89**, 156104 (2002)
114. W. Steurer, A. Apfolter, M. Koch, W.E. Ernst, B. Holst, E. Søndergård, J. Manson, Phys. Rev. Lett. **99**, 0355031 (2007)
115. J.S. Becker, R.D. Brown, D.R. Killelea, H. Yuan, S.J. Sibener, Proc. Natl. Acad. Sci. USA **108**, 977 (2011)

116. M.A. Freedman, J.S. Becker, A.W. Rosenbaum, S.J. Sibener, J. Chem. Phys. **129**, 044906 (2008)
117. M.A. Freedman, A.W. Rosenbaum, S.J. Sibener, Phys. Rev. B **75**, 113410 (2007)
118. W.K. Burton, N. Cabrera, F.C. Frank, Phil. Trans. R. Soc. (London) A **243** (1951)
119. A. Trayanov, A.C. Levi, E. Tosatti, Surf. Sci. **233**, 184 (1990)
120. J. Villain, D.R. Grempel, J. Lapujoulade, J. Phys. F, Met. Phys. **15**, 809 (1985)
121. B. Salanon, F. Fabre, J. Lapujoulade, W. Selke, Phys. Rev. B **38**, 7385 (1988)
122. J. Lapoujoulade, Surf. Sci. Rep. **20**, 191 (1994)
123. H.-J. Ernst, R. Folkerts, L. Schwenger, Phys. Rev. B **52**, 8461 (1995)
124. P. Zeppenfeld, K. Kern, R. David, G. Comsa, Phys. Rev. Lett. **62**, 63 (1989)
125. G. Bracco, C. Malò, C.J. Moses, R. Tatarek, Surf. Sci. **287/288**, 871 (1993)
126. G. Bracco, L. Pedemonte, R. Tatarek, Phys. Rev. B **54**, 10385 (1996)
127. L. Pedemonte, G. Bracco, R. Tatarek, Surf. Sci. **377–379**, 524 (1997)
128. D. Cvetko, A. Lausi, A. Morgante, F. Tommasini K.C. Prince, Surf. Sci. **269/270**, 68 (1992)
129. J. Sprösser, B. Salanon, J. Lapujoulade, Europhys. Lett. **16**, 283 (1991)
130. D. Cvetko, V. De Renzi, L. Floreano, A. Morgante, F. Tommasini, Solid State Commun. **91**, 539 (1994)
131. B. Poelsema, L.K. Verheij, G. Comsa, Phys. Rev. Lett. **51**, 2410 (1983)
132. A. Glebov, J.R. Manson, J.G. Skofronick, J.P. Toennies, Phys. Rev. Lett. **78**, 1508 (1997)
133. A. Glebov, J.P. Toennies, J.G. Skofronick, J.R. Manson, Phys. Rev. B **58**, 10012 (1998)
134. A.M. Lahee, J.R. Manson, J.P. Toennies, Ch. Wöll, J. Chem. Phys. **86**, 7194 (1987)
135. J.W.M. Frenken, Surf. Sci. **211–212**, 21 (1989)
136. J. Ellis, J.P. Toennies, Phys. Rev. Lett. **70**, 2118 (1993)
137. L. Pedemonte, R. Tatarek, G. Bracco, Phys. Rev. B **66**, 045414 (2002)
138. R.M. Bradley, J.M.E. Harper, J. Vac. Sci. Technol. A **6**, 2390 (1988)
139. L. Pedemonte, G. Bracco, C. Boragno, F. Buatier de Mongeot, U. Valbusa, Phys. Rev. B **68**, 115431 (2003)
140. G. Bracco, D. Cavanna, Phys. Rev. B **76**, 033411 (2007)
141. P.C. Dastoor, W. Allison, Surf. Sci. **433–435**, 99 (1999)
142. D. Green, D.A. Maclaren, W. Allison, P.C. Dastoor, J. Phys. D, Appl. Phys. **35**, 3216 (2002)
143. P.C. Dastoor, W. Allison, Phys. Rev. B **67**, 245403 (2003)
144. P. Bedrossian, B. Poelsema, G. Rosenfeld, L.C. Jorritsma, N.N. Lipkin, G. Comsa, Surf. Sci. **334**, 1 (1995)
145. S. Terreni, P. Cantini, M. Canepa, L. Mattera, Phys. Rev. B **56**, 6490 (1997)
146. D. Cvetko, V. De Renzi, L. Floreano, A. Morgante, M. Peloi, F. Tommasini, V. Chab, K.C. Prince, Surf. Sci. Lett. **323**, L305 (1995)
147. D. Cvetko, V. De Renzi, L. Floreano, A. Morgante, M. Peloi, F. Tommasini, V. Chab, K.C. Prince, Phys. Rev. B **51**, 17957 (1995)
148. D. Cvetko, V. De Renzi, L. Floreano, A. Morgante, M. Peloi, F. Tommasini, V. Chab, K.C. Prince, Phys. Rev. B **52**, 14941 (1995)
149. L. Floreano, D. Cvetko, F. Bruno, G. Bavdek, A. Cossaro, R. Gotter, A. Verdini, A. Morgante, Prog. Surf. Sci. **72**, 135 (2003)
150. B.J. Hinch, C. Koziol, J.P. Toennies, G. Zhang, Europhys. Lett. **10**, 341 (1989)
151. A. Crottini, D. Cvetko, L. Floreano, R. Gotter, A. Morgante, F. Tommasini, Phys. Rev. Lett. **79**, 1527 (1997)
152. A. Schiffrin, A. Riemann, W. Auwaerter, Y. Pennec, A. Weber-Bargioni, D. Cvetko, A. Cossaro, A. Morgante, J.V. Barth, PNAS **104**, 5279 (2007)
153. A. Schiffrin, J. Reichert, Y. Pennec, W. Auwaerter, A. Weber-Bargioni, M. Marschall, M. Dell'Angela, D. Cvetko, G. Bavdek, A. Cossaro, A. Morgante, J.V. Barth, J. Phys. Chem. C **113**, 12101 (2009)
154. C. Sanchez-Sanchez, V. Lanzilotto, C. Gonzalez, A. Verdini, P.L. de Andres, L. Floreano, M.F. Lopez, J.A. Martin-Gago, Chem. Eur. J. **18**, 7382 (2012)
155. S. Iannotta, T. Toccoli, F. Biasioli, A. Boschetti, M. Ferrari, Appl. Phys. Lett. **76**, 1845 (2000)

156. Y. Wu, T. Toccoli, N. Koch, E. Iacob, A. Pallaoro, P. Rudolf, S. Iannotta, Phys. Rev. Lett. **98**, 076601 (2007)
157. B.S. Zhao, G. Meijer, W. Schöllkopf, Science **331**, 892 (2011)
158. V. Druzhinina, M. DeKieviet, Phys. Rev. Lett. **91**, 19302 (2003)
159. N. Bundaleski, H. Khemliche, P. Soulisse, P. Roncin, Phys. Rev. Lett. **101**, 177601 (2008)
160. M. Busch, A. Schüller, S. Wethekam, H. Winter, Surf. Sci. Lett. **603**, 23 (2009)
161. M. Marynowski, W. Franzen, M. El-Batanouny, Phys. Rev. B **60**, 6053 (1999)
162. H. Morgner, Adv. At. Mol. Opt. Phys. **42**, 387 (2000)
163. D. Farìas, K.H. Rieder, Rep. Prog. Phys. **61**(12), 1575 (1998)
164. E. Hulpke (ed.), *Helium Atom Scattering from Surfaces*. Springer Series in Surface Science, vol. 27 (1992)
165. J.P. Toennies, Appl. Phys. **3**, 91 (1974)
166. A.P. Graham, Surf. Sci. Rep. **49**, 115 (2003)
167. W. Kress, F.W. de Wette (eds.), *Surface Phonons*. Springer Series in Surface Sciences, vol. 21 (Springer, Berlin, 1991)
168. J.P. Toennies, Experimental determination of surface phonons by helium atom and electron loss spectroscopy, in *Surface Phonons*, ed. by W. Kress, F.W. de Wette. Springer Series in Surface Sciences, vol. 21 (Springer, Berlin, 1991)
169. K.H. Rieder, N. Garcia, Phys. Rev. Lett. **49**, 43 (1982)
170. D.A. MacLaren, Ph.D. Thesis, University of Cambridge, 2002
171. K. Schönhammer, O. Gunnarsson, Surf. Sci. **117**, 53 (1982)
172. G. Iadonisi, A.C. Levi, Int. J. Quant. Chem. **26**, 823 (1984)
173. Bernd Kohler, P. Ruggerone, M. Scheffler, Surf. Sci. **368**, 213 (1996)
174. J.E. Lennard-Jones, A.F. Devonshire, Nature **137**, 1069 (1936)
175. A. Franchini, G. Santoro, J. Phys. Condens. Matter **14**, 5881 (2002)
176. V. Celli, N. Garcia, J. Hutchison, Surf. Sci. **87**, 112 (1979)
177. L. Mattera, C. Salvo, S. Terreni, F. Tommasini, Surf. Sci. **97**, 158 (1980)
178. G. Bracco, G. Scoles, J. Chem. Phys. **119**, 6277 (2003)
179. N.W. Ashcroft, N.D. Mermin, *Solid State Physics* (Harcourt, Brace & Co, New York, 1976)
180. J.M. Moix, E. Pollak, W. Allison, J. Chem. Phys. **134**, 024319 (2011)
181. U. Garibaldi, A.C. Levi, R. Spadacini, G.E. Tommei, Surf. Sci. **48**, 649 (1975)
182. N. García, J. Chem. Phys. **67**, 897 (1977)
183. G. Armand, J.R. Manson, Phys. Rev. B **18**, 6510 (1978)
184. J.L. Beeby, J. Phys. C, Solid State Phys. **5**, 3438 (1972)
185. N. Esbjerg, J.K. Nørskov, Phys. Rev. Lett. **45**, 807 (1980)
186. M. Manninen, J.K. Nørskov, M.J. Puska, C. Umrigar, Phys. Rev. B **29**, 2314 (1984)
187. P. Cortona, M.G. Dondi, F. Tommasini, Surf. Sci. **261**, L35–L38 (1992)
188. P. Cortona, M.G. Dondi, A. Lausi, F. Tommasini, Surf. Sci. **276**, 333 (1992)
189. K.T. Tang, J.P. Toennies, J. Chem. Phys. **80**, 3726 (1984)
190. M. Petersen, S. Wilke, P. Ruggerone, B. Kohler, M. Scheffler, Phys. Rev. Lett. **76**, 995 (1996)
191. M.I. Trioni, G. Fratesi, S. Achilli, G.P. Brivio, J. Phys. Condens. Matter **21**, 264003 (2009)
192. G. Benedek, G. Brusdeylins, R.B. Doak, J.G. Skofronick, J.P. Toennies, Phys. Rev. B **28**, 2104 (1983)
193. G. Lilienkamp, J.P. Toennies, J. Chem. Phys. **78**, 5210 (1983)
194. G. Benedek, L. Miglio, Z. Phys. B, Condens. Matter **50**, 93 (1983)
195. D. Evans, V. Celli, G. Benedek, J.P. Toennies, R.B. Doak, Phys. Rev. Lett. **50**, 1854 (1983)
196. G. Benedek, P.M. Echenique, J.P. Toennies, F. Traeger, J. Phys. Condens. Matter **22**, 304016 (2010)
197. N. Cabrera, V. Celli, R. Manson, Phys. Rev. Lett. **22**, 346 (1969)
198. V. Bortolani, A. Franchini, N. Garcia, F. Nizzoli, G. Santoro, Phys. Rev. B **28**, 7358 (1983)
199. V. Bortolani, A. Franchini, F. Nizzoli, G. Santoro, Phys. Rev. Lett. **52**, 429 (1984)
200. D. Eichenauer, J.P. Toennies, J. Chem. Phys. **85**, 532 (1986)
201. R.E. Allen, G.P. Aldredge, F.W. de Wette, Phys. Rev. B **4**, 1648 (1971)
202. R.E. Allen, G.P. Aldredge, F.W. de Wette, Phys. Rev. B **4**, 1661 (1971)

203. G.P. Aldredge, R.E. Allen, F.W. de Wette, Phys. Rev. B **4**, 1682 (1971)
204. G. Benedek, Phys. Status Solidi **58**, 661 (1973)
205. G. Benedek, Surf. Sci. **61**, 603 (1976)
206. M. Born, K. Huang, *Dynamical Theory of Crystal Lattices* (Oxford University Press, Oxford, 1954)
207. B.G. Dick Jr., A.W. Overhauser, Phys. Rev. **112**, 90 (1958)
208. G. Benedek, G.P. Brivio, L. Miglio, V.R. Velasco, Phys. Rev. B **26**, 497 (1982)
209. C.S. Jayanthi, H. Bilz, W. Kress, G. Benedek, Phys. Rev. Lett. **59**, 795 (1987)
210. B.M. Hall, M.H. Mohamed, L.L. Kesmodel, D.L. Mills, Phys. Rev. B **38**, 5856 (1988)
211. C. Kaden, P. Ruggerone, J.P. Toennies, G. Zhang, G. Benedek, Phys. Rev. B **46**, 13509 (1992)
212. G. Santoro, A. Franchini, V. Bortolani, D.L. Mills, R.F. Wallis, Surf. Sci. **478**, 99 (2001)
213. V. Chis, B. Hellsing, G. Benedek, M. Bernasconi, E.V. Chulkov, J.P. Toennies, Phys. Rev. Lett. **101**, 206102 (2008)
214. I.Yu. Sklyadneva, G. Benedek, E.V. Chulkov, P.M. Echenique, R. Heid, K.-P. Bohnen, J.P. Toennies, Phys. Rev. Lett. **107**, 095502 (2011)
215. S. Baroni, S. de Gironcoli, A. Dal Corso, P. Giannozzi, Rev. Mod. Phys. **73**, 515 (2001)
216. G. Benedek, M. Bernasconi, V. Chis, E. Chulkov, P.M. Echenique, B. Hellsing, J.P. Toennies, J. Phys. Condens. Matter **22**, 084020 (2010)
217. B. Gumhalter, J. Phys. Condens. Matter **24**, 104015 (2012)
218. A.C. Levi, H. Suhl, Surf. Sci. **88**, 221 (1979)
219. V. Bortolani, A.C. Levi, Riv. Nuovo Cimento **9**, 1 (1986)
220. G. Armand, J. Lapujoulade, Y. Lejay, Surf. Sci. **63**, 143 (1977)
221. A. Šiber, B. Gumhalter, Ch. Wöll, J. Phys. Condens. Matter **14**, 5913 (2002)
222. W. Allison, R.F. Willis, M. Cardillo, Phys. Rev. B **23**, 6824 (1981)
223. E.A. Akhadov, T.W. Trelenberg, S.A. Safron, J.G. Skofronick, D.H. Van Winkle, F.A. Flaherty, W. Theis, Phys. Rev. B **67**, 113406 (2003)
224. G. Scoles (ed.), *Atomic and Molecular Beam Methods*, vol. 1 (1988)
225. H. Pauly, *Atom, Molecule, and Cluster Beams 1* (Springer, Berlin, 2000), pp. 105–111. Chap. 1
226. L. Pedemonte, G. Bracco, R. Tatarek, Phys. Rev. A **59**, 3084 (1999)
227. L. Pedemonte, G. Bracco, J. Chem. Phys. **119**, 1433 (2003)
228. H.C.W. Beijerinck, N.F. Verster, Physica C **111**, 327 (1981)
229. D.P. DePonte, S.D. Kevan, F.S. Patton, Rev. Sci. Instrum. **77**, 055107 (2006)
230. T. Reisinger, G. Bracco, S. Rehbein, G. Schmahl, W.E. Ernst, B. Holst, J. Phys. Chem. A **111**, 12620 (2007)
231. URL www.beamdynamicsinc.com. For example from the company Beam Dynamics, Inc, www.beamdynamicsinc.com
232. V.L. Hirschy, J.P. Aldridge, Rev. Sci. Instrum. **42**, 381 (1971)
233. W. Christen, T. Krause, B. Kobin, K. Rademann, J. Phys. Chem. A **115**, 6997 (2011)
234. J. Wang, V.S. Shamamian, B.R. Thomas, J.M. Wilkinson, J. Riley, C.F. Giese, W.R. Gentry, Phys. Rev. Lett. **60**, 696 (1988)
235. L. Pedemonte, A. Gussoni, R. Tatarek, G. Bracco, Rev. Sci. Instrum. **73**, 4257 (2002)
236. J.R. Buckland, J.L. Folkerts, R. Balsod, W. Allison, Meas. Sci. Technol. **8**, 933 (1997)
237. K. Kern, R. David, G. Comsa, Rev. Sci. Instrum. **56**, 369 (1985)
238. M.P. Grams, A.M. Cook, J.H. Tuerner, R.B. Doak, J. Phys. D **39**, 930 (2006)
239. N.H. de Leeuw, F.M. Higgins, S.C. Parker, J. Phys. Chem. B **103**, 1270 (1999)
240. G.M. Riganese, A.D. Vita, J.C. Charlier, X. Gonze, R. Car, Phys. Rev. B **61**, 13250 (2000)
241. O.C. Wells, Scanning **8**, 120 (1986)
242. M. Koch, S. Rehbein, G. Schmahl, T. Reisinger, G. Bracco, W.E. Ernst, B. Holst, J. Microsc. **229**(1), 1 (2008)
243. S. Rehbein, R. Doak, R. Grisenti, G. Schmahl, J. Toennies, C. Wöll, Microelectron. Eng. **53**(1–4), 685 (2000)
244. T. Reisinger, S. Eder, M. Greve, H. Smith, B. Holst, Microelectron. Eng. **87**, 1011 (2010)

245. P. Witham, E. Sanchez, Rev. Sci. Instrum. **82**(10), 103705 (2011)
246. R.B. Doak, R.E. Grisenti, S. Rehbein, G. Schmahl, J.P. Toennies, C. Wöll, Phys. Rev. Lett. **83**, 4229 (1999)
247. O. Carnal, M. Sigel, T. Sleator, H. Takuma, J. Mlynek, Phys. Rev. Lett. **67**, 3231 (1991)
248. S. Eder, T. Reisinger, M.M. Greve, G. Bracco, B. Holst, New J. Phys. **14**, 073014 (2012)
249. T. Reisinger, B. Holst, J. Vac. Sci. Technol. B **26**, 2374 (2008)
250. B. Holst, W. Allison, Nature **390**, 244 (1997)
251. K. Fladischer et al., New J. Phys. **12**, 033018 (2010)
252. P. Sutter, M. Minniti, P. Albrecht, D. Farias, R. Miranda, E. Sutter, Appl. Phys. Lett. **99**, 211907 (2011)
253. A.R. Alderwick, A.P. Jardine, H. Hedgeland, D.A. MacLaren, W. Allison, J. Ellis, Rev. Sci. Instrum. **79**, 123301 (2008)
254. F.S. Patton, D.P. Deponte, G.S. Elliott, S.D. Kevan, Phys. Rev. Lett. **97**, 013202 (2006)
255. R.B. Doak, Y. Ekinci, B. Holst, J.P. Toennies, T. Al-Kassab, A. Heinrich, Rev. Sci. Instrum. **75**, 405 (2004)
256. R.B. Doak, J. Phys. Condens. Matter **16**, 283 (2004)
257. K.M. O'Donnell, A. Fahy, M. Barr, W. Allison, P.C. Dastoor, Phys. Rev. B **85**, 113404 (2012)

Chapter 13
The Helium Spin-Echo Method

Andrew Jardine

Abstract Helium spin-echo is a recently developed technique which enables surface dynamical processes, particularly diffusion, to be studied with atomic-scale precision over picosecond timescales. It yields a measurement of surface correlation with time, making possible a complete statistical description of the underlying dynamics. We describe the background to and implementation of the technique, followed by its application to adsorbate transport measurements. Finally, we look towards a wide range of future applications.

13.1 Introduction

The dynamics of atoms and molecules on surfaces are crucial to a wide range of fundamental and technological processes, ranging from catalysis and chemical reaction to the growth of novel materials and self assembly of nanostructures [1–3]. Broadly speaking, atomic scale dynamics can be divided into *periodic* or *vibrational* processes and *aperiodic* or *diffusive* processes. A wealth of spectroscopic techniques exist for studying the properties of vibrational modes [4], however studying aperiodic processes, such as adsorbate transport, is considerably more difficult.

In general, at sufficiently low temperatures adsorbates reside at well defined adsorption sites. They remain in equilibrium, although their energy fluctuates due to coupling to the substrate energy bath. At higher temperatures, larger fluctuations mean they occasionally gain sufficient energy to traverse the surrounding energy barrier and diffuse from one site to another. Once activated, jumps take place on a picosecond timescale. Such fast processes are experimentally challenging and practically all existing adsorbate diffusion data has been obtained using techniques that are sensitive over much longer intervals. Figure 13.1 shows a graphical comparison of the length and timescales over which a selection of diffusion measurement techniques are sensitive, using data from several existing reviews. Real-space imaging methods, such as scanning tunnelling microscopy (STM) or field ion microscopy

A. Jardine (✉)
The Cavendish Laboratory, JJ Thomson Avenue, Cambridge, CB3 0HE, UK
e-mail: apj24@cam.ac.uk

G. Bracco, B. Holst (eds.), *Surface Science Techniques*,
Springer Series in Surface Sciences 51, DOI 10.1007/978-3-642-34243-1_13,

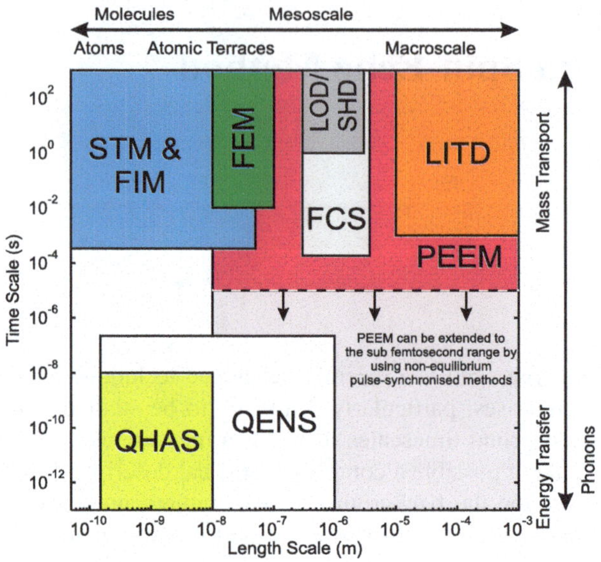

Fig. 13.1 Graphical comparison of techniques for studying surface diffusion, adapted from [5]. Diffusion processes can be measured using different techniques over short lengths and times, or equally over longer lengths and times. The figure is constructed using data from [1, 2] with additional information on STM [6], field emission microscopy (FEM) [7], fluctuation correlation spectroscopy (FCS) [8] linear optical diffraction (LOD)/second harmonic diffraction (SHD) [9] and PEEM [10–12]. Quasi-elastic neutron scattering (QENS) [13] is poorly surface sensitive but included for comparison. QHAS is the only surface technique which simultaneously allows microscopic length-scales and picosecond to nanosecond timescales to be studied, while the system is in true thermal equilibrium

(FIM) provide atomic spatial resolution, but are usually limited to millisecond framing rates. Other techniques such as photoemission electron microscopy (PEEM) can be extended to faster times, but cannot then provide atomic-scale spatial information. In addition, most techniques cannot characterise the *mechanism* of the adsorbate motion, such as the distance travelled during a single jump event. The mechanism of atomic scale of motion is directly related to fundamental characteristics such as the interaction potential around the adsorbate and the rate of energy exchange with the surface.

Helium spin-echo (HeSE) is a recently developed experimental technique which allows all of these important issues to be addressed. Broadly speaking, it is a form of helium atom scattering (HAS) experiment which makes possible a wide range 'ultra-high' resolution studies. It is particularly useful for studying adsorbate transport and diffusion using the quasi-elastic helium atom scattering (QHAS) methodology. QHAS experiments provide a unique combination of atomic-scale spatial information and picosecond temporal information and have opened an experimental window onto a previously unstudied physical regime.

13.1.1 Quasi-elastic HAS: A Unique Probe of Surface Dynamics

When helium atoms are incident on a surface, they can diffract from its periodic structure, scatter diffusely from random adsorbates, or can exchange energy with mobile surface atoms, such as surface phonons or diffusing adsorbates [14, 15]. Excitation of a vibration corresponds to transfer of a well defined quanta of energy between the incident helium and the surface mode. These are usually modest energy changes which can be measured by timing the interval required for scattered helium atoms to travel along a well defined distance, i.e. by using the time-of-flight (TOF) method.

When a helium atom scatters from a particle *diffusing* over a surface, it exchanges a small, non-quantised, amount of energy with the mobile species, in a process analogous to Doppler broadening. Averaging over the beam results in a spread of these *quasi-elastic* exchanges, forming an energy *broadening* around zero energy transfer. Quasi-elastic helium atom scattering (QHAS) experiments aim to characterise the underlying surface motion by measuring the shape and magnitude of the energy broadening as a function of scattering geometry, temperature, coverage, etc.

Quasi-elastic scattering experiments have long been carried out using neutrons to study bulk dynamics [13]. The first QHAS studies were published in 1988 by Frenken, Toennies & Wöll [16], which studied pre-melting of a lead surface. They used TOF methods to measure the energy transfer spectrum and this pioneering work was quickly followed by a series of similar studies [17]. However, the quasi-elastic broadening is usually extremely small. In TOF experiments it is masked by the intrinsic spread of energies in the helium beam, which is usually about 0.3 meV. Only a few systems exhibit sufficiently fast diffusion that quasi-elastic broadenings can be distinguished using TOF [17].

In HeSE experiments, nuclear spin-manipulation is used to measure only the *change* in energy of the helium atoms when the scatter. Hence, tiny quasi-elastic changes can be distinguished clearly, making a vastly wider range of QHAS measurements possible. HeSE experiments are also particularly appropriate for transport experiments as they provide the time Fourier transform of the energy transfer spectrum, which is a time dependent surface-correlation function, giving a rather natural description for aperiodic processes. The major benefits of QHAS can be summarised as,

- the length and timescales over which QHAS is sensitive are unique amongst the array of surface techniques available today (see Fig. 13.1);
- both rates and the detailed mechanism of motion can be distinguished, giving unprecedented dynamical information;
- neutral helium atoms are chemically inert and of sufficiently low energy (typically around 10 meV) that sample damage is almost never a concern (important when studying delicate systems, such as thiolates or water);
- the giant scattering cross section of defects [14] makes the technique remarkably sensitive to low coverages (down to at least 1 % of a monolayer), and to small species such as hydrogen.

The reason that QHAS can address both short timescales and atomic lengthscales, is that it is a scattering technique, which yields reciprocal-space information. The price is that interpretation of experimental data is somewhat more involved.

13.1.2 The Historical Development of HeSE

The spin-echo technique was conceived by Mezei in the 1970's and was quickly applied to neutron instruments at the ILL [13]. The essential principle is to use the nuclear spin on each neutron as an individual timer, so that only the energy *changes* on scattering are measured. Spin-echo measurements thus reject the resolution limiting spread of energies in the incident beam. The method was first applied to helium-3 atoms by deKieviet and co-workers in Heidelberg in the mid 1990's [18, 19]. The experiments demonstrated HeSE was a practical proposition and offered an energy resolution in the neV range, ideal for a wide range of novel work. Between 1999 and 2004, a second instrument was developed in Cambridge [20, 21], with a particular focus on adsorbate transport measurements [21]. At the time of writing, a third instrument is being developed at Technion, Israel. Together these instruments promise the formation of an active and productive research community.

13.2 Theory Background

13.2.1 Principles of Helium Scattering

The interaction between a beam of thermal helium atoms and a surface results from a combination of long range attraction, due to van der Waals interaction, and short range repulsion between electrons. A schematic potential is shown in Fig. 13.2, including conventional symbols. As the incident atoms (A) have low energies (typically about 10 meV for HeSE experiments), they scatter from the outermost electrons. The beam can reflect or diffract elastically (B), can exchange energy with the surface (C), and when there is an appreciable attractive well, resonant scattering is possible (D). We generally expect the potential to be corrugated. However, on close packed metal surfaces, electron delocalisation leads to the corrugation being smeared out. These surfaces appear perfectly flat to helium atoms and no appreciable diffraction can be seen. Surface defects, such as adsorbates, scatter the incoming wave in all directions, so reducing the apparent specular intensity with increasing coverage. The effect is routinely exploited to monitor the coverage of adsorbed species.

In order to interpret HeSE data, a quantitative relationship is required between scattered intensities and the underlying motion. Fully realistic calculations require a quantum-mechanically accurate scattering method, combined with a realistic interaction potential, both of which are complex problems. A range of approximate

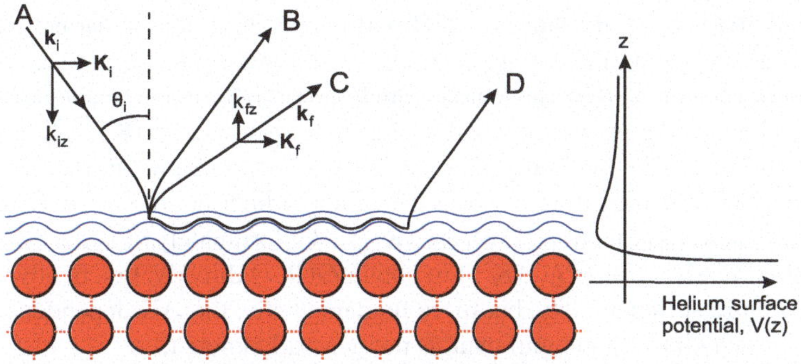

Fig. 13.2 Schematic helium-surface interaction potential, adapted from [5]. The potential follows the surface periodicity and has the general z-dependence shown. Incoming helium atoms (*A*) can scatter (or diffract) elastically (*B*), scatter inelastically by exchanging energy with the substrate (*C*) or can enter selective adsorption resonance (*D*) where they are transiently trapped in the potential well

methods have therefore been developed [14]. Fortunately, for the interpretation of dynamical data, accurate intensities are not required and for dynamical analysis the simplest kinematic approximation has been shown to work remarkably well [22]. The two fundamental parameters required are the surface-parallel momentum transfer during scattering, $\hbar\Delta\mathbf{K}$ and the energy transfer, $\hbar\Delta\omega$ (capital letters are used to indicate quantities in the surface plane). The magnitude of $\Delta\mathbf{K}$ is determined by the total scattering angle of the instrument, θ_{SD}, and the angle of incidence of the beam onto the sample, θ_i,

$$\Delta K = k_f \sin(\theta_{SD} - \theta_i) - k_i \sin(\theta_i), \tag{13.1}$$

while its direction is given by the azimuthal orientation of the sample. The energy transfer is measured during an experiment. The kinematic approximation assumes the scattered intensity, I_s, originates from an array of independent (mobile) scattering centres, so can be written as a product,

$$I_s(\Delta\mathbf{K}, \Delta\omega) = S(\Delta\mathbf{K}, \Delta\omega) \cdot \left|F(\Delta\mathbf{K}, \Delta\omega)\right|^2. \tag{13.2}$$

$F(\Delta\mathbf{K}, \Delta\omega)$ is an (amplitude) form factor, related to the shape of each species, while $S(\Delta\mathbf{K}, \Delta\omega)$ is an (intensity) structure factor which depends on the positions of the scatterers. van Hove showed that the dynamic structure factor, $S(\Delta\mathbf{K}, \Delta\omega)$, is related to a pair correlation function, $G(\mathbf{R}, t)$, by a double Fourier transform in both space and time [23],

$$G(\mathbf{R}, t) \overset{\textit{Spatial FT}}{\Longleftrightarrow} I(\Delta\mathbf{K}, t) \overset{\textit{Temporal FT}}{\Longleftrightarrow} S(\Delta\mathbf{K}, \Delta\omega), \tag{13.3}$$

so giving a remarkably simple relationship between scattering and surface dynamics. The 'van Hove pair correlation function', $G(\mathbf{R}, t)$, in principle provides a complete description of adsorbate dynamics and can be interpreted classically as the probability of finding an atom at $(\mathbf{R}, t)$ providing there was an atom at $(0, 0)$. It can also be subdivided into the sum of a self part (same atom) and a distinct part (different atom). Providing the energy dependence of the form factor is weak over the range of interest and can be neglected, which is normally the case as quasi-elastic energy exchanges are generally small, (13.2) indicates that $S(\Delta\mathbf{K}, \Delta\omega)$ is proportional to the observable quantity in an ideal TOF experiment. In contrast, HeSE experiments yield a measurement of polarisation, which is proportional to the intermediate scattering function (ISF), $I(\Delta\mathbf{K}, t)$.[1] Unfortunately, it is not possible to invert either $S(\Delta\mathbf{K}, \Delta\omega)$ or $I(\Delta\mathbf{K}, t)$ to obtain $G(\mathbf{R}, t)$ directly. Practically it is difficult to acquire sufficient data, but more fundamentally, the $\Delta\mathbf{K}$ dependence of the form factor, which is not insignificant, is not generally known.

13.2.2 The Spin-Echo Technique

The spin-echo principle is illustrated schematically in Fig. 13.3. A thermal beam of unpolarised helium-3 is produced in a supersonic expansion, **A**, resulting in a typical energy spread of 5–20 %. The beam is passed through a spin-polariser, **C** (a hexapole magnet followed by a dipole magnet) which transmits only one nuclear spin-polarisation of the beam and aligns the outgoing polarised atoms in a direction perpendicular to the axis. The beam is directed through a precession solenoid field, **D**, where the classical spin vector, **S**, undergoes precession according to

$$\frac{d\mathbf{S}}{dt} = \gamma \mathbf{S} \times \mathbf{B}, \tag{13.4}$$

where γ is the gyromagnetic ratio for helium-3 ($\gamma/2\pi = 32.43$ MHz/T). **B** is oriented along the direction of the beam, so the spins precess about that direction. The exact precession angle depends on the velocity of each individual atom concerned. The encoded beam is incident on a sample surface, **E**, from which it scatters. Some of the scattered atoms pass down the second beamline and through a second reverse precession field, **F**, before entering a spin-analyser, **G** (a second hexapole magnet). The analyser transmits only one selected polarisation into the detector, usually the cosine or 'x' component, corresponding to the analyser and polariser being aligned.

The total precession angle depends on the time, T, the helium atom spends within the field and is given by

$$\phi = \gamma \int_0^T B\,dt = \frac{\gamma}{v} \int_0^L B\,dl, \tag{13.5}$$

[1] Polarisation is proportional to the ISF when the energy changes on scattering are small compared with the mean beam energy. Strictly, there is a Fourier relationship between wavelength and magnetic field integral, so in the general case a non-linear scaling also needs to be applied, as described in Sect. 13.2.5.

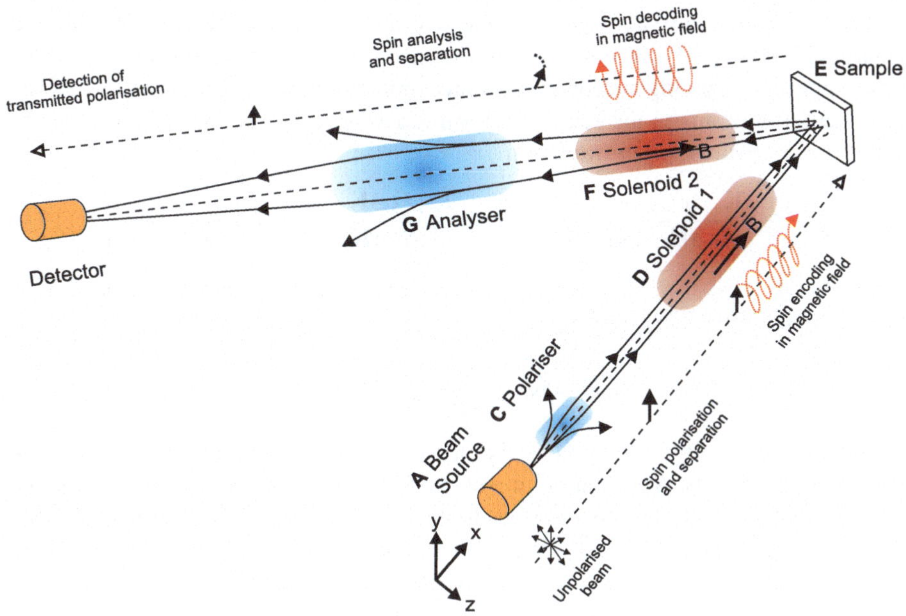

Fig. 13.3 Schematic of the spin-echo technique, adapted from [5]. The ^{3}He beam (*A*) is passed through a nuclear spin polariser (*C*), then into a solenoid (*D*), where the spins precess about the beam axis. The beam then scatters from a sample (*E*), and part of the scattered distribution passes through an identical reversed solenoid (*F*), spin-analyser (*G*) and into a helium detector. If the helium atoms experience an energy change during scattering, the final spin-direction will change, leading to a decrease in the beam averaged polarisation

where v is the velocity of atoms in either coil and L is the length of the field. The total phase accumulated in both coils is

$$\phi = \phi_1 + \phi_2 = \frac{\gamma}{v_1} \int_0^L B_1 \, dl + \frac{\gamma}{v_2} \int_0^L B_2 \, dl. \tag{13.6}$$

If the fields are equal and opposite, which is the usual condition for 'standard' spin-echo experiments, we obtain

$$\phi = \gamma \left(\frac{1}{v_1} - \frac{1}{v_2} \right) \int_0^L B \, dl \approx \gamma \frac{(v_2 - v_1)}{v_1^2} \int_0^L B \, dl, \tag{13.7}$$

which is valid when the changes in velocity are small, compared with the mean beam velocity. Since $\hbar \Delta\omega = \frac{1}{2} m v_2^2 - \frac{1}{2} m v_1^2 \approx m v_1 (v_2 - v_1)$, we obtain

$$\phi = \frac{\gamma}{m v^3} \hbar \Delta\omega \int_0^L B \, dl, \tag{13.8}$$

where we now take v to be the mean velocity of atoms in the beam, which indicates that ϕ provides a first order approximation to the energy change experienced by an individual helium atom.

Real measurements average over the distribution of energies in the scattered beam. The spin analyser transmits atoms according to the cosine of their classical spin-phase, giving the beam averaged polarisation, $P_x = \langle \cos \phi \rangle$. Since the distribution of atoms in the scattered beam is given by $S(\Delta\mathbf{K}, \Delta\omega)$, P_x can be expressed as a cosine Fourier transform,

$$P_x = \frac{\int S(\Delta\mathbf{K}, \Delta\omega)\cos(\Delta\omega t)\, d\Delta\omega}{\int S(\Delta\mathbf{K}, \Delta\omega)\, d\Delta\omega} = \frac{Re[I(\Delta\mathbf{K}, t)]}{S(\Delta\mathbf{K})} = Re\left[\frac{I(\Delta\mathbf{K}, t)}{I(\Delta\mathbf{K}, 0)}\right] \quad (13.9)$$

where $I(\Delta\mathbf{K}, t)$ is the ISF described earlier. The quantity, t, usually referred to as the 'spin-echo time', is given by

$$t = t_{SE} = \frac{\gamma\hbar}{mv^3}\int_0^L B\, dl. \quad (13.10)$$

t_{SE} depends on the field integral of the instrument and is therefore proportional to the solenoid currents. A similar argument shows that measurement of the orthogonal polarisation, $P_y = \langle \sin \phi \rangle$, corresponds to the imaginary part of the normalised ISF, $P_y = Im[I(\Delta\mathbf{K}, t)/I(\Delta\mathbf{K}, 0)]$. Hence, a 'standard' spin-echo measurement, with equal currents in both coils, yields the complete normalised ISF.

13.2.3 The ISF as a Correlation Function

At first, the ISF seems a rather inpenetrable description of surface dynamics. However, the relationship in (13.3) helps provide some physical insight. $G(\mathbf{R}, t)$ provides a description of adsorbate motion; at its simplest (taking just the self part), it gives the likelyhood of an adsorbate having moved by a distance $\mathbf{R}$, in time t. Since the ISF is the spatial Fourier transform of $G(\mathbf{R}, t)$, the ISF is also a correlation function (representing the self-similarity of the adsorbate configuration with time), but expressed in reciprocal space. Since $\Delta\mathbf{K}$ bears a reciprocal relationship with periods in real space ($2\pi/\Delta\mathbf{K}$ for a unit cell with orthogonal basis vectors) it is useful and intuitive to consider the ISF as representing apparent surface correlation, on the lengthscale and direction given by $\Delta\mathbf{K}$, over the time interval, t.

Figure 13.4(a) shows the form of a typical HeSE ISF. Adsorbate diffusion leads to a (generally exponential) decay, which in both cases shown corresponds to adsorbate correlation being lost with a characteristic time of 200–300 arb. units. The ISF does not decay fully in either case, indicating some correlation remains after large times, for example due to scattering from static species, or confinement on the lengthscale characteristic of the measurement. Periodic processes, such as adsorbate vibrations, lead to oscillations in the ISF which decay with the lifetime of the mode. Figure 13.4(b) shows the energy domain spectrum, obtained by Fourier transforming the ISF. Here, the exponential decay relates to a Lorentzian broadening of the elastic peak, accompanied by a sharp elastic signal and peaks at finite energy transfer.

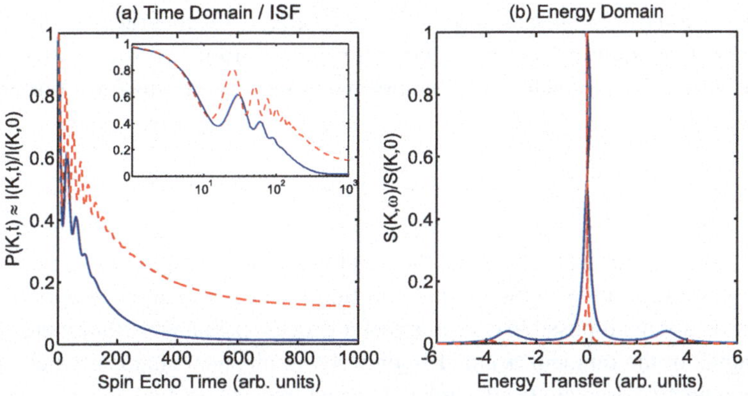

Fig. 13.4 Illustration of the typical form of a HeSE measurement in (**a**) the time domain (ISF) and (**b**) Fourier transformed to the energy domain. The *inset* shows the data with $\log(t)$ to cover a wide time range. Diffusion corresponds to an exponential decay in the ISF or a Lorentzian quasi-elastic broadening in energy. Periodic motion leads to finite energy transfer modes, corresponding to oscillations in the ISF. An elastically scattered component of the beam, due to scattering from immobile species, corresponds to a finite background in the ISF, or a sharp peak at $\Delta\omega = 0$

13.2.4 Semi-classical Formalism & Correlation Interpretation

An alternative semi-classical formalism of spin-echo has been described by Gaehler [24], which provides an intuitive understanding of the origin of "spin-echo time". Qualitatively, the nuclear spins within the beam are polarised perpendicular to the beamline, as before, and are transferred into the solenoid field. Quantum mechanically, the nuclear spin must align parallel or antiparallel with the field, so the initial polarisation can be written as a superposition. The energies of these two components are split by the field, so one travels faster and reaches the sample first, followed shortly by the other. The temporal separation between the two components at the sample corresponds to the spin-echo time. After scattering, the components are recombined and if the surface does not change inbetween, the original polarisation is regained. However, if the surface changes, coherence between the scattered components is lost and the resulting polarisation reflects the corresponding loss of correlation [5].

13.2.5 Wavelength Transfer Matrix & Tilted Projection Formalism

In Sect. 13.2.2 we described a 'standard' spin-echo experiment. More generally, the spin-phase accumulated in either solenoid, ϕ, is proportional to the wavelength,

$$\phi = \frac{\gamma m \lambda}{2\pi h} \int_0^L B\, dl = C\lambda I \tag{13.11}$$

where $C = \gamma m B_{eff}/2\pi h$ and B_{eff} is the magnetic field integral per unit current. Hence, for two identical coils the complex polarisation measured after passing through both coils is given, without approximation, by the double Fourier transform,

$$P(I_1, I_2) \propto \iint I(\lambda_1, \lambda_2) \exp(2\pi i C \lambda_1 I_1 + 2\pi i C \lambda_2 I_2)\, d\lambda_1\, d\lambda_2. \qquad (13.12)$$

$I(\lambda_1, \lambda_2)$ is a matrix that represents the mapping of wavelengths λ_1 in the first coil (before scattering) to λ_2 in the second coil (after scattering) and contains complete information about all possible energy transfer processes that are 'illuminated' by the wavelengths in the incident beam. The underlying physical characteristics (e.g. the width of a particular excitation) can be obtained from $I(\lambda_1, \lambda_2)$.

$I(\lambda_1, \lambda_2)$ can be obtained by measuring $P(I_1, I_2)$ over a sufficiently wide range of currents and Fourier transforming. However the process is extremely time consuming and has only recently been demonstrated [25]. A much faster alternative is the 'tilted-projection' approach described by Alexandrowicz [26, 27]. Essentially, any 1D measurement with a fixed ratio between I_1 and I_2 corresponds to a projection of the $I(\lambda_1, \lambda_2)$ matrix onto an equivalent axis. Hence a series of 1D measurements may be tuned, by selecting specific current ratios, to give an ultra-high energy resolution measurement of a particular mode. Selection of appropriate current ratios and determination of the actual energy-resolution requires knowledge of beam-velocity distribution, angular spread and dispersion characteristics of the mode being examined [26].

The most important projections correspond to (i) equal currents in both coils and to (ii) current in one coil only. The former is the standard spin-echo condition, which is only sensitive to energy *changes* on scattering. Conversely, case (ii) enables the *absolute* energy distribution in the beam to be quantified, providing the same information as in a TOF experiment and an essential day-to-day calibration.

13.3 Experimental Setup and Method

Figure 13.5 shows a schematic of the Cambridge HeSE instrument [21], which illustrates the key elements in any HeSE system. The beam is produced in gas recycling beam source using a supersonic nozzle expansion [28] and is directed into the scattering chamber. Here, the beam is incident on the sample surface, mounted on a precision manipulator. In the Cambridge instrument the manipulator allows 6-axes of adjustment (translation and rotation in all directions), to accurately direct scattered atoms with particular $\Delta\mathbf{K}$ down the outgoing beamline, to the detector. Magnetic components are positioned along the beamlines to implement the nuclear spin-manipulation.

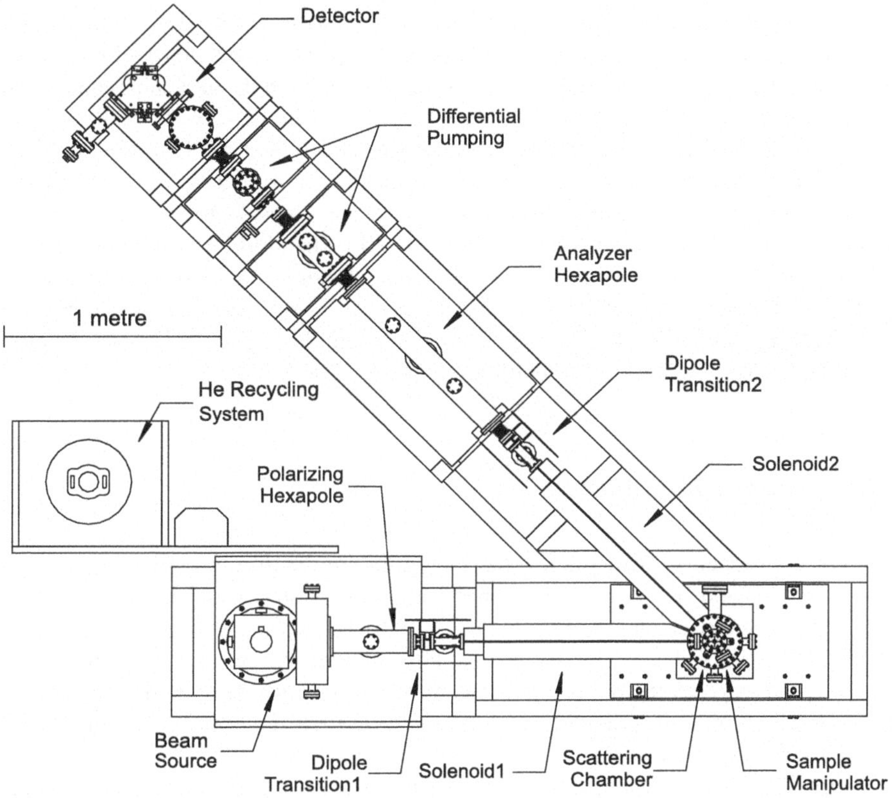

Fig. 13.5 Top-down schematic of the Cambridge HeSE instrument, adapted from [21]. The gas-recycling beam source is on the *bottom left*, and directs the beam through the magnetic components, onto the sample inside the scattering chamber (*bottom right*). Scattered atoms pass along the outgoing beamline to a mass-spectrometer detector

13.3.1 General Characteristics

The 45° scattering angle was chosen to give large surface-parallel momentum transfer, $\Delta\mathbf{K}$, from the momentum in the beam, $\mathbf{k}_i$, while the fixed geometry simplifies construction of the apparatus and gives stable, consistent behaviour. The nominal beam energy is a crucial parameter. The polariser and analyser are designed to work correctly with a single beam energy, and at other energies the polarised intensity is weaker. The Cambridge instrument is designed to operate with $E = 5k_B T/2 = 8$ meV, which represents a compromise between resolution (enabling a wide range of systems to be studied) and momentum transfer (enabling the mechanism of adsorbate motion to be determined). With an 8 meV beam (37 K nozzle) spin-echo times of 680 ps can be reached, at $\Delta\mathbf{K}$ greater than 4 Å^{-1}. Measurements of vibrational modes generally benefit from better excitation with higher energies, typically 12–15 meV, even though the intensity is weaker.

The central region of the instrument, between polariser and analyser, is carefully magnetically shielded to avoid stray fields (including the earth's field) which would otherwise lead to beam depolarisation. A further, 'Helmholz-like' magnetic field, perpendicular to the plane of the instrument, is applied in the vicinity of the sample to give an additional 45° precession and compensate for the total scattering angle.

All HeSE detectors are high sensitivity mass-spectrometers, tuned to detect mass-3. Helium is particularly difficult to ionise, and typical 'good-quality' detectors usually only have an efficiency of about 1×10^{-5}. In addition, residual hydrogen in the vacuum system and the natural abundance of deuterium leads to a significant background due to HD, which also has a mass of 3 amu. HeSE experiments are almost always signal limited, and there are considerable ongoing efforts to develop highly sensitive, low background helium detectors [29, 30].

13.3.2 Spin-Manipulation

The key components for magnetic spin-manipulation are the spin-polariser, precession solenoids and spin-analyser. Strongly inhomogeneous magnetic fields are used for the spin-polariser and analyser [18, 21], which exert a force on each ^{3}He nuclear spin given by $\mathbf{F} = \pm\mu\nabla|\mathbf{B}|$, so spatially separate the two polarisations as the beam passes through them. Hexapole magnets are typically used as $|\mathbf{B}|$ then varies quadratically with radius, leading to true geometric focusing of one spin-component and defocussing of the other.

In the Cambridge instrument, the polarising hexapole [31] is designed to focus one spin-component, diverging from the source, onto the sample, substantially improving beam intensity. The other spin-component is defocused and mostly hits the internal walls of the polarising magnet. For use with intense molecular beams the magnet must allow sufficient gas conductance around the polepieces for the defocused gas to escape before it blocks the beamline. However, since the nuclear spin-interaction is extremely weak, making a sufficiently strong and accurate magnet with suitable gas conductance was a substantial challenge ($\sim$1.1 T at a polepiece radius of 1 mm, 300 mm long, assembled to $\sim$10 μm precision), and formed a crucial step in the development of the apparatus.

The analyser magnet is designed to focus one polarisation of the diverging atoms scattered from the sample into the detector. After scattering the beam is much less intense, so gaps between polepieces are not required. Hence, the Halbach method of hexapole construction can be used, giving slightly stronger magnetic fields for the same internal bore [32].

On leaving the polarising hexapole (and after entering the analyser), the nuclear spins are polarised with respect to the local hexapole field direction. Hence, the polariser is followed by an additional dipole field to orient the spins in a uniform spatial direction. A similar field is required immediately before the spins enter the analyser, and the relative orientation of these select the real component of the polarisation (dipole fields aligned) and the imaginary component (dipole fields rotated by $\pi/2$).[2] The rate of field change in these transition regions is carefully adjusted so that the spins follow the field direction between the hexapole and dipole, but see an abrupt change between the dipole and solenoid, so begin precessing [21].

In the Cambridge instrument, the precession solenoids are conventional copper windings with a cosine winding pattern to minimise aberrations [21]. They represent a compromise between energy resolution, intensity, thermal management and mechanical rigidity. Longer solenoids give greater field integrals, hence higher resolution (longer spin-echo times), but are less mechanically rigid. In addition, the scattered intensity drops off with the inverse square of the outgoing solenoid length. Conversely, thicker windings give stronger fields but more substantial aberrations and make adequate cooling more difficult. The solenoids are 0.75 m long and consist of 15 double layers of 2×1 mm copper windings on a 32 mm former. The windings are bonded with boron-nitride loaded epoxy to improve thermal conductivity and are cooled with inner and outer water jackets, enabling continuous operation at currents of up to 8 A, giving a peak field of $\sim$0.15 T. The field integrals calibrate the instrument and can be obtained precisely either by analytic means, or by experimental calibration using a monochromating crystal such as LiF [21].

13.3.3 Using HeSE to Study Adsorbate Dynamics

The main application of HeSE is the study of adsorbate transport on surfaces. The underlying framework for analysis comes from the principles established by van Hove [23], the key being the Fourier relationship between the correlation function, $G(\mathbf{R}, t)$, and the intermediate scattering function, $I(\Delta\mathbf{K}, t)$, summarised by (13.3). The underlying principle of *experimental* HeSE analysis is to identify the nature of the motion by finding the simplest surface dynamical model which reproduces the measurements of $I(\Delta\mathbf{K}, t)$, along with its temperature and coverage variation. To do so, we make use of the *shape* of $I(\Delta\mathbf{K}, t)$.

For a perfectly static surface there is no change in its correlation with time, so the normalised ISF corresponds to a line at unity. If particles on the surface are diffusing, the overall level of $I(\Delta\mathbf{K}, t)$ decays with t, although it will only decay to zero if the entire scattered intensity comes from mobile species. For simple forms of random motion, such as hopping or Brownian motion, the adsorbate correlations decay exponentially, leading to the general form

$$I(\Delta\mathbf{K}, t) = A\exp(-\alpha t). \tag{13.13}$$

α is a decay constant, often referred to as the "dephasing rate", which characterises the rate of motion in a particular measurement. The variation of α with $\Delta\mathbf{K}$ is then of particular interest; its absolute value provides the rate of motion, while its variation with $\Delta\mathbf{K}$ corresponds to the mechanism.

---

[2]Practically, it is easier to select the imaginary component by using an additional coil to add an extra $\pi/2$ rotation just after the polariser.

Generally, $I(\Delta\mathbf{K}, t)$ contains contributions from several processes, not just adsorbate diffusion (e.g. substrate phonons) and practically, these complicate the fitting of experimental measurements. Ideally, the exponential decay is well separated in time from other processes, in which case fitting is straightforward. A three parameter fit, of the form $a\exp(-bt) + c$, is standard (representing decay amplitude, decay constant, residual value), but can be reduced to two parameters, with a subsequent improvement in fit reliability, if the exponent decays sufficiently quickly that the residual polarisation can be unambiguously identified.

In some cases the timescales are not well separated. Combined models encompassing vibration and diffusion have been proposed (e.g. [33]) but at present there is no standard approach to tackling this problem. Fourier transforming and scaling the data to the energy domain can be particularly useful (a time-domain exponential dephasing rate of $\alpha = 1\ \text{ns}^{-1}$ corresponds to an energy-domain Lorentzian FWHM of 1.32 μeV). In the energy domain, vibrational modes often appear more clearly separated from the quasi-elastic peak, enabling either Fourier filtering or direct energy domain fitting, as in the case of Cs/Cu(001) [34].

13.3.4 HeSE Signatures for Simple Forms of Motion

Analytic forms of $I(\Delta\mathbf{K}, t)$ exist for simple, idealised types of adsorbate motion. Generally, these can be obtained by constructing $G(\mathbf{R}, t)$ for the desired motion, then Fourier transforming to give the ISF [5]. Here, we state the key results.

The simplest form of motion is continuous Brownian motion. In this case, the ISF decays exponentially with time,

$$I(\Delta\mathbf{K}, t) = \exp\bigl(-\Delta\mathbf{K}^2 D|t|\bigr). \tag{13.14}$$

and gives a dephasing rate, $\alpha(\Delta\mathbf{K})$, which varies quadratically with momentum transfer,

$$\alpha(\Delta\mathbf{K}) = D\Delta\mathbf{K}^2, \tag{13.15}$$

as illustrated in Fig. 13.6. In the limit of large length scales, i.e. measurements at sufficiently small $\Delta\mathbf{K}$, all diffusion must conform to this macroscopic limit. D is the tracer diffusion constant for the motion [3], which is related to the adsorbate-substrate energy exchange rate (friction), η, through the Einstein model of Brownian motion, $D = k_B T/(m\eta)$.

When the corrugation of the substrate is important, we expect discrete hops between adsorption sites. In this case, Chudley & Elliot showed [35] that an exponentially decaying ISF results [15], where the dephasing rate is given by the periodic function,

$$\alpha(\Delta\mathbf{K}) = 2\sum_j \nu_j \sin^2\left(\frac{\Delta\mathbf{K}\cdot\mathbf{j}}{2}\right). \tag{13.16}$$

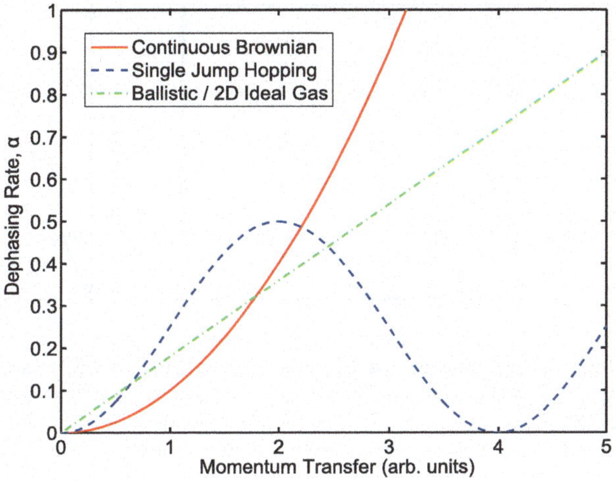

Fig. 13.6 Comparison of the expected variation of dephasing rate, α, with momentum transfer for several simple forms of motion. Continuous Brownian motion leads to a quadratic dependence, hopping results in a sinusoid, while ballistic (ideal-gas) like behaviour leads to a linear dependence

The summation is over the possible jump vectors, $\mathbf{j}$, weighted by the specific hopping rate for that jump, ν_j. For single jumps it gives the sinusoidal form illustrated by the dashed blue lines in Fig. 13.6.

Finally, when the corrugation of the substrate is small and the coupling to the substrate is also weak, we expect ballistic 2D gas like motion. In this case, the ISF takes a Gaussian form [5, 36], the width of which varies linearly with $\Delta\mathbf{K}$, as shown by the dot-dashed green line in Fig. 13.6.

13.3.5 Signatures for More Complex Forms of Motion

A number of other forms of motion have been identified in HeSE data and modelled. We do not describe these exhaustively here, but mention the key characteristics and provide references for further information. Generally, ISFs for independent motions may be combined by multiplication [13],

$$I(\Delta\mathbf{K},t) = I_1(\Delta\mathbf{K},t) \cdot I_2(\Delta\mathbf{K},t) \tag{13.17}$$

which enables more sophisticated models to be constructed.

Motion on Non-Bravais Lattices The jump model described in (13.16) applies to hopping motion between sites on a Bravais lattice. When hops can occur between sites which do not form a simple lattice (i.e. more than one lattice point in the basis) then more sophisticated modelling is required. Analytic forms analogous to the Chudley & Elliot formula have been established by Tuddenham et al. [37], based on a procedure established for interpretation of neutron scattering data. Several models have been given, but the one likely to be most frequently applied is for adsorption and hopping between sites which form a hexagonal network. Examples include the hollow sites on an f.c.c. (111) surface and the carbon atoms in a graphene sheet.

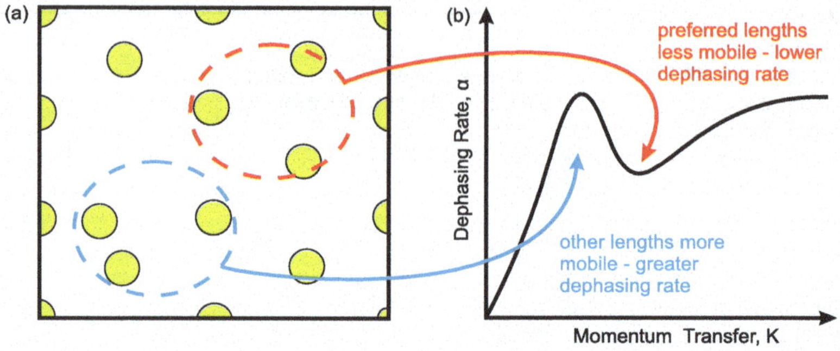

Fig. 13.7 (**a**) Illustrates a snapshot of the dynamic quasi-hexagonal structure formed by strongly repelling adsorbates. (**b**) Shows the corresponding peak and dip structure in a plot of $\alpha(\Delta\mathbf{K})$. Structures on the preferred hexagonal lengthscale are more stable, leading to a dip in the dephasing rate. Other structures are less stable, leading to increased dephasing rates

For jumps between these sites, the ISF is expected to consist of two exponentially decaying terms, whose amplitude and decay constants follow a particular relationship.

Correlated Motion When strong forces are present between adsorbates, the motion of one particle influences others nearby. A measurement of this *correlated* motion provides information about the corresponding adsorbate-adsorbate interaction potential. Strongly repelling adsorbates tend to order themselves into a quasi-hexagonal structure, as illustrated in Fig. 13.7(a). The increased stability of such structures lead to lower dephasing rates at the associated values of momentum transfer, i.e. $\Delta K = 4\pi/\sqrt{3}a_{hex}$, which is also the location of the corresponding diffraction ring. The increased stability forms a dip in $\alpha(\Delta\mathbf{K})$ (shown in Fig. 13.7(b)), known as a *de Gennes narrowing*, as it also corresponds to a narrowing of the quasi-elastic broadening in $S(\Delta\mathbf{K}, \Delta\omega)$. On other length scales, the corresponding structures are less stable, leading to an increased dephasing rate at other values of ΔK. Measurements of the magnitude and coverage dependence of the peak and dip features, in conjunction with MD simulations, enable the form of the inter-adsorbate potential to be examined [38, 39].

Confined Diffusion & Intracell Motion The models above describe unbounded diffusive motion. In the case of confined motion (e.g. between steps or other forms of nanoscale confinement) the surface configuration does not become completely uncorrelated at large times. Hence, the ISF is expected to decay to a finite value and its form is not a simple exponential. The level to which the ISF decays depends on the lengthscale of the confinement and the momentum transfer (i.e. associated lengthscale) of the measurement, and a simple quantitative relationship is given in [27, 38]. In neutron scattering studies, confinement has been distinguished through the residual level to which the ISF decays over large times [13], and these models can in principle be applied to HeSE. However, in HeSE studies the main difficulty is distinguishing this level from contributions due to elastic scattering from other elements of the surface. To date there have been few studies in confined regimes, and these have been mainly interpreted through simulation.

Rotational Motion Individual diffusing atoms and small symmetric molecules are generally well approximated by point scatterers (an extended, but symmetric adsorbate cross section has little effect [22]). However, many adsorbed species can reorient around their point of attachment to the surface. Such reorientation is a prerequisite for processes such as self-assembly, and can occur at much lower temperatures than translation. Aperiodic rotational motion, such as jump rotation, is a special form of confined diffusion and can also be observed using quasi-elastic scattering [40, 41]. A rotational ISF consists of the sum of several exponential decays, depending on the relevant geometry. The decay constant of each exponential term is constant with ΔK, but each prefactor varies, so result in a non-exponential decaying form which varies in shape with momentum transfer [41].

13.3.6 Complimentary Tools: MC and MD Numerical Simulations

Although analytic models provide the ISF for typical forms of adsorbate motion, most real systems do not conform to idealised forms. While they are valuable in developing a qualitative understanding, numerical simulations are widely used for quantitative analysis. There are two components, generation of adsorbate trajectories and simulation of quasi-elastic scattering. The aim is to produce a 'virtual experiment' which mimics the real system by capturing the essential physics within the simplest possible model.

Molecular dynamics (MD) simulations are normally used for QHAS interpretation, however Monte-Carlo (MC) simulations are also useful. In MC simulations, adsorbates sit on discrete lattice sites and make activated jumps based on a discrete jump propagation routine such as the Metropolis algorithm. MC simulations run very quickly and are useful to study geometric effects, such as the availability of adsorption sites. However the MC trajectory quantisation in both space and time is not representative of real experiments. MD methods aim to realistically simulate adsorbate motion by continuously integrating the classical equations of motion governing the system. Both slab [42] and Langevin approaches [38, 39, 43–45] have been applied to QHAS interpretation. Slab calculations build a surface geometry from several layers of mobile atoms, all of which interact with each other through explicitly defined potentials. In principle, they can give realistic phonon interactions with the substrate. The Langevin approach, which has been much more widely used, considers the interaction between the adsorbate and substrate through an effective adiabatic or 'frozen' potential, V. The generalised Langevin equation of motion is given by

$$m\ddot{\mathbf{R}}_i = \nabla V(\mathbf{R}_i) - m\eta\dot{\mathbf{R}}_i + \xi_i(t) + \sum_{i\neq j} U(\mathbf{R}_i - \mathbf{R}_j). \tag{13.18}$$

η is a frictional coupling constant, which, along with the (white noise) random impulse term, ξ, describes energy transfer to and from the surface (ξ is scaled from η according to the fluctuation dissipation theory [46], and is not an independent free parameter). The final term is added to allow for pairwise forces between adsorbates, U. Generally, Langevin simulations run much more quickly than slab calculations. The approximation works well providing the adsorbate particles are not too light, and the adsorbate–substrate interaction is not too stiff [3].

Once n adsorbate trajectories, $\mathbf{R}_i(t)$, have been obtained, the ISF may be straightforwardly constructed using the kinematic approximation. A time dependent scattered amplitude is given by

$$A(\Delta\mathbf{K},t)=\sum_{i=1}^{n}\exp\big(-i\,\Delta\mathbf{K}.\mathbf{R}_i(t)\big),$$

which is related to the ISF by autocorrelation. By Fourier transforming $A(\Delta\mathbf{K},t)$ with respect to time, then multiplying the result by its complex conjugate, we form the dynamic structure factor,

$$S(\Delta\mathbf{K},\Delta\omega)=\left|\int A(\Delta\mathbf{K},t)\exp(-i\,\Delta\omega t)\,dt\right|^2,$$

which can then be inverse transformed using (13.3) to give $I(\Delta\mathbf{K},t)$. The simulated ISF can then be compared with experiment, enabling the parameterisation of the interaction potential and value of η, to be refined. Usually, the experimental and simulated dephasing rates are compared, rather than the raw ISFs, which therefore require fitting of exponential decays to both ISFs. Typically simulated and experimental data is treated identically, to minimise systematic misfitting effects and give a fair comparison.

13.4 Applications of the Technique

A wide range of systems have now been studied using HeSE. In this section we describe a selection of experiments to illustrate the scope and limitations of the technique.

13.4.1 Self-diffusion and Adsorbate-Substrate Potentials

The earliest and conceptually most straightforward HeSE studies correspond to measurements of the diffusion rate and mechanism of CO molecules. Figure 13.8 shows a plot of the dephasing rates obtained for CO diffusing on Cu(001) [47] and is typical of processed HeSE data; each data point corresponds to an exponential fit to an individual measurement of the ISF. The two sides of the graph show data along $\langle 110\rangle$ and $\langle 100\rangle$. The general form clearly indicates a jump diffusion mechanism, following (13.16), while the small deviation from a single sinusoid indicates

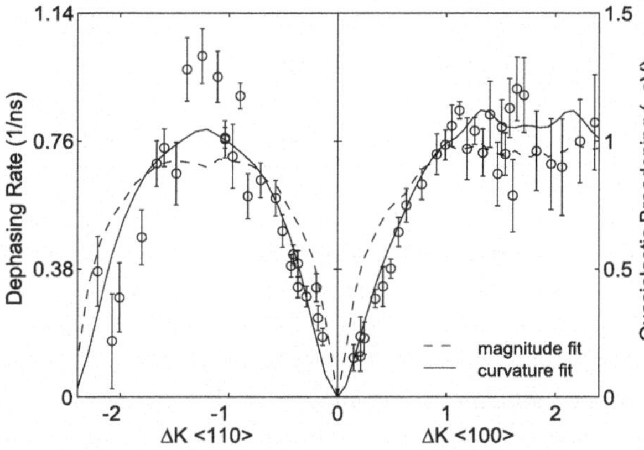

Fig. 13.8 Plot showing typical HeSE results for mobile CO on Cu(001), adapted from [47]. The data is shown as a dephasing rate, α, varying with momentum transfer, $\Delta\mathbf{K}$, which provides a characteristic signature of the motion and was taken for a CO coverage of 0.1 ML at 190 K. The *left* and *right* sides shows measurements along $\langle 110\rangle$ and $\langle 100\rangle$ directions respectively. The *lines* show two molecular dynamics fits to the data, which are described in the text

a distribution of jump lengths are present. The magnitude of $\alpha(\Delta\mathbf{K})$ gives the absolute transport rate. The temperature dependence of α can be used to estimate the rate limiting potential barrier, using the standard Arrhenius procedure.[3] However, since on (100) surfaces all measurements contain projected contributions from both $\langle 100\rangle$ and $\langle 110\rangle$, only the lowest barrier can be obtained regardless of measurement direction. Instead, it can be shown using (13.16) that since the peak values in the $\alpha(\Delta\mathbf{K})$ plots are approximately equal, the potential is roughly isotropic.

Practically, CO/Cu(001) diffuses relatively slowly, and in order to achieve measurable rates temperatures of around 190 K were required. In turn, these necessitated an overpressure of $\sim 2\times 10^{-6}$ mbar to stabilise a dynamic equilibrium coverage of 0.1 ML. Balancing the desorption characteristics and diffusion rates of a species in this way is a common requirement in HeSE experiments, as the rate of surface motion must correspond to the temporal measurement window. Generally, a maximum spin-echo time of ~ 1 ns means hopping rates of at least 10^9 s^{-1} are necessary.

To obtain a complete representation of the system, including accurate values for the adiabatic barriers to diffusion, the system was modelled using the Langevin MD approach. The basic form of the adsorbate-substrate potential was

$$V_s(x,y)=\frac{1}{4}E_H\left(1-\cos\frac{2\pi x}{a}\right)\left(1-\cos\frac{2\pi y}{a}\right)+\frac{1}{2}E_B\left(1-\cos\frac{2\pi x}{a}\cos\frac{2\pi y}{a}\right),$$

which allows the diffusion barriers, E_H and E_B, along the atop and bridge directions to be varied independently. Although such a potential is often satisfactory, for the

[3] Effective activation energies, derived from an Arrhenius plot of α with $1/T$, usually underestimate the true adiabatic barrier. Accurate values can be obtained using MD simulations [48].

CO/Cu(100) analysis further refinements were included. A different potential was used to represent the region around the adsorption site,

$$V_a(x, y) = \frac{1}{2}k(x^2 + y^2)[1 + u(x^2 + y^2)],$$

and the two were combined using $V(x, y) = \gamma V_a + (1 - \gamma)V_y$, with $\gamma = \exp -\alpha((x/a)^2 + (y/b)^2)^\beta$. Values for k and u were given by a previous study of the frustrated vibrational mode of CO [49], while $\alpha = 1000$ and $\beta = 4$ were chosen to give a smooth transition. The remaining three parameters, E_H, E_B and η were adjusted by running MD simulations and optimising, giving $E_H = 115 \pm 20$ meV, $E_B = 135 \pm 20$ meV and $\eta = 1/13$ THz. The fit to the data ('magnitude fit' in Fig. 13.8) revealed a significant discrepancy around $\Delta K = 0.3\ \text{Å}^{-1}$, indicating that the simulations substantially overestimate the number of multiple jumps occurring. A simple way of reducing the multiple jumps was to increase the friction, although this also increases the overall rate of motion. The 'curvature fit' in Fig. 13.8 illustrates this effect, but has also been scaled down by a factor of ~ 5 to lie on top of the data, for comparison. It was proposed that the true mechanism for reducing the long jump distribution is either position dependent friction, or more likely is related to the true 6-d geometry of the system, for example by coupling between frustrated rotation and translation [47].

The optimisation procedure described above is a rather general one and has been applied to a number of systems including several alkali metals on Cu(001) [34, 38, 50], CO/Pt(111) [39] and propane/Pt(111) [40]. Often interactions between adsorbates also need to be included, but the same principles apply. These experimentally optimised results for the potential and friction provide a valuable benchmark and validation of theoretical models, including density functional theory potentials and frictional coupling models.

13.4.2 *Diffusion on Non-Bravais Lattices*

As outlined in Sect. 13.3.5, species hopping between sites which do not form a simple Bravais lattice require a more sophisticated analysis, and the ISF is expected to consist of multiple exponentially decaying terms. The effect has recently been observed in the case of the fivefold symmetric cyclopentadienyl species (Cp) moving on a Cu(111) surface [51]. Cp jumps between *both* the f.c.c. and h.c.p. hollow sites and there are generally two exponential decays in the ISF [37], although along $\langle 110 \rangle$ one term has zero amplitude. Typical data, along with analytic fits are shown in Fig. 13.9(a). Two components are visible along $\langle 11\bar{2} \rangle$, but only one along $\langle 110 \rangle$, as expected, which is a convenient signature for this form of motion. The two dephasing rate components (the exponential decay rates from the ISF) are plotted in the usual way in Fig. 13.9(b) and form two branches. The relative position and shape of these branches enable the lifetime of the species in each site to be compared, corresponding to the relative heights of the potential. The Cp data is in excellent agreement with the form expected for degenerate f.c.c. and h.c.p. sites. Further

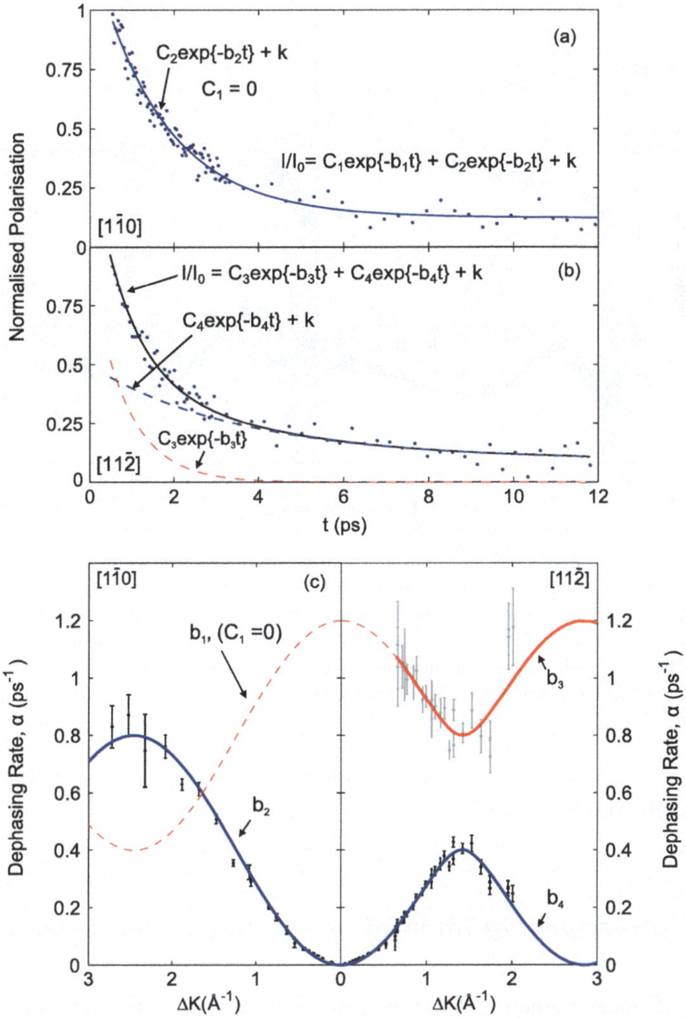

Fig. 13.9 Experimental ISFs for Cp on Cu(111) at room temperature, measured at 1.9 Å^{-1} along the (**a**) $\langle 110 \rangle$ and (**b**) $\langle 11\bar{2} \rangle$ directions, adapted from [51]; (**c**) shows the decay constants extracted from the exponential fits in (**a**) and (**b**), along with the expected form for hopping between f.c.c. and h.c.p. hollow sites

analysis revealed a potential corrugation of 40 ± 3 meV and that the adsorbate-substrate friction is high, $\eta = 2.5 \pm 0.5$ ps^{-1} [51]. Together these give a remarkably high mobility given the strength of the ionic bonding between Cp and the surface, which was attributed to the structural mismatch between adsorbate and substrate geometries. Adsorbate interactions are also remarkably weak, attributed to the cushion effect [51]. Together, these characteristics suggest that films grown from ionic

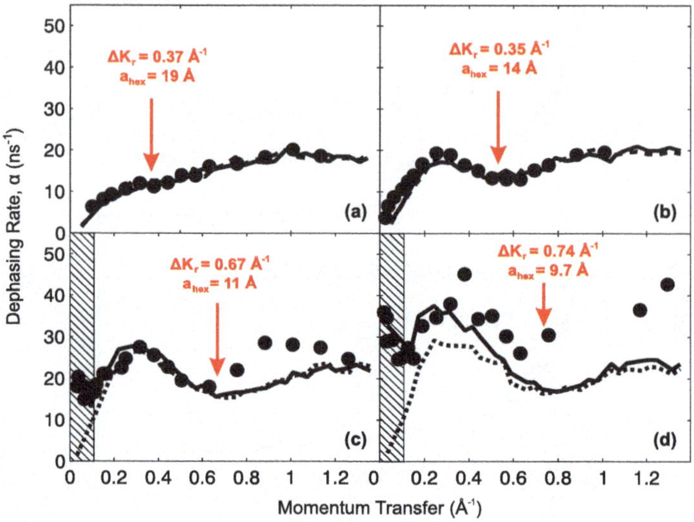

Fig. 13.10 Plot of the dephasing rates extracted from HeSE measurements of Na on Cu(001) at 155 K along the ⟨110⟩ direction, adapted from [38]. Panels (**a**) to (**d**) correspond to coverages of 0.02, 0.04, 0.065 and 0.08 ML respectively. Lines are for MD simulations including dipole-dipole adsorbate repulsion. The peak and dip structure (*red arrows*) emerges with increasing coverage, which characterises repulsive correlated motion. The hatched area corresponds to additional motion perpendicular to the surface, described in the text

molecular adsorbates could provide a valuable combination of new functionality and exceptional thermal stability.

13.4.3 Measurement of Inter-adsorbate Interaction Potentials

Several HeSE measurements have now provided information about forces between adsorbates, using the 'de Gennes narrowing' features that arise from correlated adsorbate motion. The first were for low to moderate coverages of Na on Cu(100) [38]. Figure 13.10 shows the corresponding $\alpha(\Delta\mathbf{K})$ characteristic as the coverage is increased from 0.02 ML to 0.08 ML. At low coverages the signature follows the sinusoid expected for jump diffusion, enabling the behaviour of an isolated Na atom to be characterised. As the coverage increases, characteristic peak and dip structures emerge (indicated by the red arrows), as described in Sect. 13.3.5. The lines show Langevin MD fits to the data using a Kohn & Lau dipole repulsion model for the interaction between adsorbates; the alkali atoms polarise on adsorption and the subsequent dipoles repel each other. The adsorbate-adsorbate interaction potential can be adjusted to fit the data, but in this case was derived from existing work function change data. The quality of the fit to the data confirms that pairwise repulsion is the dominant effect in this system. Similar effects have been observed with other

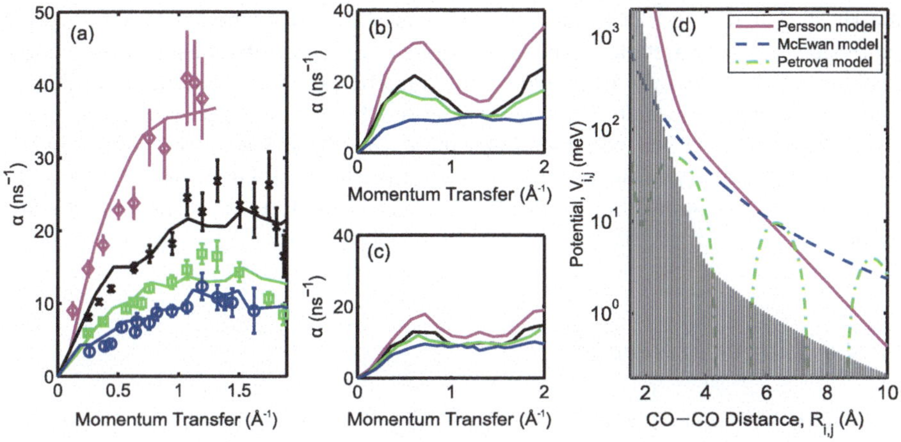

Fig. 13.11 Composite figure showing measurements and analysis of CO transport on Pt(111), adapted from [39]. (**a**) shows plots of the dephasing rate for CO moving on Pt(111) at coverages of 0.065, 0.165, 0.22 and 0.3 ML at 340 K along the $\langle 11\bar{2}\rangle$ azimuth along with the MD simulation fit. (**b**) and (**c**) show MD simulations for the incompatable CO–CO force models proposed by Persson [53] and Petrova [54], respectively. (**d**) compares several CO–CO force models with the approximate range which is compatable with the HeSE measurements (*shaded area*)

alkalis [34, 50], indicating that the behaviour is rather general. For the Na system, at the highest coverages studied an additional characteristic appears, marked by the hatched region in Fig. 13.10(c) and (d). Here, fluctuations in *local* coverage lead to deformation of the electron distribution around each Na atom, which cause apparent motion perpendicular to the surface [38, 52]. The solid lines in Fig. 13.10 show an MD simulation which includes this effect.

Quite different behaviour was observed in the case of CO on Pt(111) [39]. Here, strong pairwise repulsion was expected between adsorbed CO molecules, as it had been used to explain the changes in the heat of adsorption and thermal desorption temperatures with increasing coverage. However, HeSE measurements with increasing coverage showed no indication of any de Gennes characteristics, indicating a substantial lack of repulsive forces. Figure 13.11(a) shows $\alpha(\Delta\mathbf{K})$ for CO coverages between 0.065 and 0.3 ML. The magnitude of the data increases, indicating that CO moves more quickly with increasing coverage, which can be modelled by reducing the corrugation of the adsorbate-substrate potential with coverage [39]. However the shape does not change, indicating the mechanism of motion does not change. In contrast, Fig. 13.11(b) and (c) show MD simulations for two different CO–CO force models in the literature. Both simulations show prenounced changes with coverage which are not present in the experimental data. The immediate conclusion was that pairwise force between CO molecules, regardless of how they are mediated, must be much weaker than expected, and thus cannot be responsible for the changes in dynamics with coverage. Figure 13.11(d) compares several proposed interaction potentials (lines) with the much lower range that does not conflict with HeSE experiments (shaded area). It is important to recognise that the HeSE data does not contradict any of the previous experimental data, only their pairwise force interpretation. The lack of pairwise forces also changes our understanding of the proximity of neighbouring adsorbates, which has important consequences for processes such as chemical reaction. To explain the previous changes in adsorption energy it was concluded that there must be a delocalised change in the adsorption chemistry of the system on a lengthscale much larger than the CO–CO spacing. Understanding the origin of this effect is likely to be a substantial challenge.

13.4.4 Larger Molecules

The HeSE technique is being progressively applied to larger adsorbate species and several extended molecules have been studied. As well as providing information about specific systems, studying larger species enables more general issues to be addressed, such as the extent that simple models may be used to model the transport of complex species and the scaling of friction with lateral size.

Measurements of benzene on graphite revealed an unexpected dynamical regime [55]. The benzene-substrate potential was found to be unusually flat, while the frictional interaction was particularly strong. The combination gives rise to an atomic-scale Brownian regime, and suggests the possibility of using the frictional coupling to direct motion in nanoscale systems. Similarly, high friction was observed in the case of cyclopentadienyl on Cu(111) [51], suggesting high friction may be a general trend for extended 'flat' species.

Larger species also mean that molecular reorientation is likely to be important in adsorbate motion. HeSE can be used to directly measure the characteristics of aperiodic rotation, without requiring excitation of the species to a different rotational energy level. The signatures of translation and rotation were recently distinguished in the motion of isolated ethanethiolate (ETSH) molecules on Cu(111) [41], a prototypical system which self-assembles to form a well ordered monolayer at higher coverages. At low temperatures ETSH diffuses slowly, but the tail of the molecule performs 6-fold rotational jumps around the sulphur atom. The activation of both processes was characterised, giving the first dynamical information about this widely studied class of species. In contrast, measurements of propane on Pt(111) showed that rotational motion was less prevalent than existing interaction potentials would suggest [40], and the general form of the necessary corrections was established.

13.4.5 Quantum Contributions & Tunnelling

For light species, quantum transport effects become prevalent at low temperatures. In particular, we expect a transition from classical 'over-barrier' motion to 'thru-barrier' tunnelling [3]. Tunnelling between adjacent unit cells is possible in both ground and excited states of the adsorbate, which are populated according to the usual Boltzmann distribution. The lowest energy states are usually most localised, leading to least overlap with adjacent unit cells and so give the lowest tunnelling rate contributions. Hence, with reducing temperature a gradual transition is expected, eventually reaching a low, temperature independent rate, when only the ground state is populated. The strength of interaction between the adsorbate and substrate determines whether the adsorbate wavefunction propagates coherently (little interaction), or is frequently scattered, so becoming incoherent. This level of coherence is reflected in the characteristics of the adsorbate motion, changing from ballistic-like to hopping-like characteristics respectively. Since both the rate and mechanism of motion can be determined using HeSE, the data can be used to test of quantum-rate theories and models of quantum-coherence.

Quantum transport has recently been observed for H atoms on Pt(111), below about 140 K [56]. In this system, the $\alpha(\Delta\mathbf{K})$ curves all showed remarkably close agreement with a single jump model, indicating moderate to high friction, and enabling the data to be reliably converted to a total hopping rate per atom, Γ. Figure 13.12 shows the temperature dependence of the rate for both H and D as an Arrhenius plot. At high temperatures we see the expected Arrhenius dependence, whereas at lower temperatures the rates are higher, due to an increasly important quantum contribution. The data was modelled using a quantum transition state theory (QTST) model of dissipative tunnelling [57]. The approach was well established in the literature, but not verified experimentally. Γ is given by

$$\Gamma = 3\frac{\omega_0}{2\pi}\Xi(T)x_b\exp\bigl(-E_a/(k_BT)\bigr) \tag{13.19}$$

which approximates the system as three 1D paths between adsorption sites. x_b is the dimensionless Krammers factor which allows for friction between the adsorbate and substrate and Ξ is a temperature dependent quantum correction term. The model assumes a parabolic potential at both the adsorption site and barrier, on which both x_b and Ξ are dependent. The parabolic curvatures, together with the friction and barrier height give a four parameter theory, which has a known scaling between isotopes. The solid lines in Fig. 13.12 show the fitted model. The agreement between experiment and theory, particularly the isotopic scaling, is remarkably good, supporting the picture of dissipative tunnelling. Although the parabolic approximation cannot represent the full detail of the system, it provides an equivalent representation which seems to capture the essential physics. In particular, the curvature of the potential at the barrier is substantially greater than around the adsorption site, which leads to a narrow barrier and substantial tunnelling rates, a feature that more sophisticated models need to reproduce. It is also particularly interesting to note that the magnitude of Ξ (lower panel), which includes both prefactor and tunnelling effects [57], is substantial even approaching room temperature.

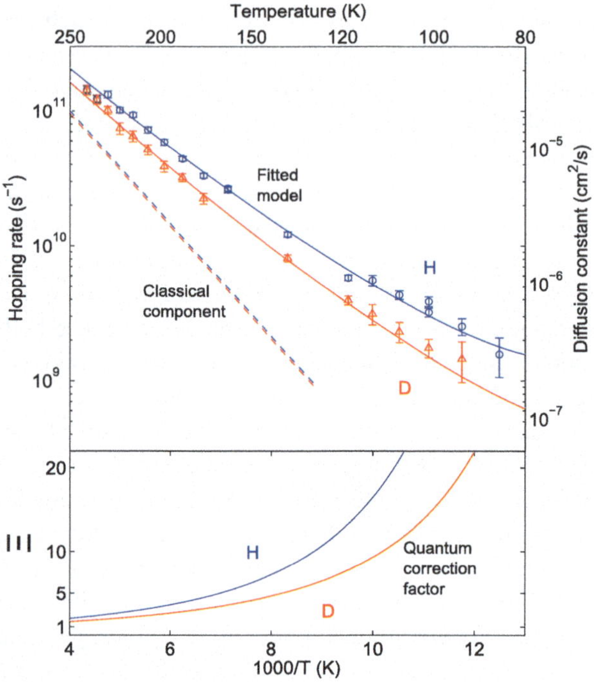

Fig. 13.12 Temperature dependence of the hopping rate of H on Pt(111), adapted from [56]. At low temperatures we see a deviation from the linear behaviour, due to an increasing proportion of activated tunnelling. The *lines* show the QTST model, described in the text. The *lower panel* shows the magnitude of the QTST quantum correction with temperature, which is substantial even at room temperature

13.4.6 Other Applications

Although we have generally focused on application to adsorbate transport, HeSE provides a generally applicable ultra-high energy-resolution scattering technique and many other types of measurement are also possible [5].

Several measurements which characterise surface vibrational modes have now been carried out. Generally, dispersion curves can be mapped out by converting the ISF to the energy domain, then applying the same analysis as with TOF methods, as illustrated by Cs/Cu(001) [34]. However, for finite energy transfers, the full resolution of HeSE is not available. The 'tilted-projection' approach (see 13.2.5) was therefore used to measure the linewidth and temperature dependence of the Cu(100) substrate modes [26]. More recently, a complete wavelength-intensity transfer-matrix has been obtained for Cu(111), by scanning currents in both solenoids and performing a 2D Fourier transform [25].

A novel form of selective adsorption resonance (SAR) measurement is are also possible. SARs cause sharp changes in the reflectivity of the sample with incident energy, which can be identified by using either one of the solenoids to energy-analyse the scattered beam with ultra-high resolution. These measurements have allowed the form of the helium-surface potential to be studied and refined with extremely high precision [20, 58].

13.5 Future Developments of the Technique

13.5.1 Instrumentation & Analysis

Substantial effort has been put into developing the HeSE technique and optimising performance at each stage. However, there remain two key factors which limit any measurement; the maximum spin-echo time available (corresponding to the energy resolution) and the available signal level. Fortunately, there are prospects for improving both.

The maximum spin-echo time is limited by the strength of the precession solenoids and their associated aberrations. At present the aberrations do not limit performance, so in the long term, higher resolution can be achieved by developing new, stronger field components. In particular, it may be possible to develop winding patterns which accept finite aberrations in order to achieve stronger field integrals. It may also be possible to use modern superconducting materials, providing flux pinning hysteresis can be adequately overcome. The maximum available spin-echo time can also be increased by reducing the energy of the He beam, albeit at the expense of momentum transfer. Many experiments do not require large momentum transfers, particularly in the case of large adsorbates or large scale nanostructured substrates, so lower beam-energies may offer a valuable compromise.

Practically all HeSE measurements benefit from increased signal, and there are ongoing efforts to improve the beam intensity and detector efficiency. Beam improvements focus on optimising the nozzle-skimmer geometry for the low temperature beam regime and promise a roughly twofold improvement in intensity. Improvements in helium detection offer much more substantial gains [29, 30], making use of the fact that spin-echo experiments do not require high temporal resolution. Large improvements in signal promise experimental runs of a few days rather than a few weeks, and will make practical many studies under low signal conditions.

However, productivity is not just limited by resolution and intensity. Relatively simple advances in instrumentation such as sample-transfer and increased automation are also likely to have a major impact on HeSE usage, and will make short collaborative experiments much more practical.

Interpretation of HeSE data from complex systems is likely to remain a consistent challenge. It will certainly be important to develop techniques which can distinguish specific aspects of motion, when multiple dynamical processes are mixed together. More sophisticated modelling is likely to include analytic and numerical components and it seems likely that collaboration across a range of experimental and theoretical disciplines will be most productive.

13.5.2 Future Applications

In the near future, a wide range of new measurements can be expected, comparable to those described in Sect. 13.4. We have already seen a much richer range of behaviour than simple theories would predict and it seems certain that future measurements on apparently 'simple' systems will bring surprises. In fact, almost every system studied to date has exhibited some novel aspect. As the breadth of available dynamical information becomes greater, we expect more comparative studies, which will begin to allow physical trends between related systems to be understood. Similarly, it seems likely that HeSE studies will become more integrated with other techniques, forming a dynamical element of more comprehensive studies rather than 'stand-alone' studies in their own right. Examples might include working towards a complete dynamical model of self-assembly, or gaining a better understanding of certain heterogeneous chemical reactions.

We are already seeing a trend amongst HeSE measurements towards larger and more complex systems. So far, this has mainly involved larger molecular species, with more internal and external degrees of freedom. Measurements on more complicated substrates will inevitably follow, ranging from simple asymmetric surfaces like f.c.c. (110), to stepped surfaces and even artificially nanostructured geometries. In the future, HeSE is likely to be used on unusually delicate systems, and surfaces which are particularly difficult to study with more conventional methods. These may include insulating surfaces, which are likely to require specialist preparation, species such as water which are susceptible to electron damage, and weakly bound species which will require low substrate temperatures.

Overall, HeSE experiments provide the first access to processes occurring on atomic length and pico- to nano-second timescales. Current experiments have barely touched on the range of novel phenomena that seem to exist in this otherwise unstudied regime. The immense range of possible experimental studies, combined with the scope for further advancement of the instrumentation and the importance of dynamical information in many areas means we expect there to be a long and productive future for this field.

References

1. J. Barth, Surf. Sci. Rep. **40**(3–5), 75 (2000)
2. R. Gomer, Rep. Prog. Phys. **53**, 917 (1990)
3. T. Ala-Nissila, R. Ferrando, S.C. Ying, Adv. Phys. **51**(3), 949 (2002)
4. D.P. Woodruff, T.A. Delchar, *Modern Techniques of Surface Science*, 2nd edn. (Cambridge University Press, Cambridge, 1994)
5. A.P. Jardine, H. Hedgeland, G. Alexandrowicz, W. Allison, J. Ellis, Prog. Surf. Sci. **84**(11–12), 323 (2009)
6. M. Rost, D. Quist, J. Frenken, Phys. Rev. Lett. **91**(2), 1 (2003)
7. A.G. Naumovets, Y.S. Vedula, Surf. Sci. Rep. **4**, 365 (1985)
8. J. Zhao, S. Granick, Macromolecules **40**(4), 1243 (2007)
9. X.D. Xiao, Y. Xie, Y. Shen, Surf. Sci. **271**(1–2), 295 (1992)
10. H. Rotermund, Surf. Sci. Rep. **29**(7–8), 265 (1997)
11. A. Kubo, N. Pontius, H. Petek, Nano Lett. **7**(2), 470 (2007)
12. A. Krasyuk, A. Oelsner, S.A. Nepijko, A. Kuksov, C. Schneider, G. Schönhense, Appl. Phys. A **76**(6), 863 (2003)
13. M. Bee, *Quasi-elastic Neutron Scattering* (IOP Publishing, Bristol, 1988)

14. D. Farias, K.H. Rieder, Rep. Prog. Phys. **61**, 1575 (1998)
15. E. Hulpke, *Helium Atom Scattering from Surfaces* (Springer, Berlin, 1992)
16. J. Frenken, J. Toennies, C. Wöll, Phys. Rev. Lett. **60**(17), 1727 (1988)
17. A.P. Graham, Surf. Sci. Rep. **49**(4–5), 115 (2003)
18. M. DeKieviet, D. Dubbers, C. Schmidt, D. Scholz, U. Spinola, Phys. Rev. Lett. **75**(10), 1919 (1995)
19. M. DeKieviet, D. Dubbers, M. Klein, C. Schmidt, M. Skrzipczyk, Surf. Sci. **377–379**(97), 1112 (1997)
20. A.P. Jardine, S. Dworski, P. Fouquet, G. Alexandrowicz, D.J. Riley, G.Y.H. Lee, J. Ellis, W. Allison, Science **304**(5678), 1790 (2004)
21. P. Fouquet, A.P. Jardine, S. Dworski, G. Alexandrowicz, W. Allison, J. Ellis, Rev. Sci. Instrum. **76**(5), 053109 (2005)
22. A.R. Alderwick, Instrumental and analysis tools for atom scattering from surfaces, Ph.D. Thesis, University of Cambridge, 2009
23. L. Van Hove, Phys. Rev. **95**(1), 249 (1954)
24. R. Gaehler, R. Golub, K. Habicht, T. Keller, J. Felber, Physica B **229**(1), 1 (1996)
25. P.R. Kole, A.P. Jardine, H. Hedgeland, G. Alexandrowicz, J. Phys. Condens. Matter **22**(30), 304018 (2010)
26. G. Alexandrowicz, A.P. Jardine, J. Phys. Condens. Matter **19**(30), 305001 (2007)
27. G. Alexandrowicz, Helium spin echo spectroscopy: measuring the dynamics of atoms, molecules and surfaces, Ph.D. Thesis, University of Cambridge, 2005
28. H. Hedgeland, P.R. Kole, W. Allison, J. Ellis, A.P. Jardine, Rev. Sci. Instrum. **80**(7), 076110 (2009)
29. M. DeKieviet, D. Dubbers, M. Klein, U. Pieles, C. Schmidt, Rev. Sci. Instrum. **71**(5), 2015 (2000)
30. A.R. Alderwick, A.P. Jardine, H. Hedgeland, D.A. MacLaren, W. Allison, J. Ellis, Rev. Sci. Instrum. **79**(12), 123301 (2008)
31. A.P. Jardine, P. Fouquet, J. Ellis, W. Allison, Rev. Sci. Instrum. **72**(10), 3834 (2001)
32. S. Dworski, G. Alexandrowicz, P. Fouquet, A.P. Jardine, W. Allison, J. Ellis, Rev. Sci. Instrum. **75**(6), 1963 (2004)
33. R. Martínez-Casado, J. Vega, A. Sanz, S. Miret-Artés, Phys. Rev. Lett. **98**(21), 1 (2007)
34. A.P. Jardine, G. Alexandrowicz, H. Hedgeland, R.D. Diehl, W. Allison, J. Ellis, J. Phys. Condens. Matter **19**(30), 305010 (2007)
35. C.T. Chudley, R.J. Elliott, Proc. Phys. Soc. **77**(2), 353 (1961)
36. J. Ellis, A. Graham, J. Toennies, Phys. Rev. Lett. **82**(25), 5072 (1999)
37. F.E. Tuddenham, H. Hedgeland, A.P. Jardine, B.A.J. Lechner, B. Hinch, W. Allison, Surf. Sci. **604**(17–18), 1459 (2010)
38. G. Alexandrowicz, A. Jardine, H. Hedgeland, W. Allison, J. Ellis, Phys. Rev. Lett. **97**(15), 1 (2006)
39. G. Alexandrowicz, P.R. Kole, E.Y.M. Lee, H. Hedgeland, R. Ferrando, A.P. Jardine, W. Allison, J. Ellis, J. Am. Chem. Soc. **130**(21), 6789 (2008)
40. A.P. Jardine, H. Hedgeland, D. Ward, Y. Xiaoqing, W. Allison, J. Ellis, G. Alexandrowicz, New J. Phys. **10**(12), 125026 (2008)
41. S. Paterson, W. Allison, H. Hedgeland, J. Ellis, A.P. Jardine, Phys. Rev. Lett. **106**, 256101 (2011)
42. J. Ellis, J. Toennies, Phys. Rev. Lett. **70**(14), 2118 (1993)
43. J. Ellis, A.P. Graham, Surf. Sci. **377–379**, 833 (1997)
44. A.P. Graham, F. Hofmann, J. Toennies, L. Chen, S. Ying, Phys. Rev. Lett. **78**(20), 3900 (1997)
45. J. Ellis, A. Graham, F. Hofmann, J. Toennies, Phys. Rev. B **63**(19), 1 (2001)
46. R. Kubo, Rep. Prog. Phys. **29**(1), 255 (1966)
47. G. Alexandrowicz, A. Jardine, P. Fouquet, S. Dworski, W. Allison, J. Ellis, Phys. Rev. Lett. **93**(15), 1 (2004)
48. A.P. Jardine, J. Ellis, W. Allison, J. Chem. Phys. **120**(18), 8724 (2004)
49. A. Graham, F. Hofmann, J.P. Toennies, J. Chem. Phys. **104**(13), 5311 (1996)

50. H. Hedgeland, P. Kole, H. Davies, A. Jardine, G. Alexandrowicz, W. Allison, J. Ellis, G. Fratesi, G. Brivio, Phys. Rev. B **80**(12), 1 (2009)
51. H. Hedgeland, B. Lechner, F. Tuddenham, A.P. Jardine, W. Allison, J. Ellis, M. Sacchi, S. Jenkins, B. Hinch, Phys. Rev. Lett. **106**(18), 186101 (2011)
52. G. Fratesi, G. Alexandrowicz, M. Trioni, G. Brivio, W. Allison, Phys. Rev. B **77**(23), 1 (2008)
53. B. Persson, Phys. Rev. B **40**(10), 7115 (1989)
54. N. Petrova, I. Yakovkin, Surf. Sci. **519**(1–2), 90 (2002)
55. H. Hedgeland, P. Fouquet, A.P. Jardine, G. Alexandrowicz, W. Allison, J. Ellis, Nat. Phys. **5**(8), 561 (2009)
56. A.P. Jardine, E. Lee, D. Ward, G. Alexandrowicz, H. Hedgeland, W. Allison, J. Ellis, E. Pollak, Phys. Rev. Lett. **105**(13) (2010)
57. P. Haenggi, P. Talkner, M. Borkovec, Rev. Mod. Phys. **62**, 251 (1990)
58. D.J. Riley, A.P. Jardine, S. Dworski, G. Alexandrowicz, P. Fouquet, J. Ellis, W. Allison, J. Chem. Phys. **126**(10), 104702 (2007)

Chapter 14
Diffraction of H_2 from Metal Surfaces

Daniel Farías, Marina Minniti, and Rodolfo Miranda

Abstract Diffraction of molecular hydrogen is a very sensitive technique to characterize the potential energy surface (PES) for hydrogen dissociative chemisorption. In this chapter we summarize recent H_2 diffraction experiments and calculations based on *ab initio* determined six-dimensional PESs performed on several single-crystal metal surfaces, ranging from non-reactive to very reactive ones, at incident energies between 20 and 200 meV. The general trends observed in experiment are discussed, as well as the validity of the approximations usually employed in the calculations. It is shown that an analysis of both H_2 elastic and rotationally inelastic diffraction intensities as a function of incident energy provides a very sensitive way to test the quality of *ab initio* determined six-dimensional PESs.

14.1 Introduction

Reactions of molecules with metal surfaces are of tremendous importance to our every day lives, as most large scale chemical processes in industry and a number of environmental protection processes involve heterogeneous catalysis in one form or another [1]. Our understanding of these rather complex reactions is based on fundamental studies of elementary reaction steps. Among these steps, perhaps the most crucial from the viewpoint of heterogeneous catalysis is the one called *dissociative*

D. Farías (✉)
Departamento de Física de la Materia Condensada and Instituto Nicolás Cabrera,
Universidad Autónoma de Madrid, Cantoblanco, 28049 Madrid, Spain
e-mail: daniel.farias@uam.es

M. Minniti
Departamento de Física de la Materia Condensada, Universidad Autónoma de Madrid,
Cantoblanco, 28049 Madrid, Spain
e-mail: marina.minniti@uam.es

R. Miranda
Departamento de Física de la Materia Condensada and Instituto Nicolás Cabrera,
Instituto Madrileño de Estudios Avanzados en Nanociencia (IMDEA-Nanociencia),
Universidad Autónoma de Madrid, Cantoblanco, 28049 Madrid, Spain
e-mail: rodolfo.miranda@uam.es

G. Bracco, B. Holst (eds.), *Surface Science Techniques*,
Springer Series in Surface Sciences 51, DOI 10.1007/978-3-642-34243-1_14,

chemisorption, in which a bond in a molecule hitting the surface is broken and the resulting fragments are bonded to the surface. The dissociation of H_2 at a metal surface constitutes a benchmark system for this step. As a consequence, the interaction of hydrogen with metal surfaces has been extensively studied, both experimentally [2–6] and theoretically [7–10]. A large number of molecular beam and associative desorption experiments have been performed, in which sticking probability curves as a function of incident conditions were measured. These experiments provided detailed information concerning the influence of incident energy and angle, vibrational state, the molecule's incident rotational state and molecular alignment on reaction [2–6]. This is certainly a quite useful piece of information, which can be used to understand how certain features of the potential energy surface (PES) control the experimental observations. However, a usual limitation of these experiments is that they provide information averaged over the whole unit cell.

A different point of view is provided by diffraction experiments. As it is well known, He-atom scattering (HAS) is a well established tool to investigate surface properties [11]. In these experiments, the positions of the different diffraction peaks provide detailed information on the surface structure, while the relative intensities of the diffraction peaks contain information on the corrugation of the particle-surface PES. Similarly, and as first pointed out by Halstead and Holloway [12], hydrogen diffraction measurements performed over a wide incident energy range should provide precise information regarding the distribution of activation barriers within the unit cell. Thus, these authors suggested that H_2 diffraction could be a promising, and maybe unique, experimental technique to gauge the H_2—surface PES within the unit cell [12]. However, these expectations have not yet been satisfied due to practical limitations in both theory and experiments.

Although the first H_2 and D_2 diffraction experiments from low-index metal surfaces have been reported in the 1980s, this line of work was not continued by the groups involved, presumably because they soon realized that the link between the experimental H_2 diffraction spectra and the PES can only be established by performing accurate dynamical calculations, which were impossible to do at that time. Though evaluation of realistic PESs for H_2 + metal surface systems is not trivial and 6D quantum dynamical calculations are computationally demanding, an exact theoretical description of H_2 scattering from first principles (within the rigid surface model) is now possible [8–10]. This possibility has renewed interest in H_2 diffraction from the experimental side and as a consequence, we have gained considerable insight into the H_2 dissociation dynamics at metal surfaces.

Diffraction of H_2 and D_2 molecular beams from surfaces is in principle very similar to He diffraction, the two major differences being: (i) the possibility of rotational-state transitions in the case of molecular scattering and (ii) the fact that diffraction competes with the reactivity channel. Concerning (i), rotationally inelastic diffraction (RID) peaks can be observed in the form of additional diffraction peaks in the angular distributions. In this process, the incident molecules convert part of their translational energy into excitation of a rotational quantum level when colliding with the surface. The position of RID peaks within an angular distribution can be obtained by combining the Bragg condition for surface diffraction with conservation of energy [13]:

$$\begin{aligned} \Delta\mathbf{K} &= \mathbf{K}_f - \mathbf{K}_i = \mathbf{G}, \\ E_f - E_i &= \Delta E_{rot}. \end{aligned} \tag{14.1}$$

In (14.1), $\mathbf{K}_f$ and $\mathbf{K}_i$ are the parallel components of the outgoing and incident wave vectors, respectively, $\mathbf{G}$ is a surface reciprocal lattice vector, E_f and E_i the final and incident beam energies and ΔE_{rot} is the rotational transition energy. For H_2 (D_2) this energy is $|\Delta E_{rot}| = 44.6$ (22.2) meV for the lowest transitions ($0 \to 2$ and $2 \to 0$) and $|\Delta E_{rot}| = 74.3$ (36.88) meV for the $1 \to 3$ and $3 \to 1$ transitions.

Concerning the reactivity channel, a quantitative measure is given by the initial sticking probability (S_0) of H_2 [3]. This probability might vary from almost zero (for non-reactive systems, like noble metals) to a value very close to one for highly reactive (also called non-activated systems) like transition metal surfaces. As a consequence, on most reactive surfaces a hydrogen overlayer can build up quickly if the experiments are done at a surface temperature below the desorption temperature of H_2 ($\sim$350 K), altering the measurements. This sets an important boundary condition to the experiments, which must be carried out at high surface temperatures, which reduces the total diffraction intensities measured to just $\sim$1 % of the incoming beam signal.

It is worth noting that the whole H_2 diffraction approach is in some sense counter intuitive, since the goal is to get information on the dissociative channel by looking into the small amount ($\sim$1 %) of diffracted H_2 molecules. The basic expectation is that the small amount of H_2 molecules surviving dissociation are carrying detailed information on the PES topology. This expectation has been confirmed by many experimental and theoretical studies of H_2 diffraction from metal surfaces. Some of them are presented in Sect. 14.4. Finally, since the classical turning point of diffracted molecules are about 2 Å away from the surface atom cores, H_2 diffraction experiments sample the PES in a region difficult to access with other experimental techniques.

14.2 Theory Background

It is beyond the scope of the present work to give a detailed description of the methods employed to calculate H_2 diffraction from metal surfaces. However, for the sake of completeness we include a brief description of the main theoretical approaches, and refer the interested reader to the proper references.

The scattering dynamics of H_2 on metal surfaces is usually described in terms of a single six-dimensional molecule-surface PES obtained from state-of-the-art electronic structure calculations [8–10]. These calculations are performed within the Born-Oppenheimer approximation, whereby electronic and phononic excitations of the surface are not considered. Inclusion of such inelastic effects in a quantum calculation modeling all six degrees of freedom of the problem is computationally too

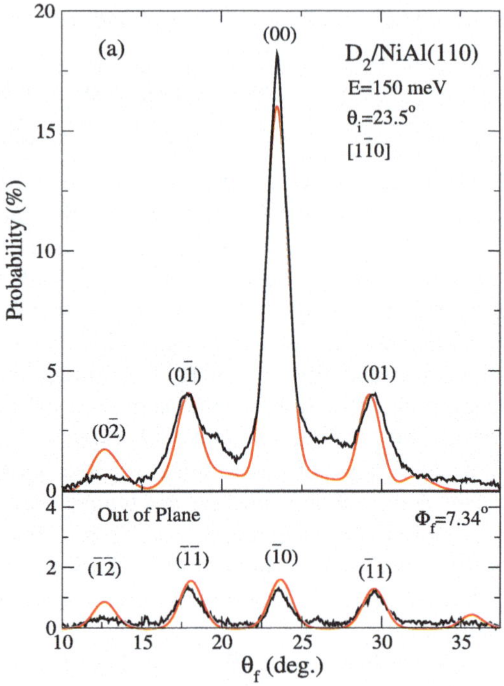

Fig. 14.1 In-plane and out-of-plane D_2 diffraction spectra for NiAl(110). *Black curves*: experiment; *red lines*: 6D quantum dynamical calculation. Theoretical calculations have been convoluted with a Gaussian function of width $\sigma = 0.7^\circ$ to account for the limited angular resolution of the measurements. Both the experimental results and the calculations have been normalized to the height of the specular peak that arises from the quantum calculations. Reproduced by permission of the IOP from [20]

expensive, and currently not possible. A few attempts have been made to include phonon inelastic processes, both in the context of quantum wave-packet [14, 15] and classical calculations [16–18]. Increasing the surface temperature leads to attenuation of diffraction intensities, due to the Debye-Waller effect, leaving the relative intensities almost unchanged. In comparisons with diffraction experiments, the effect of phonons can be taken into account by extrapolating the experimental results to a surface temperature of 0 K, using the Debye-Waller model. This justifies to a large extent the frozen surface approximation employed in the dynamical calculations. The Born-Oppenheimer approximation can be justified on the basis of recent experimental results for H_2/Pt(111) [19], which are described in more detail in Sect. 14.4.

The methods used to describe the dynamics of the H_2 + surface interaction have been described in detail elsewhere. In particular, the PES of H_2/Pd(111) [21], H_2/Pt(111) [22], H_2/NiAl(110) [23], H_2/Pd(110) [24] and H_2/Ru(0001) [25] have been determined by interpolation of *ab initio* DFT/GGA data (GGA stands for generalized gradient approximation) using the corrugation reducing procedure (CRP) [26]. The CRP has been shown to provide a precision better than 30 meV in the dynamically relevant regions for several H_2 + metal systems [21–23].

Figure 14.1 shows a comparison between experiment and 6D quantum calculations for D_2 diffraction from the NiAl(110) surfaces. The good agreement obtained for this systems, and for others discussed in more detail in Sect. 14.4, demonstrate that quantum dynamics calculations treating all molecular degrees of freedom can accurately predict diffraction patterns for hydrogen scattering from reactive metal surfaces. Inclusion of all molecular degrees of freedom is essential to account for the competition between dissociative and non-dissociative channels. It also shows that out-of-plane diffraction measurements are crucial to test the details of the PES in a wide region of space. These results support the use of 6D DFT theories to properly describe the H_2-surface interaction even in regions far away from the surface.

14.3 Experimental Setup and Method

14.3.1 General Requirements

The two books edited by Scoles [27, 28] give an excellent overview of the experimental demands for diffractive scattering from surfaces. Generally speaking, the apparatus used for He and H_2 diffraction experiments can be classified according to their scattering geometries in two different groups, whose basic features are schematically represented in Fig. 14.2. In the Surface Science Laboratory at the Universidad Autónoma de Madrid (LASUAM) there are available one of each class of these machines.

To the first class belong the so-called 'fixed-angle' systems (upper panel), in which the angle between incident and outgoing reflected beams is fixed, and angular distributions of diffracted particles are measured by continuous rotation of the crystal. In this way, the incident angle θ_i changes in time during data acquisition while the sum of incident and scattered beam angles $(\theta_i + \theta_f)$ remains constant. In the second kind of system (lower panel), the detector is mounted in the UHV scattering chamber, and is able to rotate around the crystal for any given incidence angle. Thus, the detector can measure diffraction beams in a large region of the reciprocal space for the same angle of incidence. In addition, in this kind of setup it is possible to place the detector just in front of the direct beam and measure its intensity. Final products of the experiments are absolute diffraction probabilities for a fixed incident angle, directly comparable with theoretical results, which represents a very crucial advantage of this setup. On the other hand, the proximity of the detector to the sample results in a quite limited angular resolution (about 1°), one order of magnitude larger than for 'fixed-angle' systems. Moreover, since the detector lies in the main scattering chamber and is not differentially pumped, the dynamical range of measurable intensities is very low, and lies typically two orders of magnitude below that of the differentially pumped 'fixed-angle' systems used in time-of-flight experiments.

To establish possible dynamical ranges of measured intensities becomes a central issue in case of H_2/D_2 diffraction since the reflectivity of reactive surfaces is strongly reduced due to dissociation of most of the molecules as they get in contact with the surface. As an example, typical values for hydrogen surface reflectivity from reactive surfaces are in the range of 0.5–1 % of the incident beam intensity. Moreover, to prevent hydrogen adsorption onto the surface during diffraction

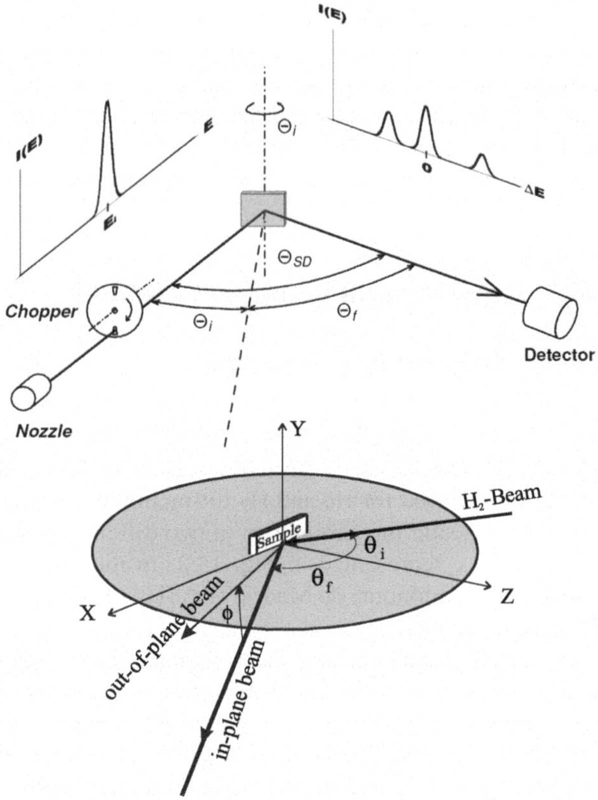

Fig. 14.2 Experimental scattering geometry for 'fixed-angle' detectors (*upper panel*) and 'rotary' detectors (*lower panel*). Trajectories of incident and outgoing beams are indicated

experiments, surface temperatures have to be kept at a value slightly higher than the molecule desorption temperature, which is in the range of 400–500 K for most hydrogen-transition metal systems.

It is possible to estimate the instantaneous coverage on the surface assuming that hydrogen desorption for the conditions used in the experiments obeys a second order desorption process. The equilibrium coverage is obtained by equaling the desorption rate to the adsorption rate. The incident beam flux is of the order of 10^{14} molecules·cm^{-2}·s^{-1}. On a Ru(0001) surface, which has an atom density $N_d = 1.57 \cdot 10^{15}$ atoms·cm^{-2}, this corresponds to an adsorption rate of 0.13 ML·s^{-1} (assuming a sticking coefficient equal to unity). The desorption rate r_{des} can be estimates as:

$$r_{des} = \nu_{des} \Theta^2 N_d e^{-\frac{E_{des}}{k_B T_s}}, \tag{14.2}$$

where Θ is the instantaneous coverage on the surface, k_B the Boltzmann constant, E_{des} the desorption energy, and ν_{des} the pre-exponential factor. From TDS experiments, the values for E_{des} and ν_{des} are known to be 1.3 eV and 3 cm^2·s^{-1}, respectively [29]. From these values, we get an instantaneous coverage of 0.02 ML at $T_s = 500$ K, which can be considered a clean Ru(0001) surface for all practical

Table 14.1 Fractional populations (in percent) of the rotational levels j_i for n-D_2 beams with stagnation conditions $P_0 d = 41.25$ Torr·cm and source temperature T_0, estimated from the effective beam rotational temperatures T_R. From [47]

E_i (meV)	T_0 (K)	T_R (K)	$j_i = 0$	$j_i = 1$	$j_i = 2$	$j_i = 3$
85	300	99	48.7	32.3	18.0	1.0
95	334	115	43.5	31.6	23.1	1.8
108	380	138	37.7	30.2	29.0	3.1
120	422	159	33.4	28.8	33.2	4.6
133	468	184	29.8	27.1	36.8	6.2
140	492	198	28.3	26.3	38.4	7.0
149	525	216	26.5	25.2	40.2	8.1

purposes. It should be noticed that the assumption of a sticking coefficient equal to one at any coverage is clearly an overestimation. The H_2 sticking coefficient at zero coverage for this surface has been experimentally determined to be in the range 0.3–0.5 for the beam energies explored in the present experiments [30]. Therefore, the estimated coverage $\Theta = 0.02$ ML should be considered as an upper limit. Increasing the surface temperature leads to a further reduction of the diffraction signal, and at the same time it increases the inelastic background due to the Debye-Waller attenuation [31]. The above cited points render H_2/D_2 diffraction measurements from reactive surfaces absolutely non-trivial, and much more challenging than He atom scattering studies.

As already mentioned, the diffraction process of H_2 and D_2 includes also the possibility of changing the molecules initial rotational state, making the determination of diffraction probabilities for molecules more complicated than for atoms. One fundamental point is, therefore, to estimate the occupation probability of the rotational levels in the incident beam. Since usually it is not possible to measure it directly, it is estimated from previous theoretical and experimental work. These investigations demonstrated that the rotational population of highly expanded molecules follow nearly a Boltzmann distribution which can be characterized by an effective rotational temperature T_R, which depends on the source temperature T_0. The relation between T_R and T_0 involves the parameter $P_0 d$, where d is the nozzle orifice diameter and P_0 the stagnation pressure [32–34]. Once the rotational temperature T_R is determined, rotational populations n_j can be obtained from a Boltzmann distribution. In most of the experiments carried out with H_2/D_2 molecules a stagnation pressure P_0 of ca. 50 bar is used. Table 14.1 gives calculated fractional populations of rotational levels for n-D_2 beams, corresponding to stagnation conditions $P_0 d = 41.25$ Torr·cm. In the case of n-H_2 beams, ~73 % of the molecules are in the $j_i = 1$ state at an incident energy of 75 meV, and ~60 % at 150 meV.

Concerning vibrations, due to the large spacing between the vibrational levels of the H_2/D_2 molecules, it can be safely stated that more than 99 % of the molecules are in the $v = 0$ state for incidence energies lower than 200 meV.

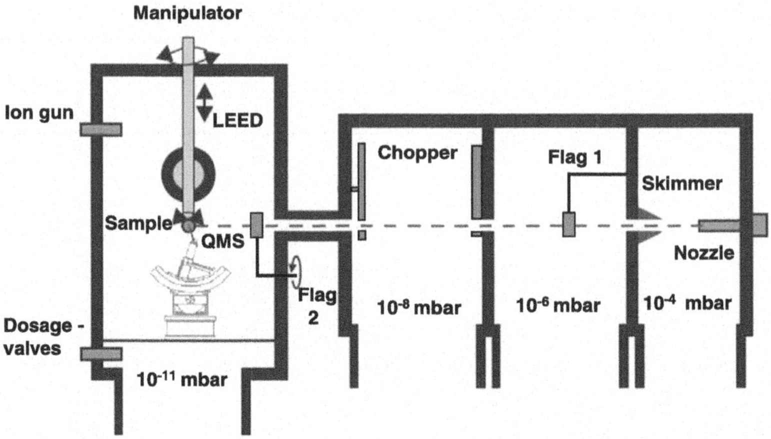

Fig. 14.3 Schematic view of the atomic and molecular beam scattering apparatus installed at LASUAM. The detector is mounted on a goniometer inside the main scattering chamber. Notice the presence of the two copper flags which allow direct H_2/D_2 sticking measurements

14.3.2 'Rotary Detector' Setup

One of the atomic and molecular beam apparatus available at LASUAM is schematically shown in Fig. 14.3. The setup consists of two separate chambers: the beam chamber and the main scattering chamber. The beam chamber is three-staged differentially pumped with one diffusion pump in each of the first two stages (5000 l/sec each) and a turbomolecular pump (350 l/sec) in the third stage. He and H_2/D_2 beams are generated by supersonic expansion through a 10 μm diameter Pt nozzle of high pressure gases (typically 40–70 bars). The nozzle tube is clamped between massive copper supports, which can be cooled to 100 K by contact with a liquid nitrogen reservoir. Moreover, the tube can be resistively heated to 800 K. As a result, the beam energy can be varied between 20–170 meV for He and 25–200 meV for H_2/D_2 beams. The energy spread is estimated to be about 1 % for He beams and ca. 10 % for H_2/D_2 beams. After expansion, the beam is collimated by a 0.5 mm skimmer and reaches the third differential pumping stage, where a rotatable disk with different aperture diameters (100–4000 μm) allows further collimation of the beam before entering the scattering chamber. In the third stage, the beam is also mechanically chopped with a magnetically coupled rotary motion feedthrough to allow phase-sensitive detection. Beam phase and frequency are registered by means of a led-photodiode couple.

The base pressure in the scattering chamber is typically 3×10^{-10} mbar, reaching 8×10^{-10} mbar with the He or H_2/D_2 beam on. This pressure increase gives rise to a continuous background in the scattering chamber which limits the signal-to-noise ratio. The signal is recovered from the background by means of a lock-in system ($f \sim 150$ Hz). The sample is mounted on a manipulator which has five degrees of freedom. Polar and azimuthal angles can be varied according to experimental requests. In addition, the crystal can be heated up to 1700 K and cooled down to 100 K.

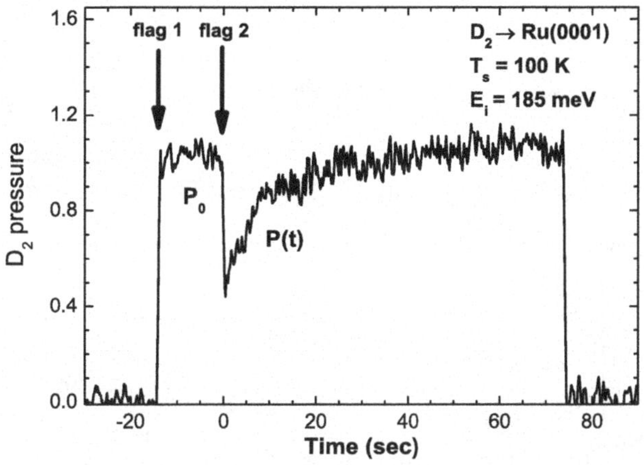

Fig. 14.4 Typical sticking measurement performed in the apparatus shown in Fig. 14.3. When the first flag is opened, the beam enters the scattering chamber, where an increase of the D_2 background partial pressure is measured by means of a quadrupole mass spectrometer. At $t = 0$, the second flag is opened, allowing D_2 molecules to stick on the Ru(0001) surface, thus causing a decrease in the background partial pressure (see text). This measurement has been performed at $\theta_i = 15°$, $E_i = 185$ meV and at a surface temperature of 100 K

For a fixed incident angle, the angular distribution of the diffracted atoms/molecules is collected with a quadrupole mass spectrometer mounted on a two-axis goniometer inside the scattering chamber, located at 5 cm from the sample. In order to improve the angular resolution, a circular slit of 1 mm diameter has been placed in front of the quadrupole ionization chamber. The detector can rotate 200° in the scattering plane, defined by the beam direction and the normal to the surface, and ±15° in the direction perpendicular to the scattering plane. Detection of both in-plane and out-of-plane intensities for fixed incident conditions gives a valuable hint for an easier interpretation of diffraction data as well as for comparison with calculations. Typical values for detection limits in this system are in the order of 10^{-3} of the incident beam intensity.

Recently, this apparatus has been equipped with two copper flags in order to be able to perform sticking measurements with the method devised by King and Wells [35]. The first copper flag is placed in the second stage of the beam generation chamber in order to stop the beam from entering the main chamber. The second copper flag is placed in the scattering chamber to prevent molecules from hitting the surface crystal (see Fig. 14.3). It is important to underline that the beam quality in terms of divergence and energy spread is the same for both diffraction and adsorption probability measurements. Indeed, with our current setup it is nowadays possible to measure He/H_2/D_2 diffraction and straightforward H_2/D_2 sticking probabilities for the same in-situ prepared sample.

Figure 14.4 shows an example of sticking measurement of D_2 molecules on the clean Ru(0001) surface performed in this apparatus. The quadrupole mass spectrometer initially measures the deuterium partial pressure due to the presence of residual

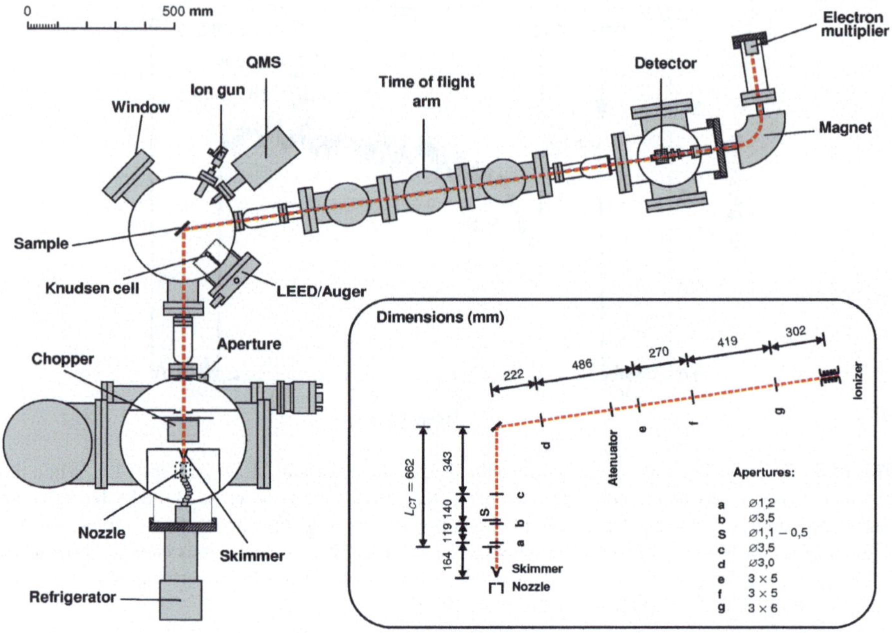

Fig. 14.5 Top-view scheme of the molecular beam time-of-flight apparatus running at the LASUAM. Reprinted with permission from [47]. Copyright (2010), American Institute of Physics

gases in the UHV chamber. By opening the first flag (first arrow in Fig. 14.4), the D_2 beam can enter the scattering chamber and strikes the second copper flag placed just in front of the sample. In this way, no sticking or adsorption on the surface is possible, and a reference deuterium background partial pressure due to the beam in the chamber (P_0) is registered. Removal of the second flag at $t = 0$ (second arrow in Fig. 14.4) allows the beam to strike the sample and causes a transient decrease in the D_2 partial pressure due to molecule sticking onto the surface. As the surface becomes covered, the transient decrease diminishes with time, until the pressure $P(t)$ reaches again P_0 when the surface is saturated. The initial sticking coefficient S_0 is then calculated as [35]:

$$S_0 = \frac{P_0 - P_{(t=0)}}{P_0}, \tag{14.3}$$

and provides an absolute value of adsorption probabilities in a relatively simple and efficient experimental way.

14.3.3 'Fixed-Angle' Setup

This setup consists of a high-resolution He atom time-of-flight spectrometer which has been transferred a few years ago to LASUAM from Peter Toennies' group in

Goettingen. Figure 14.5 shows a top-view of the entire system, which basically consists of three main units: the source chamber, the scattering chamber and the time-of-flight drift tube. Monochromatic He and H_2/D_2 beams are generated using ultrahigh purity gases at high pressures (typically up to 150 bar for He and ~50 bar for D_2) which are expanded into vacuum through a 10 μm diameter Pt nozzle. In order to avoid nozzle clogging, the gas passes through a liquid nitrogen cooled filter before reaching the nozzle. The beam energy can be varied in the range of ~20–150 meV by changing the nozzle temperature with a precision of ±1 K. The velocity spread of the incident beam is typically 1 % for He atoms and ca. 10 % for H_2/D_2 beams. To extract the central part of the supersonic free jet, a conical skimmer is used. Before colliding with the sample, the beam passes through a differentially pumped stage, where the angular divergence of the beam can be selected by a mobile aperture.

The UHV scattering chamber is equipped with Low Energy Electron Diffraction/Auger Electron Spectroscopy (LEED/AES) optics, an ion gun and a mass spectrometer for surface preparation and characterization. A standard manipulator allows the sample to be cooled by liquid N_2 and heated by electron bombardment to 1500 K. Diffraction angular distributions are measured by continuously rotating the crystal around an axis normal to the plane defined by the incident and outgoing beams. Therefore, the incidence angle is continuously varied during a measurement, and this angle is related to the corresponding final angle through $\theta_i + \theta_f = \theta_{SD}$. After scattering with a fixed angle $\theta_{SD} = 105.4°$, particles travel through three differentially pumped stages along the 1.7 m long time-of-flight drift tube before reaching the detector, where they are ionized by electron bombardment. The ions are selected by a home-made mass spectrometer and collected by an electron multiplier. Due to the strong reduction of background signal, as consequence of the large crystal-detector distance, this system makes it possible to measure diffraction intensities of the order of 10^{-5} of the incoming beam, while the angular resolution is determined by the detector acceptance angle, and is about 0.1°.

14.4 Applications of the Technique

14.4.1 General Trends

In general, larger diffraction intensities are measured with H_2 beams as compared to He beams, and a two-dimensional corrugation is detected, even from surfaces which present a quasi one-dimensional corrugation when measured with He-diffraction (like the fcc(110) ones). This is a consequence of the larger polarizability of H_2 as compared to He, which gives rise to a stronger attractive interaction [36]. As a result, larger potential well depths D are usually measured by H_2 beams in selective adsorption resonances experiments. Typical values reported for H_2 beams are in the range 30–60 meV, compared with 5–10 meV for He beams (see [11] and references therein).

Fig. 14.6 In-plane ($\phi_f = 0°$, *black curves*) and out-of-plane (*red* and *green curves*) H_2 diffraction spectra for NiAl(110), Pt(111) and Pd(111). Reproduced by permission of the IOP from [20]

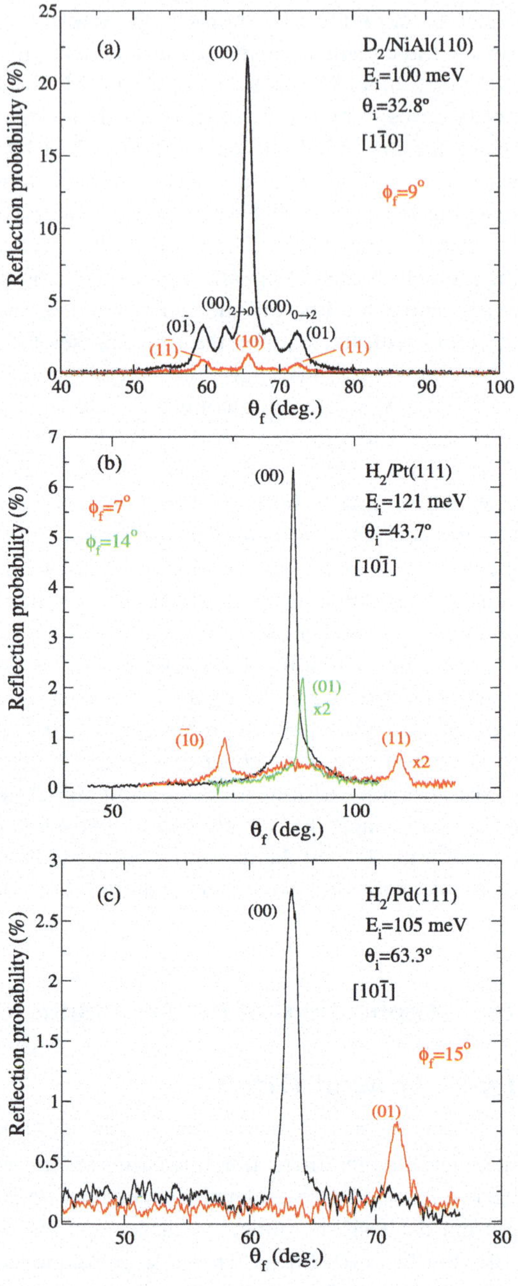

Figure 14.6 shows experimental H_2/D_2 diffraction spectra obtained with (a) NiAl(110), (b) Pt(111), and (c) Pd(111) surfaces under otherwise similar incidence conditions. The total reflectivity of the three surfaces are quite different, and goes

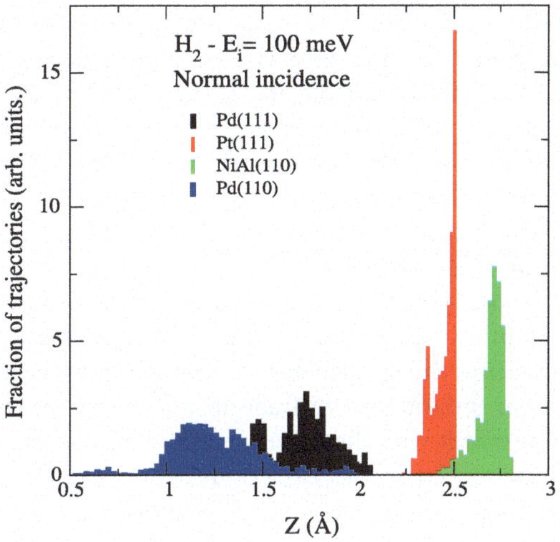

Fig. 14.7 Closest approach distance for H_2 molecules scattered from NiAl(110), Pt(111), Pd(111) and Pd(110), for $E_i = 100$ meV and normal incidence. Reproduced by permission of the IOP from [20]

from ~1 for NiAl(110) to ~0.3 for Pd(111). These spectra have been measured using a rotary detector setup, and illustrate a very significant result: out-of-plane diffraction is important, and its importance increases with increasing surface reactivity, i.e. when going from NiAl(110) and Pt(111) to Pd(111). This observation seems to be a general trend for H_2 diffraction experiments, and has been confirmed also by recent measurements in other systems, like H_2/Ru(0001) [37]. These results show clearly that out-of-plane diffraction is much more important than previously assumed in most H_2 diffraction experiments.

The importance of out-of-plane diffraction has been actually predicted in the first six-dimensional quantum calculations of H_2 diffraction, reported by Gross and coworkers for the H_2/Pd(100) system [38], a prediction which was confirmed by a similar study on H_2/Pt(111) performed in the group of Kroes [39]. The relative amount of the ratio of out-of-plane/in-plane diffraction intensities depends on impact energy and incidence angle. This is the consequence of a dynamical effect associated with grazing incidence. A model has been proposed [40], based on the periodicity of the potential along the incidence direction. According to this model, momentum change along the incidence direction is of second order while momentum change along the transverse direction is of first order. This means that, along the incidence direction, any acceleration (slowing down) is compensated by a slowing down (acceleration), while the effect of the transverse force is cumulative [40]. Therefore, pronounced out-of-plane diffraction is expected, even for conditions near grazing incidence, as is the case for the H_2/Pd(111) data shown in Fig. 14.6c.

More insight into the H_2 scattering dynamics can be obtained by performing classical trajectory calculations, in which the 6D PES is determined from DFT calculations. In this way, one gets a very good estimation of the classical turning points of H_2 at a given energy. Figure 14.7 shows the distribution of the closest approach distance to the surface of H_2 molecules scattered from NiAl(110), Pt(111), Pd(111) and Pd(110) [20]. The results correspond to $E_i = 100$ meV and normal incidence.

The consequences of the different reactivities of the three surfaces on the scattering dynamics is clear. For H_2/NiAl(110), i.e. the least reactive system, the molecules are reflected far from the surface, at Z ~ 2.7 Å. The corresponding PES presents a repulsive behavior in the entrance channel, and molecules impinging the surface with a perpendicular energy of 100 meV find a slightly corrugated hard-wall-like potential in that region. As a consequence, specular reflection is dominant on this surface (the situation resembles very much the case of He-scattering), and diffraction is relatively weak.

For H_2/Pt(111), the closest approach distance of reflected molecules is Z ~ 2.4 Å, slightly smaller than in the case of NiAl(110). It is expected that molecules reflected closer to the surface are sampling a more corrugated region of the PES, leading to more pronounced diffraction intensities as compared to the specular one. This is consistent with the experimental data shown in Fig. 14.6b. This effect is even more pronounced in the case of Pd surfaces. On Pd(111), for instance, H_2 molecules are reflected between ~1.5 Å and 2 Å, where the PES corrugation is very strong. Accordingly, there is a suppression of the specular channel and, therefore, diffraction becomes relatively more important. On Pd(110), the classical turning points are located even closer to the surface, between 1–1.5 Å. In this case, however, H_2 diffraction is not observed due to the strong influence of dynamical trapping on diffracted molecules [41].

14.4.2 H_2/Pt(111)

As mentioned in Sect. 14.2, the six-dimensional scattering dynamics calculations for H_2 + metal systems are performed within the Born-Oppenheimer approximation. By assuming this approximation (often referred to as *adiabatic approximation*), the possibility of electron-hole pair excitations is removed from the model, and the full scattering process is assumed to take place on the ground state PES. However, the ability of electronically adiabatic theory to describe reactive scattering of molecules from metal surfaces has been questioned by recent observations of electrons produced following chemical reactions or scattering of some molecules with metal surfaces [42, 43].

A recent combined experimental and theoretical work shows that despite this clear evidence of Born-Oppenheimer breakdown in some systems, one is not precluded from using the electronically adiabatic theory in all cases. In effect, the work reported by Kroes and coworkers [19] shows unambiguously that theory can accurately describe both reaction and diffractive scattering of H_2 from Pt(111) within an electronically adiabatic picture. The quantum dynamical calculations reported included the motion in all six degrees of freedom of H_2, and were based on DFT PES. The diffraction experiments were performed for fixed angles of incidence, whereby both H_2 in-plane and out-of-plane diffraction spectra were recorded along the two main symmetry directions of the Pt(111) surface. The diffraction experiments covered the range of incident energies (up to 15 kJ/mol) relevant to heterogeneous catalysis. Indeed, for the most important catalytic process involving H_2 and

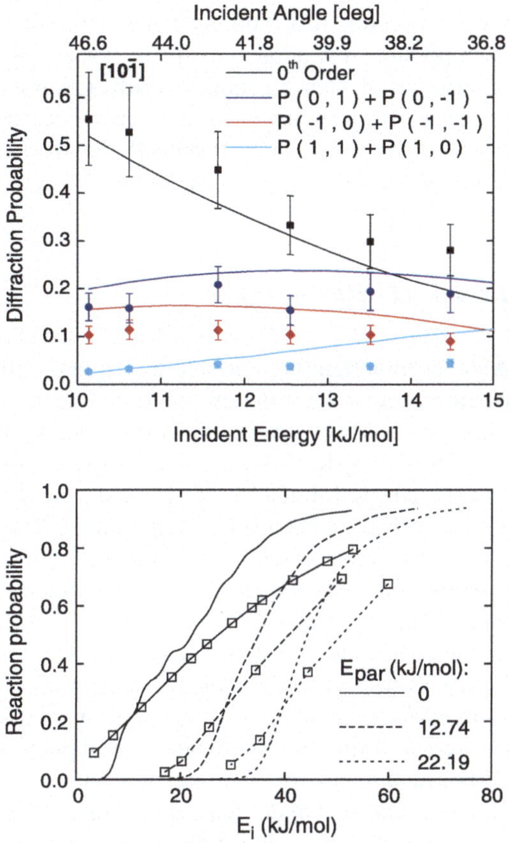

Fig. 14.8 *Top*: Experimentally determined H_2 diffraction probabilities (*symbols*) are compared with computed diffraction probabilities (*curves*), for specular scattering (*black*) and several first order out-of-plane (*colored curves* and *symbols*) diffractive scattering transitions [19]. The results are for incidence along the $[10\bar{1}]$ azimuth. Probabilities for symmetry equivalent transitions were summed. Error bars represent 68 % confidence intervals. *Bottom*: Theoretical reaction probabilities computed for normal and off-normal incidence along the $[11\bar{2}]$ direction compared to experimental results (*squares*) from Luntz et al. [44]. From [19]. Reprinted with permission from AAAS

Pt (reforming of gasoline), the average collision energy of H_2 with the a Pt surface is $2\,kT = 12.5$ kJ/mol [1]. The incident beam intensity was measured and used to normalize scattered beam intensities with respect to the incident beam, thereby yielding absolute diffraction probabilities.

The results of this work are summarized in Fig. 14.8, which shows the comparison of both experimental absolute diffraction and reaction probabilities with the theoretical results as a function of incident energy [19]. In the case of diffraction (top panel), both the energy dependence and the relative values of the diffraction probabilities are well reproduced by the theory. The agreement along the $[11\bar{2}]$ incidence direction (not shown) is likewise very good. It is worth emphasizing the importance of measuring out-of-plane diffraction, which represents roughly 50 % of the total diffraction intensity. It is only in this way that the full extent of agreement between experiment and theory can be established. Concerning reaction probabilities (bottom panel), excellent agreement is also obtained for both normal and off-normal incidence. In particular, the variation of the reaction threshold with $E_{par} = E_i \sin^2 \theta_i$ is well described and there is good overall quantitative agreement. Note that, even if one could argue that an electronically adiabatic theory could reproduce experiment for reaction due to fortuitous cancellation of errors in the PES and errors coming from possible non-adiabatic effects, it seems as extremely unlikely that such a good description of both reaction and diffraction experiments could be achieved in the presence of any significant non-adiabatic energy dissipation channel. Therefore, this study suggests that an electronically adiabatic theory is sufficient to describe scattering and reaction of H_2 from metal surfaces.

14.4.3 H_2/Ru(0001)

Another important issue to address from the theoretical point of view concerns the performance of the different functionals usually employed to calculate the 6D PES. This issue has been investigated in a recent work [45], in which experimental diffraction data for H_2/Ru(0001) were compared with six-dimensional quantum dynamics calculations performed by using two different DFT-based potential energy surfaces. These PESs have been calculated with the PW91 and the RPBE functionals, which are the two most widely used functionals employing the generalized gradient approximation.[1] The H_2 diffraction experiments on Ru(0001) were carried out in a variable angle setup at $E_{par} = E_i \sin^2 \theta_i = 35$ meV. The intensities extracted from angular distributions (along the two main symmetry directions) were firstly normalized with respect to the incident beam, and then extrapolated to 0 K applying the Debye-Waller model. Absolute experimental intensities were then compared with theoretical diffraction probabilities computed using PESs based on the PW91 and RPBE functionals for H_2 molecules with $j = 0, 1$. These results were weighted to simulate a normal hydrogen cold beam, i.e. 25 % with $j = 0$ and 75 % with $j = 1$.

Figure 14.9 shows the comparison of experimental with theoretical results corresponding to the $[10\bar{1}0]$ symmetry direction. Similar results (not shown here) were obtained along the $[11\bar{2}0]$ symmetry direction. As we can see, the comparison between experimental results and theoretical calculations using the RPBE functional is rather unsatisfactory. For instance, the general behavior of experimental intensities as a function of incident energy is not reproduced by the calculations. While all diffraction probabilities at first increase with E_i in the calculations, an overall decreasing or constant behavior with increasing incident energy is observed in the experiment. Also, in the calculations the (-10) diffraction peak is the most intense one, except for high E_i, whereas in the experiments the (01) diffraction peak is the most intense one. We can also see that the ratio of first order to specular diffraction intensities from simulations is much smaller than observed experimentally, which points to a too low corrugation amplitude for the PES calculated using the RPBE functional. These results clearly show that the PES calculated using the RPBE functional does not give a good description of the H_2 diffraction experiments from Ru(0001). In contrast, we see that much better agreement is obtained with the

[1] Calculations were also performed using the so called MIX functional, which is obtained from a weighted averaged of both PW91 and RPBE functionals. For further details see [45].

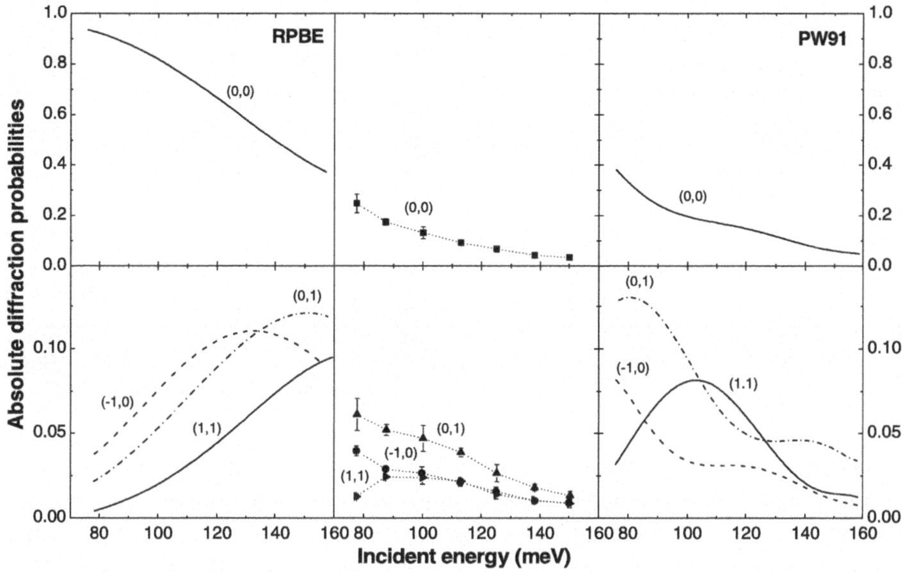

Fig. 14.9 Total diffraction probabilities calculated with the RPBE and PW91 PES, compared with experimental results measured along the [10$\bar{1}$0] symmetry direction. Both theory and experiment correspond to a fixed initial $E_{par} = 35$ meV. Adapted from [45]

quantum calculations based on the PW91 PES. In particular, both the absolute value of diffraction intensities and the general trends with incident energy are much better reproduced by the calculations, although it is worth pointing out that the agreement obtained here with a PW91 PES for H_2/Ru(0001) is not as good as the one obtained for H_2/Pt(111) using a Becke-Perdew PES (Ref. [19], shown in Fig. 14.8).

The corresponding results for reaction probabilities obtained with the two functionals are shown in Fig. 14.10. To arrive at a reliable comparison, the computed mono-energetic reaction probabilities were convoluted with the velocity distributions characterizing the experimental beams [30]. As we can see, the PW91 reaction probabilities are too large compared to the experimental sticking probabilities for the entire range of collision energies studied, whereas the RPBE reaction probabilities are too low in the energy range explored by diffraction experiments (below 200 meV). The results summarized in Figs. 14.9 and 14.10 show that, although a preference for one or the other functional cannot be given based on the reaction probabilities, the diffraction data tend to favor the PW91 functional, especially at energies below 0.15 eV. However, it is also clear that neither the PW91 nor the RPBE functionals accurately describe reaction of H_2 on Ru(0001) over the entire energy range investigated.

Although the PW91 functional overestimates the reaction probability, it yields a fairly good description of diffractive scattering of H_2 from Ru(0001). Since diffraction intensities are mainly determined by the geometrical corrugation along the unit cell, we can conclude that a much better description of the geometrical corrugation

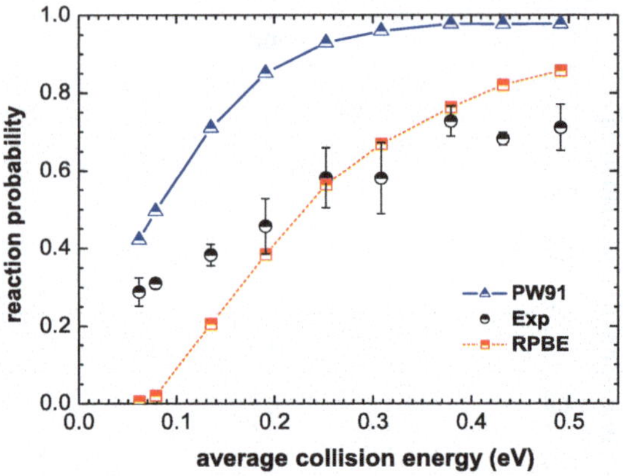

Fig. 14.10 Reaction probabilities obtained from the PW91 and RPBE PES are compared to experiment [30] for normal incident ($\nu = 0$, $j = 0$) H_2 on Ru(0001) as a function of the average collision energy. Adapted from [45]

of the unit cell is given by the PW91 PES than by the RPBE PES, for the energies explored in the experiments. These results become even more interesting if we remind that RPBE PESs usually provide more accurate molecule-surface bonding energies than the ones obtained with PW91 [46]. Thus, we can conclude that the PW91 functional describes better the PES region around 2 Å away from the surface atom nuclei, which corresponds to the region at which H_2 diffraction occurs.

14.4.4 D_2/NiAl(110)

The results already presented for H_2 /Pt(111) (see Sect. 14.4.2) suggest that quantum dynamics calculations performed within the Born-Oppenheimer approximation and using DFT-based 6D PESs, provide an excellent description of diffraction spectra. One may wonder how good is the performance of these calculations reproducing rotationally inelastic diffraction (RID) transitions, which usually represent a small amount of the total diffracted intensity (ca. 10–20 %, depending on the incident conditions). This issue has been recently addressed by a combined theoretical and experimental work on the scattering of D_2 from NiAl(110) [48], which is summarized in this section. The choice of the D_2/NiAl(110) system is justified because, owing to its minimum reaction barrier of 0.3 eV [49], the H_2/D_2 dissociation probabilities can be considered negligible in the energy range covered in the experiment (20–150 meV). Thus, low temperature targets can be used in the experiments, which minimizes Debye-Waller attenuation. In addition, previous classical and quantum dynamics calculations for this system have led to a good description of both dissociative adsorption and elastic diffraction of D_2 [49, 50], which gave confidence on the quality of the available PES to perform dynamical calculations.

The experiments have been performed using a time-of-flight setup, with fixed angle geometry. The high angular resolution of this machine, combined with its large dynamical range, make it the ideal tool to resolve the different RID peaks. It is worth emphasizing that this setup complicates significantly the theoretical analysis, since

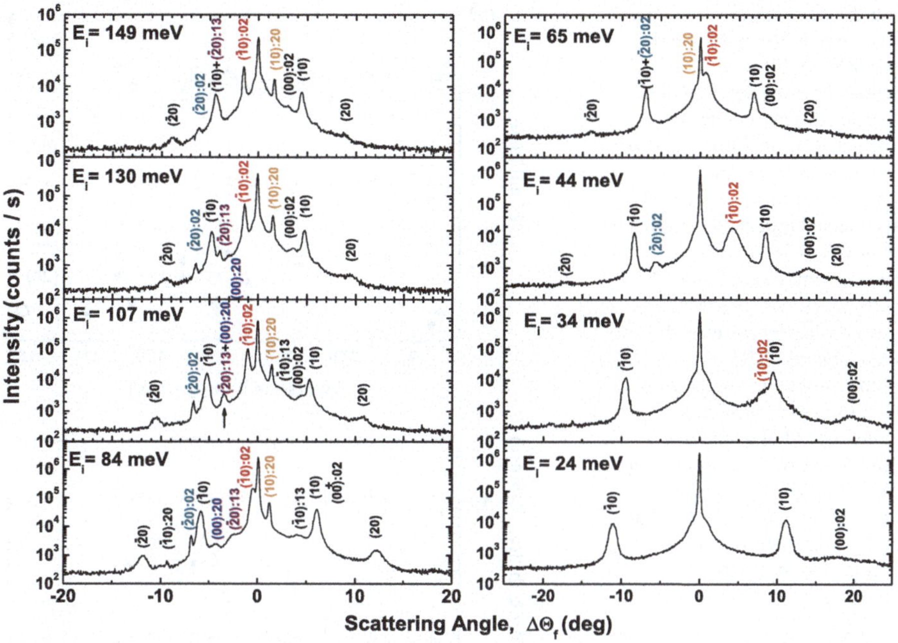

Fig. 14.11 Angular distributions for D_2/NiAl(110) along the [$1\bar{1}0$] azimuth measured at different incident energies E_i. The surface temperature is $T_S = 100$ K. Reprinted with permission from [47]. Copyright (2010), American Institute of Physics

the incidence angle is continuously varied during the measurement of diffraction spectra. Therefore, different angles of incidence must be used for each observed diffraction peak at a given incidence energy. The calculations reported in this study provide also the first example of a systematic study performed for this kind of setup, since such calculations have been prohibitively expensive until very recently due to computational limitations [48].

Figure 14.11 shows a series of angular distributions of D_2 scattered from NiAl(110) at incidence energies between 24 and 149 meV [47]. For energies below 24 meV only elastic diffraction can be seen. These spectra show a dominant specular peak, as expected for low corrugated surfaces [11]. Several RID and diffraction peaks are visible in the spectrum. The RID peaks are labeled as $(mn) : j_i j_f$, whereby (mn) denote the diffraction peak involved in the rotational transition, and j_i and j_f the initial and final rotational state, respectively. At $E_i = 24$ meV the $(\bar{1}0)$:02 peaks is kinematically allowed and can be observed around $\Delta\Theta_f = 18°$. The peaks $(\bar{2}0)$:02, $(\bar{1}0)$:02 and (00):02, involving ($j_i = 0 \rightarrow j_f = 2$) rotational transitions can be observed for $E_i > 80$ meV. In this energy range, rotational transitions ($j_i = 1 \rightarrow j_f = 3$) are also possible and, in particular, the $(\bar{2}0)$:13 channel is visible over the whole energy range. In the range [100–115] meV, there is an overlap of the $(\bar{2}0)$:13 peak, whose final scattering angle is the same (within experimental resolution) with the (00):20 peak. This is indicated by an arrow in the spectrum

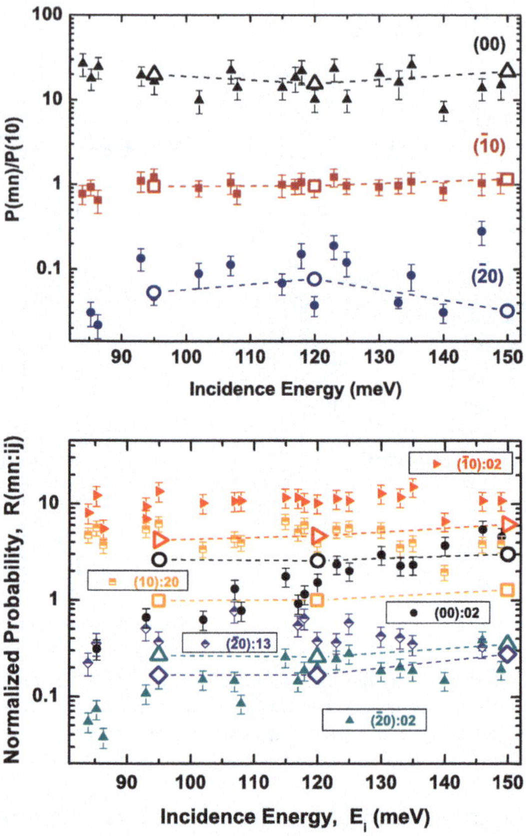

Fig. 14.12 *Top*: Experimentally determined normalized diffraction probabilities (*solid symbols*) compared with quantum diffraction probabilities (*open symbols*) for specular scattering (*black*), first (*red*) and second order diffraction (*blue*). *Bottom*: Experimentally determined normalized RID probabilities (*solid symbols*) compared with quantum RID probabilities (*open symbols*). Results are for incidence along the $[1\bar{1}0]$ symmetry direction. Error bars represent 68 % confidence intervals. *Dashed lines* are only a guide to the eye. Reproduced by permission of the PCCP Owner Societies from [48]

recorded at $E_i = 107$ meV. In such cases, time-of-flight measurements allowed to separate the contribution of each channel to the total intensity gathered at that final angle.

Figure 14.12 shows the experimentally normalized (with respect to the first order diffraction peak) determined diffraction (top panel) and RID probabilities (bottom panel) as a function of incident energy. Also shown are the results of quantum dynamics calculations performed using a 6D PES. It is worth emphasizing that, due to the experimental fixed-angle geometry used and to the j_i population of the D_2 beam, a total of 78 wave packet propagation had to be carried out. Concerning the elastic channel (top panel), we see an excellent agreement between experiment and theory. Both the relative intensities as well as their behavior as a function of incident energy are well reproduced by the calculations. In the case of the RID peak intensities, however, significant quantitative and qualitative discrepancies can be observed between theory and experiment (bottom panel). For instance, the experimentally determined intensity of the (00):02 peak increases with energy by a factor of ten in the explored energy range, whereas the corresponding theoretical intensity remains more or less at the same value in the same energy range. Furthermore, the relative intensity of the RID peaks associated with transitions from rotational states with $j_i > 0$ are not correctly described by quantum calculations. For instance, theory predicts that the (10):20 peak is more intense than the (00):02 one in the whole range of incidence energies explored, while experiment predicts the opposite below 130 meV. The same holds for less intense features: while theory predicts that the $(\bar{2}0)$:02 peak is more intense than the $(\bar{2}0)$:13 one for all incidence energies, the opposite ratio is observed in the experiment. Finally, note that although the relative RID peak intensities corresponding to $0 \rightarrow 2$ transitions are correctly reproduced by the theory, the quantitative agreement with the experimental data is clearly less satisfactory than for the elastic diffraction channels.

The observed discrepancies between experimental and theoretical results may be caused by any of the approximations made in the calculations: (i) the frozen surface approximation, (ii) the Born-Oppenheimer approximation, and (iii) inaccuracies in the PES employed in the dynamics calculations. The first point can be ruled out, because the experiment shows that the effect of thermal attenuation is comparable in all peaks seen in the diffraction spectra. In other words, the relative intensities of both elastic and RID peaks barely depend on the surface temperature. A violation of the Born-Oppenheimer approximation is also discarded by the authors of this work, mainly because previous calculations performed within the Born-Oppenheimer approximation were able to successfully account for dissociative adsorption [49] and elastic diffraction [50] in the same system. In addition, there is actually no evidence suggesting that RID processes should be more influenced by non-adiabatic effects than elastic diffraction processes.

Therefore, the discrepancies between experimental and theoretical results shown in Fig. 14.12 were attributed to inaccuracies in the PES employed in the calculations. These inaccuracies can be due either to the interpolation process or to the DFT functional used in the calculations. The fact that the PES used in this study has led to a very good description of dissociative adsorption and elastic diffraction [49, 50] strongly suggests that RID peaks are far more sensitive to subtle details of the DFT-based PES. Thus, the main conclusion of this study is that an accurate evaluation of RID intensities requires PESs that are more accurate than the existing ones. If this can be done within the framework of DFT is something that requires a thorough systematic theoretical investigations.

14.5 Future Developments

We have reviewed recent experimental and theoretical work on diffraction of H_2 and D_2 from metal surfaces. Experiments performed on Pd(111), Pt(111), NiAl(110) and Ru(0001) demonstrate that out-of-plane diffraction is important for H_2 + metal systems, and its importance increases with increasing reactivity. State-of-the-art 6D quantum dynamics calculations were performed within the rigid surface model of electronically adiabatic molecule-surface scattering. In this approach, the ground state electronic PES is based on DFT calculations using the GGA to the exchange-correlation energy [8–10]. The systematic studies reported in recent years show that such calculations provide, generally speaking, a very good description of diffraction of H_2 and D_2 from metal surfaces.

In spite of the progress made, there are still some open questions that deserve further investigations. A good example is provided by the rotationally inelastic diffraction (RID) peaks. The detailed study performed for D_2/NiAl(110) [48] clearly shows that present theory does not yield yet an accurate description of RID transitions. The discrepancies between experimental and theoretical results are very likely due to inaccuracies in the PES employed in the calculations. Similarly, the results reported for H_2/Ru(0001) show that, although the PW91 functional gives a better description of diffractive scattering than the RPBE one, the agreement obtained is not as good as the one obtained for H_2/Pt(111) using a Becke-Perdew PES. These results are expected to stimulate further developments of the theory in order to get a more accurate description of reaction, elastic diffraction and RID transition from metal surfaces.

From the experimental point of view, perhaps one of the most challenging and interesting future developments could be the study of more complex systems, like the surfaces of thin films or multicomponent alloys, with the aim of getting surfaces with dissociation properties different from those of transition metals. The combination of laser techniques with H_2 diffraction would allow, for instance, the measurement of diffraction spectra for vibrationally excited molecules, making it possible the comparison with calculations performed with H_2 molecules beyond the ground state. This would allow to test some interesting predictions of the theory, for instance that $\nu = 0$ and $\nu = 1$ H_2 react at different surface sites on Cu(001) [51].

In summary, H_2 diffraction measurements represent a quite sensitive technique to gauge the 6D H_2-surface PES within the unit cell. It samples the PES in a region around 2 Å away from the surface atom nuclei, which corresponds to the classical turning points of diffracted H_2 molecules. Therefore, the accuracy of PESs determined by state-of-the-art calculations can be tested in a region of interest also for other molecule/surface interactions, a kind of information difficult to get with other experimental techniques.

Acknowledgements We thank K.H. Rieder and J.P. Toennies for the donation of the scattering apparatus installed in our laboratory. The authors appreciate support from the Ministerio de Educación y Ciencia through projects "CONSOLIDER en Nanociencia Molecular" (CSD 2007-00010) and FIS 2010-18847, and from Comunidad de Madrid through the program NANOBIOMAGNET.

References

1. I. Chorkendorff, J.W. Niemantsverdriet, *Concepts of Modern Catalysis and Kinetics*, Student edn. (Wiley-VCH, Weinheim, 2003), p. 452
2. C.T. Rettner, M.N.R. Ashfold (eds.), *Dynamics of Gas-Surface Interactions* (Royal Society of Chemistry, London, 1991)
3. K.D. Rendulic, Surf. Sci. **272**, 34 (1992)
4. A. Hodgson, Prog. Surf. Sci. **63**, 1 (2002)
5. M.F. Bertino, D. Farías, J. Phys. C **14**, 6037 (2002)

6. G.O. Sitz, Rep. Prog. Phys. **65**, 1165 (2002)
7. G.R. Darling, S. Holloway, Rep. Prog. Phys. **58**, 1595 (1995)
8. A. Gross, Surf. Sci. Rep. **32**, 291 (1998)
9. G.J. Kroes, Prog. Surf. Sci. **60**, 1 (1999)
10. G.J. Kroes, M.F. Somers, J. Theor. Comput. Chem. **4**, 493 (2005)
11. D. Farías, K.H. Rieder, Rep. Prog. Phys. **61**, 1575 (1998)
12. D. Halstead, S. Holloway, J. Chem. Phys. **88**, 7197 (1988)
13. L. Mattera, in *Atomic and Molecular Beam Methods*, vol. 2, ed. by G. Scoles (Oxford University Press, London, 1992), pp. 366–383
14. Z.S. Wang, G.R. Darling, S. Holloway, Phys. Rev. Lett. **87**, 226102 (2001)
15. Z.S. Wang, G.R. Darling, S. Holloway, J. Chem. Phys. **120**, 2923 (2004)
16. G. Wahnström, A.B. Lee, J. Strömquist, J. Chem. Phys. **105**, 326 (1996)
17. H.F. Busnengo, W. Dong, P. Sautet, A. Salin, Phys. Rev. Lett. **87**, 127601 (2001)
18. N. Pineau, H.F. Busnengo, J.C. Rayez, A. Salin, J. Chem. Phys. **122**, 214705 (2005)
19. P. Nieto, E. Pijper, D. Barredo, G. Laurent, R.A. Olsen, E.J. Baerends, G.J. Kroes, D. Farías, Science **312**, 86 (2006)
20. D. Farías, H.F. Busnengo, F. Martín, J. Phys. C **19**, 305003 (2007)
21. H.F. Busnengo, C. Crespos, W. Dong, J.C. Rayez, A. Salin, J. Chem. Phys. **116**, 9005 (2002)
22. R. Olsen, H.F. Busnengo, A. Salin, M.F. Somers, G.J. Kroes, E.J. Baerends, J. Chem. Phys. **116**, 3841 (2002)
23. P. Rivière, H.F. Busnengo, F. Martín, J. Chem. Phys. **123**, 074705 (2005)
24. M.A. Di Césare, H.F. Busnengo, W. Dong, A. Salin, J. Chem. Phys. **118**, 11226 (2003)
25. M. Luppi, R.A. Olsen, E.J. Baerends, Phys. Chem. Chem. Phys. **8**, 688 (2006)
26. H.F. Busnengo, A. Salin, W. Dong, J. Chem. Phys. **112**, 7641 (2000)
27. G. Scoles (ed.), *Atomic and Molecular Beam Methods*, vol. 1 (Oxford University Press, New York, 1988)
28. G. Scoles (ed.), *Atomic and Molecular Beam Methods*, vol. 2 (Oxford University Press, New York, 1992)
29. P. Feulner, D. Menzel, Surf. Sci. **154**, 465 (1985)
30. I.M.N. Groot, H. Ueta, M.J.T.C. van der Niet, A.W. Kleyn, L.B.F. Juurlink, J. Chem. Phys. **127**, 244701 (2007)
31. J.R. Manson, in *Helium Atom Scattering from Surfaces*, ed. by E. Hulpke (Springer, Berlin, 1992), pp. 195–200
32. J.E. Pollard, D.J. Trevor, Y.T. Lee, D.A. Shirley, J. Chem. Phys. **77**, 4818 (1982)
33. K. Kern, R. David, G. Comsa, J. Chem. Phys. **82**, 5673 (1985)
34. M. Faubel, F.A. Gianturco, F. Ragnetti, L.Y. Rusin, F. Sondermann, U. Tappe, J.P. Toennies, J. Chem. Phys. **101**, 8800 (1994)
35. D.A. King, M.G. Wells, Proc. R. Soc. Lond. Ser. A **339**, 245 (1974)
36. A. Liebsch, J. Harris, Surf. Sci. **130**, L349 (1983)
37. P. Nieto, D. Farías, R. Miranda, M. Luppi, E.J. Baerends, M.F. Somers, M.J.T.C. van der Niet, R.A. Olsen, G.J. Kroes, Phys. Chem. Chem. Phys. **13**, 8583 (2011)
38. A. Gross, M. Scheffler, Chem. Phys. Lett. **263**, 567 (1996)
39. E. Pijper, G.J. Kroes, R.A. Olsen, E.J. Baerends, J. Chem. Phys. **116**, 9435 (2002)
40. D. Farías, C. Díaz, P. Nieto, A. Salin, F. Martín, Chem. Phys. Lett. **390**, 250 (2004)
41. D. Barredo, G. Laurent, C. Díaz, P. Nieto, H.F. Busnengo, A. Salin, D. Farías, F. Martín, J. Chem. Phys. **125**, 051101 (2006)
42. B. Gergen, H. Nienhaus, W.H. Weinberg, E.W. McFarland, Science **294**, 2521 (2001)
43. J.D. White, J. Chen, D. Matsiev, D.J. Auerbach, A.M. Wodtke, Nature **433**, 503 (2005)
44. A.C. Luntz, J.K. Brown, M.D. Williams, J. Chem. Phys. **93**, 5240 (1990)
45. P. Nieto, D. Farías, R. Miranda, M. Luppi, E.J. Baerends, M.F. Somers, M.J.T.C. van der Niet, R.A. Olsen, G.J. Kroes, Phys. Chem. Chem. Phys. **13**, 8583 (2011)
46. B. Hammer, L.B. Hansen, J.K. Nørskov, Phys. Rev. B **59**, 7413 (1999)
47. D. Barredo, G. Laurent, P. Nieto, D. Farías, R. Miranda, J. Chem. Phys. **133**, 124702 (2010)

48. G. Laurent, D. Barredo, D. Farías, P. Nieto, R. Miranda, C. Díaz, P. Rivière, M.F. Somers, F. Martín, Phys. Chem. Chem. Phys. **12**, 14501 (2010)
49. P. Rivière, M.F. Somers, G.J. Kroes, F. Martín, Phys. Rev. B **73**, 205417 (2005)
50. D. Farías, C. Díaz, P. Rivière, H.F. Busnengo, P. Nieto, M.F. Somers, G.J. Kroes, A. Salin, F. Martín, Phys. Rev. Lett. **93**, 246104 (2004)
51. M.F. Somers, D.A. McCormack, G.J. Kroes, R.A. Olsen, E.J. Baerends, R.C. Mowrey, J. Chem. Phys. **117**, 6673 (2002)

Part V
Charged Particle Techniques

Chapter 15
Low Energy Ion Scattering and Recoiling Spectroscopy in Surface Science

Vladimir A. Esaulov

Abstract This chapter presents an overview of low energy ion scattering spectroscopy and time of flight scattering and recoiling spectroscopy for the study of the structure and composition of surfaces as well as of fundamental ion surface interaction processes. The emphasis is on basic concepts regarding scattering, energy losses and charge transfer phenomena and experimental aspects involved in measurements of ion energies and time of flight measurements of scattered atoms. Some examples are provided to illustrate the type of information that can be obtained.

15.1 Introduction

The study of surfaces by scattering of low energy ions provides a powerful means of obtaining detailed information on surface structure, composition, location of adsorbates, kinetics of adsorption processes, as well as some fundamental interaction processes. When a beam of ions hits a solid surface part of the projectiles will be scattered back into the vacuum after collisions with target atoms of the top few layers. The projectiles loose energy by transfer of energy to the screened nucleus and to electrons via plasmon, electron hole and inner shell excitations. Measurement of the energy of the backscattered particles can be then used to identify the mass of these atoms in situations when the electronic losses are small enough. In many cases the target atoms "recoil" into the solid, but sometimes are scattered out as "fast recoils" and can be monitored. We address here two techniques which differ in that one analyses projectile ions (and sometimes atoms formed in neutralization processes) backscattered from the solid and in the other one also fast scattered recoil atoms are analyzed. The techniques we discuss are thus Low Energy Ion Scattering spectroscopy (LEIS or ISS) and Time of Flight Scattering and Recoiling Spectroscopy (TOF-SARS) or also Direct Recoil Spectroscopy (DRS). In both cases we consider fairly low primary ion energies in the range of 100 eV

V.A. Esaulov (✉)
Institut des Sciences Moléculaires d'Orsay, CNRS and Université Paris-Sud, UMR 8214, Bâtiment 351, 91405 Orsay Cedex, France
e-mail: vladimir.esaulov@u-psud.fr

G. Bracco, B. Holst (eds.), *Surface Science Techniques*, Springer Series in Surface Sciences 51, DOI 10.1007/978-3-642-34243-1_15, to 10 keV as opposed to the cases of Medium Energy Ion Scattering (MEIS) in the range several hundred keV, and High Energy Ion Scattering (HEIS) at energies of several MeV, also termed as the Rutherford Backscattering Spectroscopy (RBS).

One of the strengths of LEIS and also one of its potential pitfalls, comes from the neutralization of ions, which can be quite severe in case of noble gas ions. At low energies incident ions which stay too long in the vicinity of the surface, leave it as neutrals and are therefore not detected when electrostatic energy analyzers (ESA) are used. LEIS has therefore a very high first atomic layer sensitivity and is thus particularly well suited to surface analysis. Indeed LEIS was proposed for elemental identification by Smith [1, 2] in 1967 and seemed to be excellent for composition determination of the surface layer. Problems can arise when the neutralization probabilities are different for different atomic species composing the surface hampering quantitative analysis, which then needs to include a description of neutralization, which is not simple. To overcome the neutralization problem one can try to increase the incident ion energies to remove these atom specific effects or the so called "matrix" effects: effects of atomic environment. Alternatively one can use less neutralizable alkali ions when using ESA [3], or revert to detection of both, ions and neutrals by a time-of-flight (TOF) method [4, 5]. In this case however, the exclusive first-layer specificity can be lost and multiple scattering events complicate the data. Multiple scattering effects have been reduced in 180° backscattering in the impact collision variant of ISS (ICISS) proposed by Lindhard [6] in 1965 and pioneered experimentally by Aono et al. [7] in 1981. ICISS with neutral particle detection (NICISS) is also employed [8].

LEIS is nicely complemented by TOF-SARS. This pulsed beam technique produces negligible damage, is extremely sensitive to top surface atoms, and in particular can detect directly all atoms including H, which is not detected by most other spectroscopies [11, 12].

In this chapter we shall give an introductory overview of low energy ion scattering and recoiling, covering main aspects, with examples of instrumentation used, and some specific examples illustrating the techniques. This is a limited review given the brevity of the chapter, but we hope to provide the reader with a reasonably complete coverage of basic concepts and experimental methods. We do not give here a review of low energy ion scattering studies, but only provide a few examples that illustrate the type of information that can be obtained. Detailed reviews [9, 10, 13] of the use of these techniques exist in the literature along with some monographs [11, 14–16] and the interested reader will be directed to these and other relevant references regarding basic concepts involved in these studies. Some other applications of ion beams such as ion beam microscopy can be found in this book and elsewhere [17–19].

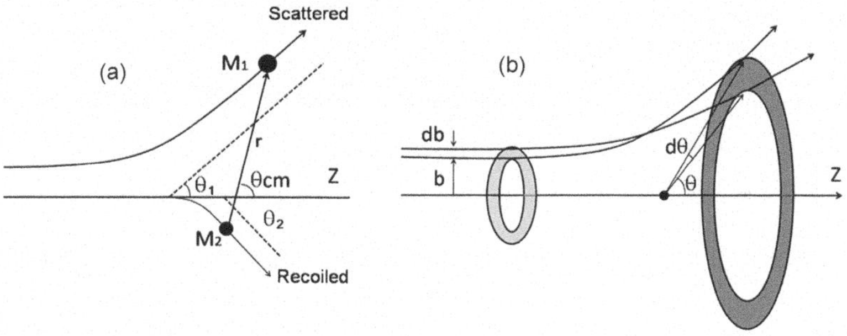

Fig. 15.1 (**a**) Scattering of particles of masses M_1 and M_2 viewed in the laboratory system, showing the center of mass and laboratory scattering and recoil angles. (**b**) Scattering of particles with impact parameters between b and $b + db$ by a central force into an angular range $2\pi\, d\theta \sin\theta$

15.2 Basic Principles

15.2.1 Energy Losses in Elastic Collisions

LEIS is conceptually a simple technique: primary ions of a known species and low keV energy impinge on a surface at a given incident angle and the energy distribution of the back scattered primary particles is measured under a fixed scattering angle. The major scattering contribution results from elastic scattering, and the trajectories of the particles can be described by one or by a sequence of two-body collisions. From energy and momentum conservation it follows that if an incident ion of energy E_0 and mass M_1 strikes a surface target atom of mass M_2 and is scattered through an angle θ_1 (Fig. 15.1), then in the laboratory frame of reference the scattered energy is given by

$$E_1 = \frac{E_0}{(1+A)^2}\left(\cos\theta_1 \pm \sqrt{A^2 - \sin^2\theta_1}\right)^2 = K E_0 \tag{15.1}$$

where $A = M_2/M_1$, the positive sign in (15.1) being for $A > 1$, while both signs are valid for $1 > A > \sin(\theta_1)$. Note the limitation on possible scattering angles for $A < 1$, i.e. $\theta_1 = \arcsin(A)$. For the case $A = 1$, $\theta_1 = 90°$.

The target atom gains energy and recoils with an energy E_2 at an angle θ_2 relative to the incident ion trajectory and

$$E_2 = E_0 \frac{4A}{(1+A)^2} \cos^2\theta_2 \tag{15.2}$$

The kinematic energy loss factor K is illustrated in Fig. 15.2. For a given scattering angle the final scattered ion energy in this binary collision is determined by the ratio of the masses of scattered particles. Consequently, the elemental analysis of the surface layer is quite straightforward and does not require a complicated analysis.

As an illustration typical LEIS spectra [20] for 500 eV He^+ scattering off a TiO_2 surface through a scattering angle of 124°, are shown in Fig. 15.3, where peaks corresponding to He scattered off the Ti and O atoms are identified. In a single

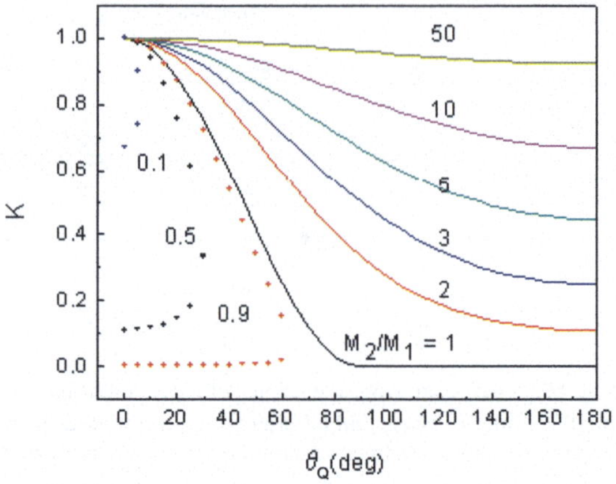

Fig. 15.2 Kinematic energy loss factor for different mass ratios

scattering event the final energies as given by (15.1) for 500 eV ions are of 387 eV and 224 eV (see also Sect. 15.4.1).

Note however that on a solid surface multiple scattering on adjacent atoms will occur and detected particles do not only correspond to single collisions. Particles that suffer multiple collisions will have a higher final energy for the same final scattering angle. Thus in the above case for e.g. consecutive scattering through 62° on two Ti atoms, the final energy would be 420 eV. Shift to lower energies may occur due to reionization of neutralized ions (see Sect. 15.2.6). Furthermore particles that penetrate into the solid, loose energy and are finally ionized in inelastic collisions (see below) and backscattered into the detector can lead to the appearance of a low energy tail. In general its intensity depends on ion energy, scattering angle and orientation of the target that determines the trajectories penetrating into the solid. This tail is particularly evident when one looks at the spectrum of neutralized particles as measured by TOF (Fig. 15.3b). We shall return to some of these points later (see Sect. 15.2.6).

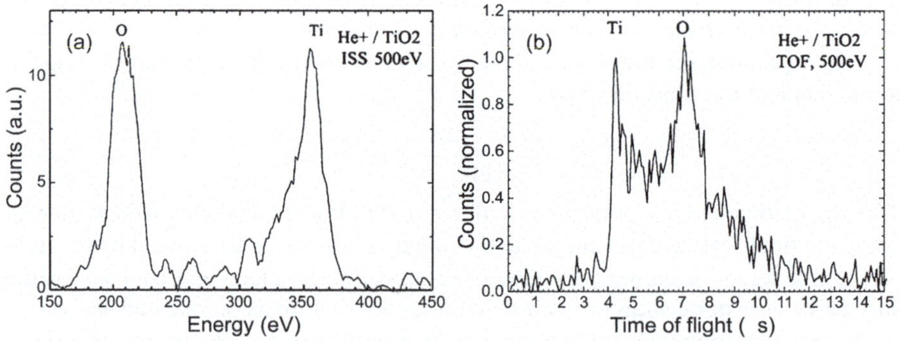

Fig. 15.3 *Left panel*: LEIS spectrum for scattering of He^+ ion on TiO_2 through a scattering angle of 124°. *Right panel*: corresponding TOF neutral spectrum [20]

The ability to resolve different masses in the LEIS spectrum is related to the energy widths of the peaks which depend on the incident ion energy spread, the energy and angular resolution of the energy analyzer and also on the thermal vibrations of the target atoms.

From (15.1) one can deduce that the resolving power (M_2/dM_2) for the case of $A > 1$, is given by:

$$\frac{M_2}{dM_2} = \frac{E_1}{dE_1}\frac{2A}{A+1}\frac{A+\sin^2\theta_1 - \cos\theta_1\sqrt{A^2-\sin^2\theta_1}}{A-\sin^2\theta_1+\cos\theta_1\sqrt{A^2-\sin^2\theta_1}} \quad (15.3)$$

The greatest mass resolution is obtained as A tends to 1 and for large scattering angles. The maximum energy transfer is obtained for 180° backscattering, but for $A = 1$ as seen above, θ_1 is limited to 90°. The choice of heavier projectiles improves resolution but at the same time in case of heavy projectiles and lighter targets the angular range of scattering is limited. Therefore for a given scattering configuration increasing the projectile mass at some point will exclude detection of the lighter species.

15.2.2 Electronic Stopping

Here it should be added that a complete description of energy losses also needs to take into account losses that can be viewed as due to a frictional force of the electrons in the solid on the moving ion. Depending upon the scattering configuration these can be small with respect to the elastic (nuclear) losses mentioned above, but in some conditions, when the trajectories of the ions on the surface of the solid or in the solid are long, they can become very large attaining several hundred eV. The "stopping" power $S(E)$ of ions in a solid is related to the energy loss per unit length dx of the ion trajectory $dE = S(E)\,dx$ and is dependent on ion velocity.

Electronic stopping of ions has attracted much attention both theoretically and experimentally [21–50], with in recent years attempts to include a proper description of the electronic structure of the solid [34–36]. Also recently the existence of a "threshold" effect of excitation of "d" electrons on the stopping of slow ions was noted [28–31] and a simple model developed. A complex aspect regards scattering through bulk monocrystals, where the electron density is very inhomogeneous [30, 31]. Similar problems exist on surfaces where regions far from the first atomic layer of the surface [37–50], where the electronic density exponentially decreases, may contribute to energy losses and also rather different types of ion trajectories can be encountered, with particles staying above the surface or probing inner layers and having straight or zigzag trajectories [37–50] etc. This leads to strong changes in energy losses. As an illustration the electron density profile at a distance of 1.283 au (atomic units) below the first atomic layer for a missing row reconstructed Au(110) surface [48, 49] is shown in Fig. 15.4. Figure 15.4 also shows examples of TOF measurements [48, 49] of energy losses of F^- ions scattered in grazing incidence

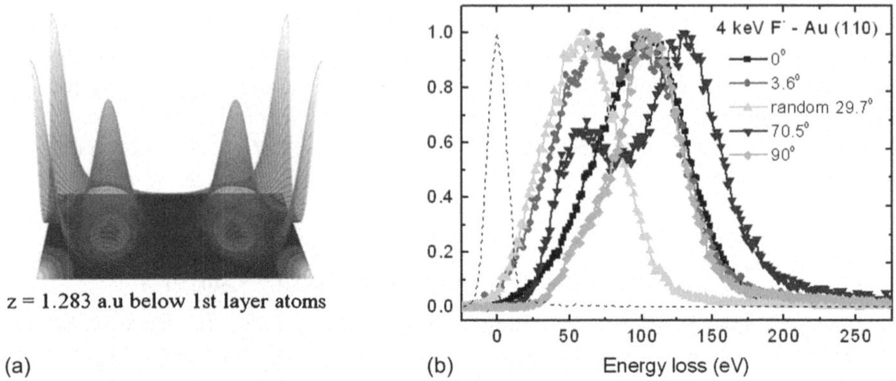

Fig. 15.4 *Left panel*: spatial charge density profiles for a reconstructed Au(110) given along the $[1\bar{1}0]$ or 0° direction. The *gray scale* density profile: low density—black (0.0031 au), high density—*white* (0.3524 au). *Right panel*: energy loss spectra of 1–4 keV F^- ions. The incident beam profile (*dashed line*) is shown for 1 and 4 keV fluorine ions

on this surface for different azimuthal orientations of the target with respect to the incident beam direction. Drastic changes can be observed as a function of azimuth.

Early descriptions of these cases treated the problem by assuming an averaged electron density [37–44]. In recent simulations [46, 48–50] the electronic density in the solid is first calculated and then included locally in calculations of stopping along the ion trajectory. As opposed to grazing scattering or for long trajectories inside a solid, these energy losses remain small in a typical LEIS experiment dealing with large angle scattering on surface atoms. On the other hand in small impact parameter collisions other inelastic processes appear and contribute to energy losses as will be mentioned below (see Sect. 15.2.6).

15.2.3 Scattering Cross Section and Interaction Potentials

Returning to the simple considerations on elastic scattering given above, it should be noted that with increasing scattering angle the ion scattering cross section decreases, so that a decrease in scattered intensity occurs. The cross section is determined by the interaction potential between the incident and projectile atom.

When an ion with nuclear charge $Z_1 e$ approaches a surface atom ($Z_2 e$) along a radius vector r, at small distances of approach, there will be a repulsive interaction between the two particles. This is usually described in terms of a screened Coulomb potential $V(r)$ with a screening function that models the deviation from a pure repulsive interaction between point charges due to the effect of the atomic electron clouds:

$$V(r) = \left(Z_1 Z_2 e^2 / 4\pi r\right) \Phi(r/a) \tag{15.4}$$

where a is a "screening length". Various approximations for the screening functions [14, 24, 51–54] are used, such as the Ziegler-Biersack-Littmark [24] (ZBL) or Molière [51, 52] etc. For instance in the Molière approximation for the Thomas-Fermi model [51, 52]:

$$\Phi(r/a) = \Phi(x) = 0.35e^{-0.3x} + 0.55e^{-1.2x} + 0.1e^{-6.0x} \tag{15.5}$$

In Firsov's [53] modeling, the screening length is:

$$a_{Firsov} = 0.4685/\left(Z_1^{0,5} + Z_2^{0,5}\right)\ (\text{Å}). \tag{15.6}$$

Using such potentials, in the center of mass reference frame of the colliding particles the scattering angle θ is given by:

$$\theta_{cm} = \pi - 2\int_{r_{min}}^{\infty} \frac{b\,dr}{r^2\sqrt{1 - \frac{V(r)}{E_{cm}} - \frac{b^2}{r^2}}} \tag{15.7}$$

where b is the impact parameter (Fig. 15.1b). From this one can calculate the differential scattering cross section $\sigma(\theta)$ into an angle θ given by

$$2\pi b\,db = -\sigma(\theta)2\pi\sin(\theta)\,d\theta = -\sigma(\theta)\,d\Omega \tag{15.8}$$

where $d\Omega$ is the solid angle.

For the simplest case of the Coulomb potential $V \sim 1/r$ we obtain the well known Rutherford scattering formula in the center-of-mass system:

$$\frac{d\sigma(E,\theta)}{d\Omega} = \frac{\left(\frac{Z_1 Z_2 e^2}{4E}\right)^2}{\sin^4\left(\frac{\theta}{2}\right)} \tag{15.9}$$

The scattering cross section for these potentials decreases rapidly with increasing scattering angle. From the point of view of LEIS this means that the number of particles scattered into large angles rapidly decreases and though from the point of view of mass resolution use of larger angles is advantageous, the detection efficiency will be low.

The choice of potentials of the type given above excludes any attractive interactions at larger distances. Usually some kind of fitting procedure to experimental data is attempted for quantitative work (see below).

15.2.4 Shadowing, Blocking and ICISS

Scattering by a repulsive atomic potential leads to the formation of a shadow [14, 55] region behind the target atoms where the incident ion can not penetrate. This is illustrated in Fig. 15.5a on the example of calculated trajectories for scattering of He^+ on Ag. An Ag atom lying behind the first scattering center, within this cone, will not result in scattering of the primary ions. One can also observe in Fig. 15.5a an increase in density of trajectories for forward scattered particles at the edge of the shadow cone. Thus since the incident flux is conserved, the shadow cone formation leads to a focusing of trajectories [11, 14, 15, 56, 57], at its edge. For the case of

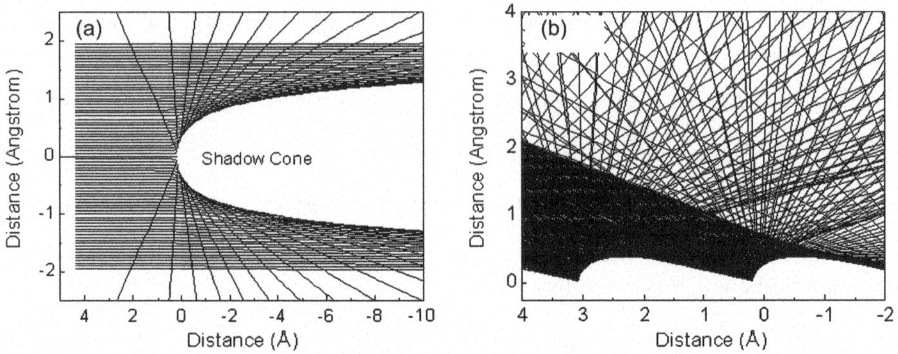

Fig. 15.5 Trajectories for 2 keV He^+ scattering on (**a**) a single Ag atom (**b**) and a string of Ag atoms illustrating the shadow cone effect and focusing. Figure (**b**) corresponds to a situation when 180° backscattering occurs

the Coulomb potential for scattering of a particle with energy E, the radius of the shadow cone at a distance d from the scattering center is given by

$$R_s = 2\left(Z_1 Z_2 e^2 d/E\right)^{0.5}$$

When one considers scattering on more than one atom several scattering events can occur. In grazing scattering on surfaces and in crystalline materials along crystal channels, shadow cones inhibit close encounters with lattice atoms. One can then observe channeling phenomenon [14, 15], in which only glancing, large impact parameter collisions occur on the crystal planes (planar channeling) or along axes of the crystal (axial channeling). Ion "steering" results in strong enhancement of scattering intensity observed along main symmetry directions in both cases of surfaces and in crystals. This is illustrated in Fig. 15.6 on the example of measured scattered intensity for the case of F^- scattering on a missing row reconstructed Au(110) crystal [50], where the well defined peaks can be related to the surface structural characteristics in specific directions as shown in the figure. Thus in this case the characteristics of the total scattered particle flux provides information on the surface structure, when studied as a function of surface azimuthal angle. A detailed discussion of general characteristics of grazing scattering may be found in a review by Winter [47]. We shall return to some other aspects of grazing scattering in Sect. 15.2.5.

In general on surfaces and in solids, depending on the relative disposition of the atoms, scattering of particles in a given direction by the first target atom, may be "blocked" by the presence of another atom along the trajectory. This leads to the formation of a "blocking cone" analogous to the shadow cone.

The shadow- and blocking-cone concepts are efficiently used for the determination of the surface structure in the so called impact collision ISS (ICISS) in close to 180° backscattering configurations. In grazing scattering on a chain of atoms primary ions are forward scattered in a series of small angle deflections by adjacent atoms. As the incidence angle increases, head on collisions can occur for a critical

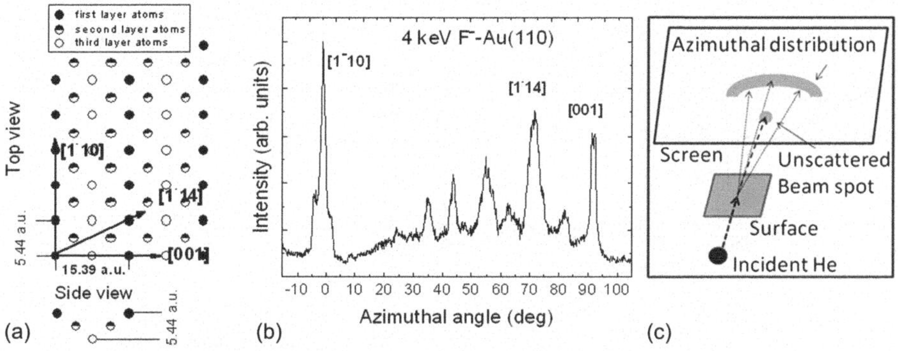

Fig. 15.6 Schematic view of (**a**) the missing row reconstructed Au(110) surface and (**b**) intensity of fluorine ions in 3.5° grazing incidence conditions for specular scattering. (**c**) Schematic diagram of grazing scattering and azimuthal distribution on a 2D detector

angle of incidence ψ_c, determined by the shadow cone size and interatomic distance d. The shadow cone radius R_s at a distance L behind the first scatterer such that the edge of the shadow cone is positioned at the next atom in Fig. 15.7 is given by $R_s = d \sin \psi_c$ and $L = d \cos \psi_c$. The focused particles at the shadow cone edge collide with the second atom at close to zero impact parameter and in backscattering, (Figs. 15.5b, 15.7a) at this critical incidence angle, a sharp increase in scattered intensity is observed. Note that in this 180° backscattering case we encounter a "double focussing" effect [13, 58] related to the initial shadow cone focusing by the first atom and subsequent blocking cone focusing by the same atom in the outgoing path (Fig. 15.7), leading to a larger enhancement of scattering intensity, with respect to other scattering configurations. This is illustrated on the example of calculated scattered intensities [58]. In practice in experiments a broadening in the intensity distribution is observed because of thermal vibrations of atoms.

When the incidence angle is further increased the ICISS spectra show contributions from shadow and blocking cone focusing on deeper layer atoms. Thus in general by varying the incidence and also the azimuthal orientation of the target important surface structural parameters can be deduced and location of adsorbates identified. An example of this will be provided later in this chapter.

In practice interpretation of ICISS data involves numerical simulations of scattering. Several computer simulation codes have been developed [24, 46, 48–50, 59–61], some of which are freely available (SRIM [24], Marlowe [60, 61], Kalypso [59]). The Kalypso code (used to calculate data in Fig. 15.5) includes a nice, instructive, visualization window, while running simulations. In general these codes rely on a binary collision approximation in which a collision between the incoming ion and a target atom is treated. This involves solving classical equations of motion using a single atom-atom interaction potential of the type mentioned above or else taking into account the summed effects of several neighboring atoms. In the so called Monte Carlo approach the parameters for the next collision is chosen randomly and the distance to the next atom is determined from the material's density.

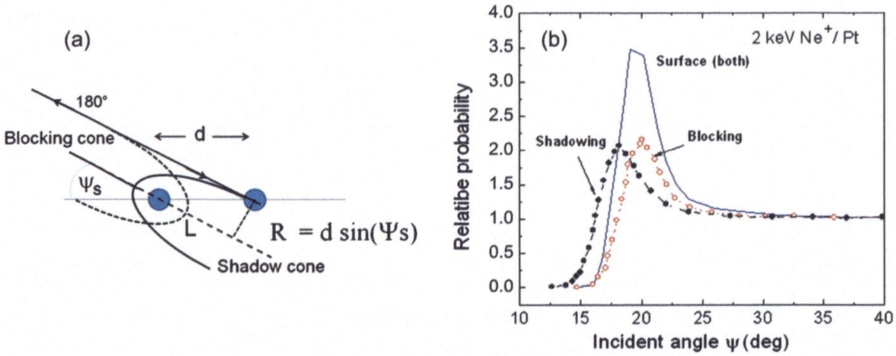

Fig. 15.7 Schematic diagram of shadow and blocking cones illustrated for backscattering. Calculated [58] surface peak (*solid line*) for 180° NICISS of 2 keV Ne^+ at Pt, due to the quasi-double–focusing effect. The *solid circle line* shows the situation for a scattering angle of less than 180°, i.e. the enhancement originating from the shadow effect (incoming path), while the *open circle* line exhibits the enhancement caused during the outgoing path (blocking effect)

This is applicable to amorphous materials (SRIM code). For crystals [46, 48–50, 59–61] (e.g., Marlowe, Kalypso) the target atomic positions are those of the given crystalline lattice and the scattered ion trajectory resulting from one collision determines the parameters for the collision on one of the next lattice atoms. Depending upon the degree of sophistication, simulations can include changes in ion trajectories because of image potential effects, ion stopping, atomic vibrations, ion induced damage, models of neutralization, beam divergence and detector acceptance (see also in Sect. 15.4.3).

15.2.5 Special Features in Grazing Scattering

While this is beyond the scope of this chapter, it is nevertheless interesting to mention some other aspects of above mentioned grazing scattering not covered in the simulations mentioned above. Experiments in which the intensity of atoms scattered grazing on the surface has been monitored on a 2D detector (Fig. 15.6) [47], show that for “random” orientation a well defined specularly reflected beam is recorded, whereas for axial channeling one observes a broad “banana shaped” or bent angular distribution and more or less pronounced peaked structures, with a central one due to focusing and side structures due to a rainbow-scattering in the collective axial interatomic potential. The azimuthal angles for these peaks (“rainbow angle”) depend on the scattering conditions, but also on the potentials seen by the projectile. It has therefore been suggested that this can be used to deduce the projectile surface potential [47].

In recent very grazing experiments [62–69], the intensity pattern of scattered He atoms was found to be different and displayed an array of clear spots superposed on a weak “banana” angular distribution (Fig. 15.8). These were attributed

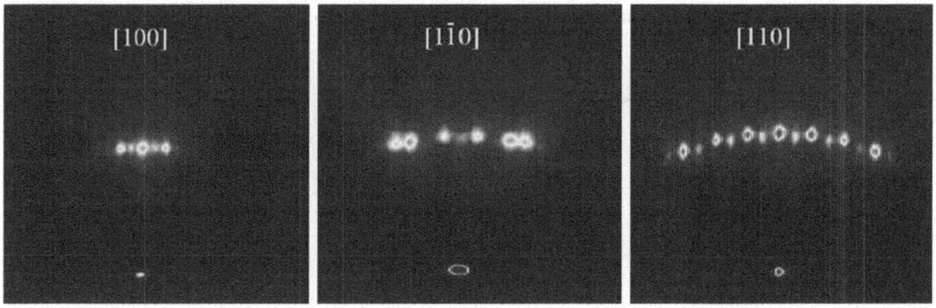

Fig. 15.8 Diffraction images [66, 67] recorded with 400 eV He on the $c2 \times 2$ reconstructed ZnSe(001), for some azimuthal directions. The *narrow spot* on the *lower part* is the direct beam

to diffraction of fast atoms, previously well known in scattering of atoms at thermal energies [70–73]. The transition from the focusing and rainbow structures to the clear diffraction regime corresponds to a sufficient lowering of perpendicular energy of projectiles, either by lowering the incident particle energy or decreasing the incident angle, which precludes excitations that could otherwise introduce dephazing. Typically diffraction was observed below 1 eV perpendicular energy for He projectiles at room temperature. Several theoretical works on describing this process have been published [62–68]. The grazing incidence fast atom diffraction (GIFAD) (for a recent review see Ref. [74]) patterns are highly surface sensitive and this led to a suggestion to use it for study of growth on surfaces.

15.2.6 Quantification and Neutralization Effects

LEIS has been quite extensively employed in surface composition analysis [9–11, 13–15]. The intensity of ions scattered from an atom A with a surface density N_A can be expressed as

$$I_A = I_0 N_A T S P_A \, d\sigma_A / d\Omega \tag{15.10}$$

where I_0 is the primary ion current, $d\sigma_A/d\Omega$ the cross section for scattering into a solid angle $d\Omega$, T is the transmission factor characterizing the spectrometer, P_A is the ion survival probability. The factor S allows for the possibility of shadowing by other atoms and surface roughness effects.

The differential cross section can be calculated using one of the interaction potentials mentioned above. The transmission function can be obtained from measurements on known surfaces. The main problem [9–11, 13–15] resides in the knowledge of the survival probability P_A. Note that P_A could in principle be concentration dependent and also be different in case of differences in the nature of the surrounding atomic species. In literature this is frequently referred to as "matrix effect" [9, 10, 13]. The situation is also complicated because depending upon collision energy, ions observed in LEIS may be surviving ions, but can also be produced by reionization of neutralized particles in small impact parameter inelastic collisions.

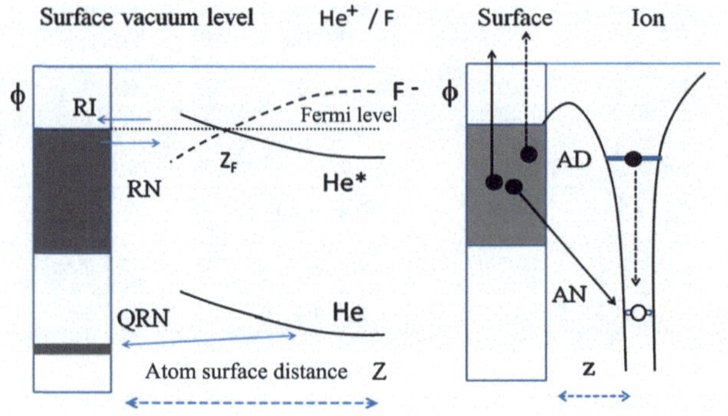

Fig. 15.9 *Left panel*: schematic diagram of the behavior of energy levels of He and F in front of a surface, indicating the respectively downward and upward shift of the levels, and resonant ionization RI, neutralization RN and quasi-resonant neutralization (QRN) processes. *Right panel*: schematic diagram of Auger neutralization (AN) and de-excitation processes (AD)

Neutralization of ions on surfaces has been the object of many studies. The different types of electron transfer phenomena that need to be considered are schematized in Fig. 15.9, which represents the relative positions of energy levels of atoms and in the solid. When an atom approaches a solid interaction leads to a broadening and shift of atomic energy levels. An "upward" or "downward" shift of energy levels for positive and negative ions respectively occurs and is given by the image potential effect [47, 75] at larger distances, Z, from the surface: $\delta E = 1/4Z$.

Resonant electron capture processes involve a transition from the occupied states of the valence band of a metal to the atomic level, at atom-surface distances, where the atomic level lies below the Fermi level [75–89]. In a static picture (ion velocity independent), for negative ions electron capture can occur, when the affinity level lies beneath the Fermi level (E_F, Fig. 15.9), for distances smaller than a critical distance Z_F and electron loss to the metal occurs for $Z > Z_F$. Inversely for the case of alkali positive ions for which the ionization potential is small, and levels are "upward" shifted, capture of electrons occurs at large distances from the surface and electron loss from the occupied atomic level to the conduction band occurs at small atom-surface distances ($Z < Z_F$). This is also the case for resonant capture into excited atomic states. This resonant electron transfer efficiency is strongly work function (ϕ) dependent, since this determines the range of distances over which the capture or loss processes occur.

This simple picture is altered in case of grazing ion surface scattering when neutralization or negative ion formation is strongly affected by dynamic so-called parallel velocity ($v_{//}$) effects [47, 75, 77–80]. This results in an effective change in the range of distances over which resonant capture and loss processes occur (for details see [77–80]).

Initially treatments of resonant charge transfer were based on simplified perturbative approaches [75]. More recently non-perturbative methods were introduced, which include complex scaling [81], coupled angular mode [78], stabilization [83, 84], self energy [84, 85], wave-packet propagation [86], and DFT calculations

[87]. These methods yield excellent agreement with experimental data [80, 87–89] on some simple systems, but some problems are still pending especially when transition metals, general cases of adsorbates and small atom surface distances [87] are involved. Recently in nanoscale systems (clusters) size effects have been observed (see Sect. 15.4.1).

Quasiresonant charge transfer. When the surface electronic structure is characterized by the presence of narrow, localized bands one can observe the occurrence of non resonant electron transfer [90, 91] between these levels and atomic states mediated by the projectile velocity. The *quasi-resonant electron capture* processes are similar to quasi-resonant charge transfer in gas-phase collisions ($A^{+/-} + B \rightarrow A + B^{+/-}$) [90, 91]. Such an "atomic-like" charge exchange process has been first proposed to explain oscillations in charge transfer in He^+ collisions with Pb [92] and more recently for N^+ scattering on NaCl [93, 94]. A recent detailed discussion of such oscillatory charge transfer processes has been given [95]. In a somewhat different context non resonant charge transfer was invoked to explain H^- formation on Si [96] and the large anion fraction in the scattering on insulators (ionic crystals) such as LiF [97, 98] and MgO [99], where it involves capture from the negative ions (F, O) in the crystal (for a review see [100]). Existence of an oscillatory behavior of the neutralization probability as a function of ion energy as in the $He^+ + Pb$ case [92] can lead to significant errors in composition analysis in LEIS and investigations over a range of energies are essential.

Auger processes (AN) usually occur when the incident particle has a core hole as e.g. for He^+ ion neutralization. An electron from the metal neutralizes the ion and the excess energy is transferred to another electron in the metal, as illustrated in Fig. 15.9b, which also shows an Auger de-excitation (AD) process. This may lead to electron emission or production of so called hot electrons. AN [101–112] of positive ions in front of metal surfaces has attracted much attention starting from the seminal works of Hagstrum [101] and more recently a number of theoretical treatments were developed for free-electron like metals [102, 103], which include more generally electronic excitations—"plasmon assisted" neutralization [104] and ion induced effects [105]. Recent theoretical treatments of AN have been applied to the description of these plasmon features observed in electron emission following AN [103, 104] and to the calculation of ion fractions in low energy scattering (LEIS) of He^+ on Al [106]. It has been found that theoretical neutralization rates reproduce quite well experimental data on He^+ neutralization on Al and Ag surfaces [107–109]. The important role of d electrons was also recently investigated [110, 111].

Effect of adsorbates. In gas surface interactions a very important problem concerns the effect of adsorbates. In the past some efforts at a theoretical modeling of the effects of impurities in promoting or poisoning of reactions has been made [113–118]. In early works, discussions of adsorbate effect have been mainly limited to consideration of work function effects [113, 118]. More recently they have been discussed in terms of site blocking and electronic effects related to changes in local density of states [114] and the electrostatic potentials [115, 116] around adsorbate sites [118]. Thus the influence of the electrostatic potentials on the energy level positions and widths of atomic states of a gas phase species approaching a surface have been discussed in some cases [116, 117]. The effect of formation of quasimolecular states corresponding to the incident particle and the adsorbate have been considered for the case of resonant electron transfer in Li scattering on a Li adsorbate [118]. An example of adsorbate induced effects in LEIS will be discussed later in this chapter.

Simple modeling. Resonant and Auger electron transfer is frequently described in an approximate framework using a rate equation approach [75]. The transition rate R is determined by the atomic level width $(\Gamma(z))$ as a function of the atom surface distance z and the population of the level is given by this width and the time spent in a given surface region. The rate R is assumed to decay exponentially away from the jellium edge:

$$R(z) = Ae^{-\alpha z} \tag{15.11}$$

The neutralization probability is then determined by the time spent by the ion near the surface in the ingoing and outgoing paths. The fraction of neutralized ions moving with a velocity v_{in} perpendicular to the surface in a time interval dt is given by

$$dN = R(z)\,dt = R(z)\,dz/v_{in} \tag{15.12}$$

After integration the survival probability in the ingoing path up to a minimum distance z_m is then:

$$P_{in} = \exp\big((A/\alpha v_{in})\exp(-\alpha z_m)\big) \tag{15.13}$$

and if $z_m = 0$

$$P_{in} = \exp\big((A/\alpha v_{in})\big) \tag{15.14}$$

For neutralization along the incoming and outgoing paths [9, 11], with a velocity v_{out}

$$P_{in} = \exp(\big(A/(\alpha v_{in} + 1/\alpha v_{out})\big) \tag{15.15}$$

This very simple expression has been used [10] to fit experimental data on survival of ions with some degree of success. However if more than one neutralization channel exists the significance of these fits is not clear. Furthermore in small impact parameter collisions reionization of atoms can occur, which complicates considerably the interpretation of neutralization data.

Inelastic processes. Experimental data on energy losses of ions and atoms as well as electron spectroscopy and photon emission data [119–129] show the existence of reionization of neutralized atoms and production of highly excited states following inner shell excitation. The interpretation of this data has been done along the lines of models used in gas phase atom-atom collisions at low energies in terms of the molecular orbital (MO) promotion [124–126] model. In this model of the atom-atom collision one follows the evolution of diabatic atomic orbitals from the separated atoms limit to the "united atom" limit at zero separation [125]. This is illustrated on the schematic MO diagram (Fig. 15.10) for Ne–Na collisions [126–128]. Strong "promotion" of the outer 4f σ orbital as the atoms approach, leads to

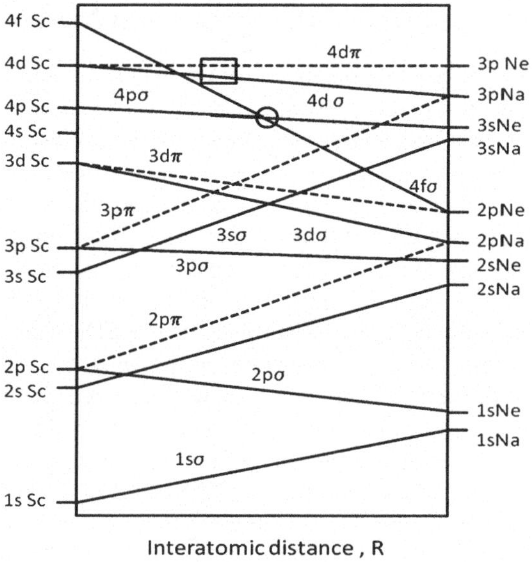

Fig. 15.10 Schematic MO diagram for the case of Ne Na collisions

the possibility of transferring electrons into higher excited states at crossings with higher lying orbitals, where electron transitions can occur (schematized by the circle). The interatomic distance at which this promotion occurs has been estimated to be about 1.45 au from gas phase collisions. Excitation processes in this description may also occur as a result of internuclear axis rotation (rotational coupling as schematized by rectangle between σ and π orbitals). In case of vacancies in inner orbitals, when energy conditions are satisfied, two electron Auger type processes in which one electron fills the inner hole and the other occupies a higher lying orbital (e.g. $2p\pi^3 4f\sigma^2 \Rightarrow 2p\pi^4 nl\sigma$) can occur [124–129]. Transition probabilities are interatomic velocity dependent [124].

Experiments in ion surface scattering show that onset of inelastic process leading to ionization, usually occurs at energies of a few hundred eV [121, 122] depending upon the atom-atom combination considered. A comparison of the dynamics of the excitation processes in gas phase and ion solid interactions may be found in [120]. In cases where a number of excitation and ionization processes occur at a given collision energy, the use of simplified approach assuming a single neutralization process, as above, is therefore obviously not justified. Note that because of neutralization processes, the detection of ions at some energy is not an indication of the onset of excitation processes at that energy.

While some authors suggest [9] that in many cases “matrix effects” are not important, in practice to make reliable statements concerning surface composition it is desirable to perform LEIS measurements over an extended energy range and also use different types of ions in an attempt to eliminate effects related to neutralization: e.g. use of He^+, Ne^+ and alkali ions. When possible a combination with time of flight analysis of scattered neutrals, as in NICISS and TOF-SARS, can be a major asset for unambiguous analysis.

15.3 Experimental Setups and Methods

The basic components of an ion scattering and time of flight spectroscopy system usually involve a primary ion beam source, mass analysis system, beam pulser in case of flight analysis, the main surface analysis system with a precision sample manipulator with provision for sample orientation and also heating/cooling. Detection and analysis systems include an electrostatic analyzer for ion energy analysis and flight tubes for TOF analysis usually equipped with multichannel plate (MCP) particle detectors and relevant electronics and data acquisition systems. In many cases these basic components are supplemented by other surface analysis techniques such as low energy electron diffraction (LEED), Auger electron spectroscopy (AES) and X-ray photoelectron spectroscopy (XPS).

Some of the simplest systems use a fixed scattering configuration in which the ion source and ESA/TOF analyzers are fixed and the sample can be rotated with respect to the incident beam direction to obtain structural information. This is for instance the case when a commercial XPS system is used for LEIS, in which the ion source is the sputter gun for cleaning the samples. Various more sophisticated systems have been developed. In the following we shall briefly describe some aspects of these LEIS and TOF systems and stop on some aspects related to electrostatic analyzers and time of flight measurements.

15.3.1 Ion Beam Lines

In general it is desirable to have an ion source with a small energy spread and for this common ion beam sources used for sputtering are not recommended. A better solution is generally a discharge source with an electron emitting filament placed in front of an anode with a hole for extracting ions (Fig. 15.11). Usually ion beams with an energy spread of 0.5 eV to a few eV in the 100 eV to 10 keV energy range can be obtained [130]. Because of a substantial gas load it is necessary to ensure differential pumping of the ion source. It is preferable to have a mass analyzer to separate spurious masses and also separate higher charge states. Common mass analyzers are Wien filters with crossed magnetic and electric fields or bending magnets. It is better to adopt a configuration in which the ion beam is not in line of sight with the analyzed sample to avoid spurious fast neutrals in the incident beam.

As an example we show a setup with such a configuration [130] in Fig. 15.11. This system was designed for both LEIS and TOF scattering spectrometry. It was equipped with two electrostatic analyzers that could rotate independently around the sample for LEIS and low energy electron spectroscopy and also with time of flight tubes for TOF measurements.

The beam line here is composed of a discharge ion source, Wien filter for mass analysis, and a quadrupole deflector for a 90° beam deflection. The ion source can produce both positive and negative ions. Although alkali ion beams can be produced

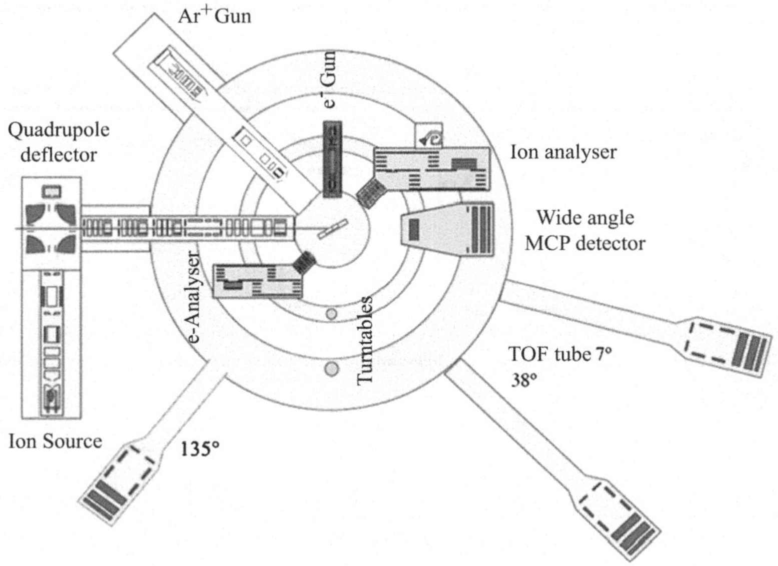

Fig. 15.11 Orsay ion scattering apparatus for LEIS, electron spectroscopy and time of flight direct recoil spectroscopy

in this type of discharge source it is more convenient to use solid emitters. In this case a simple modification of the source allows to replace the heated filament by the alkali cathode [131]. After mass selection the deflected ion beam then passes through a series of guiding deflection plates and lenses before reaching the sample. A set of dedicated deflectors are used for beam pulsing for time of flight analysis.

The sample is usually held on a three axis goniometer allowing XYZ displacements and different rotations of the sample. The incidence and azimuthal orientation as well as tilt angle of the sample can then be varied.

In case of insulating samples such as oxide cathodes that have been investigated by some groups [9], sample charging may occur upon incidence of the ion beam. To avoid this one can make use of a neutralizing electron beam. When dealing with rough insulating samples it has been found useful to direct the neutralizing electron beam on the sample close to the incident ion beam direction so as to cover the same irradiated area.

15.3.2 Energy Loss Measurements Using Electrostatic Analyzers

A complete discussion of design of electrostatic analyzers is beyond the scope of this chapter. There exist a large number of works dealing with electrostatic analyzers and here we just recall some key points. A nice review is given in [132]. Energy analysis of scattered ions entering the analyzer at a predefined angle with respect to

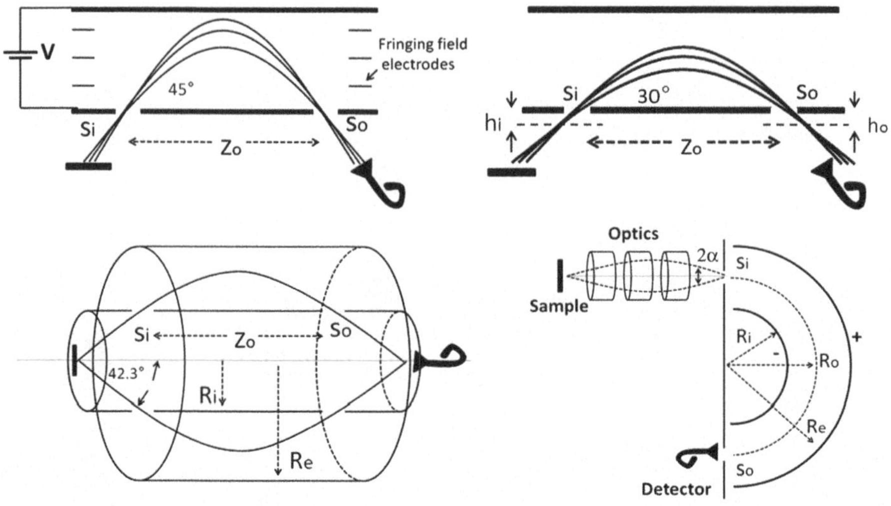

Fig. 15.12 Schematic diagrams of some common electrostatic analyzers

its electrodes, through an entrance aperture or slit, is performed by applying appropriate potentials to the electrodes that compose it, so that trajectories of ions with different energies are spatially separated. Energy selection is performed by placing an exit slit through which only ions of a given energy can pass. Various analyzer electrode geometries have been used and are distinguished by differences in their optical properties, that determine the energy resolution and transmission of the analyzer. Some common designs are illustrated in Fig. 15.12.

Amongst the most common (Fig. 15.12) are e.g. the simplest parallel plate analyzer (PPA) used in the afore mentioned Orsay system, the cylindrical mirror analyzer (CMA), the 127° cylindrical sector analyzer (SDA, not shown), the hemispherical analyzer (HSA) and the less common toroidal analyzers.

The energy of ions that pass through an analyzer or "pass" energy is related to the potential difference (V) that is applied at the analyzer electrodes [132]. For the simplest parallel plate analyzer, assuming that the source of electrons is at a distance h_i from the entrance aperture and the image at a distance h_o (Fig. 15.12), for ions entering the analyzer at an angle θ and passing through it in the plane of deflection [133]:

$$Z = (h_i + h_o)\cot(\theta) + 2h(E/V)\sin(2\theta) \qquad (15.16)$$

Particles entering at slightly different angles can also be focused at the same exit point. Indeed in this equation the first term allows one the freedom to design the PPA with second order focusing, since two of the three variables: θ, $h_i + h_o$ and h_a are independent, i.e. choose these so that

$$dZ/d\theta = 0 \quad \text{and} \quad d^2Z/d\theta = 0 \qquad (15.17)$$

Second order focusing can thus be obtained in this design for finite $h_1 + h_2$, for an entrance angle of 30°. With $h_1 + h_2 = 0$ first order focusing is obtained for

Table 15.1 Value of the constants in (15.20) for different types of analyzers

Analyzer	A	B	C	n
45° PPA	$2/z_0$	2	1	2
30° PPA	$3/z_0$	9.2	1	3
127° CDA	$2/R_0$	4/3	1	2
180° HSA	$1/R_0$	1	0	2
42° CMA	$2.2/z_0$	5.55	0	3

$\theta = 45°$. A greater degree of flexibility with regard to focusing exists for the CMA [132, 133].

For the 127° cylindrical sector analyzer (CDA):

$$E = V/2\log(R_e/R_i) \tag{15.18}$$

where R_e and R_i are the radii of the external and inner electrodes.

In case of the hemispherical 180° deflector analyzer (HSA) this is:

$$E = V/\big((R_e/R_i) - (R_i/R_e)\big) \tag{15.19}$$

The energy resolution of the analyzer is defined as the ratio of the full width at half maximum (dE) of the energy spread after particles pass through the analyzer to the pass energy E_0. It is related to the characteristic dimensions of the analyzer, characterized by a constant (A) and the spatial distribution of the particles at its entrance. In terms of widths of entrance and exit slits (S_i, S_o) and entrance angles in and out of the plane of deflection α and β, the resolution is:

$$dE/E_0 = A(S_i/2 + S_o/2) + B\alpha^n + C\beta^2 \tag{15.20}$$

These constants are given in Table 15.1 for some analyzers, where R_0 is the radius of curvature of the central trajectory and z_0 is the distance between the entrance and exit slits.

Since the energy resolution (dE) is proportional to the pass energy, lowering E_0 leads to an improvement in resolution. This is usually achieved by use of a well designed entrance electron optics. The entrance optics also serves to enhance the collection efficiency and sensitivity of analyzer.

In electrostatic analysis in general two modes of acquiring energy spectra are used. In the constant relative resolution (CRR) mode, where dE/E_0 is kept constant, one changes or sweeps the voltages across the analyzer plates and particles enter the analyzer at their initial energies. The pass energy is thus swept over the range of the acquired spectrum. In the second mode the constant absolute resolution (CAR) mode (dE constant) the particles are decelerated or accelerated to adapt their energy to the prefixed pass energy.

The electron optics plays a crucial role in the functioning of the analyzer in both modes, in that it affects the focusing of the beam at the entrance slit and hence varies its entrance conditions at the entrance slit S_i, affecting the angles in (15.20). It is one of the most complex parts of design of the analyzer. Usually it is easier to design the entrance optics for the CRR mode since the transmission conditions are the same between the sample and analyzer. However in this mode the absolute resolution worsens drastically at high energies, while the transmission decreases at low energies. The CAR mode is very attractive in that the resolution remains constant, but it implies a very careful design of the optics so as not to deteriorate the transmission of the analyzer. Also a badly designed optics will affect the focusing of the particles at the entrance of the analyzer and hence the resolution (see (15.20)). Note also that the transmission of the analyzer is energy dependent.

The design that will function adequately over a wide range of particle energies above and below the pass energy usually requires a so called "zoom" optics, which maintains focusing conditions and in which voltages across different optics elements are progressively readjusted as a function of incident energy. This aspect is particularly important in electron spectroscopy when wide energy ranges are swept like in UPS at low pass energies and somewhat less in LEIS applications.

We mention briefly some other aspects of analyzer design, which affect its operation. One important aspect relates to the boundary or fringing fields near the ends of the electrodes and at the entrance apertures. In case of the PPA the beam enters through an equipotential plate. In this case if the aperture is large grids are placed across it to minimize distortion. Additionally guard electrodes are placed at the ends of the electrodes to render the electric field in the analyzer more homogeneous. In other cases like those of the CDA or HSA the situation is more complex because the particles enter the analyzer through an equipotential plate located perpendicular with respect to the electrodes. This affects the trajectories of the particles in the analyzer and various solutions have been proposed. Depending upon spectrometer type these vary from placement of extra electrodes across the entrance to changing slightly the placement of the entrance and exit aperture plates that render the field more homogeneous. It has also been proposed [134–136] to displace slightly these apertures from the central trajectory radius.

Another perturbing factor in some cases can be that particles with energies much higher than the pass energy are backscattered from the external electrode of the analyzer and result in unwanted features such as "wings" in the spectrum. This can be countered by using a tandem spectrometer design, in which two analyzers are placed sequentially. Alternatively by use of an exit aperture in the external electrode that is designed in a manner not to affect the field in the analyzer. In some cases this is advantageously used to allow neutral particles to pass through for TOF analysis.

One final comment concerns particle detection. In many modern analyzers the exit slit is replaced by a wide aperture and the single detector is replaced by several CEMs or by MCPs allowing simultaneous acquisition of a spectrum over a more or less wide range of energies. The analyzer then deflects the ions in such a way that ions of medium pass energy E_0 are focused in the middle of the detector, while ions of lower and higher energy are focused at, respectively, larger and smaller distances. This accelerates substantially data acquisition minimizing damage to the sample by the incident ion beam. However this requires proper handling of the fringing field effects mentioned above. In case of the PPA this is particularly easy because the MCP detector is located beyond an equipotential exit plate where a guarding grid can be easily placed allowing easy use of a large detector.

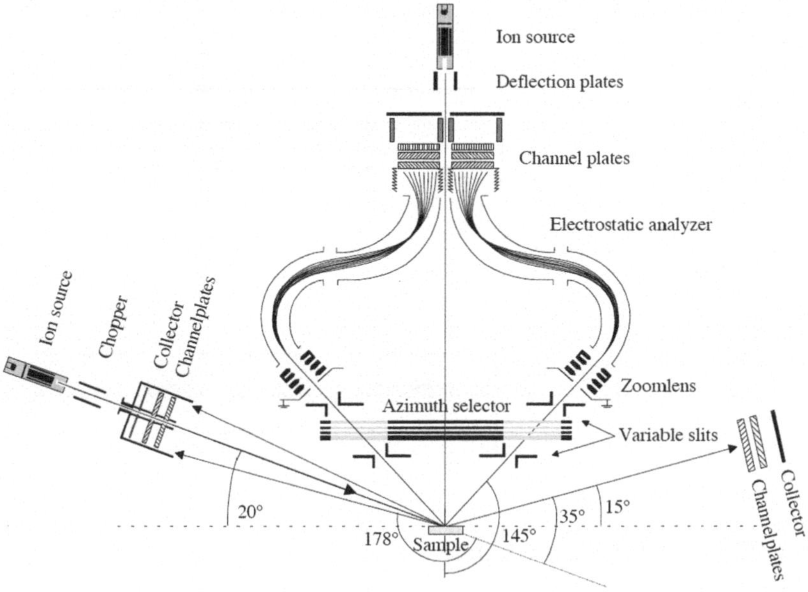

Fig. 15.13 EARISS system combined with a secondary ion gun for grazing incidence bombardment for time of flight scattering and recoil spectroscopy. Reproduced with permission from [138]

Comparisons of functioning of various analyzers and their ranking according to different criteria may be found in [132]. Many commercial systems today use the HSA with sophisticated optics and multichannel detection systems. We will stop here on a different toroidal analyzer that was proposed by Brongersma et al. [137] to provide the possibility of simultaneous energy and angle resolved analysis and a very high sensitivity, thus reducing surface damage. The energy and angle resolved ion scattering spectrometer (EARISS) uses both a special analyzer and a two-dimensional detection system.

The analyzer is illustrated in Fig. 15.13. The incident ions used for analysis pass along the analyzer axis entering the system through a central hole in the detector assembly. The analyzer is capable of imaging at the same time ions of different energies backscattered from a target into a cone with a scattering angle around 145° and an almost full (320°) azimuthal angle. The backscattered ions are transmitted through the analyzer and focused onto a two-dimensional, position-sensitive detector. Both the energy of the ions and their azimuthal angle can thus be determined independently. The scattering angle is defined by means of two circular apertures. Beyond the apertures the ions pass a zoom lens which focuses the ions at the entrance of the analyzer. The detector consists effectively of 128 detectors in radial distance. Besides focusing by ions at the entrance slit of the analyzer, the zoom lens is also capable of accelerating or decelerating the ions and thereby changing the pass-energy E_p of the analyzer. In this analyzer the angular resolution is influenced by the fact that ions may move along a spiraling path. This is avoided by an appropriately placed "azimuth selector".

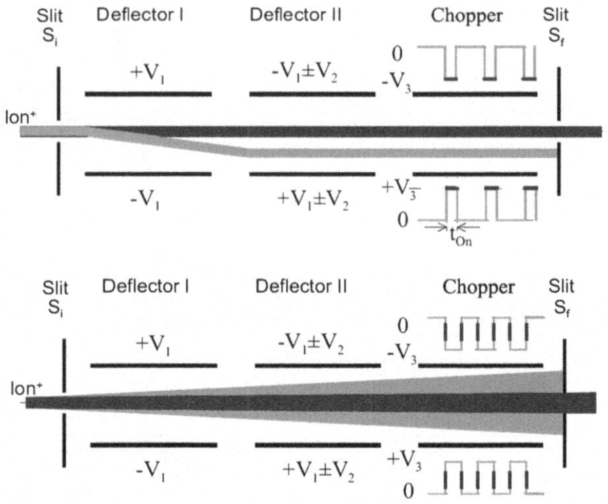

Fig. 15.14 Ion beam pulsing schemes

15.3.3 Time of Flight Systems

For time of flight analysis [139–151] the beam chopping system is usually composed of two slits placed at fairly large distance and sets of deflection plates between them. The incident beam defined by the entrance slit is deflected by a chopping pulse applied to a pair of deflection plates and is "swept" across the exit slit. Extra sets of plates are used to pre-deflect the beam to one side and then sweep across the exit slit with the deflecting pulse. When a very good time resolution is not necessary a short pulse of a few nanoseconds can be used to position a pre-deflected beam in front of the exit slit of the pulsing system.

The two types of beam pulsing schemes: pulse corresponding to the edges of the chopping voltage pulse or during the application of the pulse are illustrated in Fig. 15.14. The time resolution of the incident pulse is determined by several factors besides the initial energy spread. The "writing speed" (w) at the exit slit S_f is given by [141]

$$w = eFD/mv_0 \tag{15.21}$$

where F is the peak to peak field between plates, D the distance between the chopper and the exit slit S_f, v_0 the beam velocity and e its charge. The exit slit width results in a pulse width given by

$$dt = S_f/w \tag{15.22}$$

For a flight length L, for which the total flight time is

$$T = L\sqrt{m/2E} \tag{15.23}$$

the energy spread will then be

$$dE' = (2mE)^{1.5}\, dt/L = 4S_f E^2/eLFD \tag{15.24}$$

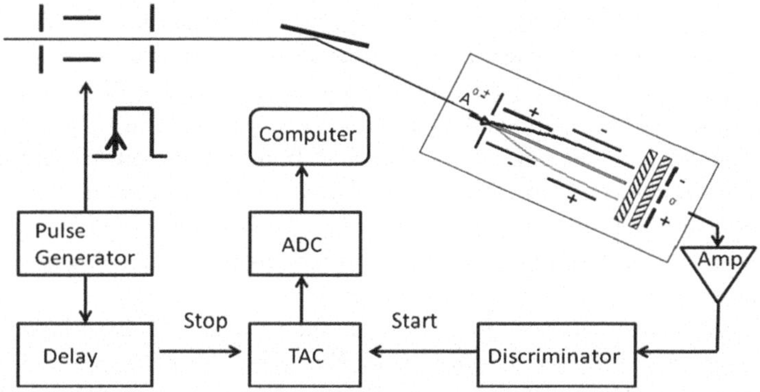

Fig. 15.15 Time of flight detection scheme [130] with three anode MCP detector schematics

While it may appear that increasing the field would improve resolution, the time variation of the pulse field introduces a spread given by [142]

$$dE'' = eFS_f \tag{15.25}$$

So that assuming that the two broadenings are independent the final energy width will be

$$dE = \sqrt{dE'^2 + dE''^2} = \sqrt{16S_f^2 E^4/(LDeF)^2 + (eFS_f)^2} \tag{15.26}$$

The minimum value of dE is obtained for $F = 2E/e(LD)^{0.5}$ and the relative energy resolution in this case will be:

$$dE/E = 2^{1.5} S_f/(DL)^{0.5} \tag{15.27}$$

An additional energy spread can appear for some scattering geometries, like in grazing scattering, when an extended target is used, because of differences in flight times of particles suffering the same energy losses, but coming from different target regions. Flight lengths covered after energy loss are then not the same. Some additional width can be introduced by "jitter" in electronics. Ion pulses of less than a nanosecond width can be obtained. This however may not be necessary in some cases considered here and frequently pulses of a few nanoseconds are used.

In the scheme described one generates two beam pulses corresponding to the up going and down going edges of the voltage pulse. The voltage pulse width has to be adapted to exclude one of these from the time window of analysis. Some authors [16] have proposed to use a secondary appropriately delayed voltage pulse applied to another set of plates to eliminate this pulse. The pulsing frequency should be such that the slowest particles from a given pulse have time to arrive at the detector before the fastest particles from the next pulse.

After scattering on the surface, the scattered or recoiled particles are detected by a detector located at the end of the flight tube as illustrated in Fig. 15.15 [130] of the Orsay setup. The flight length used was 2.2 meters for incident ion energies below 6 keV. This setup had three TOF tubes placed for detection of particles scattered through 7°, 38° and 135° for use in different types of measurements. The detectors are MCPs. In some other systems channel electron multipliers are used. There are several advantages of using MCPs. First, from a timing point of view, the arrival time of particles is better defined than on the cone of a channeltron. Secondly multichannel detection schemes can be implemented. In the Orsay setup described above this was used in a very simple manner to separate neutrals from positive and negative ions by using a set of deflection plates placed before the MCPs (Fig. 15.15). In some systems an acceleration or deceleration tube using biasable entrance and exit meshes is used to separate ions from neutrals.

Note that when using time of flight tubes as shown in Fig. 15.11, it is preferable to use an "entrance" collimator aperture to the tube and some intermediate diaphragms to avoid particle scattering on tube walls. Furthermore, in case of study of low energy ions, it can be necessary to provide magnetic shielding to avoid deflection from the earth's magnetic field.

Time of flight spectra are acquired using a time to digital converter (TDC) or time to amplitude converter (TAC) followed by an analog to digital converter (ADC). In the Orsay TOF scheme the scattered particle reaching the detector generates a start pulse in a time to amplitude converter. The stop pulse is generated by an appropriately delayed trigger pulse of the voltage pulse generator (Fig. 15.15). This inverse scheme is used because not all trigger pulses from the pulse generator would correspond to a particle reaching the detector, avoiding dead time in the TAC.

The energy distribution can be obtained from the time of flight distribution $F(t)$ from the expression:

$$N(E) = F(t)t^3/ML^2 \tag{15.28}$$

Variations of such TOF systems include setups in which measurements can be performed for a continuous range of scattering angles and also which couple time-of-flight techniques and a position-sensitive detector to capture both velocity- and spatially resolved patterns of scattered and recoiled atoms simultaneously. These are briefly outlined in the following. Systems with continuously variable scattering angles and with long time of flight tubes, with lengths from 2.5 m, have been implemented only for gas phase collisions [139]. This was achieved in one case by using a small diameter scattering chamber and a flight tube attached to a bellows. Alternatively a system with a rotating ion source has been implemented [140, 143] (with L up to 7.5 m). On the other hand, Rabalais et al. [145] and Grizzi et al. [149] have constructed special setups with large sector shaped chambers which allow a continuous rotation of the detector around the sample to allow time of flight scattering and recoiling measurements over large angular ranges from e.g. 0° up to 165° [145]. The Bariloche [149] setup is schematized in Fig. 15.16. In this system in which continuous movement is implemented by using a detector mounted on a rotating arm, a fairly short flight length is used. Additional longer time of flight tubes can be added at some fixed angles for higher resolution measurements.

Some of the above mentioned time of flight systems [146, 149] can be used in grazing scattering configuration, with measurements as a function of surface

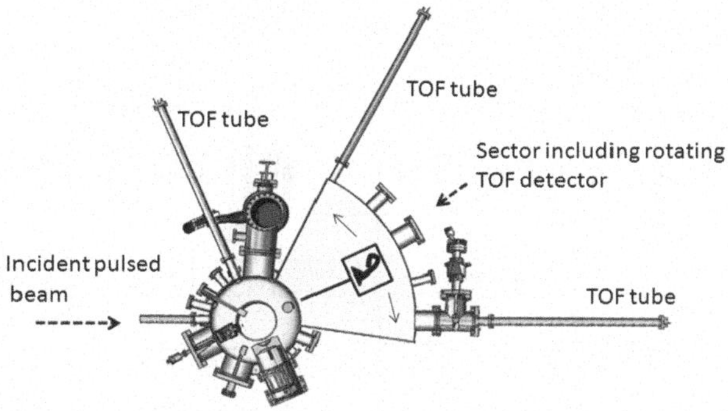

Fig. 15.16 Schematic diagram of the Bariloche [149] time of flight system with the sector shaped TOF chamber

azimuthal angle, to obtain surface structural information or information on electron transfer processes as has been mentioned above.

A versatile system in which a large position-sensitive detector is used so that time of flight spectra of spatially resolved scattered and recoiled atoms are simultaneously acquired, has been developed by Rabalais et al. [150, 151] and the technique called SARIS: Scattering and Recoiling Imaging Spectrometry. In the SARIS setup the velocities of the scattered and recoiled ions plus fast neutrals are analyzed by measuring their flight times from the sample to a large 75 mm × 95 mm position-sensitive MCP detector. In this system the distance of the MCP to sample can be varied in a 10–60 cm range and the MCP can be rotated around the sample in the scattering plane ($\phi = 0$–$80°$) and out of plane ($\theta = 0$–$160°$). In this way particles leaving the target in a range of almost 2π steradians (sr) can be collected.

Aono et al. developed a CAICISS [152–155] (coaxial impact-collision ion scattering spectroscopy) setup for backscattering ICISS, with time of flight analysis [154] and charge separation in which the neutral and ion TOF spectra are separately acquired. CAICISS [152, 153] is designed to detect particles which are backscattered at 180° by employing an MCP with a hole in the center, through which passes the incident ion beam and the MCP serves to detect backscattered particles. The charge separation is performed by an acceleration tube placed along the backscattered particle path (Fig. 15.17). A variant of these systems was recently developed for producing pulsed neutral Ne beams [155] for analysis of insulators, by introducing an ion neutralization cell in the CAICISS ion beam path.

Finally an interesting development has been the use of ion scattering and time of flight analysis [156] for monitoring growth in a thin film deposition system, under high pressure environment conditions, where techniques like AES and XPS could not be used.

15.4 Applications of Ion Scattering

In this section we shall briefly describe in more detail a few results that illustrate the type of information that can be obtained by these methods. Extensive bibliographies may be found in various reviews [9, 10, 13].

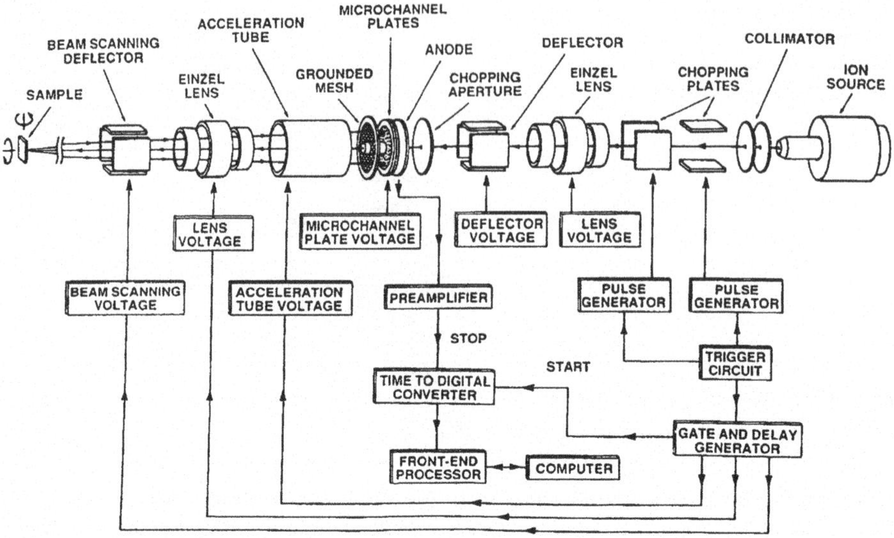

Fig. 15.17 CAICIS system of Aono et al. [152, 153]. Reproduced with permission

15.4.1 Study of Growth of Metal Clusters on TiO_2

LEIS can be used to study growth of overlayers of atoms on a surface and distinguish between island formation or layer by layer growth. As an example Fig. 15.18a shows LEIS spectra [20] for initial stages of growth of Ag on TiO_2. The initial TiO_2 substrate spectrum shows two peaks corresponding to He scattering on Ti and oxygen. Upon Ag evaporation an extra peak appears at lower energy losses corresponding to scattering on the heavier Ag atoms. Figure 15.19 shows a plot of the relative changes in intensity of the peaks as a function of Ag evaporation in equivalent monolayers as determined by a reference measurement using a quartz microbalance. These integrated intensities correspond to specular scattering conditions and are normalized to the area of the spectrum corresponding to Ag, Ti and O peaks (not including the area between the Ti and O peaks, which must correspond to multiple scattering events). In performing this plot a correction related to differences in the backscattering cross section of He on Ag, Ti and O atoms was introduced. The corresponding calculated backscattering coefficients, are 0.0018 (O), 0.0062 (Ti) and 0.011 (Ag) at 1 keV and reflect the higher backscattered yield on Ag than on other atoms.

The relative intensity of the Ag peak increases rapidly and then tends to a broad plateau in the range shown. Simultaneously, the intensity of the Ti and O peaks decreases. The more rapid initial increase of the Ag intensity is attributed to initial 2D growth followed rapidly by 3D growth of clusters. A similar picture emerges from TOF measurements of backscattered He atoms for 500 eV He^+ incidence (Fig. 15.18). In the clean TiO_2 spectrum the peak at about 8.4 μs is due to scattering on Ti, while scattering on oxygen should lead to a structure at about 11 μs. As opposed to the ion spectra, the spectrum for neutrals is broad reflecting multiple

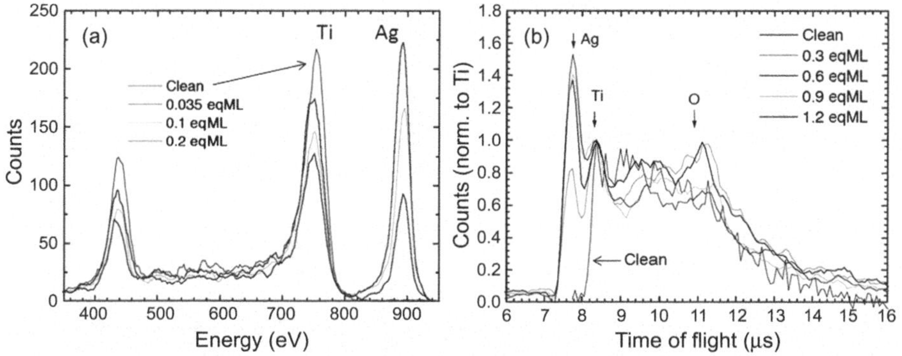

Fig. 15.18 LEIS and TOF spectra for He ion scattering on a TiO_2 surface as a function of Ag deposition

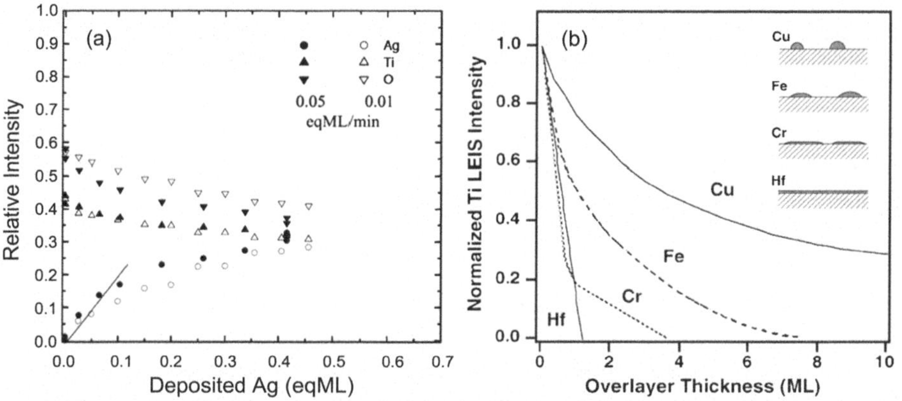

Fig. 15.19 Integrated intensities [20] of the He peaks corresponding to Ag, Ti and O as a function of the amount of deposited Ag (in eqML) for two rates of deposition: 0.05 eqML/min (*full symbols*) and 0.01 eqML/min (*hollow symbols*). (**b**) Trend of attenuation of Ti LEIS signal [157] for four metal overlayers and schematic drawing (*inset*) of initial stages of growth: Cu (3D cluster), Fe (flat island), Cr(2D island followed by 3D growth) and Hf (continuous overlayer)

scattering and scattering in inner layers. A weak structure due to oxygen can be discerned. When Ag is evaporated one observes the appearance of a peak on the shorter flight time side of the spectrum, which grows rapidly with Ag evaporation.

Some other results of LEIS studies [157] of growth of various other metals on TiO_2 are summarized in Fig. 15.19b with a cartoon of the proposed growth modes [157]. For Hf the rapid linear decrease of the Ti signal is interpreted as due to a monolayer formation. In case of Cr it was suggested that 2D growth followed by 3D growth occurs, while 3D growth is suggested for Cu.

Here ion scattering was thus efficiently used to identify cluster growth. When combined with TOF with charge separation for ion and neutral particle spectra, this has enabled studies of neutralization as a function of cluster size [158, 159], where

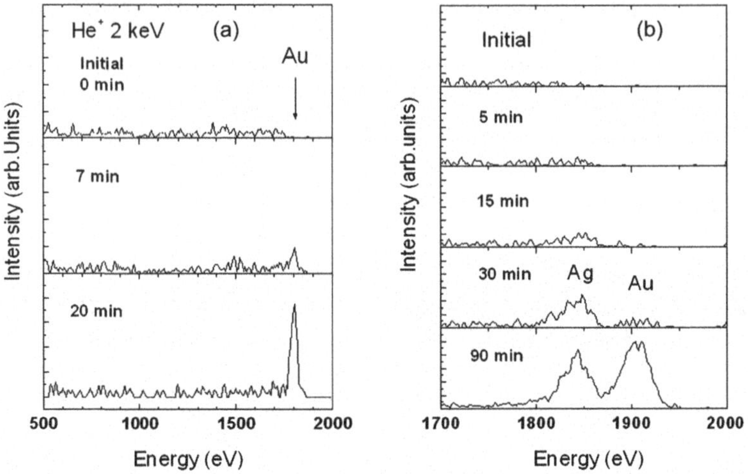

Fig. 15.20 He^+ LEIS spectra of (**a**) a cysteine monolayer on Au and (**b**) after Ag evaporation on a cysteine monolayer as a function of time. Evaporation rate was about 0.1 monolayer/minute

strong differences from Li neutralization on bulk surfaces [87, 160, 161] were found for small Au clusters. This possibility should also be kept in mind when taking neutralization into account in LEIS on e.g. catalysers.

15.4.2 Study of Metal Deposition on an Organic Layer

In molecular electronics in which a device is formed by sandwiching a molecular layer between metal electrodes, an important issue is the proper formation of the molecule metal layer contact. Frequently the molecular layer is formed by first growth of a molecular film on an appropriate substrate and then evaporation of metal to form a top contact. There are several important problems that need to be addressed here amongst which the obvious question of what actually happens when the metal is deposited: does it all stay on the surface or does some of it penetrate into the molecular layer? This can be nicely investigated by LEIS [162].

As an example we show some data from a study of Ag evaporative deposition onto an ordered L-cysteine [163] self assembled monolayer (SAM) on gold. In a first step the SAM is formed [164] by adsorption of molecules from a liquid solution onto a clean Au surface, by attachment of sulfur to gold. Ag is then evaporated in vacuum.

The data for He^+ LEIS results on the SAM layer, corresponding to a 100 nA/cm^2 2 keV He^+ ion scattered through 120°, is shown in Fig. 15.20a. On a clean Au surface one would observe a well defined He^+ ion backscattering peak due to scattering on Au atoms, but initially the 2 keV He scattering spectra do not show any clearly identifiable structures, presumably because of strong neutralization of He in the thin organic layer and the underlying gold surface. It is only after the ion beam bombards

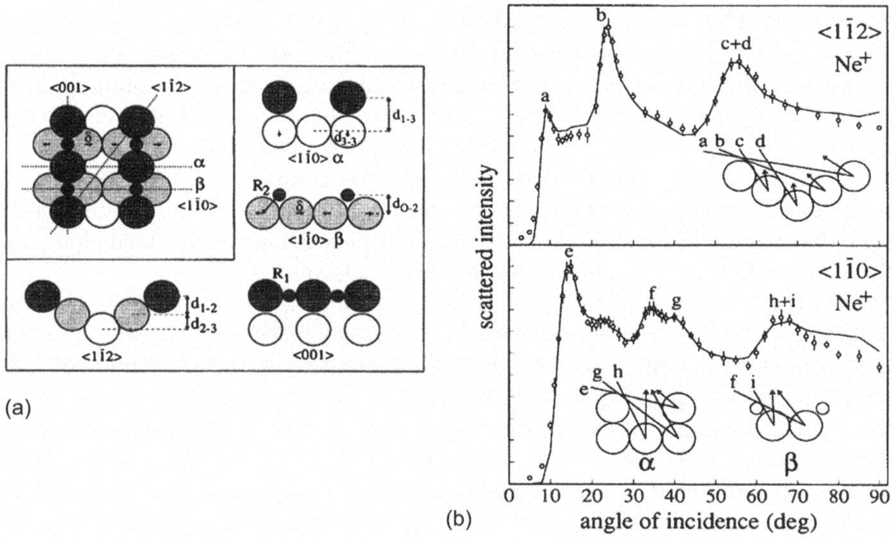

Fig. 15.21 *Left panel*: top and side view of the O(2 × 1)–Ag(110) surface in the missing row reconstruction. Atoms: Ag (*large circle*) and O (*small circles*). *Right panel*: 2.9 keV Ne^+ scattering on a 2 × 1 reconstructed Ag surface with oxygen. Letters identify focusing peaks due to trajectories shown in the *insets*. The curve is the best fit calculation. Reproduced with permission from [166]

the surface for some time, resulting in sputtering of the molecular layer, that the Au surface became visible in LEIS.

Subsequently a similar SAM was exposed to Ag evaporation and an equivalent of 5 monolayers of Ag were deposited. If the Ag atoms were located on top of the SAM at the SAM-vacuum interface, one would see a peak in the LEIS spectrum due to He^+ backscattering on Ag. However (Fig. 15.20b) initially no He^+ backscattering on Ag was observed, but only after the surface was bombarded by He^+ for about 15 min. The Ag peak intensity subsequently increased and then decreased again due to removal of Ag by sputtering and revealing the Au atoms below.

This LEIS experiment thus showed that under the evaporation conditions used, Ag atoms do not interact sufficiently with the cysteine active groups to bind them on "top" of the cysteine SAM and Ag penetrated below the SAM and formed a layer above Au.

15.4.3 Study of O(2 × 1)–Ag(110) Missing-Row Reconstruction by ICISS

Oxygen adsorption on an Ag(110) surface leads to surface reconstruction and formation of well-ordered phases [165] characterized by a (n × 1) periodicity with $2 < n < 8$. Oxygen atoms adsorb in the long bridge site leading to the formation of –Ag–O–Ag– rows along the (001) azimuthal direction. The structure of the

$O(2 \times 1)$–Ag(110) phase was studied by ICISS using He^+ and Ne^+ ion scattering [166]. Figure 15.21a shows a schematic diagram of the surface relevant to this study.

Scattered ion intensities are determined for different crystal orientations as a function of incidence angle. Figure 15.21b shows the angular distributions of the intensities of backscattered Ne along several crystal directions. This ICISS distribution displays a series of structures. These are due to shadowing and blocking effects. Because Ne is heavier than O, no backscattering of Ne off O atoms is observable and the structural information on location of O atoms can not be obtained along the (001) direction where O atoms are located in the added rows.

The structures in these figures can be explained in terms of ion trajectories shown in the insets of the figures where the surface atoms are placed in a missing row configuration. One can easily conclude that a non reconstructed surface would not lead to the same structures in Fig. 15.21b. Indeed if an extra Ag atom were placed in the first layer at the center of the inset of Fig. 15.21b then the trajectories causing the focusing peaks labeled "a" and "b" would be impossible due to shadowing by this atom. An unreconstructed surface would give rise to a single peak at an intermediate angular position between "a" and "b". The same considerations hold for peaks labeled "e" and "g". The extra unlabeled structure visible at 23° in the $\langle 1\bar{1}0 \rangle$ direction, was assigned by the authors to some ion beam damage of the surface.

In order to give a more quantitative description of the data, the authors used a Monte Carlo simulation of ion trajectories. In this code the ion surface scattering was treated as a sequence of two body classical interactions: here a Thomas Fermi Molière Firsov potential was used. The model incorporates thermal motion of crystal atoms and includes a simple description of ion neutralization and also of ion damage. In the calculations the structural and neutralization parameters are varied to obtain the best fit to the experimental spectra. The results of these simulations are compared with experiment in Fig. 15.21b. For details of the model and a more complete discussion the reader is referred to the original paper [166].

The results of this modeling allow to pinpoint the position of the oxygen atoms and give the first to second and second to third interlayer spacings. They also indicate a slight lateral displacement of the second layer Ag atoms towards the missing row and a vertical buckling of the third layer. As a comparison of this data to SEXAFS (surface extended X-ray absorption fine structure) measurements [166] the distances R_1 and R_2 between the O atoms and its nearest neighbors were found to be $R_1 = 2.045 \pm 0.005$ Å (2.05 ± 0.03 Å from SEXAFS) and $R_2 = 2.23 \pm 0.08$ Å (2.21 ± 0.03 Å from SEXAFS). Thus ion scattering provides an excellent means to study surface structure.

15.4.4 Study of Growth of Organic Layer on a Au(111) Surface by TOF SARS

In this section we consider an example of a study of the growth of a self assembled monolayer (SAM) of a dithiol (e.g. SH–$(CH_2)_n$–SH) molecule by TOF-SARS. Dithiol self-assembly has attracted attention for their use as linkers between two

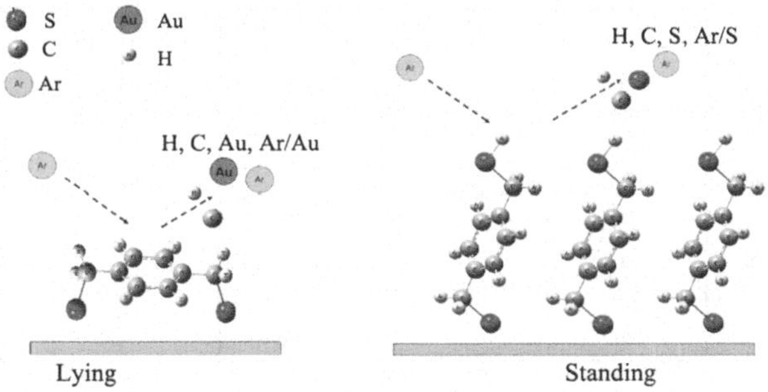

Fig. 15.22 Schematic diagram of lying down and standing up dithiol SAM configurations as investigated by TOF DRS

metallic entities in molecular electronics applications or for building complex metal-organic heterostructures. An important issue [167–169] here is to know if the dithiol SAM is in fact formed of standing up molecules with one sulfur attached to the substrate and the other sulfur free at the SAM-air interface or if one deals with a layer with both sulfur ends bonded to the substrate.

This was investigated [169] by TOF DRS for the case of alkane and aromatic dithiol SAM's on gold, along with the kinetics of their adsorption and desorption (see also [170]). The scattering geometries corresponding to the two SAM phases are schematized in Fig. 15.22.

We consider the case of 1–4 Benzenedimthanethiol (BDMT) adsorption on a Au(111) crystal [168, 169]. The clean Au(111) crystal is exposed in vacuum to BDMT vapors. In TOF-SARS, the sample is bombarded by a pulsed beam of keV Ar^+ ions. Figure 15.23a shows TOF SARS spectra for clean Au(111) and as a function of exposure to BDMT.

The spectrum for the clean Au consists of peaks corresponding to Ar scattered from Au and recoiled Au atoms. In case of a hydrocarbon polluted surface one would observe structures related to faster C and H at the left side of the Ar scattering peak, whereas the spectra in Fig. 15.23 indicate a clean surface with a sensitivity of about 1 % ML. When BDMT is introduced, the spectra initially show peaks corresponding to direct recoil (DR) H and C in addition to the scattered Ar and Au. The H and C recoil peaks are superimposed on a broader structure coming from surface recoils, i.e., recoil atoms going first towards the surface and then reflected back (inset in Fig. 15.23a). With increasing exposures the scattered Ar peak decreases and becomes broader with a long tail due to multiple scattering.

For low exposures (<100 L) the TOF spectra suggest an initially lying down phase, where both S atoms lie closer to the surface than the C atoms and are thus shadowed by C and not observed. Ar ions can penetrate through this thin layer and hence a contribution of scattered Ar and Au recoils is observed in the spectra. Above 10 kL, a higher attenuation of the Au DR and Ar scattering on Au is observed, and structures (5° incidence) assigned to recoiled S and scattering of Ar on S (Ar–S) appear. The latter can occur only if S is exposed in the film-vacuum interface. This finding was consistent with XPS data (see [169] and references therein).

These S related structures are a fingerprint for the initiation of the standing up phase, with S atoms lying at the SAM/vacuum interface. The formation of a thick

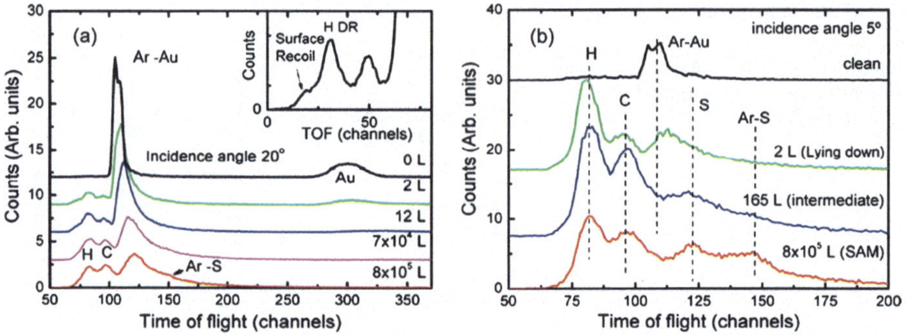

Fig. 15.23 Time of flight scattering and recoil spectra for Ar incident on a Au surface exposed to increasing doses of BDMT, for two incidence angles with respect to the surface plane. Scattering angle 45°

BDMT layer shields off the Au surface and at grazing incidence, the scattering and recoiling events take place at the outer part of the SAM. At intermediate exposures, the layer must be mixed and so some Ar scattering through the thin lying down regions is observed. At very high exposures, of the order of a mega Langmuir, the intensity of both peaks corresponding to S is large and their growth saturates. The H and C recoil peaks change in their relative intensity and both become broader (due to recoiling from different heights). Concomitantly the recoiling gold peak strongly decreases in intensity and at the highest exposures is no longer observed (Fig. 15.23b). These observations suggest the disappearance of large domains of lying down phase and formation of a standing up phase of BDMT on gold.

Thus in this study the transition of the SAM from the lying down to standing phase could be followed and from these spectra the kinetics of adsorption and also heating induced desorption was deduced. Note again the efficient detection of hydrogen in TOF-SARS, that is not detectable in a number of other techniques like Auger and X-ray photoelectron spectroscopy.

15.4.5 Chlorine Adsorption on Ag and Electron Transfer Effects. A TOF and LEIS Study

It is instructive to consider an example where electron transfer processes are clearly identifiable in the backscattering of various ions and affect interpretation of LEIS data. Figure 15.24 shows TOF spectra for Ar scattering on clean Ag(111) and after exposure to varying doses of Cl_2 [171]. It is known that chlorine decomposes on Ag, leading to atomic Cl chemisorption and formation of a Cl overlayer for exposure below 1 L and for higher exposures a rapid formation of AgCl islands occurs (see [171]). In Fig. 15.24a, the spectrum for clean Ag presents only one peak corresponding to backscattered Ar on Ag. On Cl_2 exposure one observes the appearance

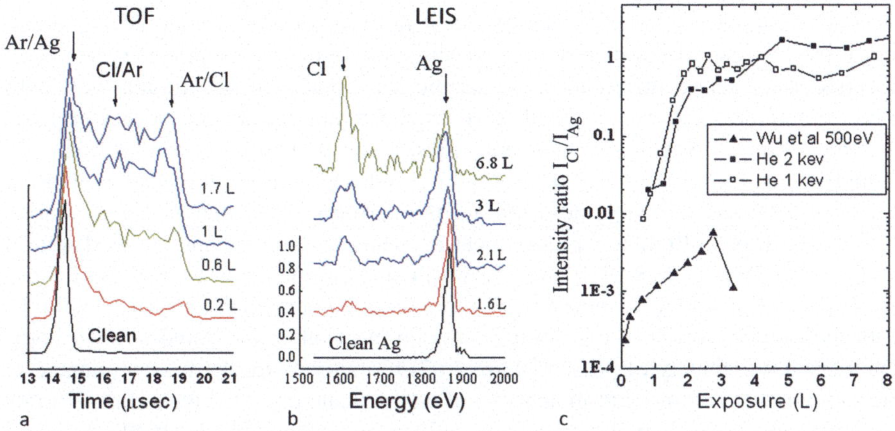

Fig. 15.24 TOF spectrum [171] for 4 keV Ar^+ scattering on an Ag(111) surface exposed to Cl as a function of Cl_2 exposure (see text), (**b**) He^+ 2 keV ISS spectrum [171] for some exposures. The scattering angle was 80° and specular conditions were used. (**c**) Intensity ratio [171] of Cl and Ag peaks deduced from the LEIS spectrum for 1 keV and 2 keV He ions and data of Wu et al. [172] at 500 eV (scattering angle 150°)

of a second structure attributable to Ar scattering on Cl atoms at 0.2 L. It then grows with higher exposures.

This information can be compared with that extracted from He ion scattering for the same exposure conditions. Figure 15.24b shows He LEIS spectra for the clean and Cl_2 exposed Ag surface. As the surface is exposed to Cl the Ag scattering peak decreases in intensity somewhat, but the peak corresponding to He scattering on Cl, remains very small until exposures above about 1.5 L when its intensity grows significantly. This effect was attributed to strong neutralization of He on the chemisorbed Cl over-layer, leading to a feeble backscattered He^+ intensity. At higher exposures formation of ionic AgCl islands leads to a different chemical environment for Cl and a lesser neutralization of He. It was assumed here that neutralization involves a quasi-resonant charge exchange process involving Cl(2s) level that is positioned differently with respect to He at different coverages, and this leads to a less favorable situation for quasi-resonant charge transfer.

Figure 15.24c shows the measured intensity ratios for Cl/Ag peaks for 1 kev and 2 keV energies and data of an earlier study [172]. It is clear that although at about 1 L Cl is clearly present on the surface, here the LEIS spectra lead to an underestimation of the amount of Cl atoms on the surface.

15.5 Concluding Remarks

Summarizing, from the examples shown above it is clear that low energy ion scattering spectroscopy provides a conceptually simple and powerful method for studying surface atomic composition, with a unique top layer sensitivity. In ICISS mode it provides a simple sensitive means of studying mass specific atomic arrangement, and provides real-space information. When supported with computer simulations accurate quantitative information on atomic locations, which compare well with other techniques, can be obtained. One of the major pitfalls in LEIS is related to effects due to Auger, resonant and quasi-resonant neutralization of ions. Although much progress in the understanding and quantitative description of these has been achieved, a reliable description for an arbitrary case may not be available. In this case, time of flight techniques measurements of scattered and recoiled neutral species (NICISS, TOF-SARS) are a very valuable complement. In TOF-SARS hydrogen detection and localization is a unique feature not available in other methods, with the rare exception of e.g. high resolution electron and infrared spectroscopy. The very low intensity of pulsed beams in TOF methods results in low damage to the sample, that is important in analysis of e.g. delicate organic layers. In LEIS this is shown to be possible in the large angle collection methods like EARISS. As noted above ion scattering has provided the means to study growth in high pressure environments. As in most other surface analytical techniques, LEIS and TOF-DRS are of course best used in conjunction with other techniques described in this book. Besides the surface analytical information, the ion scattering systems described in this chapter, provide a powerful means of studying basic phenomena like stopping of ions and electron transfer processes, important in gas surface interactions and other surface analytical techniques like secondary ion mass spectroscopy.

Acknowledgements The author would like to thank Dr. Gianangelo Bracco and Dr. Jorge Valdes for comments on the manuscript.

References

1. D.P. Smith, J. Appl. Phys. **18**, 340 (1967)
2. D.P. Smith, Surf. Sci. **25**, 171 (1971)
3. S.H. Overbury, W. Heiland, D.M. Zehner, S. Datz, R.S. Thoe, Surf. Sci. **109**, 239 (1981)
4. T.M. Buck, Y.S. Chen, G.H. Wheatley, W.F. van der Weg, Surf. Sci. **47**, 244 (1975)
5. M. Draxler, S.N. Markin, S.N. Ermolov, K. Schmid, C. Hesch, A. Poschacher, R. Gruber, M. Bergsmann, P. Bauer, Vacuum **20**, 39 (2004)
6. J. Lindhard, K. Dan, Vidensk. Selsk. Mat. Fys. Medd. **34**(14) (1965)
7. M. Aono, C. Oshima, S. Zaima, S. Otani, Y. Ishizawa, Jpn. J. Appl. Phys. **20**(1), 829 (1981)
8. H. Niehus, Surf. Sci. **166**, L107 (1986)
9. H.H. Brongersma, M. Draxler, M. de Ridder, P. Bauer, Surf. Sci. Rep. **62**, 63–109 (2007)
10. L.C.A. van den Oetelaar, H.E. van Benthem, J.H.J.M. Helwegen, P.J.A. Stapel, H.H. Brongersma, Surf. Interface Anal. **26**, 537–548 (1998)
11. J.W. Rabalais, *Principles and Applications of Ion Scattering Spectrometry* (Wiley-Interscience, New York, 2003)
12. B.J.J. Koeleman, S.T. Dezwart, A.L. Boers, B. Poelsema, L.K. Verhey, Nucl. Instrum. Methods B **218**, 225 (1983)
13. H. Niehus, W. Heiland, E. Taglauer, Surf. Sci. Rep. **17**, 213–303 (1993)
14. E.S. Mashkova, V.A. Molchanov, *Medium-Energy Ion Reflection from Solids* (North-Holland, Amsterdam, 1985)
15. L.C. Feldman, J.W. Mayer, *Fundamentals of Surface and Thin Film Analysis* (North-Holland, New York, 1986)

16. R. Smith (ed.), *Atomic & Ion Collisions in Solids and at Surfaces: Theory, Simulation and Applications* (Cambridge University Press, Cambridge, 1997)
17. D.J. Maas, R. van Gastel, Helium ion microscopy, in *Surface Science Techniques*, ed. by G. Bracco, B. Holst (Springer, Belin, 2013) (this book)
18. S. Sijbrandij, J. Notte, L. Scipioni, C. Huynh, C. Sanford, J. Vac. Sci. Technol. B **28**, 73 (2010)
19. S. Kostinski, N. Yao, J. Appl. Phys. **109**, 064311 (2011)
20. A.R. Canário, E.A. Sanchez, Yu. Bandurin, V.A. Esaulov, Surf. Sci. **547**, L887 (2003)
21. J.M. Lindhard, H.E. Scharff, K. Schiott, Dans. Vidensk. Selsk. Mat. Fys. Medd. **33**(14) (1963)
22. N.R. Arista, Interaction of charged particles with solid and surfaces. NATO Adv. Study Inst., Ser. B: Physics **271**, 489 (1991) (ed. by K. Suresh)
23. T.L. Ferrel, R.H. Ritchie, Phys. Rev. B **16**, 115 (1977)
24. J.F. Ziegler, J.P. Biersack, U. Littmark, *The Stopping and Range of Ions in Solids*, vol. 1 (Pergamon, New York, 1985)
25. P. Sigmund, *Stopping of Heavy Ions: A Theoretical Approach* (Springer, Berlin, 2004)
26. P.M. Echenique, R.M. Nieminen, R.H. Ritchie, Solid State Commun. **37**, 779 (1981)
27. M.J. Puska, R.M. Nieminen, Phys. Rev. B **27**, 6191 (1983)
28. J.E. Valdés, P. Vargas, N.R. Arista, Phys. Rev. A **53**, 1638 (1996)
29. J.E. Valdés, P. Vargas, N.R. Arista, Phys. Rev. A **56**, 4781 (1997)
30. J.E. Valdés, C. Parra, J. Díaz-Valdés, C.D. Denton, C. Agurto, F. Ortega, N.R. Arista, P. Vargas, Phys. Rev. A **68**, 064901 (2003)
31. E.A. Figueroa, E.D. Cantero, J.C. Eckardt, G.H. Lantschner, J.E. Valdés, N.R. Arista, Phys. Rev. A **75**, 010901(R) (2007)
32. A. Garcia-Lekue, J.M. Pitarke, Phys. Rev. B **64**, 35423 (2001)
33. M. Alducin, V.M. Silkin, J.I. Juaristi, E.V. Chulkov, Phys. Rev. A **67**, 032903 (2003)
34. J.E. Miraglia, M.S. Gravielle, Phys. Rev. A **67**, 062901 (2003)
35. S. Lederer, H. Winter, Phys. Rev. A **73**, 054901 (2006)
36. M.S. Gravielle, M. Alducin, J.I. Juaristi, V.M. Silkin, A **76**, 044901 (2007)
37. A. Niehof, W. Heiland, Nucl. Instrum. Methods B **48**, 306 (1990)
38. A. Närmann, W. Heiland, R. Monreal, F. Flores, P.M. Echenique, Phys. Rev. B **44**, 1991 (2003)
39. A. Robin, W. Heiland, J. Jensen, J. Juaristi, A. Arnau, Phys. Rev. A **64**, 52901 (2001)
40. A. Robin, J. Jensen, D. Osterman, W. Heiland, Nucl. Instrum. Methods B **193**, 568 (2002)
41. K. Kimura, M. Hasegawa, M. Mannami, Phys. Rev. B **36**, 7 (1987)
42. H. Winter, C. Auth, A. Mertens, A. Kirste, M.J. Steiner, Europhys. Lett. **41**, 437 (1998)
43. H. Winter, J.I. Juaristi, I. Nagy, A. Arnau, P.M. Echenique, Phys. Rev. B **67**, 245401 (2003)
44. A. Arnau, P.M. Echenique, Phys. Scr. **677**, T49 (1993)
45. L. Guillemot, E. Sánchez, V.A. Esaulov, Nucl. Instrum. Methods B **212**, 20 (2003)
46. J.E. Valdés, P. Vargas, L. Guillemot, V.A. Esaulov, Nucl. Instrum. Methods B **256**, 81 (2007)
47. H. Winter, Phys. Rep. **367**, 387 (2002)
48. J.E. Valdés, P. Vargas, C. Celedon, E. Sanchez, L. Guillemot, V.A. Esaulov, Phys. Rev. A **78**, 32902 (2008)
49. J.E. Valdés, P. Vargas, L. Guillemot, V.A. Esaulov, Nucl. Instrum. Methods B **256**, 81 (2007)
50. L. Chen, J.E. Valdés, P. Vargas, V.A. Esaulov, J. Phys. Condens. Matter **22**, 345005 (2010)
51. G. Moliere, Z. Naturforsch. **2a**, 133 (1974)
52. S.A. Cruz, E.V. Alonso, R.P. Walker, D.J. Martin, D.G. Armour, Nucl. Instrum. Methods **194**, 659 (1982)
53. B. Firsov, Sov. Phys. JETP **33**, 696 (1957)
54. J. Lindhard, M. Scharff, H.E. Schiott, K. Dan, Vidensk. Selsk. Mat. Fys. Medd. **33**(14) (1963)
55. J. Lindhard, K. Dan, Vidensk. Selsk. Mat. Fys. Medd. **34**(14) (1965)
56. A.G.J. de Wit, R.P.N. Bronckers, J.M. Fluit, Surf. Sci. **82**, 177 (1979)
57. H. Niehus, G. Comsa, Surf. Sci. **152/153**, 93 (1985)

58. R. Spitzl, H. Niehus, G. Comsa, Rev. Sci. Instrum. **61**, 760 (1990)
59. M. Karolewski, Nuc. Inst. Met. Phys. Res. **230**, 402 (2005)
60. M. Robinson, Nucl. Instrum. Methods Phys. Res., Sect. B, Beam Interact. Mater. Atoms **67**, 396 (1992)
61. M. Robinson, I. Torrens, Phys. Rev. B **9**(12), 5008 (1974)
62. P. Rousseau, H. Khemliche, A.G. Borisov, P. Roncin, Phys. Rev. Lett. **98**, 016104 (2007)
63. J.R. Manson, H. Khemliche, P. Roncin, Phys. Rev. B **78**, 155408 (2008)
64. A. Schüller, S. Wethekam, H. Winter, Phys. Rev. Lett. **98**, 016103 (2007)
65. A. Schüller, H. Winter, Phys. Rev. Lett. **100**, 097602 (2008)
66. H. Khemliche, A.G. Borisov, P. Roncin, V.H. Etgens, F. Finocchi, Appl. Phys. Lett. **95**, 151901 (2009)
67. N. Bundaleski, H. Khemliche, P. Soulisse, P. Roncin, Phys. Rev. Lett. **101**, 177601 (2008)
68. F. Aigner, N. Simonovic, B. Solleder, L. Wirtz, J. Burgdörfer, Phys. Rev. Lett. **101**, 253201 (2008)
69. A. Schüller, S. Wethekam, D. Blauth, H. Winter, F. Aigner, N. Simonovi, B. Solleder, J. Burgdörfer, L. Wirtz, Phys. Rev. A **82**, 062902 (2010)
70. I. Estermann, O. Stern, Z. Phys. **61**, 95 (1930)
71. A. Graham, Surf. Sci. Rep. **49**, 115 (2003)
72. G. Witte, J. Braun, A. Lock, J.P. Toennies, Phys. Rev. B **52**, 2165 (1995)
73. F. Hofinann, J.R. Manson, J.P. Toennies, J. Chem. Phys. **101**, 10155 (1994)
74. H. Winter, A. Schüller, Prog. Surf. Sci. **86**, 169 (2011)
75. J. Los, J.J. Geerlings, Phys. Rep. **190**, 133 (1990)
76. P. Nordlander, Phys. Rev. B **46**, 2584 (1992)
77. J.N.M. van Wunnick, R. Brako, K. Makoshi, D.M. Newns, Surf. Sci. **126**, 618 (1983)
78. A.G. Borisov, D. Teillet-Billy, J.P. Gauyacq, Phys. Rev. Lett. **68**, 2842 (1992)
79. F. Wyputta, R. Zimny, H. Winter, Nucl. Instrum. Methods B **58**, 379 (1991)
80. M. Maazouz, A. Borisov, V.A. Esaulov, J.P. Gauyacq, L. Guillemot, S. Lacombe, D. Teillet-Billy, Phys. Rev. B **55**, 13896 (1997)
81. P. Nordlander, J.C. Tully, Phys. Rev. Lett. **61**, 990 (1988)
82. S.A. Deutscher, X. Yang, J. Burgdoerfer, Nucl. Inst. Met. B **336** (1995)
83. F. Martin, M.F. Politis, Surf. Sci. **356**, 247 (1996)
84. J. Burgdoerfer, E. Kupfer, H. Gabriel, Phys. Rev. A **35**, 4963 (1987)
85. P. Kürpick, U. Thumm, Phys. Rev. A **35**, 1487 (1996)
86. V.A. Ermoshin, A.K. Kazansky, Phys. Lett. **68**, 2842 (1996)
87. L. Chen, J. Shen, J. Jia, T. Kandasamy, K. Bobrov, L. Guillemot, J.D. Fuhr, M.-L. Martiarena, V.A. Esaulov, Phys. Rev. A **84**, 052901 (2011)
88. S. Ustaze, L. Guillemot, V.A. Esaulov, P. Nordlander, D. Langreth, Surf. Sci. **415**, L1027 (1998)
89. N. Lorente, A. Borisov, J.P. Gauyacq, Surf. Sci. **429**, 46 (1999)
90. Yu.N. Demkov, Sov. Phys. JETP **18**, 138 (1964)
91. R.E. Olson, Phys. Rev. A **6**, 1822 (1972)
92. R.L. Erickson, D.P. Smith, Phys. Rev. Lett. **34**, 297 (1975)
93. D.V. Ledyankin, I. Urazgildin, V. Yurasova, Nucl. Instrum. Methods B **48**, 585 (1990)
94. D.V. Ledyankin, I. Urazgildin, V. Yurasova, Sov. Phys. JETP **67**, 2442 (1988)
95. H. Shao, D.C. Langreth, P. Nordlander, Phys. Rev. B **49**, 13948 (1994)
96. M. Maazouz, L. Guillemot, V. Esaulov, J.D. O'Connor, Surf. Sci. **398**, 49 (1998)
97. C. Auth, A.G. Borisov, H. Winter, Phys. Rev. Lett. **75**, 2292 (1995)
98. A.G. Borisov, V. Sidis, H. Winter, Phys. Rev. Lett. **77**, 1893 (1996)
99. S. Ustaze, R. Verucchi, L. Guillemot, V.A. Esaulov, Phys. Rev. Lett. **79**, 3256 (1997)
100. A.G. Borisov, V.A. Esaulov, J. Phys. Condens. Matter **12**, R177 (2000)
101. H.D. Hagstrum, Phys. Rev. **96**, 336 (1954)
102. T. Fonden, A. Zwartkruis, Phys. Rev. B **48**, 15603 (1993)
103. N. Lorente, R. Monreal, Surf. Sci. **370**, 324 (1997)
104. S.M. Ritzau, R. Baragiola, R.C. Monreal, Phys. Rev. B **59**, 15506 (1999)

105. M.A. Cazalilla, N. Lorente, R. Diez Muiño, J.-P. Gauyacq, D. Teillet-Billy, P.M. Echenique, Phys. Rev. B **58**, 13991 (1998)
106. N.P. Wang, E.A. García, R. Monreal, F. Flores, E.C. Goldberg, H.H. Brongersma, P. Bauer, Phys. Rev. A **64**, 012901 (2001)
107. S. Wethekam, A. Mertetens, H. Winter, Phys. Rev. Lett. **90**, 37602 (2003)
108. R.C. Monreal, L. Guillemot, V.A. Esaulov, J. Phys, Condens. Matter **15**, 1165 (2003)
109. Yu. Bandurin, V.A. Esaulov, L. Guillemot, R.C. Monreal, Phys. Rev. Lett. **92**, 17601 (2004)
110. D. Valdes, J.M. Blanco, V.A. Esaulov, R.C. Monreal, Phys. Rev. Lett. **97**, 04761 (2006)
111. D. Valdes, J.M. Blanco, V.A. Esaulov, R.C. Monreal, Phys. Rev. B **75**, 165404 (2007)
112. B. van Someren, P.A. Zeijlmans van Emmichoven, A. Niehaus, Phys. Rev. A **61**, 022902 (2000)
113. P.J. Feibelman, D.R. Hamman, Phys. Rev. Lett. **52**, 61 (1984)
114. J.K. Norskov, S. Holloway, N.D. Lang, Surf. Sci. **137**, 65 (1984)
115. N.D. Lang, S. Holloway, J.K. Norskov, Surf. Sci. **150**, 24 (1985)
116. P. Nordlander, N.D. Lang, Phys. Rev. B **44**, 13681 (1991)
117. A.G. Borisov, G. Makhmetov, D. Teillet-Billy, J.P. Gauyacq, Surf. Sci. Lett. **350**, L205 (1996)
118. J.P. Gauyacq, A.G. Borisov, J. Phys. Condens. Matter **10**, 6585 (1998)
119. G.E. Zampieri, F. Meier, R. Baragiola, Phys. Rev. A **29**, 116 (1984)
120. S. Lacombe, V.A. Esaulov, L. Guillemot, O. Grizzi, M. Maazouz, N. Mandarino, V.N. Tuan, J. Phys. Condens. Matter **7**, L261 (1995)
121. R. Souda, M. Aono, Nucl. Instrum. Methods Phys. Res., Sect. B, Beam Interact. Mater. Atoms **15**, 114 (1986)
122. T.W. Thomas, H. Neumann, A.W. Czanderna, J.R. Pitts, Surf. Sci. **175**, L737 (1986)
123. S. Kasi, H. Kang, C.S. Sass, J.W. Rabalais, Surf. Sci. Rep. **10**, 1 (1989)
124. E.E. Nikitin, S.Ya. Umanskii, *Theory of Slow Atomic Collisions* (Springer, Berlin, 1984)
125. M. Barat, W. Lichten, Phys. Rev. A **6**, 211 (1972)
126. J. Ostgaard Olsen, T. Andersen, M. Barat, Ch. Courbin-Gaussorgues, V. Sidis, J. Pommier, J. Agusti, N. Andersen, A. Russek, Phys. Rev. A **19**, 1457 (1979)
127. G.E. Zampieri, F. Meier, R. Baragiola, Phys. Rev. A **29**, 116 (1984)
128. S. Lacombe, L. Guillemot, M. Maazouz, N. Mandarino, E. Sanchez, V.A. Esaulov, Surf. Sci. **410**, 70–81 (1998)
129. L. Guillemot, S. Lacombe, M. Maazouz, T.V. Ngoc, V.A. Esaulov, E. Sanchez, Yu. Bandurin, A. Daschenko, V. Drobnich, Surf. Sci. **365**, 353 (1996)
130. V. Esaulov, L. Guillemot, O. Grizzi, M. Huels, S. Lacombe, T. Vu Ngoc, Rev. Sci. Instrum. **67**, 1 (1996)
131. O. Heinz, R.T. Reaves, Rev. Sci. Instrum. **39**, 1229 (1968)
132. D. Roy, J.D. Carette, *Electron Spectroscopy for Surface Analysis*, ed. by H. Ibach (Springer, Berlin, 1977)
133. J.S. Risley, Rev. Sci. Instrum. **43**, 95 (1972)
134. T.J.M. Zouros, E.P. Benis, Appl. Phys. Lett. **86**, 094105 (2005)
135. T.J.M. Zouros, E.P. Benis, J. Electron Spectrosc. Relat. Phenom. **125**, 221–248 (2002)
136. E.P. Benis, T.J.M. Zouros, J. Electron Spectrosc. Relat. Phenom. **163**, 28 (2008)
137. G.J.A. Hellings, H. Ottevanger, S.W. Boelens, C.L.C.M. Knibbeler, H.H. Brongersma, Surf. Sci. **162**, 913–920 (1985)
138. B. Moest, PhD Thesis, Technische Universiteit Eindhoven, 2004, ISBN 90-386-1855-7
139. V. Esaulov, J.P. Gauyacq, D. Dhuicq, J. Phys. B **11**, 1049 (1978)
140. J.C. Brenot, J. Pommier, D. Dhuicq, M. Barat, J. Phys. B **8**, 448 (1975)
141. T.K. Fowler, W.M. Good, Nucl. Instrum. Methods **7**, 245 (1960)
142. C.D. Moak, W.M. Good, R.F. King, J.W. Johnson, H.E. Banta, J. Judish, W.H. du Preez, Rev. Sci. Instrum. **35**, 672 (1964)
143. J.C. Brenot, D. Dhuicq, J.P. Gauyacq, J. Pommier, V. Sidis, M. Barat, E. Pollack, Phys. Rev. A **11**, 1245 (1975)
144. R. Morgenstern, M. Barat, D.C. Lorents, J. Phys. B **6**, L330 (1973)

145. O. Grizzi, M. Shi, H. Bu, J.W. Rabalais, Rev. Sci. Instrum. **61**, 740 (1990)
146. V.A. Esaulov, O. Grizzi, L. Guillemot, M. Huels, S. Lacombe, V. Ngoc Tuan, Rev. Sci. Instrum. **67**, 135 (1996)
147. J.W. Rabalais, J.A. Schultz, R. Kumar, Nucl. Instrum. Methods **218**, 719 (1983)
148. B.J.J. Koeleman, S.T. de Zwart, A.L. Boers, B. Poelsema, L.K. Verheij, Nucl. Instrum. Methods **218**, 225 (1983)
149. O. Grizzi, E. Sanchez, E. Gayone, L. Alarcon, Private communication
150. J. Yao, C. Kim, J.W. Rabalais, Rev. Sci. Instrum. **69**, 306 (1998)
151. C. Kim, A. Al-Bayati, J.W. Rabalais, Rev. Sci. Instrum. **69**, 1289 (1998)
152. M. Katayama, E. Nomura, N. Kanekama, H. Soejima, M. Aono, Nucl. Instrum. Methods B **33**, 857 (1988)
153. M. Aono, M. Katayama, E. Nomura, T. Chasse, D. Choi, M. Kato, Nucl. Instrum. Methods B **37/38**, 264 (1989)
154. I. Kamiya, M. Katayamaa, E. Nomura, M. Aono, Surface Science **242** (1991)
155. K. Umezawa, J. Surf. Sci. Nanotech. **8**, 194–196 (2010)
156. J. Im, A.R. Krauss, Y. Lin, J.A. Schultz, O.H. Auciello, D.M. Gruen, R.P. Chang, Nucl. Instrum. Methods Phys. Res. **118**, 772 (1996)
157. U. Diebold, J.M. Pan, T. Madey, Surf. Sci. **333**, 845 (1995)
158. A.R. Canario-Bettencourt, V.A. Esaulov, J. Chem. Phys. **124**, 224710 (2006)
159. G.F. Liu, Z. Sroubek, P. Kamakar, J. Yarmoff, J. Chem. Phys. **125**, 54715 (2006)
160. A.R. Canario, A.G. Borisov, J.P. Gauyacq, V.A. Esaulov, Phys. Rev. B **71**, 121401 (2005)
161. A.R. Canario, T. Kravtchuk, V.A. Esaulov, New J. Phys. **8**, 227 (2006)
162. G.C. Herdt, A.R. Czanderna, Surf. Sci. **297**, L109 (1993)
163. G. Gonella, S. Terreni, D. Cvetko, A. Cossaro, L. Mattera, O. Cavalleri, R. Rolandi, A. Morgante, L. Floreano, M. Canepa, J. Phys. Chem. B **109**, 18003-9 (2005)
164. M. Prato, O. Cavalleri, M. Canepa, A.R. Canario, V.A. Esaulov, in *8th European Conference on Molecular Electronics*, Bologna, 2008 (unpublished)
165. M. Canepa, P. Cantini, L. Mattera, S. Terreni, F. Valdenazzi, Phys. Scr. **41**, 226 (1992)
166. M. Canepa, P. Cantini, F. Fossa, L. Mattera, S. Terreni, Phys. Rev. B **47**, 15823 (1993)
167. H. Hamoudi, Z.A. Guo, M. Prato, C. Dablemont, W.Q. Zheng, B. Bourguignon, M. Canepa, V.A. Esaulov, Phys. Chem. Chem. Phys. **10**, 6836 (2008)
168. H. Hamoudi, M. Prato, C. Dablemont, O. Cavalleri, M. Canepa, V.A. Esaulov, Langmuir **26**, 7242–7247 (2010)
169. L.S. Alarcon, L. Chen, V.A. Esaulov, J.E. Gayone, E.A. Sanchez, O. Grizzi, J. Phys. Chem. C **114**, 19993–19999 (2010)
170. L.M. Rodrõguez, J.E. Gayone, E.A. Sanchez, O. Grizzi, B. Blum, R.C. Salvarezza, J. Phys. Chem. B **110**, 7095 (2006)
171. M. Casagrande, S. Lacombe, L. Guillemot, V.A. Esaulov, Surf. Sci. **445**, L36–L40 (2000)
172. K. Wu, D. Wang, J. Deng, X. Wei, Y. Cao, M. Zei, R. Zhai, X. Guo, Surf. Sci. **285**, L522 (1993)

Chapter 16
Helium Ion Microscopy

Diederik J. Maas and Raoul van Gastel

Abstract The realization of a practical helium gas field ionization source (GFIS) enabled helium ion microscopy (HIM) as a new technique to image and modify materials and microstructures. After a brief overview of most common ultra-microscopy techniques (TEM, SEM, Gallium FIB) and HIM, we introduce the interaction fundamentals of helium ions with matter. A key element of that interaction is that the resulting signals for imaging, nanofabrication and analysis, i.e. the secondary electrons and backscattered ions, are to a very high degree localized around the incidence point of the helium beam. This simple fact allows the helium ion microscope to enable a new and unique view of surfaces and provide a new method for material modification. We highlight several applications for imaging and nanofabrication using the sub-nanometer sized helium probe of the HIM.

16.1 Introduction

Microscopic techniques are widely used in the materials and life sciences for the imaging of surfaces and interfaces. Ultra-microscopy is the sub-field of microscopy that studies samples at extreme magnifications, yielding images with a field of view of less than ten microns. Over the last decades, many instrumentation advancements have fuelled the rapid rise of this sub-field of microscopy [1]. It was partly initiated by the introduction of novel optical techniques to achieve "super-resolution" and coincided with the introduction of a new generation of charged particle optics microscopes. For example, the resolving power of electron microscopes has been improved by using aberration correctors [2], better detectors and brighter sources [3]. Ion microscopy has benefited from novel sources and better electrostatic lenses.

D.J. Maas (✉)
TNO Nano-instrumentation, P.O. Box 155, 2600AD Delft, The Netherlands
e-mail: diederik.maas@tno.nl

R. van Gastel
Physics of Interfaces and Nanomaterials, MESA+ Institute for Nanotechnology, University of Twente, P.O. Box 217, 7500AE Enschede, The Netherlands
e-mail: R.vanGastel@utwente.nl

G. Bracco, B. Holst (eds.), *Surface Science Techniques*,
Springer Series in Surface Sciences 51, DOI 10.1007/978-3-642-34243-1_16,

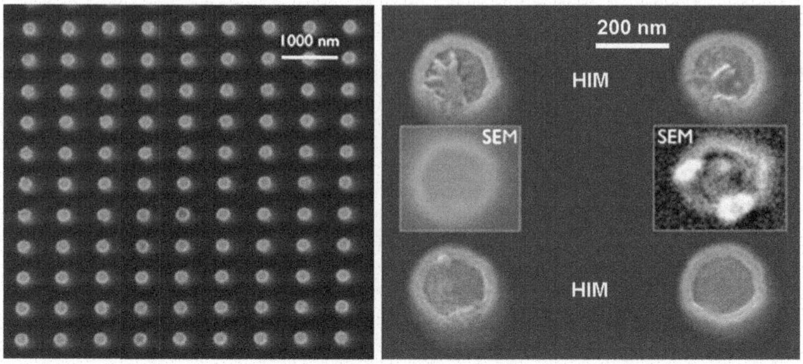

Fig. 16.1 HIM (*left panel*) and HIM plus SEM micrographs (*insets in right panel*) of copper posts with bi-layered TaN/Ta plating and an Au-top coating in a CMOS-based biosensor [4]. The HIM and SEM micrographs show different posts in the same array. Inspection with HIM at 25 keV and 0.5 pA provides clear material contrast and reveals many more details of the electrode surface than with SEM [5]. The electrodes are embedded in an insulating matrix and therefore electrically floating. When irradiated with a focused electron beam, these electrodes will charge in an unpredictable manner, thus inhibiting reliable and reproducible SEM inspection. Here, sample charging was either minimized by using a primary beam energy of 200 eV (SEM) or completely avoided by using an electron flood gun (HIM). Sample and SEM micrographs are courtesy of NXP Semiconductors. Reproduced with permission from Scanning **34**, 90 (2012) [5]. Copyright 2012 Wiley

Both fields benefit from the cleaner vacuum that is available nowadays. The latter is crucial when studying the shape and composition of surfaces, see e.g. Fig. 16.1, since beam-induced dissociation of adsorbed molecules may contaminate or even alter the sample before a reliable image can be collected.

Microscopy traditionally aims to provide information on the shape and/or composition of a sample. The first requires accurate determination of surface structure and morphology, ideally by an accurate measurement of the position of the sample atoms. The second requires elemental identification of the atomic species that the sample consists of. Almost a century ago, Louis de Broglie suggested the use of (sub-)atomic particles for the imaging of sub-300 nm structures [6]. Experimental verification of de Broglie's idea by Davisson and Germer [7] and the proof that electrons can be focused with magnetic lenses [8], inspired Max Knoll and Ernst Ruska to conceive the Electron Microscope (EM) [9]. The EM was the first ultramicroscopy instrument that attempted to achieve both feats. In their landmark 1931 paper [9], Knoll and Ruska suggested several different modes in which an electron microscope can be operated, transmission or reflection. They even contemplated the use of ions for imaging applications. Soon after, it was realized that such intensely focused particle beams can also modify a sample [10]. If well controlled, these modifications can be used for nanofabrication purposes [11]. In his review paper on nanofabrication applications of electron beams [12], Pfeiffer states *"Writing pattern features directly on silicon wafers using modified SEMs started during the 1960s [13]. Electron beam pattern generator tools became the technology of choice for the mask making industry in the late 1970s [14]."*.

The remainder of this chapter presents a brief history of ultra-microscopy, followed by the principles of the interaction of 25–35 keV helium ions with matter, the operation of the GFIS source and the ion optical column, and the principles of imaging and sample modification. We will highlight the capabilities of the HIM with several imaging and patterning applications.

16.1.1 Brief History of Ultra Microscopy

The earliest form of electron microscopy that was widely employed for materials characterization was transmission electron microscopy (TEM). The first commercial apparatus was built by Ruska and Von Borries in collaboration with Siemens in 1939. The resolution of TEM is primarily limited by the spherical aberration of the lens system. In modern day instruments that are equipped with aberration correctors, resolution in the sub-Ångstrom range is routinely achieved and the signature of individual atoms can be observed. Chemical identification of these atoms is possible by analyzing the energy of the electrons that have scattered off the target or by detecting the X-rays that are generated concomitantly. Electron energy loss spectroscopy (EELS) can provide detailed information on the chemical bonds in the target, whereas energy dispersive X-ray spectroscopy (EDS) offers fast and easy elemental analysis. A major limitation for TEM imaging is that the best resolution is only achieved on ultrathin samples with a thickness $\leq$50 nm. The preparation of such samples requires special skills and equipment like a focused ion beam (FIB), is time consuming, invasive, and destroys the sample at least partially.

Scanning electron microscopy (SEM) was the next form of microscopy that was used to characterize the sub-micron structure of materials. Although introduced in 1935 [15], it became more widely employed after dedicated SEMs were constructed in Cambridge in the fifties [16], inspired by the pioneering work in Germany and the USA [17, 18]. The resolution of SEM was initially limited by the probe size. As the probe size improved over time, the practical resolution became limited by the interaction volume of the incident electrons. State-of-the-art field emission gun (FEG) SEMs obtain a spatial resolution better than 1 nm on certain samples [19]. Materials characterization with SEM is also routinely available through energy dispersive X-ray spectroscopy (EDS) detectors.

Historically, the first instrument to achieve both the goals of atomic resolution and chemical identification was the field ion microscope (FIM), developed by Erwin Müller in 1951 [20]. In FIM, a high voltage is applied to a sharp metallic tip. The environment of the source tip consists mainly of a noble gas, e.g. helium or neon. The metal needle is kept at cryogenic temperatures to cover the surface with a thin layer of condensed noble gas. The enhancement of the electric field at the tip of the metal field-ionizes the adsorbed gas atoms and accelerates the ions towards a screen. The extremely high magnification of up to $10^6\times$ enables the visualization of individual atoms at the top of the tip, see e.g. Fig. 16.10. Chemical identification can be achieved either through the image contrast [21] or through the addition of a mass spectrometer [22]. FIM is limited in its field of view to the very end of the tip that is used to ionize the atoms of the imaging gas. Nevertheless, it has been successfully applied to study prototypical processes like adatom diffusion, island formation and field evaporation. A significant drawback of FIM, hampering its practical application to a wide range of samples, is that the sample is also the source of the particle beam.

The cathode lens microscope using low energy electrons, or Low Energy Electron Microscopy (LEEM), was the second form of electron microscopy to achieve atomic resolution, albeit exclusively in the direction perpendicular to the surface [23]. Invented by Bauer in 1962 [24], LEEM generates diffraction and interference contrast in a geometry that is very close to the one used in the original Davisson-Germer experiment [7]. It can image surface areas many microns in size and is able to chemically discriminate between surface regions through X-Ray PhotoElectron Emission Microscopy (XPEEM), but lacks the lateral resolution to laterally resolve individual atoms.

The last substantial new development in microscopic techniques came with the introduction of the Scanning Tunneling Microscope (STM) [25] and its derivatives. The use of a metallic tip as a sensor for forces and currents when it is in near contact with a surface, enabled the imaging of surfaces with atomic resolution. The spectroscopic capabilities of the STM for elastic [26] and inelastic [27] processes have enabled the characterization of the chemical identity and vibrational state of surface atoms and even their direct manipulation. The scanning motion of the tip raises some constraints for using STM, i.e. limited scan speed and hence a small field of view (FOV) and slow navigation over the sample. For present state-of-the-art instruments [28], the scan size for real-time, video rate measurements is limited to fields of view similar to those seen in FIM.

Until 1984, when liquid metal ion sources (LMIS) started to appear, the use of Scanning Ion Microscopy (SIM) for surface imaging [29] was impeded by the absence of a sufficiently bright ion source [30]. The operation of a SIM closely resembles that of a SEM. A significant difference with SEM is that in SIM the ions also sputter the sample and typically end up implanted in the sample. Focused ion beam microscopes have found their largest application area in sputtering and ion beam induced chemistry [31]. For example, in failure analysis of semiconductor devices the gallium FIB is widely employed for e.g. the preparation of thin TEM lamella from bulk specimen as well as for circuit editing to debug or even repair a failing chip [32].

Ever since the inception of HIM in the original papers by Knoll and Ruska [9], its advent remained an illusion because of two practical problems: the lack of ion optics that provide practical focal lengths without needing hundreds of Ampères of current, and the absence of a source with a sufficiently high brightness. The first problem was eventually solved through the development of manufacturable electrostatic lenses with low-aberrations [31]. Liquid metal ion sources, shown in Fig. 16.2(a), that made possible the development of FIB instruments, were however simply not suitable for noble gases. The breakthrough that made possible high-resolution HIM [33] was the realization of a practical He gas field ion source (GFIS)

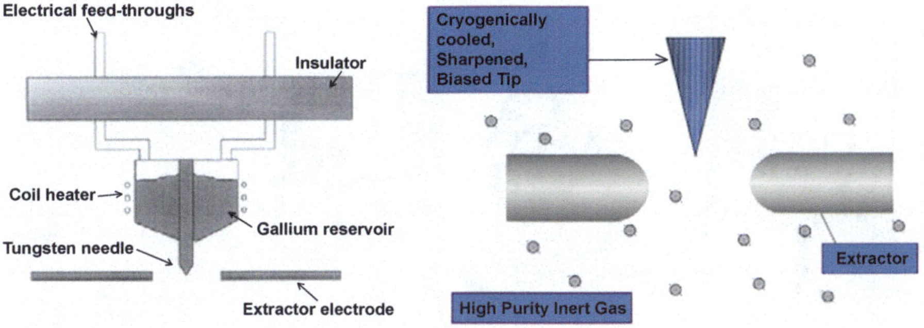

Fig. 16.2 The LMIS and GFIS are practical high-brightness ion sources for gallium and helium, respectively. *Left panel*: schematic of conventional liquid metal ion source (LMIS) as typically used in commercially available gallium focused ion beam systems (Ga-FIB). The source temperature is in the 1700–1800 K range, resulting in the wetting of the sharp Tungsten emitter with a thin liquid film of the desired ion species. *Right panel*: schematic of the gas field ion source (GFIS), the novel source that enabled the development of the helium ion microscope (HIM). The GFIS operates at a temperature between 50–70 K, emitting ionized noble gas atoms (like helium or neon) from the ultra sharp tungsten emitter. The tip of the emitter typically consists of three atoms, each delivering an ultra-bright beamlet of helium ions. *Left panel* reproduced with permission from [31]. Copyright 2005 Springer. Right panel courtesy Carl Zeiss NTS

by Ward and co-workers in 2004 [34]. This was soon followed by the implementation of this source in a commercially available instrument. Figure 16.3 displays the Carl Zeiss OrionPlus™ HIM in the TNO Van Leeuwenhoek Laboratory in Delft, one of the first instruments installed in 2008. The He GFIS is stable and delivers sufficient beam current to ensure realistic image acquisition and pattern formation times.

16.2 Theory and Background

This section presents the fundamentals of the interaction of helium ions with solids and reviews the equipment advancements that enabled the development of HIM: the gas field ion source and the ion optical column.

16.2.1 Helium Ion Interaction with the Target

After striking the sample, the incident ion beam generates a range of particles in the interaction volume:

- Secondary electrons
- Backscattered He
- Photons
- Sputtered specimen atoms

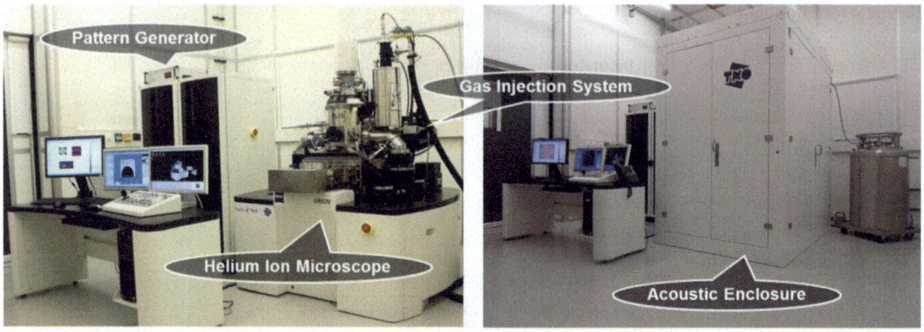

Fig. 16.3 The Carl Zeiss OrionPlus™ helium ion microscope of the TNO Van Leeuwenhoek Laboratory, a nanofacility in Delft (NL). The instrument is placed in an acoustic enclosure that suppresses room acoustic noise by at least 12 dB. The HIM is equipped with a Raith pattern generator and an Omniprobe gas injection system, which enables nanofabrication research

The interpretation of HIM images formed by secondary electrons (SE) and Rutherford backscattered ions (RBI) requires a basic understanding of the ion and electron scattering processes that occur once the primary helium ion has entered the sample. Moreover, the advancement of nanofabrication with helium ions also calls for a good understanding of the interactions that occur during the stopping of a helium ion in a target. These interactions determine the desired and undesired material modifications that can enable nanofabrication or disable imaging. Compared to electrons and gallium, the fundamental differences of the interaction of helium with the target define the HIM technique. During the propagation of ions and electrons in a solid, the direction and velocity are changed by elastic and inelastic collisions. Figure 16.4 illustrates how an elastic collision alters the particle direction, whereas an inelastic collision also slows particles down. There are several energy loss channels, e.g. direct creation of a secondary electron through valence or core ionization, or collective excitation of the solid (plasmons, excitons or phonons) [35]. When the kinetic energy of a secondary electron is sufficiently high, a cascade of SEs can be induced. As originally suggested by Salow [36], the escape of primary and secondary particles can be treated as a diffusion process. The average distance that a particle travels between the interaction events is the (in)elastic mean free path (MFP) [37]. The MFP is proportional to the inverse of the (in)elastic scattering cross section. The length of the MFP in a specific material is a function of the mass and energy of the primary particle, since these determine which channels for energy and momentum transfer are available. Along its trajectory in the target, a helium ion will undergo multiple collisions until it has lost all its kinetic energy. The total energy lost per unit distance traveled is called the stopping power, $S = S_e + S_n = -\frac{dE}{dz}$. The average electronic and nuclear interaction is described in terms of the electronic and nuclear stopping power, S_e and S_n, respectively. Already in 1915 Bohr had shown that the electronic and nuclear stopping powers are to a large extent independent of each other [38]. Ziegler, Biersack and Littmark [39] implemented the stopping model in the public domain simulation software packages TRIM and SRIM [40]. The electronic stopping is dominant in most materials for helium ions with an energy above typically

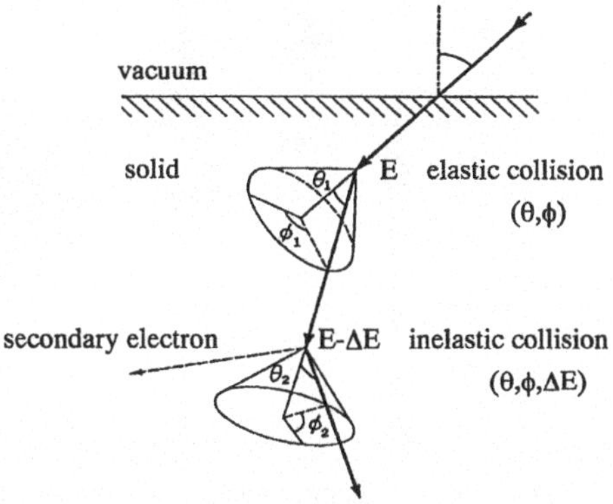

Fig. 16.4 The trajectory of a fast incident particle in a solid is altered through collisions with the target material. Elastic collisions with target atom nuclei change mainly the direction of the incident particle. Inelastic collisions with the electron cloud of the target atom slow the primary particle down and may produce secondary electrons. Figure courtesy of K. Ohya (University of Tokushima)

3–10 keV [41, 42]. Due to the mass difference, a collision with an electron hardly changes the direction of the helium ion. As a consequence, for a typical regular helium beam energy of 25 keV there is relatively little lateral scattering in the first few tens of nanometers of the target. Figure 16.5 shows the interaction volume of 5, 10, 30 and 100 keV He^+ ions in aluminum as computed with the SRIM Monte Carlo package [43]. The nuclear stopping power of helium in aluminum is negligible for incident particle energies above typically 3 keV. Indeed Fig. 16.5 illustrates how the incident energetic ions first travel forward to a significant depth, mainly losing their energy through inelastic electronic collisions. Only after the ion energy has decreased substantially, nuclear collisions cause substantial lateral scattering of the helium in the sample.

16.2.2 Secondary Electron Generation

Secondary electrons (SEs) are the most abundant of all generated particles. SEs are created along the entire trajectory of the primary particle. For imaging with SEs, this is relevant, since only those SEs can be detected that are created within at most several times the MFP range of electrons from the surface *and* that have sufficient energy to escape from the sample into the vacuum. In contrast to imaging, for lithograpic applications all SEs along the trajectory are relevant if they have sufficient energy to activate the required bond-breaking or cross-linking reaction.

Two interaction processes are responsible for the creation of secondary electrons in a collision of an helium ion with the target material [44]. Kinetic emission (KE) occurs through direct collisions that involve a substantial amount of momentum transfer to an outer shell electron and is only likely to occur in the keV ion energy range where the velocity of the primary ion matches that of an electron at the Fermi energy [45]. In HIM there are three different possible collision partners that can generate KE: primary He ions, high energy secondary electrons, and recoiled specimen atoms. Potential emission (PE) involves the neutralization of the incident positive ion via electron tunneling. The energy that is released in the neutralization is transferred to another electron in the solid, which is then emitted. For singly-charged ions

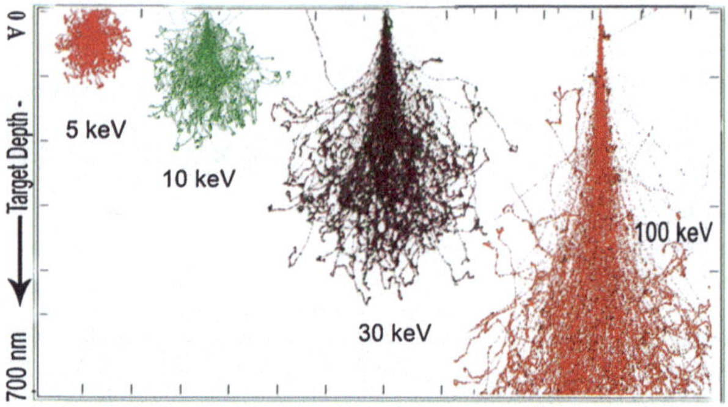

Fig. 16.5 The interaction volume of helium ions in aluminum at 5, 10, 30 and 100 keV beam energy (from SRIM modeling). With increasing energy, the target penetration depth increases while the iSE signal is better localized around the point of incidence. Reproduced with permission from J. Appl. Phys. **104** (2008), 063504 [43]. Copyright 2008 American Institute of Physics

with energies of a few keV or less, the PE process via Auger neutralization plays a major role in electron emission [46]. Figure 16.6 displays the SE yield from gold as a function of the energy of the primary particle. Panel (a) shows that for helium ions with less than 8 keV beam energy helium PE generates the most SEs. For imaging, the helium ions typically have between 20 and 35 keV beam energy, and mainly KE generates the SE signal.

The use of He ions for the imaging of surfaces with SEs provides two important benefits when compared to electron- or heavy-ion-based techniques. First, in HIM the spectrum of emitted secondary electrons peaks at rather low energies [47] compared to SEM and the backscattering of He ions is weak. This implies that the SEs are mostly of a SE1 nature, i.e. they are mainly generated by incoming ions, not by backscattered ions. The portion of the signal that results from direct beam-sample interactions is typically larger than in electron based techniques and has the potential to yield clearer images. A second benefit is the overall higher SE yield, see e.g. Ref. [48]. Not only is the SE yield for He ions substantially higher than that for electron irradiation at similar energies, it also increases with primary energy, as is illustrated in Fig. 16.6 for a Au sample. The detection of SEs in a HIM is done using an Everhart-Thornley (ET) type detector [49]. Images formed by detecting the yield of secondary electrons mainly give surface topographic information, with qualitative materials contrast, see e.g. Fig. 16.1. Variation of the angle of incidence can be used to influence the surface sensitivity and the SE yield [50].

Figure 16.7 shows a comparison of the interaction volume of 30 keV gallium and helium ions and electrons in silicon, zoomed in to the first 20 nm of the sample surface. For gallium ions the nuclear stopping power is much larger than the electronic stopping power for most beam energies. As a consequence, immediately below the point of impact on the target, the interaction volume broadens laterally to approximately 15 nm at the SE escape depth (arbitrarily chosen at a typical value of 10 nm). Hence, the spatial resolution in gallium microscopy is limited to typically 5 nm. For lower Ga beam energies, e.g. 5 keV, the balance between S_n and S_e is

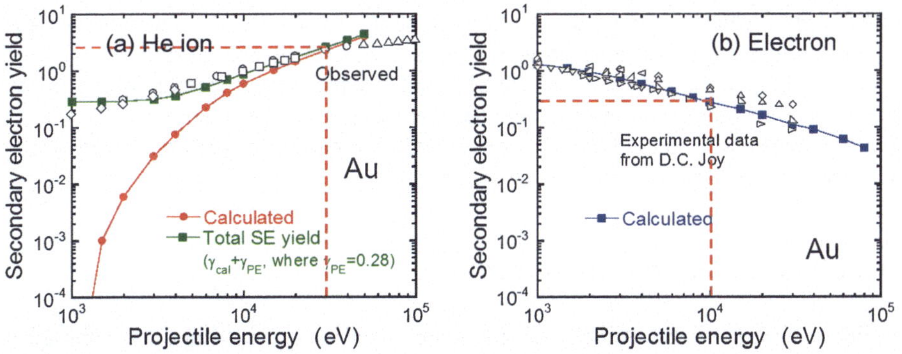

Fig. 16.6 Secondary electron yield (experimental data and Monte Carlo (MC) simulations) from a gold sample as a function of incident particle energy. Panel (**a**) is for helium ions, panel (**b**) is for electrons. The MC simulations for helium ions show that above 8 keV most iSEs stem from kinetic electron emission. The iSE yield is generally larger than the eSE yield, except for beam energies below a few keV, as is illustrated here for Au [46]. Figure courtesy of K. Ohya (University of Tokushima)

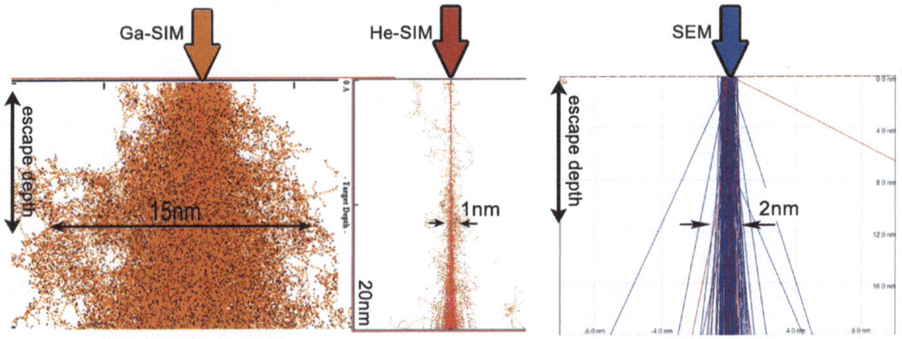

Fig. 16.7 Comparison of the interaction volume of gallium (*left*) and helium (*center*) ions and electrons (*right*) in the top 20 nm of infinitely thick silicon at 30 keV beam energy. The trajectory of the primary particles is indicated in *red*, that of (multiply) backscattered particles in a different color. The beam width is 0 nm when entering the sample. The SE escape depth is taken to be approximately 10 nm. At that depth the gallium beam has broadened to 15 nm, the helium beam to only 1 nm and the electron beam to 2 nm. Reproduced with permission from J. Appl. Phys. **104** (2008), 063504 [43]. Copyright 2008 American Institute of Physics

more favorable for high resolution imaging. The center panel of Fig. 16.7 shows that for helium ions the interaction radius is less than 1 nm even at the full SE escape depth of 10 nm. For electrons, the interaction radius is sharply peaked at 2 nm, with a certain amount of strongly laterally scattered SEs that carry lower spatial resolution information of the sample.

In conclusion, the helium ions generate a very local and strong SE signal during their trajectory in the target. Of these SEs only those that are generated in the first few nm will escape the sample and can be detected. The combination of these interaction attributes make helium ion microscopy well suited for extremely high-resolution imaging of surfaces, as well as local and dense nanopatterning. A word of caution is in place to mention that the SE yield is sensitive to the crystallinity of the target through strong variations in the SE yield as a result of channeling. The latter needs to be properly accounted for when imaging or patterning crystalline materials, as we will discuss further in Sect. 16.4.3.

16.2.3 Backscattered He Imaging

Helium ion microscopy further distinguishes itself from existing EM techniques through its alternate modes to form images. Similar to backscattered electron (BSE) imaging in SEM, microchannel plates (MCP) [51] enable the detection of backscattered He and can be used to inspect a specimen. The resulting images constitute a new form of microscopy that gives both chemical and structural contrast. In the initial HIM literature it is referred to as the Rutherford backscattered ion (RBI) mode of imaging. In a classic Rutherford backscattering spectroscopy (RBS) experiment a beam of high energy ions is used to illuminate a specimen. The spectroscopic information that is obtained resides both in the number of backscattered ions as well as in their scatter angle and recoil energy [52]. The energy of the backscattered He ions is proportional to the incident energy and the ratio of the mass of the sample atoms to the mass of the He ions. The stopping power of the specimen also allows for limited probing of depth information in an RBS-like fashion. With its sub-nanometer sized probe, HIM therefore constitutes a technique that performs a spatially resolved low energy ion scattering (LEIS) analysis on a nanometer length scale. A practical limitation is posed by the fluence that is possible in a single pixel before modification of the sample occurs.

For imaging using backscattered ions, it is important to realize that both the backscattered ion fraction and the ion MFP, and hence the chance to leave the sample and reach the RBI detector, is a function of the ion energy and depends on the depth in the sample where the backscatter event occurs. Figure 16.8 illustrates the information that resides in both types of images. The SE image in panel (a) provides detailed surface information with very strong edge contrast, akin to what is seen in SEM SE images [54]. Images recorded in RBI mode reveal more elemental contrast that can be used to identify the chemical composition of different sample regions. Furthermore, they show different topographic features that generally come closer to the true specimen morphology, albeit at lower lateral resolution. The MCP detects backscattered particles with an energy above a detection limit set by the read-out electronics. To first order it probes the scattering cross-section. The scattering cross-section for He off a given target is given by:

$$\frac{d\sigma}{d\Omega} = \left(\frac{Z_1 Z_2 e^2}{4E_0}\right)^2 \frac{1}{(\sin(\theta/2))^4} \tag{16.1}$$

σ is the scattering cross section, Ω is the solid angle, Z_1 is the atomic number of the impinging ion, Z_2 the atomic number of the target atom, E_0 the primary energy, and θ the scattering angle. As a rule of thumb we learn from this equation that

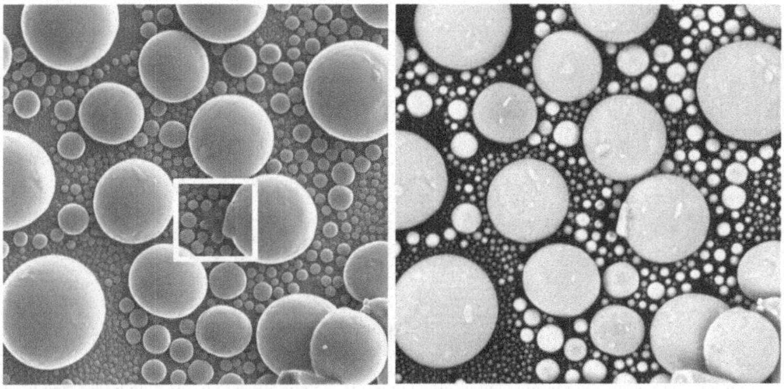

(a) (b)

Fig. 16.8 5 μm FOV images of Sn spheres on a SiO_2 background. *Panel* (**a**) shows an image recorded in SE mode, i.e. using secondary electrons, whilst *panel* (**b**) shows the same region imaged in RBI mode using backscattered He. Some Sn spheres were damaged due to sputtering (*highlighted area in left panel*) during earlier imaging at a 16 times higher magnification. The trade-off between image quality and sample damage is discussed in detail in [53]. Figure courtesy of Carl Zeiss NTS

heavier elements will appear brighter in RBI images as a result of a higher scattering cross-section. The contrast for lighter elements may be difficult to explain using (16.1) and a simple kinematic approximation, because the energy dependent gain of the microchannel plate detector [55] convolves with the energy spectrum of the backscattered He. Because the recoiled He originates from a relatively large sample interaction volume, the resolution of HIM in RBI mode is less than in SE mode and the elemental information dominates the images, as is confirmed in Fig. 16.8.

16.2.4 Backscattered He Spectroscopy

The capabilities of a HIM can be extended even further for the purpose of ion spectroscopy when it is equipped with an *energy sensitive* backscattered helium detector. The operational energy range falls in between the common LEIS and medium energy ion scattering (MEIS) spectroscopies. Helium ions backscattered from the sample may undergo several scattering events and be neutralized in the process. Multiple scatterings lead to peak broadening and increased complexity in interpreting the signal. In most situations, the neutralized helium fraction overwhelms the ion fraction, however the ratio can be strongly energy dependent. In practice the energy of backscattered He and neutral particles is detected using a Si drift detector [56]. The energy of the backscattered He follows from the kinematics of the ion-sample scattering process [52]. For a more elaborate description we refer to Chap. 15 of this book and Chap. 2 of the book by Rabalais [57]. The relevant theory will be briefly summarized here.

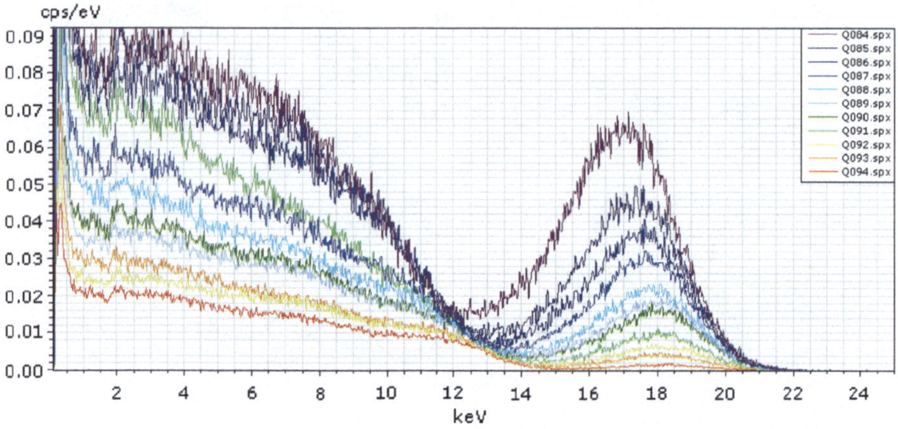

Fig. 16.9 Backscattered He spectra of a ZrO_2 film deposited on SiO_2 using an ALD process, recorded at a primary ion beam energy of 25 keV. The plotted data ranges from 1 to 40 ALD cycles. Backscattered helium beam can be used to detect differences between a monolayer and a few monolayers. Reproduced with permission from J. Vac. Sci. Technol. B **28** (2010), 73 [56]. Copyright 2010 American Vacuum Society

Rather than taking many-body interactions into account, in ion scattering it suffices to consider only pair-wise interactions of the projectile with atoms in the target to reasonably high accuracy, and combine that with the stopping power. Thus, projectile-sample interactions can largely be determined using the kinematics of binary collisions. Binary collisions are governed by conservation of momentum and conservation of energy. The conservation laws lead to the equation

$$E_1 = E_0\left(\frac{M_1}{M_1 + M_2}\right)^2\left(\cos\theta \pm \sqrt{\left(\frac{M_2}{M_1}\right)^2 - \sin^2\theta}\right)^2 = K E_0 \qquad (16.2)$$

in which E_0 and E_1 are the kinetic energy of the projectile before and after the collision respectively, M_1 and M_2 are the masses of the projectile and target particle, respectively and θ is the scattering angle.

The lateral resolution for helium ion spectroscopy (HIS) is inherently limited by sample damage (balance between signal-to-noise and target atoms sputtered away). Sputtering limits the practical pixel respectively voxel size to typically 1 µm respectively 100 nm on (almost) all materials [58]. Therefore the principles of the technique are best demonstrated on planar samples. Figure 16.9 shows the evolution of ion spectra that were recorded on a Si substrate that was repeatedly exposed to Atomic Layer Deposition (ALD) pulses of Zr and O_2. As a result of its heavier mass, the primary peak for He scattering of Zr is shifted to the right in the spectrum with respect to that of Si. With an increasing number of ALD pulses the height of the peak is observed to increase and it can be used to accurately measure the Zr coverage. We note that the HIS application works well only for the analysis of heavier elements residing on substrates that consist of mainly lower Z elements.

16.2.5 Other Modes of Operation

Other particles can be used for different operational modes to generate HIM images. The most viable of these is ionoluminescence (IL), which uses photons that are generated at by the ion beam and harvests them in a manner similar to cathodoluminescence. Although initial studies are beginning to appear [59], the number of instruments equipped with luminescence detection capabilities is still rather limited. The cross-section for the generation of X-rays by ~30 keV He ions is so small that it does not yield a usable signal.

16.2.6 Sample Damage and Nanofabrication

An important, often deleterious but sometimes beneficial [60], side effect of helium ion irradiation for imaging is damage to the sample. At the surface, charging, sputtering and cross-linking of adsorbates (e.g. carbohydrates) may occur, which alters the sample surface and hence the SE signal. In the sample, charging, charge transfer, implantation, dislocation, amorphization and even bubble formation may occur, depending on the areal dose, ion energy and substrate composition and temperature. The degradation of the silicon lattice as a function of areal dose has been extensively studied [61], and is illustrated in this chapter in Figs. 16.22, 16.24 and 16.27. The trade off between signal collection and damage to the inspected sample has been intensively studied by Castaldo et al. [53]. They repeatedly irradiated the same Sn spheres on a carbon substrate and recorded the size and shape change of the object. Similar modifications are shown in the highlighted area of the left panel of Fig. 16.8; smaller spheres shrink and disappear in the end due to sputtering while larger spheres are deformed by re-deposition of sputtered target atoms. Thus Castaldo et al. could quantify the helium fluence for which the minimum uncertainty in the size measurement of the spheres is achieved.

As much as these effects may hamper imaging, they can be conveniently applied for nanofabrication purposes. The ideal probe for nanofabrication with a scanning beam is infinitely sharp and bright and consists of particles that have an extremely localized interaction with the target substrate, resist or precursor molecules that need to be modified without causing damage to the irradiated materials. With these properties, nanostructures of arbitrary shape can be made at high speed. Unfortunately, such a probe does not exist.

The helium ion probe that a HIM provides does have several attributes that make it stand out in at least some aspects as compared to the commonly used charged particle nanofabrication tools like the gallium ion FIB and electron beam lithography (EBL) tools. These advances originate from both the unique interaction mechanism of helium ions with materials and the extreme brightness of the GFIS source [34]. Some drawbacks of the HIM are the risk of damage to the target substrate due to helium implantation at very high areal dose [61], and a relatively low beam current of at most 25 pA that limits the speed of the nanofabrication process compared to massively parallel EBL tools that have been or are being developed [12]. Nanofabrication with a MeV hydrogen ion beam has similar benefits and drawbacks [62]. Moreover, such a system is not commercially available, requires significantly more operator attention and has a substantially larger footprint than the HIM.

Nanopatterning with the HIM comes in three flavors: either by Direct Write (DW) [63–65] with the ions in the sample, by exposure of a resist (lithography) [66, 67], or as a result of local beam-induced chemical reactions of precursor molecules, adsorbed on a sample surface [68, 69]. For all these nanofabrication methods, a pattern generator is required to deliver the right helium ion dose to the right sites on the sample surface. For scanning helium ion beam lithography (SHIBL), the sample is covered by an ion-sensitive resist layer and a development step is required to convert the latent image into a nanostructure. Helium ion beam induced deposition and etch (He-IBID and He-IBIE) require the delivery of precursor molecules on the target substrate from a gas injection system (GIS). The exact shape of the deposit or etch depends on a number of factors like sample composition, precursor molecule supply, pattern scan settings and the intensity of the helium ion beam. Together they determine the interaction of the helium ions with the target atoms to create a nano-sculpture. Section 16.4.5 highlights the present state-of-the-art nanofabrication capabilities of the HIM.

16.3 Experimental Setup

16.3.1 Ion Source

The enabling technology of HIM is the high performance helium ion source that is based on Müller's FIM design [20] and is illustrated in Fig. 16.2(b). A high-voltage biased atomically sharp needle held at liquid nitrogen temperatures field ionizes adsorbed helium atoms at its tip. The field gradient projects the positively charged helium ions towards the specimen. If an aperture is placed in the center of the ion-optical column below the source, the ion beam can be collimated, focused and scanned in the same manner as an electron beam. This arrangement represents an ion source with a reduced brightness (expressed in $\mathrm{A\,cm^{-2}\,sr^{-1}\,V^{-1}}$) that is comparable to that of a cold field emission electron source with an effective source size of about 0.5 nm. Further shaping of the tip through annealing in O_2 [70, 71] and field evaporation [72, 73] allows for a significant enhancement of the source brightness. In that process the top layer of the tip is formed into a stable structure of just three atoms. The brightest emitting atom is chosen as the actual source by tilting the source such that only this beamlet passes through a beam limiting aperture, see Fig. 16.12. A FIM image of the actual configuration that is used in the field ionization gun is shown in Fig. 16.10. About 140 pA of ion current from a single atom is available in this mode, corresponding to a current density of about 0.5–1.0 $\mathrm{\mu A\,sr^{-1}}$ [74]. The atomic trimer arrangement of the field ionization gun is capable of high performance. In theory

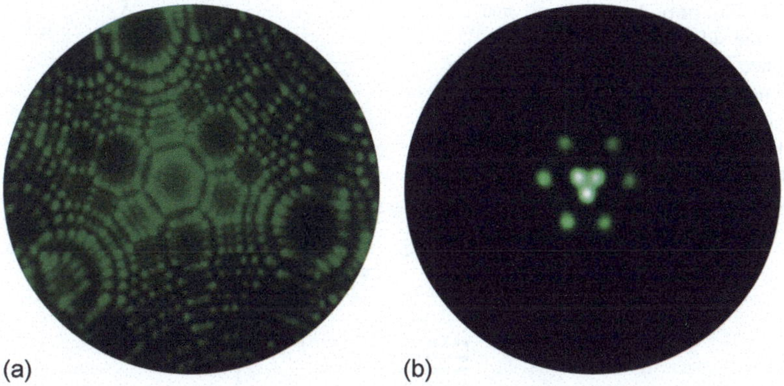

Fig. 16.10 (**a**) FIM image showing the distribution of ions originating from a standard FIM tip. (**b**) The distribution of ions from a shaped tip with the atomic trimer configuration in the last plane. The current coming from the source originates almost exclusively from the trimer. Figures courtesy of Carl Zeiss NTS

a monomer configuration will lead to even brighter sources, however the thermal stability of such a configuration is significantly less than that of the trimer configuration leading to unpractically short source lifetimes. The field ionization process in the helium gas keeps the tip clean and operation for periods typically well in excess of two weeks is possible before a tip reforming operation is necessary.

16.3.2 Probe Size

As discussed in Sect. 16.2.1, one of the areas where helium ion microscopy distinguishes itself from electron beam based techniques is in the interaction volume. The differences between ions and electrons when they interact with matter will lead to a shift in the ratio of secondary electrons produced directly by the impinging ion beam (SE1) and those produced by backscattered ions (SE2) or through other mechanisms (SE3, etc.) [54]. The lateral deposition of the primary energy occurs at much greater depths than in the typical bubble-shaped interaction volume of an electron beam of similar primary energy. This implies that the size of surface features that can be resolved when secondary electrons are used for imaging, is directly determined by the size of the probe and is affected comparatively little by SE2 contributions caused by backscattered ions [43, 75–77]. The ultimate probe size d_p that can be projected on the surface is determined by several factors and is given by [78]

$$d_p = \sqrt{d_{So}^2 + d_s^2 + d_c^2 + d_d^2} \tag{16.3}$$

$$d_{So} = M \cdot d_g \tag{16.4}$$

$$d_s = 0.5 C_S \alpha_i^3 \tag{16.5}$$

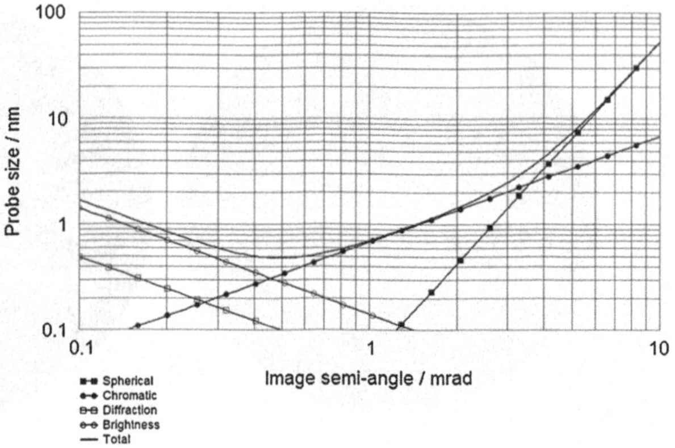

Fig. 16.11 Plot of probe size versus image side convergence angle for 25 keV landing energy, 6 mm working distance, and 0.5 pA beam current. Reproduced with permission from Nucl. Instrum. Meth. A **645** (2011), 96–101 [74]. Copyright 2010 Elsevier

$$d_c = C_C \frac{\Delta E}{E} \alpha_i \tag{16.6}$$

$$d_d = 0.6 \frac{\lambda}{\alpha_i} \tag{16.7}$$

where the four different contributions are for the demagnified source, spherical aberrations, chromatic aberrations and diffraction errors, respectively. α_i is the convergence angle of the beam. Because the emitter tip is cold and helium has only a single stable ionization state, the chromatic spread ΔE of the emitted beam is as low as 1 eV. When the system is operating at a typical primary energy of between 20 and 35 keV, chromatic aberration, which is proportional to $\Delta E/E$, is small and is further controlled by using a beam convergence angle α_i below 1 mrad in size. The wavelength λ of the ion is of the order of picometers, so diffraction is not important, and spherical aberration, which is proportional to α_i^3, is negligible because of the small convergence angle. Consequently the ion-optical performance is dominated by the demagnified source size, and (sub-)nanometer probes containing picoAmpères of current can be produced. Figure 16.11 shows the optimum probe size that can be achieved for typical operating parameters of a HIM instrument.

16.3.3 Ion Optical Column

The layout of the beam optics of a He ion microscope closely resembles that of a SEM and is shown in Fig. 16.12. The FIM image of the trimer is sent through a pair of lenses that project it directly onto the specimen. An image crossover occurs between the two lenses. The position of the crossover and a beam-limiting aperture are used to control the balance between the size of the spot that ultimately strikes the sample and the intensity of the spot. Beam stigmation, scanning motion, offset and angle are controlled by quadrupoles near lens 1 and octopoles near lens 2. This configuration can be used to acquire images of both the sample as well as the source. In the former case the beam is scanned by the octopoles near lens 2, in the latter the

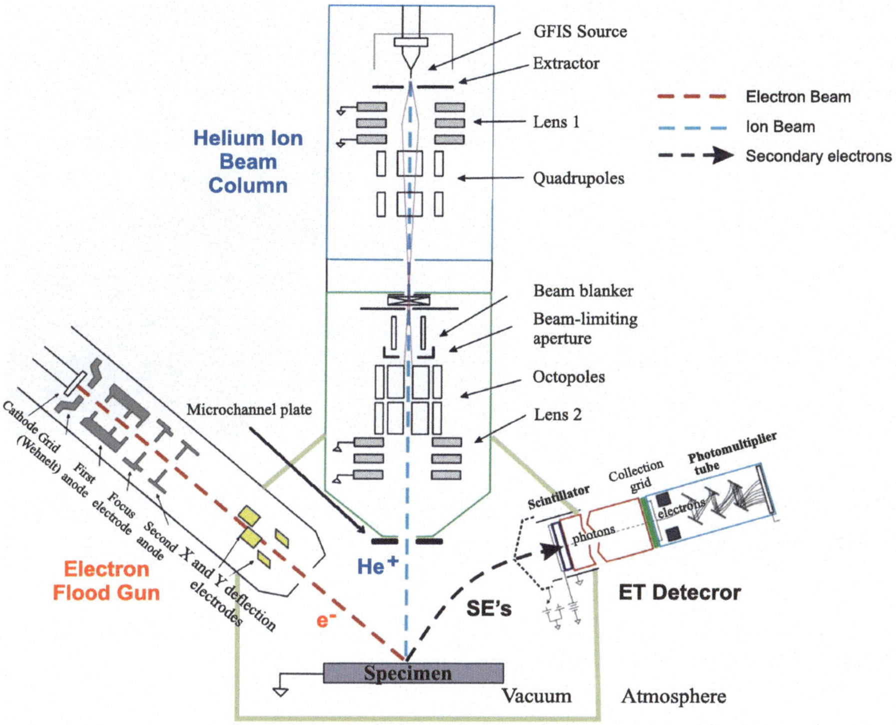

Fig. 16.12 Layout of the ion-optical column of a He ion microscope. More details on the optical performance of the HIM are presented in [74]. Figure courtesy of D. Pickard (National University of Singapore)

quadrupole at lens 1 is used to scan the FIM image of the trimer over the beam-limiting aperture to form an image like that shown in Fig. 16.10.

16.3.4 Image Resolution

The ultimate resolution that can be achieved in a HIM is achieved in SE mode and is sample dependent. Edge resolutions of 0.30 nm are routinely achieved using a 25–75 % edge criterium on a graphite sample, illustrating the stability and contrast potential of the technique [74]. Figure 16.13 shows a SE image of an asbestos fiber that was used to qualify the resolution of a HIM instrument [79]. The measured 25–75 % edge resolution is 0.24 nm. The smallest surface feature, as opposed to edge feature, that can be resolved depends on the strength of the ion-sample interaction. To illustrate the performance of a HIM instrument, we show a SE image in Fig. 16.14 of a flat bovine liver catalyse crystal [80] lattice with a periodicity of ~7 nm [81] that was resolved in a UHV environment. The visualization of individual enzymes

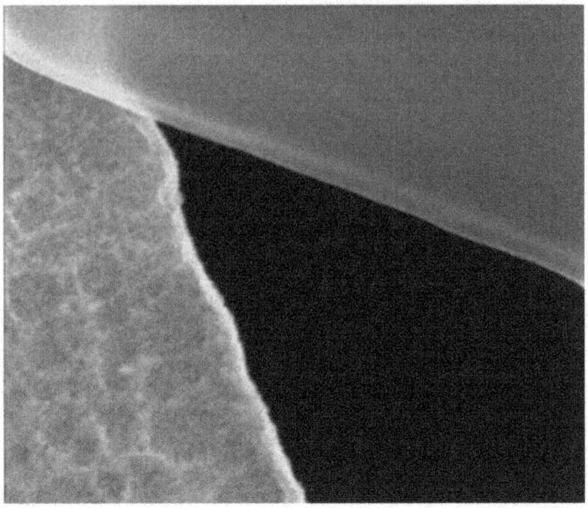

Fig. 16.13 200 nm FOV SE image of an asbestos fiber on a holey carbon film. The edge resolution that is measured on the fiber is 0.24 nm. Figure courtesy of Carl Zeiss NTS [79]

in these HIM images indicates a resolution in the low nm regime for the imaging of features on a flat surface.

16.3.5 Vacuum Performance

Before we focus on several applications of the HIM technique, we devote a final word to vacuum requirements for HIM. Figure 16.15 illustrates the influence of the beam on a pristine polycrystalline gold film under typical conditions used for imaging. The images show a gradual reduction in secondary yield and a corresponding

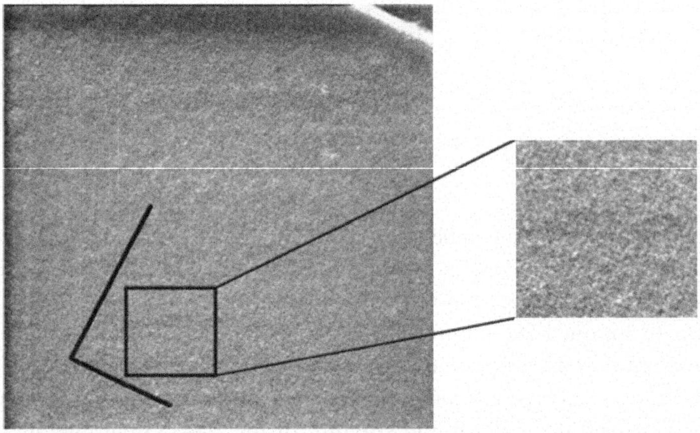

Fig. 16.14 1 μm FOV SE image of a catalyze crystal. The square symmetry of the crystal lattice with a periodicity of ~7 nm is directly visualized by the HIM. Image was recorded with the University of Twente UHV HIM

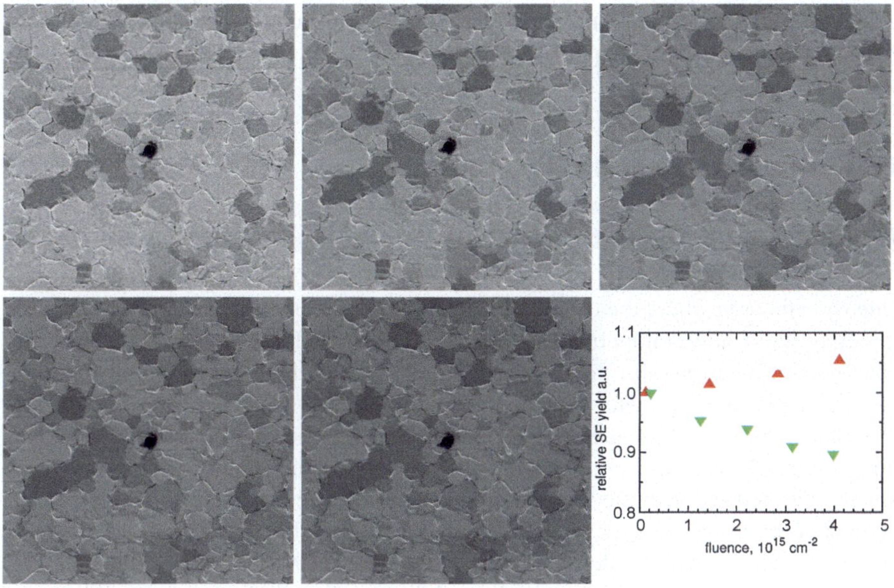

Fig. 16.15 20 μm FOV SE images of polycrystalline gold films for 1 (*top left*), 12, 23, 34, and 46 (*lower center*) exposures of 6.2×10^{12} ions cm^{-2}, respectively, recorded at a background pressure of 2×10^{-6} Torr. The image shows clear grain contrast due to differences in channeling. As the sample exposure to the He beam is increased, the image contrast decreases on an absolute scale, visualized by a darker appearance of the images. The graph shows the change in SE yield (*green*) as a function of ion fluence. For comparison, the SE yield that was recorded in an UHV environment is shown in *red*. The increase in intensity under UHV conditions is the result of sputtering/roughening of the Au surface with increased fluence

reduction in contrast. Next to the intrinsic SE yield variations due to differences in channeling for the differently oriented grains, the contrast in SE images can be primarily interpreted in terms of differences in work function. We assign the loss of signal after repeatedly imaging the same area to the buildup of a carbon layer that is formed as the impinging beam cracks residual hydrocarbons that are present in the vacuum of the instrument [10]. The buildup of these deposits was a severe problem in several of the earlier instruments and is an intrinsic problem for samples where the interaction with the beam leads to localized degassing.[1] To mitigate these effects, a dedicated UHV version of the commercially available Carl Zeiss Orion was designed and installed at the University of Twente, resulting in the enhanced instrument performance that is illustrated by the red data points in Fig. 16.15 [82]. The increased SE yield is attributed to a higher surface roughness, and correspondingly lower work function, as a result of sputtering. Several of the design features of

[1] We attribute the large variation in measured SE yields for a major part to differences in the surface cleanliness, sputtering and roughening in these experiments.

that system have now been incorporated in the Orion Plus system marketed by Carl Zeiss, further improving its performance.

16.4 Applications of the Technique

16.4.1 Imaging of Non-conductive Samples

One specific area where the use of a He ion beam offers significant added value is in the imaging of non-conductive specimen. Figure 16.1 demonstrates the efficacy of charge neutralization with the flood gun: the HIM clearly resolves the morphology of electrically floating metal posts in an isolating matrix [5]. There are fundamental differences between how electron and ion beams interact with matter when examining materials such as polymer resists, oxides, and other non-conductors. A negatively charged incident electron beam can charge materials either positive or negative depending on the energy of the beam relative to the "crossover" energies E_1 and E_2 at which each incident electron on average results in one emitted electron, thereby ensuring a state of dynamic charge balance. E_2 energies are typically around 1 keV and so the majority of SEM investigations on non-conductive samples are targeted to this energy range in order to minimize charge induced artifacts in the image [54]. The only alternative for imaging would be to apply a metallic coating to the sample. For an incident beam of positive ions the combined effect of the high SE yield and the injection of the positively charged incident ions always results in the surface of a non-conductor charging positive. At all available ion energies there is therefore the opportunity to minimize charging through the use of a low energy electron flood gun, shown in Fig. 16.12. The ion energy can always be chosen to optimize signal and contrast in the images, whilst maintaining a neutral charge on the sample. This makes HIM especially well suited for the imaging of soft, non-conductive matter [83]. This capability, as well as the large depth of field that is achieved with the GFIS source, is illustrated in Fig. 16.16 where a human pancreatic cell is imaged.

16.4.2 Critical Dimension Metrology on EUV Resist with SEM and HIM

Extreme ultraviolet (EUV) chemically amplified resist (CAR) is difficult to image in SEM due to surface charging, significant shrinkage during imaging, and beam-induced contamination. Nevertheless, the critical dimension (CD) metrology accuracy and precision requirements for future nodes on the ITRS roadmap [84] impose significant imaging challenges on scanning probe microscopes. Optical CD and CD-SEM instruments are today's tools for metrology, but they have systematic inaccuracies [85, 86]. For this reason, TNO and ASML have investigated the systematic

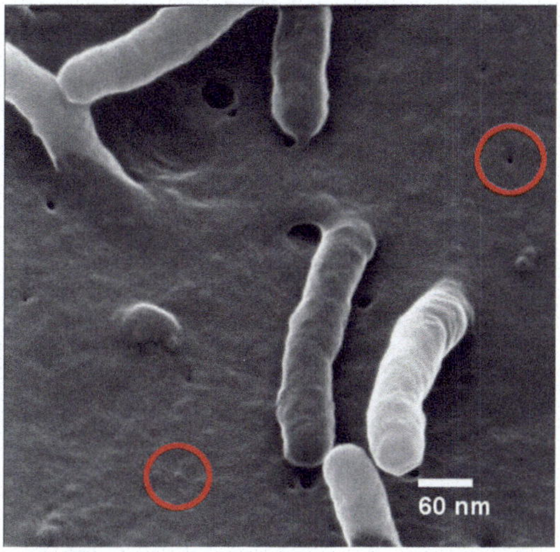

Fig. 16.16 600 nm FOV SE image of a human pancreatic cell. The image was recorded without a conductive coating. The large depth of field ensures that all features within the field of view are in focus. The surface detail is sufficient to see evidence of membrane processes, indicated by the *red circles*. Figure courtesy of P. Walther (University of Ulm)

errors of HIM and SEM [87]. In particular, we acquired SEM and HIM images of a dense line pattern at 27 nm half-pitch in EUV resist, written with ASML's EUV lithography demo tool. Figure 16.17 shows examples of raw SEM and HIM images. The SEM image and the left HIM image are recorded with 63 primary particles per pixel of 1 nm^2. Clearly, the HIM image provides a better signal-to-noise ratio. This improvement is due to the 2 to 3 times larger SE yield for helium ions than for electrons. The rightmost HIM image in Fig. 16.17 was recorded with 25 ions per pixel,

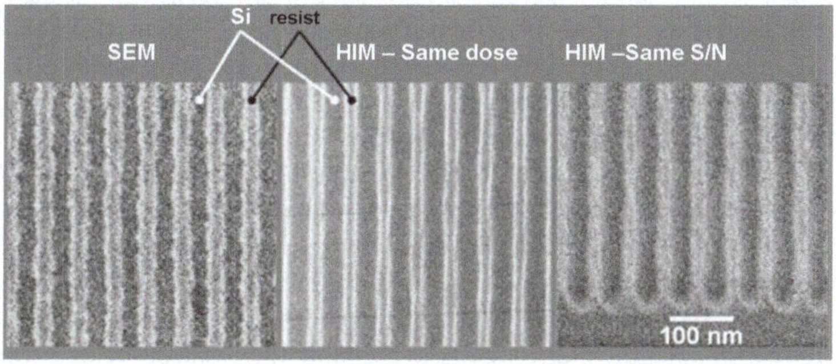

Fig. 16.17 Images of dense lines and spaces at 27 nm half pitch in EUV CAR resist. *Left panel*: Hitachi CG4000 CD-SEM image as used by ASML for metrology (500 eV beam energy and 5 pA beam current). *Center panel*: HIM image recorded at the same pixel size and dose as the CD-SEM image (25 keV and 0.5 pA). Owing to the 2 to 3 times larger SE yield per primary particle, the signal-to-noise ratio (S/N) of this image is significantly higher. *Right panel*: HIM image recorded at same pixel size at three times lower dose. This image has a comparable signal-to-noise ratio as the CD-SEM image. Nevertheless it shows significantly thinner and smoother resist lines, whereas the full pitch is the same for all images [87]. Specimen and SEM images courtesy of ASML. Reproduced with permission from MRS Proceedings **1354** (2011), ii03-03 [77]. Copyright 2011 Cambridge University Press

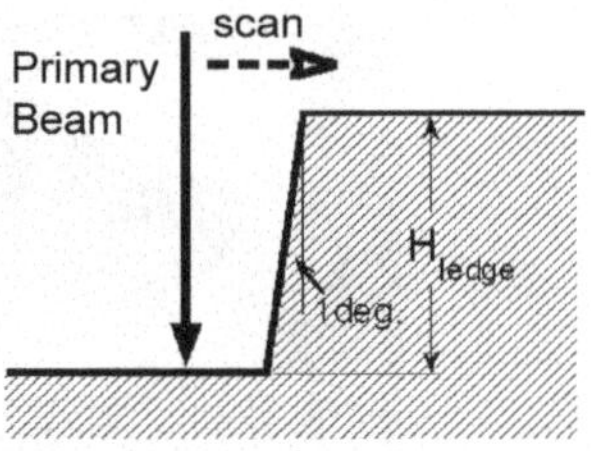

Fig. 16.18 Model of the step structure used for the MC simulation on SE contrast in SEM, Ga-FIB and HIM. Simulations are performed for a step height H_{ledge} of 10 nm and 1 μm. The primary beams are scanned on the surfaces and the line profile of the SE intensity is calculated. Reproduced with permission from J. Electron Microsc. 53 (2004), 229 [88]. Copyright 2004 Japanese Society of Microscopy

corresponding to an areal dose of 1×10^{16} ions/cm^2. As a result, the signal-to-noise ratio is more similar to that of the SEM image. Remarkably, the HIM images reveal a significantly lower resist line width (20, respectively 27 nm) than CD-SEM. The line-edge roughness (LER) is also lower in the HIM. Repeatedly imaging the same area did not change the LER, thus the observed difference is not due to resist edge smoothening by ion sputtering. Despite the significant differences in the measured CD and LER, it cannot be decided from this experiment which metrology tool is the more accurate.

A few years before these experiments could be performed, Ohya et al. modeled the topographic contrast in the SE signal in SEM, Ga-FIB and HIM [88, 89] using Monte Carlo simulations. Figure 16.18 shows the topography of one of the structures they used: a step in the substrate with a wall angle of 1° and a height of 10 nm or 1 μm. The MC simulations were used to create pseudo-SE images of the topography of the Si target, see Fig. 16.19. When compared to the SEM and FIB SE signals, the HIM SE signal peak at the step is the highest, while the peak width is the smallest for both step heights. This supports the experimental observation of Fig. 16.17 that a line edge is more clearly distinguished by helium ions.

The experimental SE images in Fig. 16.17 show that CD metrology with the HIM is at least as precise as CD-SEM. However, additional benefits are expected. Firstly, the different interaction mechanism allows for higher resolution imaging. Second, the lower interaction volume causes smaller systematic errors in CD measurements, resulting in improved topology contrast. A potential limitation is that the areal dose that is required for accurate metrology is so high that the Si substrate deforms during the recording of the image, thus inhibiting accurate metrology [61].

16.4.3 Imaging of Self-assembled Monolayers

To further illustrate the imaging capabilities of HIM, we demonstrate its application to patterns of self assembled monolayers (SAM). The SAMs were created by gas phase silanization through a PDMS stamp. Two types of molecules were

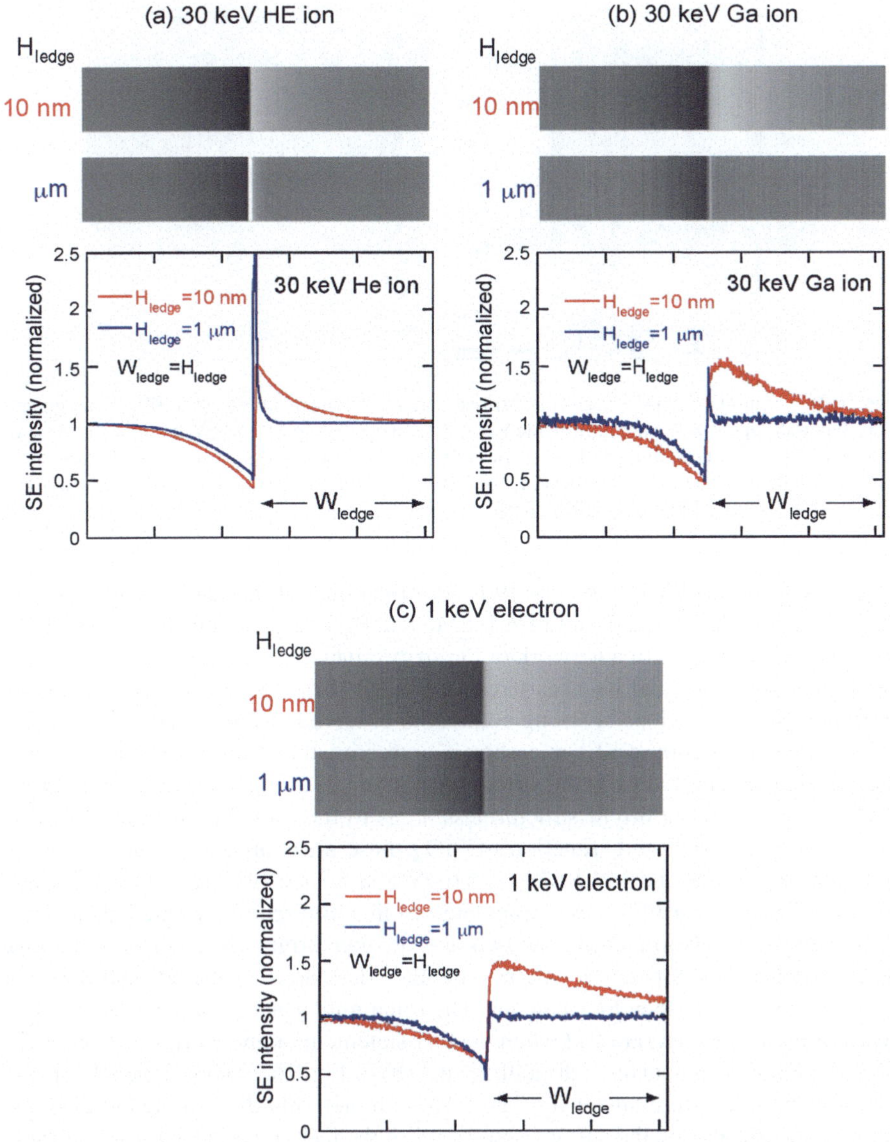

Fig. 16.19 Pseudo-SE images as computed from the MC calculations of the SE signal of a charged particle probe (having zero size) that is scanned over a step height of 10 nm (*red lines*) and 1 μm (*blue lines*) in a Si substrate. The contrast of the pseudo-SE images saturates at an intensity of two. Figure courtesy of K. Ohya [89]

used to create 4 μm wide orthogonal stripes. The molecules that were used were (3-Mercaptopropyl) trimethoxysilane (MS:$C_6H_{16}O_3SSi$) and Triethoxy1H,1H,2H, 2Htridecafluoron-octylsilane (PFS: $C_{14}H_{19}F_{13}O_3Si$). They were applied on the

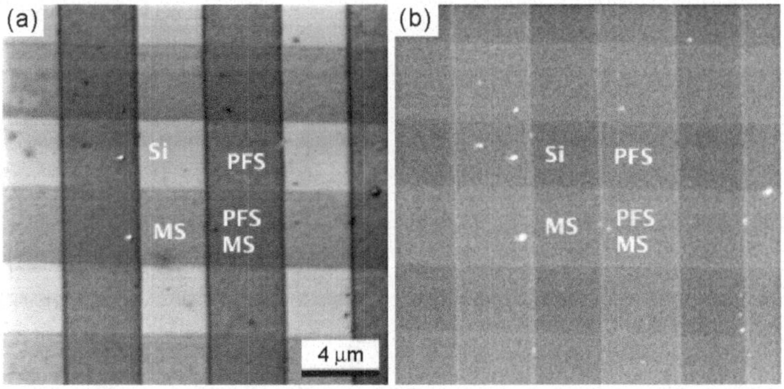

Fig. 16.20 20 µm FoV HIM images of a striped pattern of organic layers on Si(001). The images were recorded under normal beam incidence at a primary energy of 15 keV with an ion dose of 2.46×10^{16} cm^{-2}. (**a**) ET image of the stripes of PFS (vertical) and MS (horizontal). The different areas and their termination are indicated. (**b**) RBI image recorded simultaneously with (**a**). The different SAMs can still be distinguished [90]

native oxide of a Si(001) substrate [90]. The thickness of the layers is equal to the length of the molecules, 7 Å and 11 Å for MS and PFS, respectively. In Fig. 16.20 we show two HIM images of a network of the striped pattern, acquired using secondary electrons (panel (a)) and backscattered He (panel (b)). Because the SEs originate from near surface regions, a strong contrast is observed in Fig. 16.20(a). The characteristic escape depth of SEs in carbon is approximately 1 nm [48], implying that the SEs that are emitted from the stripes have interacted almost exclusively with the SAMs. We assign the bright structureless areas to the SiO_2/Si substrate. Because of the relatively low work function of SiO_2 these areas appear bright. The work function of the PFS and MS SAMs is 6.6 eV and 5.3 eV [91], respectively, where the work function for PFS was extrapolated from a different fluorinated alkanethiol [92]. This allows us to identify the two intermediate bright areas below and above the Si patches as MS patches. The intermediate dark areas to the left and the right of the Si patches are then PFS patches. The remaining square is a patch where both types of molecules overlap and where a clear statement on the work function or expected contrast mechanism is difficult. Panel (b) of Fig. 16.20 shows the RBI image that was recorded simultaneously. The SAMs are not only discernible but also discriminable, illustrating the surface sensitivity of the technique in RBI mode as well. In addition, details like increased brightness at the edges of the stripes, resulting from the silanization process are clearly visible.

Although the contrast in the RBI image is not expected, it can be explained on the basis of ion channeling mechanisms that occur in the crystalline substrate. Different to the ET images, where SEs are generated in near surface regions, the backscattering of helium as recorded in MCP images is a bulk effect, see e.g. Chap. 15 and [52, 56, 76]. For a layer of heavy elements on a lighter substrate one expects an increased backscatter yield because the heavier element has a higher cross section and will therefore add to the backscatter yield. The adlayer also decreases the energy

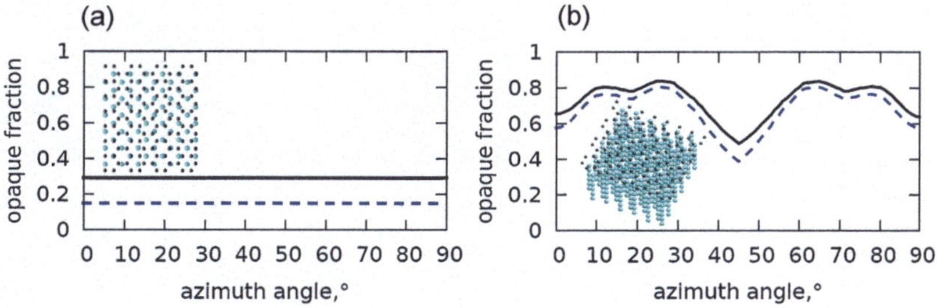

Fig. 16.21 Simulation of dechanneling contrast for clean and carbon covered Si. The graphs show the opaque fraction of the projected crystal lattice. *Blue dashed lines* are obtained for a clean Si(001) crystal *while* the *black lines* are obtained with a thin carbon layer added. (**a**) Normal incidence. The opaque projected area fraction is 15 % and 29 % for the clean and carbon covered surface, respectively. There is no azimuthal dependence for this angle of incidence. (**b**) The same calculation but for a 10° sample tilt. The average opaque projected area fractions are 68 % and 73 % for the clean and carbon covered surface, respectively. A clear dependence on the angle of incidence exists [93]

of the primary beam thus increasing the backscatter yield and reducing the range at the expense of a slightly increased lateral range in the substrate volume covered by the heavier adlayer. For the present case where a light carbon adlayer covers a heavier silicon substrate neither argument plays a role and SRIM calculations in fact indicate that no such contrast should be present in the RBI images, purely because of the small volume and weak interaction of the helium ions with the overlayers.

The absence of contrast in the RBI images that SRIM predicts is however a direct consequence of the fact that it does not take into account the crystalline structure of the specimen. The polar angle of the incident He beam is critical for the contrast in RBI images. In Fig. 16.21 we show the result of calculations of the opaque area fraction for a Si(001) crystal. The graphs show the opaque fraction of the crystal, which is proportional to the backscatter yield. For normal beam incidence plotted in Fig. 16.21(a), 15 % of the area is blocked (blue dashed line). Adding a single carbon overlayer increases the opaque fraction to 29 % (dark solid line). At normal incidence, this corresponds to an increase in the blocked fraction by 66 %, independent of the azimuth angle. Tilting the incident beam with respect to the surface normal increases the overall back scatter probability, but reduces the expected contrast ratio between a clean Si crystal and one that is covered by a single adlayer. Because the images in Fig. 16.20 are recorded under normal beam incidence along a channeling direction, we take full advantage of this dechanneling contrast in RBI images [93].

16.4.4 TEM Lamella Preparation by DW Helium Ion Beam Milling

Because of the low mass of the helium ion, sputtering by helium milling is relatively slow. Depending on the material, beam energy and local angle of incidence,

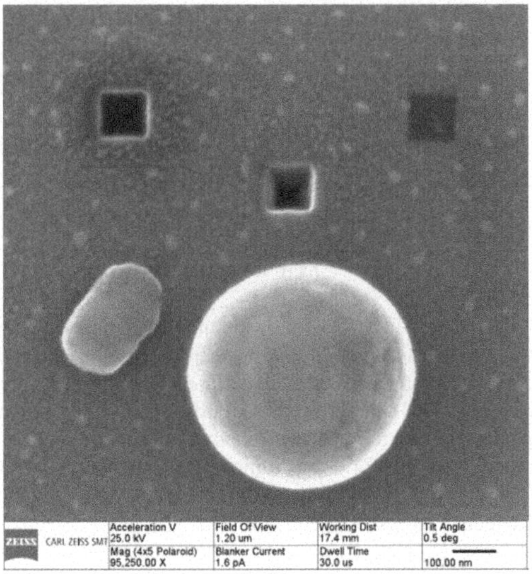

Fig. 16.22 Results of DW helium cutting a 300 nm thick Si lamella at room temperature (*lower two bubbles*) and at 700 °C (*upper three square holes*) [95]. Figure courtesy of M. Rudneva (Delft University of Technology)

the milling rate varies between 0.01 and 1 atom per ion [94]. Only at near-grazing angles of incidence can the rate exceed unity. A nice example of the cutting of materials with nanometer precision is the preparation of ultrathin TEM lamellae. Figure 16.22 shows how direct write (DW) sputtering with helium ions on Si produces either bubbles (lower half) or well-defined squares, depending on the substrate temperature [77, 95]. Previous studies show that if the dose is high enough (1.6×10^{17} ions/cm^2), Si within the ion propagation path becomes completely amorphous and small nanobubbles are formed in the amorphous region. With increased dose both the density and the diameter of the bubbles increase. Due to the poor solubility of He in amorphous Si at room temperature, the implanted He is all trapped subsurface and causes a deformation of the amorphous surface layer into a balloon-like structure [61, 96]. At an elevated temperature of 700 °C, the solubility and mobility of He in Si is increased sufficiently to avoid the nanobubble formation [97]. As a result, the shape of the holes in the Si are very similar to the design pattern in size and shape, with hardly any visible damage in the surrounding Si. Apparently, milling with the He^+ beam does not cause lateral damage or, if damage does occur, the crystal lattice is restored during or after the irradiation.

Figure 16.23 shows HIM and TEM images of $Cu_xBi_2Se_3$, a high-temperature superconducting material. The TEM lamella was created by mechanical crushing of bulk $Cu_xBi_2Se_3$ material. The fragments were mixed in a liquid and dropped on a holey carbon film that was supported by a golf grid. The sample was loaded in the HIM and was selected a suitably-sized fragment for thinning by sputtering with helium. To unravel the charge transport mechanism, knowledge of the crystal structure of $Cu_xBi_2Se_3$ is required. TEM images may resolve this structure, yet require thin samples (lamellas) that are oriented along the two symmetry planes. However, production of these lamellas is not easy, since $Cu_xBi_2Se_3$ is brittle along [001] planes and difficult to cleave in the [hk0] plane. We have thinned a flake

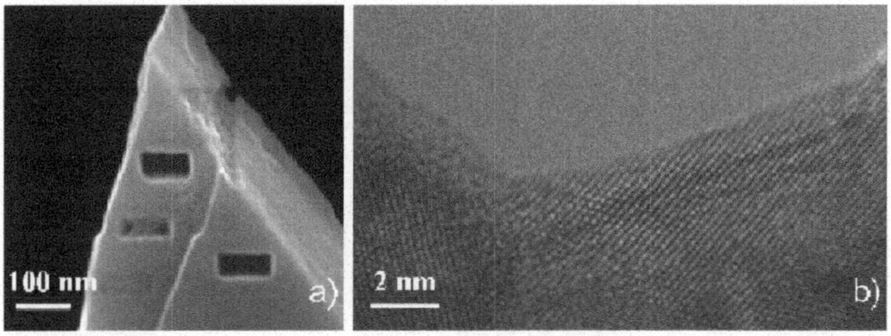

Fig. 16.23 Example of direct write milling using helium ions in $Cu_xBi_2Se_3$ crystal by HIM and TEM images. The $Cu_xBi_2Se_3$ is crystalline up to the edge, indicating that little damage is inflicted during the sample thinning with the He ion beam. Reproduced with permission from MRS Proceedings **1354** (2011), ii03-03 [77]. Copyright 2011 Cambridge University Press

of $Cu_xBi_2Se_3$ by locally irradiating the sample with helium ions. The HIM image in the left panel of Fig. 16.23 shows the sputtering effect on the flake for four different ion doses. A slight incision is made at the lowest dose. At a slightly increased helium dose a partial cut through the flake occurs, while at the two highest doses a complete hole was drilled through the target. The structure and properties of $Cu_xBi_2Se_3$ as function of the copper concentration are investigated in Ref. [98]. Our High Resolution TEM image shows a hexagonal pattern up to the HIM-made edge of the sample, as can be seen in the right panel of Fig. 16.23. The HRTEM image shows that the lamella is crystalline to the edge, indicating a very gentle milling in the HIM, possibly combined with some re-crystallization after or during the milling. Alternative TEM lamella preparation methods like gallium ion milling and ultra-microtomy showed many surface deterioration related artifacts and thus HIM offers a very useful complementary addition to existing sample preparation possibilities.

As a final illustration of DW milling, we show 100×100 nm^2 through holes that were created in silicon lamellae at room temperature. Figure 16.24 shows HIM images and electron diffraction patterns from the illuminated area for 100 nm and 300 nm thick samples. For 100 nm Si samples no damage is observed around the cut while for Si samples with a thickness of 300 nm big bubble-like structures were easily created in the area of modification. It is evident from diffraction patterns obtained in TEM that in the case of room temperature milling, some lattice deformation is observed for 300 nm thick samples. As a result, no cutting of thick Si samples is possible at room temperature. To prevent the bubble formation, out diffusion of He from the sample should be enhanced. This can be done by, for example, increasing the temperature of the sample during the cutting process [95].

Other recent experiments have shown that DW milling with helium ions can produce exceptionally narrow and high-aspect ratio cuts and vias [99, 100]. High-precision milling of graphene has been demonstrated by Bell et al. [63] and Pickard et al. [64].

Fig. 16.24 Result of HIM DW cutting of 100 nm (*left panel*) and 300 nm (*right panel*) thick Si lamellas at room temperature. For the 100 nm thick lamella, hardly any damage is observed and square holes with dimension close to the design pattern are obtained. Severe damage is inflicted in the 300 nm thick lamella, since the final stopping of 25 keV helium ions in Si occurs typically at a depth of 150–300 nm [95]. Figures courtesy of M. Rudneva (Delft University of Technology)

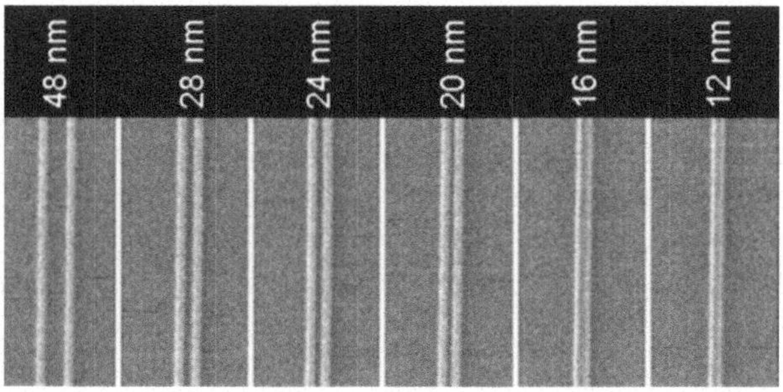

Fig. 16.25 Pt line pairs (*numbers* above the line pairs are the nominal pitch) produced with He-IBID for 4.8 nC/cm line dose. For these 13 nm wide lines this corresponds to 6×10^{16} ions/cm^2. Reproduced with permission from Microscopy Today **5** (2011), 22 [107]. Copyright 2011 Microscopy Society of America

16.4.5 Helium Ion Beam Induced Processing (He-IBIP)

The efficacy of nanofabrication with helium ions can be strongly enhanced when a continuously flowing precursor gas is introduced into the vacuum. Activation of the precursor molecules on the substrate by an incident helium ion triggers local chemistry for either additive (by beam-driven precursor deposition) or subtractive (by beam-catalyzed etching) materials processing. The enhanced fabrication efficacy of beam induced chemistry mitigates the major drawbacks of DW: the low speed and the required high dose, which may cause too much collateral damage. Circuit edit and mask repair are potential industrial applications that require well-controlled deposition of free-form nanostructures [101–103].

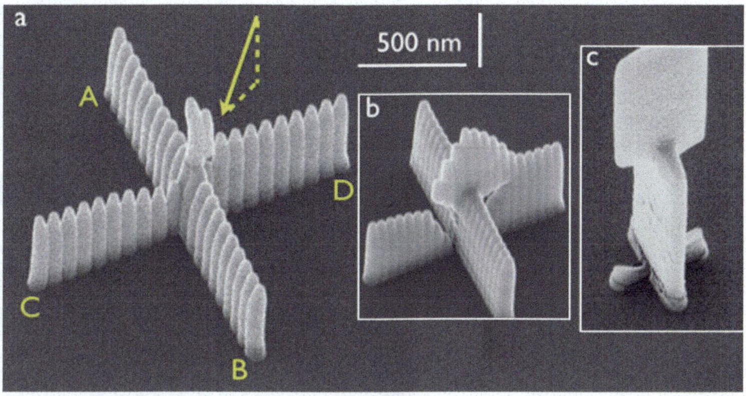

Fig. 16.26 Crosses, from two successively grown rows of 21 pillars at 75, 50 and 20 nm pitch. The AB rows were grown first. The *arrow* indicates the direction of the flux of MeC_pPtMe_3 precursor molecules from the gas injection needle. The beam energy was 25 keV, the current 0.2 pA, and the dwell time 20 μs. Reproduced with permission from Scanning **34**, 90 (2012) [5]. Copyright 2012 Wiley

Since the introduction of HIM, significant experimental progress has been made in helium ion beam induced deposition (He-IBID) [68, 69, 104]. Furthermore, several modeling or simulations studies have appeared [69, 105, 106]. Here we present some examples from experiments performed at the HIM in Delft by Scipioni et al., Alkemade et al., and Drezner et al. [5, 107, 108]. Using a Design Of Experiment [109] approach, Scipioni et al., investigated the influence of several experimental factors on the deposit size [107]. He-IBID process development is complex: a standard beam-induced deposition process using the system described below calls for the control of over ten different parameters. Several of these are used to set the flow of the reactant, several are required for defining the particle beam, and yet another set of parameters call out the routine by which the beam is scanned over the pattern of interest. The wide scatter of reported EBID and IBID outcomes in the review article by Utke indicates the complexity of the problem [110]. Scipioni et al. aimed for finding the minimum obtainable line width and gap width between line pairs of deposited platinum lines and to give a predictive formulation of the same. Figure 16.25 shows the result for the optimized deposition process, demonstrating a minimum gap width of 6 nm between two 13 nm wide Pt He-IBID lines. Remarkably, the deposited lines are well separated over their full length of 1 μm, even at the 6 nm gap width, demonstrating the strong ability of the helium ion microscope to produce fine features very reproducibly.

For optimal deposit attributes a well-controlled balance between the ion flux and the precursor molecule supply is pivotal. Reproducible results are achieved when either the depletion of the precursor molecules is mostly avoided (e.g. by fast scanning the ion beam many times over the same area) or completed (e.g. by using high ion beam currents and/or low precursor gas pressure). Alkemade et al. performed

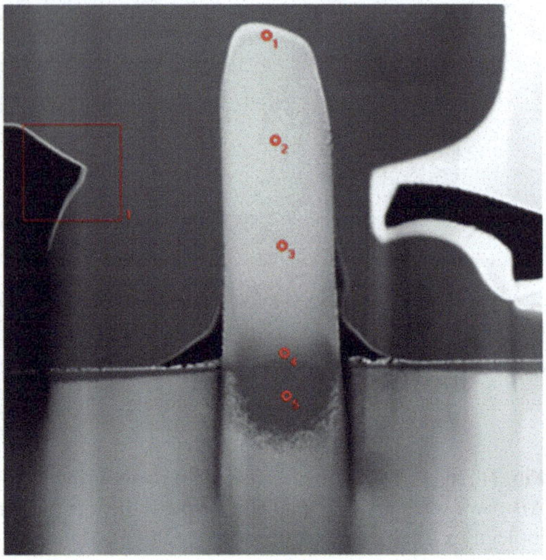

Fig. 16.27 *Left panel*: TEM cross section of a Pt He-IBID grown pillar in TNO's HIM. *Right panel*: the composition of the pillar as measured using EDS. The maximum Pt content amounts to 41 %, the highest measured metal content in a He-IBID grown structure. Figure courtesy of Intel [108]

a He-IBID experiment that nicely illustrates how a small change in the pitch between consecutively grown pillars in a pattern of two crossing lines causes drastic changes in the shape of the final deposit, see Fig. 16.26. The scan pattern was as follows: first 21 pillars were grown between points A and B, then 20 pillars between C and D, see Fig. 16.26(a). The pattern was repeated many times, resulting in an ion dose per pillar of 6 pC. The effect of the pillars along the line AB on the local precursor supply during the pillar growth along CD is different at each pitch. At 75 nm pitch, the pillars along AB form a semi-open fence. The direction of the precursor gas flow is indicated by the arrow in Fig. 16.26(a). Due to the fence, the pillar growth along the line CD is disturbed. The gas supply is enhanced before the fence, raising the pillars towards the crossing, and suppressed after the fence, reducing the pillar height. At 50 and 20 nm pitches, the fence along AB is increasingly closed, thus altering the gas supply during the deposition along the line CD. The shape of the resulting final deposits are shown in Fig. 16.26(b) and (c). Noteworthy is also the absence of overspray at the foot of the pillars. This is due to the very localized interactions of the helium beam at the sample surface.

The He-IBID cross experiments demonstrate the current precision with which extended 3D structures can be grown. It is comparable to the precision for EBID, but much better than for Ga-IBID [110]. Since He^+ ion beam milling is relatively weak, we expect that complicated structures can be grown more easily by He-IBID than by Ga-IBID.

Next to measuring the morphology of the deposit, determining the composition of the deposit is important for optimizing the deposition process. In the first He-IBID experiment [68], a maximum purity of about 20 % was demonstrated. Recently, Drezner et al. grew large Pt-rich pillars at a much lower beam current, scanning the beam much faster while the dose was applied by many repetitive exposures [108]. As a consequence, the precursor layer was much less depleted by the primary ions, thereby increasing the yield to 0.3–0.9 $\mu m^3/nC$, which is similar to the yield reported for conventional Ga based FIBs [110]. During the He-IBID deposition, the vacuum chamber background pressure rose from 4×10^{-7} to 7×10^{-6} Torr. Deposition time was 1000 s, applied to a $200 \times 200\ nm^2$ drawn box. A Raith Elphy Plus™ pattern generator was used to control beam steering, blanking and the scanning parameters: 2 μs dwell time and 4 nm pixel spacing in both x and y directions. The deposition consisted of a total number of 200,000 scans. The areal dose is 1.6×10^{18} ions/cm^2 for 0.1 pA ion current. Figure 16.27 shows a TEM image of a cross-section of a Pt He-IBID grown pillar. The sample was covered by 0.5 μm FIB tungsten deposition prior to the TEM lamella preparation by an FEI Strata 400 dualbeam FIB. Damage to the Si by such a high dose is visible under the pillar. The silicon is amorphized and the original flat surface is deformed. Energy dispersive spectroscopy (EDS) analysis was performed in the TEM, using an EDAX Genesis EDS system, on 5 points spanning the pillar height, where one point was measured below the silicon surface. The maximum measured Pt content amounts to 41 %, which is to date the highest measured metal content of a He-IBID-grown structure. Such high metal content of a deposit will very likely result in better material properties such as metal resistivity. The benefits and limitations of nanofabrication using He-IBID and helium ion beam lithography are discussed in-depth in two recent book chapters [111, 112].

16.5 Summary

Helium ion microscopy is a novel high-resolution imaging and nanofabrication technique with several rather remarkable capabilities. The unique interaction of the helium ions with the sample can be applied for imaging, patterning and modification of the target, and thus adds to the century-old field of ultra-microscopy.

For imaging, HIM unlocks a new application space, especially for imaging sensitive (non-conductive) surfaces with sub-nm resolution. While generating many secondary particles that can be used for imaging, the interaction of the primary beam with the sample can cause damage to the target. This is not unlike many other scanning microscopes that have a nanometer resolving power. Fortunately, the sample damage often occurs at much higher doses than those that are required for imaging, lithography and beam chemistry. High-resolution imagery with chemical contrast can be routinely produced with secondary electrons and backscattered helium.

The helium ion microscope has a strong ability to reproducibly fabricate fine and dense features. The low backscattering cross-section of the primary particles enables ultra-fine and -dense patterning using beam induced processes, and in resist. Further optimization of SHIBL and He-IBIP involves the consideration, evaluation and control over many experimental factors to reproducibly fabricate nanostructures with the correct desired properties, such as e.g. critical dimension, resistivity and compositional purity. Last but not least, the resolution of nanopatterning with

SHIBL and He-IBIP is not limited by the size of the ion probe but by the intrinsic length scale of the interactions of the ions and the secondary electrons with the resist, the precursor molecules and the substrate.

Helium ion microscopy is now a commercially available technique. The helium ion microscope has developed into a practical instrument that reproducibly yields new results. At present the largest challenge for scientists, engineers and marketers is in identifying useful applications that could not be achieved in the past, when HIM did not exist yet. A similar challenge will be raised once the neon ion microscope, currently under development [74], is released to the market. It is to be expected that the combination of a bright focused probe and the specifics of the interaction of neon ions with samples will once again open up a new and unique range of applications for nano-imaging and -fabrication [113].

Acknowledgements The significant investment by the Dutch NanoNed program and the Dutch Foundation for Technical Sciences (STW) is to be acknowledged for the HIM facilities at TNO and the University of Twente, respectively. ASML has financially supported the CD metrology benchmark of SEM with HIM. The HREM application is supported by NIMIC. Carl Zeiss NTS has financially and scientifically (Larry Scipioni, Bill Thompson, Colin Sanford, Louise Barriss, Lewis Stern and David Ferranti) supported the UHV instrument development program and the He-IBID experiments. Diederik Maas is grateful to Susan Ketelaars for her patience and wisdom, and acknowledges Emile van Veldhoven, Hans van der Veer, David Nijkerk, Pieter van Beek, Larry Scipioni, John Notte IV, Bill Thompson, Lewis Stern, Sybren Sijbrandij, Emile van der Drift, Vadim Sidorkin, Ping Chen, Anja van Langen-Suurling, Hozan Miro, Emma Koster, Maria Rudneva, Henny Zandbergen and, last but certainly not least, Paul Alkemade for their pleasant and close collaboration in exploring HIM applications and instrumentation. Raoul van Gastel acknowledges Gregor Hlawacek, Vasilisa Veligura, Al Lysse, Clarke Fenner, and in particular Bene Poelsema for his scientific guidance and never-ending enthusiasm in sharing his long-standing experience in ion scattering.

References

1. A.H. Zewail, Micrographia of the twenty-first century: from camera obscura to 4D microscopy. Philos. Trans. R. Soc. A **368**, 1191–1204 (2010)
2. P.W. Hawkes, Aberration correction past and present. Philos. Trans. R. Soc. A **367**, 3637–3664 (2009)
3. L.W. Swanson, G.A. Schwind, Review of the ZrO/W Schottky cathode, in *Handbook of Charged Particle Optics*, 2nd edn., ed. by J. Orloff (CRC Press, Boca Raton, 2008). ISBN 9781420045543
4. F.P. Widdershoven et al., CMOS biosensor platform, in *Proc. of 2010 IEEE International Electron Device Meeting (IEDM)*, pp. 816–819
5. P.F.A. Alkemade, E.M. Koster, E. van Veldhoven, D.J. Maas, Imaging and nanofabrication with the helium ion microscope of the Van Leeuwenhoek Laboratory in Delft. Scanning **34**, 90–100 (2012)
6. L. de Broglie, Recherches sur la théorie des quanta, Thèse de doctorat soutenue à Paris le 25 novembre 1924, Annales de Physique (10e série) III (1925), 22. Reproduced in: L. de Broglie, *Recherches sur la théorie des quanta* (Fondation Louis de Broglie, Paris, 1992)
7. C. Davisson, L.H. Germer, Diffraction of electrons by a crystal of nickel. Phys. Rev. **30**, 705–740 (1927)

8. H. Busch, Über die Wirkungsweise der Konzentrierungsspule bei der Braunschen Röhre (On the mode of action of the concentrating coil in the Braun tube). Arch. Elektrotechnik **18**, 583–594 (1927)
9. M. Knoll, E. Ruska, Das elektronenmikroskop. Z. Phys. **78**, 318–339 (1931)
10. R.L. Stewart, Insulating films formed under electron and ion bombardment. Phys. Rev. **45**, 488–490 (1934)
11. R.P. Feynman, There's plenty of room at the bottom (data storage). Caltech Eng. Sci. **23**, 22–36 (1960), lecture in 1959, reprinted in J. Microelectromech. S. **1**, 60–66 (1992). See also: http://www.zyvex.com/nanotech/feynman.html
12. H.C. Pfeiffer, New prospects for electron beams as tools for semiconductor lithography. Proc. SPIE **7378**, 737802 (2009)
13. R.F.M. Thornley, M. Hatzakis, Electron optical fabrication of solid-state devices, in *Record of 9th Symp. Electron, Ion, and Laser Beam Technol*, vol. 94, ed. by L. Martin (1967). San Francisco Press
14. J.P. Ballantyne, Mask fabrication by electron-beam lithography, in *Electron-Beam Technology in Microelectronic Fabrication*. ed. by G.R. Brewer (Academic Press, San Diego, 1980), pp. 259–307
15. M. Knoll, Aufladepotentiel und Sekundäremission Elektronenbestrahlter Körper. Z. Tech. Phys. **16**, 467–475 (1935)
16. D. McMullan, An improved scanning electron microscope for opaque specimens. Proc. Inst. Electr. Eng. **100 Part II**, 245–259 (1953)
17. M. von Ardenne, Das Elektronen-Rastermikroskop, Praktische Ausführung. Z. Tech. Phys. **19**, 407–416 (1938)
18. V.A. Zworykin, J. Hillier, R.L. Snyder, A scanning electron microscope. ASTM Bull. **117**, 15–23 (1942)
19. A.E. Vladár, M.T. Postek, B. Ming, On the sub-nanometer resolution of scanning electron and helium ion microscopes. Micros. Today **3**, 6–13 (2009)
20. E.W. Müller, Das Feldionenmikroskop. Z. Phys. **131**, 136–142 (1951)
21. T.T. Tsong, E.W. Müller, The field ion microscopical image of an ordered platinum-cobalt alloy. Appl. Phys. Lett. **9**, 7–10 (1966)
22. E.W. Müller, J.A. Panitz, S.B. McLane, The atom-probe field ion microscope. Rev. Sci. Instrum. **39**, 83–86 (1968)
23. W. Telieps, E. Bauer, An analytical reflection and emission UHV surface electron microscope. Ultramicroscopy **17**, 57–66 (1985)
24. E. Bauer, Low energy electron reflection microscopy, in *Fifth Intern. Congress for Electron Microscopy*, vol. 1, ed. by S.S. Breese Jr. (Academic Press, New York, 1962), p. D-11
25. G. Binnig, H. Rohrer, Scanning tunneling microscopy—from birth to adolescence. Rev. Mod. Phys. **59**, 615–625 (1987)
26. J. Tersoff, D.R. Hamann, Theory of the scanning tunneling microscope. Phys. Rev. B **31**, 805–813 (1985)
27. B.C. Stipe, M. Rezaei, W. Ho, Single-molecule vibrational spectroscopy and microscopy. Science **280**, 1732–1735 (1998)
28. M.J. Rost, L. Crama, P. Schakel, E. van Tol, G.B.E.M. Van Velzen-Williams, C.F. Overgauw, H. ter Horst, H. Dekker, B. Okhuijsen, M. Seynen, A. Vijftigschild, P. Han, A.J. Katan, K. Schoots, R. Schumm, W. van Loo, T.H. Oosterkamp, J.W.M. Frenken, Scanning probe microscopes go video rate and beyond. Rev. Sci. Instrum. **76**, 053710 (2005)
29. R. Levi-Setti, G. Crow, Y.L. Wang, High spatial resolution SIMS with the UC-HRL scanning ion microprobe. J. Phys. C **9**, 197–205 (1984)
30. J.B. Jergenson, Liquid metal ion source, U.S. Patent #4318029, 1982
31. L.A. Giannuzzi, F.A. Stevie, *Introduction to Focused Ion Beams: Instrumentation, Theory, Techniques and Practice* (Springer, Berlin, 2004). ISBN 978-0-387-23116-7
32. R.J. Ross, C. Boit, D. Staab (eds.), *Microelectonics Failure Analysis Desk Reference* (2011). ISBN 9781615037254, ASMI

33. E.W. Ward, J. Notte, N.P. Economou, Helium ion microscope: a new tool for nanoscale microscopy and metrology. J. Vac. Sci. Technol. B **24**, 2871–2874 (2006)
34. E.W. Ward, Atomic level ion source and method of manufacture and operation, U.S. Patent #7368727, 2004
35. B. Ziaja, R.A. London, J. Hajdu, Ionization by impact electrons in solids: electron mean free path fitted over a wide energy range. J. Appl. Phys. **99**, 033514 (2006)
36. H. Salow, Sekundarelektronen-emission. Phys. Z. **41**, 434–436 (1940)
37. M.P. Seah, W.A. Dench, Quantitative electron spectroscopy of surfaces: a standard data base for electron inelastic mean free paths in solids. Surf. Interface Anal. **1**, 2 (1979)
38. N. Bohr, LX. On the decrease of velocity of swiftly moving electrified particles in passing through matter. Philos. Mag. **30**, 581–612 (1915)
39. J.F. Ziegler, J.P. Biersack, U. Littmark, *The Stopping and Range of Ions in Matter* (Pergamon, New York, 1985). ISBN 0080220533
40. http://www.srim.org/
41. Y. Wang, M. Nastasi (eds.), *Handbook of Modern Ion Beam Materials Analysis*, 2nd edn. (Cambridge University Press, Cambridge, 2010)
42. A. Nomura, S. Kiyono, Stopping powers of copper, silver and gold for protons and helium ions of low energy. J. Phys. D, Appl. Phys. **8**, 1551–1559 (1975)
43. D. Cohen-Tanugi, N. Yao, Superior imaging resolution in scanning helium-ion microscopy: a look at beam-sample interactions. J. Appl. Phys. **104**, 063504 (2008)
44. P. Varga, H. Winter, in *Particle Induced Electron Emission II* (Springer, Berlin, 1992), p. 149
45. J.C. Tucek, S.G. Walton, R.L. Champion, Secondary-electron and negative-ion emission from Al: effect of oxygen coverage. Phys. Rev. B **53**, 14127–14134 (1996)
46. K. Ohya, K. Inai, A. Nisawa, A. Itoh, Emission statistics of X-ray induced photoelectrons and its comparison with electron- and ion-induced electron emissions. Nucl. Instrum. Methods B **266**, 541–548 (2008)
47. Y.V. Petrov, O.F. Vyvenko, A.S. Bondarenko, Scanning helium ion microscope: distribution of secondary electrons and ion channeling. Surf. Investig.: X-Ray Synchrotron Neutron Tech. **4**, 792–795 (2010)
48. R. Ramachandra, B. Griffin, D. Joy, A model of secondary electron imaging in the helium ion scanning microscope. Ultramicroscopy **109**, 748–757 (2009)
49. T.E. Everhart, R.F.M. Thornley, Wide-band detector for micro-microampere low-energy electron currents. J. Sci. Instrum. **37**, 246–247 (1960)
50. V. Castaldo, J. Withagen, C. Hagen, P. Kruit, E. van Veldhoven, Angular dependence of the ion-induced secondary electron emission for He^+ and Ga^+ beams. Micros. Microanal. **17**, 624–636 (2011)
51. J.L. Wiza, Microchannel plate detectors. Nucl. Instrum. Methods **162**, 587–601 (1979)
52. S. Kostinski, N. Yao, Rutherford backscattering oscillation in scanning helium-ion microscopy. J. Appl. Phys. **109**, 064311 (2011)
53. V. Castaldo, C.W. Hagen, P. Kruit, E. van Veldhoven, D.J. Maas, On the influence of the sputtering in determining the resolution of a scanning ion microscope. J. Vac. Sci. Technol. B **27**(6), 3196–3202 (2009)
54. L. Reimer, *Scanning Electron Microscopy: Physics of Image Formation and Microanalysis*, 2nd edn. Springer Series in Optical Sciences, vol. 45 (Springer, Berlin, 1998)
55. M. Hellsing, L. Karlsson, H.-O. Andrè, H. Nordèn, Performance of a microchannel plate ion detector in the energy range 3–25 keV. J. Phys. E, Sci. Instrum. **18**, 920–925 (1985)
56. S. Sijbrandij, J. Notte, L. Scipioni, C. Huynh, C. Sanford, Analysis and metrology with a focused helium ion beam. J. Vac. Sci. Technol. B **28**, 73–77 (2010)
57. J.W. Rabalais, *Principles and Applications of Ion Scattering Spectroscopy* (Wiley-Interscience, New York, 2003)
58. T. Wirtz, N. Vanhove, L. Pillatsch, D. Dowsett, S. Sijbrandij, J. Notte, Towards secondary ion mass spectrometry on the helium ion microscope: an experimental and simulation based feasibility study with He^+ and Ne^+ bombardment. Appl. Phys. Lett. **101**, 041601 (2012). doi:10.1063/1.4739240

59. S. Ogawa, T. Iijima, S. Awata, S. Kakinuma, S. Komatani, T. Kanayama, Helium ion microscope characterization for Cu/low-k interconnects—SE imaging and focused helium ion beam luminescence detection, in *Proceedings of the Interconnect Technology Conference and 2011 Materials for Advanced Metallization (IITC/MAM)* (IEEE International, New York, 2011), pp. 1–3
60. J.H. Franken, M. Hoeijmakers, R. Lavrijsen, J.T. Kohlhepp, H.J.M. Swagten, B. Koopmans, E. van Veldhoven, D.J. Maas, Precise control of domain wall injection and pinning using helium and gallium focused ion beams. J. Appl. Phys. **109**, 07D504 (2011)
61. R. Livengood, S. Tan, Y. Greenzweig, J. Notte, S. McVey, Subsurface damage from helium ions as a function of dose, beam energy, and dose rate. J. Vac. Sci. Technol. B **27**, 3244–3249 (2009)
62. J.A. van Kan, A.A. Bettiol, Proton beam writing: a new 3D nanolithographic technique, in *Ion Beams in Nanoscience and Technology, Particle Acceleration and Detection*, ed. by R. Hellborg et al., pp. 297–310
63. D.C. Bell, M.C. Lemme, L.A. Stern, C.M. Marcus, Precision material modification and patterning with he ions. J. Vac. Sci. Technol. B **27**, 2755–2758 (2009)
64. D. Pickard, L. Scipioni, *Graphene Nano-ribbon Patterning in the Orion Plus*. Zeiss Application Note, October 2009
65. D.J. Maas et al., Nanofabrication with a helium ion microscope. Proc. SPIE **7638**, 763814 (2010)
66. V. Sidorkin, E. van Veldhoven, E. van der Drift, P. Alkemade, H. Salemink, D. Maas, Sub-10-nm nanolithography with a scanning helium beam. J. Vac. Sci. Technol. B **27**, L18–L21 (2009)
67. D. Winston, J. Ferrera, L. Battistella, A.E. Vladár, K.K. Berggren, Modeling the point-spread function in helium-ion lithography. Scanning **34**, 121–128 (2012)
68. C.A. Sanford, L. Stern, L. Barriss, L. Farkas, M. DiManna, R. Mello, D.J. Maas, P.F.A. Alkemade, Beam induced deposition of platinum using a helium ion microscope. J. Vac. Sci. Technol. B **27**, 2660–2667 (2009)
69. P. Chen, E. van Veldhoven, C.A. Sanford, H.W.M. Salemink, D.J. Maas, D.A. Smith, P.D. Rack, P.F.A. Alkemade, Nanopillar growth by focused helium ion-beam-induced deposition. Nanotechnology **21**, 455302 (2010)
70. A. Szczepkowicz, The kinetics of hill-and-valley faceting of oxygen-covered tungsten. Surf. Sci. **606**, 202–208 (2012)
71. A. Szczepkowicz, Oxygen-covered tungsten crystal shape: time effects, equilibrium, surface energy and the edge-rounding. Surf. Sci. **605**, 1719–1725 (2011)
72. T.T. Tsong, Direct observation of interactions between individual atoms on tungsten surfaces. Phys. Rev. B **6**, 417–426 (1972)
73. J.L. Pitters, R. Urban, R.A. Wolkow, Creation and recovery of a W(111) single atom gas field ion source. J. Chem. Phys. **136**, 154704 (2012)
74. R. Hill, F.H.M. Faridur Rahman, Advances in helium ion microscopy. Nucl. Instrum. Methods A **645**, 96–101 (2011)
75. M.T. Postek, A. Vladár, C. Archie, B. Ming, Review of current progress in nanometrology with the helium ion microscope. Meas. Sci. Technol. **22**, 024004 (2011)
76. S. Sijbrandij, J. Notte, C. Sanford, R. Hill, Analysis of subsurface beam spread and its impact on the image resolution of the helium ion microscope. J. Vac. Sci. Technol. B **28**, C6F6–C6F9 (2010)
77. D.J. Maas, E. van der Drift, E. van Veldhoven, J. Meessen, M. Rudneva, P.F.A. Alkemade, Nano-engineering with a focused helium ion beam. MRS Proc. **1354** (2011) mrss11-1354-ii03-03
78. J.E. Barth, P. Kruit, Addition of different contributions to the charged particle probe size. Optik **101**, 101 (1996)
79. Zeiss Orion® Plus—product brochure, August 2011, by Carl Zeiss SMT, Oberkochen
80. W.A. Schroeder, J.R. Shelton, J.B. Shelton, B. Roberson, G. Apell, The amino acid sequence of bovine liver catalase: a preliminary report. Arch. Biochem. Biophys. **131**, 653–655 (1969)

81. R. Luftig, An accurate measurement of the catalase crystal period and its use as an internal marker for electron microscopy. J. Ultra. Mol. Struct. R. **20**, 91–102 (1967)
82. R. van Gastel, L. Barriss, C. Sanford, G. Hlawacek, L. Scipioni, A.P. Merkle, D. Voci, C. Fenner, H.J.W. Zandvliet, B. Poelsema, Design and performance of a near ultra high vacuum helium ion microscope. Micros. Microanal. **17**(S2), 928–929 (2011)
83. S.A. Boden, A. Asadollahbaik, H.N. Rutt, D.M. Bagnall, Helium ion microscopy of lepidoptera scales. Scanning **34**, 107–120 (2012)
84. http://www.itrs.net/
85. B. Bunday et al., Unified advanced optical critical dimension (OCD) scatterometry specification for sub-65 nm technology (2009 version), ISMI Tech. Transfer, 2009, docID#: 04114596F-ENG
86. B. Bunday et al., Unified advanced optical critical dimension scanning electron microscopes (CD-SEM) specification for sub-65 nm technology (2010 version), ISMI Tech. Transfer, 2010, docID#: 04114595G-ENG
87. D.J. Maas, E. van Veldhoven, J. Meessen, EUV resist metrology, some benefits and limitations, in *SPIE2011-7971* (no proceedings published)
88. K. Ohya, T. Ishitani, MonteCarlo simulations of topographic contrast in scanning ion microscope. J. Electron Microsc. **53**, 229–235 (2004)
89. T. Yamanaka, K. Inai, K. Ohya, T. Ishitani, Simulation of secondary electron emission in helium ion microscope for overcut and undercut line-edge patterns. Proc. SPIE **7272**, 72722L (2009)
90. A. George, M. Knez, G. Hlawacek, D. Hagedoorn, H.H.J. Verputten, R. van Gastel, J.E. ten Elshof, Patterning of organosilane molecular thin films from gas phase and its applications: fabrication of multifunctional surfaces and large area molecular templates for site selective material deposition. Langmuir **28**, 3045–3052 (2012)
91. J. Singh, J.E. Whitten, Adsorption of 3-mercaptopropyltrimethoxysilane on silicon oxide surfaces and adsorbate interaction with thermally deposited gold. J. Phys. Chem. C **112**, 19088–19096 (2008)
92. K.-Y. Wu, S.-Y. Yu, Y.-T. Tao, Continuous modulation of electrode work function with mixed self-assembled monolayers and its effect in charge injection. Langmuir **25**, 6232–6238 (2009)
93. G. Hlawacek, V. Veligura, S. Lorbek, T.F. Mocking, A. George, R. van Gastel, H.J.W. Zandvliet, B. Poelsema, Imaging ultrathin layers with helium ion microscopy—utilizing the channeling contrast mechanism. Beilstein J. Nanotechnol. **3**, 507–512 (2012). doi:10.3762/bjnano.3.58
94. W. Eckstein, in *Behrisch: Sputtering by Particle Bombardment* (Springer, Berlin, 2007)
95. M. Rudneva, E. van Veldhoven, S. Malladi, D.J. Maas, H.W. Zandbergen, HIM as a sculpting tool for nanosamples, in *MRS Symposium Proc.* (2012)
96. D.L. da Silva, M.J. Mörschbächer, P.F.P. Fichtner, E. Oliviero, M. Behar, Formation of bubbles and extended defects in He implanted (1 0 0) Si at elevated temperatures. Nucl. Instrum. Methods B **219–220**, 713–717 (2004)
97. P. Jung, Diffusion of implanted helium in si and SiO_2. Nucl. Instrum. Methods **91**, 362–365 (1994)
98. Y.S. Hor, A.J. Williams, J.G. Checkelsky, P. Roushan, J. Seo, Q. Xu, H.W. Zandbergen, A. Yazdani, N.P. Ong, R.J. Cava, Superconductivity in $Cu_xBi_2Se_3$ and its implications for pairing in the undoped topological insulator. Phys. Rev. Lett. **104**, 057001 (2010)
99. J. Yang, D.C. Ferranti, L.A. Stern, C.A. Sanford, J. Huang, Z. Ren, L.-C. Qin, A.R. Hall, Rapid and precise scanning helium microscope milling of solid-state nanopores for biomolecule detection. Nanotechnology **22**, 285310 (2011)
100. L. Scipioni, D.C. Ferranti, V.S. Smentkowski, R.A. Potyrailo, Fabrication and initial characterization of ultrahigh aspect ratio vias in gold using the helium ion microscope. J. Vac. Sci. Technol. B **28**, C6P18–C6P23 (2010)

101. F. Aramaki, T. Ogawa, O. Matsuda, T. Kozakai, Y. Sugiyama, H. Obaa, A. Yasaka, T. Amano, H. Shigemura, O. Suga, Development of new FIB technology for EUVL mask repair. Proc. SPIE **7969**, 79691C (2011)
102. S. Tan, R. Livengood, D. Shima, J. Notte, S. McVey, Gas field ion source and liquid metal ion source charged particle material interaction study for semiconductor nanomachining applications. J. Vac. Sci. Technol. B **28**, C6F15–C6F21 (2010)
103. EDFAS Desk Reference Committee (ed.), *Microelectronics Failure Analysis Desk Reference*, 6th edn. (ASM International, Materials Park, 2011). ISBN: 978-1-61503-725-4
104. S.A. Boden, Z. Moktadir, D.M. Bagnall, H. Mizuta, H.N. Rutt, Focused helium ion beam milling and deposition. Microelectron. Eng. **88**, 2452–2455 (2011)
105. P.F.A. Alkemade, P. Chen, E. van Veldhoven, D. Maas, Model for nanopillar growth by focused helium ion-beam-induced deposition. J. Vac. Sci. Technol. B **28**, C6F22–C6F25 (2010)
106. D.A. Smith, D.C. Joy, P.D. Rack, Monte Carlo simulation of focused helium ion beam induced deposition. Nanotechnology **21**, 175302 (2010)
107. L. Scipioni, C. Sanford, E. van Veldhoven, D. Maas, A design-of-experiments approach to characterizing beam-induced deposition in the helium ion microscope. Micros. Today **5**, 22–26 (2011)
108. Y. Drezner, Y. Greenzweig, D. Fishman, E. van Veldhoven, D.J. Maas, A. Raveh, R.H. Livengood, Structural characterization of He ion microscope platinum deposition and sub-surface silicon damage. J. Vac. Sci. Technol. B **30**, 041210 (2012). http://dx.doi.org/10.1116/1.4732074
109. D. Montgomery, *Design and Analysis of Experiments* (Wiley, New York, 1991)
110. I. Utke, P. Hoffmann, J. Melngailis, Gas-assisted focused electron beam and ion beam processing and fabrication. J. Vac. Sci. Technol. B **26**, 1197–1276 (2008)
111. E. van der Drift, D.J. Maas, Helium ion lithography, in *Nanofabrication, Techniques and Principles*, ed. by M. Stepanova, S. Dew (Springer, Berlin, 2012). ISBN 978-3-7091-0423-1
112. P.F.A. Alkemade, E. van Veldhoven, Deposition, milling, and etching with a focused helium ion beam, in *Nanofabrication, Techniques and Principles*, ed. by M. Stepanova, S. Dew (Springer, Berlin, 2012). ISBN 978-3-7091-0423-1
113. S. Tan, R. Livengood, D. Shima, J. Notte, S. McVey, Nanomachining with a focused neon beam: a preliminary investigation for semiconductor circuit editing and failure analysis. J. Vac. Sci. Technol. B **29**, 06F604 (2011)

Chapter 17
High Resolution Electron Energy Loss Spectroscopy (HREELS): A Sensitive and Versatile Surface Tool

Luca Vattuone, Letizia Savio, and Mario Rocca

Abstract We review the principles of high resolution electron energy loss spectroscopy (HREELS) and of its extensions to time, momentum and spin resolved domains. The principles of the construction design and of some results are reported and discussed. In particular, we review the different scattering mechanisms by showing paradigmatic examples of their application. Advantages and shortcomings of HREELS with respect to competing techniques are also addressed.

17.1 Introduction

The investigation of surface processes is quite demanding from the point of view of the necessary sensitivity and requires probes whose interaction with the specimen is limited to the outermost atomic layers. This may be achieved with well established bulk probes, like neutrons and X-Rays, either by enhancing the surface to volume ratio of the sample, e.g. by investigating materials in form of powders, or by shooting the probe particle beam at grazing angles from the surface normal exploiting the total reflection phenomenon (occurring below a few degrees for X-Rays). The former approach suffers of very defective and poorly defined surfaces. The latter implies the use of very large and uniform samples, which may not be easily available, and/or of beams of very high brilliance present only at large scale facilities.

Alternatively, one can use probes whose interaction cross section with matter is so large to prevent sub-surface penetration. This is the case e.g. for thermal energy

L. Vattuone (✉) · M. Rocca
Dipartimento di Fisica dell'Università di Genova, Via Dodecaneso 33, 16146, Genoa, Italy
e-mail: vattuone@fisica.unige.it

M. Rocca
e-mail: rocca@fisica.unige.it

L. Savio
CNR-IMEM Unità operativa di Genova, c/o Dipartimento di Fisica Università di Genova, Via Dodecaneso 33, 16146 Genoa, Italy
e-mail: savio@fisica.unige.it

G. Bracco, B. Holst (eds.), *Surface Science Techniques*,
Springer Series in Surface Sciences 51, DOI 10.1007/978-3-642-34243-1_17,

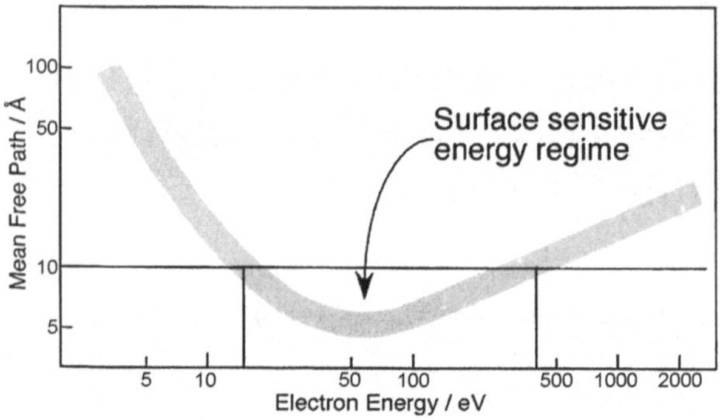

Fig. 17.1 Electron mean free path vs. kinetic energy. Taken from http://www.physics.uwo.ca/~lgonchar/courses/p9812/Chapter9_ElectronProbes.pdf

atoms (usually rare gases and in particular He since it is light, does not perturb the system by chemical interaction and He supersonic beams can be produced with a narrow velocity distribution) and low energy ions and electrons. The latter represent the most versatile probe particle since they are easy to produce, accelerate, focus, energy select, and detect. They have moreover a magnetic moment which makes them sensitive to surface magnetism. Finally their De Broglie wavelength becomes comparable with the atomic lattice spacing above 20 eV. The energy dependence of the electron mean free path, λ, through a material is reported in Fig. 17.1. It is a universal curve, largely independent of chemical composition. It initially decreases with kinetic energy, E_0, reaching a minimum of half a nm around 50 eV and then slowly increases with E_0. In the range from 10 eV to 200 eV, the penetration depth is limited to a few atomic layers but, at grazing incidence or emission, surface sensitivity is assured up to several thousand eV. The high energy behavior corresponds to the growing cross section in the forward direction for fast projectiles. The explanation for the low energy dependence is more subtle as one has to consider that the quanta of the collective modes of the electron gas (plasmons) are in the tens eV range and that the phase space available for single particle excitations shrinks with decreasing energy. The efficacy of electron energy loss mechanisms, responsible for the removal of the electrons from the beam, decreases thus more and more when moving towards lower kinetic energies.

The favorable matching of electron wavelength and lattice spacing and the high surface sensitivity made of Low Energy Electron Diffraction (LEED) a well established tool for the investigation of surface crystallography and the principal technique for a rapid inspection of surface symmetry and of surface order. Elastic backscattering from the surface lattice involves only a minority of the impinging electrons, the rest being either inelastically scattered or ending up as sample current after penetrating into the bulk of the crystal. The intensity of the specular beam (to which the dipole scattering intensity is proportional) is largest at low E_0 and at grazing incidence, conditions for which only one or just a few diffraction channels are open. The reflectivity increases, moreover, dramatically below a few eV because of quantum scattering off the attractive surface potential barrier. A high reflectivity is achieved also when the electron penetration into the bulk is hampered by energy gaps in the bulk electronic structure. The reflected beam can then come up to 30 % of the incident current [1].

The relatively high cross section of the electron-atom interaction limits the environment in which electron beams can be employed to ultra high vacuum (UHV). This limitation is not too severe since vacuum is anyhow necessary to keep the investigated surfaces in a well defined state for the time needed to perform most experiments. The electron charge implies that the experimental set up must be screened from the earth magnetic field to avoid an unwanted deflection of the electron trajectories. Such screening must be the more accurate the higher the desired energy resolution. Also the work function of the electrodes of the spectrometer must be highly uniform to avoid inhomogeneous electric fields which may cause a broadening of the velocity distribution in the beam. This is achieved by coating all surfaces with a uniform graphite film. One major drawback of electrons is the necessity of avoid surface charging. Insulator surfaces can therefore be investigated only when prepared in form of ultrathin films deposited on conducting substrates [3, 4] or when the residual conductivity associated to defects is sufficiently large.

The present chapter deals with the fundamentals and with some examples of applications of High Resolution Electron Energy Loss Spectroscopy (HREELS). At the usual cathode temperature of some 2000 K, the width of the thermal energy distribution of the emitted electrons, $k_B T$, is in the hundreds of meV. Such resolution is thus readily available without energy filtering. HREELS corresponds to the realm of a few meV full width at half maximum (FWHM) which can only be achieved by energy selecting the beam by an electrostatic device operated at pass energies in the sub-eV range. The monochromated beam is eventually accelerated to the desired energy.

When the beam strikes the surface the electrons may be either scattered elastically or inelastically or they may be resonantly captured by adsorbates thus increasing the probability to suffer an energy loss. In order to evidence inelastic events the electron beam has to be monochromatized and the scattered electrons analyzed in energy. At low kinetic energy of the impinging electrons and with the sample at, or below, room temperature loss and gain processes in the tens of meV range correspond mainly to creation and annihilation, respectively, of single vibrational quanta, such as surface phonons, and of more localized adsorbate modes. Electronic single particle transitions (i.e. interband transitions) and collective electronic excitations (plasmons and surface plasmons) occur generally in the eV range. Excitation of spin waves on magnetic samples and of Cooper pairs on superconductive materials is also possible. The inelastic scattering process may be mediated either by the long range Coulomb fields (dipole scattering) or occur at the impact with the surface atoms (impact scattering). Electron capture amplifies the latter mechanism and is often the driving force in electron induced desorption or dissociation of adsorbed species.

Dipole scattering relies on the same mechanism as the competing method of infrared absorption spectroscopy (IRAS). With respect to the latter technique HREELS has the advantages of:

(1) exploring a larger portion of reciprocal space not being limited by the very small momentum of the photons;
(2) allowing for small energy losses, while in IRAS photons below some 400 cm^{-1} are difficult to employ being absorbed by the windows separating the IR source from the UHV chamber;
(3) having a sensitivity to adsorbate coverage down to the 10^{-3} ML since the energy loss signal is recovered from regions with very low background, while in IRAS the absorbance corresponds to a decrease of an otherwise large reflectance. This remains true also after the improvement of IRAS sensitivity given by the introduction of the Fourier transform method (FT-IRAS).

On the other hand, IRAS is superior with respect to the achievable resolution (1 cm^{-1} against 1 meV of HREELS (1 meV = 8.0651 cm^{-1})) and works also at atmospheric pressure. The latter quality made IR spectroscopy more and more popular for performing experiments spanning from controlled UHV to more realistic conditions. IRAS is, moreover, much more apt for plug and play set ups, while a long training is needed to form HREELS specialists.

The impact scattering mechanism has analogy to the one operating in He atom scattering (HAS) since the energy exchange takes place during the contact of the probe particle with the surface. Given the short collision distance no limitation is present for the exchanged momentum. With respect to HAS, which presents an even higher surface sensitivity due to the inability of He atoms to overcome the very first surface layer, electrons have the advantage to be easily removed by absorption into the walls of the vacuum vessel saving heavy and expensive pumping systems. He atom beams may, on the other hand, reach energy resolutions in the sub meV range and even neV losses could be appreciated in the recently developed spin echo spectrometers [5, 6]). However, these deficiencies of HREELS are more than compensated by the higher kinetic energy which removes any limit to the energy loss value and by their other characteristics making it an economically affordable and flexible tool, definitively attractive to investigate a wide set of phenomena involving vibrational, electronic and magnetic excitations. Last but not least the success of HREELS was pushed by the production of reliable commercial spectrometers.

The present chapter is organized as follows: in the next paragraph we shall quickly revise the possible scattering mechanisms of electrons off a surface, namely dipole, impact and resonance scattering. We shall report here only the selection rules and the final formulas for the cross section under usually applicable approximations, while the reader is addressed to the original papers and reviews for a detailed mathematical treatment [7, 8]. The third paragraph is devoted to a schematic description of the design of last generation High Resolution Electron Energy Loss Spectrometers and to their developments. In particular, we shall mention Time Resolved HREELS (TR-HREELS), momentum resolved HREELS

(ELS-LEED), and Spin Polarized HREELS (SPEELS). The fourth paragraph will present selected examples of HREELS experiments aimed at the study of surface chemical reactions, at measuring surface phonon dispersion or adsorbate vibrational modes (to characterize adsorption and chemical reactions), surface plasmon dispersion, surface spin waves in magnetic materials and interband transitions. In the fifth and last paragraph the possible future developments of HREELS will be outlined.

17.2 Inelastic Scattering Mechanisms

The scattering of electrons off surfaces is a complex phenomenon and can be treated at different levels of approximation and complexity [9]. The energy loss mechanisms are generally classified in two limits referred to as dipole and impact scattering.

17.2.1 Dipole Scattering

Dipole scattering is mediated by the long range electric fields associated to the electron charge. When moving at a velocity **v** and for a distance d from the surface such fields have Fourier components extending up to a cut off frequency $\omega_c = \frac{v}{d}$ [10]. Excitation of surface and adsorbate modes through such fields can thus take place already when the electron is still far away from the surface plane. Indeed a 100 meV vibration can be excited by an electron with a kinetic energy of 1 eV already at a distance of 100 Å. Given the relatively large interaction distance neighboring oscillators will be excited nearly in phase. This scattering mechanism is therefore confined to small momentum transfers, i.e. it occurs in a narrow cone (the dipole cone) around the elastic diffraction channels as schematized in Fig. 17.2B. A typical energy loss spectrum is reported in Fig. 17.2A. The different peaks correspond to the excitation of single vibrational quanta and allow to infer the chemisorption state of the adsorbate.

The dipole mechanism is characterized by the following properties: (a) only modes causing a change of the dipole moment will be detected. In other words, the dynamical dipole moment associated with the vibrational transition must be nonzero; (b) since the Coulomb interaction is quite long range and lasts for a relatively long time (some 10^{-14} s) the microscopic details of the interaction potential are not needed to describe it.

Theoretical treatments were developed using either a purely classical picture (the electron is treated as a classical particle moving along a trajectory and interacting with the classical electric field of the surface excitations), or a semiclassical approach (the electron is treated classically while surface excitations are treated quantum mechanically [13]), or a complete quantum mechanical description [14–16].

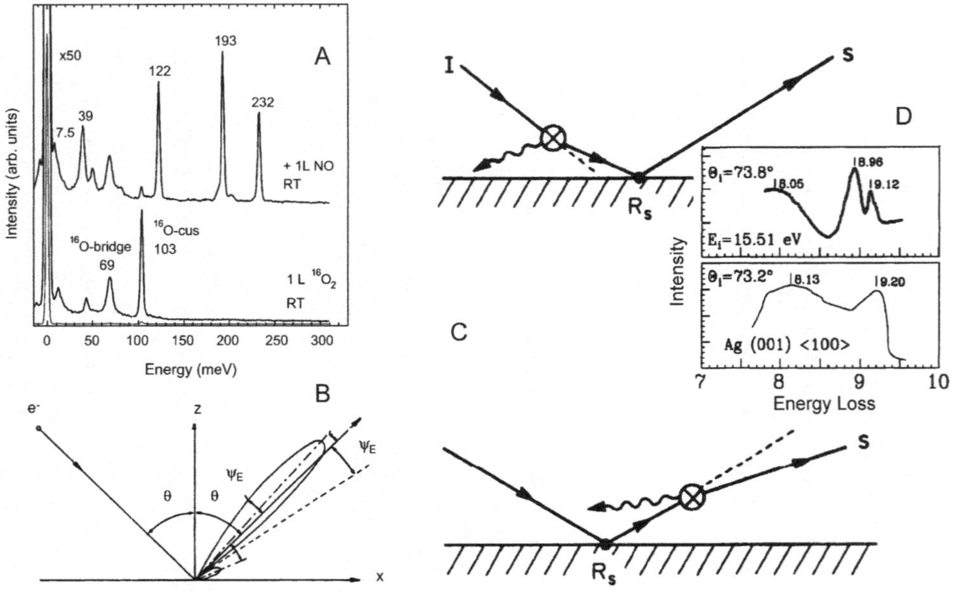

Fig. 17.2 (**A**) Typical dipole scattering spectrum of an adsorbed molecule obtained with a state of the art commercial spectrometer. The energy loss intensity is enlarged by a factor of 50 with respect to the one of the specular elastic peak. The losses are due to NO adsorbed on RuO_2. Reprinted figure with permission from Ref. [11] Copyright (2009) by Elsevier. (**B**) Schematic pictures of the angular dependence of the cross section for dipole scattering at low kinetic energy. Contrary to impact scattering, which is distributed over the whole solid angle, it is peaked along the specular direction. Reprinted figure with permission from Ref. [9] Copyright (1987) by Institute of Physics Publishing. (**C**) The two most relevant dipole excitation processes: (I) the excitation of the energy loss occurs on the incoming trajectory and (II) on the outgoing trajectory. The two processes are coherent and give rise to interference. The relevant reflection coefficients $R_s(E_0)$ may thereby differ in amplitude and phase. Reprinted figure with permission from Ref. [12] Copyright (1991) by the American Physical Society. (**D**) False loss structures mimicked by LEED Fine Structures in processes of type I of panel (**C**). The spectrum is reported vs scattered electron kinetic energy. The corresponding dependence for the elastic reflectivity, measured for the specular beam, is shown for comparison in the *lower panel*. Reprinted figure with permission from Ref. [1] Copyright (1994) by the American Physical Society

In the semi-classical theory, the probability that an electron loses an energy $\hbar\omega$ to create a surface excitation at a crystal temperature $T = 0$ K is given by [9]:

$$P(\omega) = \frac{4e^2}{\pi^2\hbar}\int_D d^2q_\parallel \frac{q_\parallel v_\perp^2}{[(\omega - q_\parallel v_\parallel)^2 + q_\parallel^2 v_\perp^2]^2} \times Im\frac{-1}{\xi_0(\mathbf{q}_\parallel, \omega) + 1} \tag{17.1}$$

The first part represents the kinematic factor. It determines the angular dependence of the cross section, peaked in an intense lobe of angular width $\psi_E = \frac{\hbar\omega}{2E_0}$ around the elastic channels. Because of the factor $q_\parallel$ the intensity goes to zero for vanishing momentum.

The last factor contains the effective dielectric function of the target ξ_0, i.e. the surface properties. The domain of integration D is determined by the angles of acceptance of monochromator and analyzer. In the semi-classical limit the surface is considered as a perfect reflector and the excitation probability equals the ratio $\frac{I_{loss}}{I_{el}}$, where I_{loss} and I_{el} are the intensity of the loss and of the elastic peak, respectively. From integration of (17.1) the following relations are derived [9], where the different dependencies are caused by the different spatial confinement of the excited modes:

$\frac{I_{loss}}{I_{el}} \propto \frac{1}{\cos\theta\sqrt{E_0}\omega^2}$ for surface phonons and plasmons (excitations decaying exponentially into the substrate;

$\frac{I_{loss}}{I_{el}} \propto \frac{1}{\cos\theta E_0\omega}$ for adsorbate modes (confined in the adlayer)

For non-metallic substrates there are no restrictions on the orientation of the observable dipoles with respect to the surface and the only requirement is the existence of a non-vanishing dynamical dipole moment. For metals, on the contrary, the image dipole will enhance the intensity of perpendicular vibrations and screen out those parallel to the surface, leading to the well known metal-surface selection rule [9]: *only molecular vibrations that give dipole changes perpendicular to a metal surface can be observed in HREELS when the scattered electrons are collected close to the specular direction.* The ultimate physical reason for this is that for metals electronic oscillations (plasma frequency) have a much higher frequency (energies of several eV) than molecular vibrations ($\hbar\omega \leq 0.6$ eV), so that the surface electrons can effectively screen parallel vibrations by giving rise to an image dipole in the opposite direction. This argument does not apply to semiconductors, characterized by 2–3 order of magnitude lower values of the plasma frequency and, *a fortiori*, to insulating samples. The metal selection rule was effectively employed to determine, e.g., the orientation of a molecular functional group with respect to a metal surface [7].

In a full quantum mechanical treatment the occurrence of different paths is included. In particular, as schematically reported in panel C of Fig. 17.2 the excitation can occur either before or after the electron impact with the surface and reflection off it. The reflection coefficient, R_s may thereby depend on the electron kinetic energy, E_0, both in amplitude and in phase. Such different channels give rise to interference and if R_s varies over the energy loss range the assumption of the proportionality of loss probability and elastic reflectivity fails. In general, such changes occur on a scale of several eV. They can, therefore, be neglected over the range of energy loss/gain of interest for vibrational spectroscopy, except for the very particular conditions corresponding to reflectivity fine structures (also called LEED fine structures, LFS). The latter are associated to reflection off the attractive image potential and can be important at low kinetic energies at the threshold conditions for the opening of the first diffraction channel besides the specular one. The assumption is, on the other hand, generally false when dealing with the larger losses associated to electronic excitations [1, 17, 18]). An example is given in panel D of Fig. 17.2. As one can see the sharp losses apparent in the spectrum reported in the upper panel occur at exactly the same kinetic energies of the scattered electron as the LFS in the specular channel (lower panel). The process is thus that the incoming electrons excite the continuum of electron-hole pairs on their way to the surface and are then reflected with different R_s (E_0) thus mimicking sharp losses. Such features can be easily distinguished from real ones since their apparent position in the spectrum changes with E_0.

For small energy losses and neglecting LFS conditions the differential cross section Σ becomes then:

$$\frac{d^2\Sigma}{d\Omega\, d\hbar\omega} = \frac{2m^2e^2v_\perp^4}{\pi\hbar^5\cos\theta_0}\frac{k_s}{k}k_i\frac{|R_I|^2P(\mathbf{q}_\parallel,\omega)}{[(\omega-q_\parallel v_\parallel)^2+q_\parallel^2v_\perp^2]^2} \tag{17.2}$$

where Ω is the solid angle and $P(\mathbf{q}_\parallel,\omega)$ incorporates the information on the physics of the surface region and, under suitable assumptions [7], is given by

$$P(\mathbf{q}_\parallel,\omega) = \frac{2\hbar q_\parallel}{\pi}\big[1+n(\omega)\big]\times Im\frac{-1}{\xi(\mathbf{q}_\parallel,\omega)+1} \tag{17.3}$$

Notably this form includes explicitly the temperature dependence of the intensity via the Bose factor $n(\omega)$. Its presence justifies why the energy gains are observed only at high temperatures and for $\hbar\omega$ values significantly lower than thermal energy k_BT.

17.2.2 Impact Scattering

Impact scattering is due to the losses arising during the collision of the electrons against the ionic cores of substrate or of adsorbate atoms. Contrary to dipolar scattering, it arises therefore from a short range interaction and is therefore not limited to small momentum transfer. In a seminal paper Ho, Willis and Plummer [19] investigated the non-dipole electron impact excitation of the vibrational modes of H adsorbed on W(100) (see Fig. 17.3) and on Mo(110) (see Fig. 17.4 and discussion in 17.4.2) and showed that for some of them the angle and energy dependence of the intensity deviates significantly from the one expected for dipole scattering. Impact scattering was not, however, systematically investigated until 1982 when the study of surface phonon spectra became a hot topic.

The theory to compute the impact scattering cross section is more demanding than for dipole scattering, since both crystal potential and multiple scattering have

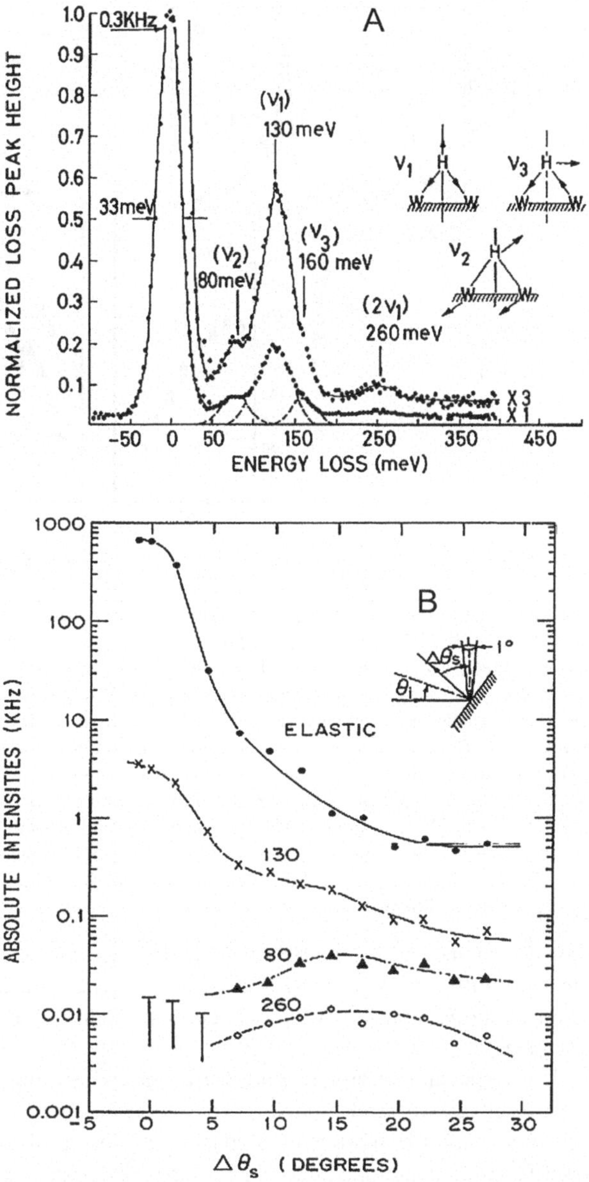

Fig. 17.3 (**A**) Energy loss spectrum recorded for out-of-specular conditions for H/W(100). The motion corresponding to the different modes is schematically reported. (**B**) Angle dependence of the absolute intensity of the modes as a function of the scattering angle θ_s. Reprinted figure with permission from Ref. [19] Copyright (1978) by the American Physical Society

to be included explicitly. The scattering probability, firstly derived by Tong, Li and Mills [22], has the form:

$$\frac{dP}{d\Omega} = \frac{mE_0}{2\pi^2\hbar^2}\frac{\cos^2\theta_s}{\cos\theta_0}S\left|M(\mathbf{k}_0, \mathbf{k}_s, \mathbf{q}_\parallel, \mathbf{u})\right|^2 \tag{17.4}$$

Where: θ_0 and θ_s are the impinging and scattering angles measured from the surface normal, respectively; E_0 is the primary kinetic energy; m is the electron mass; S is

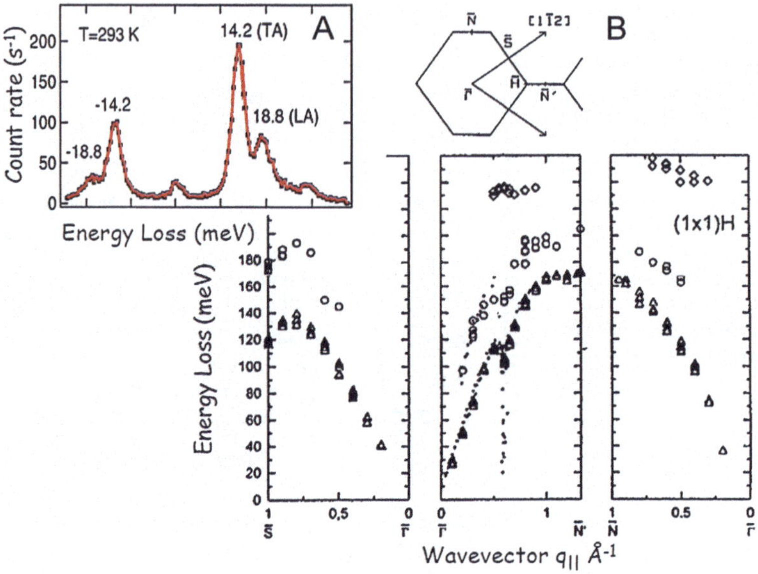

Fig. 17.4 (**A**) Typical energy loss spectrum recorded for out-of-specular conditions for Mo(110)(1 × 1) H. Thanks to the improved resolution (cf. Fig. 17.3, 1.5 meV against 33 meV) surface phonon losses are now well resolved. They correspond to the excitation of the Rayleigh wave (TA) and of the Longitudinal resonance (LA) at a wave vector transfer of 0.9 $Å^{-1}$ along the ⟨100⟩ crystallographic direction ($\bar{\Gamma} - \bar{H}$, see *inset* with the two dimensional surface Brillouin zone). The ratio between energy gains and energy losses is determined by the Bose factor. (**B**) Collection of the measured electron energy losses vs momentum transfer (*large symbols*) and comparison with HAS (*dots*). The Kohn phonon anomaly is evident in the *central panel* and is excited both by HREELS and HAS. The deep cone corresponds, on the contrary, to the excitation of electron hole pairs which are best excited with HAS. Reprinted figures with permission from Ref. [20] Copyright (2004) by the American Physical Society and from Ref. [21] Copyright (1997) by the American Physical Society

the surface area hit by the beam; M is the multiple scattering matrix element for an electron with initial wave vector $\mathbf{k}_0$ into a state with final wave vector $\mathbf{k}_s$ exciting or annihilating a phonon of wave vector $\mathbf{q}_{\|}$ (parallel to the surface) and polarization vector $\mathbf{u}$.

It is important to remark that, for impact scattering, there is no relation between elastic reflectivity and inelastic cross section, since both are determined independently by multiple scattering. Consequently normalization of the experimental data with respect to the specular elastic intensity makes no sense. The $E_0 \cos^2\theta_s$ factor in (17.4) implies that the impact cross section increases with kinetic energy and decreases with θ_i and θ_s. The actual dependence of the cross section on scattering energy and geometry is, however, so strongly modulated that no forecast for the most favorable experimental conditions is possible without calculating M. In general, since the growing cross section is more than counterbalanced by the worsening of the HREEL spectrometer performance with E_0 the best conditions for observing phonons have to be searched between 20 eV and 50 eV. This is well exemplified in Fig. 17.5, which reports the oscillatory behavior of the impact scattering cross

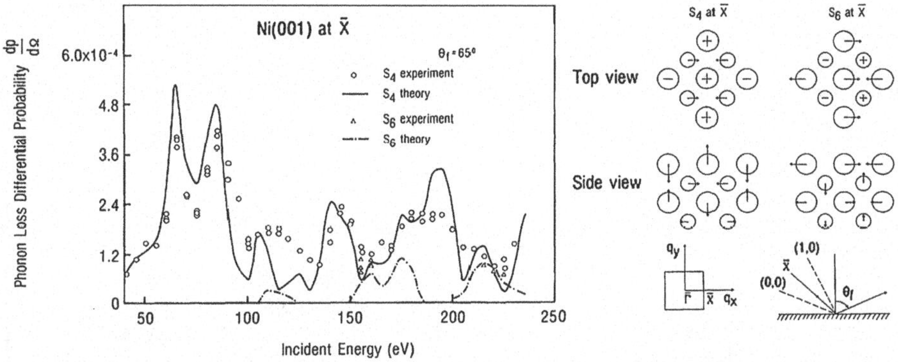

Fig. 17.5 *Left panel*: Comparison between theoretical (*lines*) and experimental (*symbols*) intensity of the S_4 and S_6 modes on Ni(100) at $\bar{X}$. *Right panel*: direction of atomic displacements in the S_4 and S_6 modes at $\bar{X}$. Up/down motion is indicated by *symbols* $+/-$, respectively; *smaller circles* denote second layer atoms. A schematic of the surface Brillouin zone and of the scattering plane is also shown. Reprinted figures with permission from Ref. [23] Copyright (1985) by the American Physical Society

section of the S_4 and S_6 phonons of Ni(100) at $\bar{X}$, as found experimentally and predicted by theory [23]. The experimental intensities have been corrected by multiplying by E_0 in order to correct them with respect to the decreasing transmittivity of the spectrometer. The two modes correspond, respectively to the counterphase motion of the nearest neighbor surface atoms in the vertical direction and along the scattering plane. Without the effect of multiple scattering the HREELS cross section for S_6 would be roughly one order of magnitude smaller than the one for S_4 and its observation would have been impossible.

Since the impact scattering cross section depends on the scalar product of the phonon eigenvector and of the transferred wave vector, $\mathbf{u} \cdot (\mathbf{k}_s - \mathbf{k}_0)$, the following selection rules follow: (a) shear polarized surface phonons (polarization parallel to the surface but perpendicular to the plane of incidence) cannot be excited for in-plane scattering events; (b) longitudinally polarized modes have vanishing cross section in the specular direction.

17.2.3 Resonance Scattering

At low kinetic energies the electrons may undergo resonant scattering. Strictly speaking it may be considered as a special kind of impact scattering with the electron remaining trapped for a while in an orbital of an adsorbate and thus increasing its energy loss probability. Resonance scattering is named after the characteristic structure of the energy dependence of the excitation cross section, which exhibits a remarkable enhancement in a limited E_0 range. Well known for gas phase scattering, resonance scattering was firstly reported for surfaces by Andersson and Davenport [24], who observed a non-dipolar behavior for OH adsorbed on Ni(100).

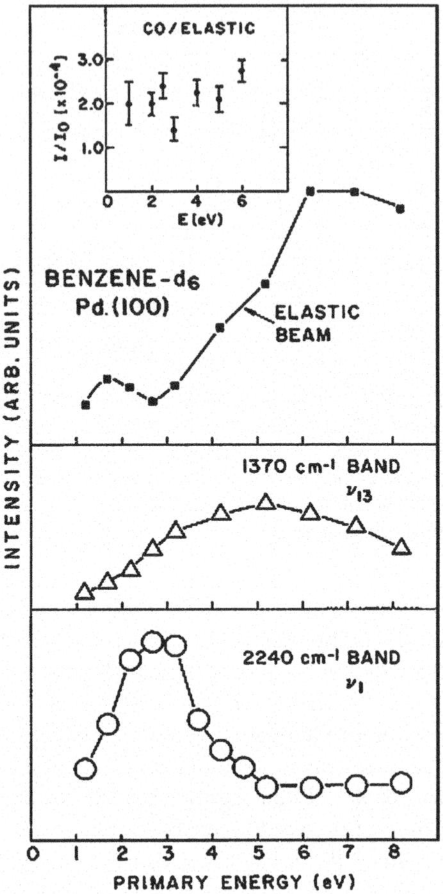

Fig. 17.6 Energy dependence of ν_1 and ν_{13} losses and of the elastic beam intensities, recorded 6° off-specular. The *inset* shows the CO loss intensity at 1820 cm^{-1}, used for calibration and normalized to the elastic beam intensity in the specular direction. Reprinted figure with permission from Ref. [30] Copyright (1984) by the American Physical Society

They proposed that the electron is temporarily trapped forming a charged ion (the so-called negative ion resonance). Two kinds of resonances are then possible:

(a) Shape resonance, if the ground state of the so-formed ion lies above the ground state of the neutral molecule;
(b) Feshbach resonance, if the ground state of the ion lies below the ground state of the neutral molecule.

A similar mechanism was later observed for both physisorbed [25–29] and chemisorbed molecules [30–34], as well as for Fuchs-Kliewer phonons of NiO(100) [35].

As an example of resonant inelastic scattering, we show in Fig. 17.6 the energy dependence of the ν_1 and ν_{13} modes and of the elastic beam for benzene chemisorbed on Pd(100). The curves are neither correlated to the specular intensity (as expected for dipole scattering) nor follow the modulation expected for impact scattering. We demand the reader to the review by Palmer and Rous [36] for further details and applications of this scattering mechanism.

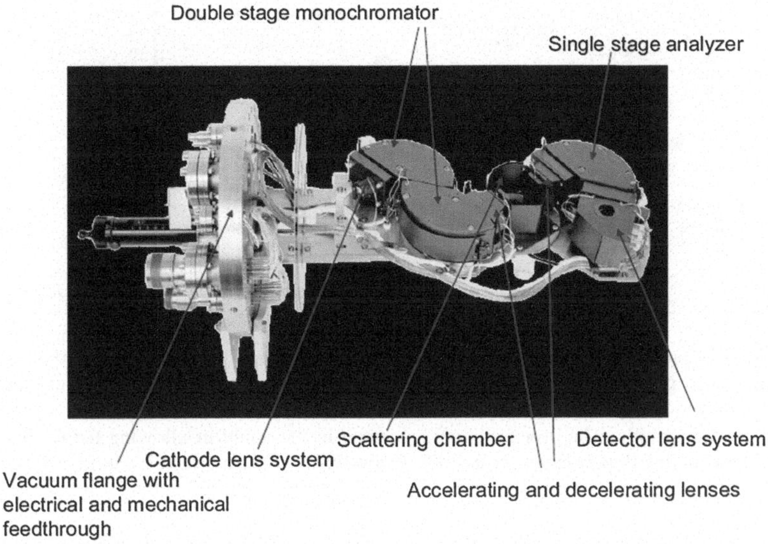

Fig. 17.7 Picture of a Delta 0.5 HR EELS Spectrometer, capable of 0.5 meV resolution in the direct beam, produced by SPECS following Ibach's design. Rielaboration of the original picture from http://www.specs.de/cms/front-content.php?idcat=128

17.3 Experimental Setups

The historical development of HREEL spectrometers is mainly correlated with significant advances in the realization of instruments combining higher monochromatic current and better resolution.

The spectrometer consists of (see Fig. 17.7): (a) a cathode lens system (A lenses) to focus the beam on the entrance slit of the premonochromator; (b) a double stage monochromator; (c) accelerating and decelerating lenses for focusing the monochromated beam onto the sample and for focusing the reflected beam onto the entrance of the energy analyzer (B lenses); (d) a single stage energy analyzer; (e) a channeltron and a detector lens system to focus the beam on its entrance (C lenses).

Monochromators and analyzers exploit the chromatic aberration of electrostatic deflectors operated at low pass energy. Among these, the cylindrical deflector analyzer (CDA) has proven to be superior to all other competing designs because, being a one dimensional device, one can use rectangular rather than circular slits. The focal property depends indeed on the width of the slit (not on its height), while the space charge is distributed over all the slit area thus reducing its pernicious effects. Moreover, with respect to the spherical deflector, the CDA has the advantage that the focal position can be finely tuned to the effective position of the exit slit by the potential applied to the plates delimiting the device in the direction normal to the deflection plane. This allows to compensate the focal position for different beam divergences caused by different space charge densities (Borsch effect) thus guaranteeing the highest current throughput for all conditions. The price one has to pay

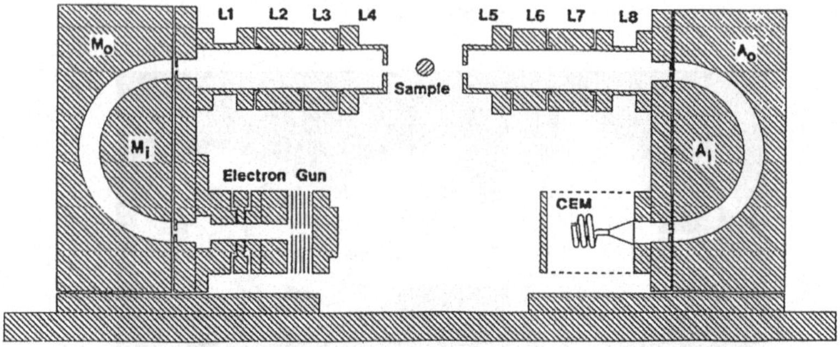

Fig. 17.8 Schematics of a dispersion compensation spectrometer. The design is based on spherical deflectors which guarantee second order focusing on the exit plane of the monochromator. The exit slits are removed so that electrons of different energies are focused simultaneously at different positions on the sample. The process is compensated in the analyzer allowing for resolutions of 20 meV. Reprinted with permission from [39]. Copyright 1991, American Institute of Physics

is the use of rectangular rather than the better developed electrostatic cylindrical lenses and to have a ribbon (rather than cylindrical) shaped beam. The monochromatic current and the brilliance of HREELS could be dramatically optimized by computer simulations of the electron trajectories in the CDA and in the lenses. This led to a toroidal rather than cylindrical shape of monochromator and analyzer. Such calculations require the solution of 3D Laplace problems aiming at the minimization of aberrations and explicitly considering fringe fields at the collimators. In the monochromator also the effect of space charge has to be taken into account in order to optimize its deflection angle [37]. Another important improvement was the introduction of an accelerating premonochromator for a coarse energy filtering since it reduces the effect of space charge. The reader is addressed to Ibach's book [8] for a more detailed discussion.

We mention that spectrometers with different design have been realized for particular applications such as time resolved (TR)-EELS, ELS-LEED and spin polarized (SP-) EELS.

TR-EELS employs the principle of dispersion compensation [38]. In a conventional HREEL setup the slits at the exit of the monochromator and at the entrance of the analyzer determine the energy resolution, by cutting off electrons with different energy and thus illuminating the sample with a monochromatic beam of limited size. In dispersion compensated spectrometers these slits are removed and the sample surface is illuminated simultaneously with electrons of different energy, focused at different positions on the sample, see Fig. 17.8. This approach greatly enhances the signal level (which is proportional to the square of the energy spread) without degrading the resolution (proportional to the spatial spread) provided the aberrations of the system are small. This spectrometer allowed to record an entire 100 point spectrum in 0.1 s [38] and to monitor the evolution of a single peak (i.e. to measure the residence time of an adsorbate on a surface) with a time resolution of 1 ms. Unfortunately, no further improvement was possible because the electron optic aberrations (large, even if the spectrometer uses a 180° spherical deflector (SDA)

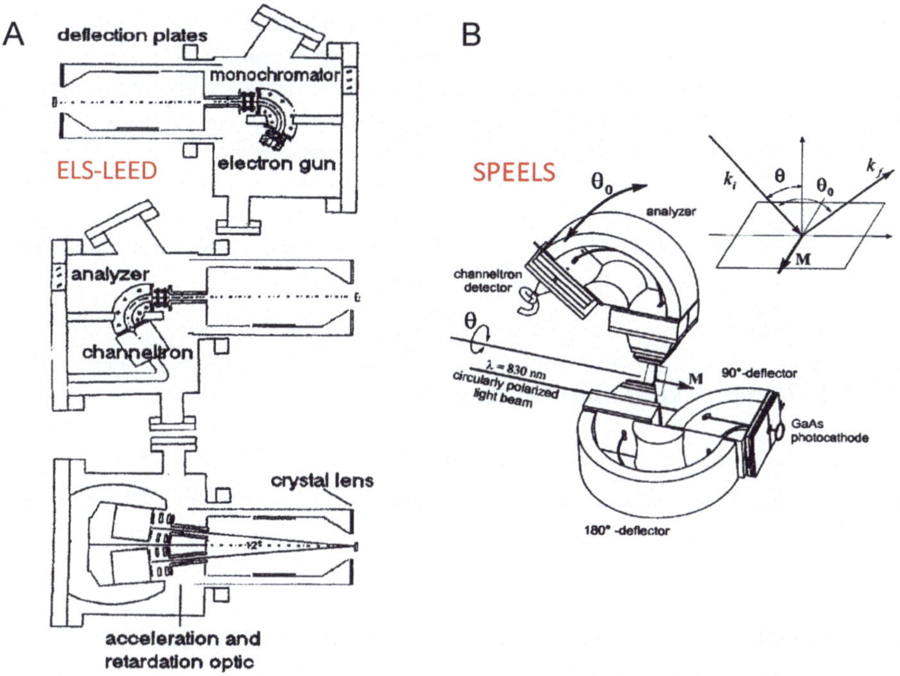

Fig. 17.9 (**A**) Schematics of an ELS-LEED spectrometer. Reprinted with permission from [41]. Copyright 1992, American Institute of Physics. (**B**) Schematics of a SPEEL spectrometer. The energy dispersive devices are based either on 90° or on 180° CDA designs to avoid spin mixing. Reprinted with permission from [51]. Copyright 2003, American Institute of Physics

design which guarantees a second order focus on its exit plane) and the rigidity of the SDA with respect to focal correction for space charge make so that the high current is obtained at an unreasonable cost in energy resolution. Although remarkable, the ms time scale remains still far from the one characteristic of most surface processes, so that this approach was not further pursued.

The ELS-LEED design is schematized in Fig. 17.9 A, B and C. This spectrometer aims at improving the momentum resolution of HREELS, usually limited to some 0.02 Å^{-1} [40] for the conventional models. In ELS-LEED the angle of incidence and of scattering are selected electronically through an octupole optic [41], as in spot profile analysis (SPA)-LEED, rather than by mechanically rotating analyzer or monochromator. The energy selection and analysis is then performed with CDA devices as in conventional HREELS. Due to the octupole deflector the electrons impinge onto the surface close to the surface normal rather than at grazing incidence. A resolution as high as 0.004 Å^{-1} was reported [42], which was exploited, e.g., to perform quite accurate measurements of surface plasmon dispersion [43, 44].

SP-EELS is employed in the investigation of magnetism at surfaces. The acronym is used to address experiments with a different degree of complexity:

indeed it refers to experiments employing a polarized primary electron beam, but the polarization analysis after scattering may be either present [45] or not [46, 47]. The polarized electron beam is obtained by using a GaAs photocathode illuminated by circularly polarized light [48]. The spin polarization of the beam is measured in a high energy Mott detector [49] or by evaluating the anisotropy of the LEED spots intensity. Suitably designed premonochromators and monochromator with 90° and 180° deflection angle, respectively allow to perform the energy selection without mixing up the spin states [50]. A scheme of the spectrometer which was realized by Ibach et al. is reported in Fig. 17.7 [51] (see also Refs. [52, 53]).

17.4 Selected Experimental Results

In this paragraph we are going to demonstrate the versatility of HREELS in surface analysis, by giving an overview on some selected results covering different subfields.

17.4.1 HREELS for the Study of Adsorbates and Surface Chemistry

The surface science literature reports thousands of papers employing HREELS to characterize adsorption at surfaces and surface reactions. The analysis of the angle and energy dependence of the vibrational losses has often been used to gain information on the adsorption site and, in particular, to discriminate between on-surface and sub-surface vibrations [54, 55]. Indeed losses whose intensity does not decrease when moving off-specular are likely to arise from impact scattering and to originate possibly from vibrations polarized parallel to the surface or involving the motion of sub-surface species. Only photo-electron diffraction, careful STM analysis or ab-initio calculations can lead to a conclusive assignment of the adsorption site, but HREELS can provide strong indications about the correctness of the starting model.

We report in the following two examples of the information provided by HREELS in the study of gas-surface interaction. Figure 17.10 compares in-specular HREEL spectra we recently recorded after dosing CO_2 on a clean and on a H-covered Ni(110) surface [56]. The modes at 50, 98, 167 and 365 meV correspond, respectively, to the ν(Ni–O), δ(OCO), ν_s(OCO) and ν(CH) vibrations of formate (HCOO) and indicate the formation of this compound as an intermediate product of the CO_2 hydrogenation reaction catalyzed by the surface. The HREEL study allowed a precise identification of the stable reaction product and, in combination with photoemission spectroscopy and ab-initio calculations, it contributed to highlight an unexpected reaction mechanism. We proved indeed that, although isolated CO_2 adsorbs onto the surface with the C atom down,

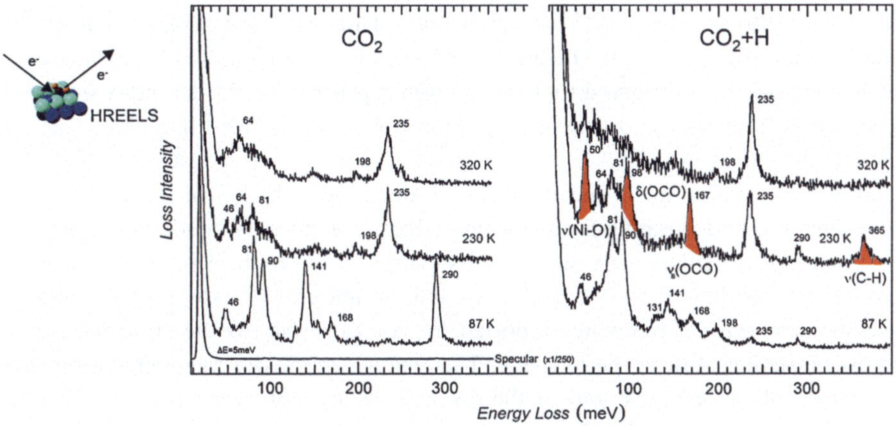

Fig. 17.10 Comparison of in-specular HREEL spectra recorded after dosing CO_2 on clean (*left*) and H covered (*right*) Ni(110). The spectra show that CO_2 reacts with H leading to formate, as demonstrated by the presence of losses at 50, 98, 167 and 365 meV. Reprinted with permission from Ref. [56]. Copyright (2008) American Chemical Society

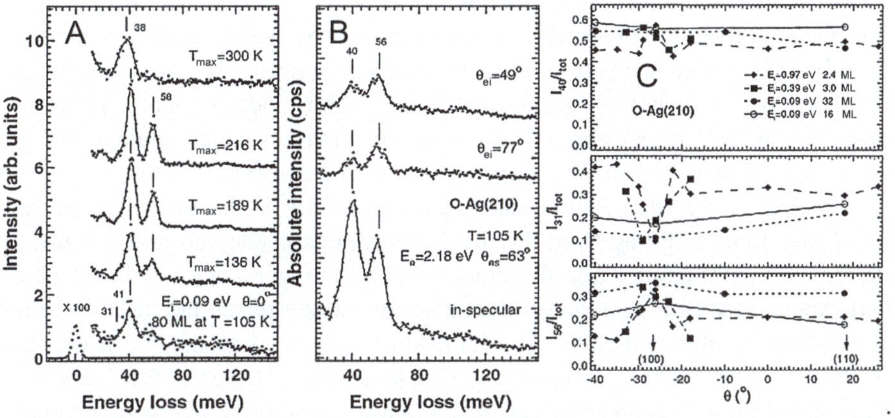

Fig. 17.11 (A) Crystal temperature dependence of the HREEL spectra. The species vibrating at 31 meV converts into the highest frequency one. (B) In-specular and off-specular spectra of oxygen adsorbed on Ag(210). The decrease of the absolute intensity when moving off-specular is much faster for the 40 meV mode, so that the 56 meV peak dominates the spectrum under these conditions. (C) Dependence of the loss intensities vs angle of incidence of the O_2 molecules. Reprinted figures with permission from Ref. [55] Copyright (2003) by the American Physical Society

when a HCO_2 complex forms, it first flips head-down binding to the surface with the two O atoms. H can thus bind to the free C end leading to HCOO formation [56].

The second example is reported in Fig. 17.11 and shows HREEL spectra recorded after dissociative O_2 adsorption on Ag(210) [55]. This is a stepped surface with one row wide (100) terraces and (110) step heights. Three losses are observed at 31, 40 and 56 meV (see panel A). The adsorption sites may be identified by comparison with the frequencies observed for O adsorption at pristine low Miller index surfaces. The lowest frequency corresponds to adatoms at (111) or (100) like planes and is identified by O at the (100) nanoterraces. The 40 meV mode must be related to a configuration similar to the one of the Ag–O added rows which form on Ag(110) and is assigned to O decorating the open step edges. The 56 meV mode is not present on pristine low Miller index surfaces. Its energy is too high to be due to adatom motion and too low to be due to the internal vibration of O_2 admolecules. Indeed it coincides with the frequency reported for AgO [57] and is therefore indicative of the formation of this type of oxide. The analysis of the different behavior of the intensities of the 40 meV and of the 56 meV peaks with respect to the electron scattering angle, reported in Fig. 17.11B confirms the different nature of their loss mechanisms. Indeed, the ratio of the 40 meV/56 meV intensities changes in favor of the higher energy mode when moving off-specular, indicating its non-dipolar nature.

As shown in Fig. 17.11 panel C the ratio of the 31 meV and of the 56 meV modes changes, moreover, as a function of the impact energies and of the angle of incidence of the O_2 molecules, as demonstrated by dosing with a supersonic molecular beam. The intensity of the 40 meV mode (upper panel) remains, on the other hand, approximately constant. Indeed, the former modes oscillate in counter-phase, the 56 meV peak being most intense when the molecules impinge at high impact energy and close to the surface normal. An adsorption path must therefore exist for which dissociation takes place very close to the surface plane and one of the atoms resulting from the dissociation process ends up in a subsurface site (56 meV vibration) while the other sticks at the step edge (40 meV vibration). Another pathway dominates when the beam strikes the surface at more grazing incidence so that both atoms end up in super surface sites, i.e. one at the step edge (40 meV) and the other on the (100) nanoterrace (31 meV). The latter may eventually diffuse to the subsurface region when mobility is activated thermally by heating the crystal. Interestingly, such mechanism is not present for Ag(410), a surface with the same open step geometry as Ag(210) but with twice as large (100) nanoterraces [58]. This finding indicates that the relaxation of the substrate atoms at the step is crucial to open the pathway leading to surface oxide formation.

When a similar experiment is repeated for Ag(511) very different frequencies are observed. This surface is characterized by 4 atom row large (100) nanoterraces, like Ag(410), but close packet rather than open step heights. The vibrational frequencies of 33 meV and 67 meV observed after dissociative oxygen adsorption are again indicative of O chemisorption and of oxide formation. In this case, however, the oxide is not AgO but rather Ag_2O [59], as unambiguously evident from the higher vibrational frequency.

17.4.2 HREELS for the Measurement of Dispersion Curves of Collective Surface Excitations: Surface Plasmons and Surface Phonons

Dispersion curves are recorded by measuring the dependence of the energy of an excitation as a function of wave vector transfer, $\mathbf{q}_{\|}$. In HREELS a monochromatic electron beam impinges onto the surface at a well defined angle with respect to the surface normal, θ_i (i.e. with a well defined impinging wave-vector $\mathbf{k}_0$) and the scattered intensity is measured at another well defined angle, θ_s, with a scattered wave vector $\mathbf{k}_s$. Independently of the energy and momentum transfer mechanism, $\mathbf{q}_{\|}$ can be unambiguously determined from the conservation of energy and momentum:

$$\frac{\hbar^2 k_0^2}{2m} = \frac{\hbar^2 k_s^2}{2m} + \hbar\omega \tag{17.5}$$

$$\mathbf{k}_{0\|} = \mathbf{k}_{s\|} + \mathbf{q}_{\|} \tag{17.6}$$

where $\hbar\omega$ is the value of the energy loss and $\mathbf{k}_0$ and $\mathbf{k}_s$ are determined by the initial energy and by the scattering geometry. Provided that only one quantum is involved in the process, $\hbar\omega$ and $\mathbf{q}_{\|}$ are linked by the dispersion relation of the excitation.

Surface Phonon Spectra

Besides Fuchs Kliewer modes [35, 62–68], corresponding to the counter-phase motion of the positively and negatively charged sublattices of crystals with at least partially ionic chemical bonds (e.g. oxides or III–V and II–VI semiconductors), only optical phonons can be excited by dipole scattering (e.g. on Si surfaces). Acoustic phonons can be probed, however, by impact scattering. Since the latter increases in importance with primary electron energy, the systematic investigation of surface phonon spectra had to wait the beginning of the '80 when spectrometers capable to work with a good energy resolution also at high impact energy and with a momentum resolution of some 0.01 Å^{-1} [69] were developed.

Although the very first surface phonon dispersion curve was reported by inelastic HAS, HREELS was used for the same purpose just shortly afterwards and promised immediately to be capable of overcoming the major shortcoming of HAS, i.e. the limited available energy in the incident probe particle limiting the energy loss range for helium atoms. HREELS proved thus to be superior to HAS for studying adsorption systems since one could then record simultaneously the dispersion of substrate and adsorbate modes. The resolution remains, however, worse than the one achievable with HAS (2 meV for out-of-specular investigations). The measurement of surface phonon dispersion curves involves recording a sequence of spectra at different scattering or incident angles under off-specular conditions. Given the small ratio of the energy losses associated to surface phonons and the impact energy, the transferred momentum is in first approximation constant over the relevant energy loss range and it is determined only by the scattering geometry [70].

The very first HREELS papers on surface phonon dispersion dealt with the Rayleigh mode of Ni(100) [69] and of Ni(100) c(2 × 2)O [71]. Most of the spectra were recorded with a primary beam energy of 180 eV to maximize the cross section and showed evidence for an adsorbate induced anomaly. Oxygen reduces thereby the interaction force between the outermost Ni planes thus reducing the Rayleigh wave frequency.

A more recent example of phonon dispersion investigation is shown in Fig. 17.4. The HREELS investigation took place at much lower impact energies [20, 21], since in the meanwhile it was recognized that optimal measurement conditions are realized by the balance of the growing inelastic cross section and the declining spectrometer performance. The investigated system, H/Mo(110), is characterized by a strong Kohn anomaly, a phonon softening at $q_{\parallel} = 2k_F$, with k_F the Fermi wave vector, caused by the nesting of the Fermi level of the adsorbate induced surface electronic state. As it is evident from the figure HREELS (● and △) and HAS (•) data coincide only for the upper branches, while the lowest one is present only in HAS. Such branch is in fact due to the formation of electron-hole pairs rather than to the excitation of surface phonons. HREELS is indeed sensitive mainly to the displacement of the ion cores, while in HAS the interaction is mediated by the electronic surface density which can therefore be excited as well. The two techniques prove therefore to be complementary for gaining the complete insight into the physics of the studied system.

Surface Plasmons

Surface plasmons mediate the surface response function to external electric fields and are therefore essential to describe photoemission and all other phenomena involving surface screening by electrons. Moreover, surface plasmons have been proposed to replace electrons and photons in the newly developed plasmonic devices for faster and more compact circuits.

In the long wavelength limit surface plasmons can be investigated by optical methods. Their dispersion, however, can only be measured by HREELS [40] or by ELS-LEED [41, 60]. In both techniques the surface plasmon dispersion curve is obtained by recording energy loss spectra for different scattering angles. Energy and momentum conservation allow then to calculate $\mathbf{q}_{\parallel}$. As an example, panel A of Fig. 17.12 reports the dispersion curves measured for the optical surface plasmon of a Ag film consisting of two dimensional (111) islands deposited on Si(111)-(7 × 7) [43]. The surface plasmon energy (i.e. the value of the energy loss) is independent of $q_{\parallel}$ for small $q_{\parallel}$, more precisely, until the parallel momentum equals $\frac{2\pi}{d}$ (where d is the island size). For larger $q_{\parallel}$ the surface plasmon propagates within the island and dispersion is observed. The Einstein oscillator like behavior at small $q_{\parallel}$ indicates therefore the occurrence of plasmon confinement inside the grains of the film. Further analysis showed that in the long wavelength limit the surface plasmon energy depends on the surface to volume ratio of the islands. Interestingly, the surface plasmons are confined also for percolated films, i.e. films showing direct

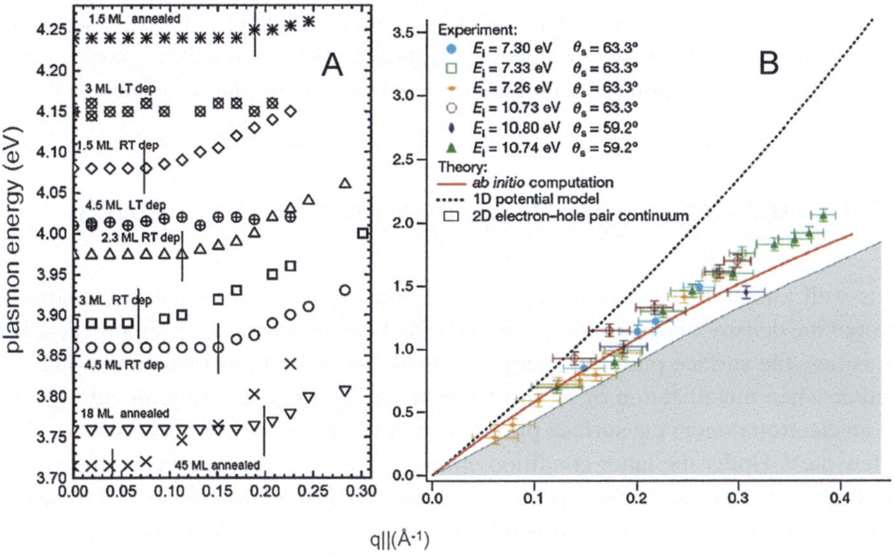

Fig. 17.12 (**A**) Dispersion of the surface plasmon as a function of $q_{\|}$ for Ag films of different thickness. The experiment was performed by ELS-LEED. The *thin lines* mark the critical $q_{\|}$ value beyond which the surface plasmon starts dispersing, which coincides with 2π times the inverse of the island diameter determined by SPA-LEED. Reprinted figure with permission from Ref. [43] Copyright (1999) by the American Physical Society. (**B**) Dispersion of the ASP for Be(1000). The different symbols correspond to the different experimental conditions reported in the legend. The *gray area* denotes the realm of intraband transitions within the Surface Shockley State. *Continuous* and *dotted lines* correspond to one and two dimensional theories. Reprinted figure with permission from Ref. [61] Copyright (2007) by Nature Publishing Group

current conductivity. The reason for this is that the anisotropy of the Ag surface plasmon dispersion with respect to crystal face and crystallographic direction impedes energy and wave vector matching at the boundary of the different grains of the film. The propagation of the surface plasmon across the borders of the islands is thus hindered even when they are electrically connected.

ELS-LEED was employed also to demonstrate the existence of the multipole plasmon mode for Ag surfaces [42, 72] taking advantage of the extreme reproducibility of the energy losses granted by the electrostatic deflection of the electron beam with the octupolar field.

The most important recent achievement on surface plasmon dispersion obtained by conventional HREELS concerns the observation of a novel type of collective electronic excitation, the acoustical surface plasmon. Such mode is associated to the excitation of the two-dimensional electron gas of Surface Shockley states (SSS) and exhibits a linear dispersion with vanishing energy in the zero wavelength limit [61, 73, 74], rather than the usual square-root dependence expected for a thin film. The dispersion is shown in panel B of Fig. 17.12 for Be(1000). The difficulty of performing such experiment lies in the small angle at which the scan curve cuts the ASP dispersion and in the very low cross section of this mode. More recently its existence was proved also for Cu(111) [74, 75] and Au(111) [73, 76]. The linear dispersion makes this mode promising for applications since it permits a distortionless propagation of a non-monochromatic signal in a future plasmonic device.

17.4.3 HREELS as a Tool to Evaluate the Density of Carriers

It is well known that the surface plasmon frequency is proportional to the square root of the density of carriers (see (17.7)) [40]. As a consequence, given the metallic densities, the surface plasmon energy is usually of several eV. In semi-conductors and/or when metallization of an insulating material is induced (e.g. by adsorption of an electron donor) the surface plasmon energy can, on the contrary, be as low as a few meV. Under the latter condition, multiple plasmon excitation can take place, leading to an increased elastic peak width easily detectable by HREELS. The temperature dependence of the such width can then be used to evaluate the carrier density [77, 78] and to monitor the metallization process [79].

Figure 17.13(A) shows the temperature dependence of the specular peak width for Si(100) (filled dots) and its fit with different models including: the carrier density determined by Bose Einstein (BE) statistics (□); BE plus intrinsic carrier plasmon thermal variations (open triangles); BE plus intrinsic carrier plus depletion layer thickness thermal variations (open diamonds); BE plus intrinsic plasmon plus topmost metallic layer plasmon variations (open dots) plasmon excitation. Only the later model describes the width at the highest T proving surface metallization. The latter is determined by the removal of the dangling bond anisotropy above 900 K caused by thermal fluctuations.

In panel (B) we report a similar physics occurring for a clean and for a H-covered ZnO($10\bar{1}0$) surface. Panels (C) and (D) report the FWHM of the elastic peak, respectively, vs. H exposure and temperature. It is evident that metallization occurs after a critical H coverage is attained, since the latter the hydrogen atoms act as electron donors. Moreover, the FWHM initially increases with temperature due to the higher probability to excite the surface plasmon mode and it reduces rapidly as soon as H desorbs.

Since in this case a broadening of the elastic peak, but no individual features were detected, the excitation responsible for it must lie below 5 meV. This result was explained by the presence of metallic (i.e. partially filled) electronic states produced by H adsorption. The carrier concentration n was then estimated by

$$\omega_p^2 = \frac{ne^2}{\epsilon_0 m^\star} \tag{17.7}$$

where $\omega_{sp} = \frac{\omega_p}{\sqrt{2}}$ is the surface plasmon frequency, ω_p the bulk plasmon frequency, $m^\star$ the effective electron mass and e its charge.

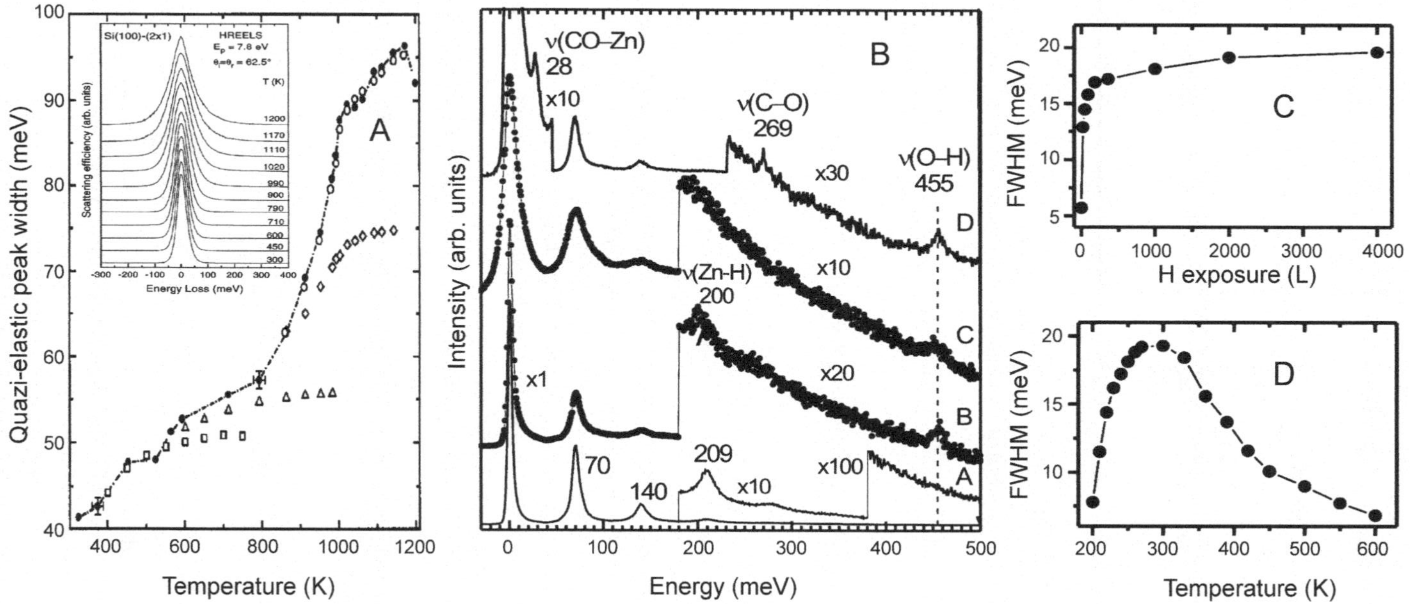

Fig. 17.13 *Left panel*: (**A**) Temperature induced broadening of the specular peak of Si(100) caused by the multiple excitation of a surface plasmon. The *inset* shows the HREEL spectra. • correspond to experimental data, other symbols and lines correspond to different theories which include from the bottom up: □ Bose-Einstein statistics, △ Bose Einstein statistics plus surface plasmon, ◇ the preceding plus depletion layer and ∘ all previous ingredients plus surface metallization occurring above 900 K because of the removal of the dangling bond anisotropy. Reprinted figure with permission from Ref. [2] Copyright (1996) by the American Physical Society. *Central* and *right panels* (**B**): spectra recorded in-specular for the clean and H-covered ZnO(10$\bar{1}$0) surface; the dependence of the quasielastic peak width upon H exposure (**C**) and on temperature (**D**) is also shown. Reprinted figure with permission from Ref. [79] Copyright (2005) by the American Physical Society

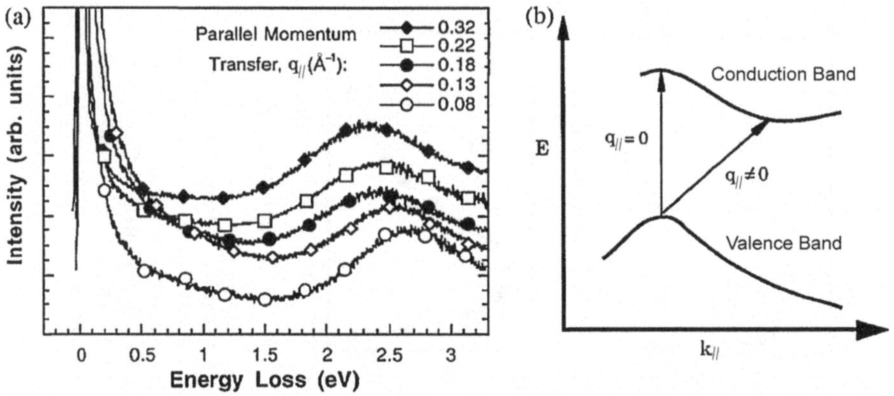

Fig. 17.14 (**A**) HREEL spectra recorded on a thin BC_2N film at different values of $q_\parallel$; (**B**) schematic illustration of interband transitions with zero and non-zero momentum change. Reprinted figure with permission from Ref. [80] Copyright (1999) by the American Physical Society

17.4.4 HREELS as a Tool to Investigate Interband Transitions

In this section we report an example of the use of HREELS to solve the apparent contradiction about the band structure of BC_2N. For this semiconductor, characterized by a tunable band gap and hence of high technological interest, a gap of 2.1 eV was inferred from photoluminescence experiments, while a value of only 1.4 eV was measured by scanning tunneling spectroscopy (STS). Thanks to HREELS measurements, and applying the energy and momentum conservation laws, it was possible to determine that the difference is due to the dependence of the band gap on $q_\parallel$, as shown by the sequence of spectra reported in Fig. 17.14(a). The experimental energy gap decreases indeed with increasing parallel momentum from $\approx$1.8 eV at $q_\parallel = 0.08\ \text{Å}^{-1}$ to 1.4 eV at $q_\parallel = 0.32\ \text{Å}^{-1}$.

The apparent contradiction between photoluminescence and STS data can thus be explained by an indirect band gap structure. Indeed, photoluminescence estimates the values of the band gap in the long wavelength limit, while STS just samples the path with the smallest gap.

17.4.5 HREELS and Superconductivity

In the previous examples HREEL spectroscopy was applied to the investigation of single crystal surfaces oriented along a well defined crystallographic direction. In case of semiconducting or insulating samples, special care is needed to avoid or to compensate for the charging of the insulating substrate, which may be achieved either by doping or by using an additional flood electron gun.

In the following we show the application of HREELS to the study of the surface of the ceramic superconductor $YBa_2Cu_3O_7$ (YBCO) [81].

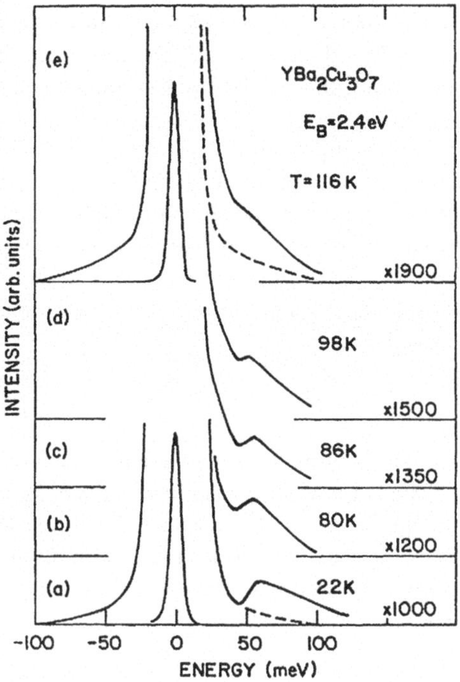

Fig. 17.15 HREEL spectra for YBCO recorded at a primary energy of 2.4 eV at different temperatures. Reprinted figure with permission from Ref. [81] Copyright (1990) by the American Physical Society

Figure 17.15 shows the evolution of the HREEL spectra vs increasing temperature. A loss feature appears at around 60 meV at the lowest T values. This energy compares well with the gap predicted by BCS theory (60 meV), but it persists also above the critical temperature of 90 K. Such dependence cannot be explained by theory and is more in line with a Bose Einstein condensation of pre-existing pairs. A similar result was reported later also for $Bi_2Sr_2CaCu_2O_8$ (BISCO) [82].

17.4.6 SP-EELS Detection of Spin Waves

The application of HREEL spectroscopy to the study of surface magnetism came only recently thanks to the development of SP-EEL spectrometers coupling energy resolution with spin polarized beams. With such a tool it was possible to confirm the theoretical prediction on the existence of well defined spin waves at short wavelengths [83]. The first clear evidence for the experimental identification of spin waves appeared seven years later [84]: broad loss bands were observed by SP-EELS and suggested to arise from spin flip scattering of the impinging electron beam induced by a magnetic sample (ultra-thin film or magnetic surface) [45, 85]. Spin flips are produced by a Stoner excitation like mechanism implying a particle-hole excitation by the impinging electron. Electron energy losses are then possible above the energies of the exchange splitting in the ferromagnetic d band.

The spin waves are expected in the same spin flip channel, but at much lower energy loss than the one characteristic of the Stoner excitation continuum. The creation of a spin wave by the incoming electron decreases the spin angular momentum of a ferromagnet by $\hbar$. To conserve total angular momentum the beam electron spin must then flip from down (the minority spin direction) to up (majority direction). Given this definition of up and down no spin wave excitation can take place for an up-spin beams and no spin annihilation for down spin beam. Thanks to this selection rule analyzing the spin distribution of the scattered beam is not necessary. Typically the loss spectrum in the spin wave region is measured both for beam polarization anti-parallel and parallel to the sample magnetization. The spin asymmetry is then defined as:

$$A = \frac{I_\downarrow - I_\uparrow}{I_\downarrow + I_\uparrow} \tag{17.8}$$

and will contain the spin wave loss feature.

The entire dispersion curve was measured for a 2 ML thick Fe film on W(110): the spin wave energies resulted to be strongly reduced with respect to the corresponding values measured for the bulk and to those predicted by theory [53]. An example of spin wave measurement is reported in Fig. 17.16 for Fe(100)p(1 $\times$ 1)O [86].

17.4.7 HREELS for the Investigation of Liquid Surfaces

Recently also liquid surfaces have attracted the attention of the surface science community [87–89]. Investigation of these systems with electron based techniques is necessarily limited to liquids with a vapor pressure low enough to allow for a reasonable mean free path of the scattered electrons. With this restriction, also HREELS can be employed. Figure 17.17 reports an example of HREEL spectra of the 1-Ethyl-3- methylimidazolium Bis (trifluoromethylsulfonyl) amide surface (EMIM in the following), recorded at different E_0 and for T = 300 K and T = 100 K [90].

Since the penetration depth of the electrons depends on primary energy [91], the similarity of the loss features detected with different electron energies at 300 K indicates that, at this T, the structure of the surface is quite similar to the bulk structure. The picture changes at T = 100 K: at low E_0 the well pronounced CF_3 and anti-symmetric O=S=O modes are almost missing while the losses around 800 cm^{-1} (due to the imidazolium-H bending modes and to the $[TF_2N]^-$ O=S=O symmetric stretch) are still present. If E_0 (and hence the penetration depth) is increased, the spectra recorded at 100 K become similar to those recorded at RT. These findings can be rationalized considering that, at low primary energy, only dipoles oscillating perpendicularly to the surface contribute to the spectrum recorded in-specular. The authors then concluded that a change of the molecular orientation of the outermost layer takes place at 100 K. Moreover, the elastic peak FWHM depends on E_0 and remarkably larger at RT than at 100 K, similarly to what observed for semiconductor surfaces with depth dependent carrier concentration [91].

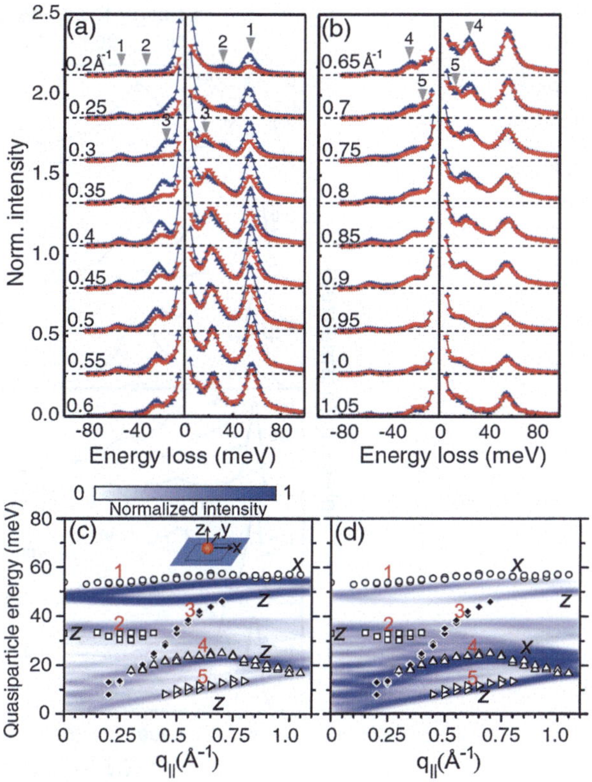

Fig. 17.16 SP-EELS spectrum recorded on Fe(001) p(1 × 1) O at different $q_{\|}$. Blue and red spectra correspond to incident electrons polarized with the spin parallel or anti-parallel to the majority electrons of the sample. Several features are visible, most of which are due to surface phonons. The feature numbered 3 shows a definitive non-Boltzmann factor like anisotropy between loss and gain peaks. It corresponds to the efficient creation (annihilation) of a surface magnon with electrons polarized in the same direction as minority (majority) electrons of the substrate. The *upper panels* report the energy loss spectra, the *lower* the calculated phonon spectra for (**c**) the O layer, (**d**) the outermost Fe layer. Reprinted figure with permission from Ref. [86] Copyright (2011) by the American Physical Society

17.5 The Future of HREELS

We deliberately limited our overview to High Resolution Electron Energy Loss Spectroscopy, thus neglecting e.g. the stunning results obtained recently recording EELS spectra with TEM allowing to identify the elemental compositions of materials [92], to map plasmonic excitations [93] and to follow electronic structural changes with unprecedented time resolution [94]. We mention these results here because they might suggest possible future breakthroughs also for High Resolution Electron Energy Loss Spectroscopy. However, on one side, vibrational spectra of a single adsorbed molecule can now be measured with low temperature STMs by Inelastic Electron Tunneling Spectroscopy (IETS) [95]. On the other side, vibra-

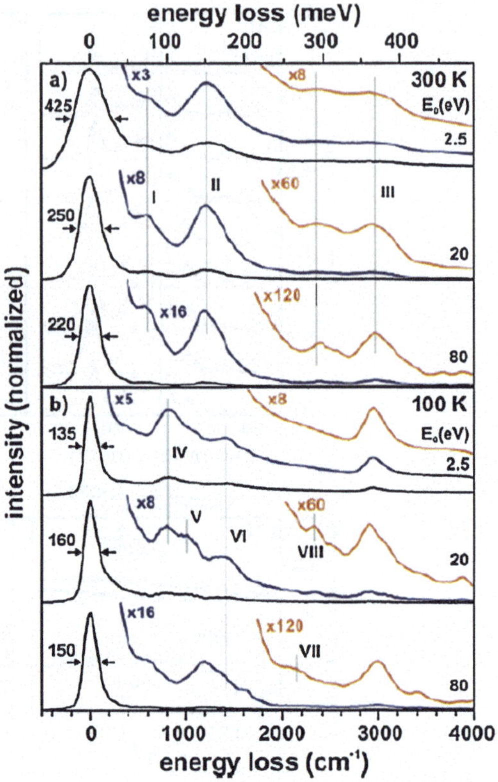

Fig. 17.17 HREELS spectra recorded in-specular at different electron energies on the surface of liquid EMIM. Reprinted with permission from [90]. Copyright 2007 American Chemical Society

tional spectra with picosecond time sensitivity have been recorded by employing, e.g., Sum Frequency Generation. HREELS will have to compete with such alternative methods which, unlike electron spectroscopy, do not need vacuum conditions as a stringent requirement and can therefore be employed also under more realistic environments. HREELS will, however, still be useful for its unique characteristic of enabling excitations with non-vanishing wave-vector, an information needed for a satisfactory description of dynamic phenomena such as surface magnons, surface plasmons and surface phonons.

References

1. M. Rocca, F. Moresco, Phys. Rev. B **50**, 18621 (1994)
2. L. Gavioli, M.G. Betti, C. Mariani, Phys. Rev. Lett. **77**, 3869 (1996)
3. Y. Wang, Z. Phys. Chem. **222**, 927 (2008)
4. H. Kuhlenbeck, G. Odörfer, R. Jaeger, G. Illing, M. Menges, Th. Mull, H.J. Freund, M. Pöhlchen, V. Stämmler, S. Witzel, C. Scharfschwerdt, K. Wennemann, T. Liedtke, M. Neumann, Phys. Rev. B **43**, 1969 (1991)
5. M. De Kieviet, D. Dubbers, Ch. Schmidt, D. Scholz, U. Spinola, Phys. Rev. Lett. **75**, 1919 (1995)

6. A.P. Jardine, H. Hedgeland, G. Alexandrowicz, W. Allison, J. Ellis, Prog. Surf. Sci. **84**, 323 (2009)
7. H. Ibach, D.L. Mills, *Electron Energy Loss Spectroscopy and Surface Vibrations* (Academic Press, San Diego, 1982)
8. H. Ibach, P.W. Hawkes, in *Electron Energy Loss Spectrometers: The Technology of High Performance*. Springer Series in Optical Sciences (1991)
9. P.A. Thiry, M. Liehr, J.J. Pireaux, R. Caudano, Phys. Scr. **35**, 368–379 (1987)
10. B.N.J. Persson, Surf. Sci. **92**, 265 (1980)
11. K. Jacobi, Y. Wang, Surf. Sci. **603**, 1600 (2009)
12. B.M. Hall, D.L. Mills, Phys. Rev. B **44**, 1202 (1991)
13. A.A. Lucas, M. Sunjić, Phys. Rev. Lett. **26**, 229 (1971)
14. E. Evans, D.L. Mills, Phys. Rev. B **5**, 4126 (1972)
15. E. Evans, D.L. Mills, Phys. Rev. B **7**, 853 (1973)
16. E. Evans, D.L. Mills, Phys. Rev. B **8**, 4004 (1973)
17. M. Rocca, U. Valbusa, Phys. Rev. Lett. **64**, 2398 (1990)
18. M. Rocca, F. Moresco, Phys. Rev. Lett. **73**, 822 (1994)
19. W. Ho, R.F. Willis, E.W. Plummer, Phys. Rev. Lett. **40**, 1463 (1978)
20. J. Kröger, S. Lehwald, H. Ibach, Phys. Rev. B **69**, 201404 (2004)
21. J. Kröger, S. Lehwald, H. Ibach, Phys. Rev. B **55**, 10895 (1997)
22. S.Y. Tong, C.H. Li, D.L. Mills, Phys. Rev. Lett. **44**, 407 (1980)
23. M.L. Xu, B.M. Hall, S.Y. Tong, M. Rocca, H. Ibach, S. Lehwald, J.E. Black, Phys. Rev. Lett. **54**, 1171 (1985)
24. S. Andersson, J.W. Davenport, Solid State Commun. **28**, 677 (1978)
25. J.E. Demuth, D. Schmeisser, Ph. Avouris, Phys. Rev. Lett. **47**, 1166 (1981)
26. L. Sanche, M. Michaud, Phys. Rev. Lett. **47**, 1008 (1981)
27. R.E. Palmer, P.J. Rous, J.L. Wilkes, R.F. Willis, Phys. Rev. Lett. **60**, 329 (1988)
28. V. Djamo, D. Teillet-Billy, J.P. Gauyacq, Phys. Rev. B **51**, 5418 (1995)
29. R. Franchy, F. Bartolucci, F. Buatier de Mongeot, F. Cemic, M. Rocca, U. Valbusa, L. Vattuone, S. Lacombe, K. Jacobi, K.B.K. Tang, R.E. Palmer, J. Villette, D. Teillet-Billy, J.P. Gauyacq, J. Phys. C **12**, 53 (2000)
30. L.L. Kesmodel, Phys. Rev. Lett. **53**, 1001 (1984)
31. T.S. Jones, N.V. Richardson, Phys. Rev. Lett. **61**, 1752 (1988)
32. H. Okuyama, H. Kato, M. Kawai, J. Yoshinobu, J. Chem. Phys. **113**, 2866 (2000)
33. C.P.A. Mulcahy, A.A. Aquino, J.J. Rogers, T.S. Jones, J. Chem. Phys. **104**, 9120 (1996)
34. J.S. Ha, S.J. Sibener, J. Chem. Phys. **98**, 10061 (1993)
35. K.W. Wulser, M.A. Langell, Phys. Rev. B **48**, 9006 (1993)
36. R.E. Palmer, P.J. Rous, Rev. Mod. Phys. **64**, 383 (1992)
37. L. Vattuone, M. Repetto, M. Rocca, Compel **11**, 85 (1992)
38. T.H. Ellis, L.H. Dubois, S.D. Kevan, M.J. Cardillo, Science **230**, 256 (1985)
39. K.J. Wu, L.D. Peterson, G.S. Elliott, S.D. Kevan, K.D. Gibson, B.J. Hinch, L.H. Dubois, Rev. Sci. Instrum. **62**, 1256 (1991)
40. M. Rocca, Surf. Sci. Rep. **22**, 1 (1995)
41. H. Claus, A. Bussenschutt, M. Henzler, Rev. Sci. Instrum. **63**, 2195 (1992)
42. V. Zielasek, N. Roonitz, M. Henzler, H. Pfnür, Phys. Rev. Lett. **96**, 196801 (2006)
43. F. Moresco, M. Rocca, T. Hildebrandt, M. Henzler, Phys. Rev. Lett. **83**, 2238 (1999)
44. H. Pfnür, T. Langer, J. Baringhaus, C. Tegenkamp, J. Phys. Condens. Matter **23**, 112204 (2011)
45. D.L. Abraham, H. Hopster, Phys. Rev. Lett. **62**, 1157 (1989)
46. J. Kirschner, D. Rebenstorff, H. Ibach, Phys. Rev. Lett. **53**, 698 (1984)
47. Y.U. Idzerda, D.M. Lind, D.A. Papaconstantopoulos, G.A. Prinz, B.T. Jonker, J.J. Krebs, Phys. Rev. Lett. **61**, 1222 (1988)
48. D.T. Pierce, R.J. Celotta, G.C. Wang, W.N. Unertl, A. Galejs, C.E. Kuyatt, S.R. Mielczarek, Rev. Sci. Instrum. **51**, 478 (1980)
49. H. Hopster, D.L. Abraham, Rev. Sci. Instrum. **59**, 49 (1988)

50. L. Vattuone, M. Rocca, Rev. Sci. Instrum. **73**, 3861 (2002)
51. H. Ibach, D. Bruchmann, R. Vollmer, M. Etzkom, P.S.A. Mumar, J. Kirschner, Rev. Sci. Instrum. **74**, 4089 (2003)
52. R. Vollmer, M. Etzkorn, P.S. Anil Kumar, H. Ibach, J. Kirschner, Phys. Rev. Lett. **91**, 147201 (2003)
53. W.X. Tang, Y. Zhang, I. Tudosa, J. Prokop, M. Etzkorn, J. Kirschner, Phys. Rev. Lett. **99**, 087202 (2007)
54. A.D. Johnson, K.J. Maynard, S.P. Daley, Q.Y. Yand, S.T. Ceyer, Phys. Rev. Lett. **67**, 927 (1991)
55. L. Vattuone, L. Savio, M. Rocca, Phys. Rev. Lett. **90**, 228302 (2003)
56. E. Vesselli, L. De Rogatis, X. Ding, A. Baraldi, L. Savio, L. Vattuone, M. Rocca, P. Fornasiero, M. Peressi, A. Baldereschi, R. Rosei, G. Comelli, J. Am. Chem. Soc. **130**, 11417 (2008)
57. B. Pettinger, X.H. Bao, I. Wilcock, M. Muhler, R. Schlögl, G. Ertl, Angew. Chem., Int. Ed. Engl. **33**, 85 (1994)
58. L. Savio, L. Vattuone, M. Rocca, Phys. Rev. Lett. **87**, 276101 (2001)
59. L. Savio, C. Giallombardo, L. Vattuone, A. Kokalj, M. Rocca, Phys. Rev. Lett. **101**, 266103 (2008)
60. T. Nagao, S. Hasegawa, Surf. Interface Anal. **30**, 488 (2000)
61. B. Diaconescu, K. Pohl, L. Vattuone, L. Savio, P. Hofmann, V.M. Silkin, J.M. Pitarke, E.V. Chulkov, P.M. Echenique, D. Farias, M. Rocca, Nature **448**, 57 (2007)
62. R. Fuchs, K.L. Kliewer, Phys. Rev. **140**, A2076 (1965)
63. X. Hou, S. Yang, G. Dong, X. Ding, X. Wang, Phys. Rev. B **35**, 8015 (1897)
64. V. De Renzi, R. Biagi, M.G. Betti, C. Mariani, Phys. Rev. B **49**, 8198 (1994)
65. K. Sakamoto, T. Suzuki, M. Harada, T. Wakita, S. Suto, A. Kasuya, Phys. Rev. B **57**, 9003 (1998)
66. P.A. Thiry, M. Liehr, J.J. Pireaux, R. Caudano, Phys. Rev. B **29**, 4824 (1984)
67. L. Savio, E. Celasco, L. Vattuone, M. Rocca, P. Senet, Phys. Rev. B **67**, 075420 (2003)
68. H. Ibach, Phys. Rev. Lett. **27**, 253 (1971)
69. S. Lehwald, J.M. Szeftel, H. Ibach, T.S. Rahman, D.L. Mills, Phys. Rev. Lett. **50**, 518 (1983)
70. M. Rocca, in *Physics of Covered Surfaces, I. Adsorbate Layers on Surfaces*, Landolt-Boernstein, vol. III/42 A2, ed. by H.P. Bonzel (2002), pp. 1–68. Chap. 4.5
71. J.M. Szeftel, S. Lehwald, H. Ibach, T.S. Rahman, J.E. Black, D.L. Mills, Phys. Rev. Lett. **51**, 268 (1983)
72. F. Moresco, M. Rocca, V. Zielasek, T. Hildebrandt, M. Henzler, Phys. Rev. B **54**, 14333 (1996)
73. S.J. Park, R.E. Palmer, Phys. Rev. Lett. **105**, 016801 (2010)
74. K. Pohl, B. Diaconescu, G. Vercelli, L. Vattuone, V.M. Silkin, E.V. Chulkov, P.M. Echenique, M. Rocca, Europhys. Lett. **90**, 57006 (2010)
75. L. Vattuone, G. Vercelli, L. Savio, M. Smerieri, M. Rocca, Plasmonics **7**, 329 (2012)
76. L. Vattuone, H. Pfnür et al. (in preparation)
77. H. Qiu, B. Meyer, Y. Wang, Ch. Wöll, Phys. Rev. Lett. **101**, 236401 (2008)
78. B.N.J. Persson, J.E. Demuth, Phys. Rev. B **30**, 5968 (1984)
79. Y. Wang, B. Meyer, M. Kunat, D. Langenberg, F. Träger, A. Birkner, Ch. Wöll, Phys. Rev. Lett. **95**, 266104 (2005)
80. Y. Chen, J.C. Barnard, R.E. Palmer, M.O. Watanabe, T. Sasaki, Phys. Rev. Lett. **83**, 2406 (1999)
81. B.N.J. Persson, J.E. Demuth, Phys. Rev. B **42**, 8057 (1990)
82. Y. Li, L. Huang, C.M. Lieber, Phys. Rev. Lett. **68**, 3240 (1992)
83. M.P. Gokhale, A. Ormeci, D.L. Mills, Phys. Rev. B **46**, 8978 (1992)
84. M. Plihal, D.L. Mills, J. Kirschner, Phys. Rev. Lett. **82**, 2579 (1999)
85. D. Venus, J. Kirschner, Phys. Rev. B **37**, 2199 (1988)
86. Y. Zhang, P.A. Ignatiev, J. Prokop, I. Tudosa, X.W. Tang, Kh. Zakeri, T.R.F. Peixoto, V.S. Stepanyuk, J. Kirschner, Phys. Rev. Lett. **106**, 127201 (2011)
87. H.P. Steinrück, Surf. Sci. **604**, 481 (2010)

88. F. Maier, J.M. Gottfried, J. Rossa, D. Gerhard, P.S. Schulz, W. Schwieger, P. Wasserscheid, H.P. Steinrück, Angew. Chem., Int. Ed. Engl. **45**, 7778 (2006)
89. B.S. Perkins, D.J. Nesbitt, Proc. Natl. Acad. Sci. **105**, 12684 (2008)
90. S. Krischok, M. Eremtchenko, M. Himmerlich, P. Lorenz, J. Uhlig, A. Neumann, R. Ottking, W.J.D. Beenken, O. Hofft, S. Bahr, V. Kempter, J.A. Schaefer, J. Phys. Chem. B **111**, 4801 (2007)
91. V.M. Polyakov, A. Elbe, J.A. Schäfer, Surf. Sci. **420**, 43 (1999)
92. K. Suenaga, M. Tence, C. Mory, C. Colliex, H. Kato, T. Okazaki, H. Shinohara, K. Hirahara, S. Bandow, S. Ijima, Science **290**, 2280 (2000)
93. J. Nelayah, M. Kociak, O. Stephan, F.J. Garcia de Abajo, M. Tence, L. Henrard, D. Taverna, I. Pastoriza-Santos, L.M. Liz-Marzan, C. Colliex, Nat. Phys. **3**, 348 (2007)
94. F. Carbone, O. Kwon, A. Zewail, Science **325**, 181 (2009)
95. B.C. Stipe, M.A. Rezaei, W. Ho, Science **280**, 1732 (1998)

Chapter 18
Low-Energy Electron Microscopy

Juan de la Figuera and Kevin F. McCarty

Abstract Low-energy electron microscopy (LEEM) images a beam of low-energy electrons that have been reflected from a sample. The technique characterizes the sample's surface in real-space with nanometer-scale lateral resolution. Through a variety of contrast mechanisms, different aspects of the surface can be imaged, including the distribution of different phases and the location of atomic steps. LEEM instrumentation can also acquire electron diffraction patterns from local regions of the surface. The ability to acquire images quickly during temperature changes, while depositing films and exposing materials to reactive gases makes LEEM extremely useful for studying dynamical processes on surfaces. New developments include aberration correction systems for improved spatial resolution and bright spin-polarized electron sources.

18.1 Introduction

Low-energy electron microscopy (LEEM [1–7]) is a technique uniquely suited to perform dynamical observations of surfaces with nanometer resolution under vacuum. Figure 18.1 schematically illustrates the technique. A parallel beam of electrons is deflected onto a sample by a prism optic (beam separator). The electrons decelerate as they approach the surface because of an electrical bias applied to the sample. Electrons backscattered from the surface are accelerated and deflected by the prism into an imaging column and finally onto an imaging detector. Apertures in the illumination and imaging columns control the size of the electron beam on the sample and the electrons that strike the detector, respectively.

In many aspects LEEM is analogous to transmission electron microscopy (TEM [8]), the standard bulk microscopy technique. For example, both techniques

J. de la Figuera (✉)
Instituto de Química-Física "Rocasolano", CSIC, Madrid, Spain 28006
e-mail: juan.delafiguera@iqfr.csic.es

K.F. McCarty
Sandia National Laboratories, Livermore, CA 94550, USA
e-mail: mccarty@sandia.gov

G. Bracco, B. Holst (eds.), *Surface Science Techniques*,
Springer Series in Surface Sciences 51, DOI 10.1007/978-3-642-34243-1_18,

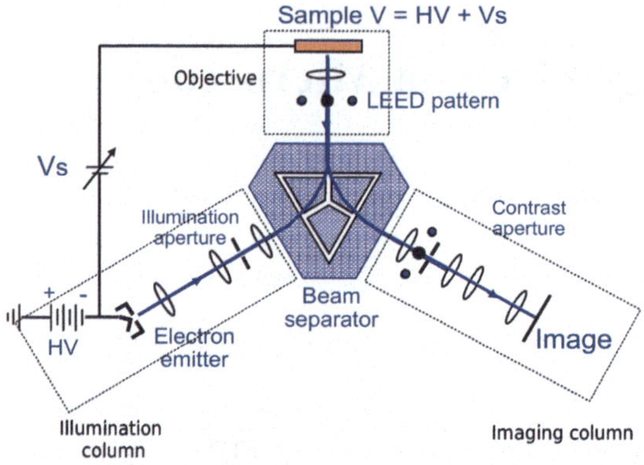

Fig. 18.1 Simplified LEEM schematic. Electrons from an emitter are formed into a parallel beam in the illumination column. After deflection by a prism optic, the beam impinges orthogonally onto a reasonably flat surface. The sample is biased to decelerate the electrons from their initial energy to a small energy, typically a few to tens of eV. Backscattered electrons are accelerated away from the sample and deflected in the opposite direction through the prism and into an imaging column. There a series of lens manipulates the electrons onto a two dimensional imaging detector. An illumination aperture defines the size of the electron beam on the sample, which is particularly useful for selected area electron diffraction. The contrast aperture in a diffraction plane selects the electrons that pass through to the detector

can acquire electron diffraction patterns from local sample areas and form images from the zeroth order electron beam (bright field imaging) or higher order diffraction beams (dark-field imaging). Although TEM can, and has been used to "look" at surfaces [9], it lacks strong contrast precisely because the operating conditions are optimized for bulk, not surface, observations. In particular, the high electron energy needed for the transmission mode gives low sensitivity to surface signals. A natural solution for observing surfaces is to employ low-energy electrons, which strongly interact with the surface. Since low-energy electrons won't penetrate through samples of practical thickness, the solution is to reflect the beam from the surface. This is the essence of LEEM. This idea was proposed before the development of "modern" surface science, with its heavy reliance on commercial technology to achieve ultrahigh vacuum (UHV). But due to technical challenges, only within the last two decades has the technique achieved its promise and become a versatile tool that allows the real *and* reciprocal space observation of surfaces during crystal growth, surface phase transitions or chemical reactions. This is achieved by using a variety of contrast mechanisms and imaging modes with nanometer-scale resolution, as explained in detail below (see Fig. 18.2). When coupled to a light source (be it a synchrotron, a laboratory-based laser or a helium discharge source) the instrumentation can provide most of the classical techniques for surface characterization: low-energy electron diffraction (LEED), scanning electron microscopy (SEM), X-ray photoelectron spectroscopy (XPS), ultraviolet photoelectron spectroscopy (UPS), etc. Use of spin-polarized electron sources or illuminated by polarized light add the capability to detect magnetic domains.

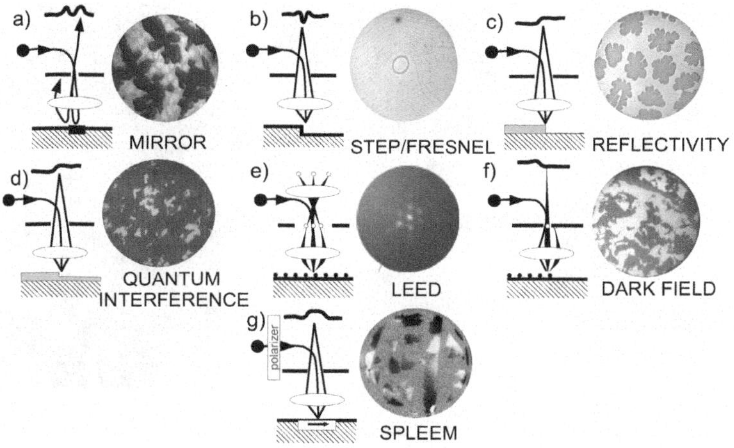

Fig. 18.2 Different working modes of a LEEM instrument. (**a**) Mirror electron microscopy of MgH_2 on Mg(0001), where the 1.25 eV electrons are reflected before striking the surface. The field of view (FOV) is 7 μm (reprinted with permission from [10]). (**b**) Phase contrast where atomic steps on a W(110) surface appear as dark lines because of interference in Fresnel diffraction (FOV is 10 μm). (**c**) Real-space image showing composition (diffraction) contrast between Cu islands and bare regions of a Ru(0001) substrate (FOV is 10 μm). (**d**) Phase contrast causing different thicknesses of Co on Ru(0001) to have different brightnesses. Average thickness is 7 ML (FOV is 10 μm). (**e**) LEED pattern of Cu/Ru(0001) showing 6-fold satellite spots around specular (zeroth order) beam. (**f**) Dark-field imaging mode using one of the satellite LEED spots in e) (FOV is 4 μm, from [11]). (**g**) Magnetic contrast in 2 ML thick islands of Co/Ru(0001) using a spin-polarized electron source, with a FOV of 4 μm. The schematics are adapted with permission from [1]

This chapter emphasizes what can and cannot be done with the current generation of LEEM instruments, with suggestions to make the most of them. We limit the scope to using electrons as the illumination source, i.e., pure LEEM. Please note, though, that all LEEMs can also perform photoemission electron microscopy (PEEM) if the appropriate light source is available [12].[1] LEEM can be used to analyze samples prepared ex-situ, as commonly done in synchrotron-based PEEM analysis. However, we particularly note that the real power of LEEM is being able to prepare materials under the carefully controlled conditions possible in an UHV-based instrument and track in real time the material's evolution. The ability to image quickly while changing temperature, annealing, exposing to gases or growing films allows for the rapid exploration of material systems. This productivity has led to a generation of LEEM users, as opposed to the LEEM "builders" of the pioneering

[1] In fact PEEM is more forgiving in terms of the samples that can be analyzed—some electrons are photoemitted normal to the surface plane even from very rough surfaces or from surfaces whose facets are not in the surface plane. In contrast, LEEM works best on relatively flat surfaces whose facets lie mainly within the surface plane. (For surfaces whose facets are not orthogonal to the electron beam, the specularly reflected electron beam is not transmitted through the imaging column under normal conditions.)

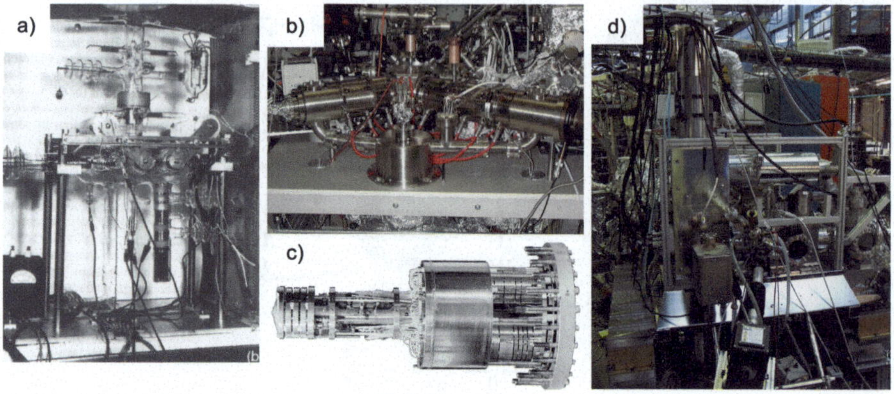

Fig. 18.3 (**a**) Glass LEEM test system, reproduced with permission from [13]. (**b**) Elmitec III instrument at Sandia National Laboratories, California. (**c**) Flange-on SPLEEM at Berkeley National Laboratories, California. (**d**) First Specs aberration-corrected LEEM/PEEM [18] at the BESSY synchrotron, Berlin

era [13]. The audience of this chapter are researchers interested in the LEEM capabilities. Hopefully some will turn into LEEM users.

Historical Development The LEEM concept, to use reflected electrons for surface crystallography, was proposed and developed by Ernst Bauer over a multidecade period of time. The first feasibility tests where conducted in the 1960s with the construction of a glass system (Fig. 18.3a) [13]. The first functioning instruments providing quality images are from the mid-1980s [14]. In contrast, the last decade of the XX century can be considered the awakening age of LEEM. The first spin polarized instrument was constructed in the beginning of the decade (see Fig. 18.3c, [15]). The founding of the first successful LEEM company (Elmitec [16]) took place in 1995. The first work on employing aberration correction also started in the same decade, with the development of the SMART LEEM [17].

Developments have recently accelerated and the technique has attracted more attention [7]. By the end of the first decade of the XXI century, we have already two commercial companies (Specs [19] entering the field in 2009 to join Elmitec [16]) with different designs (see Fig. 18.3b and d). Both offer instruments with aberration correction optics, electron energy analyzers and resolutions that go below the 10 nm mark and approach the nm level. New developments are under way, such as a cryogenic LEEM using helium cooling and an aberration-corrected instrument with a spin-polarized electron source.

18.2 Background

Using reflected electrons to image a surface is a simple idea. But it raises the obvious questions of whether a surface would have enough electron reflectivity to allow for an imaging technique, what mechanisms would provide contrast and what would be the achievable resolution [3, 13].

Imaging with Backscattered Electrons A large flux of electrons are needed to image a surface with reflected electrons in a non-scanning mode. The experience from low-energy electron diffraction (LEED) from single crystal surfaces indicated that a few percent of incident electrons should be backscattered at select energies [20]. Nevertheless LEED experience can be misleading since the technique traditionally uses electron energies above 50 eV due to experimental limitations [21]. But LEEM instruments work well at lower electron energies, where most of the contrast effects are strongest.

If the sample is at a voltage slightly more negative than that of the electron source, the entire electron beam is reflected just above the specimen surface. In this imaging mode, mirror electron microscopy (MEM [22]), the slow moving electrons are sensitive to the spatial and temporal variations in the electric field. In MEM there is obviously no scarcity of electrons for imaging. As those electrons do not actually reach the sample and they are very sensitive to changes in the electric field, their interpretation is not straightforward [23] but it can be used to reduce damage from the electron beam [10]. But even at higher energies, the reflectivity of a crystalline substrate is very high. Figure 18.4a shows experimental reflectivity, i.e., the reflected intensity normalized to the incident intensity, obtained from LEEM images as a function of electron energy. The reflected ratio of electrons is still very high (up to 20 %) at electron energies above the vacuum level, i.e., where the reflectivity falls below unity. This high reflectivity at low energies was theoretically predicted early on [3, 13], fueling the development of LEEM.

Contrast Modes The interaction of low-energy electrons with a substrate can give rise to a variety of contrast mechanisms. We have already presented the mirror mode (Fig. 18.2a), where contrast arises from differences in the electric field at the surface of the sample. Other contrast mechanisms also allow imaging: (1) the spatial distribution of different phases (Fig. 18.2c), (2) the thickness in atomic layers of films on a substrate (Fig. 18.2d), (3) the location of atomic steps on crystalline surfaces (Fig. 18.2b), and (4) the spatial distribution of differently oriented magnetic domains (Fig. 18.2g). The detailed reflected electron intensity at higher energies depends, among other factors, on the band structure of the substrate. To understand the relation of reflectivity and band structure it is instructive to consider the electron reflection under the single scattering approximation [28]. In such case, the reflected intensity would be directly related to the unoccupied band structure of the surface: a high reflectivity would occur when the density of available unoccupied electron states is low, and vice versa. Of course, multiple scattering is very important at LEEM energies, so it has to be taken into account using similar formalisms to those employed in LEED [28]. But even in the absence of detailed understanding of a particular system, it is expected that the band structure, and thus the electron reflectivity, from different materials should differ. In consequence, an energy-dependent contrast should arise from composition differences (see Fig. 18.4a). Whether two

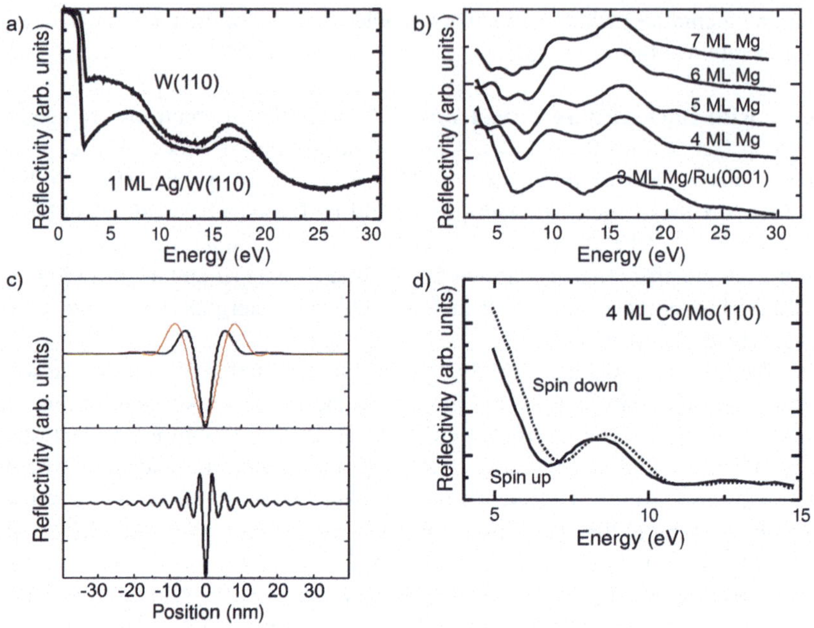

Fig. 18.4 Origin of image contrast in LEEM. (**a**) Electron reflectivity vs incoming electron energy for W(110), with and without a layer of Ag on top (reprinted with permission from [24]); (**b**) electron reflectivity as a function of film thickness for films of different thickness on Ru(0001), reprinted with permission from [25]; (**c**) real-space profile expected from Fresnel diffraction across a step with different aberrations, reprinted with permission from [26]; (**d**) electron reflectivity as a function of the incoming electron beam spin orientation relative to the sample magnetization in a 4 ML Co film on Mo(110), reprinted with permission from [27]

different elemental compositions give a useful contrast must be determined experimentally.

Another common origin of contrast in LEEM instruments is electron interference effects (phase contrast). For example, if a thin film has a sharp interface with its substrate, interference can occur between electrons reflected from the film/substrate interface and from the film surface. This gives rise to a Fabry-Perot effect that makes the electron intensity change as a function of electron energy and film thickness. Clearly, these changes in reflectivity can be used to detect film areas of different thickness, as shown for Mg films in Fig. 18.4b. Choosing appropriate electron energies allows imaging the formation of new film layers during layer-by-layer growth.

A similar effect allows detecting atomic steps: Fresnel diffraction [26, 29, 30] from electrons reflected at the lower and upper terraces around a substrate step gives rise to an oscillatory pattern of the reflected intensity around the step location (Fig. 18.4c, bottom). In non-aberration corrected instruments, the aberrations mask the fine detail of the Fresnel diffraction pattern so a step is typically seen as a single line (Fig. 18.4c, top). By observing the changes with focus and electron wavelength (i.e., energy) the up-down or down-up sense of the step can be determined [29].

In aberration corrected instruments, the fringe pattern around step edges has been observed [31]. Steps in LEEM have been detected on metal, semiconductor, nitride and oxide surfaces, and the easy observation of steps remains a crucial advantage of LEEM over other surface-sensitive electron microscopies.

Diffraction is another source of image contrast in LEEM. To improve contrast, LEEM images are usually acquired using a “contrast” aperture, which limits the diffracted beams that contribute to the observed image to the specular beam. This corresponds to the bright-field imaging mode in TEM [8]. By using deflectors in the beam separator, the illuminating or reflected beam can be tilted so real-space images can be formed from other diffraction beams (see Fig. 18.2f). If the surface has domains that give different diffraction patterns, each domain can be imaged in turn by selecting diffraction spots unique to each corresponding diffraction pattern (see Fig. 18.2f), i.e., by dark-field imaging.

18.2.1 Magnetic Contrast

Magnetic contrast can be obtained using electrons due to interactions of the illuminating electron beam either with (1) the magnetic field outside a ferromagnetic sample through the Lorentz force, or (2) by a direct interaction with the spin of the sample’s electrons. In the latter case, exchange scattering or spin-orbit interactions can produce a dependence of the reflected intensity on the electron beam spin direction, and thus, provide magnetic contrast related to the magnetization of the sample, as shown in Fig. 18.4d. Either the source can be spin polarized or the spin polarization of the electrons can be analyzed after interaction with the sample.

The first method (Lorentz force) is implemented in LEEM by tilting the illumination beam and using incident electron energies close to mirror mode. Stripe domain patterns in a perpendicular magnetized film [32] and more recently bit patterns in magnetic recording media have been imaged [33]. But the most common method is to use a spin-polarized electron source (spin-polarized LEEM, or SPLEEM [34, 35]), where exchange scattering at the sample provides a contrast that depends on the scalar product of the spin direction and the sample magnetization. The magnetic contrast can be as high as 20 % of the topographic contrast for selected substrates or films, although it is often much smaller. To remove non-magnetic contrast from the images, it is common to acquire LEEM images with opposite illuminating electron-beam spin polarizations, denoted as I_+, I_-. Then the asymmetry ratio A, corrected by the electron beam spin polarization P, is calculated pixel by pixel:

$$A = \frac{1}{P}\frac{I_+ - I_-}{I_+ + I_-}$$

By selecting the spin direction in each of three orthogonal directions, the local magnetization direction can be determined with nanometer resolution.

Dark-field imaging is also usable for magnetic imaging in antiferromagnetic domains, even without a spin polarized incident beam. An antiferromagnetic structure gives rise to a magnetic unit cell that is larger than the material's structural unit cell. Thus, it gives additionally diffracted beams of pure magnetic origin [36]. Using the magnetic diffracted beams as a source of dark-field contrast allows imaging antiferromagnetic domains in real-space. This method has recently been used on NiO [37]. The advantage of using spin-polarized electrons would be to determine additionally the spin-direction of the antiferromagnetic structure while also easily verifying the magnetic origin of the observed contrast.

18.2.2 Resolution

In contrast to TEMs where the actual resolution is orders of magnitude larger than the electron wavelength, in LEEM the experimental resolution is not far from this limit. The achieved resolution of an electron optics system can be written in a simplified description as the quadratic addition of several terms, the most important ones being the diffraction limit, the chromatic aberration and the spherical aberration. The latter two correspond to the same aberrations in light optics: considering a lens with a given focus length, spherical aberration arises because the focus depends on the radial distance to the center of the lens, while chromatic aberration corresponds to the dependence of the focal length on wavelength, i.e., electron energy [38]. Nearly all of the aberrations of a LEEM instrument arise in the objective lens [2, 38, 39], and unlike in TEM [8], a substantial part arises from chromatic aberration. Spherical aberrations are typically minimized using a contrast aperture centered around the specular beam. (A contrast aperture in PEEM also improves spatial resolution but reduces the signal available for imaging.) The chromatic aberration in LEEM is reduced by decreasing the energy spread of the electron source. For example, the current generation of instruments reaches 10 nm resolution with thermionic emission electron sources and 4 nm resolution with field emission sources, with aberration coefficients for chromatic aberrations of ≈ 150 m and for spherical aberrations of ≈ 0.5 m (at the image side) [39, 40].

The resolution of a LEEM is improved if the aberrations are reduced. But unlike in light optics, aberration correction in electron optics is quite recent. In LEEM instruments, the correction is performed by electron mirrors. Current instruments can remove the first order chromatic and spherical aberration and achieve 2 nm lateral resolution [18, 31]. Further improvements are anticipated.

18.3 Experimental Setup and Methods

From the technical point of view, a LEEM has much in common with a traditional TEM. Both use a series of lenses for manipulating the electrons that illuminate the sample and those that have interacted with the sample. Also, both form real-space images or diffraction patterns of the electrons that have interacted with the sample.

But LEEM works in reflection mode instead of in transmission, and at very low energies. Currently there are two successful commercial instrument designs [16, 19], plus a few unique, researcher-built systems.

A LEEM system (Fig. 18.1) consists of an illumination system, a beam separator, an objective lens system, and an imaging column. Additional lenses allow imaging the diffraction plane (back-focal plane) of the objective lens, while apertures can limit the illuminated area on the sample (i.e., the imaged area) and the angular spread in reciprocal space of the imaged electrons. While lenses and apertures are well-known elements of other electron microscopes (SEMs and TEMs), the beam separator is unique to LEEM. The objective lens is shared with other types of emission microscopes.

Lens System and General Considerations At the heart of any LEEM there are several electrostatic and magnetic lenses [43]. Electron lenses are well-developed and used in a variety of charged-particle instrumentation. The preference of electrostatic [44] vs magnetostatic lenses is mostly a question of convenience. While magnetic lens have smaller in aberrations, they are usually more bulky and difficult to cool down unless the coils are outside vacuum. They also rotate the electron spin and the image they form, although this effect can be avoided using self-canceling doublets.

Within the LEEM instrument the electrons have energies in the range of 10–20 keV. The objective lens decelerates the electrons to a few eV close to the sample. Decelerating the electrons requires either that the electrons travel at a high positive potential relative to the laboratory electrical ground, with the electron source grounded and the sample at a few eV, or that the electrons travel close to electrical ground and the electron gun and sample are at 10–20 kV negative potential. Obviously, having the sample close to ground potential is convenient from the experimental point of view. But it is not compatible with magnetic lenses, which have their central tube section at ground. Electrostatic lenses do not have this limitation.

The most popular commercial designs (see Fig. 18.5a, c) are dedicated LEEM instruments that employ magnetic lenses and have the electron source and sample at high voltage. In these instruments the illuminating and imaging sections and the objective and sample sections are housed in separate (but connected) vacuum chambers. The magnetic coils surround lengths of thin vacuum tubing at ground potential. As the connection between the sample area and the columns is through a small opening (several mm), differentially pumping permits imaging under pressures in the 10^{-4} mbar range while keeping the detector and electron emitter at UHV.

There are also fully electrostatic instruments (with the exception of the beam separator, see Fig. 18.5b) that have the electron source at ground and the sample close to ground. The use of electrostatic lenses makes for compact instruments that can be mounted on a standard flange (e.g., a 6″ CF flange) of a UHV chamber. Such systems comprise one commercial design (Elmitec IV [16]) and a few research systems [42, 45]. As all the electrostatic lenses are in the sample vacuum chamber, this type of LEEM usually is more demanding in terms of background pressure.

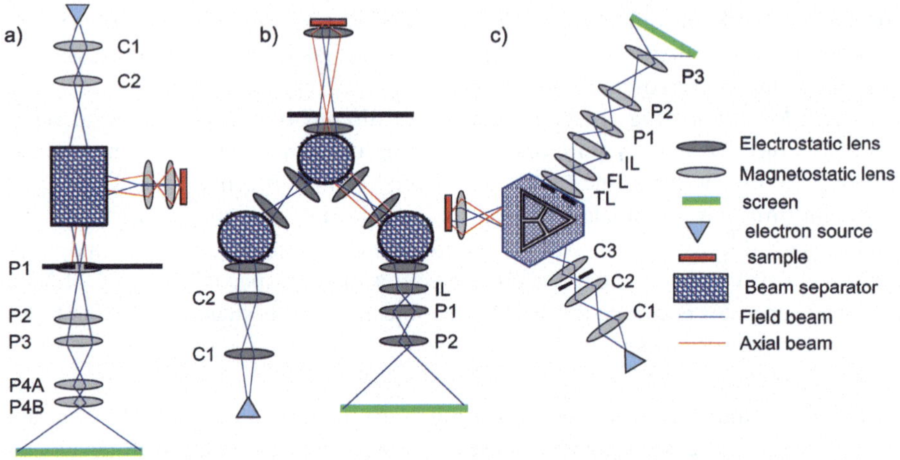

Fig. 18.5 (**a**) Specs/IBM LEEM design, reproduced with permission from [41]. (**b**) SPLEEM on-a-flange design, reproduced with permission from [42]. (**c**) Elmitec III design [16]

Illumination Sources and Condenser Optics Three types of electron sources are used in current LEEM instruments: thermionic, field emission and photoemission sources. The most popular thermionic source is based on a LaB_6 single crystal emitter. It is inexpensive, has a long lifetime, and is quite sturdy in UHV environments [46]. The main limitations is the energy spread of about 0.6 eV and the brightness.

More recently, cold field emission (CFE) sources are gaining in popularity. Although the LEEM market is too small to fuel electron source development, the TEM and SEM markets have provided for easy-to-use and reliable sources. The energy spread for cold field emission is ∼0.25 eV, significantly improving resolution.

Finally, GaAs sources are the only spin-polarized electron sources presently used for magnetic imaging (i.e., SPLEEM), both in commercial systems [16] and in custom-built instruments [15]. The energy spread is even smaller than cold field emission sources (0.11 eV). A GaAs electron source [47] consists of a GaAs cathode conditioned to negative electron affinity (NEA), which emits electrons when illuminated by circularly polarized light from a diode laser. Driving the GaAs cathode to NEA condition requires an appropriate coverage of cesium and oxygen. The need to properly prepare the GaAs surface (cleaning it, and then depositing Cs and oxygen) together with the requirement of extreme vacuum (well below 10^{-10} mbar) for reasonable lifetimes of the prepared cathode make this electron source quite demanding.[2] The traditional design suffers from the large size of the illuminating laser beam, which consequently provides a brightness lower by about 10^3 $A \times cm^{-2}\,sr^{-1}$ than a thermionic LaB_6 source. Recent designs focus the laser beam to a micrometer area and provide brightness comparable with thermal sources [48]. The spin

[2]If properly tuned, cathode preparation is routine and lifetimes are several days.

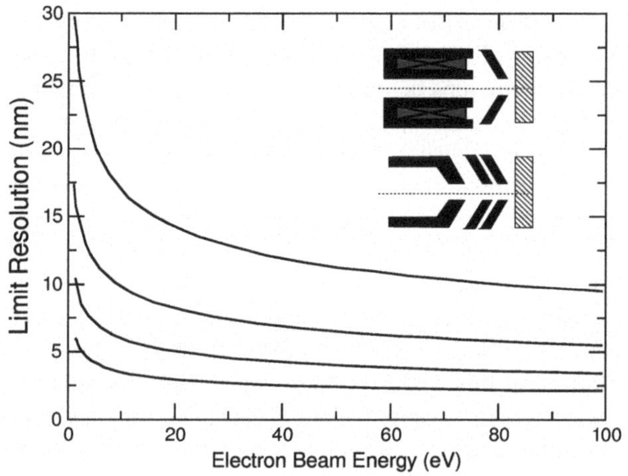

Fig. 18.6 Resolution limit as a function of the electron beam energy at the sample for a given energy spread (ΔE) and acceleration voltage (E), adapted with permission from [39]. The curves shown the resolution from top to bottom for $\Delta E = 2$ eV and $E = 5$ kV, $\Delta E = 2$ eV and $E = 15$ kV, $\Delta E = 0.25$ eV and $E = 5$ kV, $\Delta E = 0.25$ eV and $E = 15$ kV respectively. The *inset* shows two popular objective configurations adapted with permission from [50]: a fully electrostatic tetrode (*lower schematic*) and an electrostatic acceleration electrode paired with a floating magnetic lens (*upper schematic*)

direction of the electrons is perpendicular to the GaAs cathode. A spin manipulator between the GaAs cathode and the condenser optics is used to turn the spin direction to any desired orientation relative to the sample [40, 49].

The condenser lenses are used to obtain a parallel beam of electrons before reaching the sample by providing a demagnified image of the source at the objective back-focal plane. Having several condenser lenses provides maximum flexibility, although some recent designs use a minimalist approach of a single condenser lens [19]. Additionally the illuminating system can contain some deflectors and/or astigmatic octupole correctors. Typically the illuminated area on the sample is 10–100 μm in diameter depending on the design and source. Apertures can reduce the area down to about 0.2 μm.

Objective The objective lens performs a dual role: for the incoming electron beam its functions are to slow the beam from 10–20 keV to the eV range and provide a parallel beam of electrons at the sample by collecting the electrons at its back-focal plane. For the electron beam reflected from the sample, the objective increases the beam energy from several eV to 10–20 keV and forms a magnified real image of the emitted electrons (typical magnifications range from 10–50×). A diffraction pattern exists at its back-focal plane. All objectives are thus composed of an accelerating lens and an additional electrostatic or magnetostatic lens for focusing (see the inset in Fig. 18.6). As shown in Fig. 18.6, the two most important factors for determining the resolution in a LEEM objective are (1) the acceleration voltage, or conversely, the field strength between the sample and the first lens, and (2) the energy spread of the electrons. In all current instruments, the field strength is about 10–20 keV/mm.[3] Several objective designs have been tried (see inset of Fig. 18.6) with an electrostatic tetrode or a magnetic lens behind the accelerating lens being the most popular (the latter has smaller aberrations [50]). Nearly all the spherical and chromatic aberrations that limit the final instrument resolution originate at this objective system. The quantification of the objective aberrations has been studied in detail [39] due to the current interest in the aberration correction of LEEM instruments.

Beam Separator The function of the beam separator is to separate the electrons going into the sample from the electrons leaving the sample. In its simplest form it consists of a square or circular magnetic dipole with two parallel pole plates excited by a coil. A limitation of this design is that a dipole field focuses the electrons that move in a plane normal to the magnetic field. This means that the focusing properties are very different in the two axis of the electron beam. A solution to this problem is to add extra lenses for the focusing function, as in the Berkeley SPLEEM [15] (Fig. 18.5b), which uses double dipoles with opposite magnetic fields in both the illuminating beam and in the imaging section together with additional electrostatic lenses in the beam separator.[4] Most of the current systems are based on a 60° deflector that uses an array of dipole fields [51] (see Fig. 18.5c). However, 90° deflectors are being increasingly used [41] (see Fig. 18.5a). They allow stacking the optics vertically, a particular advantage for aberration correction, where an additional beam separator is needed [18]. They are constructed from a central (square) dipole magnet, surrounded by one [41] or more rings [33] at different magnetic potentials, or by additional round electrostatic lenses [52].

Imaging Optics and Imaging Systems The imaging column contains several lenses arranged so a field of view between about 1–100 μm can be imaged. After the last imaging lens, the electron current is amplified by microchannel plates before impinging on a phosphor screen. A digital camera outside of the vacuum chamber records images of the screen.

The channel-plate-based detectors are delicate and typically do not have a uniform response across their area. The main operational danger of a LEEM lies in the microchannel plates: excessive electron flux can damage the plates and the phosphor, giving rise to dead areas in the detector. Moreover, runaway events in the detector can cause damage even when no electron beam is present. When switching from real-space mode to a diffraction pattern, care has to be exercised not to exceed the maximum current of the channel plates. Controlling the image intensity safely is essential for the current detectors. In some instruments, the preferred way to control intensity is changing the illuminating beam current, while other instruments change

[3]One fact in LEEM that is often overlooked is that the sample always experiences such an applied field. As the resolution degrades mildly with acceleration voltage [39] (see Fig. 18.6), the electrical field might be somewhat reduced for delicate samples.

[4]The opposite magnetic fields cancel the spin rotation of an individual dipole so that an electron transiting the complete beam separator does not change spin direction.

the channel plate voltage. Direct imaging sensors without channel plates are being tested but they have smaller active areas and cannot be baked at the temperatures typically used to achieve UHV conditions [53].

Energy Filters Additional capability comes from the ability to filter the electrons emitted from the surface by their energy. In its simplest form, LEEM makes images from elastically reflected electrons. Any inelastically emitted electrons are then an unwanted background in the images and in LEED patterns. But the inelastic electrons also contain information, as do electrons in photoemission. An energy filter extracts information from these electrons.

Two types of energy filters are being used. One type decelerates and then bends the electron beam, which disperses it by energy. A recent design takes advantage of the energy dispersion within the beam separator itself to provide energy discrimination without extra optical elements other than apertures [41]. The magnetic field of the beam separator gives a chromatic dispersion of about 5 μm/eV [54], useable but much smaller than that of the first type of filter [1].

Two modes of operation are common. In the first, the dispersive plane of the energy filter is imaged onto the detector. Converting image intensity versus position into a count rate vs. energy gives a traditional spectrum. An aperture in the beam separator is used to select the analysis region on the sample, analogous to selected-area LEED. In the second mode, an aperture in the dispersive plane is used to select a given energy bandpass. The passed electrons are then formed into an energy filtered image. Changing the sample potential changes the electron energy that passes through the filter's aperture to the detector.

In a pure LEEM instrument (no light source), a filter enables electron energy loss spectroscopy (EELS) [55, 56] and Auger spectroscopy. Setting the filter energy to a given loss or Auger peak gives spatial maps of phase or elemental composition. When combined with light sources, the energy filter allows photoemission spectroscopy and microscopy, as is most powerfully practiced at synchrotrons. However, intense He discharge lamps and ultraviolet lasers are enabling photoemission spectromicroscopy in laboratory environments [57].

Aberration Correction Systems As mentioned before, simple electrostatic or magnetostatic lenses always have the same sign of spherical and chromatic aberrations. To increase the resolution and the acceptance angle (throughput) of the instrument, the first order aberrations have to be corrected. A successful approach is through the use of electron mirrors [38], which have opposite chromatic and spherical aberrations than lenses. Such an electron mirror can be realized with a regular electrostatic lens but with a much higher voltage, so the incoming electrons are reflected back. The main drawback of using an electron mirror is that the electrons going towards the mirror and the electrons reflected have to be separated. This means that electron mirrors had to wait until the development and familiarity with beam separators, which are specific to LEEM. With the present beam separator expertise, the adoption of electron mirrors for aberration correction is finally taking place. The current designs use either a special four-way beam separator, or stacked beams separators of the 60° [16, 40] or 90° type [18].

Methods to measure the aberrations and adjust the mirror voltages to correct for the first order aberrations all while changing other instrument parameters such as the electron energy are under development [31, 39].

18.4 Applications

LEEM instruments have been applied to a variety of surface science phenomena. Here we highlight some applications where LEEM instruments have added new insights. Most LEEM experiments share the same profile: first an atomically clean surface must be obtained, so either sputtering or annealing cycles are performed. Already at the cleaning stage, the surface state should be checked often (or continuously if possible) in LEEM. The full power is realized when all processes such as exposure to molecular or atomic fluxes are done under LEEM observation. Unless preparation procedures are very cumbersome, several experimental runs can be performed per day. This allows for rapid exploration of the experimental parameter space. LEEM instruments have been used to study metals, semiconductors or compounds such as oxides or nitrides. In most areas, the breath of the experiments has been limited more by instrument availability rather than by the technique itself.

18.4.1 Observation of Clean Surfaces

The high-resolution observation of metal and semiconductor surfaces [14], locating the atomic surface steps, was a hallmark in LEEM evolution. As mentioned before, atomic steps are imaged by Fresnel diffraction. Although the bulk of the LEEM work has been performed on metal and semiconductor surfaces, more complex systems such as binary oxides [58–60], nitrides [61–63] and quasi-crystals [64, 65] have been explored (see Fig. 18.7a–d).

The surface morphology can be directly imaged by locating the atomic steps. As LEEM can select a field of view between a few microns to up to tens of microns, it is possible to explore large areas of a surface. This often provides a dual role. On one hand, some particular types of surface areas can be located, either large flat terraces, step bunches or features created by, for example, a focused ion beam (FIB) [66]. On the other hand, sampling can reveal the average state of the surface, information that can guide experiments using complimentary techniques. Easily exploring parts of the surface millimeters apart serves to avoid the common "tunnel view" of STM where a restricted area of the surface is assumed to be representative of the full surface.

Besides steps or dislocations, which themselves can give rise to steps [62], stacking faults and twins are defects that occur in films and surfaces. Both defect types are easily located in LEEM through a combination of dark-field and bright field imaging [11, 67, 68]. In the case of two structures related by twinning, the diffraction pattern of one twin is rotated by 180° from the other's pattern. Selecting a diffracted

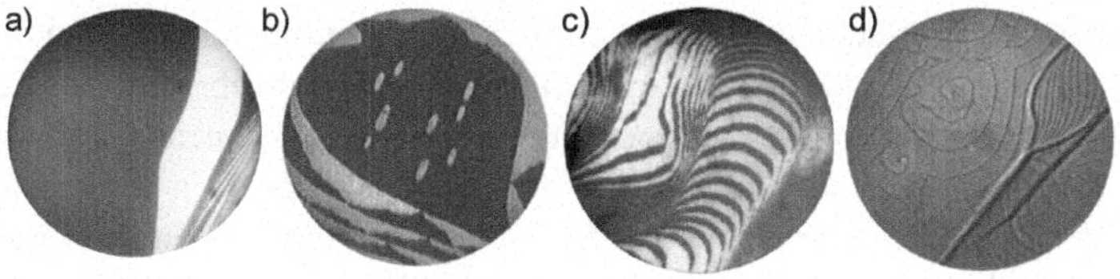

Fig. 18.7 LEEM images of different surfaces. (**a**) Dark-field image using an first order diffraction spot of Ru(0001), reproduced with permission from [69]. The field of view of 10 μm. (**b**) Dark-field image of Si(100) acquired with a diffraction spot from one of the domains of the 1 × 2 reconstruction, reproduced with permission from [71]. The field of view is 5.5 μm. (**c**) Bright field image of TiO_2(110) reprinted with permission from [58]. The field of view of 10 μm. Diffraction contrast causes the 1 × 2 reconstruction to be dark compared to the bulk-like 1 × 1 phase. (**d**) TiN(111), reprinted with permission from [63]. The field of view is 5.6 μm

spot in dark-field imaging from one twin will give the real-space distribution of that twin. Bright field imaging can be used to determine whether the dark-field image is a pure twin or a stacking fault. In the former case, no bright field contrast would be possible by symmetry. In the latter, the stacking fault can be considered a change in the unit cell of the last layers and would be seen in bright-field [68]. Sometimes the structure of the surface is such that crystallographically inequivalent but energetically equivalent terminations are possible. Examples include hcp [69] and wurzite [70] structures. In both structures the local sequence of layers in the direction perpendicular to the basal plane is such that consecutive atomic steps give rise to a rotation by 180° of the local environment, i.e., the crystal symmetry group has a screw symmetry operation. Then, a dark-field image from a first-order diffracted spot gives an image with alternatively white and dark terraces separated by atomic steps (see Fig. 18.7a).

18.4.2 Surface Dynamics: From Adatoms to Phase Transitions

The evolution of surfaces has an important role in nanoscience. Such evolution is determined by thermodynamic and kinetic effects. The ability of LEEM to locate steps in real time makes it a powerful technique for observing the evolution of surfaces under controlled conditions.

Steps As LEEM can easily image single-layer atomic steps, a popular focus has been step dynamics, both in semiconductors [72, 73] or metals [74, 75]. Although the emphasis has been on simple surfaces by extracting accurate properties for comparison with theory, an obvious evolution of the technique is to expand it to more complex surfaces such as quasi-crystals [65]. Through the interpretation of the step motion, both properties intrinsic to the steps (such as line tension) and adatom-step interactions can be determined.

A related line of work has been examining the evolution, not of more or less linear steps, but of closed steps forming islands or pits [76]. By measuring in detail the evolution of a non-equilibrium structure, such as an island array on a substrate terrace or a mound of islands, the evolution towards equilibrium can be directly compared with model simulations incorporating the mass carriers and their interaction with steps. When a set of differently sized islands on a terrace undergoes Ostwald ripening, the dependence of the island line tension on curvature makes smaller islands decay and larger islands grow [71]. The particular way the islands area changes with time gives information about the limiting process that the mass carriers experience: either diffusion along the terraces or attachment/detachment from step edges. But also the role of bulk diffusion has been highlighted in experiments studying the evolution of Pt(111) [77]. In the more complicated case of NiAl(110), separating the effects of bulk vs surface diffusion [78] required taking advantage LEEM's ability to image while oscillating the sample temperature. Another option in LEEM is the ability to observe a surface at very high temperatures, needed for refractory metals such as Rh(001) [79]. Adsorbates can also be deposited in real time during island decay, helping solve long standing questions about the role of impurities in surface evolution. The influence of sulfur on Cu island decay has thus been explained in terms of the formation of S-Cu units that serve as additional and very efficient surface mass carriers [80]. A novel development is the ability to deposit ions while imaging. The adatom concentration can be enhanced or depleted, depending on the ion energy [81, 82]. In this way, the phase space under which surface evolution can be observed is greatly expanded.

Thermal Adatoms Thermal adatoms on a crystal surface equilibrate with sinks such as steps. The concentration and distribution of the adatoms can be deduced from the detailed time evolution of the steps. This information can also be determined by studying island nucleation [83]. But to determine directly the concentration of adatoms is complicated as thermal adatom concentrations are often very low, and the adatoms themselves can move large distances, precluding the use of STM-related techniques at any but the lowest temperatures, where the surface evolution is nearly frozen. LEEM can be used to determine both the concentration and distribution of thermal adatoms on a surface by monitoring changes in the surface reflectivity. Adatoms cause the incident beam to be scattered diffusely, reducing its reflected intensity. Therefore in a bright field image a decrease in intensity can be related to an increase of adatom concentration. The method has been applied to metal [24] and carbon adatoms [84], and the sensitivity easily reaches concentrations of 10^{-4} ML.

Phase Transitions LEEM's rapid imaging gives a powerful approach to observe surface phase transitions, which are typically driven by changing temperature [85]. Depending on the phases involved, different contrast mechanisms can be used in imaging. Unlike using reciprocal space techniques to probe phase transitions, a real space view of the surface allows determining whether there is phase coexistence. An early success of LEEM was the real time observation of the transition between the Si(111) 7×7 and 1×1 surface structures [86]. The detection of phase coexistence showed directly that the transition was first order. The subtleties of the 7×7–1×1

phase transition have been studied in detail by LEEM, shedding light on the role of bulk diffusion [87] and the extended temperature range of the phase coexistence [88]. The richness of the system has motivated recent studies on the impact of the phase transition on island decay [89].

Another classic surface phase transition is the roughing transition. Although not as spectacular as Si's 1×1 to 7×7 transition, the roughening transition of Si/Ge alloys has been characterized by LEEM [90]. Additional phase transitions have been studied on other semiconductors [91] or metal/semiconductor surfaces [92]. Phase transitions in metals and oxides have received much less attention [59]. It is a glaring omission that well-known metal reconstruction phase transitions have not yet been studied by LEEM.

18.4.3 Thin Film Growth: From Self-assembly to Film Dewetting

One field where a real-time view of a process is most rewarding is thin-film growth. There are a limited number of techniques that allow real-time monitoring of thin-film growth by molecular beam epitaxy (MBE), pulsed laser deposition (PLD) or reactive MBE. The most used technique is reflection high-energy electron diffraction (RHEED), which only provides reciprocal-space information averaged over a macroscopic area. Consider the growth mode of a film by step-flow, i.e., a growth mode in which the deposited material adds to at the preexisting steps and the film grows in strict layer-by-layer mode. Step-flow gives rise to films as flat as their substrate. But as no new steps are created on the growing film, RHEED is completely blind to the number of layers deposited. LEEM is the only general purpose surface science technique that allows real-time *and* real-space monitoring of the surface growth front in PLD and MBE. For example, in Fig. 18.8 we show two examples of thin-film layer-by-layer growth, either by step-flow (Fig. 18.8a) or by island nucleation (Fig. 18.8b). The capabilities of LEEM to explore growth have been used to monitor metal growth both on metals [25, 66, 68, 93–97] and semiconductors [98, 99], semiconductor growth [100, 101], graphene growth both on metal [84, 102–110] and on SiC [111–113], and oxide growth [114–120]. The particular role of dislocations in thin-film growth has been studied in semiconductors [121], metals [62, 67, 122], and nitrides [62].

The surface sensitivity of LEEM is usually stated to arise from the short inelastic mean free path of low-energy electrons. However, at typical imaging energies, the mean free paths can be surprisingly long. This implies that changing the focus allows the buried interface between the film and the substrate to be easily imaged [125]. As heating to high temperatures under observation in LEEM is straightforward, sublimation of thin films can be tracked [126]. At lower temperatures, the dewetting process by which a continuous film breaks up [127] or a thinner film roughens to form large 3-dimensional islands (mesas) has been studied in metals [128]. Thanks to LEEM observations, the latter process has been used as part of a method [129] to produce large flat areas in metallic crystals by sputtering a mesa-containing thin film [130].

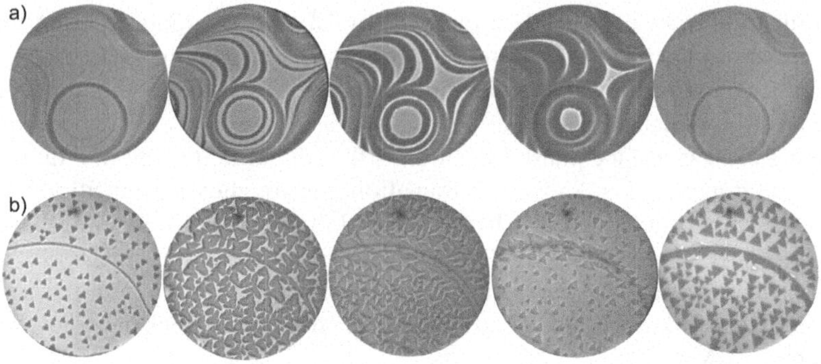

Fig. 18.8 Thin-film growth. (**a**) Step-flow in the growth of 1 ML Cr on W(110), adapted with permission from [123]. (**b**) Layer-by-layer island nucleation growth in the growth of the first three layers of Co on Ru(0001), adapted with permission from [124]. The field of view of each sequence is, respectively, 7 and 10 μm

Furthermore, the capabilities of LEEM allow not only for monitoring the growth front of the film, but for a detailed structural characterization of the film itself. By combining the different working modes of a LEEM, the interlayer spacings, the stacking sequence of the film layers or the presence of stacking faults can be resolved for each film thickness [25, 68, 94, 97].

In the same way as TEM characterizes the structure of a bulk film, LEEM can determine the microstructure of films a few atoms thick. As a complex example [11], Cu on Ru(0001) has a thickness dependent misfit dislocation network. For films two monolayers thick, the dislocation network consists of sets of parallel dislocations. Three equivalent rotational domains occur, with the dislocations aligned along the three equivalent directions of the substrate. Even if the dislocation spacing in the domains is below the resolution limit of a non-aberration corrected LEEM, each rotational domain can still be imaged in dark-field using its unique diffraction spots (see Fig. 18.9b, and composite image in Fig. 18.9c). But in addition to the rotational domains, there are also domains within the film that differ in their stacking sequence. This additional microstructure can be imaged in bright field using an appropriate energy (see Fig. 18.9d). In this way, the evolution and the interplay between the different types of film defects can be determined. In this particular example, it was found that the healing of stacking domains was impeded by the threading dislocations present at the boundaries of rotational domains.

Self-assembly of Submonolayer Films The quest for ordered arrays of nanoscale objects has popularized the study of self-organized patterns in submonolayer film growth. Monolayer-height islands deform the substrate, giving rise to an elastic interaction between islands. The competition between the (short-range) boundary cost of creating island edges is then compensated by the (long-range) elastic interaction between island edges. The resulting self-assembly gives rise to striped patterns on uniaxial systems, while isotropic systems should exhibit a transition between a bubble phase (round islands in an hexagonal pattern), a stripe phase and an

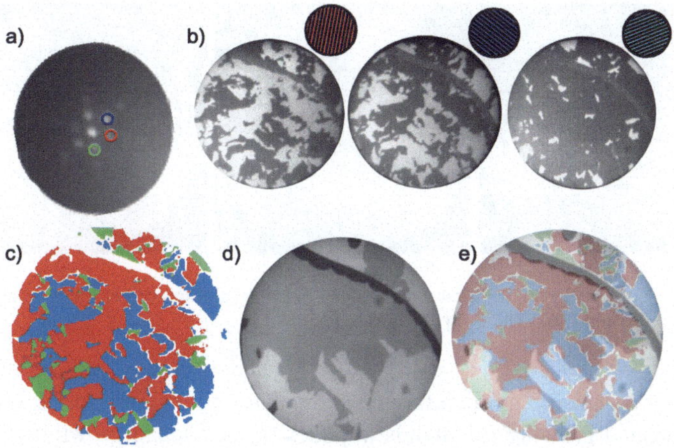

Fig. 18.9 Microstructure of a 2 ML Cu film on Ru(0001), adapted with permission from [11]. (**a**) LEED pattern arising from the superposition of three rotational domains. (**b**) LEEM dark-field images from three different diffraction spots. The *bright regions* map the spatial distribution of the three rotational orientations. (**c**) Composite image of the dark-field images in (**b**). (**d**) LEEM bright-field image showing the distribution of the two different stacking sequences (*light* and *medium gray*). (**e**) Complete microstructure of the film showing both rotational and stacking domains. The field of view is 5 μm

inverted bubble phase. The origin of the patterns lies thus in the thermodynamics of the system. They have been observed for metals [131–134] and semiconductors [135–137]. Pb islands on Cu(111) can be considered a model system for stress domain self-assembly [138–142]. Pb on Cu forms a disordered alloy until a critical Pb concentration is reached. Then, in addition to the disordered alloy, islands of pure Pb nucleate on the surface. As the coverage of Pb is increased, the shape of the Pb islands changes from round islands, to stripes, and then to inverted droplets, as predicted by theory and shown experimentally in Fig. 18.10. The pattern periodicity depends exponentially on the ratio of boundary cost (step energy) and elastic interaction. In the Pb/Cu system, detailed studies of the fluctuations of the stripe phase allowed an accurate and independent measure of the boundary energy. The long range elastic interaction was measured by studying fluctuations of the island positions in the droplet phase. The predicted periodicity matched the experimentally observed value.

Alloys Alloying in thin films or surfaces can be followed in LEEM. But as LEEM cannot image the atomic scale, quantifying the alloying is usually indirect. NiAl bulk alloys have been studied by following changes in the surface morphology [143, 144]. At surfaces, alloying and dealloying on refractory metals have been detected by the morphology of the surface [145] or more directly by the observation of a specific alloy-related LEED pattern [146, 147]. The effects on island growth and motion can be striking. In Sn on Cu(111) the formation of an alloy actually propelled the Sn islands across the substrate leaving a trail of surface alloy [148]. In

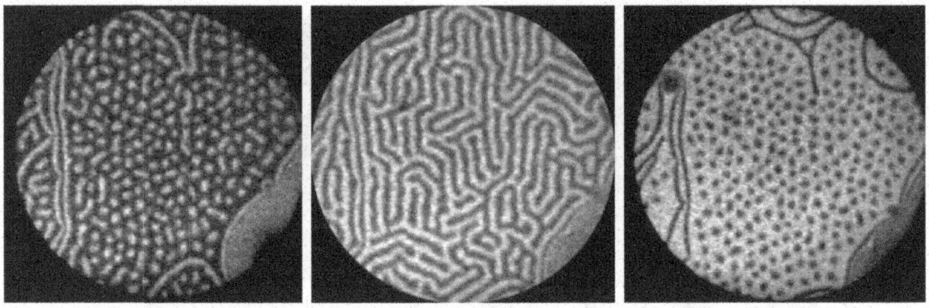

Fig. 18.10 Self-organization of Pb islands on Cu(111) as a function of the Pb coverage, reprinted with permission from [139]. The field of view is 1.8 μm

Pd/Ru(0001), at specific temperatures and deposition rates, the formation of a surface alloy around growing Pd islands impedes further island growth except at particular points. This produces distinctive labyrinthine-shaped islands, whose growth was followed in real time [96].

Magnetic Domain Studies We consider here imaging magnetic domains in ferromagnetic films or surfaces (studies of spin-dependent electron reflectivity are discussed in the VLEED section below). The number of SPLEEM instruments at the time of writing this chapter is just four, so the span of the different material combinations tested is extremely limited. Ferromagnetic elements that have been shown to provide usable contrast in SPLEEM include all the 3*d* magnetic transition metals (Fe, Co, Ni), and the rare earth Tb. In compound form, magnetite islands also give SPLEEM contrast [120]. Other compounds and alloys await experimental studies.

A straightforward SPLEEM application is to determine the magnetic domain structure of surfaces and islands [149–154]. In this way, finite-size effects on the magnetization patterns can be determined, such as the transition between single domain and vortex states in 3-dimensional magnetic islands [155], or the magnetization patterns in elongated islands [156, 157]. The Curie temperature is determined by measuring the temperature where magnetic contrast is lost. But the important point is that the Curie temperature can be measured locally, so it is easy to check the effect of thickness [124] or the local environment [158].

Since SPLEEM can measure each component of the magnetization, it can perform 3-dimensional magnetometry [159] by changing the spin-orientation of the illuminating electron beam in orthogonal directions. It then becomes a powerful method to explore spin-reorientation transitions [124, 152, 159–163]. For example, consider the magnetization of Co on Ru(0001). The growth can be interrupted at intermediate coverages between 1–2 ML and 2–3 ML, and the magnetization along two in-plane directions and the out-of-plane direction can be measured in sequence. As Fig. 18.11 shows, the 1 ML and 3 ML areas are only magnetized in-plane, unlike 2 ML areas, which are magnetized out-of-plane. Thus, there are consecutive spin-reorientation transitions as a function of thickness. Further studies followed in real time how the capping the Co with noble metals also produce additional

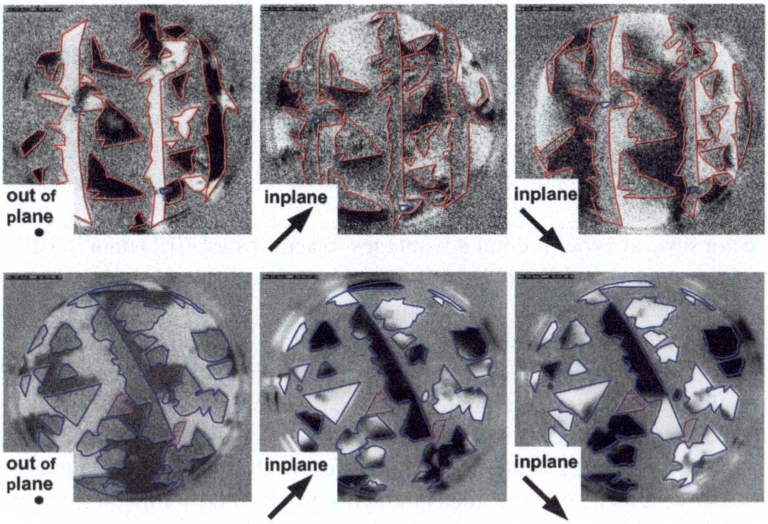

Fig. 18.11 3-D magnetometry in Co/Ru(0001) ultra-thin films, reproduced with permission from [124]. *Top row*: SPLEEM images of the same film with 2 ML Co thick islands on a continuous 1 ML Co layer. Each SPLEEM image shows magnetic contrast along the direction indicated in the *left-lower corner*. *Bottom row*: SPLEEM sequence acquired on a film with 3 ML Co thick islands on a continuous 2 ML Co layer. FOV is 2.8 μm

spin-reorientation transitions [162]. Another theme is the study of the coupling between magnetic layers. But unlike measuring magnetization using PEEM techniques, SPLEEM lacks chemical sensitivity, making the extraction of information about the magnetization of individual layers more involved.

18.4.4 Determining Surface Crystallography and Electronic Structure

Low-energy electron diffraction (LEED), the measurement of diffraction patterns with reflected electrons is the standard surface science crystallographic tool [21]. LEEM is an excellent technique to measure LEED. This can be done both in real space (with a contrast aperture, so only the specular beam is measured) or in reciprocal space and with spin-polarized electrons in SPLEEM instruments. We discuss separately the diffraction information of the "traditional" LEED energy range (50–400 eV) and that of VLEED (0–50 eV). The distinction serves to underline the different characteristics of the information obtained in each energy range.

"Traditional" LEED The "traditional" LEED allows determining surface structures comprising up to a hundred atoms with picometer accuracies by an iterative comparison of the intensity of diffracted beams vs electron energy with multiple scattering calculations from a model structure [20, 28]. In the 50–400 eV energy range the electron mean free path is small (a few atomic layers), giving it extreme surface sensitivity. The energies are high enough that the band structure of the surface can be disregarded, with the calculations employing a muffin-tin approximation. In this range, the usual collection method in LEEM is to image the back-focal plane of the objective, i.e., the diffraction pattern of the electrons leaving the sample surface.

There are several experimental advantages to acquiring LEED data in LEEM. As the illumination and the reflected electrons paths are separated by a beam separator, there is no problem measuring the specular beam or its closely spaced satellites under normal incidence conditions. The sample manipulator and electron optics allow the incident beam to be aligned precisely normal to the surface. The optical (light) path from the sample is also separate from the electron path, so measurements can be performed even when the sample is hot. (In a regular LEED instrument, light from the hot sample or the heater apparatus gives a substantial background intensity on the detector screen, making measurements at elevated temperatures difficult.) As all LEEMs include high gain microchannel plates for imaging, they can acquire LEED patterns at very low electron beam intensities. The position of the diffracted spots on the detector does not change with incoming electron energy because the diffracted electrons are deflected during their acceleration away from the surface in a way that exactly cancels the change in initial emission angle with energy [3]. Thus, there is no need for sophisticated tracking of diffracted beams with energy while measuring their intensity. Furthermore, the ability to change the power of the lens in the imaging column means that a LEEM instrument can serve as a reasonable spot-profile analysis LEED system (SPA-LEED [164]). But the reason why LEED in LEEM stands apart is the ability to select the area from which the LEED information is obtained by using the real-space imaging mode of LEEM. First a region is characterized by real-space imaging. Then a particular region is selected for LEED analysis. Diffraction from regions of interest as small as $\frac{1}{4}$ μm in diameter can be routinely obtained.

Nevertheless there are some limitations acquiring LEED in LEEM. One is that the size of the reciprocal space that can be observed without large distortions is limited to about the second Brillouin zone. The total energy range for quantitative structural analysis can thus be limited, but this can be offset by the ability to measure beams that would be domain-averaged in a regular diffractometer, such as from a 1×2 reconstruction or an hcp substrate [69]. Even under optimum conditions residual distortions in the optics prevent measuring spot positions with accuracies better than a few percent of the first Brilluoin zone, inferior to a dedicated SPA-LEED instrument. The transmission of electrons through the system might be different from a standard LEED diffractometer, although the common usage of the Pendry R factor in structural calculations ameliorates this problem.

The observation of LEED patterns in LEEM started with the first successful instruments [2]. But surprisingly there are only a few studies that have used "traditional" LEED IV calculations coupled to LEED IV measurements in LEEM, as presented in Fig. 18.12 for Cr/W(110) [123]. Both in clean substrates [69] and thin

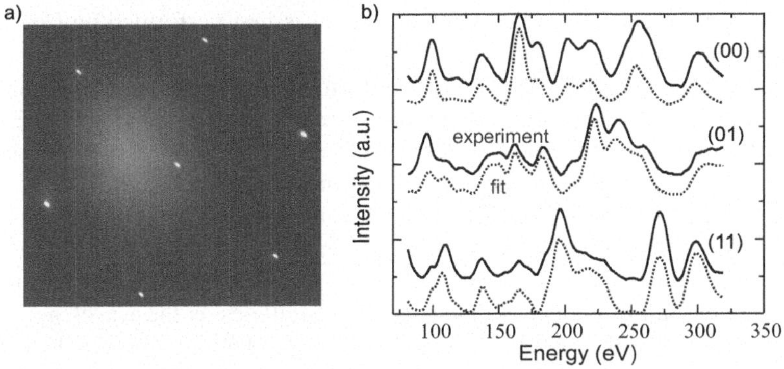

Fig. 18.12 (**a**) LEED pattern of W(110). (**b**) LEED IV measured in LEEM, and multiple scattering fit of the same W(110) surface, reproduced with permission from [123]

films [68, 97], an advantage is the ability to acquire the LEED IV from a single terrace of the substrate or film removing any possible effect of steps, which are not usually included in the multiple scattering calculations. Other than metal surfaces, to date an oxygen reconstruction on Ag [165] has been solved.

In a SPLEEM system, spin-polarized LEED can be measured. Nevertheless, to our knowledge no experimental dataset of SP-LEED has been measured in LEEM.

VLEED In VLEED (0–50 eV) the Ewald sphere is so small that usually only the specular beam or closely-spaced satellites are present. The mean free path can be quite large, extending through films tens of layers thick, and the effect of the band structure on the electron reflectivity cannot be disregarded. The traditional multiple-scattering LEED codes can fail due to convergence limitations, and also due to an incorrect treatment of the dependence of the inner potential with energy. One advantage is that the wavelength of the electrons changes more rapidly at low energies, so the information content is very high.

Besides measuring VLEED in the diffraction mode, where typically only the specular beam is observed, real-space imaging can be used instead. This gives a method to measure the diffraction information of the specular beam with nanometer resolution. This approach has been termed LEEM IV to highlight that the zeroth-order diffraction information is acquired from real-space images. The surface alloy formation of Pd on Cu(100) has been explored in this way [166, 167] by a pixel-by-pixel fitting of the VLEED IV curves to a multiple scattering calculation. The spatially resolved structure gave detailed information about alloy formation around steps. In the same way, monolayers and bilayers of graphene on Ru(0001) have been distinguished [102].

More traditional VLEED determinations can be performed by comparing the experimental data to a calculated unoccupied band structure, including the effect of anisotropic scattering. In this case, the available theoretical calculations [168–170] lag considerably behind the experimental data available. VLEED can be performed with spin-polarized electrons in an SPLEEM, from which information about the spin-split band structure can be obtained. This was first performed on Co/W(110) [171].

In some cases, comparison of VLEED information with other techniques is possible [172]. As a fingerprinting technique using reference spectra, VLEED has sometimes been used to follow the distribution and evolution of several adsorbates on surfaces, such as carbon monoxide, hydrogen or oxygen [173, 174].

The most popular use of VLEED, though, is the measurement of quantum interference peaks (QIP) in flat films. As mentioned above, the electrons reflected at the film interface and at the film surface interfere, giving rise to maxima and minima of the reflected intensity. The Reflectivity then oscillates as the wavelength of the electrons or the film thickness is changed. Although QIPs are observable at higher energies, the mean free path is then smaller so the QIPs are vanishingly small above 50 eV. A requirement for the observation of QIPs is an abrupt buried interface with high reflectivity. The high reflectivity can arise from a symmetry gap in the substrate electronic structure or a large potential step between film and substrate.

QIPs in VLEED are routinely employed to characterize thickness in ultra-thin films. QIPs have been observed on many different combinations of metal films and substrates [66, 67, 93, 94, 122, 128, 131, 171, 175–179]. It has also been observed in metal/semiconductor combinations [125] and oxide/metal [180] systems.

The QIPs can also be fit to multiple scattering calculations to obtain interplanar spacings in a similar way to regular LEED calculations [175, 177]. Another use of QIPs is determining a material's bulk band structure. The phase accumulation model (PAM, [181, 182]), used to interpret oscillations in the intensity of the photoelectrons emitted from thin films, is also applicable to QIPs. PAM indicates that a continuous electron band in the bulk is reduced to a series of discrete quantum well resonances in a thin film along the direction perpendicular to the interface, with the number of states related to the number of layers. The resonances correspond to conditions of constructive interference for electrons with a given wavelength, taking into account the phase change ϕ of the electron upon reflection at the film/substrate interface and at the film surface:

$$2kd_{film} + \phi = 2n\pi$$

where k is the electron wave vector, d_{film} is the film thickness and n is an integer. As the phase accumulation only depends on the energy, i.e., $\phi(E)$, measuring the thickness d_{osc} between reflectivity maxima at a given energy provides the electron wave vector as $k = \frac{\pi}{d_{osc}}$, allowing a parameter-free measurement of the dispersion of the unoccupied electron band. A limitation of the method is that it assumes that the electronic structure of the thin film is the same (except for the quantization of the levels) as for the bulk material. If the film is strained the assumption will not hold. Also the first few layers are expected to have a different electronic structure due to interactions with the substrate. Bulk structure determinations have been made in this way for Ag [on W(110) and on Fe(100)] [94, 177] and Cu [on Co(100)] [179].

Using a SPLEEM, the spin-dependent electron reflectivity of buried non-magnetic/magnetic interfaces can be studied. From the applied point of view, the study of spin-dependent reflectivity has suggested the design of novel spin-analyzers

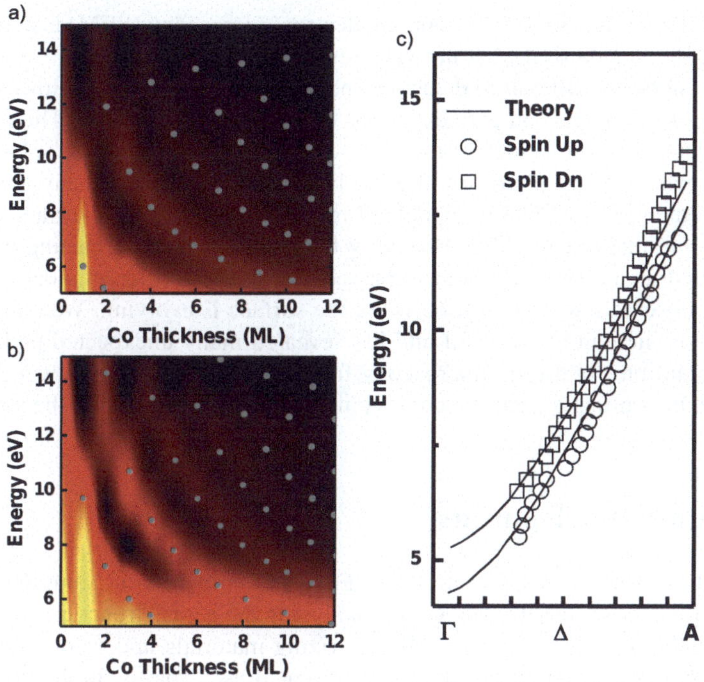

Fig. 18.13 Determining the spin-resolved unoccupied band structure, reprinted with permission from [27]. Electron reflectivity for (**a**) spin-down electrons and (**b**) spin-up electrons shows layer resolved quantum interference peaks. (**c**) Spin-split energy bands retrieved from the quantum interference peaks

[183]. The spin-dependent reflectivity of buried interfaces has been measured in MgO/Fe [180] and Cu/Co(100) [179]. The unoccupied bulk band structure has been measured for Fe [on W(110)] [178] and Co [on Mo(110)] [27] (shown in Fig. 18.13).

18.4.5 Advantages and Limitations vs Other Surface Imaging Techniques

In this section we briefly compare and contrast LEEM to other experimental techniques. The scanning probe microscopies of STM and AFM can routinely achieve atomic resolution, unlike LEEM. However, LEEM has decided advantages in imaging speed and the ease of imaging while changing the surface's state by varying, for example, temperature. X-ray magnetic circular dichroism using PEEM, XMCD-PEEM, characterizes magnetism with elemental specificity, unlike SPLEEM. But the latter technique can perform 3D magnetometry and typically is faster. Surface magneto-optic Kerr effect microscopy, SMOKE, can be performed with applied magnetic fields, unlike SPLEEM. But SPLEEM has much better resolution.

Finally the lateral resolution of scanning electron microscopy (SEM) is comparable to LEEM. But standard SEM is not very sensitive to the surface. For example, single atomic steps are difficult to detect by secondary electrons [184]. However SEM with an immersion lens can give enhanced surface sensitivity [185]. The impact of this technique is still to be determined.

In summary, LEEM is not usually the best technique to make any single type of measurement. But LEEM is capable of many different types of surface analysis, all within one instrument. Also the ability for fast real-spacing imaging combined with selected-area diffraction, is unsurpassed. But the real differentiating capability is making observations in real-time while the surface is evolving. Watching a process occur is incredibly powerful and has revealed many unexpected pathways of surface evolution. Real-time microscopy frequently determines pathways without ambiguity. In contrast, forensic analysis only sees the end result, not the path.

18.5 Future Developments

Two trends in surface science are clear. First, the emphasis on studying increasingly complicated materials. The goal is analyzing the actual "functional" materials used in real applications. Second, characterizing materials under conditions similar to those experienced during their technological use. That is, in the presence of fluids, electric fields and at low and high temperatures, etc. Instrumentation development already in progress and envisioned will allow LEEM to participate fully in these two trends. For example, new schemes of sample cooling will allow the study of materials such as high-temperature superconductors in their superconducting state. New LEEM designs will also be able to operate at much higher pressures. Achieving the same pressures (several millibar) as in near-ambient pressure X-ray photoelectron spectroscopy [186] has technical challenges. However, even at lower pressures, "near-ambient pressure" LEEM/PEEM will be a valuable microscopic compliment to XPS. LEEM-based techniques will increasingly use the spectromicroscopies enabled by imaging energy filters. Electron energy loss and Auger spectromicroscopies need only an electron source. Improved laboratory-based light sources, including focused discharge lamps and ultraviolet lasers, will bring the power of photoemission spectromicroscopy to laboratory studies.

Acknowledgements JdlF acknowledges support from the Spanish Ministry of Science and Innovation through Project No. MAT2009-14578-C03. KFM acknowledges support from the Office of Basic Energy Sciences, Division of Materials Sciences and Engineering of the US DOE under Contract No. DE-AC04-94AL85000.

References

1. L. Veneklasen, Rev. Sci. Instrum. **63**, 5513 (1992)
2. E. Bauer, Rep. Prog. Phys. **57**, 895 (1994)

3. E. Bauer, Surf. Rev. Lett. **5**(6), 1275 (1998)
4. R. Tromp, IBM J. Res. Dev. **44**(4), 503 (2000)
5. R. Phaneuf, A. Schmid, Phys. Today **56**, 50 (2003)
6. N. Rougemaille, A. Schmid, Eur. Phys. J. Appl. Phys. **50**, 20101 (2010)
7. M.S. Altman, J. Phys. Condens. Matter **22**, 084017 (2010)
8. D.B. Williams, C.B. Carter, *Transmission Electron Microscopy: A Textbook for Materials Science*, 2nd edn. (Springer, Berlin, 2009)
9. R.M. Tromp, F.M. Ross, Annu. Rev. Mater. Sci. **30**, 431 (2000)
10. T. Herranz, K. McCarty, B. Santos, M. Monti, J. de la Figuera, Chem. Mater. **22**, 1291 (2010)
11. F. El Gabaly, W. Ling, K. McCarty, J. de la Figuera, Science **308**, 1303 (2005)
12. T. Schmidt, S. Heun, J. Slezak, J. Diaz, K. Prince, G. Lilienkamp, E. Bauer, Surf. Rev. Lett. **5**, 1287 (1998)
13. E. Bauer, Surf. Sci. **299–300**, 102 (1994)
14. W. Telieps, E. Bauer, Ultramicroscopy **17**, 57 (1985)
15. K. Grzelakowski, T. Duden, E. Bauer, H. Poppa, S. Chiang, IEEE Trans. Magn. **30**, 4500 (1994)
16. Elmitec Elektronenmikroskopie GmbH
17. R. Fink, M. Weiss, E. Umbach, D. Preikszas, H. Rose, R. Spehr, P. Hartel, W. Engel, R. Degenhardt, R. Wichtendahl, H. Kuhlenbeck, W. Erlebach, K. Ihmann, R. Schlogl, H. Freund, A. Bradshaw, G. Lilienkamp, T. Schmidt, E. Bauer, G. Benner, J. Electron Spectrosc. Relat. Phenom. **84**, 231 (1997)
18. R. Tromp, J. Hannon, A. Ellis, W. Wan, A. Berghaus, O. Schaff, Ultramicroscopy **110**, 852 (2010)
19. Specs Surface Nanoanalysis Gmbh
20. M. Van Hove, S. Tong, *Surface Crystallography by LEED* (Springer, Berlin, 1979)
21. L.J. Clarke, *Surface Crystallography: An Introduction to Low Energy Electron Diffraction* (Wiley, New York, 1985)
22. S. Nepijko, N. Sedov, Adv. Imaging Electron Phys. **102**, 273 (1998)
23. S.M. Kennedy, C.X. Zheng, W.X. Tang, D.M. Paganin, D.E. Jesson, Proc. R. Soc. A **466**, 2857 (2010)
24. J. de la Figuera, N. Bartelt, K. McCarty, Surf. Sci. **600**, 4062 (2006)
25. T. Herranz, B. Santos, K. McCarty, J. de la Figuera, Surf. Sci. **605**, 903 (2011)
26. A.B. Pang, T. Müller, M. Altman, E. Bauer, J. Phys. Condens. Matter **21**(31), 314006 (2009)
27. J.S. Park, A. Quesada, Y. Meng, J. Li, E. Jin, H. Son, A. Tan, J. Wu, C. Hwang, H.W. Zhao, A.K. Schmid, Z.Q. Qiu, Phys. Rev. B **83**, 113405 (2011)
28. J. Pendry, *Low Energy Electron Diffraction: The Theory and Its Application to Determination of Surface Structure* (Academic Press, San Diego, 1974)
29. W. Chung, M. Altman, Ultramicroscopy **74**, 237 (1998)
30. S. Kennedy, N. Schofield, D. Paganin, D. Jesson, Surf. Rev. Lett. **16**, 855 (2009)
31. T. Schmidt, H. Marchetto, P. Levesque, U. Groh, F. Maier, D. Preikszas, P. Hartel, R. Spehr, G. Lilienkamp, W. Engel, R. Fink, E. Bauer, H. Rose, E. Umbach, H. Freund, Ultramicroscopy **110**, 1358 (2010)
32. K.N. Maffitt, Rev. Sci. Instrum. **39**, 259 (1968)
33. M. Mankos, D. Adler, L. Veneklasen, E. Munro, Phys. Proc. **1**, 485 (2008)
34. E. Bauer, in *Modern Techniques for Characterizing Magnetic Materials*, ed. by Y. Zhu (Springer, Berlin, 2005), pp. 361–382
35. E. Bauer, in *Magnetic Microscopy of Nanostructures*, ed. by H. Hopster, H.P. Oepen (Springer, Berlin, 2005), pp. 111–136
36. P.W. Palmberg, R.E. DeWames, L.A. Vredevoe, Phys. Rev. Lett. **21**, 682 (1968)
37. K.S.R. Menon, S. Mandal, J. Das, T.O. Mentes, M.A. Niño, A. Locatelli, R. Belkhou, Phys. Rev. B **84**, 132402 (2011). doi:10.1103/PhysRevB.84.132402
38. G. Rempfer, D. Desloge, W. Skoczylas, O. Griffith, Microsc. Microanal. **3**, 14 (1997)
39. R. Tromp, Ultramicroscopy **111**, 273 (2011)
40. K. Tsuno, T. Yasue, T. Koshikawa, Appl. Surf. Sci. **256**, 1035 (2009)

41. R. Tromp, Y. Fujikawa, J. Hannon, A. Ellis, A. Berghaus, O. Schaff, J. Phys. Condens. Matter **21**(31), 314007 (2009)
42. K. Grzelakowski, E. Bauer, Rev. Sci. Instrum. **67**, 742 (1996)
43. P. Grivet, *Electron Optics*, revised edn. (Pergamon, Elmsford, 1965)
44. G.F. Rempfer, J. Appl. Phys. **57**, 2385 (1985)
45. P. Adamec, E. Bauer, B. Lencova, Rev. Sci. Instrum. **69**, 3583 (1998)
46. Kimball Physics, Inc, Tech. Bull. (1991)
47. D.T. Pierce, R.J. Celotta, G. Wang, W.N. Unertl, A. Galejs, C.E. Kuyatt, S.R. Mielczarek, Rev. Sci. Instrum. **51**, 478 (1980)
48. X. Jin, N. Yamamoto, Y. Nakagawa, A. Mano, T. Kato, M. Tanioku, T. Ujihara, Y. Takeda, S. Okumi, M. Yamamoto, T. Nakanishi, T. Saka, H. Horinaka, T. Kato, T. Yasue, T. Koshikawa, Appl. Phys. Express **1**, 045002 (2008)
49. T. Duden, E. Bauer, Rev. Sci. Instrum. **66**, 2861 (1995)
50. J. Chmelik, L. Veneklasen, G. Marx, Optik **83**(5), 155 (1989)
51. V. Kolarik, M. Mankos, L. Veneklasen, Optik **87**, 1 (1991)
52. W. Wan, J. Feng, H. Padmore, Nucl. Instrum. Methods Phys. Res. **564**, 537 (2006)
53. R. van Gastel, I. Sikharulidze, S. Schramm, J. Abrahams, B. Poelsema, R. Tromp, S. van der Molen, Ultramicroscopy **110**, 33 (2009)
54. L. Veneklasen, Ultramicroscopy **36**, 76 (1991)
55. Y. Fujikawa, T. Sakurai, R.M. Tromp, Phys. Rev. Lett. **100**, 126803 (2008)
56. J. Sun, J.B. Hannon, R.M. Tromp, P. Johari, A.A. Bol, V.B. Shenoy, K. Pohl, ACS Nano **4**, 7073 (2010)
57. Y. Fujikawa, T. Sakurai, R.M. Tromp, Phys. Rev. B **79**, 121401 (2009)
58. K. McCarty, N. Bartelt, Surf. Sci. **540**, 157 (2003)
59. K. McCarty, N. Bartelt, Surf. Sci. **527**, L203 (2003)
60. K. McCarty, N. Bartelt, Phys. Rev. Lett. **90**, 046104 (2003)
61. S. Kodambaka, N. Israeli, J. Bareño, W. Swiech, K. Ohmori, I. Petrov, J.E. Greene, Surf. Sci. **560**, 53 (2004)
62. S. Kodambaka, S. Khare, W. Swiech, K. Ohmori, I. Petrov, J. Greene, Nature **429**, 49 (2004)
63. S. Kodambaka, J. Bareno, S. Khare, W. Swieach, I. Petrov, J. Greene, J. Appl. Phys. **98**, 034901 (2005)
64. B. Unal, Y. Sato, K. McCarty, N. Bartelt, T. Duden, C. Jenks, A. Schmid, P. Thiel, J. Vac. Sci. Technol. A **27**, 1249 (2009)
65. Y. Sato, B. Unal, T. Lograsso, P. Thiel, A. Schmid, T. Duden, N. Bartelt, K. McCarty, Phys. Rev. B **81**, 161406(R) (2010)
66. K. McCarty, Nano Lett. **6**, 858 (2006)
67. W. Ling, N. Bartelt, K. McCarty, C. Carter, Phys. Rev. Lett. **95**, 166105 (2005)
68. F. El Gabaly, J. Puerta, C. Klein, A. Saa, A. Schmid, K. McCarty, J. Cerda, J. de la Figuera, New J. Phys. **9**, 80 (2007)
69. J. de la Figuera, J. Puerta, J. Cerda, F. El Gabaly, K. McCarty, Surf. Sci. **600**, L105 (2006)
70. J. Maxson, N. Perkins, D. Savage, A. Woll, L. Zhang, T. Kuech, M. Lagally, Surf. Sci. **464**, 217 (2000)
71. N. Bartelt, W. Theis, R. Tromp, Phys. Rev. B **54**, 11741 (1996)
72. M. Mundshau, E. Bauer, W. Telieps, W. Swiech, Surf. Sci. **223**, 413 (1989)
73. N. Bartelt, R. Tromp, E. Williams, Phys. Rev. Lett. **73**, 1656 (1994)
74. M. Ondrejcek, W. Swiech, M. Rajappan, C. Flynn, Phys. Rev. B **72**, 085422 (2005)
75. M. Ondrejcek, M. Rajappan, W. Swiech, C. Flynn, Surf. Sci. **574**, 111 (2005)
76. S. Kodambaka, S. Khare, I. Petrov, J. Greene, Surf. Sci. Rep. **60**, 55 (2006)
77. B. Poelsema, J.B. Hannon, N.C. Bartelt, G.L. Kellogg, Appl. Phys. Lett. **84**, 2551 (2004)
78. K. McCarty, J. Nobel, N. Bartelt, Nature **412**, 622 (2001)
79. G. Kellogg, N. Bartelt, Surf. Sci. **577**, 151 (2005)
80. W. Ling, N. Bartelt, K. Pohl, J. de la Figuera, R. Hwang, K. McCarty, Phys. Rev. Lett. **93**, 166101 (2004)

81. M. Ondrejcek, W. Swiech, I. Petrov, M. Rajappan, C.P. Flynn, Microsc. Res. Tech. **72**, 197 (2009)
82. W. Swiech, M. Ondrejcek, C. Flynn, J. Phys. Condens. Matter **21**, 314021 (2009)
83. R.M. Tromp, M. Mankos, Phys. Rev. Lett. **81**, 1050 (1998)
84. E. Loginova, N. Bartelt, P. Feibelman, K. McCarty, New J. Phys. **10**, 093026 (2008)
85. J. Hannon, R. Tromp, Annu. Rev. Mater. Res. **33**, 263 (2003)
86. W. Telieps, E. Bauer, Surf. Sci. **162**, 163 (1985)
87. J. Hannon, H. Hibino, N. Bartelt, B. Swartzentruber, T. Ogino, G. Kellogg, Nature **405**, 552 (2000)
88. J. Hannon, F. Heringdorf, J. Tersoff, R. Tromp, Phys. Rev. Lett. **86**, 4871 (2001)
89. H. Hibino, Y. Watanabe, C. Hu, I.S.T. Tsong, Phys. Rev. B **72**, 245424 (2005)
90. J. Maxson, D. Savage, F. Liu, R. Tromp, M. Reuter, M. Lagally, Phys. Rev. Lett. **85**, 2152 (2000)
91. R. Tromp, W. Theis, N. Bartelt, Phys. Rev. Lett. **77**, 2522 (1996)
92. Y. Sato, S. Chiang, Surf. Sci. **603**, 2300 (2009)
93. M. Mundschau, E. Bauer, W. Swiech, J. Appl. Phys. **65**, 581 (1989)
94. K. Man, Z. Qiu, M. Altman, Phys. Rev. Lett. **93**, 236104 (2004)
95. K. Man, Q. Guo, M. Altman, Surf. Sci. **600**, 1060 (2006)
96. N. Rougemaille, F. El Gabaly, R. Stumpf, A. Schmid, K. Thürmer, N. Bartelt, J. de la Figuera, Phys. Rev. Lett. **99**, 106101 (2007)
97. B. Santos, J. Puerta, J. Cerda, T. Herranz, K. McCarty, J. de la Figuera, New J. Phys. **12**, 023023 (2010)
98. K. Man, M. Tringides, M. Loy, M. Altman, Phys. Rev. Lett. **101**, 226102 (2008)
99. T. Schmidt, E. Bauer, Phys. Rev. B **62**, 15815 (2000)
100. J. Hannon, J. Tersoff, M. Reuter, R. Tromp, Phys. Rev. Lett. **89**, 266103 (2002)
101. R. Tromp, J. Hannon, Surf. Rev. Lett. **9**, 1565 (2002)
102. P.W. Sutter, J. Flege, E.A. Sutter, Nat. Mater. **7**, 406 (2008)
103. E. Loginova, S. Nie, K. Thurmer, N. Bartelt, K. McCarty, Phys. Rev. B **80**, 085430 (2009)
104. E. Loginova, N. Bartelt, P. Feibelman, K. McCarty, New J. Phys. **11**, 063046 (2009)
105. E. Starodub, S. Maier, I. Stass, N. Bartelt, P. Feibelman, M. Salmeron, K. McCarty, Phys. Rev. B **80**, 235422 (2009)
106. K. McCarty, P. Feibelman, E. Loginova, N. Bartelt, Carbon **47**, 1806 (2009)
107. Y. Murata, E. Starodub, B. Kappes, C. Ciobanu, N. Bartelt, K. McCarty, S. Kodambaka, Appl. Phys. Lett. **97**, 143114 (2010)
108. E. Starodub, N. Bartelt, K. McCarty, J. Phys. Chem. C **114**, 5134 (2010)
109. J. Wofford, S. Nie, K. McCarty, N. Bartelt, O. Dubon, Nano Lett. **10**, 4890 (2010)
110. S. Nie, A. Walter, N. Bartelt, E. Starodub, A. Bostwick, E. Rotenberg, K. McCarty, ACS Nano **5**, 2298 (2011)
111. T. Ohta, F.E. Gabaly, A. Bostwick, J.L. McChesney, K.V. Emtsev, A.K. Schmid, T. Seyller, K. Horn, E. Rotenberg, New J. Phys. **10**, 023034 (2008)
112. R. Tromp, J. Hannon, Phys. Rev. Lett. **102**, 106104 (2009)
113. D. Siegel, S. Zhou, F.E. Gabaly, A. Schmid, K. McCarty, A. Lanzara, Phys. Rev. B **80**, 241407(R) (2009)
114. K. McCarty, Surf. Sci. **543**, 185 (2003)
115. K. McCarty, N. Bartelt, J. Cryst. Growth **270**, 691 (2004)
116. J. Pierce, K. McCarty, Phys. Rev. B **71**, 125428 (2005)
117. K. McCarty, J. Pierce, C. Carter, Appl. Phys. Lett. **88**, 141902 (2006)
118. J. Pierce, N. Bartelt, R. Stumpf, K. McCarty, Phys. Rev. B **77**, 195438 (2008)
119. J. Flege, J. Hrbek, P. Sutter, Phys. Rev. B **78**, 165407 (2008)
120. B. Santos, E. Loginova, A. Mascaraque, A. Schmid, K. McCarty, J. de la Figuera, J. Phys. Condens. Matter **21**, 314011 (2009)
121. J.B. Hannon, Science **313**, 1266 (2006)
122. W. Ling, J. Hamilton, K. Thurmer, G. Thayer, J. de la Figuera, R. Hwang, C. Carter, N. Bartelt, K. McCarty, Surf. Sci. **600**, 1735 (2006)

123. B. Santos, J. Puerta, J. Cerda, R. Stumpf, K. von Bergmann, R. Wiesendanger, M. Bode, K. McCarty, J. de la Figuera, New J. Phys. **10**, 013005 (2008)
124. F. El Gabaly, S. Gallego, M.C. Muñoz, L. Szunyogh, P. Weinberger, C. Klein, A. Schmid, K. McCarty, J. de la Figuera, Phys. Rev. Lett. **96**, 147202 (2006)
125. R. Tromp, A. Vandergon, F. Legoues, M. Reuter, Phys. Rev. Lett. **71**, 3299 (1993)
126. S. Tang, S. Kodambaka, W. Swiech, I. Petrov, C. Flynn, T. Chiang, Phys. Rev. Lett. **96**, 126106 (2006)
127. K. McCarty, J. Hamilton, Y. Sato, A. Saa, R. Stumpf, J. de la Figuera, K. Thürmer, F. Jones, A. Schmid, A. Talin, N. Bartelt, New J. Phys. **11**, 043001 (2009)
128. W. Ling, T. Giessel, K. Thürmer, R. Hwang, N. Bartelt, K. McCarty, Surf. Sci. **570**, L297 (2004)
129. F.E. Gabaly, A.K. Schmid, Method for preparing ultraflat, atomically perfect areas on large regions of a crystal surface by heteroepitaxy deposition, 2009
130. F.E. Gabaly, N.C. Bartelt, A.K. Schmid, J. Phys. Condens. Matter **21**, 314019 (2009)
131. L. Aballe, A. Barinov, A. Locatelli, T. Mentes, M. Kiskinova, Phys. Rev. B **75**, 115411 (2007)
132. A. Locatelli, L. Aballe, T. Mentes, F. Guo, M. Kiskinova, Surf. Sci. **601**, 4663 (2007)
133. J. de la Figuera, F. Leonard, N. Bartelt, R. Stumpf, K. McCarty, Phys. Rev. Lett. **100**, 186102 (2008)
134. T.O. Mentes, A. Locatelli, L. Aballe, E. Bauer, Phys. Rev. Lett. **101**, 085701 (2008)
135. J. Hannon, J. Tersoff, R. Tromp, Science **295**, 299 (2002)
136. G. Thayer, J. Hannon, R. Tromp, Nat. Mater. **3**, 95 (2004)
137. N. Medhekar, V. Shenoy, J. Hannon, R. Tromp, Phys. Rev. Lett. **99**, 156102 (2007)
138. R. Plass, J. Last, N. Bartelt, G. Kellogg, Nature **412**, 875 (2001)
139. R. Plass, N. Bartelt, G. Kellogg, J. Phys. Condens. Matter **14**, 4227 (2002)
140. R. van Gastel, R. Plass, N. Bartelt, G. Kellogg, Phys. Rev. Lett. **91**, 055503 (2003)
141. R. van Gastel, N. Bartelt, P. Feibelman, F. Leonard, G. Kellogg, Phys. Rev. B **70**, 245413 (2004)
142. R. van Gastel, N. Bartelt, G. Kellogg, Phys. Rev. Lett. **96**, 036106 (2006)
143. K. McCarty, J. Nobel, N. Bartelt, Phys. Rev. B **71**, 085421 (2005)
144. P. Hou, K. McCarty, Scr. Mater. **54**, 937 (2006)
145. D. Wu, W.K. Lau, Z.Q. He, Y.J. Feng, M.S. Altman, C.T. Chan, Phys. Rev. B **62**, 8366 (2000)
146. K. Man, Y. Feng, M. Altman, Phys. Rev. B **74**, 085420 (2006)
147. K. Man, Y. Feng, C. Chan, M. Altman, Surf. Sci. **601**, L95 (2007)
148. A.K. Schmid, N.C. Bartelt, R.Q. Hwang, Science **290**, 1561 (2000)
149. T. Duden, E. Bauer, Phys. Rev. Lett. **77**, 2308 (1996)
150. E. Bauer, T. Duden, R. Zdyb, J. Phys. D **35**, 2327 (2002)
151. K. Man, R. Zdyb, S. Huang, T. Leung, C. Chan, E. Bauer, M. Altman, Phys. Rev. B **67**, 184402 (2003)
152. R. Zdyb, E. Bauer, Phys. Rev. B **67**, 134420 (2003)
153. R. Zdyb, A. Locatelli, S. Heun, S. Cherifi, R. Belkhou, E. Bauer, Surf. Interface Anal. **37**, 239 (2005)
154. E. Bauer, R. Belkhou, S. Cherifi, R. Hertel, S. Heun, A. Locatelli, A. Pavlovska, R. Zdyb, N. Agarwal, H. Wang, Surf. Interface Anal. **38**, 1622 (2006)
155. H. Ding, A. Schmid, D. Li, K. Guslienko, S. Bader, Phys. Rev. Lett. **94**, 157202 (2005)
156. N. Rougemaille, A. Schmid, J. Appl. Phys. **99**, 08S502 (2006)
157. R. Zdyb, A. Pavlovska, M. Jalochowski, E. Bauer, Surf. Sci. **600**, 1586 (2006)
158. R. Zdyb, E. Bauer, Phys. Rev. Lett. **100**, 155704 (2008)
159. C. Klein, R. Ramchal, M. Farle, A. Schmid, Surf. Interface Anal. **38**, 1550 (2006)
160. T. Duden, E. Bauer, Phys. Rev. B **59**, 468 (1999)
161. C. Klein, R. Ramchal, A. Schmid, M. Farle, Phys. Rev. B **75**, 193405 (2007)
162. F. El Gabaly, K. McCarty, A. Schmid, J. de la Figuera, M.C. Muñoz, L. Szunyogh, P. Weinberger, S. Gallego, New J. Phys. **10**, 073024 (2008)
163. R. Zdyb, T. Mentes, A. Locatelli, M. Nino, E. Bauer, Phys. Rev. B **80**, 184425 (2009)

164. M.H. von Hoegen, Z. Kristallogr. **214**, 591 (1999)
165. R. Reichelt, S. Günther, J. Wintterlin, W. Moritz, L. Aballe, T.O. Mentes, J. Chem. Phys. **127**, 134706 (2007)
166. J. Hannon, J. Sun, K. Pohl, G. Kellogg, Phys. Rev. Lett. **96**, 246103 (2006)
167. J. Sun, J. Hannon, G. Kellogg, K. Pohl, Phys. Rev. B **76**, 205414 (2007)
168. M. Lindroos, H. Pfnür, D. Menzel, Phys. Rev. B **33**, 6684 (1986)
169. M. Lindroos, H. Pfnür, P. Feulner, D. Menzel, Surf. Sci. **180**, 237 (1987)
170. I. Bartos, M.A.V. Hove, M.S. Altman, Surf. Sci. **352–354**, 660 (1996)
171. T. Scheunemann, R. Feder, J. Henk, E. Bauer, T. Duden, H. Pinkvos, H. Poppa, K. Wurm, Solid State Commun. **104**, 787 (1997)
172. R.C. Jaklevic, L.C. Davis, Phys. Rev. B **26**, 5391 (1982)
173. A.K. Schmid, W. Swiech, C.S. Rastomjee, B. Rausenberger, W. Engel, E. Zeitler, A.M. Bradshaw, Surf. Sci. **331–333**, 225 (1995)
174. C.M. Yim, K.L. Man, X. Xiao, M.S. Altman, Phys. Rev. B **78**, 155439 (2008)
175. H. Poon, S. Tong, W. Chung, M. Altman, Surf. Rev. Lett. **5**, 1143 (1998)
176. M. Altman, W. Chung, Z. He, H. Poon, S. Tong, Appl. Surf. Sci. **169–170**, 82 (2001)
177. M.S. Altman, J. Phys. Condens. Matter **17**, S1305 (2005)
178. R. Zdyb, E. Bauer, Phys. Rev. Lett. **88**, 166403 (2002)
179. Y.Z. Wu, A.K. Schmid, M.S. Altman, X.F. Jin, Z.Q. Qiu, Phys. Rev. Lett. **94**, 027201 (2005)
180. Y. Wu, A. Schmid, Z. Qiu, Phys. Rev. Lett. **97**, 217205 (2006)
181. J.E. Ortega, F.J. Himpsel, G.J. Mankey, R.F. Willis, Phys. Rev. B **47**, 1540 (1993)
182. Z.Q. Qiu, N.V. Smith, J. Phys. Condens. Matter **14**, R169 (2002)
183. C. Jozwiak, J. Graf, G. Lebedev, N. Andresen, A. Schmid, A. Fedorov, F. El Gabaly, W. Wan, A. Lanzara, Z. Hussain, Rev. Sci. Instrum. **81**, 053904 (2010)
184. Y. Homma, M. Tomita, T. Hayashi, Ultramicroscopy **52**, 187 (1993)
185. I. Müllerová, K. Matsuda, P. Hrncirík, L. Frank, Surf. Sci. **601**, 4768 (2007)
186. F.F. Tao, M. Salmeron, Science **331**, 171 (2011)

Part VI
Scanning Probe Techniques

Chapter 19
Scanning Tunneling Microscopy

Ada Della Pia and Giovanni Costantini

Abstract The Nobel Prize-awarded invention of the scanning tunneling microscope (STM) has profoundly revolutionized contemporary science and technology. The STM has enabled individual atoms and molecules to be imaged, probed and handled with an unprecedented precision, thereby essentially contributing to our current understanding of the world at the nanoscale. Together with its offspring, the atomic force microscope (AFM), the STM is considered as the main innovation behind the birth of nanotechnology. This chapter is an elementary introduction to STM and to its most recent uses. Topics include a basic treatment of the underlying theory of tunneling, the description of the most commonly used experimental setups, a survey of the atomic-scale spectroscopic capabilities (scanning tunneling spectroscopy, STS) and an overview of atomic manipulation experiments.

19.1 Introduction

If a bias voltage is applied to two electrodes a few tenths of a nm apart, a current flows between them even though they are not in contact. This is due to the quantum mechanical process of electron tunneling. The resulting current depends exponentially on the electrode separation s so that even minute, subatomic variations in s produce measurable current changes. In 1981 Gerd Binnig and Heinrich Rohrer at IBM in Zürich realized that by shaping one of the electrodes as a sharp tip and scanning it across the surface of the other (Fig. 19.1), quantum tunneling can be used to build a microscope with ultra high spatial resolution [1]. Moreover, since the current depends also on the electronic properties of the electrodes, they recognized that this microscope has the ability to probe the electronic density of states of surfaces at the atomic scale. A few years later, Don Eigler at IBM in Almaden, showed that, due to the extremely localized interaction between tip and sample, it is even possible to use

A. Della Pia · G. Costantini (✉)
Department of Chemistry, The University of Warwick, Gibbet Hill Road, Coventry CV4 7AL, UK
e-mail: g.costantini@warwick.ac.uk

A. Della Pia
e-mail: A.della-pia@warwick.ac.uk

G. Bracco, B. Holst (eds.), *Surface Science Techniques*,
Springer Series in Surface Sciences 51, DOI 10.1007/978-3-642-34243-1_19,

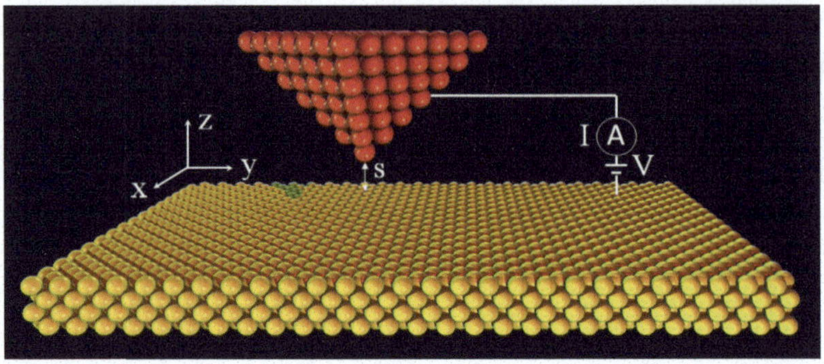

Fig. 19.1 Schematic representation of a STM. Tip and sample are held at a distance s of a few tenths of a nm and a bias voltage V, up to a few Volts, is applied between them. The resulting tunneling current I is recorded while the tip is moved across the surface. The coordinate system is also shown

this instrument to manipulate individual atoms, to position them at arbitrary locations and therefore to build artificial structures atom-by-atom [2]. This remarkable achievement brought to reality the visionary predictions made by Richard Feynman in his famous 1959 lecture "*There's plenty of room at the bottom*" [3].

The construction of this instrument, dubbed the scanning tunneling microscope (STM), was awarded the 1986 Nobel Prize in Physics and has revolutionized contemporary science and technology. The STM has enabled individual atoms and molecules to be imaged, probed and handled with an unprecedented precision, thereby essentially contributing to our current understanding of the world at the nanoscale. Together with its offspring, the atomic force microscope (AFM), the STM is considered as the main innovation behind the birth of nanotechnology.

This chapter is an elementary introduction to STM and to its most recent uses. It starts with a discussion of the physical principles and processes at the basis of STM in Sect. 19.2. This is followed by a description of the experimental setup and the technical requirements needed for actually operating such an instrument in Sect. 19.3. Section 19.4 is dedicated to the most frequent use of STM, namely imaging of surfaces, while Sect. 19.5 gives an account of its spectroscopic possibilities. Finally, Sect. 19.6 surveys the atomic manipulation capabilities of STM.

19.2 Theory of Tunneling

Figure 19.2 is a schematic representation of the energy landscape experienced by an electron when moving along the z axis of a metallic-substrate/insulator/metallic-tip tunneling junction. Usually the tip and the sample are not made of the same material and are therefore characterized by different work functions, ϕ_T and ϕ_S, respectively. At equilibrium, the two metals have a common Fermi level, resulting in an electric

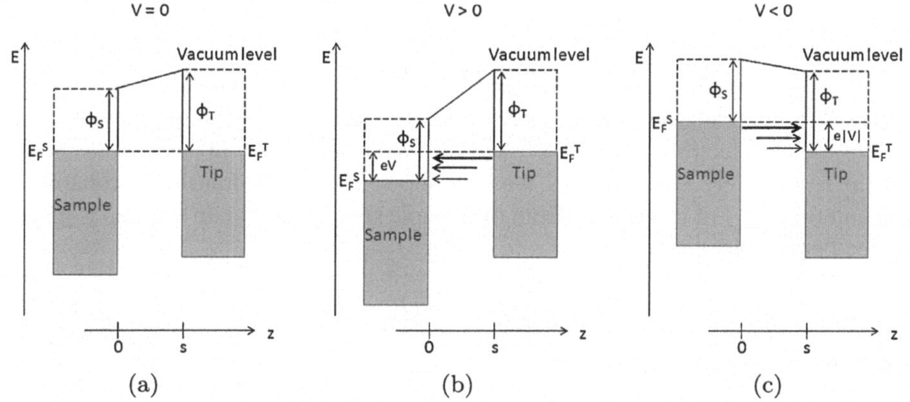

Fig. 19.2 Energy potential perpendicular to the surface plane for an electron in a tip-vacuum-sample junction. z is the surface normal direction, and s is the tip-sample distance. The *gray boxes* represent the Fermi-Dirac distribution at 0 K. $\phi_{T,S}$ and $E_F^{T,S}$ are the work functions and the Fermi levels of tip and sample, respectively. (**a**) Tip and sample in electrical equilibrium: a trapezoidal potential barrier is created. (**b**) Positive sample bias: the electrons tunnel from occupied states of the tip into unoccupied states of the sample. The thickness and the length of the *arrows* indicate the exponentially decreasing probability that an electron with the corresponding energy tunnels through the barrier. (**c**) Negative sample bias

field being established across the gap region and in different local vacuum levels, depending on the difference $\phi_T - \phi_S$ (Fig. 19.2(a)). Since the work functions in metals are of the order of several eV, the potential in the gap region is typically much higher than the thermal energy kT and thus acts as a barrier for sample and tip electrons. A classical particle cannot penetrate into any region where the potential energy is greater than its total energy because this requires a negative kinetic energy. However, this is possible for quantum mechanical objects such as electrons which are described by delocalized wave functions. This phenomenon goes under the name of *quantum tunneling*. In an unpolarized tip-sample junction the electrons can tunnel from the tip to the sample and vice versa, but there is no net tunneling current. On the contrary, if a voltage V is applied between sample and tip,[1] the Fermi level of the former is shifted by $-eV$ and a net tunneling current occurs, whose direction depends on the sign of V (Figs. 19.2(b) and (c)).

The tunneling current can be evaluated by following the time-dependent perturbation approach developed by Bardeen [4, 5]. The basic idea is to consider the isolated sample and tip as the unperturbed system described by the stationary one-particle Schrödinger equations:

$$(\mathcal{T} + \mathcal{U}_S)\psi_\mu = E_\mu \psi_\mu \tag{19.1}$$

[1]Here the convention is adopted to take the tip as a reference since experimentally the voltage is often applied to the sample while the tip is grounded. If V is the bias voltage, the energy for an electron in the sample will change by $-eV$, i.e. it will decrease for positive values of V.

and

$$(\mathcal{T} + \mathcal{U}_T)\chi_\nu = E_\nu \chi_\nu, \tag{19.2}$$

where $\mathcal{T}$ is the electron kinetic energy and ψ_μ and χ_ν are the unperturbed wave functions. The electron potentials $\mathcal{U}_S$ and $\mathcal{U}_T$ are non-zero only in the sample and in the tip, respectively. Based on this, it can be shown [5] that the transition probability per unit time $w_{\mu\nu}$ of an electron from the sample state ψ_μ to the tip state χ_ν is given by the Fermi golden rule:

$$w_{\mu\nu} = \frac{2\pi}{\hbar}|M_{\mu\nu}|^2 \delta(E_\nu - E_\mu), \tag{19.3}$$

where the matrix element is:

$$M_{\mu\nu} = \int \chi_\nu^*(\boldsymbol{x})\mathcal{U}_T(\boldsymbol{x})\psi_\mu(\boldsymbol{x})\,d^3\boldsymbol{x}. \tag{19.4}$$

The δ function in (19.3) implies that the electrons can tunnel only between levels with equal energy, that is, (19.3) accounts only for an elastic tunneling process.[2] The total current is obtained by summing $w_{\mu\nu}$ over all the possible tip and sample states and by multiplying this by the electron charge e. The sum over the states can be changed into an energy integral by considering the density of states (DOS) $\rho(E)$:

$$\sum_{\mu,\nu} \to 2\int f(\epsilon)\rho(\epsilon)\,d\epsilon. \tag{19.5}$$

The factor 2 accounts for the spin degeneracy while f, the Fermi-Dirac distribution function, takes into consideration Pauli's exclusion principle and the electronic state population at finite temperatures. As a consequence, the total current can be written as:

$$I = \frac{4\pi e}{\hbar}\int_{-\infty}^{\infty} \left[f_T\left(E_F^T - eV + \epsilon\right) - f_S\left(E_F^S + \epsilon\right)\right] \times \rho_T\left(E_F^T - eV + \epsilon\right)\rho_S\left(E_F^S + \epsilon\right)|M|^2\,d\epsilon, \tag{19.6}$$

where E_F is the Fermi energy and the indexes T and S refer to the tip and the sample, respectively. Expression (19.6) already accounts for the movement of electrons from the sample to the tip and vice versa.

Several approximations can be made to simplify (19.6) and to obtain an analytically manageable expression for I. If the thermal energy $k_B T \ll eV$, the Fermi-Dirac distributions can be approximated by step functions and the total current reduces to:[3]

$$I = \frac{4\pi e}{\hbar}\int_0^{eV} \rho_T\left(E_F^T - eV + \epsilon\right)\rho_S\left(E_F^S + \epsilon\right)|M|^2\,d\epsilon. \tag{19.7}$$

[2] The case of an inelastic tunneling process will be considered in Sect. 19.5.7.

[3] Expression (19.7) is valid only for $V > 0$. For $V < 0$ the integrand remains identical but the integration limits become $-e|V|$ and 0.

In this case only electrons with an energy differing from E_F by less than eV can participate to the tunneling current. This can be directly seen in Fig. 19.2(b) for the case of positive sample bias: tip electrons whose energy is lower than $E_F^T - eV$ cannot move because of Pauli's exclusion principle, while there are no electrons at energies higher than E_F^T.

The main problem in determining expression (19.6) is, however, the calculation of the tunneling matrix elements M since this requires a knowledge of the sample and the tip wave functions, which can be very complicated. On the other hand, for relatively small bias voltages (in the ±2 V range), Lang [6] showed that a satisfactory approximation is obtained by considering that the tunneling current flows only between the last atom of a perfectly sharp tip and the surface atom directly underneath. In this case $|M|^2$ is given by a simple one-dimensional WKB tunneling probability. In the WKB approximation [7], the probability $D(\epsilon)$ that an electron with energy ϵ tunnels through a potential barrier $U(z)$ of arbitrary shape is expressed as:

$$D(\epsilon) = \exp\left\{-\frac{2}{\hbar}\int_0^s \left[2m\big(U(z)-\epsilon\big)\right]^{\frac{1}{2}} dz\right\}. \tag{19.8}$$

This semiclassical approximation is applicable if $\epsilon \ll U$ which is in general satisfied in the case of metal samples where the work function is of the order of several eV. In order to obtain a simple analytical expression for D, the trapezoidal potential barrier of a biased tip-sample junction (see Figs. 19.2(b) and (c)) is further approximated with a square barrier of average height

$$\phi_{\mathit{eff}}(V) = (\phi_T + \phi_S + eV)/2. \tag{19.9}$$

By using this, expression (19.8) becomes:

$$D(\epsilon, V, s) = \exp\left[-2s\sqrt{\frac{2m}{\hbar^2}\left(\frac{\phi_T+\phi_S}{2}+\frac{eV}{2}-\epsilon\right)}\right] = \exp(-2ks), \tag{19.10}$$

where

$$k = \sqrt{\frac{2m}{\hbar^2}(\phi_{\mathit{eff}}-\epsilon)}. \tag{19.11}$$

The inverse decay length $2k$ dictates how sensitive the tunneling current is on variations of the tip-sample separation s and is therefore essential in determining the spatial resolution of STM. In order to evaluate k, it must be noted that electrons closest to the Fermi level experience the lowest potential barrier and are therefore characterized by an exponentially larger tunneling probability (see Figs. 19.2(b) and (c)). Thus, in a first approximation, it can be assumed that only these electrons contribute to the tunneling current which, for positive bias, is equivalent to set $\epsilon \approx eV$ in (19.11). Moreover, if the bias is much smaller than the work functions, eV can be neglected in (19.9), resulting in

$$k \cong \frac{\sqrt{m(\phi_T+\phi_S)}}{\hbar} = 5.1\sqrt{\frac{\phi_T+\phi_S}{2}}\ \text{nm}^{-1}, \tag{19.12}$$

where the work functions are expressed in eV. Using typical numbers for metallic work functions, the value of the inverse decay length $2k$ in (19.10) becomes of the order of 20 nm^{-1}. Therefore, variations in s of 0.1 nm correspond to one order of magnitude changes in the tunneling probability and, as a consequence, in the measured current. This very high sensitivity provides the STM with a vertical resolution in the picometer regime.

The lateral resolution of STM depends on how different points of the tip contribute to the total tunneling current. By considering a spherical tip shape with radius R, most of the current originates from the central position since this is closest to the surface. A point laterally displaced by Δx from the tip center is $\Delta z \approx \frac{\Delta x^2}{2R}$ further away from the substrate (higher order Δx terms are neglected in this evaluation). As a consequence, with respect to the tip center, the corresponding tunneling probability is reduced by a factor:

$$\exp\left(-2k\frac{\Delta x^2}{2R}\right). \tag{19.13}$$

By considering a tip radius $R \approx 1$ nm, the current changes by one order of magnitude for variations $\Delta x = 0.3$ nm. The actual lateral resolution is typically smaller than this upper limit and can reach down to fractions of an Å. Its specific value however depends on the precise shape of the tip which is unknown a priori. These values, together with the vertical resolution discussed above, lie at the basis of the STM atomic imaging capabilities.

Finally, by following the approximation by Lang [6], the tunneling probability (19.10) can be substituted for the tunneling matrix $|M|^2$ in (19.7) and the total tunneling current becomes:

$$I = \frac{4\pi e}{\hbar} \int_0^{eV} \rho_T\left(E_F^T - eV + \epsilon\right)\rho_S\left(E_F^S + \epsilon\right)e^{-2ks}\,d\epsilon. \tag{19.14}$$

Therefore, for a fixed lateral position of the tip above the sample, the tunneling current I depends on the tip-sample distance s, the applied voltage V and the tip and sample density of states ρ_T and ρ_S, respectively.

It should be noted that, due to the 1D approximation [6], the densities of states ρ_T and ρ_S in (19.14) do not refer to the whole tip and sample but only to the last atom of the tip and to the atom of the sample which is located directly underneath. For this reason ρ_S does depend on the lateral position of the tip (x, y) and will be referred to as local density of states (LDOS).

19.3 Experimental Setup

As seen in the previous section, variations of 0.1 nm in s induce changes in the tunneling probability of one order of magnitude. The exponential dependence in (19.10) lies at the heart of the ultimate spatial resolution of STM but places stringent requirements on the precision by which s must be controlled, as well as on the suppression of vibrational noise and thermal drift. Moreover, typical tunneling currents are in the 0.01–10 nA range, requiring high gain and low noise electronic components. The following subsections are dedicated to a general overview of technologies and methods employed to meet these specifications.

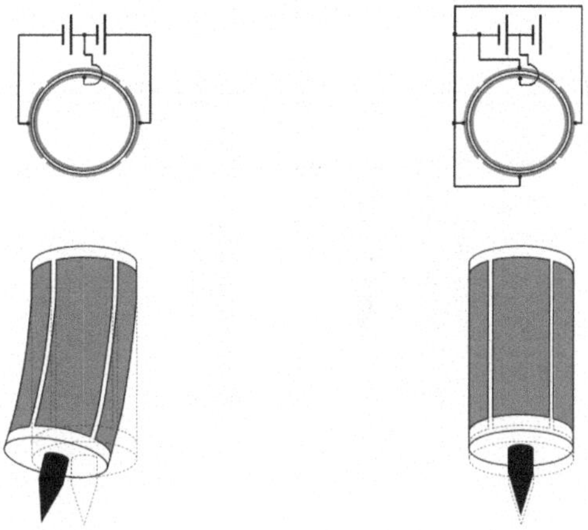

(a) (b)

Fig. 19.3 Piezoelectric scanner tube. (**a**) Schematic representation of the lateral tip displacement and (**b**) of the vertical retraction and corresponding electrical wiring (*upper row*)

19.3.1 Scanner and Coarse Positioner

The operation of a STM requires extremely fine movements of the tip relative to the sample which are realized by the use of piezoelectric ceramic actuators (*scanners*). These expand or retract depending on the voltage difference applied to their ends and, in a first approximation, the voltage-expansion relation is linear with a proportionality factor (piezo constant) usually of few nanometer/Volt. The main requirements for a good scanner are: high mechanical resonance frequencies, so as to minimize noise vibrations in the frequency region where the feedback electronics operates (see Sect. 19.3.2); high scan speeds; high spatial resolution; decoupling between x, y and z motions; minimal hysteresis and creep; low thermal drift. Although several types of STM scanner have been developed, including the bar or tube tripod, the unimorph disk and the bimorph [8], the most frequently used is a single piezoelectric tube whose outer surface is divided into four electrode sections of equal area (Fig. 19.3). By applying opposite voltages between the inner electrode and opposite sections of the outer electrode, the tube bends and a lateral displacement is obtained (Fig. 19.3(a)). The z motion is realized by polarizing with the same voltage the inner electrode in respect to all four outer electrodes (Fig. 19.3(b)). By applying several hundred Volts to the scanner, lateral scan widths up to 10 μm and vertical ones up to 1 μm can be obtained, while retaining typical lateral and vertical resolutions of 0.1 nm and 0.01 nm, respectively.

While scanning is typically done by one individual piezoelectric element, larger displacements up to several millimeters are needed to bring the tip in close proximity to the sample, to move it to different regions of the surface or to exchange samples or tips. These are achieved by mounting the scanner onto a coarse position device. Several designs have been developed to this aim including:

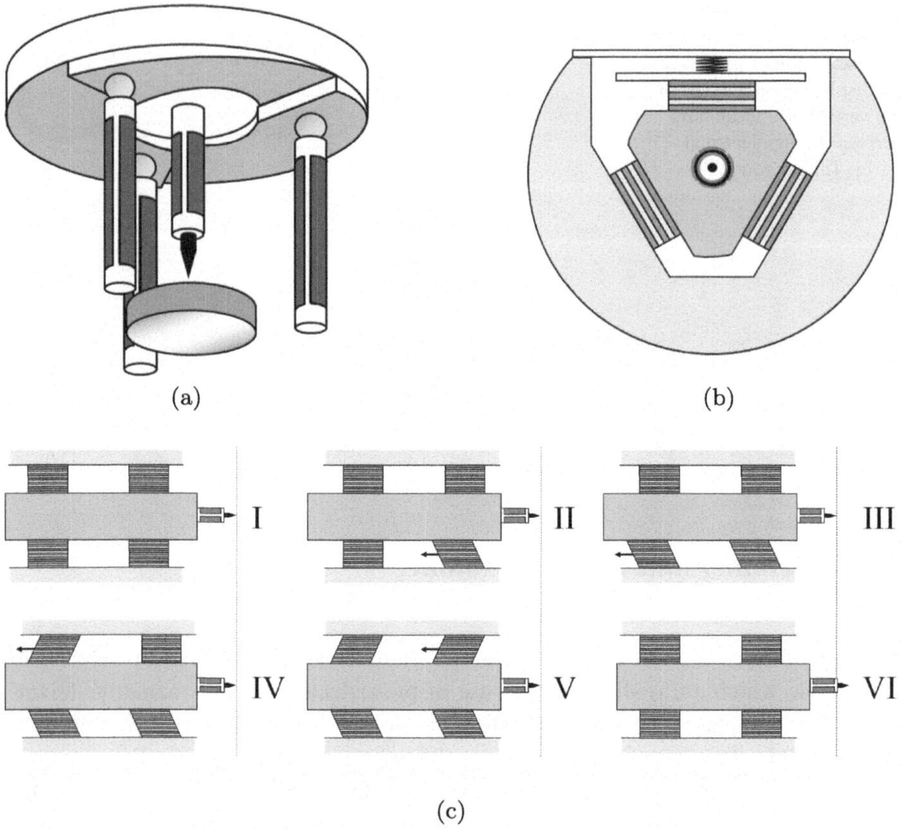

Fig. 19.4 (**a**) Schematic representation of the Besocke-beetle scanner. (**b**) Section view of the Pan motor as viewed from the tip and (**c**) schematics of its working principle

- Micrometer screws driven either manually or by a stepper motor. Mechanical coupling and thermal drift due to the heat generated by the mechanism limit the function of these devices.
- Piezoelectric walkers like the louse used in the first STM [9] or the inchworm [10], composed by two piezoelectric feet and a piezoelectric body. The movement is obtained by alternating the clamping of the first foot, the contraction or expansion of the body and the clamping of the second foot.
- Magnetic walkers where the movement is obtained by applying voltage pulses to a coil with a permanent magnet inside.
- Piezoelectric driven stick-slip motors, as the Besocke-beetle [11] or the Pan motor [12]. Together with the inchworm walker, these two latter are the most frequently used in modern commercial instruments.

The Besocke-beetle consists of a base plate with three 120° sections of a shallow ramp which is supported, through ball bearings, by three outer piezo tubes (Fig. 19.4(a)). A further piezo tube which carries the tip and acts as scanner is located in the center of the plate. Coarse lateral tip displacements are realized by using the three outer piezoelectric tubes and operating them in a linear stick-slip motion with respect to the base plate [11]. Coarse vertical displacements are achieved by inducing a rotational stick-slip motion of the piezos: because of the ramps, the rotation is coupled with a vertical motion, similar to the effect of a screw thread. A STM head based on the Besocke design has the advantage of being thermally compensated since the coarse positioning device and the scanner are made of the same components and can thus be expected to have the same thermal expansion coefficients. However, its drawbacks are a substantial bulkiness and the reduced overall coarse z travel (typically less than 1 mm).

In the motor developed by S.H. Pan and coworkers [12] the tip tube scanner is mounted onto a prism-shaped sapphire (Fig. 19.4(b)). A spring mechanism holds the prism inside a ceramic body by firmly pressing it against six piezo stacks which are used to perform coarse z movements (Fig. 19.4(b)). One step forward of the prism is achieved by sequentially applying six sharp high-voltage steps to the individual piezo stacks so that these bend backwards one at a time without moving the prism which is hold by the remaining ones (slip motion, steps II to V in Fig. 19.4(c)). In the final step, all voltages are slowly reduced to zero so that the piezo stacks simultaneously relax back to their neutral positions and, through friction, move the prism forward (stick motion, step VI in Fig. 19.4(c)). The Pan motor has a very long z travel (typically up to 10 mm) and, being particularly compact and of reduced size, it is the solution of choice for STMs operating at mK temperatures and in high magnetic fields.

19.3.2 Electronics and Control System

The voltage signals driving the piezoelectric actuators and their temporal succession and duration are generated by an electronic control system. The electronics are also used to bias the tunneling junction, to record the tunneling current and to generate the STM images. In most of the modern instruments these tasks are digitally implemented by a computer interfaced with digital to analog (DAC) and analog to digital (ADC) converters. The tunneling current is amplified by a high gain I–V converter (10^8–10^{10} V/A) usually positioned in close proximity of the tip, so as to reduce possible sources of electronic interference. This signal is then acquired by an ADC and processed by the control system. DACs are used to apply the bias voltage (from a few mV to a few V) between tip and sample and, in conjunction with high voltage amplifiers, to polarize the piezo elements. A feedback loop is integrated into the control system and is activated during the frequently used *constant current* imaging mode (see Sect. 19.4). By acting on the z motion of the scanner, the feedback varies s to keep the tunneling current constant. This is controlled by a proportional-integral and derivative (PID) filter whose parameters can be set by the operator.

Finally, a lock-in amplifier is often used in scanning tunneling spectroscopy (STS) experiments for improving the signal-to-noise ratio and for directly isolating

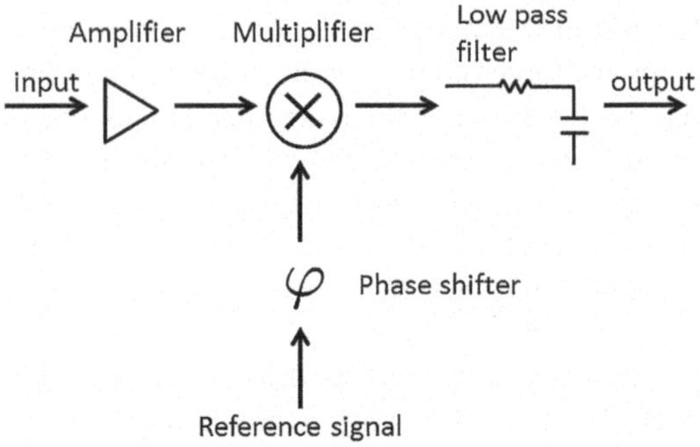

Fig. 19.5 Schematic block diagram of a lock-in amplifier

the first or second order derivative of the current with respect to the voltage. These are directly related to the electronic properties of the samples (see Sect. 19.5). In the I(V) spectroscopy technique (see Sect. 19.5.1) a small high-frequency sinusoidal signal $V_m = V_0 \sin(\omega t + \varphi_0)$ is summed to the bias voltage V applied between tip and sample (typical values of V_0 are between 1 and 100 mV, while ω is usually a few kHz). Since the amplitude of the sinusoidal signal V_0 is small, the tunneling current resulting from (19.14) can be expanded in a power series:

$$I\big(V + V_0 \sin(\omega t + \varphi_0)\big) = I(V) + \frac{dI(V)}{dV} V_0 \sin(\omega t + \varphi_0) + \frac{d^2 I(V)}{dV^2} \frac{V_0^2}{2} \sin^2(\omega t + \varphi_0) + \cdots. \qquad (19.15)$$

In a lock-in amplifier, schematized in Fig. 19.5, the amplified current is multiplied by a sinusoidal reference signal V_{ref} and then integrated over a time longer than its period. All the components in expression (19.15) with a frequency different from the reference signal are filtered out by this last step since the integral of their product with V_{ref} is averaged to 0. As a consequence, if V_{ref} has the same frequency as the modulation signal, the only remaining part is a DC signal proportional to the amplitude of the ω-component in (19.15), i.e. to $(dI/dV)\cos(\varphi - \varphi_0)$, where φ is the phase of the reference signal. By adjusting φ so that $\varphi - \varphi_0 = 0$, the output of the lock-in amplifier is maximized and the differential conductivity is obtained at the voltage V. Similarly, if V_{ref} is chosen to have a frequency 2ω, only the second harmonic component in (19.15) survives the integration step, resulting in a signal proportional to d^2I/dV^2.

19.3.3 Tip

Sharp metal tips with a low aspect ratio are essential to optimize the resolution of the STM images and to minimize flexural vibrations, respectively. Ideally, in order to obtain atomically resolved topographies and accurate spectroscopic measurements, the tip should be terminated by a single atom. In this case, because of the strong dependence on the tip-sample separation (see Sect. 19.2), most of the tunneling current

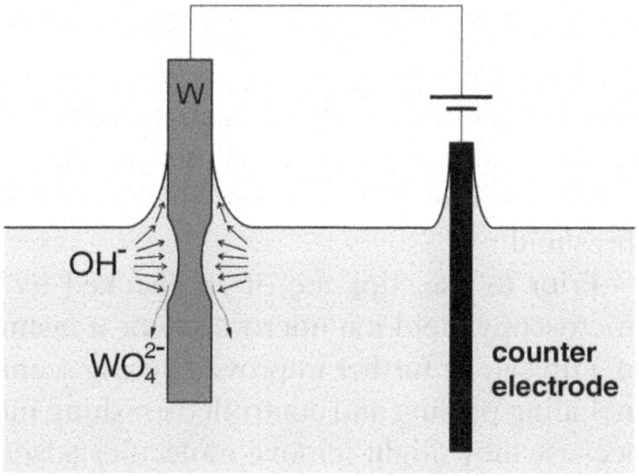

Fig. 19.6 Schematics of the electrochemical cell used for the etching of W tips in a strong alkaline solution

would originate from the last atom whose position and local DOS would precisely determine the tunneling conditions. In practice however, it is almost impossible to determine the exact atomic configuration of the tip and the actual current is often due to a number of different atoms. This is still compatible with good tunneling conditions as long as these atoms are sufficiently localized (in order to avoid "multiple tip effects") and if their structural and chemical state remains constant during scanning. An overview of the possible artifacts in STM imaging due to the tip shape can be found in [13].

The most commonly used methods to produce STM tips are to manually cut or to electrochemically etch thin wires of platinum-iridium and tungsten, respectively. These materials are chosen because of their hardness, in order to prevent tips becoming irreversible damaged after an accidental crash.[4] Due to their chemical inertness, Pt-Ir tips are often used to scan in air on atomically flat surfaces without the need of any further processing. However, they typically have inconsistent radii, while etched W tips are characterized by a more reproducible shape. These latter have the drawback that a surface oxide up to 20 nm thick is formed during etching or exposure to air [14]. For this reason W tips are primarily employed in ultra high vacuum (UHV) where the oxide layer can be removed through ion sputtering and annealing cycles.

The most used etching method is the so-called DC drop-off where the wire is immersed in a concentrated aqueous alkaline solution and the etching occurs largely at the air-electrolyte interface (Fig. 19.6). A constant positive voltage is applied between the wire (anode) and a counter electrode (cathode) inducing two reactions: the water reduction in bubbles of hydrogen gas and OH^- ions at the cathode, and the oxidative dissolution of the solid W anode to soluble tungstate ions [15]. The shape of the meniscus limits the diffusion of the OH^- ions and therefore reduces their concentration in its upper part. This slows down the etching of the wire in this region. On the other hand, also the part of the wire below the meniscus is etched at a slower rate because it is shielded by the downwards flow of tungstate ions (Fig. 19.6). The combination of these two effects leads to the typical hyperboloid

[4]Other metallic elements and even semiconductor materials have been used as tips for specific STM applications.

shape of a DC etched tip. When the weight of the immersed portion of the wire exceeds its tensile strength, the wire breaks and the actual tip is formed. In order to avoid a tip blunting, it is essential that the etching stops immediately after the wire breaks. This can be achieved by using an electronic control system that records the etching current and switches off the voltage if the current diminishes below a set threshold.

Prior to use, tips are often checked by optical microscopy, scanning electron microscopy, field ion microscopy or transmission electron microscopy. The quality of a tip can be further improved during scanning by using "*tip forming*" procedures, including pulsing and controlled crashing into metal surfaces. These processes work because they might remove molecules adsorbed on the tip or coat the tip itself with atoms of the metallic substrate, thereby producing a more stable tip apex.

If STM is performed in polar liquids, electrochemical processes might generate Faradaic or non-Faradaic currents which can be of the same order of magnitude or even larger than the tunneling current. In order to minimize these effects, the tip, except for its very apex, must be coated with an insulating material [16].

19.3.4 Vibration Isolation

A low level of mechanical noise is an essential requirement for any type of scanning probe microscopy. For this reason the core of a STM, where the tip-sample junction is located, is always equipped with one or several types of vibration damping systems. These can be stacks of metal plates separated by elastic spacers, suspension springs or eddy current dampers composed of copper elements and permanent magnets. The low frequency components of mechanical noise (< 10 Hz), which are the most difficult to eliminate, are minimized by building a small and rigid STM with a high resonance frequency. Depending on the overall size and weight of the microscope, further noise damping strategies can be adopted. Smaller, typically ambient conditions STMs, can be placed on metal or granite slabs suspended by springs or bungee cords or floating on pneumatic isolators. Sometimes, piezo-driven, feedback-controlled active vibration suppressors are also combined with passive systems. Larger versions of pneumatic isolators and active damping are used to float the frames and the chambers of big UHV STMs. The laboratory where a STM instrument is located also plays an essential role for its performance. Ground floor rooms are always preferred since they minimize low-frequency natural building oscillations, which can be very difficult to counteract. High resolution instruments are sometimes placed on large concrete blocks which are separated from the rest of the laboratory floor and rest either on a sand bed, an elastomer barrier or on second-stage pneumatic isolators [8]. Moreover, they are also often surrounded by an acoustically insulating box. All these systems essentially act as low-pass mechanical filters whose effectiveness improves with decreasing cutoff frequencies, i.e. with increasing mass and decreasing rigidity. For this reason, the body of a STM is typically a relatively heavy block of metal and the frames, slabs and vacuum chambers supporting or containing the microscope often have a considerable weight.

19.3.5 Setups for Different Environments and Temperatures

Different types of STMs have been developed that can operate in various environments such as air, inert atmosphere (N_2, Ar), vacuum, high pressure, liquid or in an electrochemical cell. The core of the different instruments is essentially the same, although the experimental chambers and setups in which they are located can vary substantially. Ambient condition STMs are typically quite compact and rigid and do not need elaborated anti-vibrational mechanisms. On the other hand, since sound waves represent a major problem, atmospheric pressure STMs are usually contained in an acoustic enclosure. A STM operating in vacuum must be hosted in a chamber with vibration-free pumps (typically ionic pumps for UHV) and must be equipped with sophisticated sample and tip manipulation mechanisms. Such systems often also have an *in-situ* surface preparation stage allowing the handling of samples without air exposure.

STM can be performed at high pressures (1–30 bar) by installing the microscope head into gas manifolds under conditions similar to those used in industrial catalytic processes. Also in this case, sample and tip manipulation and preparation stages are mandatory parts of the system. Since these types of studies are typically performed at elevated temperatures (up to 600 K) and in the presence of highly reactive gases, the metallic parts of the STM scanner and of the chamber are often gold plated, the volume of the STM chamber is kept as small as possible and the tip material is chosen to be inert toward the employed gases [17]. Moreover, low voltages are used for polarizing the piezos in order to avoid gas discharges at intermediate pressures (10^{-3}–10 mbar) and shields are added to protect the STM from the deposition of conductive materials which could create electrical shorts.

STM at the liquid/solid interface and electrochemical STM (EC-STM) need the tip and sample to be inside a liquid cell which, in turn, may be placed in a humidity-controlled atmosphere. In the case of low vapor pressure liquids, the STM can be simply operated under ambient conditions by dipping the tip into a liquid droplet deposited on the sample. A special coating must be applied to the tip when working with polar liquids (see Sect. 19.3.3).

STM can also be performed at different temperatures (in vacuum or controlled atmosphere systems): variable temperature STM (VT-STM) able to cover the 5–1400 K range, low temperature STM (LT-STM) operating at 77 K or 5 K and even milli-Kelvin STM instruments are currently available. A VT-STM is typically used to study thermally activated processes such as diffusion and growth, phase transitions, etc. These systems have sample heating and cooling stages which can be operated in a combined way so as to achieve a very precise temperature stabilization. Resistive heating is normally employed to increase the temperature, while both flow and bath cryostats with liquid nitrogen or helium as cryogenic fluids are used to reduce it. Continuous flow cryostats offer a high flexibility in temperature but are characterized by lower thermal stability, by inherent mechanical vibrations and do not easily attain temperatures below 20 K. Bath cryostats are more stable, are able to reach lower temperatures but are often also much bulkier (e.g., in order to limit the He consumption rate, a liquid He cryostat is composed by a double-stage vessel with an outer liquid nitrogen mantle). For most of these instruments the variable temperature capabilities refer to the possibility of choosing different (fixed) temperatures at which the microscope is run. However, few systems endowed with high performance position tracking and drift compensating capabilities allow a "true" variable temperature operation whereby the same surface area can be imaged with atomic resolution while its temperature is changed. By choosing optimized designs for the piezo scanners and the electronic feedback, video-rate instruments have been developed able to record several tens of images per second and thereby to follow dynamic surface processes such as molecular mobility and assembly in real time [18, 19].

LT-STMs are operated at a fixed temperature and are typically inserted inside double stage cryostats which significantly complicates the tip and sample access. However, these instruments are extremely stable with a very low thermal drift (of the order of tenths of a nm per hour) and are therefore the best choice for STS and manipulation experiments (see Sects. 19.5 and 19.6). Milli-Kelvin STMs enable temperatures to be reached where extremely interesting magnetic, quantum Hall physics and superconductivity phenomena occur. Moreover, the thermal broadening of electronic features is strongly reduced, which is required for high resolution measurements. These systems operate based on the Joule-Thomson evaporative cooling of liquid ^{3}He to temperatures of about 300 mK or liquid ^{3}He and ^{4}He mixtures below 10 mK. The STM heads can be further placed inside large-bore superconducting magnets (at present up to 15 Tesla), allowing the low temperature and high magnetic field conditions necessary to access superconductive phase transitions or to detect single spin flip processes.

19.4 STM Imaging

Since the first STM images of the surfaces of $CaIrSn_4$ and Au [1] were published back in 1982, STM has been used to analyze a wide range of materials: clean and adsorbate covered metal surfaces, semiconductors, superconductors, thin insulating layers, small and large organic molecules, individual atoms, liquid-solid interfaces, magnetic layers and surfaces, quasicrystals, polymers, biomolecules, nanoclusters and carbon nanotubes. Imaging is the most frequent application of STM used to determine the structural properties of substrates and their reconstructions, the presence of defects, sites of adsorption for adatoms and molecules and the symmetry and periodicity of adsorbate superstructures.

STM images are generated by recording the tunneling current as a function of the tip position while the tip is scanned across the sample surface. This can be done in two different ways which define the two main STM imaging modes:

- *Constant height mode*. The z section of the piezo scanner is kept fixed while the tip is moved over the substrate at a constant bias voltage (Fig. 19.7(a)). Variations of the tip-sample distance due to the surface topography produce a corresponding variation of the tunneling current which is recorded point-by-point and used to build the STM gray-level image. This mode is employed only in small areas of

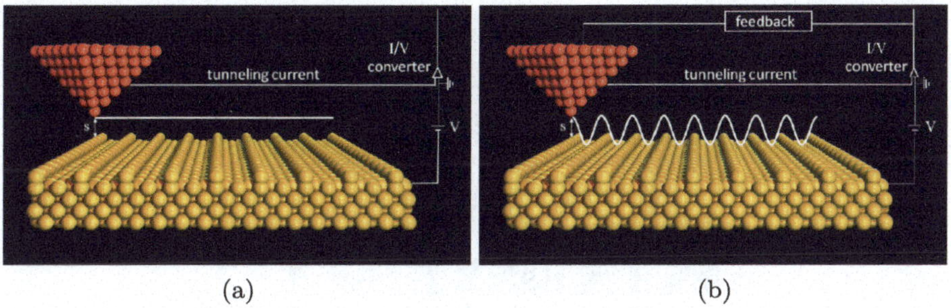

Fig. 19.7 (**a**) Constant height and (**b**) constant current imaging modes. The *thick lines* represent the trajectory followed by the tip, $s(x, y)$

extremely flat surfaces, where the probability of crashing into protrusions such as steps or defects is relatively small. Very high scanning speeds can be used because of the absence of a feedback control.

- *Constant current mode*. While the x and y sections of the piezo scanner are used to laterally move the tip across the surface, the z section is driven by the electronic feedback so as to maintain a constant tunneling current (Fig. 19.7(b)). The corresponding z-voltage applied to the scanner (feedback signal) is recorded point-by-point and used to build the STM gray-level image. This mode can be employed for any type of surface topography and is therefore the most frequently used.

Since the constant height mode is applied to atomically flat surfaces with sub-Å height variations, the exponential $I - s$ relation derived from (19.14) can be approximated by a linear dependence. As a consequence, constant height STM images are a good representation of flat surfaces. On the other hand, for less planar substrates, the constant current mode must be used which directly reproduces the surface height due to the linear voltage-extension relation of piezoelectric materials. However, even constant current images are a reliable representation of the "true" surface topography only if the sample LDOS does not vary across the scanned area. If this is not the case, a constant current profile corresponds to a complex convolution of topographical and electronic features which can be particularly relevant for surfaces covered with adsorbates.

For example, electronegative atoms like oxygen and carbon can, counter intuitively, appear as depressions in STM images when adsorbed onto metal surfaces [20, 21]. This can be understood by considering the approximation of (19.14) for small voltages and a constant tip density of states:

$$I \propto \rho_S(E_F)\exp(-2ks). \tag{19.16}$$

The chemisorption of electronegative atoms involves a charge transfer from the substrate which is accompanied by the screening of the metal conduction electrons. Both effects yield a decrease of the density of states near the Fermi level compared to the bare metal. As a consequence, in order to keep the product on the right hand side of (19.16) constant, the tip-sample distance s must be reduced, therefore

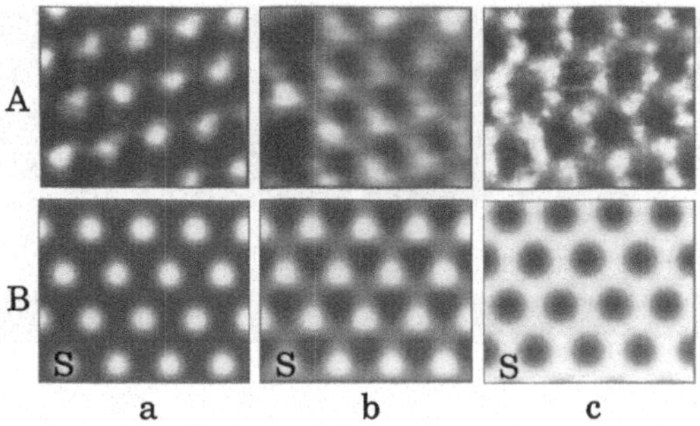

Fig. 19.8 (**A**) STM images for the p(2 × 2) overlayer of sulfur on Re(0001). The sulfur atoms are imaged in three different ways: as circles (**a**), triangles (**b**) or Y shapes (**c**). (**B**) Electron scattering quantum chemical calculations of the STM images with a tip ending in one sulfur atom (**a**), one platinum atom (**b**) or three platinum atoms (**c**). The *letter* S indicates the position of a sulfur atom. Adapted from [24]

producing the observed depression at the position of the adsorbates. A similar phenomenon explains the STM images of acetylene on Cu(001) surfaces, which appears as a dumbbell-shaped depression (see Fig. 19.15(a)) [22].

By using the full transfer Hamiltonian approach, Tersoff and Hamann [23] found that an alternative expression for (19.16) is

$$I \propto \sum_{\nu} \left|\psi_{\nu}(\boldsymbol{r_0})\right|^2 \delta\left(E_{\nu} - E_F^S\right) = \rho_S\left(\boldsymbol{r_0}, E_F^S\right). \tag{19.17}$$

The right hand side of (19.17) represents the number of sample electrons per unit volume and per unity energy calculated at the location of the tip's last atom $\boldsymbol{r_0} = (x, y, s)$ and at the sample Fermi energy. $\rho_S(\boldsymbol{r}, E)$ is often evaluated by means of density functional theory calculations (DFT) and expression (19.17) is used to construct simulated STM constant current images to be directly compared with the experimental ones.

A further critical factor in the STM visualization of adsorbates is the electronic configuration of the tip which is determined by its chemical and structural state. A prominent example of this influence is given by the p(2 × 2) layer of sulfur on Re(0001) [24]. Depending on the tip termination, the sulfur atoms can appear as circles, triangles, or Y-shaped structures in a honeycomb arrangement (upper row in Fig. 19.8). By comparison with theoretical simulations, it was determined that these visualization modes are due to a tip terminating in a single contaminant atom (e.g. S, C), in a single metallic atom (Pt, Rh, Re) or in a trimer of metallic atoms, respectively (lower row in Fig. 19.8) [24].

19.5 Scanning Tunneling Spectroscopy

Besides complicating the interpretation of STM images, the dependence of the tunneling current on the sample DOS also offers the unique opportunity of probing the electronic characteristics of surfaces with high spatial resolution. Due to the spatial localization of the tunneling current (see Sect. 19.2), STS enables to determine the electronic properties of individual atoms and molecules in relation to their structure, bonding and local environment. By changing the polarity of the bias voltage, STS gives access to both the occupied and the unoccupied states of the sample. In this sense, it is often considered as complementary to ultraviolet photoemission spectroscopy (UPS), inverse photoemission spectroscopy (IPS) and electron energy loss spectroscopy (EELS). While in these latter techniques the signal is averaged over a large surface area (between 0.1 and 2 mm in diameter), STS has the advantage of an ultimate spatial resolution. Moreover, in the case of organic samples, it also avoids possible damaging generated by intense photon or electron beam irradiation. On the other hand, STS does not provide direct chemical information and tip artifacts can strongly influence the spectroscopic data.

Having fixed the tip lateral position, the tunneling current I is a function of the applied bias voltage V and the tip-sample separation s only, the precise relation being established by (19.14). In a STS experiment the relation between two of these three parameters is measured while the remaining one is kept constant.

19.5.1 $I(V)$ Spectroscopy

$I(V)$ spectroscopy, where the tunneling current is measured as a function of the bias voltage for a constant tip-sample separation, is the most widely used technique because it provides indications about the DOS of the sample. However, its extraction from an $I(V)$ curve is a complex procedure since the DOS of the tip and the voltage dependence of the tunneling matrix elements M are usually unknown. As already explained in Sect. 19.2, $|M|^2$ can be approximated for small bias by the WKB tunneling probability D, so to obtain (19.14). By differentiating (19.14) at constant tip-sample distance s, the differential conductance for a positive sample bias is obtained as:

$$\begin{aligned}\frac{dI(V)}{dV} &\propto \rho_T(E_F^T)\rho_S(E_F^S+eV)D(eV,V,s)\\ &+\int_0^{eV}\rho_T(E_F^T-eV+\epsilon)\rho_S(E_F^S+\epsilon)\frac{\partial D(\epsilon,V,s)}{\partial V}\,d\epsilon\\ &+\int_0^{eV}\frac{d\rho_T(E_F^T-eV+\epsilon)}{dV}\rho_S(E_F^S+\epsilon)D(\epsilon,V,s)\,d\epsilon. \qquad (19.18)\end{aligned}$$

The quantity of interest, i.e. the sample LDOS ρ_S, is contained in the first term of (19.18), while the other two terms arise from the voltage dependence of the transmission coefficient D and of the tip DOS ρ_T, respectively. By assuming a constant ρ_T, the third term vanishes. For small bias voltages, the second term provides a smoothly varying background signal since D is a monotonic function of V. Hence, it is often neglected and expression (19.18) is simplified to:

$$\frac{dI(V)}{dV} \propto \rho_S(E_F^S+eV)D(eV,V,s). \qquad (19.19)$$

The monotonous dependence of D on V warrants that any possible structure in dI/dV is due to the voltage dependence of ρ_S. As a consequence, under the assumptions of constant ρ_T and small voltages, the differential conductance can be considered as a good representation of the sample density of states LDOS at the position of the tip.

The voltage dependence of D can however not be neglected for higher biases. In this case, the normalized differential conductivity $(dI/dV)/(I/V)$ is typically a better representation of ρ_S [25]. In fact, both the numerator (dI/dV) and the denominator (I/V) depend on the tunneling probability D and thus, in a first approximation, D cancels out in the ratio. Even if this procedure lacks a strict mathematic foundation, a good agreement has been found between the sample LDOS and the normalized differential conductivity, especially in the case of semiconductors [26]. However, this method cannot be used for wide gap (> 0.5 eV) semiconductors because, close to the band edge, the current approaches zero faster than dI/dV, leading to a divergence in the normalized differential conductivity [27]. A better way of dealing with higher bias values will be presented in Sect. 19.5.2.

The assumption of a constant tip DOS might be a good approximation for a metallic tip but does not hold for tips with adsorbed molecules, either intentionally picked up in a manipulation experiment (see Sect. 19.6) or accidentally transferred during scanning. On the other hand, although usually unknown, the tip DOS remains often unchanged over several subsequential measurements and therefore contributes in the same way to $I(V)$ spectra acquired at different positions on the sample. This allows to develop background subtraction schemes which rely on "calibrating" the tip on a clean spot of the surface and subsequently using this to remove tip-induced features from spectra acquired with the same tip on the objects of interest [28].

$I(V)$ spectroscopy is inherently characterized by an asymmetry in the sensitivity to occupied and unoccupied sample electronic states, the latter being far better visualized than the former. This is due to the transmission through the tunneling barrier which implies that features in ρ_S are enhanced at positive voltages, but are attenuated and difficult to recognize at negative ones. The asymmetry can be easily understood by remembering that, in (19.14), the exponential dependence of D from the bias voltage implies that most of the current is due to electrons tunneling close to the Fermi level. Figure 19.9 illustrates this effect for the case of a featureless ρ_T. At positive voltages (Fig. 19.9(a)), electrons tunnel mainly from the Fermi level of the tip into empty states of the sample. By increasing the voltage, new empty levels of the sample become accessible and, as an increase in V also reduces the tunneling barrier (see (19.10) and Fig. 19.9(b)), the corresponding peaks will be enhanced in the dI/dV spectrum. At a negative bias (Fig. 19.9(c)), the electrons tunnel from the sample E_F to empty tip levels. By decreasing the voltage, (Fig. 19.9(d)), the sample electrons at E_F tunnel into different empty tip states at higher energy but, due to the featureless ρ_T, this does not produce any new structure in the dI/dV spectrum. Any extra peak can only be due to lower-lying filled states of the sample which, however, experience a much larger tunneling barrier and are therefore strongly attenuated (Fig. 19.9(d)).

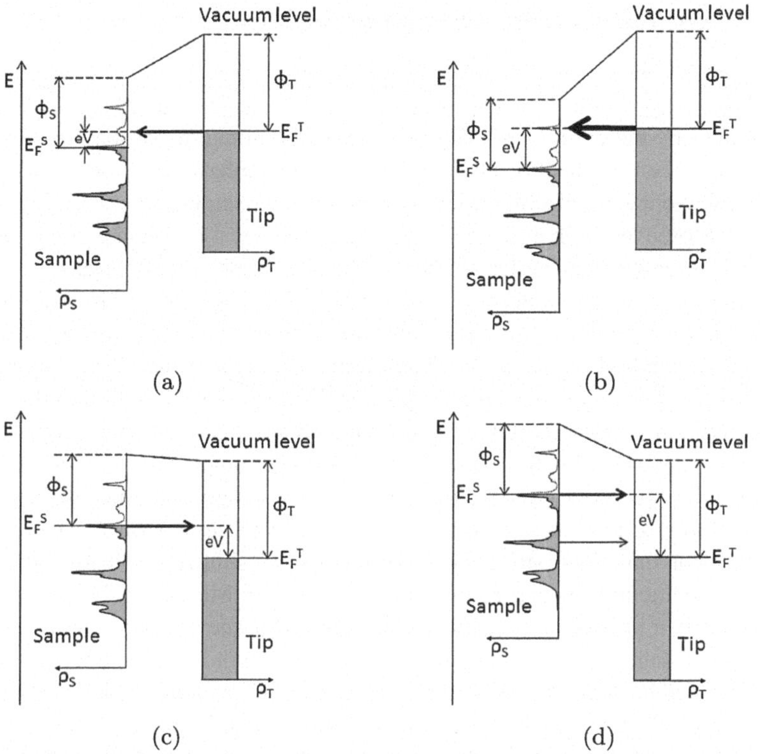

Fig. 19.9 Asymmetry of $I(V)$ spectra in the sensitivity to occupied and unoccupied sample electronic states (see text)

In order to measure an $I(V)$ curve at constant distance, the tip is placed at a certain position of the surface, the feedback is turned off, i.e. the tip-sample distance s is kept fixed, and the voltage is ramped while recording the tunneling current. The slope of the obtained $I(V)$ curve corresponds to the differential conductance between the tip and the sample and, through (19.19), to the sample LDOS. The tip-sample distance s is defined by the so-called *stabilization values*, i.e. by the values of the bias voltage and the tunneling current set before the feedback is turned off. The maximum time available for recording an $I(V)$ spectrum is determined by the actual time that the vertical and lateral position of the tip can be kept fixed with respect to the substrate. These depend on the overall stability of the system and should always be checked before performing a STS measurement. The differential conductivity can be calculated by numerically differentiating the measured $I(V)$ curve. However, this procedure typically results in very noisy conductance values. This problem is managed by measuring directly the differential conductance by means of the lock-in technique as described in Sect. 19.3.2 and Sect. 19.5.2.

19.5.2 Constant-Current Tunneling Spectra

The voltage range that can be explored in an $I(V)$ spectrum at constant s is essentially limited by the exponential dependence of the tunneling probability D on the bias voltage, see (19.10). For changes of V comparable to the work function, the tunneling current strongly increases thereby quickly saturating the high gain I–V converter. Moreover, high current values might become a problem when investigating adsorbed organic molecules which can get altered or even damaged.

An alternative way to determine the sample LDOS which avoids some of these problems is represented by so-called constant-current tunneling spectra where the ramp in V is applied while the feedback regulation is activated. This implies that the control electronics changes the tip-sample distance s as a function of the voltage so as to maintain a constant tunneling current. This procedure avoids any saturation of I and therefore benefits from an increased dynamic range with respect to constant distance $I(V)$ spectroscopy. However, it cannot be applied to spectral regions of low tunneling current, such as bias voltages close to zero or in the band gap of semiconductor materials, since the feedback would simply crash the tip into the sample in the attempt to keep I constant (see (19.14)). Moreover, the use of a lock-in amplifier is a necessary requirement in this technique and not simply a way to improve the signal-to-noise ratio. In fact, the differential conductance can clearly not be obtained by numerically differentiating the $I(V)$ signal, which is constant.

The sample LDOS can however be determined if a high-frequency sinusoidal modulation is added to the slowly varying bias voltage. In fact, if the modulation frequency is higher than the cut-off frequency of the feedback loop, the feedback does not respond to it and reacts only to the quasi-DC variations of the bias. On the other hand, as seen in Sect. 19.3.2, the lock-in amplifier outputs a signal proportional to the differential conductance. Since this occurs for any value of the bias voltage in the ramp, a dI/dV spectrum is effectively generated.

For larger values of the bias voltage V, the second term in (19.18) cannot be neglected any more and, for the case of a constant ρ_T, the correct expression for the differential conductance becomes:

$$\frac{dI[s(V),V]}{dV} \propto \rho_S\big(E_F^S+eV\big)D\big(eV,V,s(V)\big) + \int_0^{eV} \rho_S\big(E_F^S+\epsilon\big)\frac{\partial D(\epsilon,V,s(V))}{\partial V}\,d\epsilon, \tag{19.20}$$

where the dependence of s on V, due to the active feedback, has been explicitly indicated. The derivative of the transmission coefficient D with respect to the bias V can be determined by recalling the 1D WKB expression of D in (19.10). In a first approximation [29], the integral in (19.20) becomes:

$$-\frac{2e\sqrt{2m}s(V)}{\hbar\sqrt{\phi}}I(V), \tag{19.21}$$

where $\overline{\Phi}$ is the average tip and sample work function. Thus (19.20) can be solved for ρ_S:

$$\rho_S\left(E_F^S + eV\right) = \frac{1}{D(eV, V, s(V))}\left\{\frac{dI[s(V), V]}{dV} + \frac{2e\sqrt{2m}s(V)}{\hbar\sqrt{\overline{\Phi}}} I(V)\right\}, \quad (19.22)$$

which is an extension of (19.19). The absolute tip-sample distance, $s(V)$, necessary to determine the sample DOS through (19.22), is typically unknown. However, the simplest way of treating this problem is to consider $s = s_0 + \Delta s(V)$, where $\Delta s(V)$ is the tip excursion which can be measured and s_0 is a reference distance whose value affects the intensity of the peaks in the dI/dV spectrum but not their positions [30]. More sophisticated algorithms have also been proposed to deconvolve the sample DOS from STS measurements based on the 1D-WKB approximation [26, 31, 32]. Alternatively, ρ_S can be directly calculated from measured $I(V)$ curves by numerically inverting (19.14). For example, this approach has been used in order to determine the LDOS of organic nanolayers deposited on a metal surface [33].

19.5.3 CITS

Any type of STS measurement can be done with the same spatial resolution of STM, therefore also on individual molecules or atoms. Moreover, STS can be used to create 2D maps of the sample LDOS with sub-nm resolution. Such measurements are particularly interesting for quantum confined electronic systems (e.g. quantum dots or quantum corrals) or for determining the shape of molecular orbitals (see Sect. 19.5.6) [34]. In current imaging tunneling spectroscopy (CITS), spectroscopic measurements are acquired together with topographical ones, thereby obtaining a much deeper characterization of the sample surface.

There are two main ways to perform CITS. For the first one, the experimental setup is the same as for constant-current tunneling spectroscopy (Sect. 19.5.2), with feedback and lock-in amplifier working simultaneously. However, instead of applying a ramp to the bias voltage, this is kept fixed at a given value V and the tip is scanned across the surface. As a consequence, if the output of the lock-in is plot as a function of the lateral tip position, a 2D map of the differential conductance at eV is generated together with a constant current STM image. The whole spectral information can be retrieved by performing several scans at different V biases. Clearly the same issues discussed in Sect. 19.5.2 arise here when trying to extract the sample LDOS from dI/dV values acquired at different $s(V)$ values.

An alternative CITS method is to record a full $I(V)$ spectrum at constant s for every pixel of a constant current image. This implies first switching off and then reactivating the feedback electronics at each point of the image and is therefore typically a quite long procedure. As a consequence, it is often more practical to take only a STM image and, using this as a reference, to define a sparser grid of points where to measure the $I(V)$ spectra. Even if the stabilization values are identical for

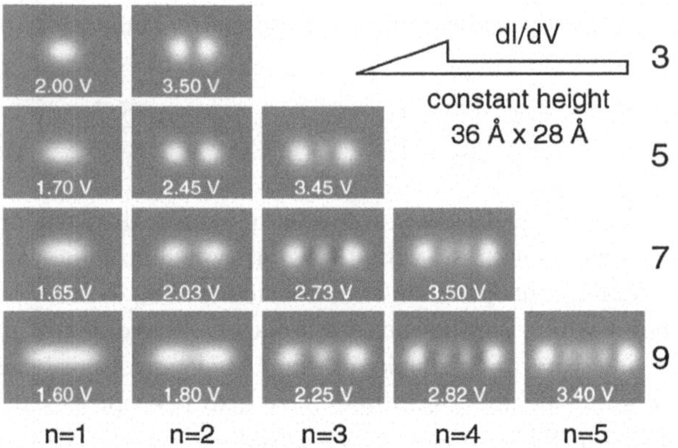

Fig. 19.10 CITS images obtained at constant distance and open feedback on copper chains on a Cu(111) surface. The images resemble the probability densities corresponding to the wave functions of electrons confined in a 1D potential well. Each *row* shows images of a Cu chain with given number of atoms indicated on the right. Each *column* refers to eigenstates with a fixed order n, specified at the *bottom of the column*. Adapted from [35]

all the $I(V)$ spectra, the corresponding tip-sample distances might be different because of topographical or electronic variations within the investigated sample area. Therefore, in some cases it might not be straightforward to relate CITS images measured in this way with 2D maps of the sample LDOS. The best method is to work with a fully disabled feedback, that is, at a true constant separation s. This type of measurements can however only be done if the surface topography is flat enough to avoid tip crashes and requires an extremely stable system and the absence of any drift. Figure 19.10 shows an example of CITS measurements done with this latter method on small chains of Cu atoms on a Cu(111) surface.

19.5.4 $s(V)$ Spectroscopy

In a $s(V)$ curve the tunnel current I is kept constant by using the feedback and the dependence of the tip-sample separation on the bias voltage is measured. As already explained in Sect. 19.5.2, this method allows a large dynamic range in V although spectral regions characterized by a low tunnel current are forbidden. In particular, $s(V)$ spectra are used to measure image-potential states whose energies lie above the vacuum barrier of the system and are therefore unaccessible to constant s spectroscopies. These bound states derive from the attractive potential existing between an electron approaching a metal surface and its positive image charge inside the metal. If their energies fall into a bulk band gap, the electrons cannot decay into bulk levels and are therefore trapped in these states. In presence of a STM tip, their position is Stark-shifted toward higher values due to the electric field between tip and sample. Image potential states can be detected as steps in a $s(V)$ curve [27, 36].

In fact, in order to keep a constant current, the tip retracts at voltages corresponding to successive image states, as the number of total states available to tunneling electrons increases.

19.5.5 *I(s)* Spectroscopy

In a $I(s)$ curve the bias voltage V is kept constant and the tunneling current is measured as a function of the tip-sample distance. This type of spectroscopy is used to determine the height of the tunneling barrier and thus, through (19.9), the local work function of the surface. In fact, according to (19.14), the current depends on s only through the exponential factor e^{-2ks}. Thus, for small values of the bias voltage,

$$\left.\frac{dI}{ds}\right|_V \approx -2kI \tag{19.23}$$

or, using (19.11),

$$\frac{d\ln(I)}{ds} = \frac{1}{I}\frac{dI(s)}{ds} = -2k = -2\frac{\sqrt{2m\phi_A}}{\hbar}, \tag{19.24}$$

where ϕ_A is the apparent local barrier height, i.e. the barrier that the electron has to overcome during the tunneling. For distances $s > 5$ Å, ϕ_A is equal to the average work function $\overline{\phi} = (\phi_S + \phi_T)/2$ which, by inverting (19.24), can be expressed as

$$\overline{\phi} \approx 0.95\left(\frac{d\ln(I)}{ds}\right)^2, \tag{19.25}$$

where $\overline{\phi}$ is expressed in eV and s in Å. On the other hand, for distances $s < 5$ Å the attractive image potential (see Sect. 19.5.4) can no longer be neglected [37] and produces a decrease in the apparent local barrier height ϕ_A which is dependent on s. As a consequence, in this case the barrier ϕ_A measured with the $I(s)$ spectroscopy cannot be directly linked to the average work function $\overline{\phi}$ [37]. Moreover, for larger values of the bias voltage, band structure effects can further complicate the interpretation of the apparent local barrier height ϕ_A [38].

In analogy to what is done for $I(V)$ spectroscopy, also $I(s)$ spectra can be measured as a function of the lateral tip position. In this case, by assuming a constant tip work function, the resulting 2D maps can be interpreted as spatial variations of the sample work function [37, 39]. Similarly to what described in Sect. 19.5.3, dI/ds can be directly measured during constant current scanning, by modulating s with a small sinusoidal signal (typical values are a few tenths of an Å) at a frequency higher than the cut-off frequency of the feedback loop and by recording the output of the lock-in amplifier tuned on the first harmonic. The local sample work function can then be obtained from (19.25). Alternatively, in analogy to CITS measurements performed on a grid (Sect. 19.5.3), a full $I(s)$ spectrum at constant V can be recorded on a grid of points defined on a reference STM image. At each point of the grid, the feedback is switched off to perform the $I(s)$ measurement and then reactivated to continue recording the topographic image. An example of this type of measurement is shown in Fig. 19.11(b) for a PVBA molecule adsorbed on Cu(111).

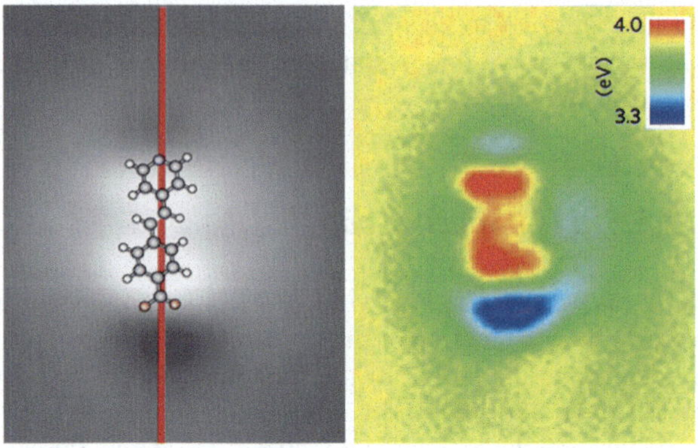

(a) (b)

Fig. 19.11 (**a**) Constant-current STM image of a PVBA molecule adsorbed on Cu(111). (**b**) 2D map of the local potential barrier. Figures adapted by permission from Macmillan Publishers Ltd: Nature Materials [39], copyright 2010

19.5.6 Orbital Mediated Tunneling Spectroscopy

In 1989 Hipps [40] observed for the first time peaks in dI/dV spectra of molecules adsorbed on a metal surface. These peaks are due to a resonant tunneling mechanism via unoccupied or occupied molecular orbitals [41] and the spectroscopy that measures these transitions is called orbital mediated tunneling spectroscopy (OMTS). The mechanism of OMTS can be understood by considering the energy landscape experienced by a tunneling electron when there is a molecule between tip and sample. For a positive sample bias (Fig. 19.12(a)), the resonance conditions are met when the Fermi level of the tip matches the energy of an empty molecular orbital. At this voltage the tunneling current increases because a new pathway becomes available for the tunneling electrons (typical changes are around 10–30 %). In fact, besides the direct tip-sample tunneling process, also a tip-molecule-sample channel opens (Fig. 19.12(a)), therefore increasing the overall tunneling probability. At voltages above the resonance, this extra path is still available although, due to the conservation of energy, it can be used only by electrons whose energy is lower than E_F^T. These must overcome a higher tunneling barrier (see (19.10) and the discussion on the asymmetry of $I(V)$ spectra in Sect. 19.5.1) implying that the contribution of the empty molecular level to the overall conductivity becomes less effective. Essentially, the contribution is maximized at the resonance itself, thus producing a peak in the dI/dV characteristics. When the molecules are decoupled from the substrate (either by other molecules in multilayers [42] or by a thin insulating film [34]) the width of the molecular resonances narrows and the contribution from the direct tunneling into substrate states gets negligible. Under these conditions, the mechanism of resonant tunneling becomes the only viable conduction path for tip electrons and not only dI/dV but also I drops at voltages above the resonance. This produces the phenomenon of negative differential resistance (NDR) which is a proof of an effective molecular decoupling (see for example Fig. 19.13(a) for a bias of about 2 V).

For negative sample biases a similar tunneling mechanism is observed although the extra channels open when occupied molecular orbitals are moved into resonance

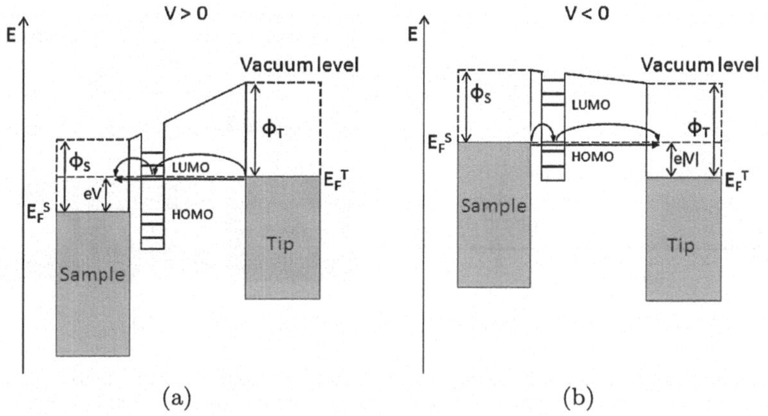

Fig. 19.12 Electron energy potential perpendicular to the surface when a molecule is located in the tip-sample gap. Resonant tunneling through unoccupied (**a**) and occupied (**b**) orbitals of an adsorbed molecule at positive and negative voltages, respectively

with the Fermi level of the tip (Fig. 19.12(b)). Finally, it should be noted that, by using a bias of a few Volts, it is possible to probe the occupied and unoccupied states of a large number of different types of molecules. However, for molecules with large energy gaps, the high voltage necessary for tunneling into the molecular states can significantly perturb the system or even break chemical bonds.

Peaks in OMTS spectra occur at voltages corresponding to resonance conditions and are due to the transition of electrons or holes through adsorbed molecules. The precise mechanism by which this happens depends on the details of the molecular electronic structure and of its coupling with both the sample and the tip. Two main scenarios can however be identified, depending on the residence time of the charge carrier onto the molecule [43]:

- Oxidation or reduction of the molecule followed by thermally induced return to the original charge state (electron hopping) [43, 44]. For a positive sample bias, an electron tunneling from the tip into an empty molecular state generates an anion which quickly relaxes into its new vibrational ground state. Successively, the extra electron tunnels from the molecule into the sample, thereby completing its conduction path. The molecule, reverted to the neutral state, relaxes back to its original ground state before a new electron restarts the cycle. Under these conditions, the rate of the reduction and oxidation processes is small enough that both the ion and the neutral molecule have time to completely relax their structure through the emission of photons or the coupling with the substrate phonons.
- Redox process occurring too rapidly for thermal relaxation to occur, such as in UPS or IPS [43]. In this case the electron residence time (10^{-15} s) is not long enough to allow the vibrational relaxation of the ion (10^{-13} s) which is thus formed in an excited state. This is similar to what happens in optical spectroscopy for a vertical transition in the Franck-Condon sense [43].

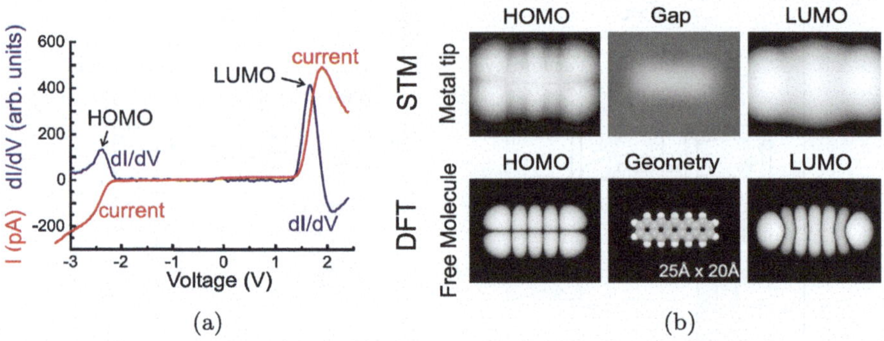

Fig. 19.13 (**a**) $I(V)$ and dI/dV spectra of a pentacene molecule adsorbed onto two layers on NaCl on Cu(111). Reprinted figure with permission from [34]. Copyright 2005 by the American Physical Society. (**b**) STM molecular orbital imaging. For a bias voltage corresponding to the HOMO and LUMO peaks in the dI/dV curve, the STM images resemble the spatial distribution of the HOMO and LUMO of the free molecule as calculated with DFT (*second row* in (**b**)). Figure adapted from [34]

The two different mechanisms lead to different positions of the observed peaks in the OMTS spectra. In fact, in the first case the measured energy corresponds to the difference between the bottoms of the potential energy surfaces of the ion and of the neutral molecule while, for a vertical transition, the difference is between an ionic excited level and the ground state of the neutral molecule.

OMTS spectroscopy of molecules that are effectively decoupled from the substrate can also be used to perform STM orbital imaging [34, 45]. As discussed above, in these conditions the tunneling current is mainly due to the molecular resonances which are energetically narrow and spatially localized. As a consequence, after the spectral position of a resonance has been identified in a dI/dV curve, a topographic image can be recorded at the corresponding bias voltage, thereby probing the spatial extension and shape of the associated molecular orbital. Due to the ultimate lateral resolution of STM, this can lead to intramolecular resolution of the orbital imaging. An example of this type of experiment is reported in Fig. 19.13 for the case of pentacene molecules adsorbed onto two layers of NaCl on Cu(111).

19.5.7 Inelastic Electron Tunneling Spectroscopy

So far we have assumed that electrons conserve their energy during the tunneling process, see (19.3). However, electrons can also tunnel inelastically between the tip and the sample exchanging energy and inducing the excitation of vibrational modes, spin-flips, magnons, plasmons, excitons, etc. either within the tunneling barrier (e.g. if there is a molecule between tip and sample) or in the electrodes themselves. These extra tunneling channels become available only above specific voltage thresholds $V^* = \hbar\omega/e$, where $\hbar\omega$ is the energy of the corresponding excitation. In fact, only

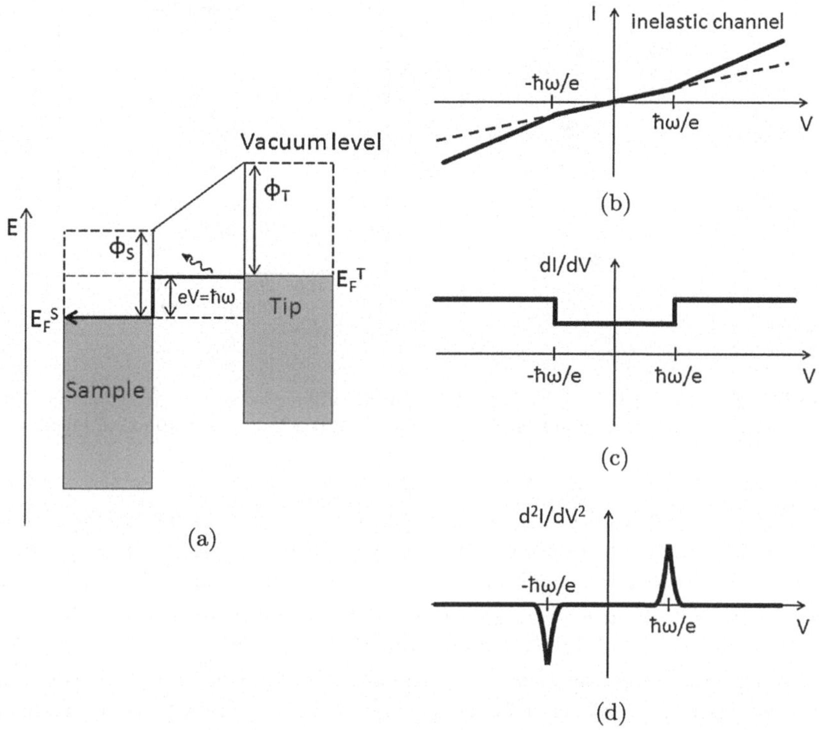

Fig. 19.14 Inelastic tunneling spectroscopy. (**a**) Schematic representation of the inelastic tunneling of electrons in a positively biased tip-vacuum-sample junction. (**b**)–(**d**) Corresponding effect on $I(V)$, dI/dV and d^2I/dV^2 spectra

beyond these values a part of the electron energy can be converted into the excitation.

The additional inelastic pathways increase the overall tunneling probability and therefore show up as discrete step-like features in the tunneling conductivity or as slope changes in $I(V)$ curves (Fig. 19.14(c) and (b), respectively). Note that this is different from the elastic tunneling case discussed in Sect. 19.5.6 since here the additional channel is available to any tunneling electron, provided its energy is larger than eV^*. However, normally, only 0.1 % of the electrons uses a vibrational inelastic channel and 5 % an electronic inelastic channel [43]. Due to this extremely low signal-to-background ratio, a lock-in technique becomes mandatory. In particular, the second harmonic component d^2I/dV^2 is measured as described in Sect. 19.3.2 since its plot versus V displays onsets of inelastic channels as peaks (Fig. 19.14(d)). This technique is called inelastic tunneling spectroscopy (IETS).

The extra inelastic channels are accessible both to electrons tunneling from the tip to the sample and to electrons following the reverse path. As a consequence, the corresponding peaks in the d^2I/dV^2 spectrum appear at the same voltage independent of sign, even if the intensities can be different due to the different density

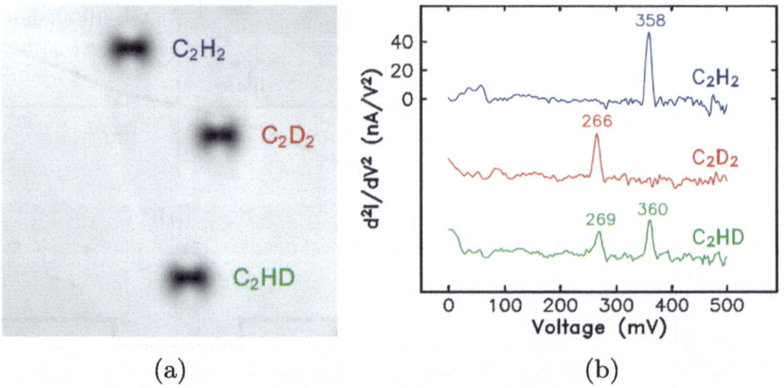

Fig. 19.15 (**a**) STM images and (**b**) STM-IETS spectra of three isotopes of acetylene on Cu(001) at 8 K. Reprinted with permission from [50]. Copyright 2002, American Institute of Physics

of states of the two electrodes. The exhibition of this symmetry is considered as the necessary fingerprint to assign specific spectral features to inelastic processes. A high spectral resolution is particularly important in IETS since the energetic separation of several inelastic processes is very small (e.g. less than 1 meV for spin flips [46], a few tenths of a meV for vibrational modes [47, 48]). As a consequence, apart from minimizing any type of instrumental broadening, IETS measurements have to be performed at cryogenic temperatures since the intrinsic line width is at least 5.4 kT due to the thermal broadening of the Fermi-Dirac distributions [8] (as a reference, 5.4 kT is equal to 0.3 meV at $T = 0.6$ K and to 140 meV at $T = 300$ K).

IETS benefits from the same spatial resolution as STM and STS and has therefore been used to measure vibrational modes of individual molecules, spin excitations of single magnetic atoms, collective plasmon excitations in 2D materials and magnons in ferromagnets on the sub-nm scale. An example of the possibilities of IETS is demonstrated in Fig. 19.15 for the vibrational excitations of single molecules. The STM image in Fig. 19.15(a) displays three apparently identical acetylene molecules. However, upon analysis of the corresponding IETS spectra (Fig. 19.15(b)), these are revealed as three differently deuterated isotopes, C_2H_2, C_2D_2 and C_2HD [49]. Similarly to other vibrational spectroscopy techniques such as infrared or Raman, IETS spectra show the isotope shifts in the C–H or C–D stretches. The essential difference of this type of measurements is the ability of doing so on individual molecules.

A different way of detecting tunneling-induced molecular vibrations by means of STM is to rely on their coupling with dynamical processes such as molecular motions. In particular, by placing the STM tip over a target molecule, a vibrational mode and an associate molecular motion can be excited by inelastically tunneling electrons if the bias voltage is increased above the corresponding threshold. By measuring the frequency of molecular diffusion events as a function of the time, the applied bias voltage and the tunneling current, it is possible to create so-called *action spectra* which reflect the vibrational spectrum of an individual molecule in a quantitative manner [51].

Optical excitations can also be revealed in an alternative way by coupling the STM with a photon detection system able to collect and analyze the luminescence stimulated by inelastic electrons [52, 53]. Such a setup has been used to characterize plasmon emission from metallic surfaces and luminescence from semiconductor quantum structures and adsorbed molecules [54, 55].

19.6 Tip-Induced Modification

Besides being an extraordinary instrument for the characterization of structural, electronic, vibrational, optical and magnetic properties of surfaces with subnanometer resolution, STM has also developed as a tool to modify and nano-engineer matter at the single molecule and atom scale.

By decreasing the distance between the tip and the sample in a controlled way, indentations can be produced in the substrate with lateral sizes down to a few nm. Nanolithography can also be performed by tunneling electrons into a layer of e-beam photoresist, thereby reaching a better resolution compared to standard electron beam lithography [56]. Many other STM-based nanopatterning and nanofabrication techniques have been developed based on a number of physical and chemical principles [57–60] including anodic oxidation, field evaporation, selective chemical vapor deposition, selective molecular desorption, electron-beam induced effects and mechanical contact. All these methods exploit the extreme lateral localization of the tunneling current and can be applied in air, liquids and vacuum.

However, the nanotechnological application that gained most attention is the ability to manipulate individual atoms and molecules on a substrate. This is possible due to a controlled use of tip-particle forces and is typically done in UHV and at low temperatures. The first atomic manipulation experiment was performed by Eigler and Schweizer in 1989 [2]. This phenomenal result fulfilled Richard Feynman's prophecy that "ultimately-in the great future-we can arrange the atoms the way we want; the very atoms, all the way down!" [3].

During a lateral manipulation experiment the tip is first placed above the particle to be moved (for example an atom) and the tunneling current is increased while keeping a constant voltage. This results in a movement of the tip toward the atom, see (19.14). If their separation is sufficiently reduced, Van der Waals forces start to come into play together with attractive and repulsive chemical interactions. When these forces equal the diffusion energy barrier, a lateral displacement of the tip can induce a movement of the atom parallel to the surface. Upon reaching the desired final position, the tip is retracted by reducing the tunneling current to the initial value, leaving the atom in the selected place. Depending on the tip-particle distance and therefore on the strength and nature of the interaction, different manipulation modes including pulling, pushing and sliding [61] were identified and used to move different types of atoms and molecules. It is worth to notice that this type of manipulation is not influenced by the value of the voltage bias.

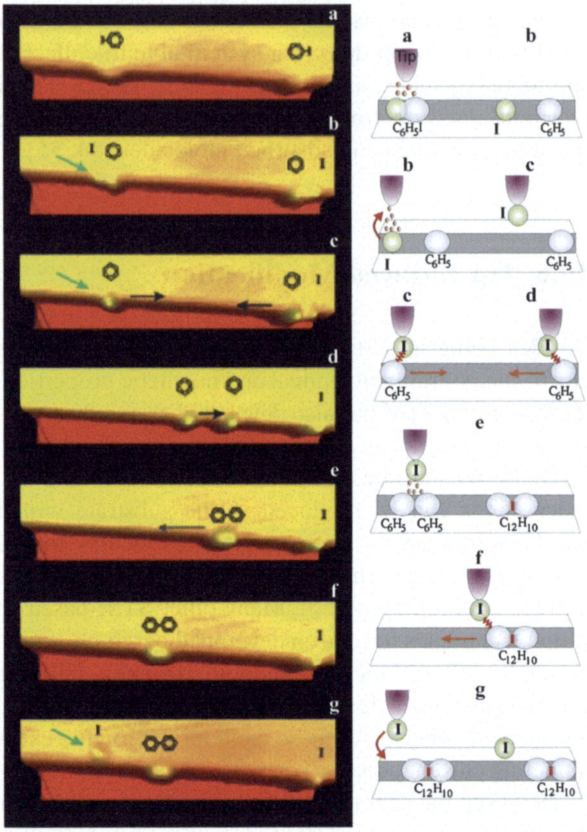

Fig. 19.16 STM images and schematics of the steps of the STM-tip induced Ullmann reaction on Cu(111). Adapted from [70]

Thanks to this technique, it was possible to fabricate artificial nanostructures such as the *quantum corral* [62] and to probe quantum mechanical effects like the quantum confinement of surface state electrons or the *quantum mirage* [63]. Lateral STM manipulation has also been used to switch between different adsorption configurations and conformations of molecules on surfaces and to modify their electronic properties in a controlled way [64].

A further application of STM manipulation is the synthesis of new molecular species based on the ability of STM to form and break chemical bonds with atomic precision. Reactants are brought close together on the surface and the actual reaction is realized by applying a voltage pulse or by exciting vibrational modes through inelastically tunneling electrons. Examples of this technique include the dissociation of diatomic molecules [65], the Ullmann reaction (Fig. 19.16) [66], the isomerization of dichlorobenzene [67] and the creation of metal-ligand complexes [68, 69].

The STM tip has also been used to perform vertical manipulations of nanoparticles where an atom (or molecule) is deliberately transfered from the surface to the tip and vice versa by using the electric field generated by the bias voltage. In contrast to the lateral manipulation, here the bonds between the surface and the atom are

(a) (b) (c)

Fig. 19.17 (**a**) Chemical structure of PTCDA. (**b**) STM and (**c**) STHM image of a PTCDA molecule adsorbed on Au(111). Adapted from [73]

broken and re-created [71]. By approaching the tip at distances of a few Å from the chosen particle, new interactions are established that reduce the atom-surface binding energy. If a voltage pulse is applied under these conditions, the resulting electric field (of the order of 10^8 V/cm) can be enough to induce the particle desorption. The vertical manipulation technique has also been used as a means to increase the lateral resolution of STM. In fact, the controlled adsorption of a specific molecule onto the tip often makes it "sharper" and can add a chemical resolution capability if the DOS of the extra molecule acts as an "energy filter" [34].

A related effect is exploited in the recently proposed scanning tunneling hydrogen microscopy (STHM) technique [73]. In STHM the experimental chamber is flooded with molecular hydrogen while the tip is scanned in constant height mode at a very close distance over the surface. H_2 can get trapped in the tip-sample junction and its rearrangement during scanning of the surface generates a new contrast mechanism based on the short-range Pauli repulsion. This is extremely sensitive to the total electron density, thereby endowing the STM with similar imaging capabilities to non-contact AFM [72] and making it able to resolve the inner structure of complex organic molecules [73] (Fig. 19.17(c)). This is not the case for standard STM which, being mainly influenced by the frontier molecular orbitals, reproduces their spatial distribution. In particular, if the molecule is adsorbed on a metallic substrate, the metal-organic interaction can lead to a substantial broadening and restructuring of the molecular orbitals which become spatially delocalized (Fig. 19.17(b)).

Acknowledgements This work was supported by EPSRC (EP/D000165/1); A. Della Pia was funded through a WPRS scholarship of the University of Warwick.

References

1. G. Binnig, H. Rohrer, C. Gerber, E. Weibel, Phys. Rev. Lett. **49**(1), 57 (1982)
2. D.M. Eigler, E.K. Schweizer, Nature **344**(6266), 524 (1990)
3. R.P. Feynman, J. Microelectromech. **1**(1), 60 (1992)
4. J. Bardeen, Phys. Rev. Lett. **6**(2), 57 (1961)
5. A.D. Gottlieb, L. Wesoloski, Nanotechnology **17**(8), R57 (2006)
6. N.D. Lang, Phys. Rev. B **34**(8), 5947 (1986)
7. L.D. Landau, E.M. Lifshitz, *Quantum Mechanics. Non-relativistic Theory* (Pergamon Press, Oxford, 1977)

8. C.J. Chen, *Introduction to Scanning Tunneling Microscopy* (Oxford University Press, New York, 2008)
9. G. Binnig, H. Rohrer, Helv. Phys. Acta **55**(6), 726 (1982)
10. A. Okumura, K. Miyamura, Y. Gohshi, J. Microsc. **152**(3), 631 (1988)
11. K. Besocke, Surf. Sci. **181**(1–2), 145 (1987)
12. S.H. Pan, E.W. Hudson, J.C. Davis, Rev. Sci. Instrum. **70**(2), 1459 (1999)
13. E. Meyer, H.J. Hug, R. Bennewitz, *Scanning Probe Microscopy: The Lab on a Tip* (Springer, Berlin, 2004)
14. I. Ekvall, E. Wahlstrom, D. Claesson, H. Olin, E. Olsson, Meas. Sci. Technol. **10**(1), 11 (1999)
15. J.P. Ibe, P.P. Bey, S.L. Brandow, R.A. Brizzolara, N.A. Burnham, D.P. Dilella, K.P. Lee, C.R.K. Marrian, R.J. Colton, J. Vac. Sci. Technol. A **8**(4), 3570 (1990)
16. L.A. Nagahara, T. Thundat, S.M. Lindsay, Rev. Sci. Instrum. **60**(10), 3128 (1989)
17. E. Laegsgaard, L. Osterlund, P. Thostrup, P.B. Rasmussen, I. Stensgaard, F. Besenbacher, Rev. Sci. Instrum. **72**(9), 3537 (2001)
18. M.J. Rost, L. Crama, P. Schakel, E. van Tol, G.B.E.M. van Velzen-Williams, C.F. Overgauw, H. ter Horst, H. Dekker, B. Okhuijsen, M. Seynen, A. Vijftigschild, P. Han, A.J. Katan, K. Schoots, R. Schumm, W. van Loo, T.H. Oosterkamp, J.W.M. Frenken, Rev. Sci. Instrum. **76**(5), 053710 (2005)
19. L. Petersen, M. Schunack, B. Schaefer, T.R. Linderoth, P.B. Rasmussen, P.T. Sprunger, E. Laegsgaard, I. Stensgaard, F. Besenbacher, Rev. Sci. Instrum. **72**(2), 1438 (2001)
20. T. Zambelli, J.V. Barth, J. Wintterlin, G. Ertl, Nature **390**(6659), 495 (1997)
21. C. Klink, L. Olesen, F. Besenbacher, I. Stensgaard, E. Laegsgaard, N.D. Lang, Phys. Rev. Lett. **71**(26), 4350 (1993)
22. F.E. Olsson, M. Persson, N. Lorente, L.J. Lauhon, W. Ho, J. Phys. Chem. B **106**(33), 8161 (2002)
23. J. Tersoff, D.R. Hamann, Phys. Rev. Lett. **50**(25), 1998 (1983)
24. P. Sautet, J. Dunphy, D.F. Ogletree, M. Salmeron, Surf. Sci. **295**(3), 347 (1993)
25. J.A. Stroscio, R.M. Feenstra, A.P. Fein, Phys. Rev. Lett. **57**(20), 2579 (1986)
26. M. Passoni, F. Donati, A.L. Bassi, C.S. Casari, C.E. Bottani, Phys. Rev. B **79**(4), 045404 (2009)
27. J.A. Stroscio, W.J. Kaiser, *Scanning Tunneling Microscopy* (Academic Press, London, 1993)
28. P. Wahl, L. Diekhoner, M.A. Schneider, K. Kern, Rev. Sci. Instrum. **79**(4), 043104 (2008)
29. B. Koslowski, C. Dietrich, A. Tschetschetkin, P. Ziemann, Phys. Rev. B **75**(3), 035421 (2007)
30. M. Ziegler, N. Neel, A. Sperl, J. Kroger, R. Berndt, Phys. Rev. B **80**(12), 125402 (2009)
31. B. Koslowski, H. Pfeifer, P. Ziemann, Phys. Rev. B **80**(16), 165419 (2009)
32. B. Naydenov, J.J. Boland, Phys. Rev. B **82**(24), 245411 (2010)
33. C. Wagner, R. Franke, T. Fritz, Phys. Rev. B **75**(23), 235432 (2007)
34. J. Repp, G. Meyer, S.M. Stojkovic, A. Gourdon, C. Joachim, Phys. Rev. Lett. **94**(2), 026803 (2005)
35. S. Folsch, P. Hyldgaard, R. Koch, K.H. Ploog, Phys. Rev. Lett. **92**(5), 056803 (2004)
36. G. Binnig, K.H. Frank, H. Fuchs, N. Garcia, B. Reihl, H. Rohrer, F. Salvan, A.R. Williams, Phys. Rev. Lett. **55**(9), 991 (1985)
37. R. Wiesendanger, *Scanning Probe Microscopy and Spectroscopy: Methods and Applications* (Cambridge University Press, Cambridge, 1994)
38. M. Becker, R. Berndt, Phys. Rev. B **81**(3), 035426 (2010)
39. L. Vitali, G. Levita, R. Ohmann, A. Comisso, A. De Vita, K. Kern, Nat. Mater. **9**(4), 320 (2010)
40. K.W. Hipps, J. Phys. Chem. **93**(16), 5958 (1989)
41. W. Mizutani, M. Shigeno, K. Kajimura, M. Ono, Ultramicroscopy **42**(a), 236 (1992)
42. M. Grobis, A. Wachowiak, R. Yamachika, M.F. Crommie, Appl. Phys. Lett. **86**(20), 204102 (2005)
43. K.W. Hipps, Scanning tunneling spectroscopy (STS), in *Handbook of Applied Solid State Spectroscopy*, ed. by D.R. Vij (Springer, New York, 2006)
44. U. Mazur, K.W. Hipps, J. Phys. Chem. B **103**(44), 9721 (1999)

45. J. Repp, G. Meyer, S. Paavilainen, F.E. Olsson, M. Persson, Science **312**(5777), 1196 (2006)
46. A.J. Heinrich, J.A. Gupta, C.P. Lutz, D.M. Eigler, Science **306**(5695), 466 (2004)
47. N. Lorente, M. Persson, Phys. Rev. Lett. **85**(14), 2997 (2000)
48. L. Vitali, M.A. Schneider, K. Kern, L. Wirtz, A. Rubio, Phys. Rev. B **69**(12), 121414 (2004)
49. B.C. Stipe, M.A. Rezaei, W. Ho, Science **280**(5370), 1732 (1998)
50. W. Ho, J. Chem. Phys. **117**(24), 11033 (2002)
51. Y. Sainoo, Y. Kim, T. Okawa, T. Komeda, H. Shigekawa, M. Kawai, Phys. Rev. Lett. **95**(24), 246102 (2005)
52. J.K. Gimzewski, B. Reihl, J.H. Coombs, R.R. Schlittler, Z. Phys. B, Condens. Matter **72**(4), 497 (1988)
53. J.G. Keizer, J.K. Garleff, P.M. Koenraad, Rev. Sci. Instrum. **80**(12), 123704 (2009)
54. F. Rossel, M. Pivetta, W.D. Schneider, Surf. Sci. Rep. **65**(5), 129 (2010)
55. P. Bharadwaj, A. Bouhelier, L. Novotny, Phys. Rev. Lett. **106**(22), 226802 (2011)
56. R.F.W. Pease, J. Vac. Sci. Technol. B **10**(1), 278 (1992)
57. H. Iwasaki, T. Yoshinobu, K. Sudoh, Nanotechnology **14**(11), R55 (2003)
58. H.J. Mamin, P.H. Guethner, D. Rugar, Phys. Rev. Lett. **65**(19), 2418 (1990)
59. C.F. Quate, Surf. Sci. **386**(1–3), 259 (1997)
60. C. Shen, M. Buck, Nanotechnology **20**(24), 245306 (2009)
61. L. Bartels, G. Meyer, K.H. Rieder, D. Velic, E. Knoesel, A. Hotzel, M. Wolf, G. Ertl, Phys. Rev. Lett. **80**(9), 2004 (1998)
62. M.F. Crommie, C.P. Lutz, D.M. Eigler, Science **262**(5131), 218 (1993)
63. H.C. Manoharan, C.P. Lutz, D.M. Eigler, Nature **403**(6769), 512 (2000)
64. F. Moresco, G. Meyer, K.H. Rieder, H. Tang, A. Gourdon, C. Joachim, Phys. Rev. Lett. **86**(4), 672 (2001)
65. B.C. Stipe, M.A. Rezaei, W. Ho, S. Gao, M. Persson, B.I. Lundqvist, Phys. Rev. Lett. **78**(23), 4410 (1997)
66. S.W. Hla, L. Bartels, G. Meyer, K.H. Rieder, Phys. Rev. Lett. **85**(13), 2777 (2000)
67. K. Morgenstern, Acc. Chem. Res. **42**(2), 213 (2009)
68. P. Liljeroth, I. Swart, S. Paavilainen, J. Repp, G. Meyer, Nano Lett. **10**(7), 2475 (2010)
69. N. Katsonis, J. Vicario, T. Kudernac, J. Visser, M.M. Pollard, B.L. Feringa, J. Am. Chem. Soc. **128**(48), 15537 (2006)
70. K.H. Rieder, G. Meyer, F. Moresco, K. Morgenstern, S.W. Hla, J. Repp, M. Alemani, L. Grill, L. Gross, M. Mehlhorn, H. Gawronski, V. Simic-Milosevich, J. Henzl, K.F. Braun, S. Foelsch, L. Bartels, in *Conference on Atoms and Molecules Near Surfaces*, ed. by L.F.J. Weiner, J. Schmiedmayer. Journal of Physics Conference Series, vol. 19 (IOP Publishing, Bristol, 2005), p. 175
71. P. Avouris, Acc. Chem. Res. **28**(3), 95 (1995)
72. L. Gross, F. Mohn, N. Moll, P. Liljeroth, G. Meyer, Science **325**(5944), 1110 (2009)
73. C. Weiss, C. Wagner, C. Kleimann, M. Rohlfing, F.S. Tautz, R. Temirov, Phys. Rev. Lett. **105**(8), 086103 (2010)

Chapter 20
Surface Characterization Using Atomic Force Microscopy (AFM) in Liquid Environments

Venetia D. Lyles, Wilson K. Serem, Jing-Jiang Yu, and Jayne C. Garno

Abstract Liquid imaging provides intrinsic advantages for AFM experiments, particularly for conducting *in situ* studies of chemical or biochemical reactions. Using liquid media has benefits for improving resolution, since the amount of force applied between the tip and sample can be reduced. Surface changes caused by immersion in different liquids can be investigated, such as for studying electrochemical reactions with different parameters of solvent polarity, pH or ion concentration. Aqueous buffers enable studies of biochemical reactions that simulate physiological conditions, with time-lapse capture of image frames at different intervals. Studies of surface changes throughout the course of self-assembly reactions have been monitored with AFM in liquid media. By injecting new molecules into the sample cell, AFM-based nanofabrication can be accomplished by nanografting protocols. Liquid environments expand the capabilities for scanning probe studies to provide insight for dynamic processes at the molecular-level.

20.1 Introduction

An intrinsic advantage of atomic force microscopy (AFM) is the flexibility to conduct experiments in different media such as in ambient air, vacuum, or in liquid environments. The capability to introduce liquids to the sample cell and acquire images in fluids broadly expands the repertoire of experimental designs, particularly for studying nanoscale changes to surfaces by capturing AFM images as reactions

V.D. Lyles · W.K. Serem · J.C. Garno (✉)
Department of Chemistry, Louisiana State University, Baton Rouge, LA 70803, USA
e-mail: jgarno@lsu.edu

V.D. Lyles
e-mail: vlyles1@lsu.edu

W.K. Serem
e-mail: wserem1@lsu.edu

J.-J. Yu
Nanomeasurements Division, Agilent Technologies, Inc., Chandler, AZ, USA
e-mail: Jing-jiang_yu@agilent.com

G. Bracco, B. Holst (eds.), *Surface Science Techniques*,
Springer Series in Surface Sciences 51, DOI 10.1007/978-3-642-34243-1_20,

proceed. For example, liquid AFM has been applied for in situ studies of the effects of different liquid media, such as with changes in pH or ion concentration [1–4]. Liquid media has been used for AFM studies with biological systems in aqueous buffers that simulate physiological conditions [3, 5], for *in situ* studies of electrochemical reactions [6–8], and for time-lapse imaging of progressive surface changes caused by chemical reactions [9–11]. In liquid media, new molecules can be introduced to the sample cell to enable AFM-based nanofabrication [12]. Accomplishing AFM experiments in liquid media also provides advantages for improving resolution, since imaging in liquid reduces or eliminates capillary and van der Waals forces between the tip and sample [13]. Liquid imaging reduces sample perturbation and minimizes or prevents damage caused by shear forces between the tip and surface. This chapter will describe state-of-the-art advances with AFM for designing creative experimental approaches for nanoscale studies in liquid media.

20.2 Theory Background

20.2.1 History of AFM

The development of scanning probe instruments during the 1980's has launched entirely new experimental approaches for nanoscale studies and has made pivotal contributions to the interdisciplinary field of nanoscience. Before 1980, there were no instruments with capabilities for visualizing surfaces at the level of individual molecules or atoms. The operating principle for AFM is based on scanning a small probe across the surface and requires positional control at the nanoscale. The precedent for scanning probe methods was developed by Heinrich Rohrer and Gerd Binnig, who were awarded the 1986 Nobel prize in physics for invention of the scanning tunneling microscope (STM) [14]. The first STM was developed at IBM Zurich Research Laboratories in Switzerland and resolved the 7×7 lattice arrangement of silicon atoms for Si(111). Imaging with STM is based primarily on electrical interactions between a conductive tip and the sample, whereas AFM imaging is reconstructed from the mechanical deflection of the probe due to physical forces. Positional control for AFM, STM and other scanning probe methods rely on the piezoelectric effect, which enables precise movement of the probe with angstrom-level control.

The first AFM instrument was introduced in 1986 by Binnig, Quate and Gerber of Stanford University in California, to measure any type of force on an atomic scale [15]. They introduced the first general-purpose AFM instrument for measuring ultra-small forces as low as 10^{-18} N. The resolution that can be obtained with AFM today is comparable to that achieved with STM [16]. Unlike STM, imaging with AFM is not restricted to conductive or semiconductor surfaces. The significance of scanning probe instruments is evidenced by diverse applications throughout the science disciplines; nanoscale imaging and measurements with scanning probe instruments provide entirely new approaches for the study of matter at small size scales.

20.2.2 Early Benchmarks for Studies Using Liquid AFM

The first AFM experiments were accomplished in air, to probe the surface of a ceramic sample of aluminum oxide (Al_2O_3), achieving lateral resolution of 30 Å and vertical resolution of less than 1 Å [15]. The first AFM experiments in liquid were reported in 1987 by Marti, Drake and Hansma, for surfaces of sodium chloride and highly oriented pyrolytic graphite (HOPG) that were covered in paraffin oil [16]. These experiments resolved the hexagonal packing arrangement of carbon atoms, achieving lateral resolution of 0.15 nm and vertical resolution of 5 pm [16]. The applicability of AFM in liquid environments for imaging biological molecules was first demonstrated by Drake et al. in 1989 [17]. Studies of chemical and biological processes in liquids have continued to progress, setting benchmarks for nanoscale resolution with investigations of corrosion, electrochemical reactions, self-assembled monolayers (SAMs), as well as deoxyribonucleic acid (DNA), proteins, and cells [18, 19].

20.3 Experimental Setup and Method

20.3.1 Basic Operating Principle of AFM

Unlike traditional optical microscopes, scanning probe instruments use a sharp probe that is scanned over the sample to acquire spatial maps of surface characteristics. The interactions between the tip and the sample are mapped to construct images at scales from microns to nanometers. The cantilever is typically made of silicon or silicon nitride with a tip radius of nanometers. The resolution of scanning probe measurements is not limited by the wavelength of light, and can achieve unprecedented nanoscale resolution. For AFM imaging, light from a diode laser is focused on the back of a cantilever and deflected toward a photodiode detector (Fig. 20.1). The photodiode detects the cantilever deflection by tracking the position of the reflected beam. An electronic feedback loop controls voltages to the piezoceramic scanner to maintain a constant force between the tip and sample. There are several designs that have been used for positional feedback with AFM. The optical deflection set-up shown in Fig. 20.1 is the most common configuration. Other approaches that have been used for monitoring tip position include optical interferometry [20], capacitive sensing [21], tuning forks [22, 23], and piezoresistive cantilevers [24].

The lattice arrangement of atoms can be visualized with AFM, providing views of molecular and atomic vacancies and adatoms [25]. The atomic lattices of substrates such as Au(111), highly oriented pyrolytic graphite (HOPG) and mica(0001) are commonly used for lateral calibration of X-Y dimensions. The heights of gold steps are used for calibration of Z dimensions. Images generated by AFM are true three-dimensional surface profiles. Samples do not require special treatments or coatings which alter their composition. Both conducting and insulating materials

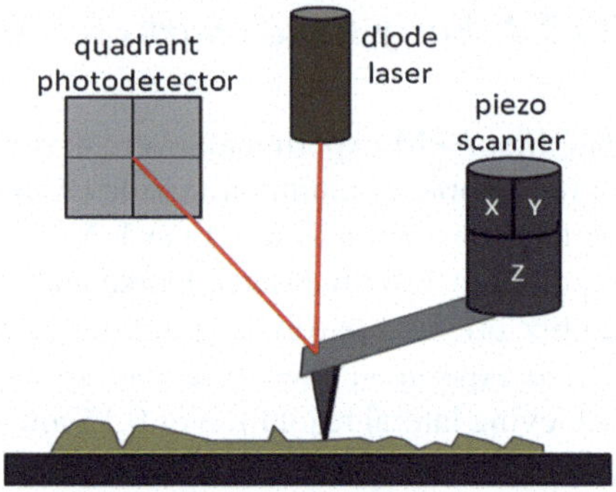

Fig. 20.1 Laser-deflection configuration used for AFM imaging

have been imaged with AFM. A vacuum environment is not required for AFM studies, and samples may be imaged in air or in liquid media.

Two basic instrument designs are possible for AFM imaging. The cantilever can be attached to the piezoscanner for scanning the tip across the sample surface; or the sample can be scanned across the probe, while the tip is held in a fixed position. There are also instrument designs which combine approaches for both tip and sample translation. The sample cell can be filled with various solvents to accomplish liquid imaging or for electrochemical AFM studies. As the tip is translated across the sample, forces between the tip and the sample cause changes in the deflection of the AFM cantilever. Depending on the instrument configuration, the types of forces that can be measured include adhesive or repulsive forces attributable to van der Waals interactions, chemical bonding, mechanical friction, electrostatic charge, or magnetic interactions. The tip can be operated in continuous direct contact with the surface (contact mode) or at a certain fixed distance from the surface (non-contact mode). For intermittent or "tapping" mode AFM (TM-AFM), the tip is driven to oscillate in and out of contact with the surface. The feedback signal for controlling the probe position with TM-AFM is obtained by maintaining a constant amplitude setting for tip deflection, rather than using a force set point. Tapping-mode experiments can be operated in liquid media, which is particularly useful for imaging fragile systems of biomolecules in buffered media.

20.3.2 Approaches for Liquid Imaging

Initial AFM liquid experiments were accomplished by simply placing a drop of liquid on the AFM tip, and focusing the laser through the liquid interface. Of course, the solvents chosen for liquid experiments should be optically transparent, and must have a relatively slow rate of evaporation, e.g. water, ethanol, butanol or hexadecane. The solvent must be compatible and nondestructive for the sample material; for example, it should not cause dissolution or corrosion of the surface. Sample cells for liquid imaging have been designed with either open or closed containment of liquids. If the liquid cell has a cover, the laser is focused through both the cover

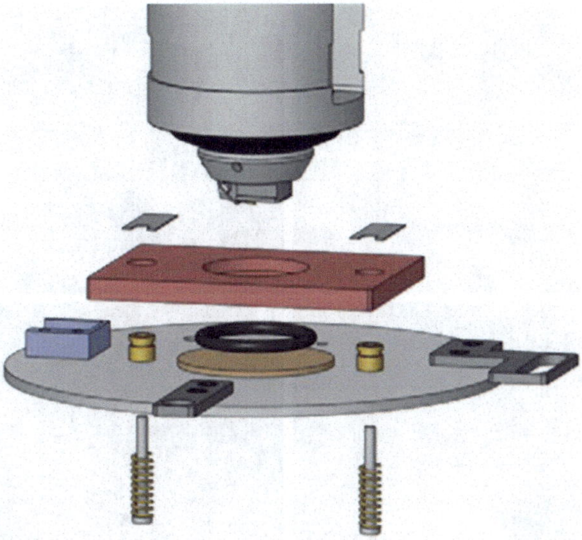

Fig. 20.2 Design of a metal sample cell assembly for AFM imaging in liquid media. (Printed with permission from Agilent Technologies, Inc.)

and liquid to reflect on the surface of the probe. Closed cells have been constructed using machined transparent materials such as glass [9, 17], quartz or plastic [26, 27]. An example design for an open liquid cell is shown in Fig. 20.2, incorporating a metal spacer sealed with an O-ring gasket to encompass a small volume of liquid. A disadvantage of the open cell design is that liquids can evaporate over time, and must be replenished at frequent intervals. Depending on the design, the volume of liquid contained in the cell ranges from 50 μL to several milliliters.

20.4 Applications of AFM in Liquids

When an AFM probe and sample are submerged in liquid media, predictable changes occur for experimental parameters due to alterations of tip-sample interactions. For example, image resolution can be improved because capillary and van der Waals forces between the tip and sample are reduced or even eliminated [13]. Depending on the viscosity, dielectric constant, conductivity, polarity or pH parameters of the liquid media, experimental conditions can be tuned to control tip-surface interactions. Liquid imaging conditions can reduce sample perturbation and will minimize or prevent damage caused by shear forces between the tip and surface. Also, liquid media can dampen vibrations, leading to reduced acoustic noise from the background. A direct comparison of the changes in vibration were shown by Kiridena et al. and Jourdan et al. who obtained frequency spectra for a free cantilever in air versus liquid media (sec-butanol) [28, 29]. When a tip was immersed in solvent, frequency spectra reveal that the natural resonance frequency of the cantilever was shifted downfield, and the damping effects of the liquid served to overall reduce the vibration of the tip.

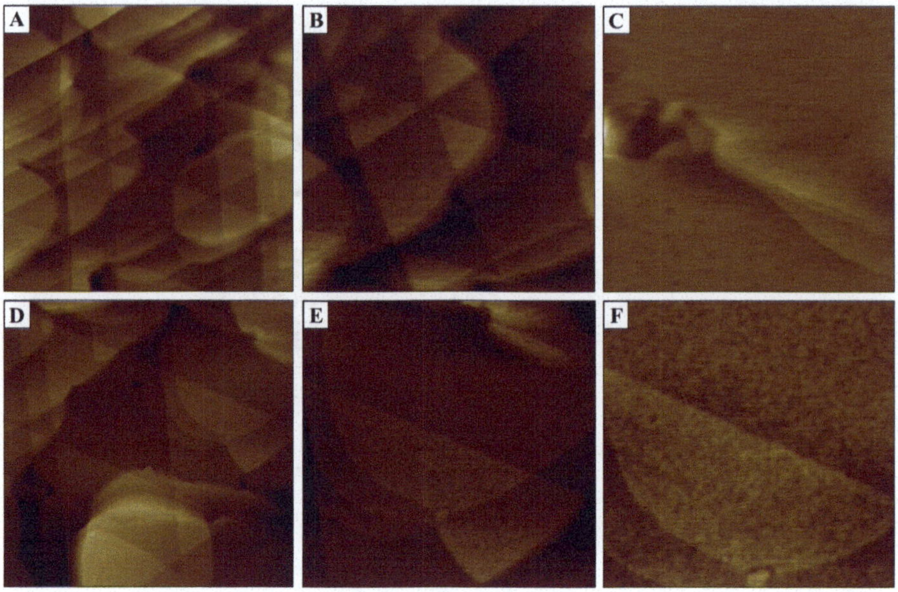

Fig. 20.3 Successive zoom-in views of the surface of a two-component monolayer for comparison of AFM resolution in air versus in liquid. Topography frames of a mixed SAM of decanethiol and tetradecanethiol imaged with contact mode in air (**A**) $1 \times 1\ \mu m^2$; (**B**) $500 \times 500\ nm^2$; (**C**) $250 \times 250\ nm^2$. Topography frames acquired for the same sample and probe in 2-butanol (**D**) $1 \times 1\ \mu m^2$; (**E**) $500 \times 500\ nm^2$; (**F**) $250 \times 250\ nm^2$

20.4.1 High Resolution AFM Imaging in Liquids

When imaging in liquids, AFM resolution can be improved to disclose subtle details of surface morphology at the molecular-level. A direct comparison of frames acquired in ambient air to views that were captured in liquid media using the same probe are shown in Fig. 20.3, for the surface of a mixed monolayer of decanethiol and tetradecanethiol (prepared overnight in 1 mM ethanolic solution, 3:1 molar ratio). For this example, the probe was first used to acquire high resolution views in ambient air, and then liquid media (sec-butanol) was injected into the sample cell. The sample is not a pure SAM, rather it is a mixture of n-alkanethiols with different chain lengths, differing by ~ 0.4 nm in height. In either air or liquid, the heights of single steps of the Au(111) substrate can be readily differentiated. However, the segregation of the different SAM domains is not resolved for the images acquired in air (Figs. 20.3A–C). The images were acquired using contact-mode AFM, and the shapes of the underlying gold terraces and step edges are clearly viewed for all of the topography frames of Fig. 20.3. The views of broad areas in Figs. 20.3A and D are quite comparable in resolution; however, as the magnification is increased the images obtained in liquid reveal differences in the nanoscale topology of the surface of the SAM. The images in ambient air (Figs. 20.3B and C) do not reveal the different domains of the decanethiol and tetradecanethiol SAMs. When the

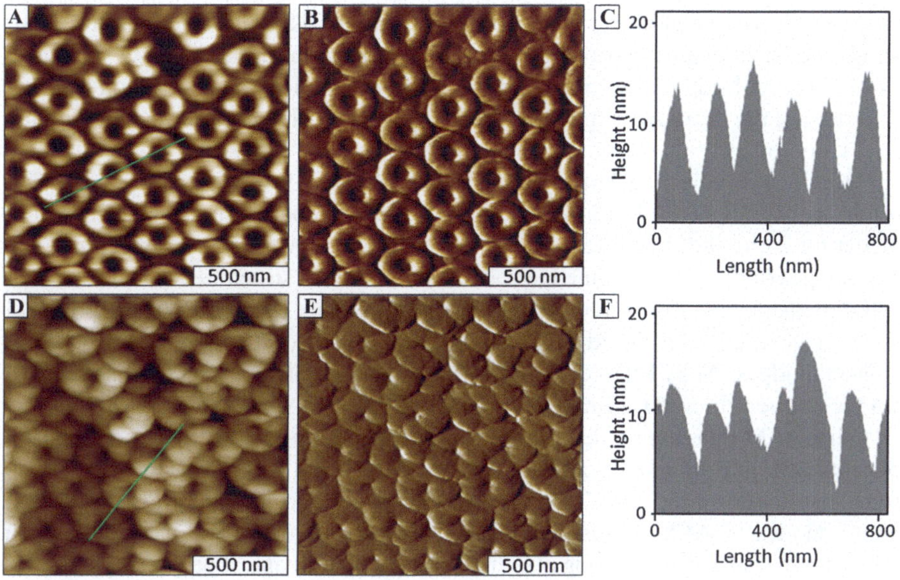

Fig. 20.4 In ethanol changed the surface morphology of OTS ring nanostructures. (**A**) Topography and (**B**) lateral force images acquired in air using for contact mode AFM in ambient air. (**C**) Cursor profile for the line in (**A**). (**D**) Topography and (**E**) lateral force images acquired in ethanol using the same probe. (**F**) Cursor profile for the line in (**D**)

images were acquired in sec-butanol (Figs. 20.3D–F) subtle changes in height are resolved throughout the flat terrace areas, as well as the details of the lacey contours of the gold step edges. Looking closely at Fig. 20.3F with a small scan size of 250 nm × 250 nm, the mottled composition of the binary mixture of SAMs is visible. When imaging in air there are attractive capillary forces operating between the tip and surface leading to overall higher forces used for imaging. Liquid media enables the use of smaller forces as well as reducing the stick-slip adhesion forces, to provide higher resolution imaging. Dynamic changes can cause AFM imaging in liquid media to be difficult, since surface features can change dimensions at the molecular-level when exposed to liquids. This is dramatically illustrated in Fig. 20.4, for multilayered ring nanostructures of octadecyltrichlorosilane (OTS). The ring nanostructures were prepared ex situ using latex particle lithography and vapor deposition [30]. Similar to the protocol for Fig. 20.3, the nanostructures were first imaged in ambient air (Figs. 20.4A–B), then liquid media (ethanol) was introduced without changing probes. An increase in the dimensions of the nanostructures is apparent in the AFM views of Figs. 20.4D–E, caused by changes in the physical size of the OTS rings. Rather dramatic changes in the lateral dimensions of the nanostructures are revealed for AFM topography frames acquired in liquid media. The width of OTS nanopatterns in dried condition measured 245 ± 16 nm. However, after exposure to ethanol, the sizes of the rings were observed to swell by tens of nm, to measure 293 ± 33 nm. The height of the nanostructures appears to have only increased slightly. The methyl-terminated OTS nanostructures consist of an alkane backbone of hydrocarbon chains with bridging links of Si–O–Si bonds, as well as non-bridged free silanol Si–OH groups which change size upon wetting. The sides of the designed ring nanostructures present a 3D interface for studying the interaction of solvents with molecular side groups.

Atomic and molecular lattices have been resolved using contact mode AFM in liquid media. High resolution images acquired using contact mode AFM in liquid media for a cleaved dolomite (104) surface were demonstrated by Pina et al. [31]. The images of dolomite obtained in water disclose lattice structures with surface unit cell dimensions in close agreement with that of the bulk structure. Views of HOPG surfaces covered with liquid parafilm oil were resolved at the atomic scale with contact mode AFM by Marti et al. [16]. Molecularly resolved views of n-alkanethiol SAMs can be achieved routinely with liquid AFM in solvent media [25, 32, 33].

Other AFM imaging modes can be accomplished in liquid media to combine the benefits of liquid imaging with dynamic protocols such as tapping mode [34–36], force modulation [28, 29, 37] or frequency modulation [38] configurations. Dynamic modes of AFM furnish additional information acquired concurrently with topography frames for mapping chemical groups or mechanical properties measurements, such as elastic response or viscoadhesion.

Atomic resolution of surfaces has been demonstrated with developments with frequency modulation AFM (FM-AFM) in liquid. A reduction of the cantilever oscillation amplitude to the range of 0.2–0.3 nm enhanced the sensitivity of the frequency signal to provide atomically resolved images of poly(p-toluenesulfonate) crystals in water [38]. Atomic features of a cleaved muscovite mica(0001) surface exhibiting large-scale corrugations and adsorbates were acquired using FM-AFM in water [39]. The cleavage plane of calcite was characterized in aqueous media using FM-AFM with atomic resolution, to resolve protruding oxygens of the carbonate groups attributed to zigzag patterns along the [421] direction of the calcite(104) surface [40]. High resolution capabilities of FM-AFM have been used to study biological surfaces in buffers, such as phospholipid/cholesterol mixed bilayers [41] and amyloid fibrils [42].

20.4.2 Measurement of Surface Forces in Liquid Media

For AFM operation with contact mode in air, there are often substantial forces present attributable to tip-surface adhesion, due to capillary forces, friction, and stick-slip adhesion, in the range of 100 nN [43]. By immersing both the tip and sample in liquid, the meniscus forces are greatly reduced [44]. Changes of the forces acting on the tip in liquid and in air are demonstrated in Fig. 20.5 with representative force-distance curves. Using the same probe, an uncoated silicon nitride tip was brought into contact with the methyl-terminated surface of a dodecanethiol SAM in air (Fig. 20.5A) and then immersed in ethanol (Fig. 20.5B). Force curves are a plot of cantilever deflection as a function of sample position along the z-axis. Forces are

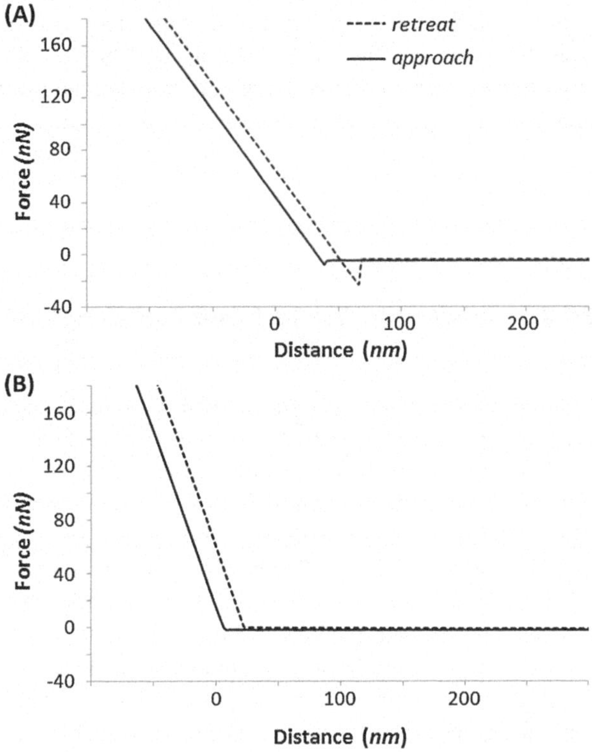

Fig. 20.5 Force versus distance curves for contact mode AFM acquired in (**A**) air and (**B**) liquid

not measured directly, but rather are calculated using the stiffness of the cantilever according Hooke's law relationship to derive values from the measured deflection of the lever. The approach-retreat cycle of typical force-distance curves obtained in air (Fig. 20.5A) and in liquid (Fig. 20.5B) show changes for the pull-off or retraction portion of the measurement. The capillary forces of attraction are substantially decreased when the tip is operated in liquid media (Fig. 20.5B).

Force measurements with AFM have emerged as a standard tool for nanoscale physical characterizations for many disciplines [45]. Molecular-level measurements of adhesion forces for biomaterials have become a significant research focus for biological AFM studies [46]. Forces can be measured with piconewton sensitivity for either specific or nonspecific protein-protein and receptor-ligand interactions, for measuring interactions between cells, and for measuring viscoelastic properties of biomaterials. For these measurements, a force versus distance curve is generated using an AFM probe that is coated with proteins such as antibodies, enzymes or desired functional groups. Imaging in liquid media can be less destructive for soft and sticky biological samples, and provides a means to study effects of pH and electrolyte concentration. The coated probe is brought into contact with the sample and then withdrawn from the surface to generate a plot of the interaction forces as a function of tip displacement. For a typical approach-retreat measurement cycle, the bending of the cantilever is monitored as the probe is brought in and out of contact with the surface. The coated probe will adhere to the sample as it is withdrawn from the surface, often with multiple pull-off points for aggregate samples. The magnitude of this adhesive force can be calculated to provide estimates of molecular bond rupture forces. Changing the pH or ionic strength of the imaging buffer is useful for studying changes for tip-protein interactions as a function of surface charge.

20.4.3 In situ Studies of Chemical Reactions with Liquid AFM

Creative experimental designs with in situ AFM have employed liquid media to view molecular changes that take place on surfaces over time during the course of chemical reactions. For example, beginning with clean substrates, the adsorption and self-organization of molecules can be viewed with time-lapse AFM images [9, 47–50], and antigen-antibody binding processes have been monitored with liquid AFM studies [51, 52]. Surface or material changes that occur with modification of pH or ionic strength can be studied with liquid AFM, in combination with electrochemistry measurements [4, 53].

An example time-lapse experiment showing step-by-step views of the unconstrained assembly of n-alkanethiol molecules onto Au(111) is shown in Fig. 20.6, for octadecanethiol (ODT) [9, 25]. Molecular-level studies were accomplished in situ by immersing freshly prepared gold substrates in a dilute solution of ODT (0.2 mM) in 2-butanol within a liquid cell. Detailed structural information of the kinetics and surface changes during adsorption is revealed as time elapsed (Figs. 20.6A–I). An adventitious adsorbate (the bright spot) and several Au(111) single atomic steps in the lower left region of the images furnish convenient landmarks as a frame of reference for in situ imaging (Fig. 20.6A). In 2-butanol, molecules of ODT initially adsorb on gold with the molecular axis of their hydrocarbon chains oriented parallel to the surface in a side-on configuration (Figs. 20.6B and C). Over time, the surface coverage increases to near saturation and a phase transition takes place to reveal taller islands composed of upright molecules with the axis of the hydrocarbon backbone oriented $\sim 30°$ from the surface normal (Figs. 20.6D and D′). Continued exposure over time produces a greater density of these islands and the growth of these nuclei until a SAM is formed with a $\sqrt{3} \times \sqrt{3}R30°$ commensurate structure. *In situ* AFM studies were conducted for different chain lengths, (C10, C22, and C40) to disclose the kinetics and mechanisms of phase transitions for n-alkanethiol self-assembly [9]. Measurements revealed that the self-assembly of longer chain length alkanethiols occurred more rapidly and produced a more complete film in comparison to shorter alkyl chains [9]. Studies were also conducted using liquid AFM for the coadsorption of mixtures of n-alkanethiols at different molecular ratios and chain lengths, to reveal differences in domain sizes and surface composition [54].

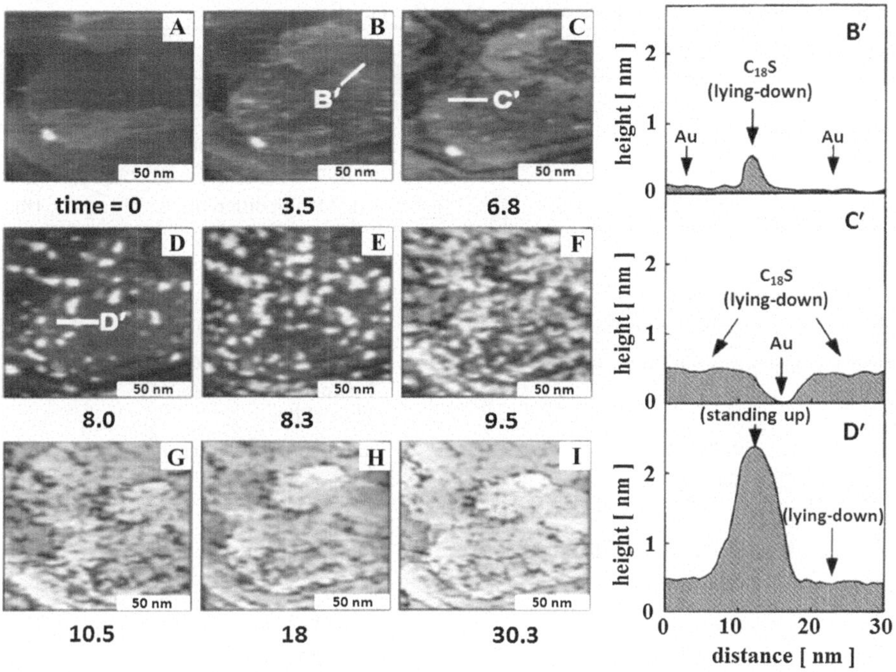

Fig. 20.6 Surface chemical reaction viewed in situ with time-lapse AFM imaging in liquid. Topography images of the self-assembly of ODT onto freshly prepared Au(111) acquired in 2-butanol. (**A–I**) Topography images acquired at different time points and (**B′**, **C′**, **D′**) corresponding height measurements of adsorbates. A small *bright spot* and *several gold* steps in the *lower left region* of the image furnish convenient landmarks for in situ imaging. (Reprinted with permission from Ref. [9])

20.4.4 Electrochemistry Studies with AFM in Liquids

A method known as scanning electrochemical microscopy (SECM) was developed to combine the AFM capabilities of mapping surfaces for electrochemical studies in a liquid cell [55]. Measurements of electron, ion and molecular transfer can be probed with SECM for applications ranging from corrosion to ion transport across cell membranes [56, 57]. Representative examples are summarized in Table 20.1. Typically, a system of three electrodes is used for electrochemical AFM studies, with the substrate serving as the working electrode [6, 11, 58]. A microfabricated probe with diameter of 10 μm or less is used as an ultramicroelectrode (UME) for SECM experiments in an electrochemical cell, while operated in non-contact mode, commonly using the shear force mode of positional feedback [7]. The resolution of SECM is on the order of 100 nm, depending on the size of area probed beneath the UME.

Processes of the underpotential deposition (UPD) of metals have been studied with electrochemical AFM [6, 8]. For these experiments, the structure of the

Table 20.1 Representative studies with SECM in liquid media

System/processes studied	Substrate(s)	Ref.
copper electrodeposition by UPD	Pt(100) and Pt(111)	[6]
polycarbonate membrane, living diatoms	glass slide, constant current mode	[7]
copper electrodeposition by UPD	Au(111) in different electrolyte solutions	[8]
track-etched polycarbonate ultrafiltration membranes	glass disk	[57]
bacterial outer membrane protein F	HOPG, MoS2, Au, Pt	[59]
electron-transfer reactions for glucose oxidase	nylon 66 membranes, hydrogel membranes, glass slides	[60]
anodization of porous alumina	aluminum in a fluid cell	[61]
galvanic electrodeposition of Pd	flame-annealed	[62]
and Pt particles	gold substrates	
corrosion of aluminum in chloride solution	aluminum alloys	[63]
kidney cells in culture medium	petri dish	[64]
films of DNA in redox solutions	Si(111) wafers	[65]
electrodeposited patterns of	gold, platinum,	[66]
conductive polymer polyaniline	carbon surfaces	
redox-active dendrimers mapped by SECM	functionalized glass substrates	[67]
Lead underpotential deposition studied in situ	Au(111)	[68]
anodic dissolution of a gold microelectrode	indium-tin-oxide	[69]
surface patterning using click chemistry by SECM	azido-functionalized glass substrates	[70]

electrodeposited layer of metal cations is mapped while the substrate is maintained at a potential that is under the equilibrium potential of the metal. Scanning probe-based patterning experiments can be accomplished with SECM; microstructures have been fabricated using the local interactions of SECM with UPD or electropolymerization reactions. Voltages are applied between the tip and substrate to enable patterning in liquid media containing reagents for designed local reactions. Atomic resolution was achieved with electrochemical AFM for the investigation of the UPD of copper on Au(111) in a sulfate solution by Manne et al. [8]. The atomically resolved structures revealed differences in lattice packing for different electrolyte solutions. Further experiments with electrochemical AFM have been recently reviewed, and applications include studies of electron transfer kinetics, electrocatalysis as well as mass/charge transfer reactions [56].

20.4.5 Nanofabrication with AFM in Liquid Media

Nanoshaving and nanografting are scanning probe-based lithography approaches which are used to rapidly and reproducibly write nanopatterns of thiol SAMs and other nanomaterials in liquid media. Commercial scanning probe instruments typically provide software to control the speed, direction, length and applied force of the scanning motion of a tip. To accomplish nanoshaving, the tip is used to scratch or scrape away adsorbates under high force for designated areas. For nanoshaving in liquid, the surrounding solvent media can facilitate dissolution of the displaced material. When the liquid cell contains dilute solutions of new molecules, nanografted patterns can be inscribed. Nanografting was first introduced in 1997 by Xu et al. and is accomplished in liquid media by applying mechanical force to an AFM probe to produce nanopatterns within a matrix monolayer [71]. Under low force (less than 1 nN), high resolution AFM characterizations of surfaces can be acquired in situ without modification of the surface. The tip becomes a tool for nanofabrication only when the force applied to the probe is increased to a certain displacement threshold. Nanografting is accomplished by force-induced displacement of molecules of a matrix SAM, followed immediately by the surface self-assembly of molecules such as n-alkanethiols from the liquid media. The molecules to be patterned are dissolved within the surrounding liquid, whereas the substrates are precoated with a protective layer to prevent nonspecific adsorption of molecules throughout areas of the surface. Various surface chemistries can be designed by choosing SAMs of different lengths and terminal groups. An example nanografted pattern prepared in ethanol media is shown in Fig. 20.7, in which the grafted molecules of 1,8-octanedithiol are taller than the surrounding matrix monolayer of hexanethiol [72]. Under self-assembly conditions of nanografting with nanoscale geometric confinement, nanopatterns of α, ω-alkanedithiols written by nanografting form upright monolayers directly with heights corresponding to a standing-up conformation to generate a surface presenting free –SH groups.

A different surface assembly mechanism takes place during nanografting due to spatial confinement to produce patterns directly with an upright conformation [25]. As the molecules of the matrix film are displaced underneath the tip, a transient microenvironment is generated exposing Au(111) for the simultaneous assembly of thiolated molecules. The assembly of nanografted organothiols bypasses the mobile "lying-down" phase due to spatial confinement between the surrounding matrix and the AFM probe; molecules from liquid media assemble immediately onto areas of the exposed substrate into a standing configuration because there is insufficient space on the surface for the molecules to assemble in a lying-down orientation [25]. Kinetic studies have demonstrated that the spatially constrained self-assembly process occurs 10-fold faster than the unconstrained adsorption of organothiol SAMs [25]. A key requirement of nanografting is that n-alkanethiols chemisorb spontaneously to metal surfaces. The speed of the AFM tip influences the composition of the monolayers formed along the writing track. A kinetic Monte Carlo model of solution and nanografted deposition of n-alkanethiols on gold surfaces was reported

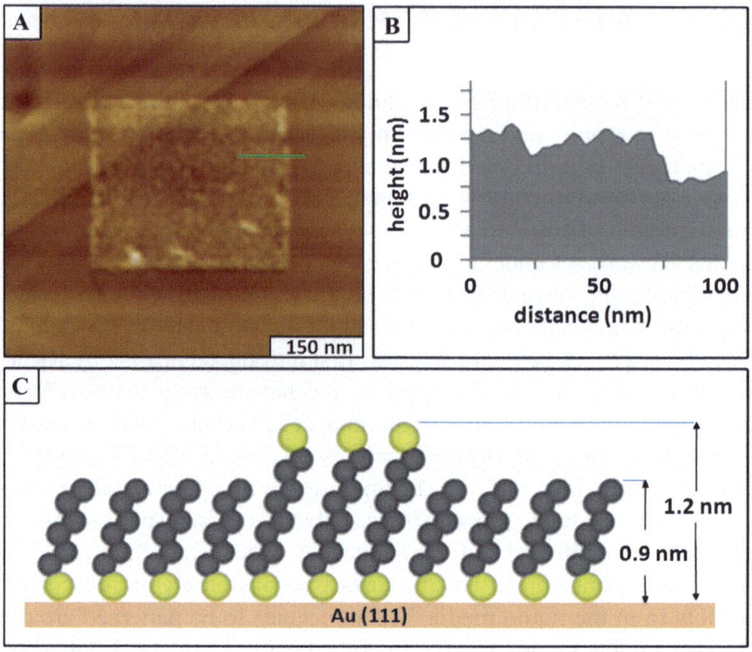

Fig. 20.7 Square nanopattern of 1,8-octanedithiol nanografted within hexanethiol SAM in ethanol. (**A**) Contact-mode topography view of the pattern acquired in ethanol; (**B**) selected cursor profile for the line drawn in (**A**); (**C**) model of molecular heights. (Reprinted with permission from Ref. [72])

by Ryu and Schatz, which corroborates experimental observations for the variation of the heterogeneity of written SAMs with the writing speed of an AFM tip [73].

Designed patterns of hexanethiol molecules nanografted within a matrix of dodecanethiol/Au(111) are displayed in Fig. 20.8, for an AFM liquid cell experiment conducted in ethanol [74]. Nine patterns of mouse ear designs that were shorter in height than the matrix SAM were prepared by inscribing concentric circles ranging from 40 to 210 nm in diameter (Fig. 20.8A). A higher magnification view of a single pattern is shown in the topography frame (Fig. 20.8B) and corresponding frictional force image (Fig. 20.8C). Imaging in liquid media enables resolution of fine details such as etch pits, scars and step edges, even after the tip was used to inscribe multiple patterns. The depth of the patterns measured 0.7 ± 0.2 nm shown by a representative cursor plot (Fig. 20.8D). The outline for writing the patterns is displayed in Fig. 20.8E which indicates the designed tip trajectory when nanografting. For the experiments of Fig. 20.8, each concentric ring was executed by tracing the AFM tip twice at an applied force of 4.8 nN.

Examples of AFM studies with nanografting are summarized in Table 20.2 to show the diverse types of molecules and liquid media used for experiments. Since nanografting protocols are accomplished in liquid media, further successive chemical steps can be developed to introduce fresh reagents for 3D fabrication of more

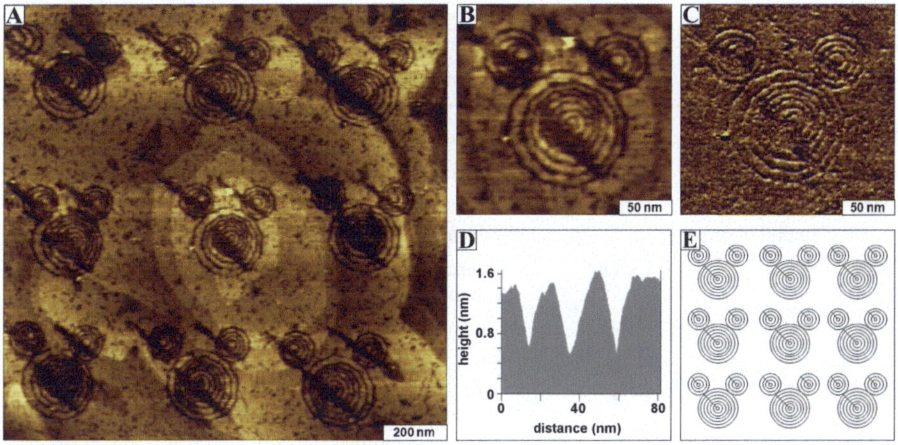

Fig. 20.8 Patterns produced by nanografting hexanethiol within a dodecanethiol monolayer in ethanol. (**A**) Topography frame of 3×3 array of ring designs; (**B**) zoom-in topography and (**C**) lateral force frames of a single pattern; (**D**) cursor profile for line in (**B**). (**E**) Schematic used for the nanopattern design. (Reprinted with permission from Ref. [74])

complex nanostructures [75, 76]. By combining nanografting and the designed spatial selectivity of SAM headgroups, in situ studies with nanografting in liquid media provides capabilities for studying surface reactions at the nanoscale. The spatial selectivity of the headgroups can be used in subsequent steps to direct the attachment of proteins [51, 77], or for molecular assembly [10].

20.4.6 AFM Studies of Biological Samples in Liquids

Since biological processes take place in aqueous environments, liquid AFM offers in situ advantages compared to electron microscopy techniques when investigating biological samples. Using a liquid cell, biomolecules can be imaged in near-physiological, buffered conditions at ambient temperatures. The highest resolution reported for biological imaging with liquid AFM is on the order of 7 Å laterally, and ∼1 Å for vertical resolution [3, 86]. In buffer solution, the pH and ionic strength of the imaging media can be adjusted to balance the van der Waals and electrostatic interactions between the tip and the sample [3]. Resolution is affected by the pH and ionic strength of the surrounding environment, which influences adhesive interactions between the tip and sample. To achieve high resolution imaging of biomolecules, Scheuring et al. describe a process for adjusting the pH and ionic strength of buffers to balance the van der Waals and electrostatic interactions between the tip and sample, at a loading force of 100 pN [3]. Conformational changes of single biomolecules have been visualized for systems of membrane proteins such as bacteriorhodopsin, porin OmpF and aquaporin-Z using contact mode AFM in

Table 20.2 Representative experimental conditions and molecules studied with nanografting

Year	Liquid media	Pattern headgroup	Nanografted molecule	Matrix film	Ref.
1997	2-butanol	CH_3	1-octadecanethiol	1-decanethiol	[71]
1999	2-butanol	CH_3	1-octadecanethiol	1-decanethiol	[32]
2002	decahydro-naphthalene	OH	11-mercapto-1-undecanal	1-octadecanethiol	[78]
2003	water	COOH	3-mercapto-1-propanoic acid	$C_{11}(EG)_6$	[79]
2003	water	OH	6-mercaptohexan-1-ol	$C_{11}(EG)_6$	[79]
2005	ethanol or 2-butanol	CH_3	Mixed n-alkanethiols 1-decanethiol and 1-octadecanethiol, 10:1	1-decanethiol and 1-octadecanethiol, 10:1	[80]
2005	2-butanol or Poly-α-olefin oil	CF_3	$CF_3(CF_2)_9(CH_2)SH$	$HOC_6SH:C_{10}SH = 2:1$	[80]
2006	ethanol	CH_3	1-octadecanethiol	decanethiol	[81]
2006	2-butanol or hexadecane	CH_3	1-octadecanethiol	$C_{10}SH:C_{18}SH = 9:1$	[81]
2007	2-butanol	SH	1-decanedithiol	1-decanethiol	[82]
2007	2-butanol	SH	biphenyl 4,4′-dithiol	1-decanethiol	[82]
2008	ethanol	COOH	11-mercapto-1-undecanoic acid	1-octadecanethiol	[83]
2008	ethanol	COOH			
2008	ethanol	CH_3	dodecanethiol	1,9-nonanedithiol	[84]
2008	ethanol	SH	1,8-octanedithiol	hexanethiol	[84]
2008	ethanol	CH_3	1-hexanethiol	thiolated biotin SAMs	[85]

buffer [3, 87–89]. Structures such as polypeptide loops, α-helices and β-strands of membrane proteins were identified within close-packed arrays of membrane proteins. Example high-resolution images of bacteriorhodopsin are shown in Fig. 20.9, for raw AFM topography frames acquired in buffer using different load forces. The circled structures are surfaces of individual extracellular proteins with a trimer center. Sequences of AFM images were acquired and analyzed to provide insight for the nanomechanics and conformational changes of membrane proteins [3].

Electrochemical AFM studies of biological samples have also been accomplished in buffer solutions for the hexagonally packed intermediate layer of ordered membrane proteins adsorbed to HOPG and other conductive supports [59, 90]. Studies of electron transfer and adsorption of myoglobin on surfaces of a graphite electrode were conducted in situ using tapping-mode AFM combined with cyclic voltammetry measurements in Tris-HCl buffer solutions [91].

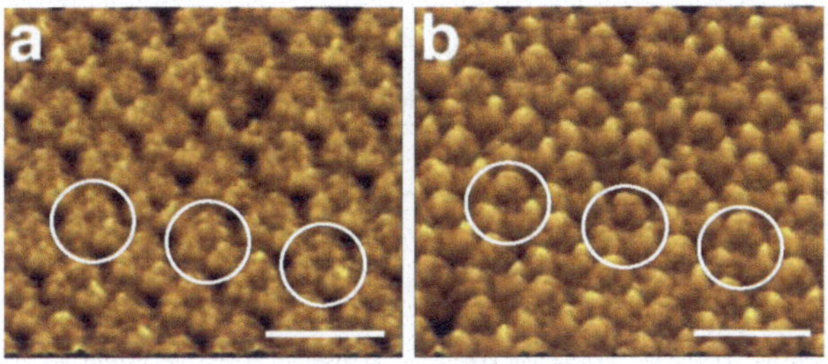

Fig. 20.9 High resolution liquid AFM images of the cytoplasmic surface of the bacteriorhodopsin membrane protein acquired using contact mode AFM. (**A**) Raw AFM topograph acquired at a loading force of ∼200 pN; (**B**) imaged at loading force of ∼100 pN. (Scale bars: 10 nm.) *Circles* indicate the trimer structure of individual extracellular proteins. (Reprinted with permission from Ref. [3])

Biological surfaces can be problematic for high resolution AFM imaging, because of the soft and sticky nature of the samples. Tapping mode has increasingly become more broadly applied and predominant for biological studies. For TM-AFM, the tip is driven to oscillate in and out of contact with the surface intermittently, which minimizes the destructive stick-slip shear forces that occur with continuous contact mode AFM. The tip does not have sufficient time to form transient interfacial bonds with the sample because it is driven to rapidly oscillate during scans across the surface. For soft, fragile samples of biomolecules such as proteins or cells, tapping mode has become the method of choice for nondestructive AFM imaging [13]. The tip is driven to oscillate by a piezoceramic actuator [92] or by an external AC electromagnetic field, as with magnetic acoustic AFM or MAC-mode [93, 94]. Table 20.3 provides examples of the range of sample types and imaging modes that have been used to investigate biological materials using AFM in liquid media.

Liquid AFM studies do not require the use of fluorescent labels to detect binding processes, instead local structural changes can be directly detected. There are advantages for AFM studies of biomolecules such as cells and proteins as compared to methods with fluorescence and electron microscopies, which require chemical treatment and stains. In fluorescent microscopy, biomolecules are tagged with fluorophores that can potentially influence the overall binding affinity [106]. Liquid AFM offers capabilities for investigations in non-denaturative environments without the requirement for fluorescent labels, providing insight into biomolecular mechanisms and binding interactions. High-resolution views of the extracellular and cytoplasmic surfaces of bacterial membrane channels involved in osmoregulation (aquaporin Z, Escherichia coli) were captured in buffer using contact-mode AFM by Scheuring et al. [88]. Structural changes of the protein surface were analyzed after samples were treated with trypsin for cleavage of the N-terminal fragment.

The ability to directly monitor dynamic changes of the conformation or association of biomolecules using time-lapse images is an inherent advantage of liquid AFM. By successively acquiring *in situ* images in liquid media, progressive changes of surfaces or materials that occur over time can be recorded. The self-assembly of collagen fibrils onto mica was tracked using time-lapse AFM by Cisneros et al.,

Table 20.3 Examples of biological samples studied in different liquid media using AFM

Biological system studied	Substrate	Media	AFM mode(s)	Ref.
outer membrane proteins: porin OmpF, aquaporin-Z, bacteriorhodopsin	mica	buffer	contact	[3]
bovine serum albumin, tobacco etch virus capsid protein, tobacco mosaic protein, specific antibodies (IgG)	–COOH monolayers	water	contact	[26]
amylin fibrils	mica	buffer	contact time-lapse AFM	[47]
collagen fibrils	mica	buffer	tapping	[48]
DNA-protein binding	mica	buffer	tapping	[49]
F-actin filaments, human amylin fibrils, nuclear pore complexes	mica	buffer	contact time-lapse AFM	[50]
Esherichia coli bacteria	glass slide	2-butanol	contact	[86]
Escherichia coli water channel, aquaporin Z	mica	buffer	contact	[88]
myoglobin	graphite	buffer	tapping	[91]
fibronectin	TiO_2	water	MAC-mode	[93]
purple membrane protein (Halobacterium salinarum)	mica	buffer	MAC-mode	[94]
Dipalmitoylphosphatidylcholine and cholesterols	mica	PBS	frequency modulation	[95]
bone marrow-derived mast cells	glass coverslips	PBS	tapping/contact	[96], [97]
double-stranded plasmid DNA	mica	*n*-propanol	contact	[98]
DNA	mica	butanol	tapping	[99]
Escherichia coli RNA polymerase	mica	buffer	tapping	[100]
G protein-coupled receptors rhodopsin and opsin	mica	buffer	contact	[101]
ATP synthase rotor proteins	mica	buffer	contact	[102]
RNA polymerase binding to DNA	mica	buffer	contact time-lapse AFM	[103]
canine kidney cells	glass coverslips	PBS	contact time-lapse AFM	[104]
DNA	mica	aqueous solutions	contact time-lapse AFM	[105]

with tapping-mode AFM experiments conducted in liquid [48]. A time-lapse sequence of AFM images for the process of the assembly of RNA polymerase and DNA into complexes was reported by Guthold et al. for experiments conducted in Hepes/MgCl2 buffer, revealing high-resolution views of protein binding to DNA [103]. Madine-Darby Canine Kidney (MDCK) cells were studied using time-lapse AFM in physiological conditions, revealing movement of fibers through the cells during imaging [104]. Studies of the structure and dynamics of DNA were conducted in aqueous solutions by Lyubchenko and Shlyakhetenko using time-lapse AFM [105]. Time-lapse images disclosed steps of the unfolding and branch migration of a Holliday junction over time. The succession of interaction events for the tumor suppressor protein p53 with a DNA fragment were investigated with time-lapse AFM using tapping mode in liquid [49]. Interactions such as association, re-association, sliding and direct binding of the protein to DNA were detected for samples attached to mica surfaces. The growth of amyloid fibrils was observed in liquid media by Goldsbury et al., evidencing bidirectional growth of amylin over time [47]. Changes were monitored over intervals of hours as the mica surface was continuously scanned.

The capabilities of in situ imaging and AFM-based nanofabrication have been combined for molecular-level studies of DNA and proteins [75, 77, 107]. Nanostructures of proteins were produced on gold substrates using nanografting with gold-binding residues such as cysteine [108], or with thiol modification of the protein molecules [109]. In a multi-step approach, nanografted patterns of SAMs can be used for site-selective adsorption of proteins. The terminal moieties of SAM nanopatterns can be designed with chemistries that avoid nonspecific protein adsorption, yet make specific interactions with targeted proteins. Examples of nanografted patterns have been reported with protein binding mediated by covalent [51, 110], electrostatic [110], and specific [111] molecular interactions. By incorporating a short thiol linker at the end of the strands, nanografted structures of single-stranded oligonucleotides of DNA as narrow as 10 nm were produced using nanografting [112–116]. Typically, for the natural adsorption of thiolated DNA on gold surfaces, the DNA molecules tend to lie down with the molecular backbone aligned parallel to the substrate. However, for patterns of DNA oligomers that are nanografted, a relatively close-packed structure with a standing conformation is produced with the molecules oriented in an upright configuration [114, 115]. Investigations of enzyme-digestion and label-free hybridization of single stranded DNA nanostructures were conducted in situ for nanografted patterns of DNA in liquid media [113–117].

20.5 Future Developments of Liquid AFM

An emerging challenge for nanoscale measurements is to capture and quantify the magnitude of structural changes for different materials in response to environmental parameters. Environmental factors such as pH, solvent polarity, ionic strength and temperature are dynamic variables which influence scanning probe experiments. With liquid imaging, in situ studies of electrochemistry, surface assembly reactions, and chemical/physical mapping of samples can be accomplished with time-lapse capture of surface changes. Combining liquid AFM with other imaging modes has made it possible to acquire information while simultaneously imaging samples, for nanoscale mapping of surface properties. Improvements to AFM instruments for probe designs and increased imaging speed will better facilitate investigations for studies of chemical/biochemical kinetics.

Future developments of liquid AFM techniques will couple other analytical approaches such as Raman or infrared spectroscopies for multifunctional instrumentation. The practical hurdles for developing such approaches are the low intensity optical signals and lack of sensitivity for such small size scales. Imaging in liquids provides a way to detect surface changes as influenced by solvents or pH effects at the nanoscale, for dynamic studies of the influence of environmental stimulus. The capabilities of liquid AFM will be important for emerging research efforts in developing stimuli-responsive materials and polymers.

Acknowledgements The authors received financial support from the National Science Foundation (DMR-0906873) and also from the Dreyfus Foundation (Camille Dreyfus Teacher-Scholar Award). Venetia D. Lyles is supported by a Fellowship from the Louisiana Board of Regents. Wilson K. Serem is an LSU doctoral candidate supported by study-leave from Masinde Muliro University, Kenya. The authors thank Dr. K. Lusker for helpful discussion.

References

1. D.J. Muller, A. Engel, Biophys. J. **73**, 1633 (1997)
2. Y.L. Lyubchenko, L.S. Shlyakhtenko, Proc. Natl. Acad. Sci. USA **94**, 496 (1997)
3. S. Scheuring, D. Müller, H. Stahlberg et al., Eur. Biophys. J. **31**(3), 172 (2002)
4. T.J. Senden, C.J. Drummond, Colloids Surf. A **94**, 29 (1995)
5. J.R. Kenseth, J.A. Harnisch, V.W. Jones et al., Langmuir **17**, 4105 (2001)
6. R.M. Rynders, R.C. Alkier, J. Electrochem. Soc. **141**, 1166 (1994)
7. Y. Lee, Z. Ding, A.J. Bard, Anal. Chem. **74**, 3634 (2002)
8. S. Manne, P.K. Hansma, J. Massie et al., Science **251**, 183 (1991)
9. S. Xu, S.J.N. Cruchon-Dupeyrat, J.C. Garno et al., J. Chem. Phys. **108**(12), 5002 (1998)
10. J.-F. Liu, S. Cruchon-Dupeyrat, J.C. Garno et al., Nano Lett. **2**, 937 (2002)
11. T.A. Land, J.J. DeYoreo, J.D. Lee, Surf. Sci. **384**, 136 (1997)
12. S. Xu, G.-Y. Liu, Langmuir **13**, 127 (1997)
13. P.K. Hansma, J.P. Cleveland, M. Radmacher et al., Appl. Phys. Lett. **64**(13), 1738 (1994)
14. G. Binnig, H. Rohrer, Helv. Phys. Acta **55**(6), 726 (1982)
15. G. Binnig, C.F. Quate, C. Gerber, Phys. Rev. Lett. **56**(9), 930 (1986)
16. O. Marti, B. Drake, P.K. Hansma, Appl. Phys. Lett. **51**(7), 484 (1987)
17. B. Drake, C.B. Prater, A.L. Weisenhorn et al., Science **243**, 1586 (1989)
18. C.A. Bippes, D.J. Muller, Rep. Prog. Phys. **74**, 086601 (2011)
19. Y.L. Lyubchenko, L.S. Shlyakhtenko, T. Ando, Methods **54**, 274 (2011)
20. R. Erlandsson, G.M. McClelland, C.M. Mate et al., J. Vac. Sci. Technol. A **6**, 266 (1988)
21. D.W. Abraham, C. Williams, J. Slinkman et al., J. Vac. Sci. Technol. B **9**, 703 (1991)
22. H. Edwards, L. Taylor, W. Duncan et al., J. Appl. Phys. **82**, 980 (1997)
23. F.J. Giessibl, Appl. Phys. Lett. **73**, 3956 (1998)

24. M. Tortonese, R.C. Barrett, C.F. Quate, Appl. Phys. Lett. **62**, 834 (1993)
25. S. Xu, P.E. Laibinis, G.Y. Liu, J. Am. Chem. Soc. **120**(36), 9356 (1998)
26. M.E. Browning-Kelley, K. Wadu-Mesthrige, V. Hari et al., Langmuir **13**, 343 (1997)
27. K. Schoenwald, Z.C. Peng, D. Noga et al., Rev. Sci. Instrum. **81**, 053704 (2010)
28. W. Kiridena, V. Jain, P.K. Kuo et al., Surf. Interface Anal. **25**, 383 (1997)
29. J.S. Jourdan, S.J. Cruchon-Dupeyrat, Y. Huan et al., Langmuir **15**, 6495 (1999)
30. J.-R. Li, J.C. Garno, Nano Lett. **8**, 1916 (2008)
31. C.M. Pina, C. Pimentel, M. Garcia-Merino, Surf. Sci. **604**(21–22), 1877
32. S. Xu, S. Miller, P.E. Laibinis et al., Langmuir **15**(21), 7244 (1999)
33. J.-J. Yu, Y. Horng Tan, X. Li et al., J. Am. Chem. Soc. **128**, 11574 (2006)
34. C.A.J. Putman, K.O. VanderWerf, B.G. De Grooth et al., Appl. Phys. Lett. **64**(18), 2454 (1994)
35. W. Han, S.M. Lindsay, T. Jing, Appl. Phys. Lett. **69**, 4111 (1996)
36. A. Noy, C.H. Sanders, D.V. Vezenov et al., Langmuir **14**, 1508 (1998)
37. W.J. Price, S.A. Leigh, S.M. Hsu et al., J. Phys. Chem. A **110**, 1382 (2006)
38. T. Fukuma, K. Kobayashi, K. Matsushige et al., Appl. Phys. Lett. **86**, 193108 (2005)
39. T. Fukuma, K. Kobayashi, K. Matsushige et al., Appl. Phys. Lett. **87**(3) (2005)
40. S. Rode, N. Oyabu, K. Kobayashi et al., Langmuir **25**(5), 2850 (2009)
41. H. Asakawa, T. Fukuma, Nanotechnology **20**, 264008 (2009)
42. T. Fukuma, Jpn. J. Appl. Phys. **48**, 08JA01 (2009)
43. A.L. Weisenhorn, P.K. Hansma, T.R. Albrecht et al., Appl. Phys. Lett. **54**(26), 2651 (1989)
44. A.L. Weisenhorn, P. Maivald, H.-J. Butt et al., Phys. Rev. B **45**, 11226 (1992)
45. H.-J. Butt, B. Cappella, M. Kappl, Surf. Sci. Rep. **59**, 1 (2005)
46. J. Zlatanova, S.M. Lindsay, S.H. Leuba, Biphys. Molec. Biol. **74**, 37 (2000)
47. C. Goldsbury, J. Kistler, U. Aebi et al., J. Mol. Biol. **285**, 33 (1999)
48. D.A. Cisneros, C. Hung, C.M. Franz et al., J. Struct. Biol. **154**, 232 (2006)
49. Y. Jiao, D.I. Cherny, G. Heim et al., J. Mol. Biol. **314**, 233 (2001)
50. M. Stolz, D. Stoffler, U. Aebi et al., J. Struct. Biol. **131**, 171 (2000)
51. K. Wadu-Mesthrge, N.A. Amro, J.C. Garno et al., Biophys. J. **80**, 1891 (2001)
52. J.N. Ngunjiri, J.C. Garno, Anal. Chem. **80**, 1361 (2008)
53. A.J. Bard, F.-R.F. Fan, D.T. Pierce et al., Science **254**, 68 (1991)
54. D.L. Bu, T.J. Mullen, G.Y. Liu, ACS Nano **4**(11), 6863 (2010)
55. A.J. Bard, F.-R.F. Fan, J. Kwak et al., Anal. Chem. **61**, 132 (1989)
56. P. Sun, F.O. Laforge, M.V. Mirkin, Phys. Chem. Chem. Phys. **9**, 802 (2007)
57. J.V. Macpherson, P.R. Unwin, Anal. Chem. **72**, 276 (2000)
58. S. Manne, P.K. Hansma, J. Massie et al., Science **251**(4990), 183 (1991)
59. P.L.T.M. Frederix, P.D. Bosshart, T. Akiyama et al., Nanotechnology **19**, 384004 (2008)
60. D.T. Pierce, P.R. Unwin, A.J. Bard, Anal. Chem. **64**, 1795 (1992)
61. H. Zhang, D. Zhang, Y. He, Microsc. Res. Tech. **66**, 126 (2005)
62. Y. Shen, M. Trauble, G. Wittstock, Phys. Chem. Chem. Phys. **10**, 3635 (2008)
63. A. Davoodi, J. Pan, C. Leygraf et al., Electrochem. Solid-State Lett. **8**, B21 (2005)
64. P.M. Diakowski, Z. Ding, Phys. Chem. Chem. Phys. **9**, 5966 (2007)
65. L.H. Lie, M.V. Mirkin, S. Hakkarainen et al., J. Electroanal. Chem. **603**, 67 (2007)
66. J. Shou, D.O. Wipf, J. Electrochem. Soc. **144**, 1202 (1997)
67. C.A. Nijhuis, J.K. Sinha, G. Wittstock et al., Langmuir **22**, 9770 (2006)
68. C.-H. Chen, N. Washburn, A.A. Gewirth, J. Phys. Chem. **97**, 9754 (1993)
69. S. Meltzer, D. Mandler, J. Electrochem. Soc. **142**, L82 (1995)
70. S.-Y. Ku, K.-T. Wong, A.J. Bard, J. Am. Chem. Soc. **130**, 2392 (2008)
71. S. Xu, G.Y. Liu, Langmuir **13**(2), 127 (1997)
72. J.-J. Yu, J.N. Ngunjiri, A.T. Kelley et al., Langmuir **24**, 11661 (2008)
73. S. Ryu, G.C. Schatz, J. Am. Chem. Soc. **128**(35), 11563 (2006)
74. J.N. Ngunjiri, A.T. Kelley, Z.M. Lejeune et al., Scanning **30**(2), 123 (2008)
75. G. Yang, J.C. Garno, G.-Y. Liu, *Scanning Probe-Based Lithography for Production of Biological and Organic Nanostructures on Surfaces (4.01)* (Elsevier, Amsterdam, 2011)

76. T. Tian, Z.M. LeJeune, W.K. Serem et al., in *Tip-Based Nanofabrication*, ed. by A.A. Tseng (Springer, New York, 2011)
77. J.N. Ngunjiri, J.-R. Li, J.C. Garno, in *Nanodevices for the Life Sciences*, ed. by C.S.S.R. Kumar (Wiley-VCH, New York, 2006)
78. J.F. Liu, S. Cruchon-Dupeyrat, J.C. Garno et al., Nano Lett. **2**(9), 937 (2002)
79. D.J. Zhou, X.Z. Wang, L. Birch et al., Langmuir **19**(25), 10557 (2003)
80. W.J. Price, P.K. Kuo, T.R. Lee et al., Langmuir **21**(18), 8422 (2005)
81. W.J. Price, S.A. Leigh, S.M. Hsu et al., J. Phys. Chem. A **110**(4), 1382 (2006)
82. J. Liang, L.G. Rosa, G. Scoles, J. Phys. Chem. C **111**(46), 17275 (2007)
83. J. Ngunjiri, J.C. Garno, Anal. Chem. **80**(5), 1361 (2008)
84. J.H. Yu, J.N. Ngunjiri, A.T. Kelley et al., Langmuir **24**(20), 11661 (2008)
85. Y.H. Tan, M. Liu, B. Nolting et al., ACS Nano **2**(11), 2374 (2008)
86. N.A. Amro, L.P. Kotra, K. Wadu-Mesthrige et al., Langmuir **16**(6), 2789 (2000)
87. A.D. Schenk, P.J.L. Werten, S. Scheuring et al., J. Mol. Biol. **350**, 278 (2005)
88. S. Scheuring, P. Ringler, M. Borgnia et al., EMBO J. **18**, 4981 (1999)
89. D.J. Muller, F.A. Schabert, G. Buldt et al., Biophys. J. **68**, 1681 (1995)
90. P.L.T.M. Frederix, M.R. Gullo, T. Akiyama et al., Nanotechnology **16**(8), 997 (2005)
91. S. Boussaad, N.J. Tao, J. Am. Chem. Soc. **121**, 4510 (1999)
92. C.A.J. Putman, K.O. Van der Werf, B.G. De Grooth et al., Appl. Phys. Lett. **64**(18), 2454 (1994)
93. S.R. Sousa, M.M. Bras, P. Moradas-Ferreira et al., Langmuir **23**, 7046 (2007)
94. F. Kienberger, C. Stroh, G. Kada et al., Ultramicroscopy **97**, 229 (2003)
95. H. Asakawa, T. Fukuma, Nanotechnology **20**(26), 7 (2009)
96. Z. Deng, V. Lulevich, F.T. Liu et al., J. Phys. Chem. B **114**(18), 5971 (2010)
97. T. Zink, Z. Deng, H. Chen et al., Ultramicroscopy **109**(1), 22 (2008)
98. H.G. Hansma, J. Vesenka, C. Siegerist et al., Science **256**(5060), 1180 (1992)
99. A.D.L. Humphris, A.N. Round, M.J. Miles, Surf. Sci. **491**(3), 468 (2001)
100. S. Kasas, N.H. Thomson, B.L. Smith et al., Biochem. **36**(3), 461 (1997)
101. Y. Liang, D. Fotiadis, S. Filipek et al., J. Biol. Chem. **278**(24), 21655 (2003)
102. D.J. Müller, N.A. Dencher, T. Meier et al., FEBS Lett. **504**(3), 219 (2001)
103. M. Guthold, M. Bezanilla, D.A. Erie et al., Proc. Natl. Acad. Sci. USA **91**, 12927 (1994)
104. C.-A. Schoenenberger, J.H. Hoh, Biophys. J. **67**, 929 (1994)
105. Y.L. Lyubchenko, L.S. Shlyakhtenko, Methods **47**, 206 (2009)
106. D.J. Müller, A. Engel, U. Matthey et al., J. Mol. Biol. **327**(5), 925 (2003)
107. M. Liu, N.A. Amro, G.-Y. Liu, Annu. Rev. Phys. Chem. **59**, 367 (2008)
108. M.A. Case, G.L. McLendon, Y. Hu et al., Nano Lett. **3**, 425 (2003)
109. Y. Hu, A. Das, M.H. Hecht et al., Langmuir **21**(20), 9103 (2005)
110. N.A. Amro, G.-y. Liu, Proc. Natl. Acad. Sci. USA **99**, 5165 (2002)
111. Y.H. Tan, M. Liu, B. Nolting et al., ACS Nano **2**, 2374 (2008)
112. P.V. Schwartz, Langmuir **17**, 5971 (2001)
113. E. Mirmomtaz, M. Castronovo, C. Grunwald et al., Nano Lett. ASAP (2008)
114. M. Liu, N.A. Amro, C.S. Chow et al., Nano Lett. **2**, 863 (2002)
115. M. Liu, G.-Y. Liu, Langmuir **21**(5), 1972 (2005)
116. M. Castronovo, S. Radovic, C. Grunwald et al., Nano Lett. **8**, 4140 (2008)
117. M. Castronovo, A. Lucesoli, P. Parisse et al., Nat. Commun. **2**, 297 (2011)

Chapter 21
Atomic Force Microscopy for Surface Imaging and Characterization of Supported Nanostructures

Franciszek Krok, Bartosz Such, Jacek J. Kolodziej, and Marek Szymonski

Abstract This chapter presents an overview of Atomic Force Microscopy (AFM) principles followed by details on AFM instrumentation. In particular, the frequency modulation method of the non-contact AFM (NC-AFM) used in ultra-high vacuum conditions is explained in detail. Then, applications of NC-AFM for an atomic-scale range characterization of semiconductor and isolator surfaces as well as supported nanostructures are introduced.

21.1 Introduction

Atomic force microscope [1] has been invented in order to circumvent two serious limitations of STM. Firstly, for STM a surface/nanostructure with enough electrical conductivity is required and, secondly, the tunneling current is strongly dependent on the local electronic structure of the sample and the probe. A scanning probe microscopy based on the concept of a profilometer with a sharp tip acting as a probe is a natural extension of the STM method [2], capable of analyzing insulating samples with high accuracy, providing that a new physical quantity will be measured instead of the tunneling current. In this context, the tip-surface interaction forces, such as a long-range electrostatic interactions, van der Waals forces, and most importantly short-range chemical forces should be taken into account. In general, those interaction forces have repulsive core dominant at short-range and a long-range attractive part. There are two regions of tip-surface distance where the forces are typically

F. Krok (✉) · B. Such · J.J. Kolodziej · M. Szymonski
Centre for Nanometer-Scale Science and Advanced Materials (NANOSAM), Institute of Physics, Jagiellonian University, ul. Reymonta 4, 30-059 Krakow, Poland
e-mail: franciszek.krok@uj.edu.pl

B. Such
e-mail: such@uj.edu.pl

J.J. Kolodziej
e-mail: jkolodz@if.uj.edu.pl

M. Szymonski
e-mail: ufszymon@uj.edu.pl

G. Bracco, B. Holst (eds.), *Surface Science Techniques*,
Springer Series in Surface Sciences 51, DOI 10.1007/978-3-642-34243-1_21,

used as the feedback signal for the operation of an atomic force microscope: in the repulsive core range and in the attractive region. In the former case, a static method of force measurement is usually utilized. However, in the attractive regime, the dependence of the force on distance is relatively weak, and in order to achieve required sensitivity (and avoid a so-called "snap to contact" phenomenon), the static force measurement is frequently replaced by dynamic methods. Already in 1986, Binnig and co-workers [1] proposed a dynamic method of investigation of tip-surface interaction forces. In general, in the AFM technique, the tip is mounted at the end of a flexible cantilever and brought into contact with the surface. While the cantilever scans the surface, the topographic features cause its deflection, which can be measured accurately. The cantilever deflection characteristics enables us to retrieve a signal, which is often called "topographic". Various aspects of the tip surface interaction relevant of operation of the atomic force microscope are discussed in detail in the first sections of this chapter (Sect. 21.2). In Sect. 21.3, the experimental instrumentation of various AFM methods, allowing measurements of particular properties of the sample surface are described. In particular, it includes the discussion of the static and the dynamic modes of AFM operation. Operation of the dynamic AFM mode within the attractive range of the interacting potential could results in accidental crash of the tip with sample surface leading to uncontrolled decoration of the tip apex which in turn could change the interaction potential and consequently the imaging conditions of the dynamic AFM [3]. Even uncontrolled but stable chemical functionalization of the tip opens new exciting possibilities for imaging with atomic resolution and at the same time sensing only a particular kind on surface atoms which interact by chemical forces with the decorated tip apex. Such possibilities has been explored in recent studies of the surface reconstruction of AIII-BV semiconductors (Sect. 21.4), and they are discussed in detail through this chapter. Moreover, AFM imaging performed with a separation of the electrostatic term from the total tip-sample interactions is used for acquiring an additional chemical sensitivity of the method as shown by the rapidly developing Kelvin force probe microscopy (KPFM) (Sect. 21.4.5). A thorough understanding of elementary processes responsible for energy dissipation at sliding interfaces requires experimental and theoretical studies of interatomic/intermolecular interactions between the respective surfaces. In particular, proper evaluation of the lateral forces between atoms of the interacting interfaces is essential for understanding frictional properties of materials. Lateral force microscopy (LFM) has already been used with success for studying friction at the atomic scale on solid surfaces. Section 21.4.6 is devoted to exploring such new prospects in studies of nanostructures which are "topographically" flat. Although the atomic force microscopy is reaching a stage of maturity, new exciting opportunities for its future development are still to be explored. Recent advances in tuning fork/qPlus sensor technology applied in low temperature AFM systems, may come closer to the goal of the ultimate tool capable of sub-atomic resolution imaging, 3D force spectroscopy combined with STM and AFM analysis of the same surface area. These new trends in AFM microscopy/spectroscopy are outlined at the final part of this chapter (Sect. 21.5).

21.2 Theory Background

All the techniques under the common name Scanning Probe Microscopy (SPM) are based on monitoring interactions between a surface and a sharp tip moved over the surface. In order to yield atomic resolution the interaction has to vary in that scale both in direction perpendicular to the surface and in the surface plane. In the prototype of all SPMs, i.e. the Scanning Tunneling Microscopy (STM), the current tunneling between two electrodes: the tip apex and the surface acts as a perfect feedback information due to its exponential dependence on the distance. In order to circumvent the limitations imposed by sample conductivity, the idea of using the force between the tip and the sample as a feedback signal was born. Hence, the Atomic Force Microscopy was invented and since the very beginning [1] recorded a remarkable string of successes [3–7]. In order to measure forces, even as low as piconewtons, suitable sensors had to be developed. The most common solution comprises of a flexible cantilever with spring constant typically from 0.01 N/m for friction experiments to as high as 2000 N/m for quartz tuning fork solutions. The cantilever bends proportionally to the force acting at the tip apex. Hence, if the amount of bending and the spring constant of the cantilever are known the total force can be evaluated. The most widespread way of the detection of cantilever deflection is based on a laser beam deflection from the cantilever top surface towards a position-sensitive photodiode. Alternative solutions, such piezoresistive or piezoelectric cantilevers are also used. Calibration of the bending is relatively easy by collecting force-distance curve on a hard material. Determination of the spring constant k is considerably more challenging and usually is the main source of uncertainty in force measurements. Various methods of spring constant determination are reviewed in the literature [8–11] while the simplest, however not yielding the highest precision is deducting k from dimensions of a rectangular cantilever assuming Young modulus E of a material: $k = \frac{Ewt^3}{4L^3}$, where L, w and t are length, width and thickness of a cantilever, respectively [6].

21.2.1 *Interaction Forces in AFM*

Forces in AFM have nonlinear distance dependence with a pronounced minimum (i.e. point of the largest attraction), a few tenths of nanometer from the surface, and a strongly repulsive core at short distances stemming from Pauli repulsion [12, 13]. The total force acting between the AFM tip and the surface is usually composed of a few components of different magnitudes and ranges (see Fig. 21.1). The long-range components (i.e. acting at the distances of nanometers), such as van der Waals and electrostatic forces usually are unable to yield atomic resolution due to their slow lateral variations. They do play important role in imaging process forming attractive background necessary for stable operation of the microscope. Especially the presence of van der Waals forces is inevitable. Its magnitude depends on the

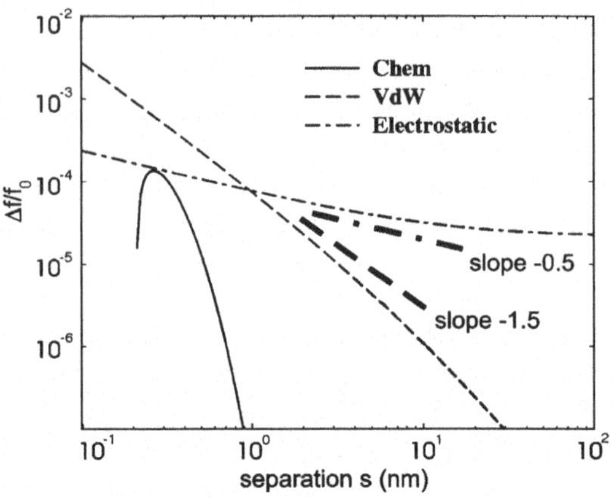

Fig. 21.1 Calculated distance dependence of frequency shift for chemical, electrostatic and van der Waals interactions according to [13]. Reprinted with permission from [13]

material of the tip and the sample and the geometry of the tip and its potential for a spherical tip can be expressed as:

$$V_{vdW} = -\frac{A_H R}{6z} \qquad (21.1)$$

where A_H is a Hamaker constant, R is the radius of the tip apex and z is the tip-surface distance, while for the more realistic tip shapes, corrections to that formula are necessary[14]. Then, the magnitude of electrostatic forces is expressed as:

$$F_{el} = -\frac{\pi \epsilon_0 R (V_{tip} - V_{surface})^2}{z}. \qquad (21.2)$$

As seen, the force depends on a squared difference of voltages of a tip and a surface [15]. Thanks to that, electrostatic force can be minimized for normal NC-AFM operation by applying an appropriate bias, and additionally can be used in our favor for imaging variations of the surface potential by using Kelvin probe force microscopy (KPFM) (see Sect. 21.4.5).

At tip-surface separations lower that 0.5 nm the dominant role is played by the chemical forces acting between the tip apex and the surface. Their exact nature (covalent, ionic etc.) is dependent on the material of which the tip and the surface are made and has to be considered for each case separately. Simplistic interpretations of protrusions as atoms in NC-AFM images can be misleading. The most typical example are (001) surfaces of alkali halides (NaCl, KBr etc.) containing ions of two kinds. Due to ionic origin of imaging interaction on such surfaces either anions or cations are imaged as protrusions. It is very difficult to determine which one is imaged as their sublattices are identical [16, 17]. That is why, the first undoubtful identification of sublattices of a ionic crystal was performed for CaF_2 for which surface geometries of Ca and F sublattices differ [18]. Other examples of NC-AFM images reflecting particular sublattices only can be found for many different systems, such as group III–V semiconductors [19] or graphite 21.4.1 [20]. The repulsive interactions are used mostly in contact mode AFM (contact AFM) in which a tip is pressed against the sample during scanning, providing the information of topography and friction. The resolution of contact AFM is limited, however, as a result of a large contact area even for very hard materials.

21.2.2 Dynamics of Cantilever in NC-AFM

Direct measurement of the tip-surface force is possible in the contact AFM mode. However, often it is required to avoid contact in order to preserve delicate structure either on the tip or on the surface. Therefore, it is necessary to place the tip at certain distance from the surface where it experiences an attractive force. However, the static cantilever is prone to 'snap to contact' event when the stiffness of the cantilever, k, becomes smaller than 'stiffness' of tip-surface interaction k_{ts}^{max}: $k < \max(\frac{\partial^2 V_{ts}}{\partial z^2}) = k_{ts}^{max}$, where V_{ts} is the potential of tip-surface interaction [21]. To circumvent that problem, a non-contact Force Microscopy (NC-AFM) was devised. In a NC-AFM, a cantilever is oscillated at its resonant frequency with an amplitude ranging from a Angstrom to a few dozens of nanometers. Than, the 'snap to contact' can be prevented when maximal tip-surface force F_{ts}^{max} is smaller than a product of cantilever spring constant k and amplitude A_0. In vacuum NC-AFM, frequency modulation (FM) detection scheme is typically used [22] while the AM detection scheme dominates in ambient and liquid conditions [23]. The cantilever is self-excited at resonant frequency by a feedback loop and the tip-surface interaction can be deduced from changes in the cantilever frequency (frequency shift, Δf). Then, as shown by Giessibl, for interactions described by inverse power law the frequency shift, Δf, is equal:

$$\Delta f = -\frac{f_0}{kA_0^2}\langle F_{ts}\big(z(t) - z_0\big)\rangle \tag{21.3}$$

where z_0 is an equilibrium position of a cantilever. Giessibl [25] showed than that formula can be reduced to

$$\Delta f = \frac{f_0}{2k}\frac{\partial^2 V_{ts}(z_{min})}{\partial z^2}, \tag{21.4}$$

if the distance of the closest approach of a tip to a surface, z_{min}, is larger than the amplitude A_0, and

$$\Delta f = -\frac{1}{\sqrt{2}\pi}\frac{f_0}{kA_0^{2/3}}\frac{C}{z_0^{n-1/2}}I_1(n) \tag{21.5}$$

for the amplitudes A_0 much larger than the closest approach distance z_{min} if the assumed force distance dependence was $F_{ts}(z) = -Cz^{-n}$ and $I_1(n) = \int_{-\infty}^{\infty}\frac{1}{(1+y^2)^n}dy$. The formula (21.4), linking the frequency shift with the force gradient appeared in the very first papers on FM NC-AFM [22] and sometimes is mistakenly used without the requirement of small oscillations. Nice comparison of the applicability range of both formulas can be found in Ref. [24]. The frequency shift is a function of distance z, spring constant k, amplitude A_0 and resonant frequency f_0 of a free cantilever. In order to separate the parameters depending on the

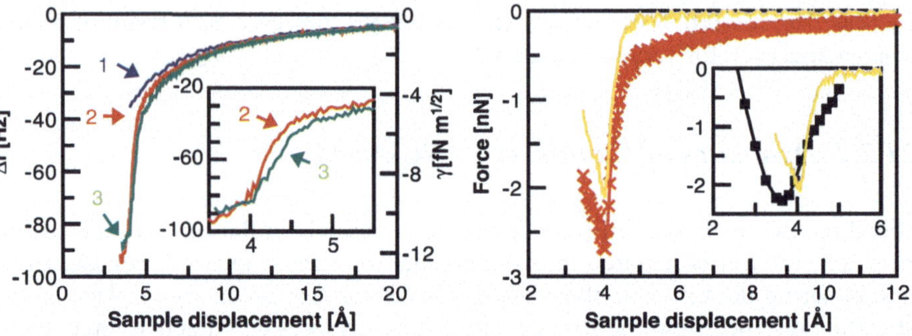

Fig. 21.2 Experimental frequency shifts (*left panel*) and forces calculated from them (*right panel*) taken over different positions (numbered *1–3*) within a cell of Si(111) 7 × 7 surface. Note in the *right panel* that short range forces (*yellow*) are separated from the total force (*red*). Reprinted with permission from [12]

interaction, i.e. distance dependence from parameters of the cantilever itself, which could allow for comparing different experiments with each other, Giessibl proposed using in analysis of large amplitude NC-AFM data 'normalized' (sometimes called also 'reduced') frequency shift [25]:

$$\gamma(z) = \frac{\Delta f k A_0^{3/2}}{f_0}. \tag{21.6}$$

21.2.3 Reverse Problem in NC-AFM

The reverse problem i.e. determining the force from the frequency shift (Fig. 21.2) is rather difficult to solve mainly due to the fact that the force is changing in a quite complicated way along the trajectory of the tip. In order to quantitatively determine the force, a full frequency vs. distance curve has to be known reaching as far as several nanometers from the surface where even long-rang forces cease to act. Quite a few numerical methods were proposed to deduce forces from such set of data [26, 27], but recently the most popular is the method proposed by Sader and Jarvis [28], who derived the following analytical expression:

$$F(z) = 2k \int_z^{\infty} \left[\left(1 + \frac{A_0^{1/2}}{8\sqrt{\pi(t-z)}} \right) \Omega(t) - \frac{A_0^{3/2}}{\sqrt{2(t-z)}} \frac{d\Omega(t)}{dt} \right] dt, \tag{21.7}$$

where $\Omega(z) = \frac{\Delta f}{f_0}$.

In NC-AFM not only conservative forces can be measured [29–31]. Since in typical, constant amplitude realization of FM NC-AFM the feedback loop is constantly putting energy into the oscillating cantilever in order to keep the amplitude constant, the dissipation power can be measured. Certainly, even far from the surface the cantilever dissipates energy. The measure of that dissipation is provided by the drive amplitude, fed constantly to the free cantilever ($A_{drive\ free}$). The energy lost during one cycle of oscillation is $E_{int} = 2\pi \frac{1/2kA_0^2}{Q}$. When a cantilever approaches the surface additional dissipation appears: more energy has to be put into the system and the drive amplitude grows (A_{drive}). Then the energy lost is given by:

$$\Delta E = 2\pi \frac{1/2kA_0^2}{Q} \left(\frac{|A_{drive}|}{|A_{drive\ free}|} - 1 \right), \tag{21.8}$$

where A_0 is the amplitude of cantilever oscillations and Q is the cantilever quality factor.

21.3 Experimental Setup and Methods

As it was pointed out in Sect. 21.2 there are two basic modes of operation of AFM in UHV conditions: the static and the dynamic mode. Both rely on the measurement of interactions between a sample surface and a sharp tip attached to the free end of a cantilever. The most advanced AFM cantilevers, whose typical lateral dimensions are of the order of 100 μm, while the thickness is less than μm, are microfabricated from silicon or silicon nitride using photolithographic techniques. The attached tips can be of conical or pyramidal shape, 3–10 μm long with apex of radius as small as 2 nm. Using AFM in UHV implies usually that the microscope is a part of a UHV system having also sample preparation capabilities. A major objective in design of such UHV system is to provide mechanical vibration isolation which guarantee stability and high resolution of the AFM measurements. Thus, usually the UHV system is designed as a rigid setup isolated from ground noise by pneumatic legs of a resonant vibration frequency of about 1 Hz. The frequencies significantly higher than 1 Hz are then not transmitted through the legs. Also, the AFM stage is often equipped with its own damping system like, for example, the one based on eddy current vibration isolation system introduced by Omicron GmbH [32]. Apart from the external mechanical noise, the detection signals of interest in AFM are still a subject of other sources of noise like the thermal noise or the intrinsic noise of the deflection sensor, all having a tremendous influence on the quality of the AFM imaging [33].

21.3.1 Contact Atomic Force Microscopy

The first mode of the AFM operation which we will discuss is the static mode, called contact-AFM, where the tip is brought into a direct contact with a sample surface. Commonly, soft cantilevers having their spring constant (k) in the range from 0.01 to 1 N/m, and with the free resonance frequency of few tens of kHz are used. Within the contact area, the tip-apex atoms experience repulsive forces due to the tip electronic orbital overlapping with those of the surface atoms. The force acting on the tip apex results in the cantilever deflection. The deflection is monitored/measured

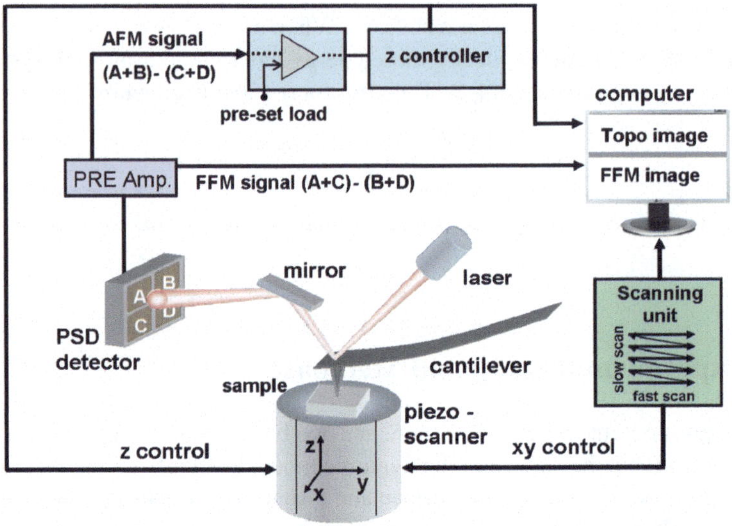

Fig. 21.3 Schematic diagram of AFM operation in contact mode allowing for surface topography measurements as well as for mapping of frictional forces. See explanation in text

by a detection system, for example, the system detecting tunneling current between the cantilever and reference electrode (as in the first AFM microscope built by Binnig [1]), electrical (piezoresistive, capacitance) systems, and the most commonly used optical (laser beam bounce, interferometer) deflection. The forces measured in this mode range from few nN down to a few tenths of nN, thus, for a cantilever of the spring constant of 1 N/m, sub-nanometer deflection, roughly corresponding to the atomic size, can be measured.

In Fig. 21.3 a simple scheme of AFM microscope, operating in the contact mode is shown. Before starting the AFM measurements, the force detection system is adjusted by directing the laser beam onto the top of the cantilever and the reflected beam is directed, by means of a mirror, into the center of a position sensitive photodetector (PSD). The PSD detector consists of four quadrants (A, B, C, D) and the differential signals from the quadrants allow to measure the vertical and torsional deflection of the cantilever. The sample is mounted on a piezo-scanner which scans the sample in the xy plane in a raster manner and move it vertically in the "z" direction (see Fig. 21.3). Upon the approach, the tip is brought into contact with the surface to reach the pre-set load force. The exerted load induces the vertical deflection of the cantilever and in turn a vertical movement of the reflected laser beam on the PSD which, corresponds to a certain difference in the signals from the top (A+B) and the bottom (C+D) parts of the photodetector, this difference is the "AFM signal" (Fig. 21.3). While the surface is scanned, the topographic features cause a change of the vertical deflection detected by the "AFM signal". Then, if the AFM is operated in a so called "height mode", the vertical sample movement is driven by a feedback loop system which keeps the tip-surface interaction constant (a constant load). The "AFM signal" from the PSD is an input signal for the feedback controller which adjusts the "z" voltage applied to the scanner to control the height of the sample and maintain the cantilever deflection constant during the scans. The scanner height variation is a direct measurement of the topographic profile of the sample surface (surface topographic image). There is also another way to perform the AFM measurement, in the so called "constant height" mode. In this mode, the sample (scanner) is moved without changing its "z" coordinate. The feedback loop is not active and the map of the "AFM signal" is a result of the measurement. This mode of operation can only be applied to very flat surface planes and for relatively small scanned areas. The acquisition time in this mode is relatively short as compared to the "constant load" mode as there is no need for the z coordinate adjustment. Also the surface measurements in the "constant height" mode are not affected by noise and/or artefacts which may be generated by the feedback loop. However, AFM operation in this mode is affected by the thermal drift, inevitable during room temperature measurements, and the sample tilts in both, the fast and slow scan directions. Also scanning across structures higher than the cantilever elastic deformation range results in the tip damage. In practice, in order to overcome these limitations, the measurements are often performed in the "quasi-constant height" mode. In such a case the feedback loop is not completely switched off, but it is set to low loop gain. Consequently, the tip-surface distance is determined by averaging of the interaction over many surface sites and stays almost constant. This allows for correcting for the vertical "z" drift of the piezo-scanner and for compensating of the sample tilts.

Apart from the vertical bending of the cantilever, while the AFM measurement is performed in the contact mode, one can measure its torsional bending (twisting) while the sample is moved in the direction perpendicular to the long axis of the cantilever. This mode of AFM operation is called lateral force microscopy (LFM), or friction force microscopy (FFM) and it allows for direct measurement of frictional forces acting between the tip and the surface. The torsional deflection of the cantilever is measured by acquiring the "FFM signal" which is the difference of the left (A+C) and right (B+D) parts of the PSD detector (Fig. 21.3). The average magnitude of the torsional deflection is proportional to the friction force sensed by the tip. For scanning over flat surface this directly determines the value of the friction coefficient. For non-flat surface morphologies, however, the interpretation of this measurement is ambiguous since the friction force can not be separated from other lateral forces in a simple way. Scans along the same line but in opposite directions may, in such a case, help to distinguish between deflections caused by the substrate slopes from the local variations of the frictional properties.

21.3.2 Non-contact Atomic Force Microscopy

In the other mode, called non-contact atomic force microscopy (NC-AFM) or dynamic force microscopy (DFM) mode, the cantilever is forced to oscillate in close proximity of the sample surface with frequency near or at its resonance frequency. In contrast to contact-AFM mode, stiffer cantilevers, with k in the order of tens of

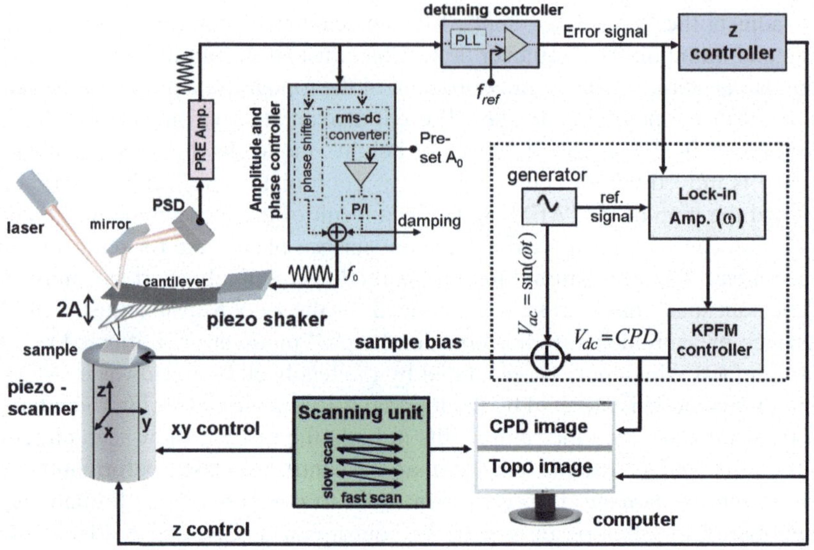

Fig. 21.4 Schematic circuit diagram of AFM operation in non-contact FM mode allowing for surface topography measurements as well as for mapping of the contact potential (the scheme of electronics within the *dashed line*). See explanation in text

N/m and, consequently, with resonance frequencies of hundreds of kHz, are used. While scanning, the total tip-sample interactions change and lead to variation of the cantilever oscillation frequency or amplitude. These changes of amplitude and frequency, with respect to the pre-defined (reference) amplitude and frequency, are used as feedback signals to track the surface topography. Two regimes (sub-modes) of operation of dynamic AFM are employed. In AM-AFM (so called amplitude modulation) mode the cantilever is excited off its eigenfrequency but still within the resonance peak and the variation of its oscillation amplitude is used as a feedback signal to acquire the surface topography. This mode is used when operating AFM in ambient or liquid environments. In the other sub-mode, i.e. FM-AFM (so called frequency modulation) mode, the amplitude of cantilever oscillation is kept constant while the change of its resonant oscillation frequency is a source of contrast. Topographic images are taken by profiling the surface with constant frequency shift i.e. keeping constant difference between the interacting cantilever resonant frequency and the resonant frequency of free cantilever. FM-AFM mode is typically used in UHV conditions.

In Fig. 21.4 a scheme of the FM-AFM mode is shown. The cantilever with eigenfrequency f_0 and spring constant k, mounted on the piezo-shaker, is forced to oscillated with amplitude A_0 by a feedback loop as a self-driven oscillator. FM-AFM mode requires operation of two feedback loops. Within the first loop, the deflection signal from PSD detector after amplification (in PRE Amp.) is fed into the "Amplitude and phase" controller. The purpose of this controller is to provide feedback such that the cantilever oscillate at constant, pre-set amplitude. Within the controller, the

PSD signal is branched into an rms-to-dc converter and a phase shifter. The phase shifter controls phase shift, ϕ, between the mechanical excitation generated by the piezo-shaker and the oscillation of the cantilever detected by PSD. The rms-to-dc converter provides a dc signal which is a measure of the oscillation amplitude. Then, this signal is compared to the pre-set value of amplitude and the amplitude error signal is generated. The amplitude error after amplification with a proportional/integral (P/I) controller, is multiplied by the phase shifted cantilever deflection signal generated in the phase shifter. The resulting signal drives the piezo-shaker. The phase shifter is adjusted in such a way that the driving signal required for establishing the desired oscillation amplitude is minimal; in such a case the phase ϕ is equal exactly 90°. The second loop controls the cantilever oscillation frequency change allowing the topography of the sample surface to be acquired. The cantilever deflection signal from PSD is fed into a detuning controller where the phase-locked loop (PLL) detector measures the oscillation frequency (f) and compares it with the predefined set-frequency f_{ref}. An output signal (the error signal) is delivered which is proportional to the difference between the input frequency f and the reference frequency f_{ref}. This error signal is a feedback signal to regulate the tip-sample distance (with the help of the z controller) maintaining a constant tip-sample interaction by minimizing $f - f_{ref}$.

21.3.3 Kelvin Probe Force Microscopy

Kelvin probe force microscopy (KPFM), realized as an extension of dynamic force microscopy technique, offers surface chemical sensitivity at the nanometer scale by means of surface contact potential difference (CPD) measurements acquired in parallel to the AFM imaging. This is done by a separation of the electrostatic term from the total tip-sample interactions. The method of the contact potential measurement is based on the original idea by Lord Kelvin who, in 1898, demonstrated a method for an experimental determination of the surface potential (the work function) using a vibrating plate capacitor [34]. In this method, the capacitor is constituted by parallel metal plates, one of which has known work function (it is called reference electrode) and the other has unknown work function (to be measured). Upon connecting the plates electrically electrons will flow from the plate with the smaller work function to the plate with the higher one until the Fermi levels are equal. Due to the periodic change of the distance between the plates, the capacitance is changing, resulting in an alternate current flowing between the plates. Then an external *dc* voltage is applied between the plates and regulated until the current is nullified. This *dc* voltage corresponds to the contact potential difference of the two materials which constitute the plates of the capacitor. In the KPFM experiment the same idea is employed but the tip and the sample surface play the roles of the vibrating capacitor 'plates'. As the cantilever is a precise force sensor, in order to compensate the CPD between the sample and cantilever instead of the current, the electrostatic tip-sample force is reduced to zero by applying a dc voltage to the sample. Thus KPFM measures the CPD with high sensitivity. Scanning allows for mapping of the surface potential with spacial resolution in nanometer-scale. To run the surface contact potential measurements with KPFM there is a need to operate a third loop as shown in Fig. 21.4. The sample is biased with sinusoidal voltage $V_{bias} = V_{ac}\sin(\omega t) + V_{dc}$. This sample bias induces oscillatory electrostatic force (see (21.2)), with frequency ω, between the tip and the sample resulting in additional oscillating component superimposed on the mechanical oscillation of the AFM tip. The time-dependent cantilever oscillation, with an angular frequency ω, induces the variation of detuning controller output signal (the error signal). A lock-in amplifier is employed to extract the ω component of the error signal. The output signal of the lock-in amplifier is directly proportional to the difference between CPD and V_{dc}. Then, the CPD value is measured by applying V_{dc} to the sample (with the help of Kelvin regulator), such that the output signal of the lock-in amplifier is nullified. This V_{dc} value is acquired for each point on the sample surface, simultaneously with the topography signal, composing a map of the work function on surface.

21.4 Applications of the Technique

21.4.1 NC-AFM of Semiconductor Surfaces

For the first time the atomic resolution using the NC-AFM technique was achieved on 7×7 Si(111) surface. It was done independently and almost concurrently by Giessibl [35] and Kitamura and Iwatsuki [36]. Results of both researches were published in the beginning of year 1995. Shortly after, further experimental and theoretical studies of the same surface followed [37–48]. The Si(111)(7×7) is thus perhaps the most thoroughly studied surface, using NC-AFM technique. Other semiconductor surfaces studied using NC-AFM included GaAs(110) [49–52], InAs(110) [53], InP(110) [54–56] GaAs(001) [19, 57, 58], InSb(001) [19, 58, 59], InAs(001) [58, 60], InP(001) [58, 61].

As far as the imaging mechanisms are concerned the studies yielded convergent results; as discussed in Sect. 21.2 the atomically resolved details are, in most cases, due to tip-surface interactions of covalent nature. In all discussed cases the tip structures were formed *ad hoc*, by intentional crashing the tip onto the investigated surface, and they remained a mystery. Therefore, the interpretation of the interactions is based on hypothetic tip structures. In experimental practice, a tip is crashed against investigated surface either intentionally or unintentionally several times until some stable configuration, having also atomic resolution capability, is reached. The control parameters of the microscope, such as Δf, bias, scan rate, feedback loop strength, etc., are then tuned to reach optimum performance. Hard crashes happens also during data acquisition and they often result in total loss of atomic resolution, large change in vdW background (in the latter case the atomic contrast may be restored by changing the detuning) and sometimes in discontinuity of the pattern, i.e. the atomic resolution is not lost but the pattern is either shifted or completely

Fig. 21.5 Si defects in the Sn layer on the Sn/Si(111)($\sqrt{3} \times \sqrt{3}$)$R30^{\circ}$ surface manipulated to create the "Si" symbol. Reprinted with permission from [62]

changes its character. Sometimes soft events are observed when the pattern continuity is preserved but its intensity changes and/or some rows appear/disappear. Such soft event is likely associated with a change in the tip front atom electric charge. This variability, however, do not disqualifies the NC-AFM technique as a tool for surface structure studies since irrespective of the nature of imaging interactions, the patterns reflect the surface lattice symmetries.

On well known surfaces, like Si(111)7 $\times$ 7 or (110) faces of AIII-BV crystals the NC-AFM has been most often tested as a new technique to reach understanding of the atomically resolved imaging or atom manipulation mechanisms. In order to illustrate the state-of-the-art of the technique we show Fig. 21.5, presenting the NC-AFM image of the Sn covered Si(111) surface [62].

Below we review few attempts to use the NC-AFM as a tool for determination of complex surface structures, unknown or questioned. In all cases the NC-AFM is performed, using Park Scientific Instruments VP2 apparatus, at room temperature. Commercial piezoresistive silicon non-contact cantilevers are used as probes. Resonant frequencies of the cantilevers used are about 200 kHz and the spring constant is 20 N/m. Cantilever oscillation amplitudes are about 10 nm and detunings in the range 5–30 Hz (see Ref. [58] for more experimental details).

Structure of the As-Rich GaAs(001) Surface

The structure is imaged with high resolution using NC-AFM in the quasi-constant-height mode by repulsive interactions [63]. Double dimer rows, reminiscent of the $\beta 2$ structure [64], running along [$1\bar{1}0$] direction are evidenced (Fig. 21.6). Frequent irregularities are seen in the double dimer rows.

The gray tone covering the smooth fields with no atomic details may be identified with the zero level of short range interactions. Relative to this zero-level tone local repulsive short-range interactions are indicated by the brighter tones. The $\beta 2$ dimers are imaged with ultimate resolution, i.e., atoms constituting the dimers are resolved—see an insert in Fig. 21.6a). More rare, *single-dimer* (the most likely $\alpha 2$), *"three-dimer"* (β) motifs, and single adatoms are also seen [63].

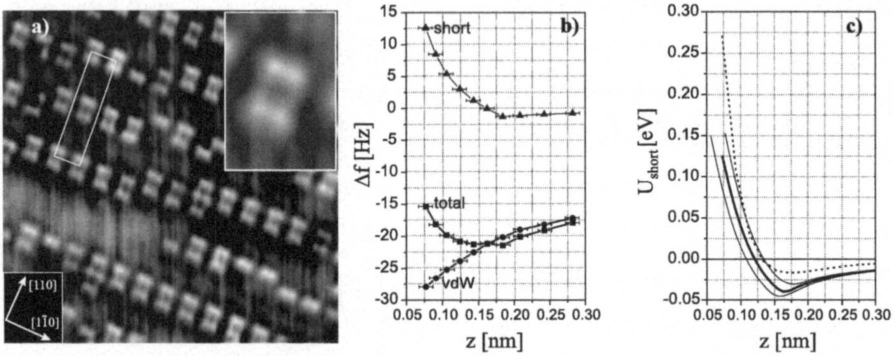

Fig. 21.6 10 nm × 10 nm NC-AFM image of 2 × 4 reconstructed GaAs(001) surface. The *insert* shows in magnification the double dimer detail. The *white rectangle* indicates the $c(2 \times 8)$ unit cell. (**b**) FM-AFM interaction curves over As top layer atoms. The "total" curve is the sum of "van der Waals" and "short" curves. (**c**) Potential of the short-range tip-surface interaction obtained from the "short" Δf curve—*solid line*. Error limits are indicated by *thin solid lines*. Krypton-krypton interaction potential is included for reference (*dotted line*) The z-scale zero is arbitrary. Reprinted with permission from [63]

It is interesting to note that in the image of Fig. 21.6 atomic features have a width of only 2.2 Å (Gaussian FWHM) which is less than the arsenic double atomic radius (2.4 Å). Such a result indicates that the imaging is provided by interactions of extremely short range. This is investigated further by continuous scanning the surface in the quasi-constant-height mode and simultaneously varying the average detuning in steps [63]. For different detunings the Δf values over crystallographically equivalent surface atoms are analyzed and the frequency shifts arising due to the short range and van der Waals interactions are derived. Finally the interaction curves are drawn (Fig. 21.6b) and the short-range interaction potential $U_{\text{short}}(z)$ is evaluated using the Sader-Jarvis algorithm [28]—see Fig. 21.6c.

The potential $U_{\text{short}}(z)$ has attractive tail and the attractive energy is of the order of the room temperature kT. The attractive interaction has rather large spatial extent as it can be deduced from the size of the darker field around the double dimer in Fig. 21.6. The decay of this weak attractive interaction is similar to the krypton-krypton interaction tail which indicates its dispersive nature. The $U_{\text{short}}(z)$ potential is steep in its repulsive part and also in this region it is comparable to the Kr-Kr interatomic potential (Fig. 21.6c). Thus, the interaction providing atomic details is the core-core repulsion. Since the core-core interactions are concentrated around atomic nucleus and have a symmetry close to the circular one, the imaging mode reached for As-rich GaAs(001) surface gives the in-plane coordinates of protruding surface atoms directly. It is interesting to note also, that no formation of chemical bonds between the tip and the surface As dimers occurs suggesting that the tip front atom is passivated arsenic atom.

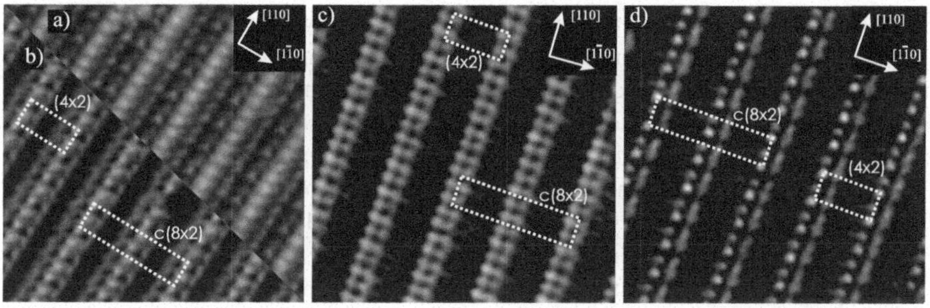

Fig. 21.7 FM-AFM pattern of the $c(8 \times 2)/(4 \times 2)$InAs(001) surface at RT. The size of the map is 10 nm × 10 nm. Parts (**a**) and (**b**) show respectively the dominant "attractive" and the dominant "repulsive" features as bright details. Unit cells of the structure are indicated by the *dotted rectangles*. Parts (**c**) and (**d**) show other "repulsive" FM-AFM patterns of the same surface, i.e., for different tips. Size of all maps is 10 nm × 10 nm

Structure of In-Rich InAs(001) Surface

The case of In-rich InAs(001) surface illustrates a variety of NC-AFM pattern that may be obtained over the same structure or in other words the chemical sensitivity of the NC-AFM technique.

Exemplary NC-AFM pattern is shown in Fig. 21.7a, b. In order to show important details of the NC-AFM map we split it into two parts showing the positive and the negative of the same pattern. The grayscales are tuned separately (by linear operations only) for each part in such a way that the dominant features are shown as clear and bright details. In fact only these dominant features on the FM-AFM patterns can be unambiguously interpreted. Other weaker features, often not having proper surface lattice symmetries, arise as a mix of details reflecting weakly interacting surface sites, accidentally shifted images contributed by secondary atoms of the tip, as well as residual differentials (due to quasi-constant-height imaging mode). The pattern is locally reconstructed either $c(8 \times 2)$ or (4×2) consistently with the LEED data [65]. The "attractive" side of the pattern is presented in Fig. 21.7a. In this figure the brighter tones denote lower cantilever frequencies, i.e. stronger attractive (or less repulsive) interactions. The pattern is composed of dominant rows of bright features arranged on 4×1 lattice. The bright features are obviously associated with identical points on the surface lattice likely with rows of atoms. In the Fig. 21.7b the "repulsive" side of the same pattern is shown and the brighter tones indicate more repulsive/less attractive interactions. In this figure one finds small circular features constituting rows of double periodicity along [110], positioned half way between neighbor attracting dominant rows. In the $c(8 \times 2)$ reconstructed regions one may find that the mirror symmetry axis is on the double-period row while the glide reflection axis is found on the attracting row. Full width at half maximum (FWHM) is estimated between 2 and 2.5 Å for the sharp repulsive features (Fig. 21.7b) and ~3.5 Å for the dominant attractive features (Fig. 21.7a).

The other example of the NC-AFM pattern for the InAs(001)$c(8 \times 2)/(4 \times 2)$ surface is presented in Fig. 21.7b. In this pattern the only meaningful information is on the repulsive end. Both the resolution and the symmetry of the pattern are excellent, and the tip is chemically selective, since features of only one specific type are seen. They form a set of double rows having double periodicity along [110]. The latter becomes clear only after careful examination as the deviations from the 4×1 symmetry are rather small (the bright features are grouped in blocks containing four features each). One can identify parts of the pattern having $c(8 \times 2)$ symmetry and find out that mirror symmetry axes are located in the middle of the double rows. The FWHM of the bright spots composing the pattern is 2.5 Å.

One more NC-AFM pattern obtained for the same surface is presented in Fig. 21.7c. On the "repulsive side" there are distinct single rows composed of sharp bright features and the period along the rows is double. The mirror symmetry axes are located on these double period rows. The rows include alternative brighter and darker sharp, atomic-like features, of FWHM equal to 2.5 Å corresponding to atoms of the same kind and in similar chemical environment located alternatively at two different levels.

To conclude, by NC-AFM study three distinctly different structural elements on the $c(8 \times 2)/(4 \times 2)$InAs(001) surface are found: namely the single period rows, the double-period rows and, the double rows, parallel to [110]. By symmetry considerations, one can find their relative positions on surface. The double rows are placed half way between the neighboring single-period rows. The double-period rows are located in the middle of double rows. These conclusions are based only on the pattern symmetries and not dependent on the tip structure or the imaging mode [60].

Structure of In-Rich InSb(001) and Ga-Rich GaAs(001) Surfaces

InSb(001)$c(8 \times 2)$ and GaAs(001)$c(8 \times 2)$ surfaces were investigated using NC-AFM [19]. Obtained NC-AFM maps of these In(Ga)-rich surfaces are shown in Figs. 21.8a–c and g. The bright tones denote here more attractive interactions. The patterns' symmetry is in all cases 4×1. For both surfaces several types of atomic rows are visible running along [001] crystallographic direction. The patterns may be interpreted taking into account that the formation of contrast in NC-AFM is due to convolution of two factors: chemical interactions and topography [19].

If certain surface structure model may be assumed it is possible to analyze NC-AFM images assuming certain "chemical" and topographic factors (depending on the atom elevation in the model) for each surface-atom type and then defining "interaction strength" coefficients by multiplication of the two factors. The "interaction strength" coefficient are then represented by gray tone to create an image. The chemical factors are interactively modified until simulated maps similar to the experimental patterns are obtained. This scheme may seem oversimplified but, since the NC-AFM tip structure is not known, even most advanced simulations would in fact produce the same results, i.e., either confirmation or rejection of the assumed model based on the pattern symmetries only.

It is found that only the interaction maps simulated for the ζ model [66] are similar to the experimental data. These maps are shown in Figs. 21.8d–f. The chemical

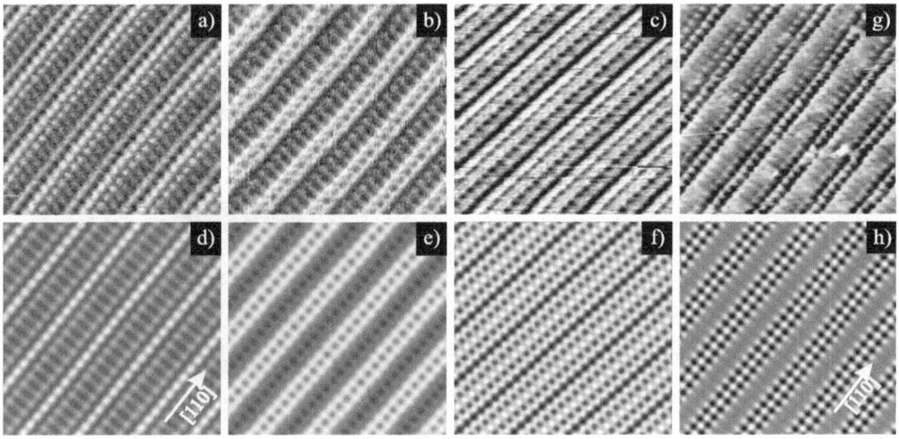

Fig. 21.8 (**a**, **b**, **c**)—NC-AFM maps of $c(8 \times 2)$ InSb surface. Maps (**a**) and (**b**) are recorded in the constant height mode and the map (**c**) is obtained in the topographic mode. (**d**, **e**, **f**)—-Simulated AFM maps. (**g**) A map of $c(8 \times 2)$ GaAs. (**h**) Simulated corresponding map. The size of all images is 10 nm × 10 nm. Reprinted with permission from [19]

factor sets obtained for maps 21.8d and 21.8e suggest particularly simple interpretation of the maps 21.8a and 21.8b i.e., the map 21.8a is an image of the In sublattice, and the map 21.8b is an image of the Sb sublattice. For the map 21.8c both sublattices contribute to the pattern. For the interaction map simulated for GaAs (Fig. 21.8h) and the experimental pattern 21.8g the interpretation is more complex: on the attractive "side" it is reflecting the Ga sublattice, however, repulsive interactions have to be set over As atoms to account for intense dark spots found in the pattern. The protruding type-1 atoms [66] are absent which correspond well to the current views on this surface structure [66, 67].

The obtained results are also consistent with theoretical calculations by Tobik et al. and Ke et al. [51, 56] who concluded that in case of covalent interaction of the silicon tip with surfaces of AIIIBV compounds three single-atomic imaging modes leading to different patterns are expected. The modes correspond to three most likely tip apexes, namely: the AIII apex, BV apex, and the Si apex. Interactions of these tips with the surface should lead to images of BV sublattice, AIII sublattice, and composition of the two sublattices respectively. The theoretical predictions correlate particularly well with experimental results for InSb for which all three AFM imaging modes are demonstrated. The result for GaAs indicate more complex tip-surface interactions that likely include electrostatic and/or core-core contributions.

21.4.2 NC-AFM of Nanostructured Insulator Surfaces

While metal or semiconductor surfaces may be studied on local scales using scanning tunneling microscopy, in case of insulators the NC-AFM have no alternative.

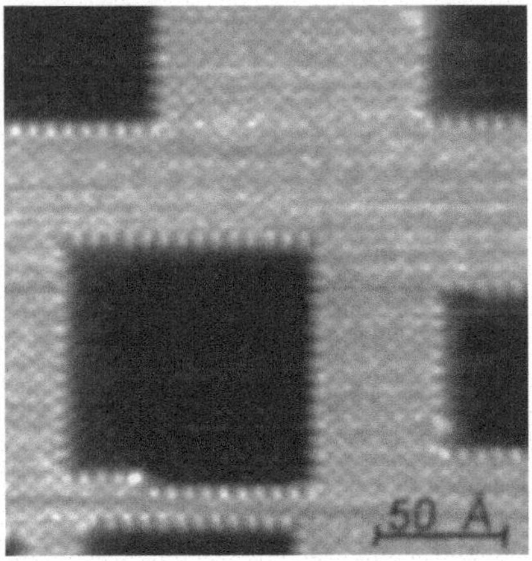

Fig. 21.9 Atomically resolved NC-AFM image of radiation-induced pit in KBr surface (23 nm × 23 nm). Reprinted with permission from [69]

Below we present two examples of DFM technique used to study insulating surfaces [68–70]. Other examples may be found in Refs. [71–75].

Electron Irradiated KBr Surface

Electron-stimulated desorption of (100)KBr surface has been investigated in vacuum with NC-AFM and mass spectroscopy. It has been found that both desorption components (K and Br) show oscillatory dependence on the electron dose with the oscillation amplitude decaying gradually [68].

The NC-AFM images of irradiated KBr surface, taken for small increments of the irradiation dose, have indicated that the desorption proceeds in a layer-by-layer regime and that one period of oscillation corresponds to removal of a single monolayer. Exemplary NC-AFM image is presented in Fig. 21.9. It has been found that K desorption signal is strictly correlated with the edge atom density on the surface. The maximum efficiency of the desorption process is found for the surface with the top layer ~50 % desorbed when the density of edge atoms reaches its maximal value at about 5 % of the total atom density in the KBr(100) monoatomic layer ($0.92 \times 10^{15}/\mathrm{cm}^2$). The minimum efficiency of the desorption process is observed for the mostly flat surface.

The NC-AFM studies supplemented with the rich literature data on point defects in alkali halides allowed to determine the following scenario for the electron stimulated desorption processes in alkali halides. First, incident electron scattering in the crystal results in an electron hole plasma in the scattering volume, and then by fast selftrapping of electrons and holes mobile Frenkel defects, i.e., H-centers and excited F*-centers are produced. These defects diffuse through the crystal independently. If the H-center reaches the surface the halogen atom is emitted but no defect on the surface is left as the H-center is essentially an interstitial atom. The F*-center can not initiate desorption unless it is trapped at the step. At the flat surface F*-centers are reflected back to the bulk where they may convert to immobile (ground) state (F). The F-centers may annihilate with the H-centers thus

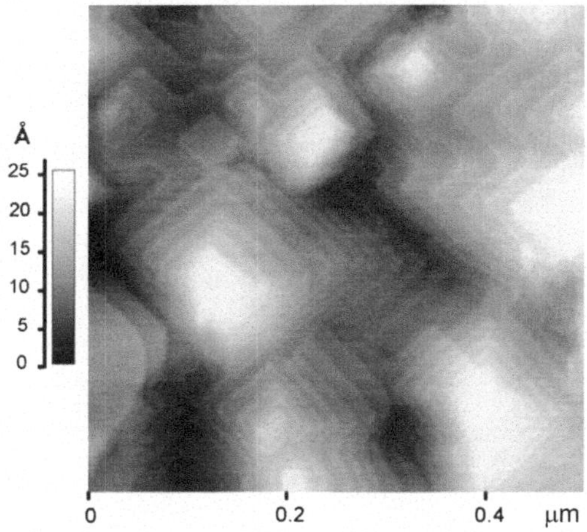

Fig. 21.10 Surface topography of the KBr film grown on InSb(001) substrate. The film thickness is 120 monolayers. The image size is 0.5 μm × 0.5 μm. Reprinted with permission from [70]

the steady state concentration of immobile F-centers is built after certain dose of irradiation. This concentration at surface proximity is controlled by the surface recombination/reflection rate for diffusing F*-centers, determined by the surface step density. A steady-state desorption process is thus controlled by the changing branching ratio between the deexcitation paths which initiate atomic emission and the paths which do not. The dumping of the oscillations is explained by the decoherence in the layer-by-layer desorption mode increasing with time. More detailed description of the investigation may be found in Refs. [68, 69].

Pyramidal Structures Formed During KBr Homoepitaxy

Multilayer KBr films growth by MBE in the [001] surface orientation show regular pyramidal structures on a mesoscale dimension. This phenomenon have been studied by NC-AFM in Ref. [70].

If alkali halides are evaporated from a moderately heated effusion cell their vapors are composed predominantly of alkali halide monomers. Thus during the homoepitaxy the film growth is controlled by diffusion of the alkali halide molecules on the substrate surface. The surface of ∼100 Å thick film, imaged using NC-AFM, is shown in Fig. 21.10. Steps on the surface are aligned along [010] and [100] directions.

Formation of terraces and islands on the surface is explained in terms of terrace-ledge-kink model [70]. An alkali halide molecule is initially adsorbed on the crystal surface at random position. Due to its high diffusivity it migrates thermally until it is caught at the ledge. Then it walks along the ledge until it is incorporated at kink. Observed terraces on the pyramidal structures have much smaller sizes than expected on the bases of KBr molecule diffusivity. Therefore barriers for transition across steps (Ehrlich-Schwoebel barriers (ESB) have to be invoked to explain the surface topography. Prior studies of adsorption systems with ESB have been carried

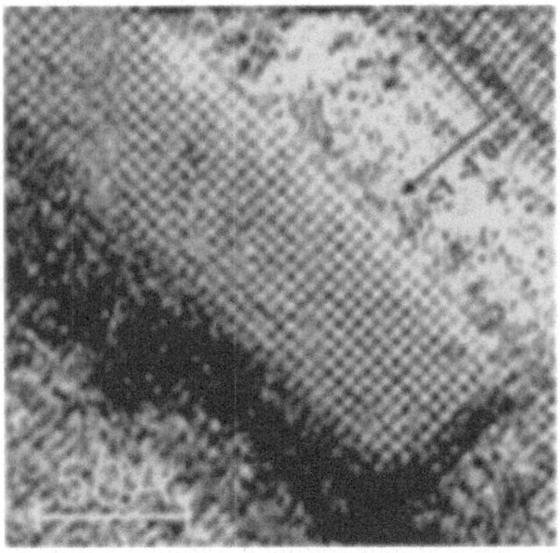

Fig. 21.11 Topographic DFM image (20 nm × 20 nm) of the Au wire on the InSb(001) surface. Reprinted with permission from [76]

out with STM and therefore limited to conducting surfaces. With DFM such studies are also possible for insulating and weakly conducting surfaces.

21.4.3 Au Islands Supported on InSb

Submonolayer deposition of Au on InSb(001) results in formation of interesting nanostructures, i.e., narrow, up to 800 nm long nanowires are observed—cf. Fig. 21.11, [76]. The nanowires are created along atomic troughs (along the [110] direction) of the reconstructed InSb(001) surface. The temperature window for the nanowire formation is relatively narrow: 600–650 K.

In Fig. 21.11 an atomically resolved image of the nanowire formed at 600 K is presented. It should be emphasized that the atomic resolution is seen both for the substrate and for the wire in the same image.

21.4.4 NC-AFM of Supported Molecules

Intensive NC-AFM studies concerning organic molecules on surfaces are inspired by rapid maturing of organic electronic technologies as well as by expectations for novel, single-molecule computing devices [78–88].

In Fig. 21.12a large InSb terrace with adsorbed 3,4,9,10-perylene-tetracarboxylic-dianhydride (PTCDA) molecules is shown [89]. As seen, on the terrace, the molecules form initially long chains along [110] crystallographic direction, while at steps the chains are shorter and more frequent. Interestingly, the chains appear bright from far (they are imaged by attractive interactions) while at closer tip-surface approach—see Figs. 21.12b–d the contrast reverses and the PTCDA molecules appear as dark features on brighter, uniform background. The molecules are preferentially oriented with their longer axis parallel to [110] and they are adsorbed in two

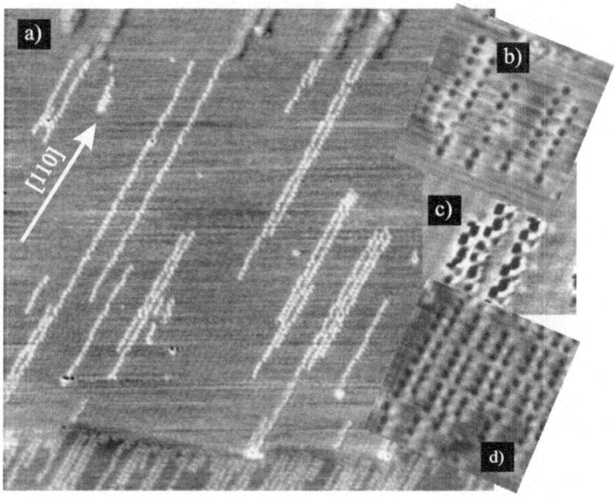

Fig. 21.12 (**a**) Overview (150 nm × 150 nm) of the InSb surface with 0.06 ML PTCDA deposited at 520 K, topographic image, (**b**, **c**, **d**) parts of the surface imaged with higher resolution. Reprinted with permission from [89]

equivalent positions shifted (by ~4 Å) perpendicular to the chain. Other, skewed orientations of PTCDA molecules are also possible—see Fig. 21.12c. At steps, where a density of molecules is larger, characteristic pairing of molecular chains is seen. Since there are no atomic details visible on the background it is claimed that the interaction over substrate is due to mesoscopic polarization forces only. Therefore the dark fields over molecule cores are ascribed to the repulsive short range interaction. Analysis of tip surface interactions evidenced that the high-resolution DFM imaging is possible through core-core repulsive interaction in the constant height mode [89]. It has been also shown that the PTCDA/InSb system is suitable for manipulation of single molecules by the DFM tip.

21.4.5 KPFM of Supported Nanostructures

The KPFM technique apart from topographic imaging of the sample surface, probes the surface electronic and chemical properties at nanometer scales on a broad range of materials [90]. Thus, using KPFM the surface areas of different chemical compositions can be identified despite they are topographically not distinguishable [91]. An example is presented in Fig. 21.13a where a typical topography of KBr islands, grown on InSb(001) surface is shown. The substrate surface is composed of large, atomically flat terraces with monatomic substrate steps oriented along the main crystallographic directions of [110] and $[1\bar{1}0]$. The KBr islands are visible as rectangular brighter features. In Fig. 21.13b, a contact potential difference image (CPD image) of KBr/InSb(001) surface, acquired simultaneously with topographic image, is shown. The dark features on the CPD map correspond to KBr islands. The map represents the voltage applied to the sample in order to compensate the CPD between the surface and the tip. Thus, a darker contrast on KBr islands (lower voltage) corresponds to decreased work function of the islands, as compared with the substrate.

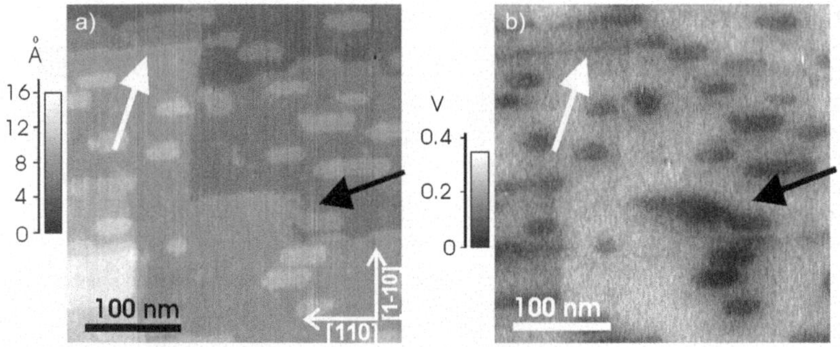

Fig. 21.13 Topography (**a**) and CPD (**b**) images of KBr islands grown on InSb(001) surface acquired with KPFM. The *black arrows* indicate the KBr island which is topographically not resolved from the substrate terrace and can be identified only with the help of CPD image. The *white arrows* highlight the substrate step-edge (**a**) with corresponding CPD signal drop in (**b**). The parameters for topography imaging: $f_0 = 111.1$ kHz, $\Delta f = -17$ Hz and for the CPD measurements: $\omega = 500$ Hz and $V_{ac} = 300$ mV. Reprinted with permission from [91]

There is one straightforward observation emerging from Figs. 21.13(a) and (b), which demonstrates the advantages of KPFM for imaging heterogeneous surfaces. Namely, there are some KBr islands (marked by black arrow in Fig. 21.13(a) that have grown up attached to the substrate terrace edge and they can be recognized only using the CPD signal but not the topographic one. Moreover, the KPFM provides more reliable topographic height of surface structures, as compared to the standard NC-AFM, in case when the surface of heterogeneous systems is imaged. Since the surface topography is acquired while keeping the total tip-sample interactions constant, the apparent topography is highly influenced by the difference of the electrostatic interaction between the probe and the imaged surface areas of different composition in a classical NC-AFM. In KPFM, by the elimination of the electrostatic interaction term, this effect is minimized [92].

Apart from CPD contrast related to chemical composition of imaged surfaces also the surface defects can be imaged/detected. In Fig. 21.13b, together with the dark features corresponding to the KBr islands, there is a decrease of the surface potential visible in a form of zigzag line (see the white arrow in the image). The zigzag line corresponds to the substrate terrace edges and it reflects the variation of electrostatic potential (ES) across monatomic steps. The variation of the ES indicates trapping of electrical charge at the step states which induce pinning of the Fermi level [93]. The measured drop of CPD (decrease of the work function) indicates that the steps are positively charged. However, the measured value of the reduction of the surface potential (about 30 mV) should not be directly assigned to the magnitude of a variation of the electrostatic potential. Most likely the true local ES variation is much higher but the measured value is highly influenced by the averaging due to a finite tip size [94].

For certain systems, KPFM can provide even a direct information on the chemical compositions of surface nanostructures [77]. In Fig. 21.14(a) the topography and (b) contact potential (CPD) images of nanostructures grown on InSb(001) surface

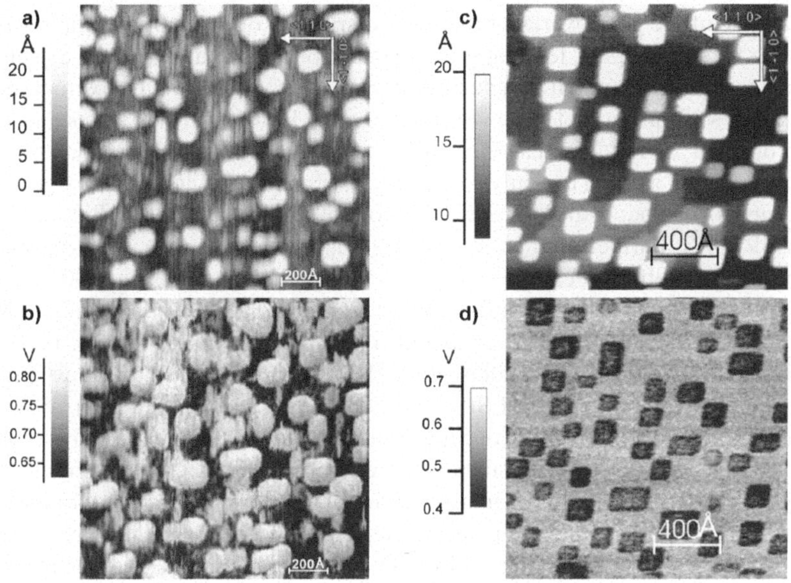

Fig. 21.14 The topography and surface potential images of Au nanostructures grown on InSb(001) surface with and without post-deposition annealing. (**a**) The topography and (**b**) corresponding CPD map of 0.2 ML of Au deposited on InSb(001) at 400 K. (**c**) Topography and (**d**) corresponding CPD mapping of the same system after post-deposition annealing at 650 K for 2 h. Reprinted with permission from [77]

after room temperature gold depositions are shown. For the experimental conditions used, gold grows predominantly in a form of rectangular islands with typical height of a few monolayers (about 2.0 nm) and with certain amount of material spread over the substrate surface (i.e., between the islands). The gold islands have higher surface potential in comparison with the substrate surface according to what, in general, is expected due to higher work function of Au with respect to a clean InSb surface. From the comparison between the topography and CPD images, it follows that the CPD map provides more details concerning the developed surface topography. Some small features that are difficult to be seen in the topography image, due to a large variation in the image Z direction, are easily recognized with the help of CPD signal. Moreover, almost the same CPD contrast for Au islands and spread material indicates that they are of the same chemical composition. Thus, KPFM is able to give the information about the chemical composition of surface morphology, provided there is some reference marker on the imaged surface.

Adsorbate-substrate reactions can also be observed with KPFM. It has been established that the post-growth annealing of the initial gold islands (shown in Fig. 21.14a) at a temperature of 600 K results in the inversion of the CPD contrast between the Au nanostructures and InSb substrate surface as observed in Fig. 21.14d. The inversion of the CPD contrast suggests a composition change of the islands due to formation of an InAu alloy phase. It seems that upon annealing in the presence of the gold overlayer there is a disruption of the cation-anion bonds on the substrate surface [95], leading to segregation of indium atoms, their dissolution in Au islands and as a result to a decreasing work function of the nanostructures.

The KPFM is also a tool for investigation of the local electronic states and transport properties of modern nanoelectronic devices. In this respect, quantum dots (QDs) are widely studied not only for better understanding of low-dimensional electron systems [96] but also due to their important in electronic and photonic applications [97, 98]. Further interest in studying nanometer-size QDs stems from their extraordinary abilities for charge storage [99], cold electron emission [100] and photoluminescence [101]. The work function of a single QD is an important parameter because it is related to the barrier height for the carrier injection into, or for the carrier ejection from the QD. Salem et al. [102] used the KPFM to investigate the change of CPD after charging of Si quantum dots. Si QDs, of 2–8 nm in size, were grown by plasma decomposition of SiH_4 on a SiO_2 substrate. The local charge injection to the dots was performed using the biased tip in the contact mode. Figure 21.15 shows the topography and CPD images of the Si QDs as-grown (a and b) and after the charging process (c and d), respectively. The charged DQs show an increased CPD, while the substrate maintains the same surface potential. The change of the dot CPD depends on the dot size, i.e., larger dots have a higher potential value. Measuring the CPD changes as a function of the dot diameter and comparing the results with the calculated charging energy of separated dots, the number of injected electrons can be evaluated. It has been found [102] that one electron could be stored in a Si QD of diameter up to 2.8 nm, whereas there can be three electrons in QDs having diameters from 4.7 to 7.4 nm.

Although the concept of CPD measurements in KPFM is based on the detection of long-range electrostatic tip-sample interactions, atomic-resolution KPFM is demonstrated on a variety of surfaces, including semiconductors [94, 103–105] and ionic crystals [106]. In atomically resolved KPFM, the measured CPD is defined as local-CPD (LCPD). The LCPD is attributed the atomic-scale contrast of CPD, due to detection of short-range, bias dependent forces acting between the tip-apex atoms and surface atoms. In case of ionic crystal surface, the tip-sample interactions resulting in atomic LCPD contrast is the dipole interaction between the surface atom and the tip apex atom [107] whereas in case of semiconductor surfaces, the short-range force is attributed to a covalent bonding interaction [108].

21.4.6 Friction Force Microscopy Measurements at Atomic-Scale

The macroscopic friction between solids is related to the interactions taking place in single, nanometer-scale asperity contacts of the sliding surfaces [109, 110]. Soon after its invention, AFM technique was applied to measure the frictional forces at the nanometer-size contact. In the FFM the tip apex sliding against the sample surface mimics such an asperity and allows the exploration of the nanometer scale contact mechanics [111]. For an atomically sharp tip, the atomic scale stick-slip behavior of

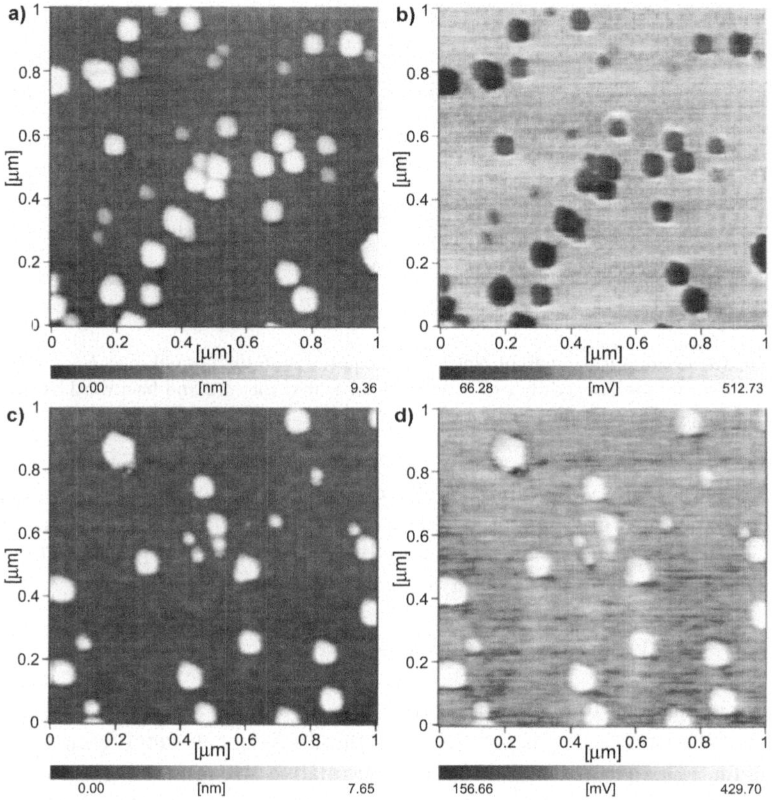

Fig. 21.15 Topographic and corresponding surface potential images for Si QDs before (**a** and **b**) and after (**c** and **d**) dots charging, respectively. Reprinted with permission from [102]

the frictional forces is observed when the tip is dragged over atomically flat terraces. A pioneering work in this field of atomic-scale friction was performed in 1987 by Mate et al. [112]. In their experiment, the tungsten tip was sliding over a graphite surface, and the lateral forces exhibited a modulation (a saw-tooth pattern) with the atomic periodicity of the graphite lattice. Furthermore, the FFM signal exhibited hysteresis between the forward and backward scan (the friction force loop).

Figure 21.16a shows a typical FFM map, with corresponding friction force traces along the indicated lines (b), acquired on NaCl(100) surface in UHV [113]. The lateral force reveals two opposite saw-tooth profiles when scanning forwards and backwards as seen in the FFM profiles (Fig. 21.16b)). The saw-tooth modulation has the periodicity of the NaCl surface lattice and is characteristic for the stick-slip process. Generally, the slope of the sticking part of the friction loop, at the turning points of tip motion, reflects the stiffness of the tip-sample contact. From the average value of the friction force, the dynamic friction coefficient can be evaluated provided that the load is known. The area enclosed in this hysteresis loop is an energy dissipated in one complete cycle due to friction. The behavior of the measured

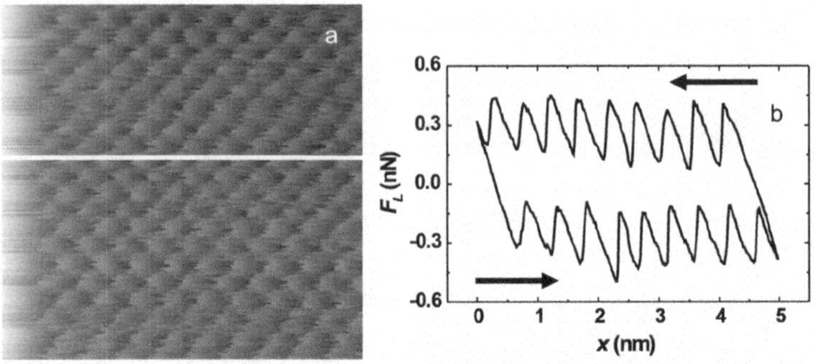

Fig. 21.16 (**a**) Friction force map of NaCl(100) acquired with the normal load $F_N = 0.65$ nN in UHV. (**b**) Friction loop formed by two scan lines measured forward and backward, respectively. Reprinted with permission from [113]

friction is well explained within Prandt-Tomlinson (P-T) model [114, 115]. In the P-T model, the FFM tip-apex, attached the cantilever assumed to be a spring with a spring constant k, is drag over a sinusoidal surface potential. This leads to a combined surface-tip potential of the form:

$$V_{total}(x,t) = -\frac{E_0}{2}\cos\left(\frac{2\pi x}{a}\right) + \frac{1}{2}k_{eff}(x - vt)^2 \qquad (21.9)$$

where E_0 is the peak-to-peak amplitude of the tip-sample potential, a is the lattice constant of the surface, k_{eff} is the effective lateral spring constant (contact stiffness) and v is the velocity of the tip support. The relative strength of the surface corrugation E_0 with respect to the potential energy stored in the spring of k_{eff} is defined by the parameter $\eta = \frac{2\pi^2 E_0}{k_{eff} a^2}$. According to the P-T model, changing the ratio of the surface potential corrugation to the lateral stiffness, the friction force can be controlled. Thus, for $\eta > 1$ the stick-slip behavior is observed whereas for values of $\eta < 1$ smooth sliding, without energy dissipation, is expected (this regime is called "superlubrilicity") [116].

Experimentally, the transition from dissipative stick-slip to ultra-low friction has been realized by introducing periodic variation of the tip-surface interactions due to a mechanical shaking of the tip while sliding on the surface ("dynamic superlubricity") [118]. Such a transition between two friction regimes is presented in Fig. 21.17a. In the upper part, the friction force map was acquired without shaking the cantilever, whereas some mechanical excitation, with frequency corresponding to the first normal resonance of the cantilever in contact, was applied in the lower part. A significant change in the frictional behavior is observed upon switching the excitation. Instead of stick-slip, the tip motion is smoothed out, and the energy dissipation accompanying the tip jumps consequently disappears (Fig. 21.17b). Such a remarkable way of reduction of the friction can, in practice, results in creation of wear-free contacts as reported recently by Lantz and co-workers [119]. They explored a system for a data storage which involves the use of AFM and thin polymer film. The performance of such system implies the use of AFM probes either to made

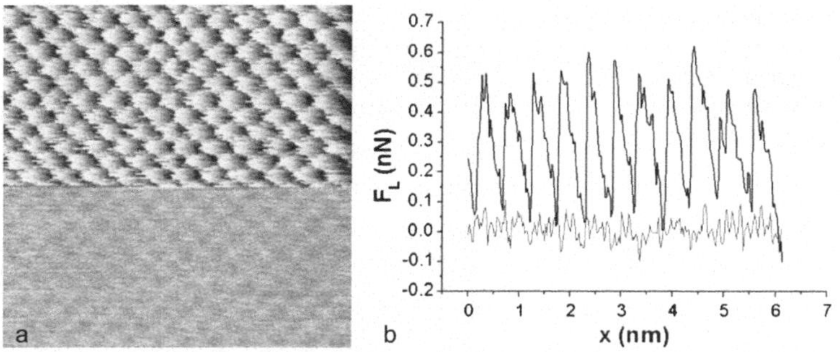

Fig. 21.17 (**a**) Friction force map acquired an atomically flat NaCl(100) surface with normal load of 2.73 nN without (the *upper part* of the image) and with mechanical shaking of the cantilever (the *lower part* of the image). (**b**) Representative scan lines from the *upper part* (*thick line*) and the *lower part* (*thin line*) of the map in (**a**). Reprinted with permission from [117]

a dense array of holes in the polymer film (a "write" procedure) or, in the inverse procedure of "read", for retrieval of the stored data. A demanding challenge for the optimum system performance is to extend the AFM tips lifetime by reducing their wear. Lantz and co-workers reported that upon the mechanical excitation of the tip, they were able to drag a silicon tip across a polymer surface for the best part of a kilometer without damaging either the tip or the surface. To get impressiveness of that performance, consider that it is an equivalent of someone climbing a ladder from Earth to Mars without wearing out the soles of their shoes [120].

21.5 Future Development of the Technique

In recent years NC-AFM related techniques have been developing very fast. One of the fields where fast progress was especially prominent is cryogenic NC-AFM. Following the footsteps of STM, for which going to low temperatures gave access to new fields of physics and chemistry, cryogenic NC-AFM is opening a range of new applications. There are some examples of highly successful NC-AFM experiments performed with the use of cantilever-based systems [12, 121–123], however the rapid growth of applications and accessibility of the field coincided with the introduction of quartz tuning forks as sensors [124, 125]. There are a few features making it a very useful sensor. Perhaps the main advantage is the way of detecting its deflection: the deformation of a prong of a tuning fork generates a current due to piezoelectric effect, which can be easily detected. Therefore, the optical detection system, quite cumbersome to work with in a bath cryostat can be shed. Additionally, a tuning fork can be equipped with a tip made of any material (for instance tungsten or Pt-Ir alloy) which can be connected to preamplifier by a separate lead allowing in principle for dual AFM/STM operation. Additionally, a tuning fork has much higher stiffness compared to typical silicon cantilevers (typically 1000–2000 N/m vs. a few N/m) allowing to operate sensor with small amplitudes—in sub-nanometer

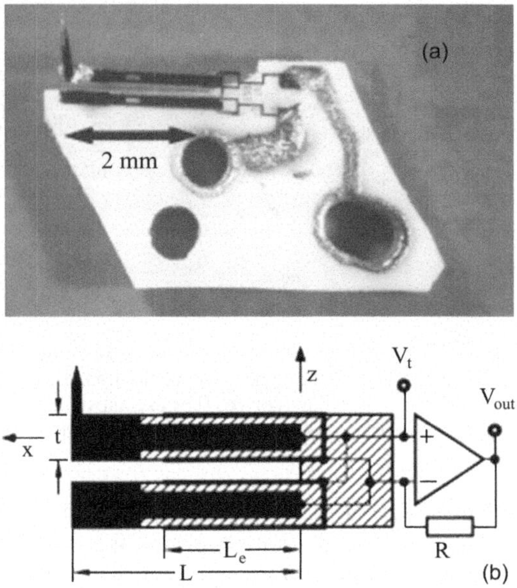

Fig. 21.18 A tuning fork-based NC-AFM sensor in qPlus configuration. Reprinted with permission from [125]

or even sub-Angstrom range that makes them particularly suitable to probe short range, chemical interaction, which ranges are of similar magnitude.

Tuning forks (Fig. 21.18) are typically used in qPlus configuration [125, 126]—in which one of the prongs is glued to a ceramic holder while another, with a tip attached at its end is oscillating freely (other geometries were, however, tested [127]). Therefore, all the methods of calculation forces from frequency shift developed for cantilevers are valid for tuning forks. Quality factors depend strongly on temperature and can well exceed 100000 at liquid helium (LHe) temperature (more typical values are 10000–50000), while it drops below 1000 at room temperature (RT). The resonant frequency of a free (non-glued) tuning fork is usually 32678 Hz, but after gluing and attaching the tip it drops to 20000–25000 Hz. There are already quite a few examples of a huge potential of that method in the literature.

The cryogenic NC-AFM has potentially extremely high resolution—exceeding even the abilities of STM (see Fig. 21.19). Gross and co-workers have shown that by adsorbing on a tip apex a single CO molecule it was possible to resolve internal structure of a planar, weakly bound organic molecule [128]. In their experiment the tip approached so close to the molecule that started to experience repulsive interactions which, in turn allowed for resolution of internal structure of aromatic backbone of a pentacene molecule. Similarly, the same group of authors have showed that the NC-AFM could be used in the experiment where a molecule unknown a priori was used (Fig. 21.20). The molecule examined was collected from deep sea sediments and the sophisticated NMR methods were used to determine its structure. However, the NMR was only capable to narrow the choice to four possible structural variants. NC-AFM images accompanied with DFT simulations allowed determination a proper structure out of four possibilities [129].

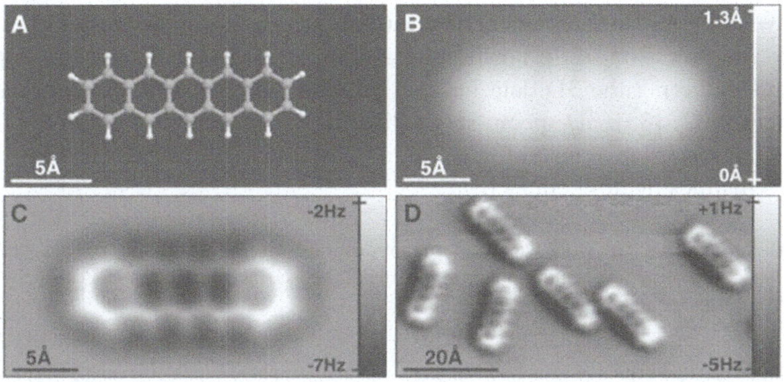

Fig. 21.19 A pentacene molecule on Cu(111) surface: (**a**) ball-and-stick model of a pentacene molecule; (**b**) STM image of a molecule; (**c**) and (**d**) constant height NC-AFM images. Reprinted with permission from [128]

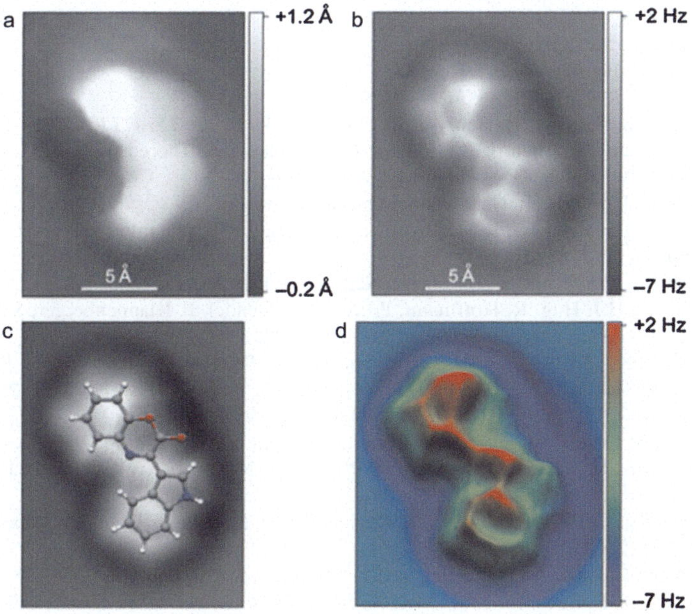

Fig. 21.20 Images of a molecule identified as cephalandole: (**a**) STM image; (**b**) NC-AFM image—row data; (**c**) the NC-AFM image with the molecular structure superimposed; (**d**) filtered NC-AFM image. Reprinted with permission from [129]

Another group of experiments shows the great potential stemming from the chance of very stable and precise positioning of a tip over the sample while measuring forces at the same time. The prominent examples are measurement of the force required for moving a Co atom on the Pt(111) surface [130] and presentation of the switching of surface dimers on a Si(001) [131]. As another result of extreme stability and precision of tip positioning at LHe temperature the spectroscopy experiments gathering the 3D data became possible without difficult atom tracking techniques [132, 133]. 3D spectroscopy, like the experiments for graphite [134] or alkali halides [135] give insight into friction properties of materials and tip-surface system dynamics in close contact.

Acknowledgements The work was supported by Joint Research Programme (PSRP—085/2010) "Molecular assemblies on semiconductors and insulating surfaces" under the Polish-Swiss Research Programme.

References

1. G. Binnig, C.F. Quate, Ch. Gerber, Phys. Rev. Lett. **56**, 930 (1986)
2. G. Binnig, H. Roehrer, Ch. Gerber, E. Weibel, Phys. Rev. Lett. **50**, 120 (1982)
3. W.A. Hofer, A.S. Foster, A.L. Shluger, Rev. Mod. Phys. **75**, 1287 (2003)
4. S. Morita, R. Wiesendanger, E. Meyer (eds.), *Noncontact Atomic Force Microscopy* (Springer, Berlin, 2002)
5. R. Garcia, R. Perez, Surf. Sci. Rep. **47**, 197 (2002)
6. F.J. Giessibl, Rev. Mod. Phys. **75**, 949 (2003)
7. C. Barth, A.S. Foster, C.R. Henry, A.L. Shluger, Adv. Mater. **23**, 477 (2011)
8. N.A. Burnham, X. Chen, C.S. Hodges, G.A. Matei, E.J. Thorenson, C.J. Roberts, M.C. Davies, S.J.B. Tendler, Nanotechnology **14**, 1 (2003)
9. C.A. Clifford, M.P. Seah, Nanotechnology **16**, 1666 (2005)
10. S.M. Cook, T.E. Schaefer, K.M. Chynoweth, M. Wigton, R.W. Simmonds, K.M. Lang, Nanotechnology **17**, 2135 (2006)
11. C.T. Gibson, D.A. Smith, C.J. Roberts, Nanotechnology **16**, 234 (2005)
12. M.A. Lantz, H.J. Hug, R. Hoffmann, P.J.A. van Schendel, P. Klappenberger, S. Martin, A. Baratoff, H.-J. Guentherodt, Science **291**, 2580 (2001)
13. M. Guggisberg, M. Bammerlin, Ch. Loppacher, O. Pfeiffer, A. Abdurixit, V. Barwich, R. Bennewitz, A. Baratoff, E. Meyer, H.-J. Guentherodt, Phys. Rev. B **61**, 11151 (2000)
14. C. Argento, R.H. French, J. Appl. Phys. **80**, 6081 (1996)
15. J. Olsson, N. Lin, V. Yakimov, R. Erlandsson, J. Appl. Phys. **84**, 4060 (1998)
16. R. Bennewitz, S. Schaer, V. Barwich, O. Pfeiffer, E. Meyer, F. Krok, B. Such, J. Kolodziej, M. Szymonski, Surf. Sci. **474**, L197 (2001)
17. M.A. Venegas de la Cerda, J. Abad, A. Madgavkar, D. Martrou, S. Gauthier, Nanotechnology **19**, 045503 (2008)
18. A.S. Foster, C. Bath, A.L. Shluger, M. Reichling, Phys. Rev. Lett. **86**, 2373 (2001)
19. J.J. Kolodziej, B. Such, M. Szymonski, F. Krok, Phys. Rev. Lett. **90**, 226101 (2003)
20. H. Hoelscher, W. Allers, U.D. Schwarz, A. Schwarz, R. Wiesendanger, Phys. Rev. B **62**, 6967 (2000)
21. N.A. Burnham, R.J. Colton, J. Vac. Sci. Technol. A **7**, 2906 (1989)
22. T.R. Albrecht, P. Gruetter, D. Horne, D. Rugar, J. Appl. Phys. **69**, 668 (1991)
23. Y. Martin, C.C. Williams, H.K. Wickramasinghe, J. Appl. Phys. **61**, 4723 (1987)
24. H. Hoelscher, U.D. Schwarz, R. Wiesendanger, Appl. Surf. Sci. **140**, 244 (1999)
25. F.J. Giessibl, Phys. Rev. B **56**, 16010 (1997)
26. U. Duerig, Appl. Phys. Lett. **76**, 1203 (2000)
27. F.J. Giessibl, Appl. Phys. Lett. **78**, 123 (2001)
28. J.E. Sader, S.P. Jarvis, Appl. Phys. Lett. **84**, 1801 (2004)
29. B. Gotsmann, C. Seidel, B. Anczykowski, H. Fuchs, Phys. Rev. B **60**, 11051 (1999)

30. B. Anczykowski, B. Gotsmann, H. Fuchs, J.P. Cleveland, V.B. Elings, Appl. Surf. Sci. **140**, 376 (1999)
31. C. Loppacher, R. Bennewitz, O. Pfeiffer, M. Guggisberg, M. Bammerlin, S. Schaer, V. Barwich, A. Baratoff, E. Meyer, Phys. Rev. B **62**, 13674 (2000)
32. http://www.omicron.de
33. J. Polesel-Maris, M.A. Venegas de la Cerda, D. Martrou, S. Gauthier, Phys. Rev. B **79**, 235401 (2009)
34. L. Kelvin, G. Fitzgerald, W. Francis, Philos. Mag. J. Sci. **46**, 82 (1898)
35. F.J. Giessibl, Science **267**, 68 (1995)
36. S. Kitamura, M. Iwatsuki, Jpn. J. Appl. Phys. **34**, L145 (1995)
37. T. Uchihashi, Y. Sugawara, T. Tsukamoto, M. Ohta, S. Morita, M. Suzuki, Phys. Rev. B **45**, 9834 (1997)
38. R. Perez, I. Stich, M.C. Payne, K. Terakura, Phys. Rev. B **58**, 10835 (1998)
39. N. Nakagiri, M. Suzuki, K. Okiguchi, H. Sugimura, Surf. Sci. **373**, 373 (1997)
40. S. Kitamura, M. Iwatsuki, Jpn. J. Appl. Phys. **3**, L668 (1996)
41. R. Luthi, E. Meyer, M. Bammerlin, A. Baratoff, T. Howald, Ch. Gerber, H.-J. Guntherod, Z. Phys. B **100**, 165 (1996)
42. F.J. Giessibl, S. Hembacher, H. Bielefeldt, J. Mannhart, Science **289**, 422 (2000)
43. M.A. Lantz, H.J. Hug, P.J.A. van Schendel, R. Hoffman, S. Martin, A. Baratoff, A. Abdurixit, H.-J. Guntherod, Ch. Gerber, Phys. Rev. Lett. **84**, 2642 (2000)
44. M. Guggisberg, O. Pfeiffer, S. Schaar, V. Barvich, M. Bammerlin, C. Loppacher, R. Bennewitz, A. Baratoff, E. Meyer, Appl. Phys. A (suppl.) **72**, S19 (2001)
45. M.A. Lantz, H.J. Hug, R. Hoffman, S. Martin, A. Baratoff, H.-J. Guntherod, Phys. Rev. B **68**, 035324 (2003)
46. R. Nishi, S. Araragi, K. Shirai, Y. Sugawara, S. Morita, Appl. Surf. Sci. **210**, 90 (2003)
47. S. Morita, Y. Sugimoto, N. Oyabu, R. Nishi, O. Custance, Y. Sugawara, M. Abe, J. Electron Microsc. **53**, 163 (2004)
48. Y. Sugimoto, T. Namikawa, K. Miki, M. Abe, S. Morita, Phys. Rev. B **77**, 195424 (2008)
49. S. Morita, M. Abe, K. Yokoyama, Y. Sugawara, J. Cryst. Growth **210**, 408 (2000)
50. N. Uehara, H. Hosoi, K. Sueoka, K. Mukasa, Nanotechnology **15**, S97 (2004)
51. S.H. Ke, T. Uda, R. Perez, I. Stich, K. Terakura, Phys. Rev. B **60**, 11631 (1999)
52. S.H. Ke, T. Uda, I. Stich, K. Terakura, Phys. Rev. B **63**, 245323 (2001)
53. A. Schwarz, W. Allers, U.D. Schwarz, R. Wiesendanger, Phys. Rev. B **62**, 13617 (2000)
54. H. Uejama, M. Ohta, Y. Sugawara, S. Morita, Jpn. J. Appl. Phys. **34**, L1086 (1995)
55. Y. Sugawara, M. Ohta, H. Uejama, S. Morita, Science **270**, 1646 (1995)
56. J. Tobik, I. Stich, K. Terakura, Phys. Rev. B **63**, 245324 (2001)
57. J.J. Kolodziej, B. Such, M. Szymonski, Phys. Rev. B **71**, 165419 (2005)
58. J.J. Kolodziej, B. Such, M. Goryl, F. Krok, P. Piatkowski, M. Szymonski, Appl. Surf. Sci. **252**, 7614 (2006)
59. G. Goryl, D. Toton, N. Tomaszewska, J.S. Prauzner-Bechcicki, L. Walczak, A. Tejeda, A. Taleb-Ibrahimi, L. Kantorovich, E.G. Michel, J.J. Kolodziej, Phys. Rev. B **82**, 165311 (2010)
60. G. Goryl, D. Toton, M. Goryl, N. Tomaszewska, J. Kolodziej, Surf. Sci. **605**, 2073 (2011)
61. B. Such, J.J. Kolodziej, F. Krok, P. Piatkowski, M. Szymonski, Surf. Sci. **600**, 2379 (2006)
62. Y. Sugimoto, P. Pou, O. Custance et al., Science **322**, 413 (2008)
63. J.J. Kolodziej, M. Goryl, J. Konior, M. Reichling, M. Szymonski, Phys. Rev. B **76**, 245314 (2007)
64. W.G. Schmidt, F. Bechstedt, Phys. Rev. B **54**, 16742 (1996)
65. C. Kendrick, G. LeLay, A. Kahn, Phys. Rev. B **54**, 17877–17883 (1996)
66. C. Kumpf, L.D. Marks, D. Ellis, D. Smilgies, E. Landemark, M. Nielsen, R. Feidenhans'l, J. Zegenhagen, O. Bunk, J.H. Zeysig, Y. Su, R.L. Johnson, Phys. Rev. Lett. **86**, 3586 (2001)
67. S.-H. Lee, W. Moritz, M. Scheffler, Phys. Rev. Lett. **85**, 3890 (2000)
68. B. Such, J. Kolodziej, P. Czuba, P. Piatkowski, P. Struski, F. Krok, M. Szymonski, Phys. Rev. Lett. **85**, 2621 (2000)

69. M. Goryl, B. Such, F. Krok, K. Meisel, J.J. Kolodziej, M. Szymonski, Surf. Sci. **593**, 147 (2005)
70. J.J. Kolodziej, B. Such, P. Czuba, F. Krok, P. Piatkowski, M. Szymonski, Surf. Sci. **506**, 12 (2002)
71. R. Bennewitz, M. Reichling, E. Matthias, Surf. Sci. **387**, 69 (1997)
72. M. Heyde, M. Sterrer, H.-P. Rust, H.-J. Freund, Appl. Phys. Lett. **87**, 083104 (2005)
73. R. Hoffmann, C. Barth, A.S. Foster, A.L. Shluger, H.J. Hug, H.-J. Guentherodt, R.M. Nieminen, M. Reichling, J. Am. Chem. Soc. **127**, 17863 (2005)
74. C. Barth, M. Reichling, Nature **414**, 54 (2001)
75. M. Reichling, C. Barth, Phys. Rev. Lett. **83**, 768 (1999)
76. M. Szymonski, M. Goryl, F. Krok, J.J. Kolodziej, F. Buatier de Mongeot, Nanotechnology **18**, 044016 (2007)
77. M. Goryl, J.J. Kolodziej, F. Krok, P. Piatkowski, B. Such, M. Szymonski, Microelectron. Eng. **81**, 394 (2005)
78. Ch. Loppacher, M. Guggisberg, O. Pfeiffer, E. Meyer, M. Bammerlin, R. Luethi, R. Schlitter, J.K. Gimzewski, H. Tang, C. Joachim, Phys. Rev. Lett. **90**, 066107 (2003)
79. L. Nony, E. Gnecco, A. Baratoff, R. Bennewitz, O. Pfeiffer, S. Maier, A. Wetzel, E. Meyer, Ch. Gerber, Nano Lett. **4**, 2185 (2004)
80. T. Kunstmann, A. Schlarb, M. Fendrich, Th. Wagner, R. Moeller, R. Hoffmann, Phys. Rev. B **71**, 121403 (2005)
81. S.A. Burke, J.M. Mativetsky, R. Hoffmann, P. Gruetter, Phys. Rev. Lett. **94**, 096102 (2005)
82. B. Such, D. Weiner, A. Schirmeisen, H. Fuchs, Appl. Phys. Lett. **89**, 093104 (2006)
83. S. Maier, L.-A. Fendt, L. Zimmerli, T. Glatzel, O. Pfeiffer, F. Diederich, E. Meyer, Small **4**, 1115 (2008)
84. F. Loske, R. Bechstein, J. Schuette, F. Ostendorf, M. Reichling, A. Kuehnle, Nanotechnology **20**, 065606 (2009)
85. K. Lammle, T. Trevethan, A. Schwarz, M. Watkins, A.L. Shluger, R. Wiesendanger, Nano Lett. **10**, 2965 (2010)
86. B. Such, T. Trevethan, T. Glatzel, S. Kawai, L. Zimmerli, E. Meyer, A.L. Shluger, C.H.M. Amijs, P. de Mendoza, A.M. Echavarren, ACS Nano **4**, 3429 (2010)
87. P. Rahe, M. Nimmrich, A. Greuling, J. Schuette, I.G. Stara, J. Rybacek, G. Huerta-Angeles, I. Stary, M. Rohlfing, A. Kuehnle, J. Phys. Chem. C **114**, 1547 (2010)
88. T. Trevethan, B. Such, T. Glatzel, S. Kawai, A.L. Shluger, E. Meyer, P. de Mendoza, A.M. Echavarren, Small **7**, 1264 (2011)
89. J.J. Kolodziej, M. Goryl, J. Konior, F. Krok, M. Szymonski, Nanotechnology **18**, 135302 (2007)
90. W. Melitz, J. Shena, A.C. Kummel, S. Lee, Surf. Sci. Rep. **66**, 1 (2011)
91. F. Krok, P. Piatkowski, J.J. Kolodziej, B. Such, P. Struski, P. Czuba, M. Szymonski, Surf. Sci. **566–568**, 63 (2004)
92. S. Sadewasser, M.Ch. Lux-Steiner, Phys. Rev. Lett. **91**, 266101 (2003)
93. Ch. Sommerhalter, Th.W. Matthes, Th. Glatzel, A. Jaeger-Waldau, Appl. Phys. Lett. **75**, 286 (1999)
94. F. Krok, K. Sajewicz, J. Konior, M. Goryl, P. Piatkowski, M. Szymonski, Phys. Rev. B **77**, 235427 (2008)
95. Y. Shapira, F. Xu, D.M. Hill, J.H. Weaver, Appl. Phys. Lett. **51**, 118 (1987)
96. H. Sakaki, Solid State Commun. **92**, 119 (1994)
97. D. Leonard, M. Krishnamurthy, C.M. Reaves, S.P. Denbaars, P.M. Petroff, Appl. Phys. Lett. **63**, 3203 (1993)
98. J.-Y. Marzin, J.-M. Gerard, A. Izrael, D. Barrier, G. Bastard, Phys. Rev. Lett. **73**, 716 (1994)
99. B.J. Hinds, T. Yamanaka, S. Oda, J. Appl. Phys. **90**, 6402 (2001)
100. K. Hishiguchi, X. Zhao, S. Oda, J. Appl. Phys. **92**, 2748 (2002)
101. K. Arai, S. Oda, Phys. Status Solidi C **0**, 1254 (2003)
102. M.A. Salem, H. Mizuta, S. Oda, Appl. Phys. Lett. **85**, 3262 (2004)
103. S. Kitamura, K. Suzuki, M. Iwatsuki, C.B. Mooney, Appl. Surf. Sci. **157**, 222 (2000)

104. K. Okamoto, Y. Sugawara, S. Morita, Appl. Surf. Sci. **188**, 381 (2002)
105. G.H. Enevoldsen, T. Glatzel, M.C. Christensen, J.V. Lauritsen, F. Besenbacher, Phys. Rev. Lett. **100**, 236104 (2008)
106. F. Bocquet, L. Nony, Ch. Loppacher, Th. Glatzel, Phys. Rev. B **78**, 035410 (2008)
107. L. Nony, A.S. Foster, F. Bocquet, Ch. Loppacher, Phys. Rev. Lett. **103**, 036802 (2009)
108. S. Sadewasser, P. Jelinek, C.-K. Fang, O. Custance, Y. Yamada, Y. Sugimoto, M. Abe, S. Morita, Phys. Rev. Lett. **103**, 266103 (2009)
109. Y. Mo, K.T. Turner, I. Szlufarska, Nature **457**, 1116 (2009)
110. O. Ben-David, S.M. Rubinstein, J. Fineberg, Nature **463**, 76 (2010)
111. E. Gnecco, R. Bennewitz, T. Gyalog, E. Meyer, J. Phys. Condens. Matter **13**, R619 (2001)
112. C. Mate, G. McClelland, R. Erlandsson, S. Chiang, Phys. Rev. Lett. **59**, 1942 (1987)
113. E. Gnecco, R. Bennewitz, T. Gyalog, Ch. Loppacher, M. Bammerlin, E. Meyer, H.-J. Guentherodt, Phys. Rev. Lett. **84**, 1172 (2000)
114. L. Prandtl, Z. Angew. Math. Mech. **8**, 85 (1928)
115. G.A. Tomlinson, Philos. Mag. **7**, 905 (1929)
116. E. Gnecco, S. Maier, E. Meyer, J. Phys. Condens. Matter **20**, 354004 (2008)
117. E. Gnecco, A. Socoliuc, S. Maier, J. Gessler, T. Glatzel, A. Baratoff, E. Meyer, Nanotechnology **20**, 025501 (2009)
118. A. Socoliuc, E. Gnecco, S. Maier, O. Pfeiffer, A. Baratoff, R. Bennewitz, E. Meyer, Science **313**, 208 (2006)
119. M.A. Lantz, D. Wiesmann, B. Gotsmann, Nat. Nanotechnol. **4**, 586 (2009)
120. E. Gnecco, Nature **461**, 178 (2009)
121. W. Allers, A. Schwarz, U.D. Schwarz, R. Wiesendanger, Rev. Sci. Instrum. **69**, 221 (1998)
122. H.J. Hug, B. Stiefel, P.J.A. van Schendel, A. Moser, S. Martin, H.-J. Guentherodt, Rev. Sci. Instrum. **70**, 3625 (1999)
123. R. Hoffmann, L.N. Kantorovich, A. Baratoff, H.J. Hug, H.-J. Guentherodt, Phys. Rev. Lett. **92**, 146103 (2004)
124. F.J. Giessibl, Appl. Phys. Lett. **73**, 3956 (1998)
125. F.J. Giessibl, Appl. Phys. Lett. **76**, 1470 (2000)
126. A. Bettac, J. Koeble, K. Winkler, B. Uder, M. Maier, A. Feltz, Nanotechnology **20**, 264009 (2009)
127. M. Heyde, M. Kulawik, H.-P. Rust, H.-J. Freund, Rev. Sci. Instrum. **75**, 2446 (2004)
128. L. Gross, F. Mohn, N. Moll, P. Lilieroth, G. Meyer, Science **325**, 1110 (2009)
129. L. Gross, F. Mohn, N. Moll, G. Meyer, R. Ebel, W.M. Abdel-Mgeed, M. Jaspars, Nat. Chem. **2**, 821 (2010)
130. M. Ternes, C.P. Lutz, C.F. Hirjibehedin, F.J. Giessibl, A.J. Heinrich, Science **319**, 1066 (2008)
131. A. Sweetman, S. Jarvis, R. Danza, J. Bamidele, S. Gangopadhyay, G.A. Shaw, L. Kantorovich, P. Moriarty, Phys. Rev. Lett. **106**, 136101 (2011)
132. M. Abe, Y. Sugimoto, O. Custance, S. Morita, Appl. Phys. Lett. **87**, 173503 (2005)
133. Y. Sugimoto, P. Pou, M. Abe, P. Jelinek, R. Perez, S. Morita, O. Custance, Nature **446**, 64 (2007)
134. B.J. Albers, T.C. Schwendemann, M.Z. Baykara, N. Pilet, M. Liebmann, E.I. Altmann, U.D. Schwarz, Nat. Nanotechnol. **4**, 307 (2009)
135. B. Such, T. Glatzel, S. Kawai, S. Koch, E. Meyer, J. Vac. Sci. Technol. B **28**, C4B1 (2010)

Index

G. Bracco, B. Holst (eds.), *Surface Science Techniques*,
Springer Series in Surface Sciences 51, DOI 10.1007/978-3-642-34243-1,

T

U

V

W

X

Y

Z

The manufacturer's authorised representative in the EU is Springer
Nature Customer Service Centre GmbH, Europaplatz 3, 69115 Heidelberg,
Germany. If you have any concerns regarding our products, please
contact ProductSafety@springernature.com

Printed and bound by CPI Group (UK) Ltd, Croydon, CR0 4YY
05/10/2026
02228123-0019